HANDBOOK OF
APPLIED ECONOMETRICS
AND STATISTICAL INFERENCE

STATISTICS: Textbooks and Monographs

D. B. Owen, Founding Editor, 1972–1991

HANDBOOK OF APPLIED ECONOMETRICS AND STATISTICAL INFERENCE

EDITED BY

AMAN ULLAH
University of California, Riverside
Riverside, California

ALAN T. K. WAN
City University of Hong Kong
Kowloon, Hong Kong

ANOOP CHATURVEDI
University of Allahabad
Allahabad, India

CRC Press
Taylor & Francis Group
Boca Raton London New York

CRC Press is an imprint of the
Taylor & Francis Group, an **informa** business

CRC Press
Taylor & Francis Group
6000 Broken Sound Parkway NW, Suite 300
Boca Raton, FL 33487-2742

First issued in paperback 2020

© 2002 by Taylor & Francis Group, LLC
CRC Press is an imprint of Taylor & Francis Group, an Informa business

No claim to original U.S. Government works

ISBN 13: 978-0-367-57867-1 (pbk)
ISBN 13: 978-0-8247-0652-4 (hbk)

Visit the Taylor & Francis Web site at
http://www.taylorandfrancis.com

and the CRC Press Web site at
http://www.crcpress.com

To the memory of Viren K. Srivastava
Prolific researcher,
Stimulating teacher,
Dear friend

Preface

This *Handbook* contains thirty-one chapters by distinguished econometricians and statisticians from many countries. It is dedicated to the memory of Professor Viren K Srivastava, a profound and innovative contributor in the fields of econometrics and statistical inference. Viren Srivastava was most recently a Professor and Chairman in the Department of Statistics at Lucknow University, India. He had taught at Banaras Hindu University and had been a visiting professor or scholar at various universities, including Western Ontario, Concordia, Monash, Australian National, New South Wales, Canterbury, and Munich. During his distinguished career, he published more than 150 research papers in various areas of statistics and econometrics (a selected list is provided). His most influential contributions are in finite sample theory of structural models and improved methods of estimation in linear models. These contributions have provided a new direction not only in econometrics and statistics but also in other areas of applied sciences. Moreover, his work on seemingly unrelated regression models, particularly his book *Seemingly Unrelated Regression Equations Models: Estimation and Inference*, coauthored with David Giles (Marcel Dekker,

Inc., 1987), has laid the foundation of much subsequent work in this area. Several topics included in this volume are directly or indirectly influenced by his work.

In recent years there have been many major developments associated with the interface between applied econometrics and statistical inference. This is true especially for censored models, panel data models, time series econometrics, Bayesian inference, and distribution theory. The common ground at the interface between statistics and econometrics is of considerable importance for researchers, practitioners, and students of both subjects, and it is also of direct interest to those working in other areas of applied sciences. The crucial importance of this interface has been reflected in several ways. For example, this was part of the motivation for the establishment of the journal *Econometric Theory* (Cambridge University Press); the *Handbook of Statistics* series (North-Holland), especially Vol. 11; the North-Holland publication *Handbook of Econometrics*, Vol. I–IV, where the emphasis is on econometric methodology; and the recent *Handbook of Applied Economic Statistics* (Marcel Dekker, Inc.), which contains contributions from applied economists and econometricians. However, there remains a considerable range of material and recent research results that are of direct interest to both of the groups under discussion here, but are scattered throughout the separate literatures.

This *Handbook* aims to disseminate significant research results in econometrics and statistics. It is a consolidated and comprehensive reference source for researchers and students whose work takes them to the interface between these two disciplines. This may lead to more collaborative research between members of the two disciplines. The major recent developments in both the applied econometrics and statistical inference techniques that have been covered are of direct interest to researchers, practitioneres, and graduate students, not only in econometrics and statistics but in other applied fields such as medicine, engineering, sociology, and psychology. The book incorporates reasonably comprehensive and up-to-date reviews of recent developments in various key areas of applied econometrics and statistical inference, and it also contains chapters that set the scene for future research in these areas. The emphasis has been on research contributions with accessibility to practitioners and graduate students.

The thirty-one chapters contained in this *Handbook* have been divided into seven major parts, viz., Statistical Inference and Sample Design, Nonparametric Estimation and Testing, Hypothesis Testing, Pretest and Biased Estimation, Time Series Analysis, Estimation and Inference in Econometric Models, and Applied Econometrics. Part I consists of five chapters dealing with issues related to parametric inference procedures and sample design. In Chapter 1, Barry Arnold, Enrique Castillo, and

José Maria Sarabia give a thorough overview of the available results on
Bayesian inference using conditionally specified priors. Some guidelines
are given for choosing the appropriate values for the priors' hyperparameters, and the results are elaborated with the aid of a numerical example.
Helge Toutenburg, Andreas Fieger, and Burkhard Schaffrin, in Chapter 2,
consider minimax estimation of regression coefficients in a linear regression
model and obtain a confidence ellipsoid based on the minimax estimator.
Chapter 3, by Pawel Pordzik and Götz Trenkler, derives necessary and
sufficient conditions for the best linear unbiased estimator of the linear
parametric function of a general linear model, and characterizes the subspace of linear parametric functions which can be estimated with full efficiency. In Chapter 4, Ahmad Parsian and Syed Kirmani extend the concepts
of unbiased estimation, invariant estimation, Bayes and minimax estimation
for the estimation problem under the asymmetric LINEX loss function.
These concepts are applied in the estimation of some specific probability
models. Subir Ghosh, in Chapter 5, gives an overview of a wide array of
issues relating to the design and implementation of sample surveys over
time, and utilizes a particular survey application as an illustration of the
ideas.

The four chapters of Part II are concerned with nonparametric estimation and testing methodologies. Ibrahim Ahmad in Chapter 6 looks at the
problem of estimating the density, distribution, and regression functions
nonparametrically when one gets only randomized responses. Several
asymptotic properties, including weak, strong, uniform, mean square, integrated mean square, and absolute error consistencies as well as asymptotic
normality, are considered in each estimation case. Multinomial choice models are the theme of Chapter 7, in which Jeff Racine proposes a new
approach to the estimation of these models that avoids the specification
of a known index function, which can be problematic in certain cases.
Radhey Singh and Xuewen Lu in Chapter 8 consider a censored nonparametric additive regression model, which admits continuous and categorical
variables in an additive manner. The concepts of marginal integration and
local linear fits are extended to nonparametric regression analysis with censoring to estimate the low dimensional components in an additive model. In
Chapter 9, Mezbahur Rahman and Aman Ullah consider a combined parametric and nonparametric regression model, which improves both the (pure)
parametric and nonparametric approaches in the sense that the combined
procedure is less biased than the parametric approach while simultaneously
reducing the magnitude of the variance that results from the non-parametric
approach. Small sample performance of the estimators is examined via a
Monte Carlo experiment.

In Part III, the problems related to hypothesis testing are addressed in three chapters. Anil Bera and Aurobindo Ghosh in Chapter 10 give a comprehensive survey of the developments in the theory of Neyman's smooth test with an emphasis on its merits, and put the case for the inclusion of this test in mainstream econometrics. Chapter 11 by Bill Farebrother outlines several methods for evaluating probabilities associated with the distribution of a quadratic form in normal variables and illustrates the proposed technique in obtaining the critical values of the lower and upper bounds of the Durbin–Watson statistics. It is well known that the Wald test for autocorrelation does not always have the most desirable properties in finite samples owing to such problems as the lack of invariance to equivalent formulations of the null hypothesis, local biasedness, and power nonmonotonicity. In Chapter 12, to overcome these problems, Max King and Kim-Leng Goh consider the use of bootstrap methods to find more appropriate critical values and modifications to the asymptotic covariance matrix of the estimates used in the test statistic. In Chapter 13, Jan Magnus studies the sensitivity properties of a "t-type" statistic based on a normal random variable with zero mean and nonscalar covariance matrix. A simple expression for the even moments of this t-type random variable is given, as are the conditions for the moments to exist.

Part IV presents a collection of papers relevant to pretest and biased estimation. In Chapter 14, David Giles considers pretest and Bayes estimation of the normal location parameter with the loss structure given by a "reflected normal" penalty function, which has the particular merit of being bounded. In Chapter 15, Akio Namba and Kazuhiro Ohtani consider a linear regression model with multivariate t errors and derive the finite sample moments and predictive mean squared error of a pretest double k-class estimator of the regression coefficients. Shalabh, in Chapter 16, considers a linear regression model with trended explanatory variable using three different formulations for the trend, viz., linear, quadratic, and exponential, and studies large sample properties of the least squares and Stein-rule estimators. Emphasizing a model involving orthogonality of explanatory variables and the noise component, Ron Mittelhammer and George Judge in Chapter 17 demonstrate a semiparametric empirical likelihood data based information theoretic (ELDBIT) estimator that has finite sample properties superior to those of the traditional competing estimators. The ELDBIT estimator exhibits robustness with respect to ill-conditioning implied by highly correlated covariates and sample outcomes from nonnormal, thicker-tailed sampling processes. Some possible extensions of the ELDBIT formulations have also been outlined.

Time series analysis forms the subject matter of Part V. Judith Giles in Chapter 18 proposes tests for two-step noncausality tests in a trivariate

VAR model when the information set contains variables that are not directly involved in the test. An issue that often arises in the approximation of an ARMA process by a pure AR process is the lack of appraisal of the quality of the approximation. John Galbraith and Victoria Zinde-Walsh address this issue in Chapter 19, emphasizing the Hilbert distance as a measure of the approximation's accuracy. Chapter 20 by Anoop Chaturvedi, Alan Wan, and Guohua Zou adds to the sparse literature on Bayesian inference on dynamic regression models, with allowance for the possible existence of nonnormal errors through the Gram–Charlier distribution. Robust in-sample volatility analysis is the substance of the contribution of Chapter 21, in which Xavier Yew, Michael McAleer, and Shiqing Ling examine the sensitivity of the estimated parameters of the GARCH and asymmetric GARCH models through recursive estimation to determine the optimal window size. In Chapter 22, Koichi Maekawa and Hiroyuki Hisamatsu consider a nonstationary SUR system and investigate the asymptotic distributions of OLS and the restricted and unrestricted SUR estimators. A cointegration test based on the SUR residuals is also proposed.

Part VI comprises five chapters focusing on estimation and inference of econometric models. In Chapter 23, Gordon Fisher and Marcel-Christian Voia consider the estimation of stochastic coefficients regression (SCR) models with missing observations. Among other things, the authors present a new geometric proof of an extended Gauss–Markov theorem. In estimating hazard functions, the negative exponential regression model is commonly used, but previous results on estimators for this model have been mostly asymptotic. Along the lines of their other ongoing research in this area, John Knight and Stephen Satchell, in Chapter 24, derive some exact properties for the log-linear least squares and maximum likelihood estimators for a negative exponential model with a constant and a dummy variable. Minimum variance unbiased estimators are also developed. In Chapter 25, Murray Smith examines various aspects of double-hurdle models, which are used frequently in demand analysis. Smith presents a thorough review of the current state of the art on this subject, and advocates the use of the copula method as the preferred technique for constructing these models. Rick Vinod in Chapter 26 discusses how the popular techniques of generalized linear models and generalized estimating equations in biometrics can be utilized in econometrics in the estimation of panel data models. Indeed, Vinod's paper spells out the crucial importance of the interface between econometrics and other areas of statistics. This section concludes with Chapter 27 in which William Griffiths, Chris Skeels, and Duangkamon Chotikapanich take up the important issue of sample size requirement in the estimation of SUR models. One broad conclusion that can be drawn from this paper is that the usually stated sample size requirements often understate the actual requirement.

The last part includes four chapters focusing on applied econometrics. The panel data model is the substance of Chapter 28, in which Aman Ullah and Kusum Mundra study the so-called immigrants home-link effect on U.S. producer trade flows via a semiparametric estimator which the authors introduce. Human development is an important issue faced by many developing countries. Having been at the forefront of this line of research, Aunurudh Nagar, in Chapter 29, along with Sudip Basu, considers estimation of human development indices and investigates the factors in determining human development. A comprehensive survey of the recent developments of structural auction models is presented in Chapter 30, in which Samita Sareen emphasizes the usefulness of Bayesian methods in the estimation and testing of these models. Market switching models are often used in business cycle research. In Chapter 31, Baldev Raj provides a thorough review of the theoretical knowledge on this subject. Raj's extensive survey includes analysis of the Markov-switching approach and generalizations to a multivariate setup with some empirical results being presented.

Needless to say, in preparing this *Handbook*, we owe a great debt to the authors of the chapters for their marvelous cooperation. Thanks are also due to the authors, who were not only devoted to their task of writing exceedingly high quality papers but had also been willing to sacrifice much time and energy to review other chapters of the volume. In this respect, we would like to thank John Galbraith, David Giles, George Judge, Max King, John Knight, Shiqing Ling, Koichi Maekawa, Jan Magnus, Ron Mittelhammer, Kazuhiro Ohtani, Jeff Racine, Radhey Singh, Chris Skeels, Murray Smith, Rick Vinod, Victoria Zinde-Walsh, and Guohua Zou. Also, Chris Carter (Hong Kong University of Science and Technology), Hikaru Hasegawa (Hokkaido University), Wai-Keong Li (City University of Hong Kong), and Nilanjana Roy (University of Victoria) have refereed several papers in the volume. Acknowledged also is the financial support for visiting appointments for Aman Ullah and Anoop Chaturvedi at the City University of Hong Kong during the summer of 1999 when the idea of bringing together the topics of this *Handbook* was first conceived. We also wish to thank Russell Dekker and Jennifer Paizzi of Marcel Dekker, Inc., for their assistance and patience with us in the process of preparing this *Handbook*, and Carolina Juarez and Alec Chan for secretarial and clerical support.

Aman Ullah
Alan T. K. Wan
Anoop Chaturvedi

Contents

Contents

xiv

Contributors

Ibrahim A. Ahmad   Department of Statistics, University of Central Florida, Orlando, Florida

Barry C. Arnold   Department of Statistics, University of California, Riverside, Riverside, California

Sudip Ranjan Basu   National Institute of Public Finance and Policy, New Delhi, India

Anil K. Bera   Department of Economics, University of Illinois at Urbana–Champaign, Champaign, Illinois

Enrique Castillo   Department of Applied Mathematics and Computational Sciences, Universities of Cantabria and Castilla-La Mancha, Santander, Spain

Anoop Chaturvedi   Department of Statistics, University of Allahabad, Allahabad, India

Duangkaman Chotikapanich   School of Economics and Finance, Curtin University of Technology, Perth, Australia

R. W. Farebrother   Department of Economic Studies, Faculty of Social Studies and Law, Victoria University of Manchester, Manchester, England

A. Fieger   Service Barometer AG, Munich, Germany

Gordon Fisher   Department of Economics, Concordia University, Montreal, Quebec, Canada

John W. Galbraith   Department of Economics, McGill University, Montreal, Quebec, Canada

Aurobindo Ghosh   Department of Economics, University of Illinois at Urbana–Champaign, Champaign, Illinois

Subir Ghosh   Department of Statistics, University of California, Riverside, Riverside, California

David E. A. Giles   Department of Economics, University of Victoria, Victoria, British Columbia, Canada

Judith A. Giles   Department of Economics, University of Victoria, Victoria, British Columbia, Canada

Kim-Leng Goh   Department of Applied Statistics, Faculty of Economics and Administration, University of Malaya, Kuala Lumpur, Malaysia

William E. Griffiths   Department of Economics, University of Melbourne, Melbourne, Australia

Hiroyuki Hisamatsu   Faculty of Economics, Kagawa University, Takamatsu, Japan

George G. Judge   Department of Agricultural Economics, University of California, Berkeley, Berkeley, California

Maxwell L. King   Faculty of Business and Economics, Monash University, Clayton, Victoria, Australia

S.N.U.A. Kirmani   Department of Mathematics, University of Northern Iowa, Cedar Falls, Iowa

John L. Knight   Department of Economics, University of Western Ontario, London, Ontario, Canada

Shiqing Ling   Department of Mathematics, Hong Kong University of Science and Technology, Clear Water Bay, Hong Kong, China

Xuewen Lu Food Research Program, Agriculture and Agri-Food Canada, Guelph, Ontario, Canada

Koichi Maekawa Department of Economics, Hiroshima University, Hiroshima, Japan

Jan R. Magnus Center for Economic Research (CentER), Tilburg, University, Tilburg, The Netherlands

Michael. McAleer Department of Economics, University of Western Australia, Nedlands, Perth, Western Australia, Australia

Ron C. Mittelhammer Department of Statistics and Agricultural Economics, Washington State University, Pullman, Washington

Kusum Mundra Department of Economics, San Diego State University, San Diego, California

A. L. Nagar National Institute of Public Finance and Policy, New Delhi, India

Akio Namba Faculty of Economics, Kobe University, Kobe, Japan

Kazuhiro Ohtani Faculty of Economics, Kobe University, Kobe, Japan

Ahmad Parsian School of Mathematical Sciences, Isfahan University of Technology, Isfahan, Iran

Pawel R. Pordzik Department of Mathematical and Statistical Methods, Agricultural University of Poznań, Poznań, Poland

Jeffrey S. Racine Department of Economics, University of South Florida, Tampa, Florida

Mezbahur Rahman Department of Mathematics and Statistics, Minnesota State University, Mankato, Minnesota

Baldev Raj School of Business and Economics, Wilfrid Laurier University, Waterloo, Ontario, Canada

José María Sarabia Department of Economics, University of Cantabria, Santander, Spain

Samita Sareen Financial Markets Division, Bank of Canada, Ontario, Canada

Stephen E. Satchell Faculty of Economics and Politics, Cambridge University, Cambridge, England

Burkhard Schaffrin   Department of Civil and Environmental Engineering and Geodetic Science, Ohio State University, Columbus, Ohio

Shalabh   Department of Statistics, Panjab University, Chandigarh, India

R. S. Singh   Department of Mathematics and Statistics, University of Guelph, Guelph, Ontario, Canada

Christopher L. Skeels   Department of Statistics and Econometrics, Australian National University, Canberra, Australia

Murray D. Smith   Department of Econometrics and Business Statistics, University of Sydney, Sydney, Australia

Helge Toutenburg   Department of Statistics, Ludwig-Maximilians–Universität München, Munich, Germany

Götz Trenkler   Department of Statistics, University of Dortmund, Dortmund, Germany

Aman Ullah   Department of Economics, University of California, Riverside, Riverside, California

H. D. Vinod   Department of Economics, Fordham University, Bronx, New York

Marcel-Christian Voia   Department of Economics, Concordia University, Montreal, Quebec, Canada

Alan T. K. Wan   Department of Management Sciences, City University of Hong Kong, Hong Kong S.A.R., China

Xavier Chee Hoong Yew   Department of Economics, Trinity College, University of Oxford, Oxford, England

Victoria Zinde-Walsh   Department of Economics, McGill University, Montreal, Quebec, Canada

Guohua Zou   Institute of System Science, Chinese Academy of Sciences, Beijing, China

Selected Publications of Professor Viren K. Srivastava

Books:

1. *Seemingly Unrelated Regression Equation Models: Estimation and Inference* (with D. E. A. Giles), Marcel-Dekker, New York, 1987.
2. *The Econometrics of Disequilibrium Models*, (with B. Rao), Greenwood Press, New York, 1990.

Research papers:

1. On the Estimation of Generalized Linear Probability Model Involving Discrete Random Variables (with A. R. Roy), *Annals of the Institute of Statistical Mathematics*, 20, 1968, 457–467.
2. The Efficiency of Estimating Seemingly Unrelated Regression Equations, *Annals of the Institute of Statistical Mathematics*, 22, 1970, 493.
3. Three-Stage Least-Squares and Generalized Double k-Class Estimators: A Mathematical Relationship, *International Economic Review*, Vol. 12, 1971, 312–316.

4. Disturbance Variance Estimation in Simultaneous Equations by k-Class Method, *Annals of the Institute of Statistical Mathematics*, 23, 1971, 437–449.
5. Disturbance Variance Estimation in Simultaneous Equations when Disturbances Are Small, *Journal of the American Statistical Association*, 67, 1972, 164–168.
6. The Bias of Generalized Double k-Class Estimators, (with A. R. Roy), *Annals of the Institute of Statistical Mathematics*, 24, 1972, 495–508.
7. The Efficiency of an Improved Method of Estimating Seemingly Unrelated Regression Equations, *Journal of Econometrics*, 1, 1973, 341–50.
8. Two-Stage and Three-Stage Least Squares Estimation of Dispersion Matrix of Disturbances in Simultaneous Equations, (with R. Tiwari), *Annals of the Institute of Statistical Mathematics*, 28, 1976, 411–428.
9. Evaluation of Expectation of Product of Stochastic Matrices, (with R. Tiwari), *Scandinavian Journal of Statistics*, 3, 1976, 135–138.
10. Optimality of Least Squares in Seemingly Unrelated Regression Equation Models (with T. D. Dwivedi), *Journal of Econometrics*, 7, 1978, 391–395.
11. Large Sample Approximations in Seemingly Unrelated Regression Equations (with S. Upadhyay), *Annals of the Institute of Statistical Mathematics*, 30, 1978, 89–96.
12. Efficiency of Two Stage and Three Stage Least Square Estimators, *Econometrics*, 46, 1978, 1495–1498.
13. Estimation of Seemingly Unrelated Regression Equations: A Brief Survey (with T. D. Dwivedi), *Journal of Econometrics*, 8, 1979, 15–32.
14. Generalized Two Stage Least Squares Estimators for Structural Equation with Both Fixed and Random Coefficients (with B. Raj and A. Ullah), *International Economic Review*, 21, 1980, 61–65.
15. Estimation of Linear Single-Equation and Simultaneous Equation Model under Stochastic Linear Constraints: An Annotated Bibliography, *International Statistical Review*, 48, 1980, 79–82.
16. Finite Sample Properties of Ridge Estimators (with T. D. Dwivedi and R. L. Hall), *Technometrics*, 22, 1980, 205–212.
17. The Efficiency of Estimating a Random Coefficient Model, (with B. Raj and S. Upadhyaya), *Journal of Econometrics*, 12, 1980, 285–299.
18. A Numerical Comparison of Exact, Large-Sample and Small-Disturbance Approximations of Properties of k-Class Estimators, (with T. D. Dwivedi, M. Belinski and R. Tiwari), *International Economic Review*, 21, 1980, 249–252.

19. Dominance of Double k-Class Estimators in Simultaneous Equations (with B. S. Agnihotri and T. D. Dwivedi), *Annals of the Institute of Statistical Mathematics*, 32, 1980, 387–392.
20. Estimation of the Inverse of Mean (with S. Bhatnagar), *Journal of Statistical Planning and Inferences*, 5, 1981, 329–334.
21. A Note on Moments of k-Class Estimators for Negative k, (with A. K. Srivastava), *Journal of Econometrics*, 21, 1983, 257–260.
22. Estimation of Linear Regression Model with Autocorrelated Disturbances (with A. Ullah, L. Magee and A. K. Srivastava), *Journal of Time Series Analysis*, 4, 1983, 127–135.
23. Properties of Shrinkage Estimators in Linear Regression when Disturbances Are not Normal, (with A. Ullah and R. Chandra), *Journal of Econometrics*, 21, 1983, 389–402.
24. Some Properties of the Distribution of an Operational Ridge Estimator (with A. Chaturvedi), *Metrika*, 30, 1983, 227–237.
25. On the Moments of Ordinary Least Squares and Instrumental Variables Estimators in a General Structural Equation, (with G. H. Hillier and T. W. Kinal), *Econometrics*, 52, 1984, 185–202.
26. Exact Finite Sample Properties of a Pre-Test Estimator in Ridge Regression (with D. E. A. Giles), *Australian Journal of Statistics*, 26, 1984, 323–336.
27. Estimation of a Random Coefficient Model under Linear Stochastic Constraints, (with B. Raj and K. Kumar), *Annals of the Institute of Statistical Mathematics*, 36, 1984, 395–401.
28. Exact Finite Sample Properties of Double k-Class Estimators in Simultaneous Equations (with T. D. Dwivedi), *Journal of Econometrics*, 23, 1984, 263–283.
29. The Sampling Distribution of Shrinkage Estimators and Their F-Ratios in the Regression Models (with A. Ullah and R. A. L. Carter), *Journal of Econometrics*, 25, 1984, 109–122.
30. Properties of Mixed Regression Estimator When Disturbances Are not Necessarily Normal (with R. Chandra), *Journal of Statistical Planning and Inference*, 11, 1985, 15–21.
31. Small-Disturbance Asymptotic Theory for Linear-Calibration Estimators, (with N. Singh), *Technometrics*, 31, 1989, 373–378.
32. Unbiased Estimation of the MSE Matrix of Stein Rule Estimators, Confidence Ellipsoids and Hypothesis Testing (with R. A. L. Carter, M. S. Srivastava and A. Ullah), *Econometric Theory*, 6, 1990, 63–74.
33. An Unbiased Estimator of the Covariance Matrix of the Mixed Regression Estimator, (with D. E. A. Giles), *Journal of the American Statistical Association*, 86, 1991, 441–444.

34. Moments of the Ratio of Quadratic Forms in Non-Normal Variables with Econometric Examples (with Aman Ullah), *Journal of Econometrics*, 62, 1994, 129–141.
35. Efficiency Properties of Feasible Generalized Least Squares Estimators in SURE Models under Non-Normal Disturbances (with Koichi Maekawa), *Journal of Econometrics*, 66, 1995, 99–121.
36. Large Sample Asymptotic Properties of the Double k-Class Estimators in Linear Regression Models (with H. D. Vinod), *Econometric Reviews*, 14, 1995, 75–100.
37. The Coefficient of Determination and Its Adjusted Version in Linear Regression Models (with Anil K. Srivastava and Aman Ullah), *Econometric Reviews*, 14, 1995, 229–240.
38. Moments of the Fuction of Non-Normal Vector with Application Econometric Estimators and Test Statistics (with A. Ullah and N. Roy), *Econometric Reviews*, 14, 1995, 459–471.
39. The Second Order Bias and Mean Squared Error of Non-linear Estimators (with P. Rilstone and A. Ullah), *Journal of Econometrics*, 75, 1996, 369–395.

HANDBOOK OF
APPLIED ECONOMETRICS
AND STATISTICAL INFERENCE

1

Bayesian Inference Using Conditionally Specified Priors

BARRY C. ARNOLD   University of California, Riverside, Riverside, California

ENRIQUE CASTILLO   Universities of Cantabria and Castilla–La Mancha, Santander, Spain

JOSÉ MARÍA SARABIA   University of Cantabria, Santander, Spain

1. INTRODUCTION

Suppose we are given a sample of size n from a normal distribution with known variance and unknown mean μ, and that, on the basis of sample values $x_1, x_2, \ldots, x_n$, we wish to make inference about μ. The Bayesian solution of this problem involves specification of an appropriate representation of our prior beliefs about μ (summarized in a prior density for μ) which will be adjusted by conditioning to obtain a relevant posterior density for μ (the conditional density of μ, given $\underline{X} = \underline{x}$). Proper informative priors are most easily justified but improper (nonintegrable) and noninformative (locally uniform) priors are often acceptable in the analysis and may be necessary when the informed scientist insists on some degree of ignorance about unknown parameters. With a one-dimensional parameter, life for the Bayesian analyst is relatively straightforward. The worst that can happen is that the analyst will need to use numerical integration techniques to normalize and to quantify measures of central tendency of the resulting posterior density.

1

Moving to higher dimensional parameter spaces immediately complicates matters. The "curse of dimensionality" begins to manifest some of its implications even in the bivariate case. The use of conditionally specified priors, as advocated in this chapter, will not in any sense eliminate the "curse" but it will ameliorate some difficulties in, say, two and three dimensions and is even practical for some higher dimensional problems.

Conjugate priors, that is to say, priors which combine analytically with the likelihood to give recognizable and analytical tractable posterior densities, have been and continue to be attractive. More properly, we should perhaps speak of priors with convenient posteriors, for their desirability hinges mostly on the form of the posterior and there is no need to insist on the prior and posterior density being of the same form (the usual definition of conjugacy). It turns out that the conditionally specified priors that are discussed in this chapter are indeed conjugate priors in the classical sense. Not only do they have convenient posteriors but also the posterior densities will be of the same form as the priors. They will prove to be more flexible than the usually recommended conjugate priors for multidimensional parameters, yet still manageable in the sense that simulation of the posteriors is easy to program and implement.

Let us return for a moment to our original problem involving a sample of size n from a normal distribution. This time, however, we will assume that both the mean and the variance are unknown: a classic setting for statistical inference. Already, in this setting, the standard conjugate prior analysis begins to appear confining. There is a generally accepted conjugate prior for (μ, τ) (the mean μ and the precision (reciprocal of the variance) τ). It will be discussed in Section 3.1, where it will be contrasted with a more flexible conditionally conjugate prior. A similar situation exists for samples from a Pareto distribution with unknown inequality and scale parameters. Here too a conditionally conjugate prior will be compared to the usual conjugate priors. Again, increased flexibility at little cost in complexity of analysis will be encountered. In order to discuss these issues, a brief review of conditional specification will be useful. It will be provided in Section 2. In Section 3 conditionally specified priors will be introduced; the normal and Pareto distributions provide representative examples here. Section 4 illustrates application of the conditionally specified prior technique to a number of classical inference problems. In Section 5 we will address the problem of assessing hyperparameters for conditionally specified priors. The closing section (Section 6) touches on the possibility of obtaining even more flexibility by considering mixtures of conditionally specified priors.

2. CONDITIONALLY SPECIFIED DISTRIBUTIONS

In efforts to describe a two-dimensional density function, it is undeniably easier to visualize conditional densities than it is to visualize marginal densities. Consequently, it may be argued that joint densities might best be determined by postulating appropriate behavior for conditional densities. For example, following Bhattacharyya [1], we might specify that the joint density of a random vector (X, Y) have every conditional density of X given $Y = y$ of the normal form (with mean and variance that might depend on y) and, in addition, have every conditional density of Y given $X = x$ of the normal form (with mean and variance which might depend on x). In other words, we seek all bivariate densities with bell-shaped cross sections (where cross sections are taken parallel to the x and y axes). The class of such densities may be represented as

$$f(x, y) = \exp\{(1, x, x^2)A(1, y, y^2)'\}, \ (x, y) \in \mathbb{R}^2 \tag{1}$$

where $A = (a_{ij})_{i,j=0}^2$ is a 3×3 matrix of parameters. Actually, a_{00} is a norming constant, a function of the other a_{ij}s chosen to make the density integrate to 1. The class (1) of densities with normal conditionals includes, of course, classical bivariate normal densities. But it includes other densities; some of which are bimodal and some trimodal!

This normal conditional example is the prototype of conditionally specified models. The more general paradigm is as follows.

Consider an ℓ_1-parameter family of densities on $\mathbb{R}$ with respect to μ_1, a measure on $\mathbb{R}$ (often a Lebesgue or counting measure), denoted by $\{f_1(x; \underline{\theta}) : \underline{\theta} \in \Theta\}$ where $\Theta \subseteq \mathbb{R}^{\ell_1}$. Consider a possible different ℓ_2-parameter family of densities on $\mathbb{R}$ with respect to μ_2 denoted by $\{f_2(y; \underline{\tau}) : \tau \in T\}$ where $T \subseteq \mathbb{R}^{\ell_2}$. We are interested in all possible bivariate distributions which have all conditionals of X given Y in the family f_1 and all conditionals of Y given X in the family f_2. Thus we demand that

$$f_{X|Y}(x|y) = f_1(x; \underline{\theta}(y)), \ \forall x \in S(X), \ y \in S(Y) \tag{2}$$

and

$$f_{Y|X}(y|x) = f_2(y; \underline{\tau}(x)), \ \forall x \in S(X), \ y \in S(Y) \tag{3}$$

Here $S(X)$ and $S(Y)$ denote, respectively, the sets of possible values of X and Y.

In order that (2) and (3) should hold there must exist marginal densities f_X and f_Y for X and Y such that

$$f_Y(y)f_1(x; \underline{\theta}(y)) = f_X(x)f_2(y; \underline{\tau}(x)), \ \ \forall x \in S(X), \ y \in S(Y) \tag{4}$$

To identify the possible bivariate densities with conditionals in the two prescribed families, we will need to solve the functional equation (4). This is not always possible. For some choices of f_1 and f_2 no solution exists except for the trivial solution with independent marginals.

One important class of examples in which the functional equation (4) is readily solvable are those in which the parametric families of densities f_1 and f_2 are both exponential families. In this case the class of all densities with conditionals in given exponential families is readily determined and is itself an exponential family of densities. First, recall the definition of an exponential family.

Definition 1 (Exponential family) *An ℓ_1-parameter family of densities $\{f_1(x; \underline{\theta}) : \underline{\theta} \in \Theta\}$, with respect to μ_1 on $S(X)$, of the form*

$$f_1(x; \underline{\theta}) = r_1(x)\beta_1(\underline{\theta}) \exp\left\{ \sum_{i=1}^{\ell_1} \theta_i q_{1i}(x) \right\} \tag{5}$$

is called an exponential family of distributions.

Here Θ is the natural parameter space and the $q_{1i}(x)$s are assumed to be linearly independent. Frequently, μ_1 is a Lebesgue measure or counting measure and often $S(X)$ is some subset of Euclidean space of finite dimension.

Note that Definition 1 contines to be meaningful if we underline x in equation (5) to emphasize the fact that x can be multidimensional.

In addition to the exponential family (5), we will let $\{f_2(y; \underline{\tau}) : \tau \in T\}$ denote another ℓ_2-parameter exponential family of densities with respect to μ_2 on $S(Y)$, of the form

$$f_2(y; \underline{\tau}) = r_2(y)\beta_2(\underline{\tau}) \exp\left\{ \sum_{j=1}^{\ell_2} \tau_j q_{2j}(y) \right\}, \tag{6}$$

where T is the natural parameter space and, as is customarily done, the $q_{2j}(y)$s are assumed to be linearly independent.

The general class of the bivariate distributions with conditionals in the two exponential families (5) and (6) is provided by the following theorem due to Arnold and Strauss [2].

Theorem 1 (Conditionals in exponential families) *Let $f(x, y)$ be a bivariate density whose conditional densities satisfy*

$$f(x|y) = f_1(x; \underline{\theta}(y)) \tag{7}$$

and

$$f(y|x) = f_2(y; \underline{\tau}(x)) \tag{8}$$

for some function $\underline{\theta}(y)$ and $\underline{\tau}(x)$ where f_1 and f_2 are defined in (5) and (6). It follows that $f(x, y)$ is of the form

$$f(x, y) = r_1(x)r_2(y)\exp\{\underline{q}^{(1)}(x)M\underline{q}^{(2)}(y)'\} \tag{9}$$

in which

$$\underline{q}^{(1)}(x) = (q_{10}(x), q_{11}(x), q_{12}(x), \dots, q_{1\ell_1}(x))$$
$$\underline{q}^{(2)}(y) = (q_{20}(y), q_{21}(y), q_{22}(y), \dots, q_{2\ell_2}(y))$$

where $q_{10}(x) = q_{20}(y) \equiv 1$ and M is a matrix of parameters of appropriate dimensions (i.e., $(\ell_1 + 1) \times (\ell_2 + 1)$) subject to the requirement that

$$\int_{S(X)} \int_{S(Y)} f(x, y) \, d\mu_1(x) \, d\mu_2(y) = 1$$

For convenience we can partition the matrix M as follows:

$$M = \begin{pmatrix} m_{00} & | & m_{01} & \cdots & m_{0\ell_2} \\ -- & + & -- & -- & -- \\ m_{10} & | & & & \\ \cdots & | & & \tilde{M} & \\ m_{\ell_1 0} & | & & & \end{pmatrix} \tag{10}$$

Note that the case of independence is included; it corresponds to the choice $\tilde{M} \equiv 0$.

Note that densities of the form (9) form an exponential family with $(\ell_1 + 1)(\ell_2 + 1) - 1$ parameters (since m_{00} is a normalizing constant determined by the other m_{ij}s).

Bhattacharyya's [1] normal conditionals density can be represented in the form (9) by suitable identification of the q_{ij}s. A second example, which will arise again in our Bayesian analysis of normal data, involves normal-gamma conditionals (see Castillo and Galambos [3]). Thus we, in this case, are interested in all bivariate distributions with $X|Y = y$ having a normal distribution for each y and with $Y|X = x$ having a gamma distribution for each x. These densities will be given by (9) with the following choices for the rs and qs.

$$\left. \begin{array}{l} r_1(x) = 1, r_2(y) = y^{-1}I(y > 0) \\ q_{11}(x) = x, q_{21}(y) = -y \\ q_{12}(x) = x^2, q_{22}(y) = \log y \end{array} \right\} \tag{11}$$

The joint density is then given by

$$f(x, y) = y^{-1} \exp\left\{(1, x, x^2)M\begin{pmatrix} 1 \\ -y \\ \log y \end{pmatrix}\right\} I(x \in \mathbb{R}, y > 0) \tag{12}$$

Certain constraints must be placed on the m_{ij}s, the elements of M, appearing in (12) and more generally in (9) to ensure integrability of the resulting density, i.e., to identify the natural parameter space. However, in a Bayesian context, where improper priors are often acceptable, no constraints need be placed on the m_{ij}s. If the joint density is given by (12), then the specific forms of the conditional distributions are as follows:

$$X|Y = y \sim N(\mu(y), \sigma^2(y)) \tag{13}$$

where

$$\mu(y) = \frac{m_{10} - m_{11}y + m_{12}\log y}{2(-m_{20} + m_{21}y - m_{22}\log y)} \tag{14}$$

$$\sigma^2(y) = \frac{1}{2}(-m_{20} + m_{21}y - m_{22}\log y)^{-1} \tag{15}$$

and

$$Y|X = x \sim \Gamma(m_{02} + m_{12}x + m_{22}x^2, m_{01} + m_{11}x + m_{21}x^2) \tag{16}$$

A typical density with normal-gamma conditionals is shown in Figure 1. It should be remarked that multiple modes can occur (as in the distribution with normal conditionals).

Another example that will be useful in a Bayesian context is one with conditionals that are beta densities (see Arnold and Strauss [2]). This corresponds to the following choices for rs and qs in (9):

$$\left. \begin{aligned} r_1(x) &= [x(1-x)]^{-1}I(0 < x < 1) \\ r_2(y) &= [y(1-y)]^{-1}I(0 < y < 1) \\ q_{11}(x) &= \log x \\ q_{21}(y) &= \log y \\ q_{12}(x) &= \log(1-x) \\ q_{22}(y) &= \log(1-y) \end{aligned} \right\} \tag{17}$$

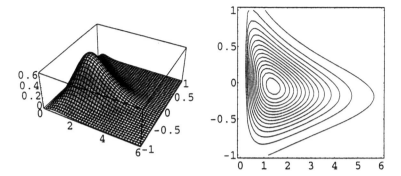

Figure 1. Example of a normal-gamma conditionals distribution with $m_{02} = -1, m_{01} = 1, m_{10} = 1.5, m_{12} = 2, m_{11} = 1, m_{20} = 3, m_{22} = m_{21} = 0$, showing (left) the probability density function and (right) the contour plot.

This yields a joint density of the form

$$
\begin{aligned}
f(x, y) = [x(1 - x)y(1 - y)]^{-1} \exp\{ & m_{11} \log x \log y + m_{12} \log x \log(1 - y) \\
+ \ & m_{21} \log(1 - x) \log y + m_{22} \log(1 - x) \log(1 - y) \\
+ \ & m_{10} \log x + m_{20} \log(1 - x) \\
+ \ & m_{01} \log y + m_{02} \log(1 - y) \\
+ \ & m_{00} \} I(0 < x, y < 1)
\end{aligned}
\tag{18}
$$

In this case the constraints on the m_{ij}s to ensure integrability of (17) are quite simple:

$$
\left.
\begin{aligned}
m_{10} > 0, \ m_{20} > 0, \ m_{01} > 0, \ m_{02} > 0 \\
m_{11} \le 0, \ m_{12} \le 0, \ m_{21} \le 0, \ m_{22} \le 0
\end{aligned}
\right\}
\tag{19}
$$

There are some non-exponential family cases in which the functional equation (4) can be solved. For example, it is possible to identify all joint densities with zero median Cauchy conditionals (see Arnold et al. [4]). They are of the form

$$
f(x, y) \propto (m_{00} + m_{10}x^2 + m_{01}y^2 + m_{11}x^2 y^2)^{-1}
\tag{20}
$$

Extensions to higher dimensions are often possible (see Arnold et al. [5]).

Assume that $\underline{X}$ is a k-dimensional random vector with coordinates $(X_1, X_2, \ldots, X_k)$. For each coordinate random variable X_i of $\underline{X}$ we define the vector $\underline{X}_{(i)}$ to be the $(k - 1)$-dimensional vector obtained from $\underline{X}$ by deleting X_i. We use the same convention for real vectors, i.e., $\underline{x}_{(i)}$ is obtained from $\underline{x}$

by deleting x_i. We concentrate on conditional specifications of the form "X_i given $\underline{X}_{(i)}$."

Consider k parametric families of densities on $\mathbb{R}$ defined by

$$\{f_i(x; \underline{\theta}_{(i)}) : \underline{\theta}_{(i)} \in \Theta_i\}, \quad i = 1, 2, \ldots, k \tag{21}$$

where $\underline{\theta}_{(i)}$ is of dimension ℓ_i and where the ith density is understood as being with respect to the measure μ_i. We are interested in k-dimensional densities that have all their conditionals in the families (21). Consequently, we require that for certain functions $\underline{\theta}_{(i)}$ we have, for $i = 1, 2, \ldots, k$,

$$f_{X_i|\underline{X}_{(i)}}(x_i|\underline{x}_{(i)}) = f_i(x_i; \underline{\theta}_{(i)}(\underline{x}_{(i)})) \tag{22}$$

If these equations are to hold, then there must exist marginal densities for the $\underline{X}_{(i)}$s such that

$$\begin{aligned} f_{\underline{X}_{(1)}}(\underline{x}_{(1)})f_1(x_1; \underline{\theta}_{(1)}(\underline{x}_{(1)})) &= f_{\underline{X}_{(2)}}(\underline{x}_{(2)})f_2(x_2; \underline{\theta}_{(2)}(\underline{x}_{(2)})) \\ \cdots &= f_{\underline{X}_{(k)}}(\underline{x}_{(k)})f_k(x_k; \underline{\theta}_{(k)}(\underline{x}_{(k)})) \end{aligned} \right\} \tag{23}$$

Sometimes the array of functional equations (23) will be solvable. Here, as in two dimensions, an assumption that the families of densities (21) are exponential families will allow for straightforward solution.

Suppose that the k families of densities $f_1, f_2, \ldots, f_k$ in (21) are $\ell_1, \ell_2, \ldots, \ell_k$ parameter exponential families of the form

$$f_i(t; \underline{\theta}_{(i)}) = r_i(t) \exp\left\{\sum_{j=0}^{\ell_i} \theta_{ij} q_{ij}(t)\right\}, \quad i = 1, 2, \ldots, k \tag{24}$$

(here θ_{ij} denotes the jth coordinate of $\underline{\theta}_{(i)}$ and, by convention, $q_{i0}(t) \equiv 1, \forall i$). We wish to identify all joint distributions for $\underline{X}$ such that (22) holds with the f_is defined as in (24) (i.e., with conditionals in the prescribed exponential families).

By taking logarithms in (23) and differencing with respect to $x_1, x_2, \ldots, x_k$ we may conclude that the joint density must be of the following form:

$$f_{\underline{X}}(\underline{x}) = \left[\prod_{i=1}^{k} r_i(x_i)\right] \exp\left\{\sum_{i_1=0}^{\ell_1} \sum_{i_2=0}^{\ell_2} \cdots \sum_{i_k=0}^{\ell_k} m_{i_1, i_2, \ldots, i_k} \left[\prod_{j=1}^{k} q_{ii_j}(x_j)\right]\right\} \tag{25}$$

The dimension of the parameter space of this exponential family is $\left[\prod_{i=1}^{k}(\ell_i + 1)\right] - 1$ since $m_{00\ldots0}$ is a function of the other m's, chosen to ensure that the density integrates to 1. Determination of the natural parameter space may be very difficult and we may well be enthusiastic about accepting non-integrable versions of (25) to avoid the necessity of identifying which ms correspond to proper densities [4].

3. CONDITIONALLY SPECIFIED PRIORS

Suppose that we have data, $\underline{X}$, whose distribution is governed by the family of densities $\{f(\underline{x}; \underline{\theta}) : \underline{\theta} \in \Theta\}$ where Θ is k-dimensional ($k > 1$). Typically, the informative prior used in this analysis is a member of a convenient family of priors (often chosen to be a conjugate family).

Definition 2 (Conjugate family) *A family $\mathcal{F}$ of priors for $\underline{\theta}$ is said to be a conjugate family if any member of $\mathcal{F}$, when combined with the likelihood of the data, leads to a posterior density which is again a member of $\mathcal{F}$.*

One approach to constructing a family of conjugate priors is to consider the possible posterior densities for $\underline{\theta}$ corresponding to all possible samples of all possible sizes from the given distribution beginning with a locally uniform prior on Θ. More often than not, a parametric extension of this class is usually considered. For example, suppose that $X_1, \ldots X_n$ are independent identically distributed random variables with possible values 1, 2, 3. Suppose that

$$P(X_i = 1) = \theta_1, \quad P(X_i = 2) = \theta_2, \quad P(X_i = 3) = 1 - \theta_1 - \theta_2$$

Beginning with a locally uniform prior for (θ_1, θ_2) and considering all possible samples of all possible sizes leads to a family of Dirichlet $(\alpha_1, \alpha_2, \alpha_3)$ posteriors where $\alpha_1, \alpha_2, \alpha_3$ are positive integers. This could be used as a conjugate prior family but it is more natural to use the augmented family of Dirichlet densities with $\alpha_1, \alpha_2, \alpha_3 \in \mathbb{R}^+$.

If we apply this approach to i.i.d. normal random variables with unknown mean μ and unknown precision (the reciprocal of the variance) τ, we are led to a conjugate prior family of the following form (see, e.g., deGroot [6]).

$$f(\mu, \tau) \propto \exp\left(a \log \tau + b\tau + c\mu\tau + d\mu^2\tau\right) \tag{26}$$

where $a > 0, b < 0, c \in \mathbb{R}, d < 0$. Densities such as (26) have a gamma marginal density for τ and a conditional distribution for μ, given τ that is normal with precision depending on τ.

It is not clear why we should force our prior to accommodate to the particular kind of dependence exhibited by (26). In this case, if τ were known, it would be natural to use a normal conjugate prior for μ. If μ were known, the appropriate conjugate prior for τ would be a gamma density. Citing ease of assessment arguments, Arnold and Press [7] advocated use of independent normal and gamma priors for μ and τ (when both are unknown). They thus proposed prior densities of the form

$$f(\mu, \tau) \propto \exp\left(a \log \tau + b\tau + c\mu + d\mu^2\right) \tag{27}$$

where $a > 0, b < 0, c \in \mathbb{R}, d < 0$.

The family (27) is not a conjugate family and in addition it, like (26), involves a specific assumption about the (lack of) dependence between prior beliefs about μ and τ.

It will be noted that densities (26) and (27), both have normal and gamma conditionals. Indeed both can be embedded in the full class of normal-gamma conditionals densities introduced earlier (see equation (12)). This family is a conjugate prior family. It is richer than (26) and (27) and it can be argued that it provides us with desirable additional flexibility at little cost since the resulting posterior densities (on which our inferential decisions will be made) will continue to have normal and gamma conditionals and thus are not difficult to deal with. This is a prototypical example of what we call a conditionally specified prior. Some practical examples, together with their corresponding hyperparameter assessments, are given in Arnold et al. [8].

If the possible densities for $\underline{X}$ are given by $\{f(\underline{x}; \underline{\theta} \in \Theta\}$ where $\Theta \subset \mathbb{R}^k, k > 1$, then specification of a joint prior for $\underline{\theta}$ involves describing a k-dimensional density. We argued in Section 2 that densities are most easily visualized in terms of conditional densities. In order to ascertain an appropriate prior density for $\underline{\theta}$ it would then seem appropriate to question the informed scientific expert regarding prior beliefs about θ_1 given specific values of the other θ_is. Then, we would ask about prior beliefs about θ_2 given specific values of $\underline{\theta}_{(2)}$ (the other θ_is), etc. One clear advantage of this approach is that we are only asking about univariate distributions, which are much easier to visualize than multivariate distributions.

Often we can still take advantage of conjugacy concepts in our effort to pin down prior beliefs using a conditional approach. Suppose that for each coordinate θ_i of $\underline{\theta}$, if the other θ_is (i.e. $\underline{\theta}_{(i)}$) were known, a convenient conjugate prior family, say $f_i(\theta_i|\underline{\alpha}), \underline{\alpha}_i \in A_i$, is available. In this notation the $\underline{\alpha}_i$s are "hyperparameters" of the conjugate prior families. If this is the case, we propose to use, as a conjugate prior family for $\underline{\theta}$, all densities which have the property that, for each i, the conditional density of θ_i, given $\underline{\theta}_i$, belongs to the family f_i. It is not difficult to verify that this *is* a conjugate prior family so that the posterior densities will also have conditionals in the prescribed families.

3.1 Exponential Families

If, in the above scenarios, each of the prior families f_i (the prior for θ_i, given $\underline{\theta}_{(i)}$) is an ℓ_i-parameter exponential family, then, from Theorem 1, the resulting conditionally conjugate prior family will itself be an exponential family.

It will have a large number of hyperparameters (namely $\prod_{i=1}^{k}(\ell_i + 1) - 1$), providing flexibility for matching informed or vague prior beliefs about $\underline{\theta}$.

Formally, if for each i a natural conjugate prior for θ_i (assuming $\underline{\theta}_{(i)}$ is known) is an ℓ_i-parameter exponential family of the form

$$f_i(\theta_i) \propto r_i(\theta_i) \exp\left[\sum_{j=1}^{\ell_i} \eta_{ij} T_{ij}(\theta_i) \right] \tag{28}$$

then a convenient family of priors for the full parameter vector $\underline{\theta}$ will be of the form

$$f(\underline{\theta}) = \left[\prod_{i=1}^{k} r_i(\theta_i) \right] \exp\left\{ \sum_{j_1=0}^{\ell_1} \sum_{j_2=0}^{\ell_2} \cdots \sum_{j_k=0}^{\ell_k} m_{j_1 j_2 \cdots j_k} \left[\prod_{i=1}^{k} T_{i,j_i}(\theta_i) \right] \right\} \tag{29}$$

where, for notational convenience, we have introduced the constant functions $T_{i0}(\theta_i) = 1$, $i = 1, 2, \ldots, k$.

This family of densities includes all densities for $\underline{\theta}$ with conditionals (for θ_i given $\theta_{(i)}$ for each i) in the given exponential families (28). The proof is based on a simple extension of Theorem 1 to the n-dimensional case.

Because each f_i is a conjugate prior for θ_i (given $\underline{\theta}_{(i)}$), it follows that $f(\underline{\theta})$, given by (29), is a conjugate prior family and that all posterior densities have the same structure as the prior. In other words, a priori and a posteriori, θ_i given $\underline{\theta}_{(i)}$ will have a density of the form (28) for each i. They provide particularly attractive examples of conditionally conjugate (conditionally specified) priors.

As we shall see, it is usually the case that the posterior hyperparameters are related to the prior hyperparameters in a simple way. Simulation of realizations from the posterior distribution corresponding to a conditionally conjugate prior will be quite straightforward using rejection or Markov Chain Monte Carlo (MCMC) simulation methods (see Tanner [9]) and, in particular, the Gibbs sampler algorithm since the simulation will only involve one-dimensional conditional distributions which are themselves exponential families. Alternatively, it is often possible to use some kind of rejection algorithm to simulate realizations from a density of the form (29).

Note that the family (29) includes the "natural" conjugate prior family (obtained by considering all possible posteriors corresponding to all possible samples, beginning with a locally uniform prior). In addition, (29) will include priors with independent marginals for the θ_is, with the density for θ_i selected from the exponential family (28), for each i. Both classes of these more commonly encountered priors can be recognized as subclasses of (29), obtained by setting some of the hyperparameters (the $m_{j_1, j_2, \ldots, j_k}$s) equal to zero.

Consider again the case in which our data consist of n independent identically distributed random variables each having a normal distribution with mean μ and precision τ (the reciprocal of the variance). The corresponding likelihood has the form

$$f_{\underline{X}}(\underline{x}; \mu, \tau) = \frac{\tau^{n/2}}{(2\pi)^{n/2}} \exp\left[-\frac{\tau}{2}\sum_{i=1}^{n}(x_i - \mu)^2\right] \qquad (30)$$

If τ is known, the conjugate prior family for μ is the normal family. If μ is known, the conjugate prior family for τ is the gamma family. We are then led to consider, as our conditionally conjugate family of prior densities for (μ, τ), the set of densities with normal-gamma conditionals given above in (12). We will rewrite this in an equivalent but more convenient form as follows:

$$f(\mu, \tau) \propto \exp\left[m_{10}\mu + m_{20}\mu^2 + m_{12}\mu \log \tau + m_{22}\mu^2 \log \tau\right]$$
$$\times \exp\left[m_{01}\tau + m_{02}\log \tau + m_{11}\mu\tau + m_{21}\mu^2\tau\right] \qquad (31)$$

For such a density we have:

1. The conditional density of μ given τ is normal with mean

    $$E(\mu|\tau) = \frac{-(m_{10} + m_{11}\tau + m_{12}\log \tau)}{2(m_{20} + m_{21}\tau + m_{22}\log \tau)} \qquad (32)$$

    and precision

    $$1/\text{var}(\mu|\tau) = -2(m_{20} + m_{21}\tau + m_{22}\log \tau) \qquad (33)$$

2. The conditional density of τ given μ is gamma with shape parameter $\alpha(\mu)$ and intensity parameter $\lambda(\mu)$, i.e.,

    $$f(\tau|\mu) \propto \tau^{\alpha(\mu)-1} e^{-\lambda(\mu)\tau} \qquad (34)$$

    with mean and variance

    $$E(\tau|\mu) = \frac{1 + m_{02} + m_{12}\mu + m_{22}\mu^2}{-(m_{01} + m_{11}\mu + m_{21}\mu^2)} \qquad (35)$$

    $$\text{var}(\tau|\mu) = \frac{1 + m_{02} + m_{12}\mu + m_{22}\mu^2}{(m_{01} + m_{11}\mu + m_{21}\mu^2)^2} \qquad (36)$$

In order to have a proper density, certain constraints must be placed on the m_{ij}s in (31) (see for example Castillo and Galambos [10]). However, if we are willing to accept improper priors we can allow each of them to range over $\mathbb{R}$.

In order to easily characterize the posterior density which will arise when a prior of the form (31) is combined with the likelihood (30), it is convenient to rewrite the likelihood as follows:

$$f_{\underline{X}}(x; \mu, \tau) = (2\pi)^{-n/2} \exp\left[\frac{n}{2}\log\tau - \frac{\sum_{i=1}^{n} x_i^2}{2}\tau + \sum_{i=1}^{n} x_i\mu\tau - \frac{n}{2}\mu^2\tau\right] \quad (37)$$

A prior of the form (31) combined with the likelihood (37) will yield a posterior density again in the family (31) with prior and posterior hyperparameters related as shown in Table 1. From Table 1 we may observe that four of the hyperparameters are unaffected by the data. They are the four hyperparameters appearing in the first factor in (31). Their influence on the prior is eventually "swamped" by the data but, by adopting the conditionally conjugate prior family, we do not force them arbitrarily to be zero as would be done if we were to use the "natural" conjugate prior (26).

Reiterating, the choice $m_{00} = m_{20} = m_{12} = m_{22} = 0$ yields the natural conjugate prior. The choice $m_{11} = m_{12} = m_{21} = m_{22} = 0$ yields priors with independent normal and gamma marginals. Thus both of the commonly proposed prior families are subsumed by (31) and in all cases we end up with a posterior with normal-gamma conditionals.

3.2 Non-exponential Families

Exponential families of priors play a very prominent role in Bayesian statistical analysis. However, there are interesting cases which fall outside the

Table 1. Adjustments in the hyperparameters in the prior family (31), combined with likelihood (37)

Parameter	Prior value	Posterior value
m_{10}	m_{10}^*	m_{10}^*
m_{20}	m_{20}^*	m_{20}^*
m_{01}	m_{01}^*	$m_{01}^* - \frac{1}{2}\sum_{i=1}^{n} x_i^2$
m_{02}	m_{02}^*	$m_{02}^* + n/2$
m_{11}	m_{11}^*	$m_{11}^* + \sum_{i=1}^{n} x_i$
m_{12}	m_{12}^*	m_{12}^*
m_{21}	m_{21}^*	$m_{21}^* - n/2$
m_{22}	m_{22}^*	m_{22}^*

exponential family framework. We will present one such example in this section. Each one must be dealt with on a case-by-case basis because there will be no analogous theorem in non-exponential family cases that is a parallel to Theorem 1 (which allowed us to clearly identify the joint densities with specified conditional densities).

Our example involves classical Pareto data. The data take the form of a sample of size n from a classical Pareto distribution with inequality parameter α and precision parameter (the reciprocal of the scale parameter) τ. The likelihood is then

$$f_{\underline{X}}(\underline{x}; \alpha, \tau) = \prod_{i=1}^{n} \tau\alpha(\tau x_i)^{-(\alpha+1)} I(\tau x_i > 1)$$

$$= \alpha^n \tau^{-n\alpha} \left(\prod_{i=1}^{n} x_i\right)^{-(\alpha+1)} I(\tau x_{1:n} > 1) \qquad (38)$$

which can be conveniently rewritten in the form

$$f_{\underline{X}}(\underline{x}; \alpha, \tau) = \exp\left[n \log \alpha - n\alpha \log \tau - \left(\sum_{i=1}^{n} \log x_i\right)(\alpha + 1)\right] I(\tau x_{1:n} > 1)$$

$$\qquad (39)$$

If τ were known, the natural conjugate prior family for α would be the gamma family. If α were known, the natural conjugate prior family for τ would be the Pareto family. We are then led to consider the conditionally conjugate prior family which will include the joint densities for (α, τ) with gamma and Pareto conditionals. It is not difficult to verify that this is a six (hyper) parameter family of priors of the form

$$f(\alpha, \tau) \propto \exp[m_{01} \log \tau + m_{21} \log \alpha \log \tau]$$
$$\times \exp[m_{10}\alpha + m_{20} \log \alpha + m_{11}\alpha \log \tau] I(\tau c > 1) \qquad (40)$$

It will be obvious that this is not an exponential family of priors. The support depends on one of the hyperparameters. In (40), the hyperparameters in the first factor are those which are unchanged in the posterior. The hyperparameters in the second factor are the ones that are affected by the data. If a density is of the form (40) is used as a prior in conjunction with the likelihood (39), it is evident that the resulting posterior density is again in the family (40). The prior and posterior hyperparameters are related in the manner shown in Table 2.

The density (40), having gamma and Pareto conditionals, is readily simulated using a Gibbs sampler approach. The family (40) includes the two most frequently suggested families of joint priors for (α, τ), namely:

Table 2. Adjustments in the parameters in the prior (40) when combined with the likelihood (39)

Parameter	Prior value	Posterior value
m_{10}	m_{10}^*	$m_{10}^* - \sum_{i=1}^n \log x_i$
m_{20}	m_{20}^*	$m_{20}^* + n$
m_{01}	m_{01}^*	m_{01}^*
m_{11}	m_{11}^*	$m_{11}^* - n$
m_{21}	m_{21}^*	m_{21}^*
c	c^*	$\min(x_{1:n}, c^*)$

1 *The "classical" conjugate prior family.* This was introduced by Lwin [11]. It corresponded to the case in which m_{01} and m_{21} were both arbitrarily set equal to 0.
2 *The independent gamma and Pareto priors.* These were suggested by Arnold and Press [7] and correspond to the choice $m_{11} = m_{21} = 0$.

4. SOME CLASSICAL PROBLEMS

4.1 The Behrens–Fisher Problem

In this setting we wish to compare the means of two or more normal populations with unknown and possibly different precisions. Thus our data consists of k independent samples from normal populations with

$$X_{ij} \sim N(\mu_i, \tau_i), \qquad i = 1, 2, \ldots, k; \; j = 1, 2, \ldots, n_i \qquad (41)$$

Our interest is in the values of the μ_is. The τ_is (the precisions) are here classic examples of (particularly pernicious) nuisance parameters. Our likelihood will involve $2k$ parameters. If all the parameters save μ_j are known, then a natural conjugate prior for μ_j would be a normal density. If all the parameters save τ_j are known, then a natural conjugate prior for τ_j will be a gamma density. Consequently, the general conditionally conjugate prior for (μ, τ) will be one in which the conditional density of each μ_j, given the other $2k - 1$ parameters, is normal and the conditional density of τ_j, given the other $2k - 1$ parameters, is of the gamma type. The resulting family of joint priors in then given by

$$f(\underline{\mu}, \underline{\tau}) = (\tau_1 \tau_2 \ldots \tau_k)^{-1} \exp \left\{ \sum_{j_1=0}^{2} \sum_{j_2=0}^{2} \cdots \sum_{j_k=0}^{2} \sum_{j'_1=0}^{2} \sum_{j'_2=0}^{2} \cdots \right.$$

$$\left. \cdots \sum_{j'_k=0}^{2} \left[m_{\underline{j}\underline{j}'} \prod_{i=1}^{k} q_{ij_i}(\mu_i) \prod_{i'=1}^{k} q'_{i'j'_{i'}}(\tau_{i'}) \right] \right\} \tag{42}$$

where

$q_{i0}(\mu_i) = 1,$

$q_{i1}(\mu_i) = \mu_i,$

$q_{i2}(\mu_i) = \mu_i^2,$

$q'_{i'0}(\tau_{i'}) = 1,$

$q'_{i'1}(\tau_{i'}) = -\tau_{i'},$

$q'_{i'2}(\tau_{i'}) = \log \tau_{i'}.$

There are thus $3^{2k} - 1$ hyperparameters. Many of these will be unaffected by the data. The traditional Bayesian prior is a conjugate prior in which only the $4k$ hyperparameters that are affected by the data are given non-zero values. An easily assessed joint prior in the family (42) would be one in which the μs are taken to be independent of the τs and independent normal priors are used for each μ_j and independent gamma priors are used for the τ_js. This kind of prior will also have $4k$ of the hyperparameters in (42) not equal to zero.

Perhaps the most commonly used joint prior in this setting is one in which the μ_js are assumed to have independent locally uniform densities and, independent of the μ_js, the τ_js (or their logarithms) are assumed to have independent locally uniform priors. All three of these types of prior will lead to posterior densities in the family (42). We can then use a Gibbs sampler algorithm to generate posterior realizations of $(\underline{\mu}, \underline{\tau})$. The approximate posterior distribution of $\sum_{i=1}^{k}(\mu_i - \hat{\mu})^2$ can be perused to identify evidence for differences among the μ_is. A specific example of this program is described in Section 5.1 below.

4.2 2 × 2 Contingency Tables

In comparing two medical treatments, we may submit n_i subject to treatment i, $i = 1, 2$ and observe the number of successes (survival to the end of the observation period) for each treatment. Thus our data consists of two independent random variables (X_1, X_2) where $X_i \sim binomial(n_i, p_i)$.

The odds ratio

$$\Psi(\underline{p}) = \frac{p_1(1-p_2)}{p_2(1-p_1)} \tag{43}$$

is of interest in this setting. A natural conjugate prior for p_1 (if p_2 is known) is a beta distribution. Analogously, a beta prior is natural for p_2 (if p_1 is known). The corresponding conditionally conjugate prior for (p_1, p_2) will have beta conditionals (cf. equation (18)) and is given by

$$\begin{aligned}
f(p_1, p_2) = {} & [p_1(1-p_1)p_2(1-p_2)]^{-1} \\
& \times \exp[m_{11} \log p_1 \log p_2 + m_{12} \log p_1 \log(1-p_2) \\
& + m_{21} \log(1-p_1) \log p_2 + m_{22} \log(1-p_1) \log(1-p_2) \\
& + m_{10} \log p_1 + m_{20} \log(1-p_1) \\
& + m_{01} \log p_2 + m_{02} \log(1-p_2) + m_{00}] \\
& \times I(0 < p_1 < 1) I(0 < p_2 < 1)
\end{aligned} \tag{44}$$

When such a prior is combined with the likelihood corresponding to the two independent binomial X_is, the posterior density is again in the family (44). Only some of the hyperparameters are affected by the data.

The usual prior in this situation involves independent beta priors for p_1 and p_2. Priors of this form are of course included as special cases in (44) but it is quite reasonable to expect non-independent prior beliefs about the efficacy of the two treatment regimes. Observe that simulated realizations from a posterior of the form (44) are readily generated using a Gibbs sampler algorithm (with beta conditionals). This permits ready simulation of the posterior distribution of the parametric function of interest (the odds ratio (43)).

4.3 Regression

The conditionally specified prior approach can also be used in other classical situations. We will describe an example involving simple linear regression, but analogous ideas can be developed in more complex settings.

Assume that we have n independent random variables $X_1. X_2, \ldots, X_n$ whose marginal distributions follow a linear regression model. Thus

$$X_i \sim N(\alpha + \beta t_i, \sigma^2), \quad i = 1, , 2, \ldots, n \tag{45}$$

where the t_is are known quantities and the parameters α, β, and σ^2 are unknown. Here $\alpha \in \mathbb{R}$, $\beta \in \mathbb{R}$ and $\sigma^2 \in \mathbb{R}^+$. Often α and β are the parameters of interest whereas σ^2 is a nuisance parameter. As we have done in previous sections, we reparameterize in terms of precision $\tau (= 1/\sigma^2)$. If β and τ were known, we would use a normal prior for α. If α and τ were known, a normal prior for β would be used. If α and β were known, a routinely used prior for τ would be a gamma distribution. Our conditional specification route would then lead to a joint prior with normal, normal, and gamma conditionals. Thus we would use

$$
\begin{aligned}
f(\alpha, \beta, \tau) \propto \exp\{ & m_{100}\alpha + m_{200}\alpha^2 + m_{010}\beta + m_{020}\beta^2 + m_{110}\alpha\beta \\
& + m_{120}\alpha\beta^2 + m_{210}\alpha^2\beta + m_{102}\alpha \log \tau + m_{012}\beta \log \tau \\
& + m_{220}\alpha^2\beta^2 + m_{202}\alpha^2 \log \tau + m_{022}\beta^2 \log \tau + m_{112}\alpha\beta \log \tau \\
& - m_{121}\alpha\beta^2\tau + m_{122}\alpha\beta^2 \log \tau - m_{211}\alpha^2\beta\tau \\
& + m_{212}\alpha^2\beta \log \tau + m_{222}\alpha^2\beta^2 \log \tau \} \\
\times \exp\{ & -m_{001}\tau + m_{002} \log \tau - m_{101}\alpha\tau - m_{011}\beta\tau - m_{201}\alpha^2\tau \\
& - m_{021}\beta^2\tau - m_{111}\alpha\beta\tau \}
\end{aligned}
\tag{46}
$$

The two factors on the right-hand side of (46) involve, respectively, hyperparameters which are not and are affected by the data. The seven hyperparameters that are affected by the data are changed in value as displayed in Table 3.

The classical Bayesian approach would set all hyperparameters in the first factor of (46) equal to zero. This 7-parameter conjugate prior might be adequate but it clearly lacks flexibility when compared with the full prior

Table 3. Adjustments in the parameters in the prior family (46), combined with the likelihood corresponding to the model (45)

Parameter	Prior value	Posterior value
m_{001}	m_{001}^*	$m_{001}^* + n/2$
m_{002}	m_{002}^*	$m_{002}^* + \dfrac{\sum_{i=1}^{n} x_i^2}{2}$
m_{011}	m_{011}^*	$m_{011}^* - \sum_{i=1}^{n} x_i t_i$
m_{021}	m_{021}^*	$m_{021}^* + \dfrac{\sum_{i=1}^{n} t_i^2}{2}$
m_{101}	m_{101}^*	$m_{101}^* - \sum_{i=1}^{n} x_i$
m_{111}	m_{111}^*	$m_{111}^* + \sum_{i=1}^{n} t_i$
m_{201}	m_{201}^*	$m_{201}^* + n/2$

family (46). It is not easy to justify the dependence structure that is implicitly assumed when using such a restrictive prior.

Another possibility would involve independent priors for α, β, and τ. Whether we pick a prior in the full family (46) or from some subfamily, we will still be able to use a simple Gibbs sampler approach to simulating realizations from the posterior density with its normal, normal, and gamma conditionals.

5. HYPERPARAMETER ASSESSMENT STRATEGIES

If we have agreed to use a conditionally specified prior such as (29), (40), etc., we will be faced with the problem of selecting suitable values of the hyperparameters to match as closely as possible the prior beliefs of our informed expert who is supplying the a priori information. It must be emphasized that use of conditionally specified priors (as is the case with use of other convenient flexible families of priors) does not imply that we believe that our expert's beliefs will precisely match some distribution in the conditionally specified family. What we hope to be the case is that the conditionally specified family will be flexible enough to contain a member which will approximate the informed expert's belief quite adequately.

In order to select a conditionally specified prior to represent our expert's prior beliefs about a multidimensional parameter $\underline{\theta}$, it will be necessary to elicit quite a bit of information.

The information provided by a human expert will typically be inconsistent. That is to say, there probably will not exist *any* distribution which matches exactly all of the many prior pieces of probabilistic information provided by the expert. What we will try to do is to find a conditionally specified prior that is, in some sense, least at variance with the given information.

There will be some arbitrariness in how we measure such discrepancies but we take comfort in the fact that eventually the data will outweigh any unwarranted prior assumptions.

Our knowledge of the one-dimensional conditional distributions involved in our conditionally specified prior will usually allow us to compute a variety of conditional moments and percentiles explicitly as functions of the hyperparameters in the conditionally specified prior. Armed with this information, we will then elicit from our informed expert the subjective evaluations of the true prior values of a selection of conditional moments and percentiles. We will usually ask for more values of conditional moments and percentiles than there are hyperparameters in the model. We reiterate that we don't expect to find a choice of hyperparameters that will match the expert's elicited moments and percentiles exactly (they probably won't

even be consistent). However we do plan to choose values of the hyperparameters to make the conditional moments and percentiles of the prior agree
as much as possible with the information provided by the expert.

We can illustrate this technique by returning to the normal example
discussed in Section 3.1. In that example we had n i.i.d. observations with
a common normal density with mean μ and precision τ. The conditionally
specified prior is of the form (31) with conditional means and variances in
terms of the hyperparameters explicitly available in equations (32)–(36).

For several different values of τ, say $\tau_1, \tau_2, \ldots, \tau_m$, we ask the informed
expert to provide his or her informed best guesses for conditional means and
variances of μ given τ. Suppose that the following information is provided
(the subscript A denotes assessed value)

$$E_A(\mu|\tau = \tau_i) = \xi_i, i = 1, 2, \ldots, m \tag{47}$$

$$var_A(\mu|\tau = \tau_i) = \eta_i, i = 1, 2, \ldots, m \tag{48}$$

Next, for several different values of μ, say $\mu_1, \mu_2, \ldots, \mu_\ell$, assessed values
are provided for conditional moments of τ given μ:

$$E_A(\tau|\mu = \mu_j) = \Psi_j, j = 1, 2, \ldots, \ell \tag{49}$$

$$var_A(\tau|\mu = \mu_j) = \xi_j, j = 1, 2, \ldots, \ell. \tag{50}$$

One approach to selecting appropriate values of the hyperparameters (to
make the assessed values (47)–(50) as close as possible to the actual conditional moments provided by (32)–(36)), is to set up an objective function of
the form

$$D(\underline{m}) = \sum_{i=1}^{m}(E(\mu|\tau = \tau_i) - \xi_i)^2$$

$$+ \sum_{i=1}^{m}(var(\mu|\tau = \tau_i) - \eta_i)^2$$

$$+ \sum_{j=1}^{\ell}(E(\tau|\mu = \mu_j) - \Psi_j)^2$$

$$+ \sum_{j=1}^{\ell}(var(\tau|\mu = \mu_j) - \xi_j)^2 \tag{51}$$

(where the conditional moments are given in (32)–(36)) and, using a reliable
optimization program, choose values of $\underline{m}$ to minimize $D(\underline{m})$ in (51). The
objective function $D(\underline{m})$ is admittedly somewhat arbitrary and a refined
version might well involve some differential weighting of the terms on the

right-hand side of (51). However, some version of this approach can be used for many of the assessment problems that will be encountered in the application of conditionally specified priors.

In the particular example at hand, a simpler alternative is available. By referring to equations (32)–(36) we can verify that if the assessed moments (47)–(50) are approximately equal to the corresponding conditional moments written as functions of $\underline{m}$, then the following array of approximate linear relations should hold:

$$\xi_i/\eta_i \approx m_{10} + m_{11}\tau_i + m_{12}\log \tau_i$$
$$-\eta_i^{-1}/2 \approx m_{20} + m_{21}\tau_i + m_{22}\log \tau_i$$
$$-\psi_j/\chi_j \approx m_{01} + m_{11}\mu_j + m_{21}\mu_j^2$$
$$(\psi_j^2/\chi_j) - 1 \approx m_{02} + m_{12}\mu_j + m_{22}\mu_j^2 \tag{52}$$

Least-squares evaluations of the hyperparameters can then be obtained using a standard regression program. Concrete examples are given in Arnold et al. [8,12,13].

In this section we have not been concerned with selecting hyperparameters that will ensure a proper (i.e., integrable) prior. For reasonable sample sizes and for priors which are not "heavily" improper the posterior distributions will normally turn out to be proper. Specific details regarding the constraints necessary to guarantee propriety of prior and/or posterior densities of conditionally specified form may be found in Arnold et al. [4] and the references cited there. In order to rationally use the Gibbs sampler for posterior simulations we will need a proper posterior density (see Hobert and Casella [14] for discussion of this topic).

5.1 An Example

Samples of nitrogen from two different sources are to be compared. Twelve samples are taken from the air and eight samples are obtained by chemical reaction in a container at standard temperature and pressure. The masses of the samples are as follows (as reported by Jeffreys [15]):

Population 1 (Air samples)					
2.31035	2.31026	2.31024	2.31012	2.31027	2.31017
2.30986	2.31010	2.31001	2.31024	2.31010	2.31028
Population 2 (Chemical samples)					
2.30143	2.29890	2.29816	2.30182	2.29869	2.29940
2.29849	2.29889				

Our modelling assumptions are that observations in population 1 are i.i.d. normal $N(\mu_1, \tau_1)$ and observations in population 2 are i.i.d. normal $N(\mu_2, \tau_2)$.

Our interest focuses on the difference between the means $\nu \overset{\Delta}{=} \mu_1 - \mu_2$. To this end, we will specify a conditionally conjugate joint prior for $(\mu_1, \mu_2, \tau_1, \tau_2)$ and update it with the given data to obtain a posterior density for $(\mu_1, \mu_2, \tau_1, \tau_2)$ that is still conditionally specified. Using Gibbs sampler simulations from this posterior density, we will be able to generate an approximation to the posterior density of the difference between the mean masses for the two types of samples. In addition we will observe the approximate posterior density of $\xi \overset{\Delta}{=} \tau_1 / \tau_2$, to determine whether or not we are in a Behrens–Fisher situation (i.e., whether $\xi = 1$).

From (42), our conditionally conjugate prior family for $(\mu_1, \mu_2, \tau_1, \tau_2)$ is given by

$$f(\mu_1, \mu_2, \tau_1, \tau_2) \propto (\tau_1 \tau_2)^{-1} \exp[m_{1000}\mu_1 + m_{0100}\mu_2 - m_{0010}\tau_1$$
$$- m_{0001}\tau_2 + \cdots + m_{2222}\mu_1^2 \mu_2^2 \log \tau_1 \log \tau_2] \qquad (53)$$

which has $3^4 - 1 = 80$ hyperparameters. Only 8 of these, namely

$$m_{0010}, \quad m_{0001}, \quad m_{0020}, \quad m_{0002}, \quad m_{1010}, \quad m_{0101}, \quad m_{2010}, \text{ and } m_{0201}$$

will be changed from prior to posterior by the likelihood of the data set. The classical Bayesian analysis of this data set would give nonzero values to some or all of these eight hyperparameters and set the remaining 72 equal to 0. Considerable additional flexibility will be provided by the full 80 hyperparameter family.

For illustrative purposes we will analyze our nitrogen data assuming diffuse prior information, i.e., we will initially set all m_{ij}s equal to zero. For comparison, reference can be made to Arnold et al. [13], where with a similar data set two alternative informative prior analyses are described, namely:

1. Independent conjugate priors for each parameter (the only nonzero ms in (53) are $m_{1000}, m_{0100}, m_{0010}, m_{0001}, m_{2000}, m_{0200}, m_{0020}$, and m_{0002}); and
2. A classical analysis that assumes that only the hyperparameters that will be affected by the data are nonzero (i.e., $m_{0010}, m_{0001}, m_{0020}, m_{0002}, m_{1010}, m_{0101}, m_{2010}$, and m_{0201}).

In the diffuse prior case, our prior is of the form

$$f(\underline{\mu}, \underline{\tau}) \propto (\tau_1 \tau_2)^{-1} \text{ if } \tau_1, \tau_2 > 0 \text{ and } -\infty < \mu_1, \mu_2 < \infty. \qquad (54)$$

The posterior distribution becomes

$$f(\underline{\mu}, \underline{\tau}|\text{data}) \propto (\tau_1\tau_2)^{-1} \exp\left(\frac{n_1}{2}\log \tau_1 + \frac{n_2}{2}\log \tau_2\right.$$

$$- \tau_1\frac{1}{2}\sum_{j=1}^{n_1} x_{1j}^2 - \tau_2\frac{1}{2}\sum_{j=1}^{n_2} x_{2j}^2 + \mu_1\tau_1\sum_{j=1}^{n_1} x_{1j}$$

$$\left. + \mu_2\tau_2\sum_{j=1}^{n_2} x_{2j} - \frac{n_1}{2}\mu_1^2\tau_1 - \frac{n_2}{2}\mu_2^2\tau_2\right) \qquad (55)$$

and the posterior conditional distributions to be used in the Gibbs sampler are

$$\mu_1|\tau_1 \sim N\left(\mu = \frac{1}{n_1}\sum_{j=1}^{n_1} x_{1j};\ \sigma^2 = \frac{1}{n_1\tau_1}\right)$$

$$\mu_2|\tau_2 \sim N\left(\mu = \frac{1}{n_2}\sum_{j=1}^{n_2} x_{2j};\ \sigma^2 = \frac{1}{n_2\tau_2}\right)$$

$$\tau_1|\mu_1 \sim \Gamma\left(\frac{n_1}{2};\ \frac{1}{2}\sum_{j=1}^{n_1} x_{1j}^2 - \mu_1\sum_{j=1}^{n_1} x_{1j} + \mu_1^2\frac{n_1}{2}\right)$$

$$\tau_2|\mu_2 \sim \Gamma\left(\frac{n_2}{2};\ \frac{1}{2}\sum_{j=1}^{n_2} x_{2j}^2 - \mu_2\sum_{j=1}^{n_2} x_{2j} + \mu_2^2\frac{n_2}{2}\right)$$

Using the data reported by Jeffreys [15], the nonzero hyperparameters in the posterior density of $(\mu_1, \mu_2, \tau_1, \tau_2)$ are

$$m_{0010} = 6$$
$$m_{0020} = -32.0213$$
$$m_{1020} = 27.722$$
$$m_{2020} = -6$$
$$m_{0001} = 4$$
$$m_{0002} = -21.1504$$
$$m_{0102} = 18.3958$$
$$m_{0202} = 4$$

Gibbs sampler simulation for the posterior densities of $v = \mu_1 - \mu_2$ and $\xi = \tau_1/\tau_2$ were based on 1300 iterations, discarding the first 300. Smooth approximate posterior densities were then constructed using kernel density estimates. The resulting posterior density approximations are displayed in Figures 2 and 3.

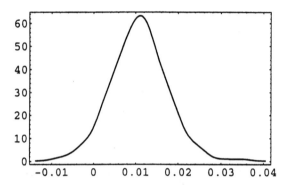

Figure 2. Diffuse priors: simulated density of $\mu_1 - \mu_2$ using the Gibbs sampler with 1000 replications and 300 starting runs.

The corresponding approximate posterior means and variances are

$E(v) = 0.01074$

$var(v) = 0.00004615$

$E(\xi) = 2.3078$

$var(\xi) = 4.3060$

From Figure 2 we see that $\mu_1 - \mu_2$ is slightly positive (air samples have more mass than chemically produced samples). The shortest 90% interval for v would not include $v = 0$. It is also evident that we really are in a

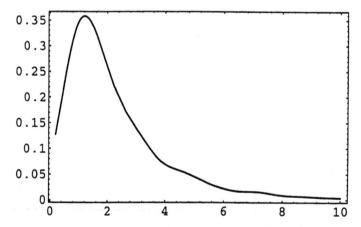

Figure 3. Diffuse priors: simulated density of τ_1/τ_2 using the Gibbs sampler with 1000 replications and 300 starting runs.

Behrens–Fisher situation since the posterior distribution of $\xi = \tau_1/\tau_2$ is centered considerably to the right of $\xi = 1$.

6. MIXTURES

Conditionally specified priors form flexible families for modeling a broad spectrum of prior beliefs. But they do have their limitations. They will often be inadequate to model multimodal prior densities. In such settings, as is commonly done in routine conjugate Bayesian analysis, resort can be made to the use of priors that are finite mixtures of conditionally conjugate priors. Conjugacy will be lost, as will the ability to use the Gibbs sampler, but, in the case of conditionally specified priors which form an exponential family (as in Section 3.1), approximate evaluation of posterior moments corresponding to finite-mixture priors will not be unusually difficult. We will pay a relatively high price for added flexibility, but if faced with definitely multimodal prior beliefs, we must pay.

7. ENVOI

For many classical data analysis situations, conditionally conjugate priors offer a tractable and more flexible alternative to the usual rather restrictive conjugate priors. They are not a panacea but certainly they merit space in the practicing Bayesian's prior distributional tool case.

REFERENCES

1. A. Bhattacharyya. On some sets of sufficient conditions leading to the normal bivariate distribution. *Sankhya*, 6:399–406, 1943.
2. B. C. Arnold and D. Strauss. Bivariate distributions with conditionals in prescribed exponential families. *Journal of the Royal Statistical Society, Series B*, 53:365–375, 1991.
3. E. Castillo and J. Galambos. Bivariate distributions with normal conditionals. In *Proceedings of the IASTED International Symposium: Simulation, Modelling and Development—SMD '87*, pp. 59–62, Cairo, 1987. ACTA Press.
4. B. C. Arnold, E. Castillo, and J. M. Sarabia. *Conditional Specification of Statistical Models*. Springer Series in Statistics, Springer Verlag, New York, 1999.

5. B. C. Arnold, E. Castillo and J. M. Sarabia. General conditional specification models. *Communications in Statistics A, Theory and Methods,* 24:1–11, 1995.

6. M. H. DeGroot. *Optimal Statistical Decisions.* McGraw-Hill, New York, 1970.

7. B. C. Arnold and S. J. Press. Bayesian estimation and prediction for Pareto data. *Journal of the American Statistical Association,* 84:1079–1084, 1989.

8. B. Arnold, E. Castillo and J. M. Sarabia. Bayesian analysis for classical distributions using conditionally specified priors. *Sankhya,* B60:228–245, 1998.

9. M. A. Tanner. *Tools for Statistical Inference. Methods for the Exploration of Posterior Distributions and Likelihood Functions.* Springer Series in Statistics, Springer Verlag, New York, 1996.

10. E. Castillo and J. Galambos. Conditional distributions and the bivariate normal distribution. *Metrika, International Journal for Theoretical and Applied Statistics,* 36(3):209–214, 1989.

11. T. Lwin. Estimating the tail of the Paretian law. *Skand. Aktuarietidskrift.,* 55:170–178, 1972.

12. B. C. Arnold, E. Castillo and J. M. Sarabia. The use of conditionally conjugate priors in the study of ratios of gamma scale parameters. *Computational Statistics and Data Analysis,* 27:125–139, 1998.

13. B. C. Arnold, E. Castillo, and J. M. Sarabia. Comparisons of means using conditionally conjugate priors. *Journal of the Indian Society of Agricultural Statistics,* 49:319–344, 1997.

14. J. P. Hobert and G. Casella. The effect of improper priors on Gibbs sampling in hierarchical linear mixed models. *Journal of the American Statistical Association,* 91:1461–1473, 1996.

15. H. S. Jeffreys. *Theory of Probability.* Oxford University Press, 1961.

2

Approximate Confidence Regions for Minimax–Linear Estimators

HELGE TOUTENBURG   Ludwig-Maximilians-Universität München, Munich, Germany

A. FIEGER   Service Barometer AG, Munich, Germany

BURKHARD SCHAFFRIN   Ohio State University, Columbus, Ohio

1. INTRODUCTION

In statistical research of the linear model there have been many attempts to provide estimators of β which use sample and prior information simultaneously. Examples are the incorporation of prior information in the form of exact or stochastic restrictions (see [1–3]) and the use of inequality restrictions which leads to the "minimax" estimation.

Minimax estimation is based on the idea that the quadratic risk function for the estimate $\hat{\beta}$ is not minimized over the entire parameter space R^L, but only over an area $B(\beta)$ that is restricted by a priori knowledge. For this, the supremum of the risk is minimized over $B(\beta)$ in relation to the estimate (minimax principle).

In many of the models used in practice, knowledge of a priori restrictions for the parameter vector β is available in a natural way. Reference [4] shows a variety of examples from the field of economics (such as input–output models), where the restrictions for the parameters are so-called workability conditions of the form $\beta_i \geq 0$ or $\beta_i \in (a_i, b_i)$ or $E(y_t|X) \leq a_t$ and, more generally,

$$A\beta \leq a$$

Minimization of $S(\beta) = (y - X\beta)'(y - X\beta)$ under inequality restrictions can be done with the simplex algorithm. Under general conditions we obtain a numerical solution. The literature deals with this problem under the generic term of inequality restricted least squares (see [5–8]). The advantage of this procedure is that a solution $\hat{\beta}$ is found that fulfills the restrictions. The disadvantage is that the statistical properties of the estimates are not easily determined and no general conclusions about superiority can be made. If all restrictions define a convex area, this area can often be enclosed in an ellipsoid of the following form:

$$B(\beta) = \{\beta : \beta'T\beta \leq k\}$$

with the origin as center point or in

$$B(\beta, \beta_0) = \{\beta : (\beta - \beta_0)'T(\beta - \beta_0) \leq k\}$$

with the center point vector β_0.

2. CONFIDENCE REGIONS ON THE BASIS OF THE OLSE

We consider the linear regression model

$$y = X\beta + \epsilon, \qquad \epsilon \sim N(0, \sigma^2 I) \tag{1}$$

with nonstochastic regressor matrix X of full column rank K. The sample size is T. The restriction of uncorrelated errors is not essential since it is easy to give the corresponding formulae for a covariance matrix $\sigma^2 W \neq \sigma^2 I$. If no further information is given, the Gauss–Markov estimator is OLSE: the ordinary least-squares estimator

$$b = (X')^{-1}X'y = S^{-1}X'y \approx N(\beta, \sigma^2 S^{-1}) \tag{2}$$

with $S = X'X$. The variance factor σ^2 is estimated by

$$s^2 = (y - Xb)'(y - Xb)(T - K)^{-1} \approx \frac{\sigma^2}{T - K}\chi^2_{T-K} \tag{3}$$

2.1 Confidence Regions for β on the Basis of b

From (2) we get

$$\sigma^{-1}S^{1/2}(b - \beta) \approx N(0, I) \tag{4}$$

whence it follows that

$$\frac{1}{K}\sigma^{-2}(b-\beta)'S(b-\beta) \approx \frac{1}{K}\chi_K^2 \tag{5}$$

As this χ^2 variable is independent of s^2, we may observe that

$$K^{-1}s^{-2}(b-\beta)'S(b-\beta) \approx F_{K,T-K} \tag{6}$$

From the central $F_{K,T-K}$ distribution we define the $(1-\alpha)$ fractile $F_{K,T-K}$ $(1-\alpha)$ by

$$P(F \leq F_{K,T-K}(1-\alpha)) = 1-\alpha \tag{7}$$

Using these results we have

$$P\left(\frac{(b-\beta)'S(b-\beta)}{s^2 K} \leq F_{K,T-k}(1-\alpha)\right) = 1-\alpha \tag{8}$$

This characterizes a simultaneous confidence region for β which is formed by the interior of the K-dimensional ellipsoid (see Figure 1)

$$\frac{1}{K}\frac{(b-\beta)'S(b-\beta)}{s^2} = F_{K,T-k}(1-\alpha) \tag{9}$$

In practice, besides the simultaneous confidence region, one may often be interested in the resulting *intervals for the components* β_i. They are deduced in the Appendix and the interval for the ith component β_i $(i = 1, \ldots, K)$ is given by

$$b_i - g_i \leq \beta_i \leq b_i + g_i \tag{10}$$

with

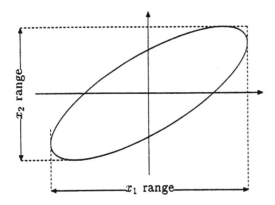

Figure 1.   Region of the components x_1 and x_2 of an ellipsoid $x'Ax = r$.

$$g_i = \sqrt{F_{K,T-K}(1-\alpha)s^2 K(S^{-1})_{ii}} \tag{11}$$

where $(S^{-1})_{ii}$ is the ith diagonal element of S^{-1} and b_i is the ith component of b. The length of the interval (10) is

$$l_i = 2g_i \tag{12}$$

The points of intersection of the ellipsoid (9) with the β_i-axes result from (A3) as

$$\beta_i = b_i \pm \sqrt{\frac{F_{K,T-K}(1-\alpha)s^2 K}{(S)_{ii}}} \tag{13}$$

In the special case of a diagonal matrix $S = X'X$ (as, e.g., in the case of orthogonal regressors), S^{-1} is diagonal and we have $(S^{-1})_{ii} = 1/(S)_{ii}$. Hence in this case the points of intersection (13) with the β_i-axes coincide with the end-points of the confidence intervals (10). However, in general we have $(S_{ii})^{-1} \le (S^{-1})_{ii}$.

3. MINIMAX–LINEAR ESTIMATION

Under the additional condition

$$\beta'B\beta \le r \tag{14}$$

with a positive definite $(K \times K)$ matrix B and a constant $r \ge 0$, the mini-max–linear estimation (MMLE) is of the form

$$b^* = (r^{-1}\sigma^2 B + S)^{-1} X'y = D^{-1}X'y \tag{15}$$

where $D = (r^{-1}\sigma^2 B + S)$; see, e.g., [9], Theorem 3.19. This estimator is biased, with bias vector

$$\lambda = \text{bias}(b^*, \beta) = E(b^*) - \beta = (D^{-1}S - B\beta)\beta = -r^{-1}\sigma^2 D^{-1}B\beta \tag{16}$$

and the covariance matrix

$$\sigma^2 V = E[(b^* - E(b^*))(b^* - E(b^*))'] = \sigma^2 D^{-1}SD^{-1} \tag{17}$$

Assuming normal distribution of ϵ, it is observed that

$$b^* - \beta \approx N(\lambda, \sigma^2 V) \tag{18}$$

$$V^{-1/2}(b - \beta) \approx N(V^{-1/2}\lambda, \sigma^2 I) \tag{19}$$

$$\sigma^{-2}(b^* - \beta)'V^{-1}(b^* - \beta) \sim \chi^2_K(\delta) \tag{20}$$

with the noncentrality parameter

$$\delta = \sigma^{-2}\lambda'V^{-1}\lambda \tag{21}$$

As the MMLE b^* (15) is dependent on the unknown parameter σ^2, this estimator is not operational.

3.1 Substitution of σ^2

We confine ourselves in the substitution to σ^2 with a positive constant c and obtain the feasible estimator (see [10])

$$b_c^* = D_c^{-1}X'y \tag{22}$$

with

$$D_c = \left(r^{-1}cB + S\right) \tag{23}$$

$$\text{bias}(b_c^*, \beta) = \left(D_c^{-1}S - I\right)\beta = -r^{-1}cD_c^{-1}B\beta = \lambda_c \tag{24}$$

$$\sigma^2 V_c = \sigma^2 D_c^{-1} S D_c^{-1} \tag{25}$$

and

$$\sigma^2(b_c^* - \beta)'V_c^{-1}(b_c^* - \beta) \sim \chi_K^2(\delta_c) \tag{26}$$

where the noncentrality parameter δ_c is given by

$$\delta_c = \sigma^{-2}\lambda_c'V_c^{-1} = \sigma^{-2}\beta'\left(SD_c^{-1} - I\right)D_cS^{-1}D_c\left(D_c^{-1}S - I\right)\beta \tag{27}$$

$$= \sigma^{-2}\beta'(S - D_c)S^{-1}(S - D_c)\beta = (\sigma^2 r^2)^{-1}c^2\beta'B'S^{-1}B\beta \tag{28}$$

We note that δ_c is unknown, too, along with the unknown $\sigma^{-1}\beta$.

The choice of c has to be made such that the feasible MMLE b_c^* is superior to the Gauss–Markov estimator b. Based on the scalar quadratic risk of an estimator $\hat{\beta}$

$$R\left(\hat{\beta}, a\right) = a'E\left[\left(\hat{\beta} - \beta\right)\left(\hat{\beta} - \beta\right)'\right]a \tag{29}$$

with a fixed $K \times 1$ vector $a \neq 0$, it holds that

$$R(b, a) \geq \sup_c\{R(b_c^*, a) : \beta'B\beta \leq r\} \tag{30}$$

if (see [2])

$$c \leq 2\sigma^2 \tag{31}$$

This (sufficient) condition follows from a general lemma on the robustness of the MMLE against misspecification of the additional restriction $\beta'B\beta \leq r$ since the substitution of σ^2 by c may be interpreted as a misspecified ellipsoid of the shape

$$\beta' B\beta \le r\sigma^2 c^{-1} \tag{32}$$

The condition (31) is practical if a lower bound for σ^2 is known:

$$\sigma_1^2 \le \sigma^2 \tag{33}$$

resulting in the choice

$$c = 2\sigma_1^2 \tag{34}$$

for c. Such a lower bound may be reclaimed using the estimator s^2 of σ^2:

$$P\left(\frac{s^2(T-K)}{\chi_{T-K}^2(1-\alpha)} \le \sigma^2\right) = 1 - \alpha \tag{35}$$

Hence one may choose $\sigma_1^2 \le s^2(T-K)/\chi_{1-\alpha}^2$ at a $1 - \alpha$ level of significance. The estimator b_c^* with $c = 2\sigma_1^2$ is called the two-stage minimax–linear estimator (2SMMLE).

4. APPROXIMATION OF THE NONCENTRAL χ^2 DISTRIBUTION

From formula (24.21) in [11], a noncentral χ^2 distribution may be approximated by a central χ^2 distribution according to

$$\chi_K^2(\delta_c) \approx a\chi_d^2 \tag{36}$$

with

$$a = \frac{K + 2\delta_c}{K + \delta_c}, \qquad d = \frac{(K + \delta_c)^2}{K + 2\delta_c} \tag{37}$$

where, due to the unknown δ_c, the factor a and the number of degrees of freedom d are also unknown.

With the approximation (36), formula (26) becomes

$$a^{-1}\sigma^{-2}(b_c^* - \beta)'V_c^{-1}(b_c^* - \beta) \approx \chi_d^2 \tag{38}$$

i.e., approximately (in the case of independence of s^2) we have

$$\frac{(b_c^* - \beta)'V_c^{-1}(b_c^* - \beta)}{ads^2} \approx F_{d,T-K} \tag{39}$$

The desired confidence region for β at the level $1 - \alpha$ is defined by the interior of the ellipsoid

$$(b_c^* - \beta)'V_c^{-1}(b_c^* - \beta) < ads^2 F_{d,T-K}(1 - \alpha) \tag{40}$$

Since δ_c, a and d are unknown, relation (40) cannot be applied directly. To overcome this problem we use the following approach, which is essentially based on an approximation.

4.1 Bounds for δ_c

We rewrite the noncentrality parameter δ_c (27) as follows. From

$$\text{bias}(b_c^*, \beta) = \lambda_c = (D_c^{-1}S - I)\beta = -r^{-1}cD_c^{-1}B\beta \tag{41}$$

we get

$$\begin{aligned}\delta_c &= \sigma^2 \lambda_c' V_c^{-1} \lambda_c = \sigma^2 r^{-2} c^2 \beta' B D_c^{-1} D_c S^{-1} D_c D_c^{-1} B\beta \\ &= \sigma^{-2} r^{-2} c^2 \beta' B S^{-1} B\beta \end{aligned} \tag{42}$$

Let $\lambda_{\min}(A)$ denote the minimal and $\lambda_{\max}(A)$ the maximal eigenvalue of a matrix A. Then it is well known that "Raleigh's inequalities"

$$0 \le \beta' B\beta \lambda_{\min}\left(B^{1/2} S^{-1} B^{1/2}\right) \le \beta' BS^{-1} B\beta \le r\lambda_{\max}\left(B^{1/2} S^{-1} B^{1/2}\right) \tag{43}$$

hold true, yielding for a general c and with the inequality (33) at first

$$0 \le \beta' B\beta \lambda_{\min}\left(B^{1/2} S^{-1} B^{1/2}\right) \le \beta' BS^{-1} B\beta \le r\lambda_{\max}\left(B^{1/2} S^{-1} B^{1/2}\right) \tag{44}$$

and for $c = 2\sigma_1^2$ especially

$$\delta_c \le 2cr^{-1}\lambda_{\max}\left(B^{1/2} S^{-1} B^{1/2}\right) = \delta_0 \tag{45}$$

Hence, the upper bound δ_0 for δ_c can be calculated for any c.

Using this inequality, we get the following upper bounds for the coefficients a and d of the approximation (36)

$$a \le \frac{K + 2\delta_0}{K + \delta_0} = a_0 \tag{46}$$

and

$$d \le \frac{(K + \delta_0)^2}{K + 2\delta_0} = d_0 \tag{47}$$

i.e.,

$$ad = K + \delta_c \le K + \delta_0 = a_0 d_0 \tag{48}$$

Replacing a and d by a_0 and d_0 respectively, the approximate confidence region for β becomes

$$\left\{\beta : (b_c^* - \beta)' V_c^{-1}(b_c^* - \beta) < (K + \delta_0) s^2 F_{d_0, T-K}(1 - \alpha)\right\} \tag{49}$$

We have $(K + \delta_c) \leq (K + \delta_0)$, but $F_{d,T-K}(1 - \alpha) \geq F_{d_0,T-K}(1 - \alpha)$, for realistic choices of α and $T - K \geq 3$. Thus the impact of changing the actual parameter to its maximal value δ_0, on the volume of the confidence region (49) used in practice, has to be analysed numerically. Simulations (see Section 5) were carried out which showed that using δ_0 instead of δ_c would increase the volume of the confidence region.

With the abbreviation

$$g_i^0 = \sqrt{F_{d_0 T-K}(1 - \alpha)s^2(K + \delta_0)(V_c)_{ii}} \tag{50}$$

it follows, from (49), that the confidence intervals for the components from β may be written as

$$KI_i = \left[b_{c_i}^* - g_i^0 \leq \beta_i \leq b_{c_i}^* + g_i^0\right] \tag{51}$$

5. PROPERTIES OF EFFICIENCY

Let us now investigate the efficiency of the proposed solution. Assume that the confidence level $1 - \alpha$ is fixed. Replacing δ_c by the least favourable value δ_0 influences the length of the confidence intervals

(a) True, but unknown confidence region (40) on the basis of δ_c.
     Length of the confidence interval:

$$2g_i^c = 2\sqrt{F_{d,T-K}(1 - \alpha)s^2(K + \delta_c)(V_c)_{ii}}$$

(b) Practical confidence region (49) on the base of δ_0.
     Length of the confidence interval according to (50):

$$2g_i^0 = 2\sqrt{F_{d_0,T-K}(1 - \alpha)s^2(K + \delta_0)(V_c)_{ii}}$$

By defining the ratio

$$\frac{\text{Length of the interval on the basis of } \delta_0}{\text{Length of the interval on the basis of } \delta_c}$$

we get (for all $i = 1, \ldots, K$) the same stretching factor

$$f = f(\delta_c, \delta_0, K, T - K) = \frac{2g_i^0}{2g_i^c} = \sqrt{\frac{(K + \delta_0)F_{d_0,T-K}(1 - \alpha)}{(K + \delta_c)F_{d,T-K}(1 - \alpha)}} \tag{52}$$

For given values of δ_c (where $\delta_c = 0.1$ and $\delta_c = 1$) for the $T - K = 10$ and $T = K = 33$, respectively, we have calculated the stretching factor in dependence of δ_0 and varying values of K (Figures 2–4). The stretching factor

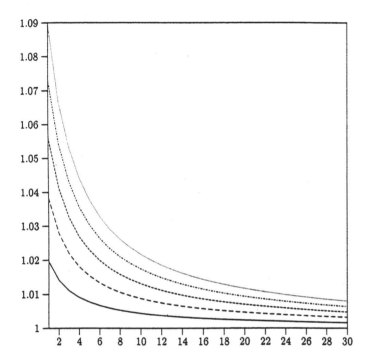

Figure 2. Stretching factor f (vertical axis) depending on K (horizontal axis). With increasing K the stretching factor decreases. Results are presented for $\delta_c = 1$, $T = 33$, and additionally varying δ_0 starting from $\delta_c + 0.1$ (solid line, step 0.1). With increasing difference $(\delta_0 - \delta_c)$ the stretching factor increases; see also Figure 4.

decreases with increasing K (number of regressors) and increases with the distance $(\delta_0 - \delta_c)$.

Another means of rating the quality of the practical confidence region (49) is to determine the equivalent confidence level $1 - \alpha$ of the true (but unknown) confidence region (40). The true confidence region is defined approximately through

$$P\left(\frac{(b_c^* - \beta)' V_c^{-1}(b_c^* - \beta)}{ads^2} \leq F_{d,T-K}(1 - \alpha)\right) = 1 - \alpha \qquad (53)$$

The replacement of δ_c by its maximum δ_0 leads to an increased confidence interval given by

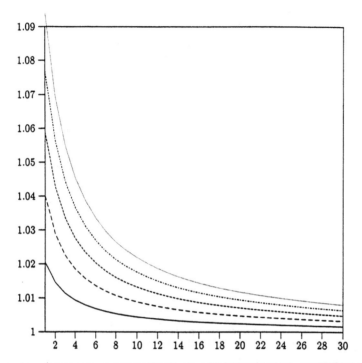

Figure 3. Stretching factor f (vertical axis) depending on K (horizontal axis). With increasing K the stretching factor decreases. Results are presented for $\delta_c = 1$, $T = K + 10$, and additionally varying δ_0 from $\delta_c + 0.1$ (solid line, step 0.1). With increasing difference $(\delta_0 - \delta_c)$ the stretching factor increases; see also Figure 4.

$$P\left(\frac{(b_c^* - \beta)'V_c^{-1}(b_c^* - \beta)}{a_0 d_0 s^2} \leq F_{d_0 T-K}(1 - \alpha)\right) = 1 - \alpha_1 \tag{54}$$

Hence, by combination of (53) and (54), we find for the true (and smaller) confidence region

$$P\left(\frac{(b_c^* - \beta)'V_c^{-1}(b_c^* - \beta)}{ads^2} \leq f^2 F_{d,T-K}(1 - \alpha)\right) = 1 - \alpha_1 \geq 1 - \alpha \tag{55}$$

with $f \geq 1$, from (52). Replacing the unknown noncentrality parameter δ_c by its maximum δ_0 results in an increase of the confidence level, as we have $\alpha_1 \leq \alpha$ Figures 5 and 6 present values of α_1 for varying values of T and K.

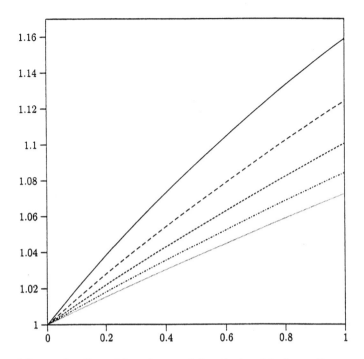

Figure 4. Stretching factor f (vertical axis) depending on the difference $(\delta_0 - \delta_c)$ (horizontal axis). With increasing difference $(\delta_0 - \delta_c)$ the stretching factor increases. Results are presented for $\delta_c = 1$, $T = 33$, and additionally varying K from 1 (solid line) to 5. With increasing K the stretching factor decreases; see also Figure 2.

As a consequence, in practice we choose a smaller confidence level of, e.g., $1 - \alpha = 0.90$ to reach a real confidence level of $1 - \alpha_1 < 1$ (also for greater distances $\delta_0 - \delta_c$). Both the stretching factor f and the amount by which the confidence level increases are increasing with $\delta_0 - \delta_c$.

$$\delta_0 - \delta_c \leq \delta_0 - \delta_u \leq \delta_0 \tag{56}$$

where, according to (42) and (43),

$$\delta_u = 4\sigma_1^2 r^{-2}\left[\lambda_{\min}\left(B^{1/2}S^{-1}B^{1/2}\right)\right]\beta'B\beta \geq 0 \tag{57}$$

turns out to be a lower bound of the true noncentrality parameter δ_c. The upper bound δ_0 is calculated for concrete models, such that it becomes possible to estimate the maximum stretch factor f and the maximal

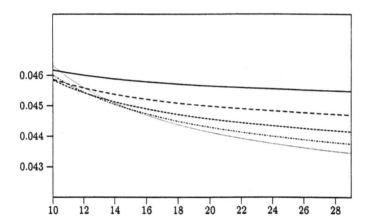

Figure 5. Confidence level α_1 (vertical axis) depending on T (horizontal axis) for $f^2 = 1.02^2$ and $\alpha = 0.05$. Additionally, varying K from 1 (solid line) to 5, α_1 decreases with increasing K.

increase of the confidence level from $1 - \alpha$ to $1 - \alpha_1$. In this way the practicability of the proposed method is given in addition to the estimation of its efficiency.

If the ellipsoid of the prior information is not centered in the origin but in a general midpoint vector $\beta_0 \neq 0$, i.e.,

$$(\beta - \beta_0)' B(\beta - \beta_0) \leq r \tag{58}$$

then the MMLE becomes

$$b^*(\beta_0) = \beta_0 + D^{-1} X'(y - X\beta_0) \tag{59}$$

with

$$\text{bias}(b^*(\beta_0), \beta) = (D^{-1} S - I)(\beta - \beta_0) \tag{60}$$

and (see (17))

$$V(b^*(\beta_0)) = V(b^*) = V \tag{61}$$

All the preceding results remain valid if we replace, for λ in (16) and δ_c in (24), the vector β by $(\beta - \beta_0)$, provided that δ in (21) and δ_c in (27) are defined with accordingly changed λ and λ_c.

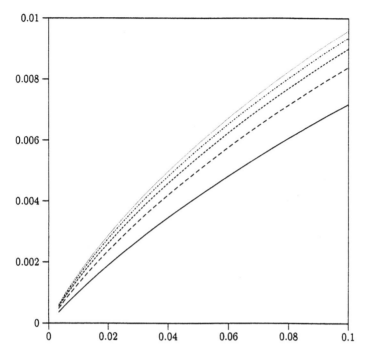

Figure 6.   Difference of the confidence level $(\alpha - \alpha_1)$ (vertical axis) depending on α (horizontal axis) for $f^2 = 1.02^2$ and $K = 3$. Additionally, varying $T = 10$ (solid line), 15, 20, 25, and 30, with increasing T the difference $\alpha - \alpha_1$ increases.

6. COMPARING THE VOLUMES

The definition of a confidence ellipsoid is based on the assumption that the unknown parameter β is covered with probability $1 - \alpha$ by the random ellipsoid. If one has the choice between alternative ellipsoids, one would choose the ellipsoid with the smallest volume. In other words, the MDE (mean dispersion error) superiority of the MMLE with respect to the Gauss–Markov estimator in the sense of (14) does not necessarily lead to a preference for the ellipsoids based on the MMLE. Hence in the following we determine the volume of both ellipsoids. The volume of the T-dimensional unit sphere

$$x'x \le 1$$

(x being a $T \times 1$ vector) is given as

$$\text{Vol}_E = \frac{\pi^{T/2}}{\Gamma\left(1 + \frac{T}{2}\right)} \tag{62}$$

For an ellipsoid $x'Ax \leq 1$ with a positive definite matrix A, the volume is

$$\text{Vol}(A) = \text{Vol}_E |A|^{-1/2} \tag{63}$$

6.1 Gauss–Markov Estimator

The confidence ellipsoid for β on the basis of the Gauss–Markov estimator b in (2) is, according to (8),

$$\frac{1}{s^2 K F_{K,T-K}(1-\alpha)}(b-\beta)'S(b-\beta) \leq 1 \tag{64}$$

so that the volume is

$$\text{Vol}(b) = \left(s^2 K F_{K,T-K}(1-\alpha)\right)^{K/2} |S|^{-1/2} \text{Vol}_E \tag{65}$$

6.2 MMLE

Based on the approximations (36) and (48), the confidence region using the MMLE b_c^* was (see (49))

$$\frac{1}{(K+\delta_0)s^2 F_{d_0,T-K}(1-\alpha)}(b_c^* - \beta)'V_c^{-1}(b_c^* - \beta) \leq 1 \tag{66}$$

and hence its volume is

$$\text{Vol}(b_c^*) = \left((K+\delta_0)s^2 F_{d_0,T-K}(1-\alpha)\right)^{K/2} |V_c^{-1}|^{-1/2} \text{Vol}_E \tag{67}$$

Comparison of both volumes gives

$$q = \frac{\text{Vol}(b_c^*)}{\text{Vol}(b)} = [f(0,\delta_0,K,T-K)]^K \frac{|X'X|^{1/2}}{|V_c^{-1}|^{1/2}} \tag{68}$$

where $f(0, \delta_0, K, T-K)$ is the maximal stretch factor (52) for the lower bound $\delta_u = 0$ of the noncentrality parameters δ_c. The case $\delta_u = 0$ corresponds to $T \to \infty$, i.e., to change from the MMLE to the Gauss–Markov estimator. The MMLE b_c has smaller variance than the Gauss–Markov estimator b:

$$0 \leq (X'X)^{-1} - V_c \quad \text{(nonnegative definite)}$$

i.e., we have

$$0 \le V_c^{-1} - X'X = 2r^{-1}cB + r^{-1}c^2 BSB = C$$

or

$$V_c^{-1} = X'X + C \quad \text{with} \quad C \ge 0 \qquad \text{(nonnegative definite)} \tag{69}$$

From (69) we may conclude that

$$|X'X| \le |V_c^{-1}|$$

and thus

$$|I + r^{-1}cBS^{-1}|^{-1} = \frac{|X'X|^{1/2}}{|V_c^{-1}|^{1/2}} \le 1 \tag{70}$$

So the relation (68) between both volumes turns out to be the product of a function $f \ge 1$ and the expression (70) which is ≤ 1.

The ratio (68) has to be investigated for a concrete model and given data, as δ_0 (and hence f) and the quality (70) are dependent on the data as well as on the strength of the additional condition.

Let $X'X = S$ and assume the condition $\beta'S\beta \le r$. Then, according to Section 4, we have

$$V_c^{-1} = \left(r^{-2}c^2 + 1 + 2r^{-1}c\right)S = \left(r^{-1}c + 1\right)^2 S$$

and

$$|V_c^{-1}| = \left(r^{-1}c + 1\right)^{2K}|S| \tag{71}$$

Analogously, from (45) with $c = 2\sigma_1^2$, we get

$$\delta_0 = 2r^{-1}c \tag{72}$$

This results in a change of the relation of the volumes (68) to

$$q = q(r^{-1}c) = \left(\frac{f(0, 2r^{-1}c, K, T - K)}{(r^{-1}c + 1)}\right)^K \tag{73}$$

7. CONCLUSIONS

In this paper we have demonstrated the use of prior information in the form of inequalities (c.f. (14)) and an approximation of the noncentral chi-square distribution to get practicable confidence regions that are based in the minimax estimator. We have computed the relationship between the true (but unknown) and the approximated confidence intervals. Furthermore, the relationship between the volumes of the confidence ellipsoids based on the OLSE and the minimax estimator has been investigated.

APPENDIX

Assume an ellipsoid

$$x'Ax = r$$

with positive definite matrix A and $1 \times T$ vector $x' = (x_1, \ldots, x_T)$. We determine the regions of the x_i components for the ellipsoid. Without loss of generality, we solve this problem for the first component x_1 only; this is equivalent to finding an extremum under linear constraints.

Let $e_1' = (1, 0, \ldots, 0)$ and μ be a Lagrange multiplier. Further, let

$$f(x) = x_1 = e_1'x$$
$$g(x) = x'Ax - r$$

and

$$F(x) = f(x) + \mu g(x)$$

Then we have to solve

$$F(x) = \text{stationary}_{x,y}$$

which leads to the necessary normal equations

$$\frac{\partial F(x)}{\partial x} = e_1 + 2\hat{\mu}Ax = 0$$

$$\frac{\partial F(x)}{\partial \mu} = x'Ax - r = 0$$

$$(A1)$$

From (A1) it follows that

$$x'e_1 + 2\hat{\mu}x'Ax = 0$$

thus we get

$$2\hat{\mu} = -\frac{x_1}{r}$$

Inserting this into (A1) gives

$$e_1 + 2\hat{\mu}Ax = e_1 - \frac{x_1}{4}Ax = 0$$

or

$$x = A^{-1}e_1\frac{r}{x_1}$$

and therefore

$$e_1'x = x_1 = e_1'A^{-1}e_1\frac{r}{x_1}$$

or

$$x_1 = \pm\sqrt{r(A^{-1})_{11}}$$

with $(A^{-1})_{11}$ as the first diagonal element of the matrix A^{-1}. In the case that the ellipsoid is not centered in the origin

$$(x - x_0)'A(x - x_0) = r$$

the regions of the x_i components become

$$x_{0_i} - \sqrt{r(A^{-1})_{ii}} \le x_i \le x_{0_i} + \sqrt{r(A^{-1})_{ii}} \tag{A2}$$

The intersection points of the ellipsoid $x'Ax = r$ with the coordinate axes follow from

$$\big(0, \ldots, x_i - x_{0,i}, 0, \ldots, 0\big)'A(0, \ldots, x_i - x_0, 0, \ldots, 0)$$

$$= (x_t - x_{0_i})^2(A)_{ii} = r$$

as

$$x_i = x_{0_i} \pm \sqrt{\frac{r}{(A)_{ii}}} \tag{A3}$$

REFERENCES

1. J. S. Chipman, M. M. Rao. The treatment of linear restrictions in regression analysis. *Econometrica*, 32, 198–209, 1964.
2. T. A. Yancey, G. G. Judge, M. E. Bock. Wallace's weak mean square error criterion for testing linear restrictions. *Econometrica*, 41, 1203–1206, 1973.
3. T. A. Yancey, G. G. Judge, M. E. Bock. A mean square error test when stochastic restrictions are used in regression. *Communications in Statistics, Part A—Theory and Methods*, 3, 755–768, 1974.
4. P. Stahlecker. *A prior Information und Minimax-Schätzung im linearen Regressionsmodell*. Athenäum, Frankfurt/Main, 1987.

5. G. G. Judge, T. Takayama. Inequality restrictions in regression analysis. *Journal of the American Statistical Association*, 66, 166–181, 1966.

6. J. M. Dufour. Nonlinear hypothesis, inequality restrictions and nonnested hypotheses: Exact simultaneous tests in linear regression. *Econometrica*, 57, 335–355, 1989.

7. J. Geweke. Exact inference in the inequality constrained normal linear regression model. *Journal of Applied Econometrics*, 1, 127–141, 1986.

8. J. J. A. Moors, J. C. van Houwelingen. Estimation of linear models with inequality restrictions. *Technical Report 121*. Tilburg University, The Netherlands, 1987.

9. C. R. Rao, H. Toutenburg. *Linear Models: Least Squares and Alternatives*. 2nd ed., Springer, 1999.

10. H. Toutenburg. *Prior Information in Linear Models*, Wiley, 1982, p. 96.

11. M. G. Kendall, A. Stuart. *The Advanced Theory of Statistics*, Griffin, 1977, p. 245.

3

On Efficiently Estimable Parametric Functionals in the General Linear Model with Nuisance Parameters

PAWEL R. PORDZIK[*]   Agricultural University of Poznań, Poznań, Poland

GÖTZ TRENKLER[†]   University of Dortmund, Dortmund, Germany

1. INTRODUCTION AND PRELIMINARIES

Consider the general linear models $\mathcal{M}_a = \{\mathbf{y}, \mathbf{W}\gamma + \mathbf{Z}\delta, \sigma^2 \mathbf{V}\}$ and $\mathcal{M} = \{\mathbf{y}, \mathbf{W}\gamma, \sigma^2 \mathbf{V}\}$, in which $\mathbf{y}$ is an observable random vector with expectation $E_a(\mathbf{y}) = \mathbf{W}\gamma + \mathbf{Z}\delta$ in the former and $E(\mathbf{y}) = \mathbf{W}\gamma$ in the latter model, and with the same dispersion matrix $D(\mathbf{y}) = \sigma^2 \mathbf{V}$ in both cases. The matrices $\mathbf{W}$, $\mathbf{Z}$, and $\mathbf{V}$ are known, each allowed to be deficient in rank, while the positive scalar σ^2 and the subvectors in $(\gamma' : \delta')'$ are unknown parameters. Thus it is assumed that the expectation vector in $\mathcal{M}_a$ consists of two parts. $\mathbf{W}\gamma$, involving main parameters, and $\mathbf{Z}\delta$, comprising nuisance parameters (wide applicability of such models in statistical practice is well known). From now on it is assumed that neither $\mathcal{C}(\mathbf{W}) \subset \mathcal{C}(\mathbf{Z})$ nor $\mathcal{C}(\mathbf{Z}) \subset \mathcal{C}(\mathbf{W})$, where $\mathcal{C}()$ denotes the column space of a matrix argument. Furthermore, for con-

---

[*]Support by Graduiertenkolleg, Department of Statistics, University of Dortmund, is gratefully acknowledged.

[†]Research was supported by Deutsche Forschungsgemeinschaft, Grant Tr 253/2–3.

sistency of the model $\mathcal{M}_a$, it is assumed that $\mathbf{y} \in \mathcal{C}(\mathbf{W} : \mathbf{Z} : \mathbf{V})$; accordingly, concerning the model $\mathcal{M}$, it is assumed that $\mathbf{y} \in \mathcal{C}(\mathbf{W} : \mathbf{V})$. In the sequel, let $\mathcal{S}_a$ and $\mathcal{S}$ denote the sets that comprise all vectors $\mathbf{y}$ satisfying the consistency condition for the model $\mathcal{M}_a$ and $\mathcal{M}$, respectively.

In this paper, the situation is considered in which there is uncertainty about inclusion of nuisance parameters in the model. Assuming that an adequate model for the experiment may be given by $\mathcal{M}$ or $\mathcal{M}_a$, the question arises on efficiency of inference induced by a wrong choice. The problem of comparing the model $\mathcal{M}$ with its augmented by nuisance parameters counterpart $\mathcal{M}_a$, closely related to the problem of evaluating efficiency of inference under uncertainty about model specification, has gained the attention of many researchers and practitioners. One of the goals of such comparisons was to characterize consequences of the presence of nuisance parameters for the precision of best linear unbiased estimator (BLUE) of a functional of main parameters. Numerous equivalent conditions were derived for linear functions $\mathbf{p}'\gamma$ to be estimated with full efficiency in the model $\mathcal{M}_a$, or, as stressed in [1], p. 350, under which "$\mathbf{p}'\gamma$ are variance-robust with respect to overspecification of the model (i.e., whose BLUEs under $\mathcal{M}_a$ retain their variances when $\mathcal{M}$ is the true model)." This aspect of evaluating the performance of $\mathcal{M}_a$ has so far been studied under additional assumptions imposed on design and dispersion matrices. For a survey of results and extensive discussion on the issue under the standard linear model with $\mathbf{V} = \mathbf{I}$, $\mathbf{I}$ stands for an identity matrix of appropriate dimension, and a weakly singular model, where $\mathcal{C}(\mathbf{W} : \mathbf{Z}) \subset \mathcal{C}(\mathbf{V})$, the reader is referred to [2] and [1], respectively.

The aim of this paper is twofold. First, another characteristic of the problem is presented; it can be viewed as an extension of the orthogonality condition introduced in the context of the analysis of variance model for two-way classification of data. Secondly, the problem is considered for the general linear model with no extra assumptions imposed on the matrices $\mathbf{W}$, $\mathbf{Z}$, and $\mathbf{V}$. In Section 2, necessary and sufficient conditions are derived for the BLUE of a parametric function $\mathbf{p}'\gamma$ in $\mathcal{M}_a$ to have the same variance as the corresponding estimator under the model $\mathcal{M}$. Furthermore, extending the approach presented in [1], the subspace of functionals $\mathbf{p}'\gamma$ which can be estimated with full efficiency under the model with nuisance parameters is characterized.

Given subspaces $\mathcal{U}$, $\mathcal{V}$, and $\mathcal{W}$ of the finite dimensional Euclidean space $\mathcal{R}^n$, let $\dim(\mathcal{U})$, $\mathcal{U} \cap \mathcal{V}$, $\mathcal{U} + \mathcal{V}$, and $\mathcal{U} \oplus \mathcal{V}$ denote the dimension of $\mathcal{U}$, the intersection, and the sum and direct sum of $\mathcal{U}$ and $\mathcal{V}$, respectively. If $\mathcal{W} \subset \mathcal{U}$, then we have

$$\mathcal{U} \cap (\mathcal{V} + \mathcal{W}) = (\mathcal{U} \cap \mathcal{V}) + \mathcal{W} \tag{1}$$

Given a real matrix $\mathbf{A}$, let $\mathbf{A}'$, $\mathbf{A}^-$ and $r(\mathbf{A})$ denote the transpose of $\mathbf{A}$, a generalized inverse of $\mathbf{A}$ and the rank of $\mathbf{A}$, respectively. By the symbol $C^\perp(\mathbf{A})$ the orthocomplement of $C(\mathbf{A})$ will be denoted, and $\mathbf{A}^\perp$ will represent any matrix such that $C^\perp(\mathbf{A}) = C(\mathbf{A}^\perp)$. Further, $\mathbf{P_A}$ and $\mathbf{Q_A}$ will stand for the orthogonal projectors onto $C(\mathbf{A})$ and $C^\perp(\mathbf{A})$, respectively. Let $\mathbf{A}$ and $\mathbf{B}$ be disjoint matrices, i.e., $C(\mathbf{A}) \cap C(\mathbf{B}) = \{\mathbf{0}\}$. By the symbol $\mathbf{P_{A|B}}$ let a projector onto $C(\mathbf{A})$ along $C(\mathbf{B})$ be denoted, cf. [3]. If $\mathbf{C}$ is a matrix satisfying the condition $C(\mathbf{C}) \subset C(\mathbf{A} : \mathbf{B})$, then we have $C(\mathbf{C}) = C(\mathbf{C}_1) \oplus C(\mathbf{C}_2)$, where a matrix $\mathbf{C}_1$ is such that

$$C(\mathbf{C}_1) = C(\mathbf{A}) \cap C(\mathbf{C} : \mathbf{B}) = C(\mathbf{P_{A|B}C}) \tag{2}$$

and $\mathbf{C}_2$ is defined accordingly. Finally, for the purposes of this paper, a result given in [4] (Theorem 3) is restated. Let $\kappa = \mathbf{K}\beta$ be a vector of parametric functions estimable in the model $\{\mathbf{y}, \mathbf{X}\beta, \sigma^2\mathbf{V}\}$, i.e., $C(\mathbf{K}') \subseteq C(\mathbf{X}')$, and let $\hat{\kappa}$ and $\hat{\kappa}_r$ denote the best linear unbiased estimator of κ under the above-mentioned and restricted model $\{\mathbf{y}, \mathbf{X}\beta | \mathbf{R}\beta = \mathbf{0}, \sigma^2\mathbf{V}\}$, respectively. Then it follows that the dispersion matrices of these two estimators are identical, $D(\hat{\kappa}) = D(\hat{\kappa}_r)$, if and only if

$$\mathbf{KGR}_1' = \mathbf{0} \tag{3}$$

where $\mathbf{R}_1$ and $\mathbf{G}$ are any matrices satisfying $C(\mathbf{R}_1') = C(\mathbf{X}') \cap C(\mathbf{R}')$ and

$$\mathbf{XGX}' = \mathbf{V} - \mathbf{VQ_x}(\mathbf{Q_xVQ_x})^-\mathbf{Q_xV} \tag{4}$$

2. RESULTS

It is well known that when augmenting a linear model with concomitants, the increase of efficiency in estimating parametric functions must come through substantial decrease in variance of observations. If it is not the case and, consequently, the inference base is overparameterized, then the variance of the estimators is inflated. It is of some interest to characterize functionals of main parameters which are variance-robust with respect to overparameterization of the model. Following the notation introduced in [1], let $\mathcal{E}_a$ denote the class of linear functions of the main parameters γ which are estimable in the model with nuisance parameters $\mathcal{M}_a = \{\mathbf{y}, \mathbf{W}\gamma + \mathbf{Z}\delta, \sigma^2\mathbf{V}\}$; that is, let

$$\mathcal{E}_a = \{\mathbf{p}'\gamma : \mathbf{p} = \mathbf{W}'\mathbf{q}, \mathbf{q} \in C^\perp(\mathbf{Z})\} \tag{5}$$

Furthermore, denote by $\mathbf{p}'\hat{\gamma}_a$ and $\mathbf{p}'\hat{\gamma}$ the BLUE of $\mathbf{p}'\gamma$ obtained under the model $\mathcal{M}_a$ and $\mathcal{M} = \{\mathbf{y}, \mathbf{W}\gamma, \sigma^2\mathbf{V}\}$, respectively. In this section, we characterize the subclass of the functionals $\mathbf{p}'\gamma$ for which the BLUE under the model

$\mathcal{M}_a$ has the same variance as the corresponding estimator under the model $\mathcal{M}$; subsequently, we denote the subclass of such functionals by $\mathcal{E}_0$; i.e.,

$$\mathcal{E}_0 = \left\{ \mathbf{p}'\gamma \in \mathcal{E}_a : Var(\mathbf{p}'\hat{\gamma}_a) = Var(\mathbf{p}'\hat{\gamma}) \right\} \tag{6}$$

First note that $\mathcal{M}$ can be considered as the restricted model $\{\mathbf{y}, \mathbf{W}\gamma + \mathbf{Z}\delta | \mathbf{Z}\delta = \mathbf{0}, \sigma^2\mathbf{V}\}$. Further, on account of Theorem 1 in [4], note that to obtain the best linear unbiased estimator of $\mathbf{p}'\gamma$ one can replace $\mathbf{Z}\delta = \mathbf{0}$ with a subset of restrictions expressed by estimable parametric functions under the model $\mathcal{M}_a$. In view of the relation $\mathcal{C}(\mathbf{0} : \mathbf{Z})' \cap \mathcal{C}(\mathbf{W} : \mathbf{Z})' = \mathbf{C}\,\mathcal{C}(\mathbf{0} : \mathbf{Q}_\mathbf{w}\mathbf{Z})'$, a representation of such restrictions takes the form $\mathbf{Q}_\mathbf{w}\mathbf{Z}\delta = \mathbf{0}$. Following this approach, and making use of the results stated in (3) and (4), necessary and sufficient conditions for every functional $\mathbf{p}'\gamma \in \mathcal{E}_a$ to be estimated with the same variance under the models $\mathcal{M}_a$ and $\mathcal{M}$ can easily be proved.

Theorem 1. The classes $\mathcal{E}_a$ and $\mathcal{E}_0$ coincide if and only if one of the following equivalent conditions holds:

$$\mathbf{P}_{\mathbf{UZ}^\perp}\mathbf{P}_{\mathbf{UW}^\perp} = \mathbf{P}_{\mathbf{UW}^\perp}\mathbf{P}_{\mathbf{UZ}^\perp} \tag{7}$$

$$\mathcal{C}(\mathbf{W} : \mathbf{Z}) \cap \mathcal{C}(\mathbf{VW}^\perp) \subset \mathcal{C}(\mathbf{Z}) \tag{8}$$

$$\mathcal{C}(\mathbf{W} : \mathbf{Z}) \cap \mathcal{C}(\mathbf{VZ}^\perp) \subset \mathcal{C}(\mathbf{W}) \tag{9}$$

where $\mathbf{U}$ is a matrix such that $\mathbf{V} = \mathbf{U}'\mathbf{U}$.

Proof. It is clear that both classes $\mathcal{E}_a$ and $\mathcal{E}_0$ coincide if and only if

$$D(\mathbf{Q}_\mathbf{Z}\mathbf{W}\hat{\gamma}) = D(\mathbf{Q}_\mathbf{Z}\mathbf{W}\hat{\gamma}_a) \tag{10}$$

Assume for a moment that $\mathbf{X} = (\mathbf{W} : \mathbf{Z})$, $\beta = (\gamma' : \delta')'$, $\mathbf{K} = \mathbf{Q}_\mathbf{Z}\mathbf{X}$ and $\mathbf{R} = \mathbf{Q}_\mathbf{W}\mathbf{X}$. Considering $\mathbf{Q}_\mathbf{Z}\mathbf{W}\hat{\gamma}_a$ and $\mathbf{Q}_\mathbf{Z}\mathbf{W}\hat{\gamma}$ as the BLUE of $\mathbf{K}\beta$ under the model $\{\mathbf{y}, \mathbf{X}\beta, \sigma^2\mathbf{V}\}$ and $\{\mathbf{y}, \mathbf{X}\beta | \mathbf{R}\beta = \mathbf{0}, \sigma^2\mathbf{V}\}$, respectively, due to (3) and (4), one can replace (10) by the condition

$$\mathbf{Q}_\mathbf{Z}\mathbf{U}'(\mathbf{I} - \mathbf{U}\mathbf{Q}_\mathbf{X}(\mathbf{Q}_\mathbf{X}\mathbf{U}'\mathbf{U}\mathbf{Q}_\mathbf{X})^-\mathbf{Q}_\mathbf{X}\mathbf{U}')\mathbf{U}\mathbf{Q}_\mathbf{W} = \mathbf{0} \tag{11}$$

Further, expressing the last equality by orthogonal projectors, one can write equivalently

$$\mathbf{P}_{\mathbf{UZ}^\perp}\mathbf{P}_{\mathbf{UW}^\perp} = \mathbf{P}_{\mathbf{UZ}^\perp}\mathbf{P}_{\mathbf{UX}^\perp}\mathbf{P}_{\mathbf{UW}^\perp} \tag{12}$$

In view of the relation $\mathcal{C}(\mathbf{UX}^\perp) \subset \mathcal{C}(\mathbf{UW}^\perp) \cap \mathcal{C}(\mathbf{UZ}^\perp)$, the right-hand side of the equality (12) simplifies to $\mathbf{P}_{\mathbf{UX}^\perp}$, thus establishing the condition (7). For the proof of (8), it is enough to express the equality (11) as

$$(\mathbf{Z}^\perp)'\mathbf{P}_{\mathbf{X}|\mathbf{VX}^\perp}\mathbf{VW}^\perp = \mathbf{0}$$

and note that, due to (2), it holds that $C(\mathbf{P}_{\mathbf{X}|\mathbf{VX}^\perp}\mathbf{VW}^\perp) = C(\mathbf{X}) \cap C(\mathbf{VW}^\perp)$. Starting with the transposition of (11), the proof of the condition (9) follows the same lines.

It has to be mentioned that the equality $\mathcal{E}_a = \mathcal{E}_0$, as shown above, being equivalent to commutativity of the orthogonal projectors $\mathbf{P}_{\mathbf{UZ}^\perp}$ and $\mathbf{P}_{\mathbf{UW}^\perp}$, is consequently equivalent to any of (A 2) through (A 46) conditions presented in [5], Theorem 1. Furthermore, it is to be noted that the dispersion equality (10) does not imply equality of the BLUEs of $\mathbf{p}'\gamma \in \mathcal{E}_0$ obtained under the models $\mathcal{M}_a$ and $\mathcal{M}$, as it does when additionally $C(\mathbf{W} : \mathbf{Z}) \subset C(\mathbf{V})$ (cf. [1] Theorem 3.3). Instead of this, it follows that (for $\mathcal{S} \neq \mathcal{S}_a$), under the conditions given in Theorem 1, the BLUE of every functional $\mathbf{p}'\gamma \in \mathcal{E}_0$ obtained in the model $\mathcal{M}_a$ continues to be the BLUE of $\mathbf{p}'\gamma$ under the model $\mathcal{M}$ (in other words, assuming that the model $\mathcal{M}$ is true, the equality $\mathbf{p}'\hat{\gamma}_a = \mathbf{p}'\hat{\gamma}$ holds almost surely, i.e., for every $\mathbf{y}$ for which the model $\mathcal{M}$ is consistent). This statement is a straightforward conclusion of (11) and Corollary 1 given in [6]. In the context of partitioned models, it seems to be an independently interesting contribution to discussion of the problem when the BLUE of every estimable parametric function under the general linear model $\{\mathbf{y}, \mathbf{X}\beta, \sigma^2\mathbf{V}\}$ continues to be its BLUE under the restricted model $\{\mathbf{y}, \mathbf{X}\beta|\mathbf{R}\beta = \mathbf{0}, \sigma^2\mathbf{V}\}$ (cf. [7, 8]).

To trace back and gain a deeper insight into the characterization of the equality $\mathcal{E}_a = \mathcal{E}_0$, first refer to a situation when the consistency conditions for the models $\mathcal{M}$ and $\mathcal{M}_a$ coincide. Let $C(\mathbf{Z}) \subset C(\mathbf{V} : \mathbf{W})$, that is, $\mathcal{S}_a = \mathcal{S}$, then the projection $\mathbf{P}_{\mathbf{VW}^\perp|\mathbf{W}}\mathbf{Z}$ is well defined and, by (2), the condition (8) can be expressed as $C(\mathbf{P}_{\mathbf{VW}^\perp|\mathbf{W}}\mathbf{Z}) \subset C(\mathbf{Z})$. Since the equality $\mathbf{P}_{\mathbf{VW}^\perp|\mathbf{W}} + \mathbf{P}_{\mathbf{W}|\mathbf{VW}^\perp} = \mathbf{I}$ holds true onto the column space of $(\mathbf{V} : \mathbf{W})$ (cf. [3]), the last inclusion can be rewritten as

$$C(\mathbf{P}_{\mathbf{W}|\mathbf{VW}^\perp}\mathbf{Z}) \subset C(\mathbf{Z}) \tag{13}$$

or, equivalently, $C(\mathbf{W}) \cap C(\mathbf{VW}^\perp : \mathbf{Z}) \subset C(\mathbf{Z})$.

Further, restricting the considerations to the weakly singular linear model $\mathcal{M}_a$, wherein $C(\mathbf{W} : \mathbf{Z}) \subset C(\mathbf{V})$, one can express the condition (13) as

$$\mathbf{P}_{\mathbf{Z}|\mathbf{VZ}^\perp}\mathbf{P}_{\mathbf{W}|\mathbf{VW}^\perp}\mathbf{Z} = \mathbf{P}_{\mathbf{W}|\mathbf{VW}^\perp}\mathbf{Z}$$

or, making use of a general representation of oblique projectors (cf. [3]), in the form stated in [1], Theorem 3.5; i.e., as

$$\mathbf{W}'\mathbf{V}^-\mathbf{P}_{\mathbf{Z}|\mathbf{VZ}^\perp}\mathbf{P}_{\mathbf{W}|\mathbf{VW}^\perp}\mathbf{Z} = \mathbf{W}'\mathbf{V} - \mathbf{Z}$$

For further discussion of $\mathcal{E}_a = \mathcal{E}_0$ under the more restrictive assumption $\mathbf{V} = \mathbf{I}$, the reader is referred to [2, 5].

In the next theorem, the subclass $\mathcal{E}_0$ of functionals $\mathbf{p}'\gamma$ that can be estimated with full efficiency under the general linear model $\mathcal{M}_a$ is characterized. This extends the corresponding result obtained in the context of the weakly singular model (cf. [1], Theorem 3.1).

Theorem 2.   The subclass $\mathcal{E}_0$ of functionals $\mathbf{p}'\gamma \in \mathcal{E}_a$ for which the BLUE under the model $\mathcal{M}_a$ has the same variance as the corresponding estimator under the model $\mathcal{M}$ is given by

$$\mathcal{E}_0 = \{\mathbf{p}'\gamma : \mathbf{p} = \mathbf{W}'\mathbf{q}, \mathbf{q} \in \mathcal{C}^\perp(\mathbf{Z} : \mathbf{VW}^\perp) \cap \mathcal{C}(\mathbf{Z} : \mathbf{V})\} \qquad (14)$$

Proof.   Let $\mathbf{p}'\gamma \in \mathcal{E}_a$, that is, $\mathbf{p} = \mathbf{W}'\mathbf{q}$ and $\mathbf{q} \in \mathcal{C}(\mathbf{Q_Z})$. By the remark preceding Theorem 1, consider $\mathbf{p}'\hat{\gamma}$ as the BLUE of $\mathbf{q}'\mathbf{W}\gamma$ in the model $\{\mathbf{y}, \mathbf{W}\gamma + \mathbf{Z}\delta | \mathbf{Q_W Z}\delta = \mathbf{0}, \sigma^2\mathbf{V}\}$. Then putting $\mathbf{X} = (\mathbf{W} : \mathbf{Z})$, $\mathbf{K} = \mathbf{q}'\mathbf{X}$, and $\mathbf{R} = \mathbf{Q_W X}$, and making use of (3) and (4), one can write the necessary and sufficient conditon for $Var(\mathbf{p}'\gamma) = Var(\mathbf{p}'\hat{\gamma}_a)$ in the form

$$\mathbf{q}'\mathbf{P}_{\mathbf{X}|\mathbf{VX}^\perp}\mathbf{VQ_W} = \mathbf{0} \qquad (15)$$

On account of the relations (2) and $\mathcal{C}(\mathbf{VX}^\perp) \subset \mathcal{C}(\mathbf{VQ_W})$, it follows that $\mathcal{C}(\mathbf{P}_{\mathbf{X}|\mathbf{VX}^\perp}\mathbf{VQ_W}) = \mathcal{C}(\mathbf{X}) \cap (\mathbf{VQ_W})$ and, consequently, the equality (15) can be expressed as

$$\mathbf{q} \in \mathcal{C}(\mathbf{Q_Z}) \cap [\mathcal{C}^\perp(\mathbf{X}) + \mathcal{C}^\perp(\mathbf{VQ_W})]$$

Thus, applying (??) and noting that all $\mathbf{q} \in \mathcal{C}^\perp(\mathbf{X})$ generate zero functionals, one obtains

$$\mathcal{E}_0 = \{\mathbf{p}'\gamma : \mathbf{p} = \mathbf{W}'\mathbf{q}, \mathbf{q} \in \mathcal{C}^\perp(\mathbf{Z} : \mathbf{VW}^\perp)\}$$

Now, to exclude functionals which are estimated with zero variance, observe that $Var(\mathbf{p}'\hat{\gamma}_a) = 0$ if and only if $\mathbf{q} \in \mathcal{C}^\perp(\mathbf{V} : \mathbf{Z})$; this follows from (2), (1), and the fact that $\mathbf{p}'\hat{\gamma}_a$, being the BLUE of $\mathbf{q}'\mathbf{X}\beta$ under the model $\{\mathbf{y}, \mathbf{X}\beta, \sigma^2\mathbf{V}\}$, can be represented in the form $\mathbf{q}'\mathbf{P}_{\mathbf{X}|\mathbf{VX}^\perp}\mathbf{y}$ (cf. [3], Theorem 3.2. Since the relation $\mathcal{C}^\perp(\mathbf{Z} : \mathbf{V}) \subset \mathcal{C}^\perp(\mathbf{Z} : \mathbf{VW}^\perp)$ together with (1) implies

$$\mathcal{C}^\perp(\mathbf{Z} : \mathbf{VW}^\perp) = \mathcal{C}^\perp(\mathbf{Z} : \mathbf{V}) \oplus \mathcal{C}(\mathbf{Z} : \mathbf{V}) \cap \mathcal{C}^\perp(\mathbf{Z} : \mathbf{VW}^\perp)$$

the proof of (14) is complete.

The dimension of the subspace $\mathcal{E}_0$ can easily be determined. Let $\mathbf{T}$ be any matrix such that $\mathcal{C}(\mathbf{T}) = \mathcal{C}(\mathbf{Z} : \mathbf{V}) \cap \mathcal{C}^\perp(\mathbf{Z} : \mathbf{VW}^\perp)$. Then, by the equality $\mathcal{C}(\mathbf{W} : \mathbf{VW}^\perp) = \mathcal{C}(\mathbf{W} : \mathbf{V})$, it follows that $\mathcal{C}(\mathbf{T}) \cap \mathcal{C}^\perp(\mathbf{W}) = \{\mathbf{0}\}$ and, consequently,

$$dim(\mathcal{E}_0) = r(\mathbf{W}'\mathbf{T}) = r(\mathbf{T})$$

Further, making use of the well-known properties of the rank of matrices, $r(\mathbf{AB}) = r(\mathbf{A}'\mathbf{AB}) = r(\mathbf{A}) - dim[\mathcal{C}(\mathbf{A}) \cap \mathcal{C}^\perp(\mathbf{AB})]$, with $\mathbf{A} = (\mathbf{Z} : \mathbf{V})$ and $\mathbf{AB} = (\mathbf{Z} : \mathbf{VW}^\perp)$, one obtains the equality $r(\mathbf{T}) = r(\mathbf{Z} : \mathbf{V}) - r(\mathbf{Z} : \mathbf{VW}^\perp)$. By $r(\mathbf{A} : \mathbf{B}) = r(\mathbf{A}) + r(\mathbf{Q_A B})$, this leads to the following conclusion.

Corollary 1. Let $\mathcal{E}_0$ be the subspace of functionals $\mathbf{p}'\gamma$ given in (14), then

$$dim(\mathcal{E}_0) = dim[\mathcal{C}(\mathbf{VZ}^\perp) \cap \mathcal{C}(\mathbf{W})]$$

3. REMARKS

The conditions stated in Theorem 1, when referred to the standard partitioned model with $\mathbf{V} = \mathbf{I}$, coincide with the orthogonality condition relevant to the analysis of variance context. Consider a two-way classification mode $\mathcal{M}_a(\mathbf{I}) = \{y, \mathbf{W}\gamma + \mathbf{Z}\delta, \sigma^2\mathbf{I}\}$ where $\mathbf{W}$ and $\mathbf{Z}$ are known design matrices for factors, and the unknown components of the vectors γ and δ represent factor effects. An experimental design, embraced by the model $\mathcal{M}_a(\mathbf{I})$, is said to be orthogonal if $\mathcal{C}(\mathbf{W}:\mathbf{Z}) \cap \mathcal{C}(\mathbf{W}^\perp) \perp \mathcal{C}(\mathbf{W}:\mathbf{Z}) \cap \mathcal{C}(\mathbf{Z}^\perp)$ (cf. [9]). As shown in [10], p.43, this condition can equivalently be expressed as

$$\mathcal{C}(\mathbf{W} : \mathbf{Z}) \cap \mathcal{C}(\mathbf{W}^\perp) \subset \mathcal{C}(\mathbf{Z}) \tag{16}$$

$$\mathcal{C}(\mathbf{W} : \mathbf{Z}) \cap \mathcal{C}(\mathbf{Z}^\perp) \subset \mathcal{C}(\mathbf{W}) \tag{17}$$

or, making use of orthogonal projectors, $\mathbf{PwPz} = \mathbf{PzPw}$. The concept of orthogonality, besides playing an important role when testing a set of nested hypotheses, allows for evaluating efficiency of a design (by which is meant its precision relative to that of an orthogonal design). For a discussion of orthogonality and efficiency concepts in block designs, the reader is referred to [11]. It is well known that one of the characteristics of orthogonality is that the BLUE of $\eta = \mathbf{W}'\mathbf{Q}_z\mathbf{W}\gamma$ does not depend on the presence of δ in the model $\mathcal{M}_a(\mathbf{I})$, or, equivalently, $D(\hat{\eta}) = D(\hat{\eta}_a)$ (cf. [5], Theorem 6). The result stated in Theorem 1 can be viewed as an extension of this characteristic to the linear model with possibly singular dispersion matrix. First, note that the role of (8) and (9) is the same as that of (16) and (17), in the sense that each of them is a necessary and sufficient condition for η to be estimated with the same dispersion matrix under the respective models with and without nuisance parameters. Further, considering the case when $\mathcal{C}(\mathbf{Z}) \subset \mathcal{C}(\mathbf{V} : \mathbf{W})$, which assures that $\mathcal{S} = \mathcal{S}_a$, the conditions of Theorem 1 are equivalent to the equality $\hat{\eta} = \hat{\eta}_a$ almost surely. Thus, as in the analysis of variance context, it provides a basis for evaluating the efficiency of other designs having the same subspaces of design matrices and the same singular dispersion matrix. The singular linear model, commonly used for analyzing categorical data, cf. [12, 13], can be mentioned here as an example of such a context to which this remark refers.

REFERENCES

1. K. Nordström, J. Fellman. Characterizations and dispersion-matrix robustness of efficiently estimable parametric functionals in linear models with nuisance parameters, *Linear Algebra Appl.* 127:341–361, 1990.

2. J. K. Baksalary. A study of the equivalence between Gauss–Markoff model and its augmentation by nuisance parameters, *Math. Operationsforsch. Statist. Ser. Statist.* 15:3–35, 1984.

3. C. R. Rao. Projectors, generalized inverses and the BLUE's, *J. Roy. Statist. Soc. Ser. B* 36:442–448, 1974.

4. J. K. Baksalary, P. R. Pordzik. Inverse-partitioned-matrix method for the general Gauss–Markov model with linear restrictions, *J. Statist. Plann. Inference* 23:133–143, 1989.

5. J. K. Baksalary. Algebraic characterizations and statistical implications of the commutativity of orthogonal projectors, *Proceedings of the Second International Tampere Conference in Statistics* (T. Pukkila and S. Puntanen, eds.), University of Tampere, 1987, pp.113–142.

6. J. K. Baksalary, P. R. Pordzik. Implied linear restrictions in the general Gauss–Markov model, *J. Statist. Plann. Inference* 30:237–248, 1992.

7. T. Mathew. A note on the best linear unbiased estimation in the restricted general linear model, *Math. Operationsforsch. Statist. Ser. Statist.* 14:3–6, 1983.

8. W. H. Yang, H. Cui, G. Sun. On best linear unbiased estimation in the restricted general linear model, *Statistics* 18:17–20, 1987.

9. J. N. Darroch, S. D. Silvey. On testing more than one hypothesis, *Ann. Math. Statist.* 34:555–567, 1963.

10. G. A. F. Seber. *The Linear Hypothesis: A General Theory*, 2nd edition, Griffin's Statistical Monographs & Courses, Charles Griffin, London, 1980.

11. T. Calínski. Balance, efficiency and orthogonality concepts in block designs, *J. Statist. Plann. Inference* 36:283–300, 1993.

12. D. G. Bonett, J. A. Woodward, P. M. Bentler. Some extensions of a linear model for categorical variables, *Biometrics* 41:745–750, 1985.

13. D. R. Thomas, J. N. K. Rao. On the analysis of categorical variables by linear models having singular covariance matrices, *Biometrics* 44:243–248, 1988.

4

Estimation under LINEX Loss Function

AHMAD PARSIAN   Isfahan University of Technology, Isfahan, Iran

S.N.U.A. KIRMANI   University of Northern Iowa, Cedar Falls, Iowa

1. INTRODUCTION

The classical decision theory approach to point estimation hinges on choice of the loss function. If θ is the estimand, $\delta(X)$ the estimator based on a random observable X, and $L(\theta, d)$ the loss incurred on estimating θ by the value d, then the performance of the estimator δ is judged by the risk function $R(\theta, \delta) = E\{L(\theta, \delta(X))\}$. Clearly, the choice of the loss function L may be crucial. It has always been recognized that the most commonly used squared error loss(SEL) function

$$L(\theta, d) = (d - \theta)^2$$

is inappropriate in many situations. If the SEL is taken as a measure of inaccuracy, then the resulting risk $R(\theta, \delta)$ is often too sensitive to the assumptions about the behavior of the tail of the probability distribution of X. The choice of SEL may be even more undesirable if it is supposed to represent a real financial loss. When overestimation and underestimation are not equally unpleasant, the symmetry in SEL as a function of $d - \theta$ becomes a burden. In practice, overestimation and underestimation of the same magnitude often have different economic consequences and the actual loss function is asymmetric. There are numerous such examples in the literature. Varian (1975) pointed out that, in real-estate valuations, underassessment results in an approximately linear loss of revenue whereas overassessment often results in appeals with attendant substantial litigation and other costs. In food-processing industries it is undesirable to overfill containers, since there is no cost recovery for the overfill. If the containers are underfilled, however, it is possible to incur a much more severe penalty arising from misrepresentation of the product's actual weight or volume, see Harris (1992). In dam construction an underestimation of the peak water level is usually much more serious than an overestimation, see Zellner (1986). Underassessment of the value of the guarantee time devalues the true quality, while it is a serious error, especially from the business point of view, to overassess the true value of guarantee time. Further, underestimation of the failure rate may result in more complaints from customers than expected. Naturally, it is more serious to underestimate failure rate than to overestimate failure rate, see Khattree (1992). Other examples may be found in Kuo and Dey (1990), Schabe (1992), Canfield (1970), and Feynman (1987). All these examples suggest that any loss function associated with estimation or prediction of such phenomena should assign a more severe penalty for overestimation than for underestimation, or vice versa.

Ferguson (1967), Zellner and Geisel (1968), Aitchison and Dunsmore (1975), Varian (1975), Berger (1980), Dyer and Keating (1980), and Cain (1991) have all felt the need to consider asymmetric alternatives to the SEL. A useful alternative to the SEL is the convex but asymmetric loss function

$$L(\theta, d) = b \exp[a(d - \theta)] - c(d - \theta) - b$$

where a, b, c are constants with $-\infty < a < \infty$, $b > 0$, and $c \neq 0$. This loss function, called the LINEX (LINear-EXponential) loss function, was proposed by Varian (1975) in the context of real-estate valuations. In addition, Klebanov (1976) derived this loss function in developing his theory of loss functions satisfying a Rao–Blackwell condition. The name LINEX is justified by the fact that this loss function rises approximately linearly on one side of zero and approximately exponentially on the other side. Zellner (1986) provided a detailed study of the LINEX loss function and initiated a good deal of interest in estimation under this loss function. The objective of the present paper is to provide a brief review of the literature on point estimation of a real-valued/vector-valued parameter when the loss function is LINEX.

The outline of this paper is as follows. The LINEX loss function and its key properties are discussed in Section 2. Section 3 is concerned with unbiased estimation under LINEX loss; general results available in the literature are surveyed and results for specific distributions listed. The same mode of presentation is adopted in Sections 4 and 5: Section 4 is devoted to invariant estimation under LINEX loss and Section 5 covers Bayes and minimax estimation under the same loss function. The LINEX loss function has received a good deal of attention in the literature but, of course, there remain a large number of unsolved problems. Some of these open problems are indicated in appropriate places.

2. LINEX LOSS FUNCTION AND ITS PROPERTIES

Thompson and Basu (1996) identified a family of loss functions $L(\Delta)$, where Δ is either the estimation error $\delta(X) - \theta$ or the relative estimation error $(\delta(X) - \theta)/\theta$, such that

- $L(0) = 0$
- $L(\Delta) > (<)L(-\Delta) > 0$, for all $\Delta > 0$
- $L(\cdot)$ is twice differentiable with $L'(0) = 0$ and $L''(\Delta) > 0$ for all $\Delta \neq 0$, and
- $0 < L'(\Delta) > (<) - L'(-\Delta) > 0$ for all $\Delta > 0$

Such loss functions are useful whenever the actual losses are nonnegative, increase with estimation error, overestimation is more (less) serious than underestimation of the same magnitude, and losses increase at a faster (slower) rate with overestimation error than with underestimation error. Considering the loss function

$$L^*(\Delta) = b\exp(a\Delta) + c\Delta + d$$

and imposing the restrictions $L^*(0) = 0$, $(L^*)'(0) = 0$, we get $d = -b$ and $c = -ab$; see Thompson and Basu (1996). The resulting loss function

$$L^*(\Delta) = b\{\exp(a\Delta) - a\Delta - 1\} \tag{2.1}$$

when considered as a function of θ and δ, is called the LINEX loss function. Here, a and b are constants with $b > 0$ so that the loss function is nonnegative. Further,

$$L'(\Delta) = ab\{\exp(a\Delta) - 1\}$$

so that $L'(\Delta) > 0$ and $L'(-\Delta) < 0$ for all $\Delta > 0$ and all $a \neq 0$. In addition,

$$L(\Delta) - L(-\Delta) = 2b\{\sinh(a\Delta) - a\Delta\}$$

and

$$L'(\Delta) + L'(-\Delta) = 2ab\{\cosh(a\Delta) - 1\}$$

Thus, if $a > (<)0$, the $L'(\Delta) > (<) - L'(-\Delta)$ and $L(\Delta) > (<)L(-\Delta)$ for $\Delta > 0$. The shape of the LINEX loss function (2.1) is determined by the constant a, and the value of b merely serves to scale the loss function. Unless mentioned otherwise, we will take $b = 1$.

In Figure 1, values of $\exp(a\Delta) - a\Delta - 1$ are plotted against Δ for selected values of a. It is seen that for $a > 0$, the curve rises almost exponentially when $\Delta > 0$ and almost linearly when $\Delta < 0$. On the other hand, for $a < 0$, the function rises almost exponentially when $\Delta < 0$ and almost linearly when $\Delta > 0$. So the sign of a reflects the direction of the asymmetry, $a > 0$ $(a < 0)$ if overestimation is more (less) serious than underestimation; and its magnitude reflects the degree of the asymmetry. An important observation is that for small values of $|a|$ the function is almost symmetric and not far from a squared error loss (SEL). Indeed, on expanding $\exp(a\Delta) \approx 1 + a\Delta + a^2\Delta^2/2$, $L(\Delta) \approx a^2\Delta^2/2$, a SEL function. Thus for small values of $|a|$, optimal estimates and predictions are not very different from those obtained with the SEL function. In fact, the results under LINEX loss are consistent with the "heuristic" that estimation under

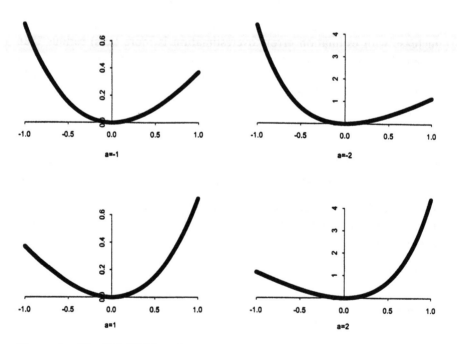

Figure 1. The LINEX loss function.

SEL corresponds to the limiting case $a \to 0$. However, when $|a|$ assumes appreciable values, optimal point estimates and predictions will be quite different from those obtained with a symmetric SEL function, see Varian (1975) and Zellner (1986).

The LINEX loss function (2.1) can be easily extended to meet the needs of multiparameter estimation and prediction. Let Δ_i be the estimation error $\delta_i(X) - \theta_i$ or the relative error $(\delta_i(X) - \theta_i)/\theta_i$ when the parameter θ_i is estimated by $\delta_i(X)$, $i = 1, \ldots, p$. One possible extension of (2.1) to the p-parameter problem is the so-called separable extended LINEX loss function given by

$$L(\Delta) = \sum_{i=1}^{p} b_i \{ \exp(a_i \Delta_i) - a_i \Delta_i - 1 \}$$

where $a_i \neq 0$, $b_i > 0$, $i = 1, \ldots, p$ and $\Delta = (\Delta_1, \ldots, \Delta_p)'$. As a function of Δ, this is convex with minimum at $\Delta = (0, \ldots, 0)$. The above loss function has been utilized in multi-parameter estimation and prediction problems, see Zellner (1986) and Parsian (1990a).

When selecting a loss function for the problem of estimating a location parameter θ, it is natural to insist on the invariance requirement $L(\theta + k, d + k) = L(\theta, d)$, where $L(\theta, d)$ is the loss incurred on estimating θ by the value d; see Lehmann and Casella (1998). The LINEX loss function (2.1) satisfies this restriction if $\Delta = \delta(X) - \theta$ but not if $\Delta = (\delta(X) - \theta)/\theta$. Similarly, the loss function (2.1) satisfies the invariance requirement $L(k\theta, kd) = L(\theta, d)$ for $\Delta = (\delta(X) - \theta)/\theta$ but not for $\Delta = \delta(X) - \theta$. Consequently, when estimating a location (scale) parameter and adopting the LINEX loss, one would take $\Delta = \delta(X) - \theta$ ($\Delta = (\delta(X) - \theta)/\theta$) in (2.1). This is the strategy in Parsian et al (1993) and Parsian and Sanjari (1993). An analogous approach may be adopted for the p-parameter problem when using the separable extended LINEX loss function. If the problem is one of estimating the location parameter μ in a location-scale family parameterized by $\theta = (\mu, \sigma)$, see Lehmann and Casella (1998), the loss function (2.1) will satisfy the invariance requirement $L((\alpha + \beta\mu, \beta\sigma), \alpha + \beta d) = L((\mu, \sigma), d)$ on taking $\Delta = (\delta(X) - \mu)/\sigma$, where $\beta > 0$.

For convenience in later discussion, let

$$\rho(t) = b \{ \exp(at) - at - 1 \}, \qquad -\infty < t < +\infty$$

where $a \neq 0$ and $b > 0$. Then, $\rho(t)$ may be described as the kernel of the LINEX loss function. In terms of $\rho(t)$, the LINEX loss function corresponding to $\Delta = \delta(X) - \theta$ is

$$L(\theta, \delta) = \rho(\delta(X) - \theta) \tag{2.2}$$

whereas the LINEX loss function corresponding to $\Delta = (\delta(X) - \theta)/\theta$ is

$$L(\theta, \delta) = \rho((\delta(X) - \theta)/\theta) \tag{2.3}$$

3. UNBIASED ESTIMATION UNDER LINEX LOSS

The accuracy, or rather inaccuracy, of an estimator δ in estimating an unknown parameter $\theta \in \Theta$ is measured by the risk function

$$R(\theta, \delta) = E_\theta\{L(\theta, \delta(X))\} \tag{3.1}$$

It is well known that, in general, estimators or predictors with uniformly minimum risk (UMR) do not exist. One way of avoiding this difficulty is to restrict the class of estimators by ruling out estimators that too strongly favor one or more values of θ at the cost of neglecting other possible values. This can be achieved by requiring the estimator to satisfy some condition enforcing a certain degree of impartiality. One such condition requires that the "bias" of estimation, sometimes called the systematic error of the estimator δ, be zero. The purpose of this section is to show that a similar theory of estimation under LINEX loss can be obtained.

The classical notion of mean-unbiasedness

$$E_\theta(\delta(X)) = \theta, \qquad \forall \theta \in \Theta \tag{3.2}$$

is "symmetric" in the estimation error $\Delta = \delta(X) - \theta$ and is inappropriate if overestimation and underestimation are not equally serious, see Andrews and Phillips (1987). Lehmann (1951) provided a valuable insight into the concept of unbiasedness. He introduced a more general concept of risk-unbiasedness which reduces to the usual notion of mean-unbiasedness (3.2) in the case of SEL; see Lehmann (1988) for a detailed exposition. Following Lehmann's definition, an estimator $\delta(X)$ of θ is said to be risk-unbiased if it satisfies

$$E_\theta\{(\theta, \delta(X))\} \le E_\theta\{L(\theta', \delta(X))\}, \qquad \forall \theta' \ne \theta \tag{3.3}$$

If the loss function is as in (2.2), then (3.3) reduces to

$$E_\theta\{\exp[a\delta(X)]\} = \exp(a\theta), \qquad \forall \theta \in \Theta \tag{3.4}$$

We say $\delta(X)$ is a LINEX-unbiased estimator (L-UE) of θ w.r.t. the loss (2.2) if it satisfies condition (3.4) and we then refer to θ as an L-estimable parameter. If (3.4) does not hold, $\delta(X)$ is said to be biased and we define its bias as (see Parsian and Sanjari 1999)

$$b(\theta) = a^{-1}\{\ln[\exp(a\theta)] - \ln E_\theta\{\exp[a\delta(X)]\}\}$$
$$= \theta - a^{-1} \ln E_\theta\{\exp[a\delta(X)]\} \tag{3.5}$$

Intuitively, condition (3.4) seems to be more appropriate in the context of LINEX loss than condition (3.2), which does not take into account the direction of the estimation error.

It is easy to verify that if

$$E_\theta[g(X)] = \exp(a\theta), \qquad \forall \theta \in \Theta \tag{3.6}$$

then $\delta(X) = a^{-1} \ln[g(X)]$ is an L-UE of θ, provided that $g(X)$ is positive with probability one. Also, if δ is an L-UE of θ, then the risk function with respect to the loss (2.2) reduces to

$$R(\theta, \delta) = -aE_\theta[\delta(X) - \theta], \qquad \forall \theta \in \Theta \tag{3.7}$$

If the loss function is as in (2.3) with $\theta = \sigma$, then (3.3) reduces to

$$E_\sigma\left\{ \delta(X) e^{a\delta(X)/\sigma} \right\} = e^a E_\sigma[\delta(X)] \tag{3.8}$$

For LINEX loss, the function $L(\theta, \delta)$ is convex in $\delta(X)$. Hence, a theory of uniformly minimum risk unbiased (UMRU) estimation under LINEX loss can be easily developed on the same lines as the classical theory of uniformly minimum variance unbiased (UMVU) estimation. In particular, it follows from the Rao–Blackwell–Lehmann–Scheffe theorem (see Lehmann and Casella1998) that, if T is a complete sufficient statistic for θ and $h(X)$ is any L-UE of $\gamma(\theta)$, then

$$\delta^*(T) = a^{-1} \ln E\{\exp[ah(X)|T\} \tag{3.9}$$

is the unique UMRU estimator (UMRUE) of $\gamma(\theta)$ under LINEX loss (2.2). Thus, in presence of a complete sufficient statistic T, if $g(T) > 0$ with probability one and

$$E_\theta[g(T)] = \exp(a\gamma(\theta)), \qquad \forall \theta \in \Theta$$

the estimator $\delta^*(T) = a^{-1} \ln[g(T)]$ is UMRUE of $\gamma(\theta)$.

It must be mentioned here that Klebanov (1976), who did not use the name LINEX, was the first to discuss L-UE and some related notions. Following up on Klebanov's work, Shafie and Noorbaloochi (1995) proved that the best LINEX-unbiased estimators dominate the corresponding UMVU estimators of the location parameter of a location family. We refer to Shafie and Noorbaloochi (1995) for proof, additional properties, and interesting insights.

Example. Let $X \sim \Gamma(\alpha, \sigma)$. That is, let X have a gamma distribution with shape parameter α and scale parameter σ so that X has mean $\alpha\sigma$ and variance $\alpha\sigma^2$. Suppose α is known, and consider the problem of estimating σ when the loss function is the LINEX loss (2.2) with $\theta = \sigma$. Define

$$H_\alpha(x) = \sum_{k=1}^{\infty} \frac{\Gamma(\alpha)}{\Gamma(\alpha + k)} \frac{x^k}{k!}$$

It is easy to see that

$$E_\sigma[H_\alpha(X)] = \exp(\sigma)$$

Since X is complete sufficient for σ, it follows that, for $a > 0$,

$$\delta^*(X) = \frac{1}{a} \ln H_\alpha(aX)$$

is UMRUE of σ. The function $H_\alpha(x)$ defined above is useful in a number of problems to be considered later.

3.1 Special Probability Models

Normal distribution

Let $X_1, \ldots, X_n$ be a random sample of size n from $N(\theta, \sigma^2)$. The problem is to find the best L-UE of θ using LINEX loss (2.2).

(a) Variance is known

It is easy to check that the UMRUE of θ is

$$\delta^*(X) = \bar{X} - \frac{a\sigma^2}{2n}$$

(b) Variance is unknown

In this case, based on the $H_\alpha(\cdot)$ function, UMRUE of θ does not exist.

(c) Estimation of variance

It is well known that sample mean and sample variance are jointly complete sufficient for (θ, σ^2) and

$$\sum (X_i - \bar{X})^2 \sim \sigma^2 \chi^2_{(n-1)}$$

Using this fact, it can be shown that

$$\frac{1}{a} \ln \left\{ H_{(n-1)/2} \left(\frac{a}{2} \sum (X_i - \bar{X})^2 \right) \right\}$$

is UMRUE of σ^2 if $a > 0$. σ^2 is not L-estimable if $a < 0$.

Poisson distribution

Let $X_1, \ldots, X_n$ be a random sample of size n from the Poisson distribution with mean θ and consider the problem of estimating θ using LINEX loss (2.2). It is easy to verify that

$$\frac{n}{a} \ln\left(1 + \frac{a}{n}\right) \bar{X}$$

is UMRUE of θ provided that $a > -n$. Obviously θ is not L-estimable if $a < -n$.

Binomial distribution

Let $X \sim B(n, p)$, where n is known. To find an L-UE of $\gamma(p)$, using (3.6), we have to find a mean-unbiased estimator of $\exp\{a\gamma(p)\}$. But it is well known, see Lehmann and Casella (1998), that $\exp\{a\gamma(p)\}$ has a mean-unbiased estimator if and only if it is a polynomial of degree at most n. Thus, L-UEs of $\gamma(p)$ exist only in very restricted cases. In particular, $\gamma(p) = p$ is not L-estimable.

Exponential distribution

Let $X_1, \ldots, X_n$ be a random sample of size n from $E(\mu, \sigma)$, where $E(\mu, \sigma)$ denotes an exponential distribution with location parameter μ and mean $\mu + \sigma$. The problem is to find the best L-UE of the unknown parameter using LINEX loss (2.2).

(a) Scale parameter σ is known

It is easy to verify that, in this case, the UMRUE of μ is

$$X_{(1)} + \frac{1}{a} \ln\left(1 - \frac{a\sigma}{n}\right)$$

provided that $a\sigma < n$. If $a\sigma > n$, the parameter μ is not L-estimable.

(b) Scale parameter σ is unknown

It is easy to verify that, in this case, the UMRUE of μ is

$$X_{(1)} + \frac{1}{a} \ln\left\{1 - \frac{a}{n(n-1)} \sum (X_i - X_{(1)})\right\}$$

provided that $a < 0$. If $a > 0$, the parameter μ is not L-estimable; see Khattree (1992) for a different discussion of this problem under the loss (2.2).

(c) Location parameter μ is known

It is easy to see that, in this case, $T = \sum(X_i - \mu)$ is complete sufficient for σ and that the best L-UE of σ is

$$\frac{1}{a}\ln[H_n(aT)]$$

provided that $a > 0$. σ is not L-estimable if $a < 0$.

(d) Location parameter μ is unknown

The problem is to estimate σ using LINEX loss, when μ is a nuisance parameter, and it is easy to see that, in this case, $T^* = \sum(X_i - X_{(1)})$ is a function of complete sufficient statistic and that the best L-UE of σ is

$$\frac{1}{a}\ln[H_{n-1}(aT^*)]$$

provided that $a > 0$. σ is not L-estimable if $a < 0$.

Two or more exponential distributions

Let $X_{11}, \ldots, X_{1n1}$ and $X_{21}, \ldots, X_{2n2}$ be independent random samples from $E(\mu_1, \sigma_1)$ and $E(\mu_2, \sigma_2)$, respectively. The problem is to find the best L-UE of the unknown parameter of interest. This problem arises in life testing and reliability when a system consists of two independent components having minimum guaranteed lives μ_1 and μ_2 and constant hazard rates $1/\sigma_1$, $1/\sigma_2$. The estimators given below are obtained and discussed in Parsian and Kirmani (2000).

(a) σ_1, σ_2 are known and $\mu_1 = \mu_2 = \mu$ but is unknown

Let $W = \min(X_{1(1)}, X_{2(1)})$, where $X_{i(1)} = \min\{X_{i1}, \ldots, X_{ini}\}$. Then it can be seen that

$$W + \frac{1}{a}\ln\left(1 - \frac{a}{\sigma}\right)$$

is the best L-UE of μ, provided that $a < \sigma$ where $\sigma = n_1\sigma_1^{-1} + n_2\sigma_2^{-1}$. Obviously, μ is not L-estimable if $a > \sigma$.

(b) $\mu_1 = \mu_2 = \mu$ and μ, σ_1, σ_2 are unknown

Parsian and Kirmani (2000) show that the best L-UE of μ in this case is

$$W + \frac{1}{a}\ln(1 - aG^{-1})$$

provided that $a < 0$. Here $G = \sum n_i(n_i - 1)T_i^{-1}$ and $T_i = \sum(X_{ij} - W)$. μ is not L-estimable if $a > 0$. This problem is discussed for the SEL case by Ghosh and Razmpour (1984).

(c) $\sigma_1 = \sigma_2 = \sigma$ and μ_1, μ_2, σ are unknown

Let $T^* = \sum\sum(X_{ij} - X_{i(1)})$. Then, for $a > 0$, the best L-UE of σ is

$$\frac{1}{a}\ln\left[H_{n_1+n_2-1}(aT^*)\right]$$

and σ is not L-estimable if $a < 0$. On the other hand it can be seen that, for $a < 0$, the best L-UE of μ_i, $i = 1, 2$, is

$$X_{i(1)} + \frac{1}{a}\ln(1 - b_i T^*)$$

where $b_i = a/n_i(n_1 + n_2 - 2)$ and μ_i, $i = 1, 2$, are not L-estimable for $a > 0$. This shows that quantile parameters, i.e., $\mu_i + b\sigma$, are not L-estimable for any b.

4. INVARIANT ESTIMATION UNDER LINEX LOSS FUNCTION

A different impartiality condition can be formulated when symmetries are present in a problem. It is then natural to require a corresponding symmetry to hold for the estimators. The location and scale parameter estimation problems are two important examples. These are invariant with respect to translation and multiplication in the sample space, respectively. This strongly suggests that the statistician should use an estimation procedure which also has the property of being invariant.

We refer to Lehmann and Casella (1998) for the necessary theory of invariant (or equivariant) estimation. Parsian et al. (1993) proved that, under the LINEX loss function (2.2), the best location-invariant estimator of a location parameter θ is

$$\delta^*(X) = \delta_o(X) - a^{-1}\ln\left\{E_{\theta=0}\left[\exp\left(a\delta_o(X)\middle| Y = y\right]\right\}\right. \tag{4.1}$$

where $\delta_o(X)$ is any location-invariant estimator of θ with finite risk and $Y = (Y_1, \ldots, Y_{n-1})$ with $Y_i = X_i - X_n$, $i = 1, \ldots, n-1$.

Notice that $\delta^*(X)$ can be written explicitly as a Pitman-type estimator

$$\delta^*(X) = a^{-1}\ln\left\{\int e^{-au}f(X - u)du \middle/ \int f(X - u)du\right\} \tag{4.2}$$

Parsian et al. (1993) showed that δ^* is, in fact, minimax under the LINEX loss function (2.2).

4.1 Special Probability Models

Normal distribution

(a) Known variance

Let $X_1, \ldots, X_n$ be a random sample of size n from $N(\theta, \sigma^2)$. The problem is to find the best location-invariant estimator of θ using LINEX loss (2.2). It is easy to see that in this case (4.2) reduces to

$$\delta^*(X) = \bar{X} - \frac{a\sigma^2}{2n}$$

This estimator is, in fact, a generalized Bayes estimator (GBE) of θ, it dominates $\bar{X}$, and it is the only minimax-admissible estimator of θ in the class of estimators of the form $c\bar{X} + d$; see Zellner (1986) , Rojo (1987), Sadooghi-Alvandi and Nematollahi (1989) and Parsian (1990b).

Exponential distribution

Let $X_1, \ldots X_n$ be a random sample of size n from $E(\mu, \sigma)$, where σ is known. The problem is to find the best location-invariant estimator of μ using LINEX loss (2.2). It is easy to see that in this case (4.2) reduces to

$$\delta^*(X) = X_{(1)} + \frac{1}{a}\ln\left(1 - \frac{a\sigma}{n}\right)$$

provided that $a\sigma < n$.

Linear model problem

Consider the normal regression model $Y \sim N_p\,(X\beta, \sigma^2 I)$, where X $(p \times q)$ is the known design matrix of rank $q(\leq p)$, $\beta(q \times 1)$ is the vector of unknown regression parameters, and $\sigma(> 0)$ is known. Then the best invariant estimator of $\theta = \lambda'\beta$ is

$$\delta^*(Y) = \lambda'\hat{\beta} - \frac{a\sigma^2}{2}\lambda'(X'X)^{-1}\lambda$$

which is minimax and GBE, where $\hat{\beta}$ is OLSE of β. Also, see Cain and Janssen (1995) for a real-estate prediction problem under the LINEX loss and Ohtani (1995), Sanjari (1997), and Wan (1999) for ridge estimation under LINEX loss.

An interesting open problem here is to prove or disprove that $\delta^*(X)$ is admissible.

Parsian and Sanjari (1993) derived the best scale-invariant estimator of a scale parameter θ under the modified LINEX loss function (2.3). However, unlike the location-invariant case, the best scale-invariant estimator has no general explicit closed form. Also, see Madi (1997).

4.2 Special Probability Models

Normal distribution

Let $X_1, \ldots, X_n$ be a random sample of size n from $N(\theta, \sigma^2)$. The problem is to find the best scale-invariant estimator of σ^2 using the modified LINEX loss (2.3).

(a) Mean known

It is easy to verify that the best scale-invariant estimator of σ^2 is (take $\theta = 0$)

$$\delta^*(X) = \left\{ \frac{1 - e^{-2a/(n+2)}}{2a} \right\} \sum X_i^2$$

(b) Mean unknown

In this case, the best scale-invariant estimator of σ^2 is

$$\delta^*(X) = \left\{ \frac{1 - e^{-2a/(n+1)}}{2a} \right\} \sum (X_i - \bar{X})^2$$

Exponential distribution

Let $X_1, \ldots, X_n$ be a random sample of size n from $E(\mu, \sigma)$. The problem is to find the best scale-invariant estimator of σ using the modified LINEX loss (2.3).

(a) Location parameter is known

In this case, the best scale-invariant estimator of σ is (take $\mu = 0$)

$$\delta^*(X) = \left\{ \frac{1 - e^{-a/(n+1)}}{a} \right\} \sum X_i$$

(b) Location parameter is known

The problem is to estimate σ using LINEX loss, when μ is a nuisance parameter, and in this case the best scale-invariant estimator is

$$\delta^*(X) = \left\{\frac{1 - e^{a/n}}{a}\right\} \sum (X_i - X_{(1)})$$

Usually, the best scale-invariant estimator is inadmissible in the presence of a nuisance parameter. Improved estimators are obtained in Parsian and Sanjari (1993). For pre-test estimators under the LINEX loss functions (2.2) and (2.3), see Srivastava and Rao (1992), Ohtani (1988, 1999), Giles and Giles (1993, 1996) and Geng and Wan (2000). Also, see Pandey (1997) and Wan and Kurumai (1999) for more on the scale parameter estimation problem.

5. BAYES AND MINIMAX ESTIMATION

As seen in the previous two sections, in many important problems it is possible to find estimators which are uniformly (in θ) best among all L-unbiased or invariant estimators. However, this approach of minimizing the risk uniformly in θ after restricting the estimators to be considered has limited applicability. An alternative and more general approach is to minimize an overall measure of the risk function associated with an estimator without restricting the estimators to be considered. As is well known in statistical theory, two natural global measures of the size of the risk associated with an estimator θ are

(1)   the average:

$$\int R(\theta, \delta)\omega(\theta)d\theta \tag{5.1}$$

for some suitably chosen weight function ω and
(2)   the maximum of the risk function:

$$\sup_{\theta \in \Theta} R(\theta, \delta) \tag{5.2}$$

Of course, minimizing (5.1) and (5.2) lead to Bayes and minimax estimators, respectively.

5.1   General Form of Bayes Estimators for LINEX Loss

In a pioneering paper, Zellner (1986) initiated the study of Bayes estimates under Varian's LINEX loss function. Throughout the rest of this paper, we will write $\theta|X$ to indicate the posterior distribution of θ. Then, the posterior risk for δ is

$$E_{\theta|X}[L(\theta, \delta(X)] = e^{a\delta(X)}E_{\theta|X}[e^{-a\theta}] - a[\delta(X) - E_{\theta|X}(\theta)] - 1 \qquad (5.3)$$

Writing $M_{\theta|X}(t) = E_{\theta|X}[e^{t\theta}]$ for the moment-generating function of the posterior distribution of θ, it is easy to see that the value of $\delta(X)$ that minimizes (5.3) is

$$\delta_B(X) = -a^{-1}\ln M_{\theta|X}(-a) \qquad (5.4)$$

provided, of course that, $M_{\theta|X}(\cdot)$ exists and is finite. We now give the Bayes estimator $\delta_B(X)$ for various specific models.

It is worth noting here that, for SEL, the notions of unbiasedness and minimizing the posterior risk are incompatible; see Noorbaloochi and Meeden (1983) as well as Blackwell and Girshick (1954). Shafie and Noorbaloochi (1995) observed the same phenomenon when the loss function is LINEX rather than SEL. They also showed that if θ is the location parameter of a location-parameter family and $\delta(X)$ is the best L-UE of θ then $\delta(X)$ is minimax relative to the LINEX loss and generalized Bayes against the improper uniform prior. Interestingly, we can also show that if $\delta(X)$ is the Bayes estimator of a parameter θ with respect to LINEX loss and some proper prior distribution, and if $\delta(X)$ is L-UE of θ with finite risk, then the Bayes risk of $\delta(X)$ must be zero.

5.2   Special Probability Models

Normal distribution

(a) Variance known

Let $X_1, \ldots, X_n$ be a random sample of size n from $N(\theta, \sigma^2)$. The natural estimator of θ, namely $\bar{X}$, is minimax and admissible relative to a variety of symmetric loss functions.

The conjugate family of priors for θ is $N(\mu, \tau^2)$, and

$$\theta|\bar{X} \sim N\left(\frac{n\tau^2\bar{X} + \sigma^2\mu}{n\tau^2 + \sigma^2}, \frac{\sigma^2\tau^2}{n\tau^2 + \sigma^2}\right)$$

Therefore, the unique Bayes estimator of θ using LINEX loss (2.2) is

$$\delta_B(X) = \frac{n\tau^2 \bar{X}}{n\tau^2 + \sigma^2} - \frac{1}{n\tau^2 + \sigma^2}\left(\frac{a\tau^2\sigma^2}{2} - \mu\sigma^2\right)$$

Notice that as $\tau^2 \to \infty$ (that is, for the diffused prior on $(-\infty, +\infty)$)

$$\delta_B(X) \to \bar{X} - \frac{a\sigma^2}{2n} \equiv \delta^*(X)$$

i.e., $\delta^*(X)$ is the limiting Bayes estimator of θ and it is UMRUE of θ. It is also GBE of θ.

Remarks

(1) $\delta^*(X)$ dominates $\bar{X}$ under LINEX loss; see Zellner (1986).
(2) Let $\delta_{c,d}(X) = c\bar{X} + d$; then it is easy to verify that $\delta_{c,d}(X)$ is admissible for θ under LINEX loss, whenever $0 \leq c < 1$ or $c = 1$ and $d = -a\sigma^2/2n$; otherwise, it is inadmissible; see Rojo (1987) and Sadooghi-Alvandi and Nematollhi (1989). Also, see Pandey and Rai (1992) and Rodrigues (1994).
(3) Taking $\lambda = \sigma^2/n\tau^2$ and $\mu = 0$, we get

$$\delta_B(X) = \frac{1}{1 + \lambda}\left(\bar{X} - \frac{a\sigma^2}{2n}\right)$$

   i.e., $\delta_B(X)$ shrinks $\delta^*(X)$ towards zero.
(4) Under LINEX loss (2.2), $\delta^*(X)$ is the only minimax admissible estimator of θ in the class of all linear estimators of the form $c\bar{X} + d$, see Parsian (1990b). Also, see Bischoff et al (1995) for minimax and Γ-minimax estimation of the normal distribution under LINEX loss function (2.2) when the parameter space is restricted.

(b) Variance unknown

Now the problem is to estimate θ under LINEX loss (2.2), when σ^2 is a nuisance parameter.

   If we replace σ^2 by the sample variance S^2 in $\delta^*(X)$ to get $\bar{X} - aS^2/2n$, the obtained estimator is, of course, no longer a Bayes estimator (it is empirical Bayes!); see Zellner (1986). We may get a better estimator of θ by replacing σ^2 by the best scale-invariant estimator of σ^2. However, a unique Bayes, hence admissible, estimator of θ is obtained when

$$\theta|r \sim N\left(\mu, (\lambda r)^{-1}\right)$$

$$r = 1/\sigma^2 \sim IG(\alpha, \beta)$$

where IG denotes the inverse Gaussian distribution. This estimator is

$$\delta_B(X) = \frac{n\bar{X} + \lambda\mu}{n+\lambda} - \frac{1}{a}\ln\left[\left(1 - \frac{a^2}{(n+\lambda)\alpha}\right)^{\frac{(n+1)}{2}}\right]\frac{K_\nu\left(\sqrt{2\gamma\left(\alpha - \frac{a^2}{n+\lambda}\right)}\right)}{k_\nu(\sqrt{2\gamma\alpha})}$$

provided that $\alpha > a^2/(n+\lambda)$, where $\nu = (n+1)/2$, $2\gamma = nS^2 + n\lambda(n+\lambda)^{-1}/$ $(\bar{X} - \mu)^2 + \alpha/\beta^2$, and $K_\nu(\cdot)$ is the modified Bessel function of the third kind; see Parsian (1990b).

An interesting open problem here is to see if one can get a minimax and admissible estimator of θ, under LINEX loss (2.2), when σ^2 is unknown. See Zou (1997) for the necessary and sufficient conditions for a linear estimator of finite population mean to be admissible in the class of all linear estimators. Also, see Bolfarine (1989) for further discussion of finite population prediction under LINEX loss function.

Poisson distribution

Let $X_1, \ldots, X_n$ be a random sample of size n from $P(\theta)$. The natural estimator of θ is $\bar{X}$ and it is minimax and admissible under the weighted SEL with weight θ^{-1}. The conjugate family of priors for θ is $\Gamma(\alpha, \beta)$, so that

$$\theta|\bar{X} \sim \Gamma(\alpha + n\bar{X}, \beta + n)$$

and the unique Bayes estimator of θ under the loss (2.2) is

$$\delta_B(X) = \left\{\frac{n}{a}\ln\left(\frac{\beta+n+a}{\beta+n}\right)\right\}\left(\bar{X} + \frac{\alpha}{n}\right)$$

where $\beta + n + a > 0$. Note that $\delta_B(X)$ can be written in the form $\delta_{c,d}(X) = c\bar{X} + d$, and $\delta_B(X) \to c^*\bar{X} + d$ as $\beta \to 0$ and $\delta_B(X) \to c^*\bar{X}$ as $\alpha, \beta \to 0$, where $c^* = n/a\ln(1 + a/n)$. Now, it can be verified that $\delta_{c,d}$ is admissible for θ using LINEX loss whenever either $a \leq -n$, $c > 0$, $d > 0$, or $a > -n$, $0 < c < c^*$, $d \geq 0$.

Further, it can be proved that $\delta_{c,d}$ is inadmissible if $c > c^*$ or $c = c^*$ and $d > 0$, because it is dominated by $c^*\bar{X}$; see Sadooghi-Alvandi (1990) and Kuo and Dey (1990). Notice that the proof of Theorem 3.2 for $c = c^*$ in Kuo and Dey (1990) is in error. Also, see Rodrigues (1998) for an application to software reliability and Wan et al (2000) for minimax and Γ-minimax estimation of the Poisson distribution under a LINEX loss function (2.2) when the parameter space is restricted.

An interesting open problem here is to prove or disprove that $c^*\bar{X}$ is admissible.

Evidently, there is a need to develop alternative methods for establishing admissibility of an estimator under LINEX loss. In particular, it would be interesting to find the LINEX analog of the "information inequality" method for proving admissibility and minimaxity when the loss function is quadratic.

Binomial distribution

Let $X \sim B(n, p)$, where n is unknown and p is known. A natural class of conjugate priors for n is $NB(\alpha, \theta)$ and

$$n|X, \alpha, \theta \sim NB(\alpha + X, \theta q)$$

where $q = 1 - p$. The unique Bayes estimator of n is

$$\delta_B(X) = c(\theta)X + \alpha(c(\theta) - 1)$$

where $c(\theta) = 1 + a^{-1} \ln\{(1 - \theta q e^{-a})/(1 - \theta q)\}$.

Now, let $\delta_{c,d}(X) = cX + d$. Then, see Sadooghi-Alvandi and Parsian (1992), it can be seen that

(1)   For any $c < 1$, $d \geq 0$, the estimator $\delta_{c,d}(X)$ is inadmissible.
(2)   For any $d \geq 0$, $c = 1$, the estimator $\delta_{c,d}(X)$ is admissible.
(3)   For any $d \geq 0$, $c > 1$, the estimator $\delta_{c,d}(X)$ is admissible if $a \leq \ln q$.
(4)   Suppose $a > \ln q$ and $c^* = a^{-1} \ln[(e^a - q)/(1 - q)]$, then $c^* > 1$ and
     (a)   if $1 < c < c^*$, then $\delta_{c,d}(X)$ is admissible;
     (b)   if $c > c^*$ or $c = c^*$ and $d > 0$, then $\delta_{c,d}(X)$ is inadmissible, being dominated by c^*X.

Finally, the question of admissibility or inadmissibility of c^*X remains an open problem.

Exponential distribution

Because of the central importance of the exponential distribution in reliability and life testing, several authors have discussed estimation of parameters from different points of view. Among them are Basu and Ebrahimi (1991), Basu and Thompson (1992), Calabria and Pulcini (1996), Khattree (1992), Mostert et al. (1998), Pandey (1997), Parsian and Sanjari (1997), Rai (1996), Thompson and Basu (1996), and Upadhyay et al (1998).

Parsian and Sanjari (1993) discussed the Bayesian estimation of the mean σ of the exponential distribution $E(0, \sigma)$ under the LINEX loss function (2.3); also, see Sanjari (1993) and Madi (1997). To describe the available results, let $X_1, \ldots, X_n$ be a random sample from $E(0, \sigma)$ and adopt the LINEX loss function (2.3) with $\theta = \sigma$. Then it is known that

(1)   The Bayes estimator of σ w.r.t. the inverse gamma prior with parameters α and η is

$$\delta_B(X) = c(\alpha) \sum X_i + d(\alpha, \eta)$$

where $c(\alpha) = (1 - e^{-a/(n+\alpha+1)})/a$, $d(\alpha, \eta) = c(\alpha)\eta$.

(2) As $\alpha \to 0$, $c(\alpha) \to c^*$, and $\delta_B(X) \to c^* \sum X_i + c^* d \equiv \delta^*(X)$, where $c^* = (1 - e^{-a/(n+1)})/a$.

(3) $\delta^*(X)$ with $d = 0$ is the best scale-invariant estimator of σ and is minimax.

(4) Let $\delta_{c,d}(X) = c \sum X_i + d$; then $\delta_{c,d}(X)$ is inadmissible if (a) $c < 0$ or $d < 0$; or (b) $c > c^*$, $d \geq 0$; or (c) $0 \leq c < c^*$, $d = 0$; and it is admissible if $0 \leq c \leq c^*$, $d \geq 0$; see Sanjari (1993).

As for Bayes estimation under LINEX loss for other distributions, Soliman (2000) has compared LINEX and quadratic Bayes estimates for the Rayleigh distribution and Pandey et al. (1996) considered the case of classical Pareto distribution. On a related note, it is crucial to have in-depth study of the LINEX Bayes estimate of the scale parameter of gamma distribution because, in several cases, the distribution of the minimal sufficient statistic is gamma. For more in this connection, see Parsian and Nematollahi (1995).

ACKNOWLEDGMENT

Ahmad Parsian's work on this paper was done during his sabbatical leave at Queen's University, Kingston, Ontario. He is thankful to the Department of Mathematics and Statistics, Queen's University, for the facilities extended. The authors are thankful to a referee for his/her careful reading of the manuscript.

REFERENCES

Aitchison, J., and Dunsmore, I. R. (1975), *Statistical Prediction Analysis.* London: Cambridge University Press.

Andrews, D. W. K., and Phillips, P. C. V. (1987), Best median-unbiased estimation in linear regression with bounded asymmetric loss functions. *J. Am. Statist. Assoc.,* **82**, 886–893.

Basu, A. P., and Ebrahimi, N. (1991), Bayesian approach to life testing and reliability estimation using asymmetric loss function. *J. Statist. Plann. Infer.,* **29**, 21–31.

Basu, A. P., and Thompson, R. D. (1992), Life testing and reliability estimation under asymmetric loss. *Surval. Anal.,* 3–9, 9–10.

Berger, J. O. (1980), *Statistical Decision Theory: Foundations, Concepts, and Methods.* New York: Academic Press.

Bischoff, W., Fieger, W., and Wulfert, S. (1995), Minimax and Γ-minimax estimation of bounded normal mean under the LINEX loss. *Statist. Decis.,* **13**, 287–298.

Blackwell, D., and Girshick, M. A. (1954), *Theory of Games and Statistical Decisions.* New York: John Wiley.

Bolfarine, H. (1989), A note on finite population prediction under asymmetric loss function. *Commun. Statist. Theor. Meth.,* **18**, 1863–1869.

Cain, M. (1991), Minimal quantile measures of predictive cost with a generalized cost of error function. *Economic Research Papers*, No. 40, University of Wales, Aberystwyth.

Cain, M., and Janssen, C. (1995), Real estate prediction under asymmetric loss function. *Ann. Inst. Statist. Math.,* **47**, 401–414.

Calabria, R., and Pulcini, G. (1996), Point estimation under asymmetric loss functions for left truncated exponential samples. *Commun. Statist. Theor. Meth.,* **25**, 585–600.

Canfield, R. V. (1970), A Bayesian approach to reliability estimation using a loss function. *IEEE Trans. Reliab., R-19*, 13–16.

Dyer, D. D., and Keating, J. P. (1980), Estimation of the guarantee time in an exponential failure model. *IEEE Trans. Reliab., R-29*, 63–65.

Ferguson, T. S. (1967), *Mathematical Statistics: A Decision Theoretic Approach.* New York: Springer-Verlag.

Feynman, R. P. (1987), Mr. Feynman goes to Washington. *Engineering and Science*, California Institute of Technology, 6–62.

Geng, W. J., and Wan, A. T. K. (2000), On the sampling performance of an inequality pre-test estimator of the regression error variance under LINEX loss. *Statist. Pap.,* **41**(4), 453–472.

Ghosh, M., and Razmpour, A. (1984), Estimation of the common location parameter of several exponentials. *Sankhyya–A,* **46**, 383–394.

Giles, J. A., and Giles, D. E. A. (1993), Preliminary-test estimation of the regression scale parameter when the loss function is asymmetric. *Commun. Statist. Theor. Meth.,* **22**, 1709–1733.

Giles, J. A., and Giles, D. E. A. (1996), Risk of a homoscedasticity pre-test estimator of the regression scale under LINEX loss. *J. Statist. Plann. Infer.*, **50**, 21–35.

Harris, T. J. (1992), Optimal controllers for nonsymmetric and nonquadratic loss functions. *Technometrics*, **34**(3), 298–306.

Khattree, R. (1992), Estimation of guarantee time and mean life after warranty for two-parameter exponential failure model. *Aust. J. Statist.*, **34**(2), 207–215.

Klebanov, L. B. (1976), A general definition of unbiasedness. *Theory Probab. Appl.*, **21**, 571–585.

Kuo, L., and Dey, D. K. (1990), On the admissibility of the linear estimators of the poisson mean using LINEX loss functions. *Statist. Decis.*, **8**, 201–210.

Lehmann, E. L. (1951), A general concept of unbiasedness. *Ann. Math. Statist.*, **22**, 578–592.

Lehmann, E. L. (1988), Unbiasedness. In *Encyclopedia of Statistical Sciences*, vol. **9**, eds. S. Kotz and N. L. Johnson. New York: John Wiley.

Lehmann, E. L., and Casella, G. (1998), *Theory of Point Estimation*. 2nd ed. New York: Springer-Verlag.

Madi, M. T. (1997), Bayes and Stein estimation under asymmetric loss functions. *Commun. Statist. Theor. Meth.*, **26**, 53–66.

Mostert, P. J., Bekker, A., and Roux, J. J. J. (1998), Bayesian analysis of survival data using the Rayleigh model and LINEX loss. *S. Afr. Statist. J.*, **32**, 19–42.

Noorbaloochi, S., and Meeden, G. (1983), Unbiasedness as the dual of being Bayes. *J. Am. Statist. Assoc.*, **78**, 619–624.

Ohtani, K. (1988), Optimal levels of significance of a pre-test in estimating the disturbance variance after the pre-test for a linear hypothesis on coefficients in a linear regression. *Econ. Lett.*, **28**, 151–156.

Ohtani, K. (1995), Generalized ridge regression estimators under the LINEX loss function. *Statist. Pap.*, **36**, 99–110.

Ohtani, K. (1999), Risk performance of a pre-test estimator for normal variance with the Stein-variance estimator under the LINEX loss function. *Statist. Pap.*, **40**(1), 75–87.

Pandey, B. N. (1997), Testimator of the scale parameter of the exponential distribution using LINEX loss function. *Commun. Statist. Theor. Meth.,* **26**(9), 2191–2202.

Pandey, B. N., and Rai, O. (1992), Bayesian estimation of mean and square of mean of normal distribution using LINEX loss function. *Commun. Statist. Theor. Meth.,* **21**, 3369–3391.

Pandey, B. N., Singh, B. P., and Mishra, C.S. (1996), Bayes estimation of shape parameter of classical Pareto distribution under the LINEX loss function. *Commun. Statist. Theor. Meth.,* **25**, 3125–3145.

Parsian, A. (1990a), On the admissibility of an estimator of a normal mean vector under a LINEX loss function. *Ann. Inst. Statist. Math.,* **42**, 657–669.

Parsian, A. (1990b), Bayes estimation using a LINEX loss function. *J. Sci. IROI,* **1**(4), 305–307.

Parsian, A., and Kirmani, S. N. U. A. (2000), Estimation of the common location and location parameters of several exponentials under asymmetric loss function. Unpublished manuscript.

Parsian, A., and Nematollahi, N. (1995), Estimation of scale parameter under entropy loss function. *J. Stat. Plann. Infer.* **52** , 77–91.

Parsian, A., and Sanjari, F. N. (1993), On the admissibility and inadmissibility of estimators of scale parameters using an asymmetric loss function. *Commun. Statist. Theor. Meth.,* **22**(10), 2877–2901.

Parsian, A., and Sanjari, F. N., (1997), Estimation of parameters of exponential distribution in the truncated space using asymmetric loss function. *Statist. Pap.,* **38**, 423–443.

Parsian, A., and Sanjari, F. N., (1999), Estimation of the mean of the selected population under asymmetric loss function. *Metrika,* **50**(2), 89–107.

Parsian, A., Sanjari, F. N., and Nematollahi, N. (1993), On the minimaxity of Pitman type estimators under a LINEX loss function. *Commun. Statist. Theor. Meth.,* **22**(1), 91–113.

Rai, O. (1996), A sometimes pool estimation of mean life under LINEX loss function. *Commun. Statist. Theor. Meth.,* **25**, 2057–2067.

Rodrigues, J. (1994), Bayesian estimation of a normal mean parameter using the LINEX loss function and robustness considerations. *Test,* **3**(2), 237–246.

Rodrigues, J. (1998), Inference for the software reliability using asymmetric loss functions: a hierarchical Bayes approach. *Commun. Statist. Theor. Meth.,* **27**(9), 2165–2171.

Rojo, J. (1987), On the admissibility of $c\bar{X} + d$ w.r.t. the LINEX loss function. *Commun. Statist. Theor. Meth.,* **16**, 3745–3748.

Sadooghi-Alvandi, S. M. (1990), Estimation of the parameter of a Poisson distribution using a LINEX loss function. *Aust. J. Statist.,* **32**(3), 393–398.

Sadooghi-Alvandi, S. M., and Nematollahi, N. (1989), On the admissibility of $c\bar{X} + d$ relative to LINEX loss function. *Commun. Statist. Theor. Meth.,* **21**(5), 1427–1439.

Sadooghi-Alvandi, S. M., and Parsian, A. (1992), Estimation of the binomial parameter n using a LINEX loss function. *Commun. Statist. Theor. Meth.,* **18**(5), 1871–1873.

Sanjari, F. N. (1993), Estimation under LINEX loss function. Ph.D Dissertation, Department of Statistics, Shiraz University, Shiraz, Iran.

Sanjari, F. N. (1997), Ridge estimation of independent normal means under LINEX loss," *Pak. J. Statist.,* **13**, 219–222.

Schabe, H. (1992), Bayes estimates under asymmetric loss. *IEEE Trans. Reliab.,* R.**40**(1), 63–67.

Shafie, K., and Noorbaloochi, S. (1995), Asymmetric unbiased estimation in location families. *Statist. Decis.,* **13**, 307–314.

Soliman, A. A. (2000), Comparison of LINEX and quadratic Bayes estimators for the Rayleigh distribution. *Commun. Statist. Theor. Meth.,* **29**(1), 95–107.

Srivastava, V. K., and Rao, B. B. (1992), Estimation of disturbance variance in linear regression models under asymmetric criterion. *J. Quant. Econ.,* **8**, 341–345

Thompson, R. D., and Basu, A. P. (1996), Asymmetric loss functions for estimating system reliability. In *Bayesian Analysis in Statistics and Econometrics in Honor of Arnold Zellner,* eds. D. A. Berry, K. M. Chaloner, and J. K. Geweke. New York: John Wiley & Sons.

Upadhyay, S. K., Agrawal, R. and Singh, U. (1998), Bayes point prediction for exponential failures under asymmetric loss function. *Aligarh J. Statist.,* **18**, 1–13.

Varian, H. R. (1975), A Bayesian approach to real estate assessment. In *Studies in Bayesian Econometrics and Statistics in Honor of Leonard J. Savage*, eds. Stephan E. Fienberg and Arnold Zellner, Amsterdam: North-Holland, pp. 195–208.

Wan, A. T. K. (1999), A note on almost unbiased generalized ridge regression estimator under asymmetric loss. *J. Statist. Comput. Simul.,* **62**, 411–421.

Wan, A. T. K. and Kurumai, H. (1999), An iterative feasible minimum mean squared estimator of the disturbance variance in linear regression under asymmetric loss. *Statist. Probab. Lett.,* **45**, 253–259.

Wan, A. T. K., Zou, G., and Lee, A. H. (2000), Minimax and Γ-minimax estimation for the Poisson distribution under the LINEX loss when the parameter space is restricted. *Statist. Probab. Lett.,* **50**, 23–32.

Zellner, A. (1986), Bayesian estimation and prediction using asymmetric loss functions. *J. Am. Statist. Assoc.,* **81**, 446–451.

Zellner, A., and Geisel, M. S. (1968), Sensitivity of control to uncertainty and form of the criterion function. In *The Future of Statistics*, ed. Donald G. Watts. New York: Academic Press, 269–289.

Zou, G. H. (1997), Admissible estimation for finite population under the LINEX loss function. *J. Statist. Plann. Infer.,* **61**, 373–384.

5
Design of Sample Surveys across Time

SUBIR GHOSH University of California, Riverside, Riverside, California

1. INTRODUCTION

Sample surveys across time are widely used in collecting social, economic, medical, and other kinds of data. The planning of such surveys is a challenging task, particularly for the reason that a population is most likely to change over time in terms of characteristics of its elements as well as in its composition. The purpose of this paper is to give an overview on different possible sample surveys across time as well as the issues involved in conducting these surveys.

For a changing population over time, the objectives may not be the same for different surveys over time and there may be several objectives even for the same survey. In developing good designs for sample surveys across time, the first step is to list the objectives. The objectives are defined in terms of desired inferences considering the changes in population characteristics and composition. Kish (1986), Duncan and Kalton (1987), Bailar (1989), and Fuller (1999) discussed possible inferences from survey data collected over time and presented sampling designs for making those inferences. Table 1 presents a list of possible inferences with examples.

In Section 2, we present different designs for drawing inferences listed in Table 1. Section 3 describes the data collected using the designs given in Section 2. Section 4 discusses the issues in survey data quality. A brief discussion on statistical inference is presented in Section 5. Section 6 describes the Current Population Survey Design for collecting data on the United States labor market conditions in the various population groups, states, and even sub-state areas. This is a real example of a sample survey for collecting the economic and social data of national importance.

2. SURVEY DESIGNS ACROSS TIME

We now present different survey designs.

Table 1. Possible inferences with examples

Inferences on	Examples
(a) Population parameters at distinct time points	Income by counties in California during 1999
(b) Net change (i.e., the change at the aggregate level)	Change in California yearly income between 1989 and 1999
(c) Various components of individual change (gross, average, and instability)	Change in yearly income of a county in California from 1989 to 1999
(d) Characteristics based on cumulative data over time	Trend in California income based on the data from 1989 to 1999
(e) Characteristics from the collected data on events occurring within a given time period	Proportion of persons who experienced a criminal victimization in the past six months
(f) Rare events based on cumulative data over time	Cumulate a sufficient number of cases of persons with uncommon chronic disease
(g) Relationships among characteristics	Relationship between crime rates and arrest rates

2.1 Cross-sectional Surveys

Cross-sectional surveys are designed to collect data at a single point in time or over a particular time period. If a population remains unchanged in terms of its characteristics and composition, then the collected information from a cross-sectional survey at a single point in time or over a particular time period is valid for other points in time or time periods. In other words, a cross-sectional survey provides all pertinent information about the population in this situation. On the other hand, if the population does change in terms of characteristics and composition, then the collected information at one time point or time period cannot be considered as valid information for other time points or time periods. As a result, the net changes, individual changes, and other objectives described in (b) and (c) of Table 1 cannot be measured by cross-sectional surveys. However, in some cross-sectional surveys, retrospective questions are asked to gather information from the past. The strengths and weaknesses of such retrospective information from a cross-sectional survey will be discussed in Section 4.

2.2 Repeated Surveys

These surveys are repeated at different points of time. The composition of population may be changing over time and thus the sample units may not overlap at different points in time. However, the population structure (geographical locations, age groups, etc.) remains the same at different time points. At each round of data collection, repeated surveys select a sample of population existing at that time. Repeated surveys may be considered as a series of cross-sectional surveys. The individual changes described in (c) of Table 1 cannot be measured by replicated surveys.

2.3 Panel Surveys

Panel surveys are designed to collect similar measurements on the same sample at different points of time. The sample units remain the same at different time points and the same variables are measured over time. The interval between rounds of data collection and the overall length of the survey can be different in panel surveys.

A special kind of panel studies, known as cohort studies, deals with sample universes, called cohorts, for selecting the samples. For another kind of panel studies, the data are collected from the same units at sampled time points.

Panel surveys permit us to measure the individual changes described in (c) of Table 1. Panel surveys are more efficient than replicated surveys in measuring the net changes described in (b) of Table 1 when the values of the characteristic of interest are correlated over time. Panel surveys collect more accumulated information on each sampled unit than replicated surveys.

In panel surveys, there are possibilities of panel losses from nonresponse and the change in population composition in terms of introduction of new population elements as time passes.

2.4 Rotating Panel Surveys

In panel surveys, samples at two different time points have complete overlap in sample units. On the other hand, for rotating panel surveys, samples at different time points have partial or no overlap in sampling units. For samples at any two consecutive time points, some sample units are dropped from the sample and some other sample units are added to the sample. Rotating panel surveys reduce the panel effect or panel conditioning and the panel loss in comparison with non-rotating panel surveys. Moreover, the introduction of new samples at different waves provides pertinent information of a changing population. Rotating panel surveys are not useful for measuring the objective (d) of Table 1. Table 2 presents a six-period rotating panel survey design.

The Yates rotation design

Yates (1949) gave the following rotation design.

Part of the sample may be replaced at each time point, the remainder being retained. If there are a number of time points, a definite scheme of replacement is followed; e.g., one-third of the sample may be replaced, each selected unit being retained (except for the first two time points) for three time points.

Table 2. A six-period rotating panel design

			Time			
Units	1	2	3	4	5	6
1	X				X	X
2	X	X				X
3	X	X	X			
4		X	X	X		
5			X	X	X	
6				X	X	X

The rotating panel given in Table 2 is in fact the Yates rotation design (Fuller 1999).

The Rao–Graham rotation design

Rao and Graham (1964) gave the following rotation pattern.

We denote the population size by N and the sample size by n. We assume that N and n are multiples of $n_2 (\geq 1)$. In the Rao–Graham rotation design, a group of n_2 units stays in the sample for r time points ($n = n_2 r$), leaves the sample for m time points, comes back into the sample for another r time points, then leaves the sample for m time points, and so on.

Table 3 presents a seven-period Rao–Graham rotation design with $N = 5$, $n = r = 2$, $n_2 = 1$, and $m = 3$. Notice that the unit $u(u = 2, 3, 4, 5)$ stays in the sample for the time points $(u - 1, u)$, leaves the sample for the time points $(u + 1, u + 2, u + 3)$, and again comes back for the time points $(u + 4, u + 5)$ and so on. Since we have seven time points, we observe this pattern completely for $u = 2$ and partially for $u = 3, 4$, and 5.

For $u = 1$, the complete pattern in fact starts from the time points $(u + 4, u + 5)$, i.e., $(5, 6)$. Also note that the unit $u = 1$ is in the sample at time point 1 and leaves the sample for the time points $(u + 1, u + 2, u + 3)$, i.e., $(2, 3, 4)$. For the rotation design in Table 2, we have $N = 6$, $n = r = 3$, $n_2 = 1$, and $m = 3$, satisfying the Rao–Graham rotation pattern described above. Since there are only six time points, the pattern can be seen partially.

One-level rotation pattern

For the one-level rotation pattern, the sample contains n units at all time points. Moreover, $(1 - \mu) n$ of the units in the sample at time t_{i-1} are retained in the sample drawn at time t_i, and the remaining μn units are replaced with the same number of new ones. At each time point, the enumerated sample reports only one period of data (Patterson 1950, Eckler 1955).

Table 3. A seven-period rotating panel design

				Time			
Units	1	2	3	4	5	6	7
1	X				X	X	X
2	X	X				X	X
3		X	X				X
4			X	X			
5				X	X		

For the rotation design in Table 2, $\mu = \frac{1}{3}$ and $n = 3$, and for the rotation design in Table 3, $\mu = \frac{1}{2}$ and $n = 2$.

Two-level rotation pattern

Sometimes it is cheaper in surveys to obtain the sample values on a sample unit simultaneously instead of at two separate times. In the two-level rotation pattern, at time t_i a new set of n sample units is drawn from the population and the associated sample values for the times t_i and t_{i-1} are recorded (Eckler 1955).

This idea can be used to generate a more than two-level rotation pattern.

2.5 Split Panel Surveys

Split panel surveys are a combination of a panel and a repeated or rotating panel survey (Kish 1983, 1986). These surveys are designed to follow a particular group of sample units for a specified period of time and to introduce new groups of sample units at each time point during the specified period. Split panel surveys are also known as supplemental panel surveys (Fuller 1999).

The simplest version of a split panel survey is a combination of a pure panel and a set of repeated independent samples. Table 4 presents an example.

A complicated version of a split panel survey is a combination of a pure panel and a rotating panel. Table 5 presents an example.

The major advantage of split panel surveys is that of sharing the benefits of pure panel surveys and repeated surveys or rotating panel surveys. For

Table 4. The simplest version of a six-period split panel survey

			Time			
Units	1	2	3	4	5	6
1	X	X	X	X	X	X
2	X					
3		X				
4			X			
5				X		
6					X	
7						X

Table 5. A complicated version of a six-period split panel survey

Units	Time 1	2	3	4	5	6
1	X	X	X	X	X	X
2	X				X	X
3	X	X				X
4	X	X	X			
5		X	X	X		
6			X	X	X	
7				X	X	X

inference (d) of Table 1 only the pure panel part is used. For inference (f) of Table 1 only the repeated survey or rotating panel survey part is used. But for the remaining inferences both parts are used. Moreover, the drawbacks of one part are taken care of by the other part. For example, the new units are included in the second part and also the second part permits us to check biases from panel effects and respondent losses in the first part.

Table 6 presents the relationship between survey designs across time given in the preceding sections and possible inferences presented in Table 1 (Duncan and Kalton 1987, Bailar 1989).

3. SURVEY DATA ACROSS TIME

We now explain the nature of survey data, called longitudinal data, collected by survey designs presented in Sections 2.1–2.5. Suppose that two variables

Table 6. Survey designs versus possible inferences in Table 1

Units	Time (a)	(b)	(c)	(d)	(e)	(f)	(g)
2.1	X				X		X
2.2	X	X			X	X	X
2.3	X	X	X	X	X		X
2.4	X	X	X		X	X	X
2.5	X	X	X	X	X	X	X

X and Y are measured on N sample units. In reality, there are possibilities of measuring a single variable as well as more than two variables.

The cross-sectional survey data are $x_1, \ldots x_N$ and $y_1, \ldots, y_N$ on N sample units. The repeated survey data are x_{it} and y_{it} for the ith sample unit at time t on two variables X and Y. The panel survey data are x_{it} and y_{it}, $i = 1, \ldots, N$, $t = 1, \ldots, T$. There are T *waves* of observations on two variables X and Y. The rotating panel survey data from the rotating panel in Table 2 are (x_{it}, y_{it}), $(x_{i(t+1)}, y_{i(t+1)})$, $(x_{i(t_+2)}, y_{i)t+2)})$, $i = t + 2$, $t = 1, 2, 3, 4$; (x_{11}, y_{11}), (x_{15}, y_{15}), (x_{16}, y_{16}), (x_{21}, y_{21}), (x_{22}, y_{22}), (x_{26}, y_{26}). The split panel survey data from the split panel in Table 5 are (x_{1t}, y_{1t}), $t = 1, \ldots, 6$; (x_{it}, y_{it}), $(x_{i(t+1)}, y_{i(t+1)})$, $(x_{i(t+2)}, y_{i)t+2)})$, $i = t + 3$, $t = 1, 2, 3, 4$; (x_{21}, y_{21}), (x_{25}, y_{25}), (x_{26}, y_{26}), (x_{31}, y_{31}), (x_{32}, y_{32}), (x_{36}, y_{36}).

4. ISSUES IN SURVEY DATA QUALITY

We now present the major issues in the data quality for surveys across time.

4.1 Panel Conditioning/Panel Effects

Conditioning means a change in response that occurs at a time point t because the respondent has had the interview waves $1, 2, \ldots, (t - 1)$. The effect may occur because the prior interviews may influence respondents' behaviors and the way the respondents answer questions. Panel conditioning is thus a reactive effect of prior interviews on current responses. Although a rotating panel design is useful for examining panel conditioning, the elimination of the effects of conditioning is much more challenging because of its confounding with the effects of other changes.

4.2 Time-in-Sample Bias

Panel conditioning introduces time-in-sample bias from respondents. But there are many other factors contributing to the bias. If a significantly higher or lower level of response is observed in the first wave than in subsequent waves, when one would expect them to be the same, then there is a presence of time-in-sample bias.

4.3 Attrition

A major source of sample dynamics is attrition from the sample. Attrition of individuals tends to be drastically reduced when all individuals of the original households are followed in the interview waves. Unfortunately, not all individuals can be found or will agree to participate in the survey interview waves. A simple check on the potential bias from later wave nonresponse can be done by comparing the responses of subsequent respondents as well as nonrespondents to questions asked on earlier waves.

4.4 Telescoping

Telescoping means the phenomenon that respondents tend to draw into the reference period events that occurred before (or after) the period. The extent of telescoping can be examined in a panel survey by determining those events reported on a given wave that had already been reported on a previous wave (Kalton et al. 1989). Internal telescoping is the tendency to shift the timing of events within the recall period.

4.5 Seam

Seam refers to a response error corresponding to the fact that two reference periods are matched together to produce a panel record. The number of transitions observed between the last month of one reference period and the first month of another is far greater than the number of transitions observed between months within the same reference period (Bailar 1989). It has been reported in many studies that the number of activities is far greater in the month closest to the interview than in the month far away from the interview.

4.6 Nonresponse

In panel surveys, the incidence of nonresponse increases as the panel wave progresses. Nonresponse at the initial wave is a result of refusals, not-at-homes, inability to participate, untraced sample units and other reasons. Moreover, nonresponse occurs at subsequent waves for the same reasons. As a rule, the overall nonresponse rates in panel surveys increase with successive waves of data collection. Consequently, the risk of bias in survey estimates increases considerably, particularly when nonrespondents systematically differ from respondents (Kalton et al. 1989).

The cumulative nature of this attrition has an effect on response rates. In a survey with a within-wave response rate of 94% that does not return to nonrespondents, the number of observations will be reduced to half of the original sample observations by the tenth wave (Presser 1989).

Unit nonresponse occurs when no data are obtained for one or more sample units. Weighting adjustments are normally used to compensate for unit nonresponse.

Item nonresponses occur when one or more items are not completed for a unit. The remaining items of the unit provide responses. Imputation is normally used for item nonresponse.

Wave nonresponses occur when one or more waves of panel data are unavailable for a unit that has provided data for at least one wave (Lepkowski 1989).

4.7 Coverage Error

Coverage error is associated with incomplete frames to define the sample population as well as the faulty interview. For example, the coverage of males could be worse than that of females in a survey. This error results in biased survey estimates.

4.8 Tracing/Tracking

Tracing (or tracking) arises in a panel survey of persons, households, and so on, when it is necessary to locate respondents who have moved. The need for tracing depends on the nature of the survey including the population of interest, the sample design, the purpose of the survey, and so on.

4.9 Quasi-Experimental Design

We consider a panel survey with one wave of data collected before a training program and another wave of data collected after the training program. This panel survey is equivalent to the widely used quasi-experimental design (Kalton 1989).

4.10 Dynamics

In panel surveys, the change over time is a challenging task to consider.

Target population dynamics

When defining the target population, the consequences of birth, death, and mobility during the life of the panel must be considered. Several alternative population definitions can be given that incorporate the target population dynamics (Goldstein 1979).

Household dynamics

During the life of a panel survey, family and household type units could be created, could disappear altogether, and undergo critical changes in type. Such changes create challenges in household and family-level analyses.

Characteristic dynamics

The relationship between a characteristic of an individual, family, or household at one time and the value of the same or another characteristic at a different time could be of interest in a panel survey. An example is the measurement of income mobility in terms of determining the proportion of people living below poverty level in one period who remain the same in another period.

Status and behavior dynamics

The status of the same person may change over time. For example, children turn into adults, heads of households, wives, parents, and grandparents over the duration of a panel survey. The same is also true for behavior of a person.

4.11 Nonsampling Errors

Two broad classes of nonsampling errors are nonobservation errors and measurement errors. Nonobservation errors may be further classified into nonresponse errors and noncoverage errors. Measurement errors may be further subdivided into response and processing errors. A major benefit of a panel survey is to measure such nonsampling errors.

4.12 Effects of Design Features on Nonsampling Errors

We now discuss four design features of a panel survey influencing the nonsampling errors (Cantor 1989, O'Muircheartaigh 1989).

Interval between waves

The longer the interval between waves, the longer the reference period for which the respondent must recall the events. Consequently, the greater is the chance for errors of recall.

In panel surveys, each wave after the first wave produces a bounded interview with the boundary data provided by the responses on the previous wave. The longer interval between waves increases the possibility of migration and hence increases the number of unbounded interviews. For example, the interviewed families or individuals who move into a housing unit after the first housing unit contact are unbounded respondents.

The longer interval between waves increases the number of people who move between interviews and have to be traced and found again. Consequently, this increases the number of people unavailable to follow up.

As the interval between waves decreases, telescoping and panel effects tend to increase.

Respondent selection

Three kinds of respondent categories are normally present in a household survey. A self-respondent does answer questions on his or her own, a proxy respondent does answer questions for another person selected in the survey, and a household respondent does answer questions for a household.

Response quality changes when household respondents change between waves or a self-respondent takes the place of a proxy-respondent.

Respondent changes create problems when retrospective questions are asked and the previous interview is used as the start-up recall period. It is difficult to eliminate the overlap between the reporter's responses.

Respondent changes between waves are also responsible for differential response patterns because of different conditioning effects.

Mode of data collection

The mode of data collection influences the initial and follow-up cooperation of respondents, the ability to effectively track survey respondents and maintain high response quality and reliability. The mode of data collection includes face-to-face interviews, telephone interviews, and self-completion questionnaires. The mixed-mode strategy, which is a combination of modes of data collection, is often used in panel surveys.

Rules for following sample persons

The sample design for the first wave is to be supplemented with the following rules for determining the method of generation of the samples for subsequent waves. The rules specify the plans for retention, dropping out of sample units from one wave to the next wave, and adding new sample units to the panel.

Longitudinal individual unit cohort design

If the rule is to follow all the sample individual units of the first wave in the samples for the subsequent waves, then it is an individual unit cohort design. The results from such a design are meaningful only to the particular cohort from which the sample is selected. For example, the sample units of individuals with a particular set of characteristics (a specific age group and gender) define a cohort.

Longitudinal individual unit attribute-based design

If the rule is to follow the sample of individuals, who are derived from a sample of addresses (households) or groups of related persons with common dwellings (families) of the first wave in the samples for the subsequent waves, then it is an individual unit attribute-based design. The analyses from such a design focus on individuals but describe attributes of families or households to which they belong.

Longitudinal aggregate unit design

If the rule is to follow the sample of aggregate units (dwelling units) of the first wave in the sample for the subsequent waves, then it is an aggregate unit design. The members of a sample aggregate unit (the occupants of a sample dwelling unit) of the first wave may change in the subsequent waves. At each wave, the estimates of characteristics of aggregate units are developed to represent the population.

Complex panel surveys are based on individual attribute-based designs and aggregate unit designs. The samples consist of groups of individuals available at a particular point in time, in contrast to individuals with a particular characteristic in surveys based on individual unit cohort designs. Complex panel surveys attempt to follow all individuals of the first wave sample for the subsequent waves but lose some individuals due to attrition, death, or movement beyond specified boundaries and add new individuals in the subsequent wave samples due to birth, marriage, adoption or cohabitation.

4.13 Weighing Issues

In panel surveys, the composition of units, households, and families changes over time. Operational rules should be defined regarding the changes that would allow the units to be considered in the subsequent waves of sample following the first wave of sample, the units to be terminated, and the units to be followed over time. When new units (individuals) join the sample in the subsequent waves of sample, the nature of retrospective questions for them should be prepared. The weighting procedures for finding unbiased estimators of the parameters of interest should be made. The adjustment of weights for reducing the variances and biases of the estimators resulting from undercoverage and nonresponse should also be calculated.

5. STATISTICAL INFERENCE

Both design-based and model-based approaches are used in drawing inferences on the parameters of interest for panel surveys. The details are available in literature (Cochran 1977, Goldstein 1979, Plewis 1985, Heckman and Singer 1985, Mason and Fienberg 1985, Binder and Hidiroglou 1988, Kasprzyk et al. 1989, Diggle et al. ,1994, Davidian and Giltinan 1995, Hand and Crowder 1996, Lindsey 1999). The issues involved in deciding one model over the others are complex. Some of these issues and others will be discussed in Chapter 9 of this volume.

6. EXAMPLE: THE CURRENT POPULATION SURVEY

The Current Population Survey (CPS) provides data on employment and earnings each month in the USA on a sample basis. The CPS also provides extensive data on the U.S. labor market conditions in the various population groups, states, and even substate areas. The CPS is sponsored jointly by the U.S. Census Bureau and the U.S. Bureau of Labor Statistics (Current Population Survey: TP 63).

The CPS is administered by the Census Bureau. The field work is conducted during the week of the 19th of the month. The questions refers to the week of the 12th of the same month. In the month of December, the survey is often conducted one week earlier because of the holiday season.

The CPS sample is a multistage stratified probability sample of approximately 56,000 housing units from 792 sample areas. The CPS sample consists of independent samples in each state and the District of Columbia. California and New York State are further divided into two substate areas: the Los Angeles–Long Beach metropolitan area and the rest of California; New York City and the rest of New York State. The CPS design consists of independent designs for the states and substate areas. The CPS sampling uses the lists of addresses from the 1990 Decennial Census of Population and Housing. These lists are updated continuously for new housing built after the 1990 census.

Sample sizes for the CPS sample are determined by reliability requirements expressed in terms of the coefficient of variation (CV).

The first stage of sampling involves dividing the United States into primary sampling units (PSUs) consisting of metropolitan area, a large county, or a group of smaller counties within a state.

The PSUs are then grouped into strata on the basis of independent information obtained from the decennial census or other sources. The strata are constructed so that they are as homogeneous as possible with respect to labor force and other social and economic characteristics that are highly correlated with unemployment. One PSU is sampled per stratum. For a self-weighting design, the strata are formed with equal population sizes. The objective of stratification is to group PSUs with similar characteristics into strata having equal population sizes. The probability of selection for each PSU in the stratum is proportional to its population as of the 1990 census. Some PSUs have sizes very close to the needed equal stratum size. Such PSUs are selected for sample with probability one, making them self-representing (SR). Each of the SR PSUs included in the sample is considered as a separate stratum.

In the second stage of sampling, a sample of housing units within the sample PSUs is drawn.

Ultimate sampling units (USUs) are geographically compact clusters of about four housing units selected during the second stage of sampling. Use of housing unit clusters lowers the travel costs for field representatives.

Sampling frames for the CPS are developed from the 1990 Decennial Census, the Building Permit Survey, and the relationship between these two sources. Four frames are created: the unit frame, the area frame, the group quarters frame, and the permit frame. A housing unit is a group of rooms or a single room occupied as a separate living quarter. A group quarters is a living quarter where residents share common facilities or receive formally authorized care. The unit frame consists of housing units in census blocks that contain a very high proportion of complete addresses and are essentially covered by building permit offices. The area frame consists of housing units and group quarters in census blocks that contain a high proportion of incomplete addresses, or are not covered by building permit offices. The group quarters frame consists of group quarters in census blocks that contain a sufficient proportion of complete addresses and are essentially covered by building permit offices. The permit frame consists of housing units built since the 1990 census, as obtained from the Building Permit Survey.

The CPS sample is designed to be self-weighting by state or substate area. A systematic sample is selected from each PSU at a sampling rate of 1 in k, where k is the within-PSU sampling interval which is equal to the product of the PSU probability of selection and the stratum sampling interval. The stratum sampling interval is normally the overall state sampling interval.

The first stage of selection of PSUs is conducted from each demographic survey involved in the 1990 redesign. Sample PSUs overlap across surveys and have different sampling intervals. To ensure housing units get selected for only one survey, the largest common geographic areas obtained when intersecting, sample PSUs in each survey are identified. These intersecting areas as well as residual areas of those PSUs, are called basic PSU components (BPCs). A CPS stratification PSU consists of one or more BPCs. For each survey, a within-PSU sample is selected from each frame within BPCs. Note that sampling by BPCs is not an additional stage of selection.

The CPS sampling is a one-time operation that involves selecting enough sample for the decade. For the CPS rotation system and the phase-in of new sample designs, 19 samples are selected. A systematic sample of USUs is selected and 18 adjacent sample USUs are identified. The group of 19 sample USUs is known as a *hit string*. The within-PSU sort is performed so that persons residing in USUs within a hit string are likely to have similar labor force characteristics. The within-PSU sample selection is performed independently by BPC and frame.

The CPS sample rotation scheme is a compromise between a permanent sample and a completely new sample each month. The data are collected using the 4-8-4 sampling design under the rotation scheme. Each month, the data are collected from the sample housing units. A housing unit is interviewed for four consecutive months and then dropped out of the sample for the next eight months, is brought back in the following four months, and then retired from the sample. Consequently, a sample housing unit is interviewed only eight times. The rotation scheme is designed so that outgoing housing units are replaced by housing units from the same hit string which have similar characteristics. Out of the two rotation groups replaced from month to month, one is in the sample for the first time and the other returns after being excluded for eight months. Thus, consecutive monthly samples have six rotation groups in common. Monthly samples one year apart have four rotation groups in common. Households are rotated in and out of the sample, improving the accuracy of the month-to-month and year-to-year change estimates. The rotation scheme ensures that in any one month, one-eighth of the housing units are interviewed for the first time, another eighth is interviewed for the second time, and so on.

When a new sample is introduced into the ongoing CPS rotation scheme, the phase-in of a new design is also practiced. Instead of discarding the old CPS sample one month and replacing it with a completely redesigned sample the next month, a gradual transition from the old sample design to the new sample design is undertaken for this phase-in scheme.

7. CONCLUSIONS

In this paper, we present an overview of survey designs across time. Issues in survey data quality are discussed in detail. A real example of a sample survey over time for collecting data on the United States labor market conditions in different population groups, states, and substate areas, is also given. This complex survey is known as the Current Population Survey and is sponsored by the U.S. Census Bureau and the U.S. Bureau of Labor Statistics.

ACKNOWLEDGMENT

The author would like to thank a referee for making comments on an earlier version of this paper.

REFERENCES

B. A. Bailer. Information needs, surveys, and measurement errors. In: D. Kasprzyk et al., eds. *Panel Surveys*. New York: Wiley, 1–24, 1989.

D. A. Binder, M. A. Hidiroglou. Sampling in time. In: P. R. Krishnaiah and C. R. Rao, eds. *Sampling, Handbook of Statistics: 6*. Amsterdam: North-Holland, 187–211, 1988.

D. Cantor. Substantive implications of longitudinal design features: the National Crime Survey as a case study. In: D. Kasprzyk et al., eds. *Panel Surveys*. New York: Wiley, 25–51, 1989.

W. G. Cochran. *Sampling Techniques*. New York: Wiley, 1977.

M. Davidian, D. M. Giltinan. *Nonlinear Models for Repeated Measurement Data*. London: Chapman & Hall/CRC, 1995.

P. J. Diggle, K.-Y. Liang, S. L. Zeger. *Analysis of Longitudinal Data*. Oxford: Clarendon Press, 1994.

G. J. Duncan, G. Kalton. Issues of design and analysis of surveys across Time. *Int. Statist. Rev.* 55:1: 97–117, 1987.

A. R. Eckler. Rotation sampling. *Ann. Math. Statist.* 26: 664–685, 1955.

W. A. Fuller. Environmental surveys over time. *J. Agric. Biol. Environ. Statist.* 4:4: 331–345, 1999.

S. Ghosh. *Asymptotics, Nonparamretics, and Time Series.* New York: Marcel Dekker, 1999a.

S. Ghosh. *Multivariate Analysis, Design of Experiments, and Survey Sampling.* New York: Marcel Dekker, 1999b.

H. Goldstein. *The Design and Analysis of Longitudinal Studies: Their Role in the Measurement of Change.* New York: Academic Press, 1979.

D. J. Hand, M. Crowder. *Practical Longitudinal Data Analysis.* London: Chapman & Hall/CRC, 1996.

J. J. Heckman, B. Singer. *Longitudinal Analysis of Labor Market Data.* Cambridge: Cambridge University Press, 1985.

G. Kalton. Modeling considerations: Discussions from a survey sampling perspective. In: D. Kasprzyk et al., eds. *Panel Surveys.* New York: Wiley, 575–585, 1989.

G. Kalton, D. Kasprzyk, D. B. McMillen. Nonsampling errors in panel surveys. In: D. Kasprzyk et al., eds. *Panel Surveys.* New York: Wiley, 249–270, 1989.

D. Kasprzyk, G. J. Duncan, G. Kalton, M. P. Singh, eds. *Panel Surveys.* New York: Wiley, 1989.

R. C. Kessler, D. F. Greenberg. *Linear Panel Analysis: Models of Quantitative Change.* New York: Academic Press, 1981.

G. King, R. O. Keohane, S. Verba. *Designing Social Inquiry: Scientific Inference in Qualitative Research.* New Jersey: Princeton University Press, 1994.

L. Kish. Data collection for details over space and time. In: T. Wright, ed. *Statistical Methods and the Improvement of Data Quality.* New York: Academic Press: 73–84, 1983.

L. Kish. Timing of surveys for public policy. *Aust. J. Statist.* 28: 1–12, 1986.

L. Kish. *Statistical Design for Research.* New York: Wiley, 1987.

J. M. Lepkowski. Treatment of wave nonresponse in panel surveys. In: D. Kasprzyk et al., eds. *Panel Surveys.* New York: Wiley, 348–374, 1989.

J. K. Lindsey. *Models for Repeated Measurements.* Oxford: Oxford University Presss, 1999.

P. C. Mahalanobis. On large scale sample surveys. *Phil. Trans. Roy. Soc. Lond. Ser. B*: 231: 329–451, 1944.

P. C. Mahalanobis. A method of fractal graphical analysis. *Econometrica* 28: 325–351, 1960.

W. M. Mason, S. E. Fienberg. *Cohort Analysis in Social Research: Beyond Identification Problem*. New York: Springer-Verlag, 1985.

C. O'Muircheartaigh. Sources of nonsampling error: Discussion. In: D. Kasprzyk et al., eds. *Panel Surveys*. New York: Wiley, 271–288, 1989.

H. D. Patterson. Sampling on successive occasions with partial replacement of units. *J. Roy. Statist. Soc. Ser. B* 12: 241–255, 1950.

I. Plewis. *Analysing Change: Measurement and Explanation using Longitudinal Data*. New York: Wiley, 1985.

S. Presser. Collection and design issues: discussion. In: D. Kasprzyk et al., eds. Panel Surveys. New York: Wiley, 75-79, 1989.

J. N. K. Rao, J. E. Graham. Rotation designs for sampling on repeated occasions. *J. Am. Statist. Assoc.* 50: 492–509, 1964.

U.S. Department of Labor Bureau of Labor Statistics and U.S. Census Bureau. *Current Population Survey*: TP63, 2000.

F. Yates. Sampling methods for censuses and surveys. London: Charles Griffin, 1949 (fourth edition, 1981).

6

Kernel Estimation in a Continuous Randomized Response Model

IBRAHIM A. AHMAD   University of Central Florida, Orlando, Florida

1. INTRODUCTION

Randomized response models were developed as a mean of coping with nonresponse in surveys, especially when the data collected are sensitive or personal as is the case in many economic, health, or social studies. The technique initiated in the work of Warner (1965) for binary data and found many applications in surveys. For a review, we refer to Chaudhuri and Mukherjee (1987). Much of this methodology's applications centered on qualitative or discrete data. However, Poole (1974) was first to give the continuous analog of this methodology in the so-called "product model." Here, the distribution of interest is that of a random variable Y. The interviewer is asked, however, to pick a random variable X, independent of Y, from a known distribution $F(x) = (x/T)^{\beta}$, $0xT, \beta \geq 1$ and only report $Z = XY$. Let $G(H)$ denote the distribution of $Y(Z)$ with probability density function (pdf) $g(h)$. One can represent the df of Y in terms of that of Z and X as follows (cf. Poole 1974):

$$G(y) = H(yT) - (yt/\beta)h(yT), \qquad y \in R^1 \tag{1.1}$$

Hence

$$g(y) = T(1 - 1/\beta)h(yT) - (yT^2/\beta)h'(yT) \tag{1.2}$$

Thus, when $\beta = 1$, this is the uniform $[0, T]$ case, the methodology requires to ask the interviewer to multiply his/her value of Y by any number chosen between 0 and T and to report only the product. Let the collected randomized response data be $Z_1, \ldots, Z_n$. We propose to estimate $G(y)$ and $g(y)$, respectively, by

$$\hat{G}_n(y) = \hat{H}_n(yT) - (yT/\beta)\hat{h}_n(yT) \tag{1.3}$$

and

$$\hat{g}_n(y) = T(1 - 1/\beta)\hat{h}_n(yT) - (yT^2/\beta)\hat{h}_n'(yT) \tag{1.4}$$

where $\hat{h}_n(z) = (na_n)^{-1} \sum_{i=1}^{n}(z - Z_i/a_n)$, $\hat{H}_n(z) = \int_{-\infty}^{z} \hat{h}_n(w)dw$, and $\hat{h}_n'(z) = (d/dz)\hat{h}_n(z)$, with $k(u)$ a known symmetric, bounded pdf such that $| u|k(u) \to 0$ as $|u| \to \infty$ and $\{a_n\}$ are positive constants such that $a_n \to 0$ as $n \to \infty$. We note that $\hat{H}_n(z) = (1/n) \sum_{i=1}^{n} K([z - Z_i]/a_n)$ with $K(u) = \int_{-\infty}^{u} k (w)dw$ and $\hat{h}_n'(z) = (na_n^2)^{-1} \sum_{i=1}^{n} k'([z - Z_i]/a_n)$. In Section 2, we study some sample properties of $\hat{g}_n(y)$ and $\hat{G}_n(y)$, including an asymptotic representation of the mean square error (mse) and its integrated mean square error (imse) and discuss its relation to the direct sampling situation. A special case of $\hat{G}_n(y)$, using the "naive" kernel, was briefly discussed by Duffy and Waterton (1989), where they obtain an approximation of the mean square error.

Since our model is equivalently put as $\ln Z = \ln X + \ln Y$, we can see that the randomized response model may be viewed as a special case of the "deconvolution problem." In its general setting, this problem relates to estimating a pdf of a random variable U (denoted by p) based on contaminated observations $W = U + V$, where V is a random variable with known pdf q. Let the pdf of W be l. Thus, $l(w) = p * q(w)$. To estimate $p(w)$ based on a sample $W_1, \ldots, W_n$ we let k be a known symmetric pdf with characteristic function ϕ_k and let $a_n = a$ be reals such that $a \to 0$ and $na \to \infty$ as $n \to \infty$. Further, let ϕ_p, ϕ_q and ϕ_l denote the characteristic functions corresponding to U, V, and W, respectively. If $\hat{\phi}_l(t) = (1/n) \sum_{j=1}^{n} e^{itW_j}$ denote the empirical characteristic function of $W_1, \ldots, W_n$, then the deconvolution estimate of $p(.)$ is

$$\hat{p}_n(w) = \frac{1}{2\pi} \int_{-\infty}^{\infty} e^{i\theta w} \left[\hat{\phi}_l(\theta)\phi_k(a\theta)/\phi_q(\theta) \right] d\theta \tag{1.5}$$

Hence, if we let

$$\hat{k}_a^*(w) = \frac{1}{2\pi} \int_{-\infty}^{\infty} e^{i\theta w} \big[\phi_k(\theta)/\phi_q(\theta/a) \big] d\theta \tag{1.6}$$

then

$$\hat{p}_n(w) = \frac{1}{na} \sum_{j=1}^{n} k_a^*\big(\tfrac{W_i - w}{a}\big) \tag{1.7}$$

This estimate was discussed by many authors including Barry and Diggle (1995), Carroll and Hall (1988), Fan (1991a, b, 1992), Diggle and Hall (1993), Liu and Taylor (1990a, b)), Masry and Rice (1992), Stefanski and Carroll (1990), and Stefanski (1990). It is clear, from this literature, that the general case has, in addition to the fact that the estimate is difficult to obtain since it requires inverting characteristic functions, two major problems. First, the explicit mean square error (or its integrated form) is difficult to obtain exactly or approximately (c.f. Stefanski 1990). Second, the best rate of convergence in this mean square error is slow (cf. Fan 1991a). While we confirm Fan's observation here, it is possible in our special case to obtain an expression for the mean square error (and its integral) analogous to the usual case and, thus, the optimal width is possible to obtain.

Note also that by specifying the convoluting density f, as done in (1.1), we are able to provide an explicit estimate without using inversions of characteristic functions, which are often difficult to obtain, thus limiting the usability of (1.5). In this spirit, Patil (1996) specializes the deconvolution problem to the case of "nested" sampling, which also enables him to provide an estimate analogous to the direct sampling case.

Next, let us discuss regression analysis when one of the two variables is sensitive. First, suppose the response is sensitive; let (U, Y) be a random vector with df $G(u,y)$ and pdf $g(u,y)$. Let $X \sim F(x) = (x/T)^\beta$, $0 \le x \le T$, and we can observe (U, Z) only where $Z = XY$. Further, let $H(u, z)$ $(h(u, z))$ denote the df (pdf) of (U, Z). Following the reasoning of Poole (1974), we have that, for all (u, y),

$$G(u, y) = H(u, yT) - (yT/\beta)H^{(0,1)}(u, yT), \tag{1.8}$$

where $H^{(0,1)}(u, y) = (\partial/\partial y)H(u, y)$. Thus, the pdf of (U, T) is

$$g(u, y) = [(1 - 1/\beta)T]h(u, yT) - (yT^2/\beta)h^{(0,1)}(u, yT) \tag{1.9}$$

where $h^{(0,1)}(u, y) = \partial h(u, y)/\partial y$. If we let $g_1(u) = \int g(u, y)dy$, then the regression $R(u) = \int g(u, y)dy/g_1(u)$ is equal to

$$R(u) = \left\{ [(1 - 1/\beta)T] \int zh(u, z)dz - (1/T\beta)z^2 h^{(1,0)}(u, z)dz \right\}/g_1(u) \tag{1.10}$$

But, since $E(Z|U) = E(X)E(Y|U) = (\beta T/\beta + 1)R(u)$, we get that

$$R(u) = [(1 + 1/\beta)/T]r(u) \tag{1.11}$$

where $r(u)$ is the regression function of Z on U, i.e. $r(u) = E(Z|U = u)$. If $(U_1, Z_1), \ldots, (U_n, Z_n)$ is a random sample from $H(u, z)$, then we can estimate $R(u)$ by

$$\hat{R}_n(u) = [(1 + 1/\beta)/T]\hat{r}_n(u) \tag{1.12}$$

where $\hat{r}_n(u) = \sum_{i=1}^{n} Z_i k([u - U_i]/a_n)/\sum_{i=1}^{n} k([u - U_i]/a_n)$. Note that $\hat{R}_n(u)$ is a constant multiple of the usual regression estimate of Nadaraya (1965), c.f. Scott (1992).

Secondly, suppose that the regressor is the sensitive data. Thus, we need to estimate $S(y) = E(U|Y = y)$. In this case $g_2(y) = [(1 - 1/\beta)/T]h_2(yT) - [yT^2/\beta]h_2'(yT)$ and thus

$$S(y) = \left\{ T(1 - 1/\beta) \int uh(u, yT)du - (yT^2/\beta) \int uh^{(0,1)}(u, yT)dy \right\}/g_2(y) \tag{1.13}$$

which can be estimated by

$$\hat{S}_n(y) = \left\{ T(1 - 1/\beta) \sum_{i=1}^{n} U_i k\left(\frac{yT - Z_i}{a_n}\right) - [yT^2/\beta] \sum_{i=1}^{n} U_i k'\left(\frac{yT - Z_i}{a_n}\right).a_n^{-1} \right\} /$$

$$na_n \hat{g}_{2n}(y), \tag{1.14}$$

where $\hat{g}_{2n}(y)$ is as given in (1.4) above.

In Section 3, we discuss the mse and imse as well as other large sample properties of both $\hat{R}_n(u)$ and $\hat{S}_n(y)$. Note that since the properties of $\hat{R}_n(u)$ follow directly from those of the usual Nadaraya regression estimate (cf. Scott (1992) for details) we shall concentrate on $\hat{S}_n(y)$. Comparison with direct sampling is also mentioned. For all estimates presented in this paper, we also show how to approximate the mean absolute error (mae) or its integrated form (imae).

2. ESTIMATING PROBABILITY DENSITY AND DISTRIBUTION FUNCTIONS

In this section, we discuss the behavior of the estimates $\hat{g}_n(u)$ and $\hat{G}_n(u)$ as given in (1.4) and (1.3), respectively. We shall also compare these properties with those of direct sampling to try to measure the amount of information loss (variability increase) due to using the randomized response methodology.

First, we start by studying $g_n(y)$.

$$\text{mse}(\hat{g}_n(y)) = T^2(1 - 1/\beta)^2\text{mse}(h_n(yT)) + (y^2T^4/\beta^2)\text{mse}(\hat{h}'_n(yT))$$
$$- 2[yT^3(1 - 1/\beta)/\beta]E[\hat{h}_n(yT) - h(yT))][\hat{h}'_n(yT) - h'(yT)]$$

$$(2.1)$$

Denoting by amse the approximate (to first-order term) form of mse we now proceed to obtain $\text{amse}(\hat{g}_n(y))$.

It is well known, c.f. Scott (1992), that

$$\text{amse}(\hat{h}(yT)) = h(yT)R(k)/na + \sigma_k^4 a^4 (h''(yT))^2/4 \tag{2.2}$$

where $R(k) = \int k^2(w)dw$. Similarly, it is not difficult to see that

$$\text{amse}(\hat{h}'_n(yT)) = h(yT)R(k')/na^3 + \sigma_k^4 a^4 (h^{(3)}(yT))^2/4 \tag{2.3}$$

Next,

$$E(\hat{h}_n(yT) - h(yT))(\hat{h}'_n(yT) - h'(yT)) = E\hat{h}_n(yT)\hat{h}'_n(yT) - E\hat{h}_n(yT)h'(yT)$$
$$- Eh(yT)\hat{h}'_n(yT) + h(yT)h'(yT)$$

$$(2.4)$$

But

$$E\hat{h}_n(yT)\hat{h}'_n(yT) = (na^3)^{-1}Ek\left(\frac{yT - Z_1}{a}\right)k'\left(\frac{yT - Z_1}{a}\right)$$
$$- \left(\frac{n-1}{n}\right)E\hat{h}_n(yT)E\hat{h}'_n(yT)$$
$$= J_1 + J_2, \text{ say}$$

Now, using $" \simeq "$ to mean the first-order approximation, we see that

$$J_1 = (na^3)^{-1}\int k\left(\frac{yT - z}{a}\right)k'\left(\frac{yT - z}{a}\right)h(z)dz \simeq (na^2)^{-1}h(yT)\int k(w)k'(w)dw$$

and

$$j_2 \simeq h(yT)h'(yT) + \frac{\sigma^2 a^2}{2}\left\{h''(yT)h'(yT) + h(yT)h^{(3)}(yT)\right\}$$
$$+ \frac{\sigma^4 a^4}{4}h''(yT)h^{(3)}(yT)$$

Thus

$$E\hat{h}_n(yT)\hat{h}_n'(yT) \simeq \frac{h(yT)}{na^2} \int k(w)k'(w)dw + h(yT)h'(yT)$$

$$+ \frac{\sigma^2 a^2}{2}\left[h''(yT)h'(yT) + h(yT)h^{(3)}(yT)\right]$$

$$+ \frac{\sigma^2 a^2}{4}h''(yT)h^{(3)}(yT)$$

Also

$$E\hat{h}_n(yT) \cdot h'(yT) \simeq h'(yT)\left\{h(yT) + \frac{\sigma^2 a^2}{2}h''(yT)\right\}$$

and

$$E\hat{h}_n'(yT) \cdot h(yT) \simeq h(yT)\left\{h'(yT) + \frac{\sigma^2 a^2}{2}h^{(3)}(yT)\right\}$$

Hence, the right-hand side of (2.4) is approximately equal to

$$\frac{h(yT)}{na^2}\int k(w)k'(w)dw + \frac{\sigma^4 a^4}{4}h''(yT)h^{(3)}(yT) \qquad (2.5)$$

But, clearly, $\int k(w)k'(w)dw = \lim_{|w|\to\infty} k^2(w) = 0$, and this shows that (2.5) is equal to

$$\frac{\sigma^4 a^4}{4}h''(yT)h^{(3)}(yT) \qquad (2.6)$$

Hence,

$$\text{amse}(\hat{g}_n(y)) = \frac{h(yT)}{na^3}R(aT(1 - 1/\beta)k - (yT^2/\beta)k')$$

$$+ \frac{\sigma^4 a^4}{4}\left\{T(1 - 1/\beta)h''(yT) - (yT^2/\beta)h^{(3)}(yT)\right\}^2 \qquad (2.7)$$

The aimse is obtained from (2.7) by integration over y.

In the special case when randomizing variable X is uniform $[0,1]$, the amse reduces to

$$\text{amse}(\hat{g}_n(y)) = \frac{y^2 h(y)}{na^3}R(k') + \frac{\sigma^4 a^4}{4}y^2\left(h^{(3)}(y)\right)^2 \qquad (2.8)$$

while its imse is

$$\text{aimse}(\hat{g}_n(x)) = \frac{R(k')EZ^2}{na^3} + \frac{\sigma^4 a^4 R(\phi_h^{(3)})}{4} \tag{2.9}$$

where $\phi_h^{(3)}(y) = yh^{(3)}(y)$. Thus, the smallest value occurs at $a^* = \{R(k')EZ^2/(\sigma^2 R(\phi_h^{(3)})n)\}^{1/7}$. This gives an order of convergence in aimse equal to $n^{-4/7}$ instead of the customary $n^{-4/5}$ for direct sampling. Thus, in randomized response sampling, one has to increase the sample size from, say, n in direct sampling to the order of $n^{7/5}$ for randomized response to achieve the same accuracy in density estimation. This confirms the results of Fan (1991a) for the general deconvolution problem.

Next we state two theorems that summarize the large sample properties of $\hat{g}_n(y)$. The first addresses pointwise behavior while the second deals with uniform consistency (both weak and strong). Proofs are only sketched.

Theorem 2.1

(i) Let yT be a continuity point of $h(.)$ and $h'(.)$ and let $na^3 \to \infty$ as $n \to \infty$; then $\hat{g}_n(y) \to g(y)$ in probability as $n \to \infty$.

(ii) If yT is a continuity point of $h(.)$ and $h'(.)$ and if, for any $\varepsilon > 0$, $\sum_{n=1}^{\infty} e^{-\varepsilon na^3} < \infty$, then $\hat{g}_n(y) \to g(y)$ with probability one as $n \to \infty$.

(iii) If $na^3 \to \infty$ and $na^7 \to 0$ as $n \to \infty$, then $\sqrt{na^3}[\hat{g}_n(y) - g(y)]$ is asymptotically normal with mean 0 and variance $\sigma^2 = h(yT)R(k')(yT^2/\beta)^2$ provided that $h^{(3)}(.)$ exists and is bounded.

Proof

(i) Since yT is obviously a continuity point of $h(.)$ and $h'(.)$, then $\hat{h}_n(yT) \to h(yT)$ and $\hat{h}_n'(yT) \to h'(yT)$ in probability as $n \to \infty$ and the result follows from the definition of $\hat{g}_n(y)$.

(ii) Using the methods of Theorem 2.3 of Singh (1981), one can show that, under the stated conditions, $\hat{h}_n(yT) \to h(yT)$ and $\hat{h}_n'(yT) \to h'(yT)$ with probability one as $n \to \infty$. Thus, the result follows.

(iii) Again, note that

$$\sqrt{na^3}[\hat{g}_n(y) - g(y)] = \sqrt{na^3}\left[\hat{h}(yT) - h(yT)\right][T(1 - 1/\beta)]$$
$$- \sqrt{na^3}\left[\hat{h}_n'(yT) - h'(yT)\right](yT^2/\beta) \tag{2.10}$$

But, clearly, $\hat{h}(yT) - h(yT) = O_p((na)^{-1}) + O(a^4)$ and, hence, the first term on the right-hand side of (2.10) is $O_p(a/\sqrt{n}) + O((na^7)^{1/2}) = o_p(1)$. On the other hand, that $\sqrt{na^3}(\hat{h}_n'(yT) - h'(yT))$ is asymptotically normal with mean 0 and variance σ^2 follows from an argument similar to that used in density estimation, cf. Parzen (1962).

Theorem 2.2. Assume that $zh'(z)$ is bounded and uniformly continuous.

(i) If $\int e^{itw}k^{(r)}(w)dw$ and $\int we^{itw}k^{(r)}(w)dw$ are absolutely integrable (in t) for $r = 0,1$, then $\sup_y |\hat{g}_n(y) - g(y)| \to 0$ in probability as $n \to \infty$, provided that $na^{2r+2} \to \infty$ as $n \to \infty$.

(ii) If $\int |d\theta^3 k^{(r)}(\theta)| < \infty$ for $r = 0$ and 1, and $s = 0$ and 1, if $\int |d(k^{(r)}(\theta))^2| < \infty$, $r = 0$ and 1, if $EZ^2 < \infty$, and if $\ln \ln n/(na^{2r+2})^{1/2} \to 0$ as $n \to \infty$, then $\sup_y |\hat{g}_n(y) - g(y)| \to 0$ with probability one as $n \to \infty$.

Proof

(i) Recall the definition of $\hat{g}_n(x)$ from (1.4). In view of the result of Parzen (1962) for weak consistency and Nadaraya (1965) for strong consistency in the direct sampling case, we need only to work with $\sup_z |z| \| \hat{h}_n^{(r)}(z) - h^{(r)}(z) \|$. First, look at $\sup_z |z| |E\hat{h}_n^{(r)}(z) - h^{(r)}(z)|$. But, writing $\Psi(u) = uh^{(r)}(u)$, we easily see that

$$\sup{}_z|z|\left|E\hat{h}_n^{(r)}(z) - h^{(r)}(z)\right| \leq \sup{}_z\left|\int \Psi(z - au)k^{(r)}(u)du - \Psi(z)\right|$$

$$+ a\sup{}_z\left|h^{(r)}(z)\right|\int|w|k(w)dw \leq Ca \tag{2.11}$$

Hence, (2.11) converges to zero as $n \to \infty$. Next, writing $\psi_n(t) = \int e^{itz} z\hat{h}_n^{(r)}(z)dz$, one can easily see that

$$\psi_n(t) = a^{-r+1}\xi_n(t)v(at) + a^{-r}\eta_n(t)\mu(at), \tag{2.12}$$

where $\xi_n(t) = (1/n)\sum_{j=1}^n$, and $\eta_n(t) = (1/n)\sum_{j=1}^n Z_j e^{itZ_j}$, $v(t) = \int we^{itw}k^{(r)}(w)dw$, and $\mu(t) = \int e^{itw}k^{(r)}(w)dw$. Hence,

$$E\psi_n(t) = \psi(t) = a^{-r+1}\xi(t)v(at) + a^{-r}\eta(t)\mu(at) \tag{2.13}$$

where $\xi(t) = E\xi_n(t)$ and $\eta(t) = E\eta_n(t)$. Thus we see that

$$z\left(\hat{h}_n^{(r)}(z) - E\hat{h}_n^{(r)}(z)\right) = \frac{1}{2\pi}\int_{-\infty}^{\infty} e^{itz}(\psi_n(t) - \psi(t))dt. \tag{2.14}$$

Hence, as in Parzen (1962), one can show that

$$E^{\frac{1}{2}}\left\{\sup{}_z|z|\left|\hat{h}_n^{(r)}(z) - E\hat{h}_n^{(r)}(z)\right|\right\}^2 \leq \frac{c}{2\pi}\left\{a^{-r}n^{-\frac{1}{2}}\int|v(t)|dt + a^{-a-1}n^{-\frac{1}{2}}\int|\mu(t)|\right\} \tag{2.15}$$

The right-hand side of (2.14) converges to 0 as $n \to \infty$. Part (i) is proved.

(ii) Let $H_n(z)$ and $H(z)$ denote the empirical and real cdf's of Z. Note that

$$z\left[\hat{h}_n^{(r)}(z) - E\hat{h}_n^{(r)}(z)\right] = (z/a^{r+1})\int k^{(r)}\left(\frac{z-w}{a}\right)d[H_n(w) - H(w)]$$

$$= I_{1n} + I_{2n}$$

where

$$I_{1n} = a^{-r}\left(\frac{z-w}{a}\right)k^{(r)}\left(\frac{z-w}{a}\right)d[H_n(w) - H(w)]$$

and

$$I_{2n} = a^{-r}\int wk^{(r)}\left(\frac{z-w}{a}\right)d[H_n(w) - H(w)]$$

Using integration by parts,

$$|I_{1n}| \leq a^{-r}\left|\int [H_n(w) - H(w)]d\left(\frac{z-w}{a}\right)k^{(4)}\left(\frac{z-w}{a}\right)\right|$$

$$\leq a^{-r-1}\sup_w |H_n(w) - H(w)|\int\left|d\theta k^{(4)}(\theta)\right| \tag{2.16}$$

$$= O_{wp1}\left(\ln\ln n/na^{2r+2}\right)^{1/2}$$

Next, using Cauchy–Schwartz inequality for signed measures,

$$|I_{2n}| \leq a^{-r}\left\{\int w^2 d(H_n(w) - H(w))\right\}^{1/2}$$

$$\left\{\int\left(k^{(r)}\left(\frac{z-w}{a}\right)\right)^2 d(H_n(w) - H(w))\right\}^{1/2} \tag{2.17}$$

$$\leq a^{-r}\left\{O_{wp1}\left(\ln\ln n/\left(n^{\frac{1}{2}}\right)\right)\right\}\cdot J_n, \quad \text{say}$$

where

$$|J_n^{-1}\sup_w|H_n(w) - H(w)|\int\left|d(k^{(r)}(\theta)^2\right| = O_{wp1}\left(\ln\ln n/\left(na^2\right)^{\frac{1}{2}}\right) \tag{2.18}$$

From (2.16) and (2.17), we see that $I_{2n} = O_{wp1}(\ln\ln n/(na^{2r+2})^{1/2})$. Thus, (ii) is proved and so is the theorem.

Next, we study the behavior of $\hat{G}_n(y)$ given in (1.3). First, note that

$$\text{mse}(\hat{G}_n(y)) = \text{mse}\left(\hat{H}_n(yT)\right) + \left(\frac{yT}{\beta}\right)^2\text{mse}\left(\hat{h}_n(yT)\right)$$

$$-\frac{2yT}{\beta}E\left(\hat{H}_n(yT) - H(yT)\right)\left(\hat{h}_n(yT) - h(yT)\right)$$

Set $z = yT$ and note that

$$\text{amse}\left(\hat{H}_n(z)\right) = \frac{H(z)(1 - H(z))}{n} - \frac{2a}{n}h(z)\int uk(u)K(u)du + \frac{\sigma^4 a^4}{4}(h'(z))^2$$

(2.19)

and

$$\text{amse}\left(\hat{h}_n(z)\right) = \frac{h(z)R(k)}{na} + \frac{\sigma^4 a^4}{4}(h''(z))^2$$

(2.20)

Let us evaluate the last term in $\text{mse}(\hat{G}_n(y))$.

$$E\hat{H}_n(z)\hat{h}_n(z) = \frac{1}{na}EK\left(\frac{z - Z_1}{a}\right)k\left(\frac{z - Z_1}{a}\right)$$

$$+ \frac{n - 1}{n}EK\left(\frac{z - Z_1}{a}\right)Ek\left(\frac{z - Z_1}{a}\right)$$

$$= J_1 + J_2, \quad \text{say}$$

$$J_2 \simeq \frac{n - 1}{n}\left\{h(z)H(z) + \frac{\sigma^2 a^2}{2}\left(h''(z) + h'(z)\right) + \frac{\sigma^4 a^4}{4}h'(z)h''(z)\right\}$$

and

$$J_1 \simeq \frac{h(z)}{2n} - \frac{ah'(z)}{n}\int uk(u)K(u)du$$

Thus

$$E\hat{H}_n(z)\hat{h}_n(z) \simeq \frac{h(z)}{2n} - \frac{ah'(z)}{n}\int uk(u)K(u)du$$

$$+ h(z)H(z) + \frac{\sigma^2 a^2}{2}\left(h'(z) + h''(z)\right) + \frac{\sigma^4 a^4}{4}h'(z)h''(z)$$

and

$$E\hat{h}_n(z)H(z) + E\hat{H}_n(z)h(z) \simeq 2h(z)H(z) + \frac{\sigma^2 a^2}{2}\left(h'(z) + h''(z)\right)$$

Hence,

$$E\left(\hat{H}_n(z) - H(z)\right)\left(E\hat{h}_n(z) - h(z)\right) \simeq \frac{h(z)}{2n}\{1 - 2H(z)\}$$

$$-\frac{a}{n}h'(z)\int uk(u)K(u)du \qquad (2.21)$$

$$+\frac{\sigma^4 a^4}{4}h'(z)h''(z)$$

Hence,

$$\mathrm{amse}\left(\hat{G}_n(y)\right) = \frac{1}{na}R(k)h(yT)\left(\frac{yT}{\beta}\right)^2$$

$$+\frac{1}{n}\left\{H(yT)(1 - H(yT)) - \frac{yT}{\beta}h(yT)(1 - 2H(yT))\right\}$$

$$+\frac{\sigma^4 a^4}{4}\left\{h'(yT) - \left(\frac{yT}{\beta}\right)h''(yT)\right\}^2$$

$$(2.22)$$

Again, the aimse is obtained from (2.20) by integration with respect to y. Note that the optimal choice of a here is of order $n^{-1/3}$. Thus, using randomized response to estimate $G(y)$ necessitates that we increase the sample size from n to an order of $n^{5/3}$ to reach the same order of approximations of *mse* or *imse* in direct sampling.

Finally, we state two results summarizing the large sample properties of $\hat{G}_n(y)$ both pointwise and uniform. Proofs of these results are similar to those of Theorems 2.1 and 2.2 above and, hence, are omitted.

Theorem 2.3

(i) Let yT be a continuity point of $h(.)$ and let $na \to \infty$ as $n \to \infty$; then $\hat{G}_n(y) \to G(y)$ in probability as $n \to \infty$.

(ii) If yT is a continuity point of $h(.)$ and if, for any $\varepsilon > 0$, $\sum_{n=1}^{\infty} e^{-\varepsilon na_n} < \infty$, then $\hat{G}_n(y) \to G(y)$ with probability one as $n \to \infty$.

(iii) If $na \to \infty$ and $na^5 \to 0$ as $n \to \infty$ and if $h''(.)$ exists and is bounded, then $\sqrt{na}(\hat{G}_n(y) - G(y))$ is asymptotically normal with mean 0 and variance $(yT/\beta)^2 h(yT)R(k)$.

Theorem 2.4. Assume that $zh(z)$ is bounded and uniformly continuous.

(i) If $\int e^{itw}k(w)dw$ and $\int e^{itw}wk(w)dw$ are absolutely integrable (in t), then $\sup_y |\hat{G}_n(y) - G(y)| \to 0$ in probability provided that $na^2 \to 0$ as $n \to \infty$.

(ii) If $|dk(\theta)| < \infty$, if $EZ^2 < \infty$, and if $\ln\ln n/(na^2) \to 0$ as $n \to \infty$, then $\sup_y |\hat{G}_n(y) - G(y)| \to 0$ with probability one as $n \to \infty$.

Before closing this section, let us address the representation of the mean absolute error (mae) of both $\hat{g}_n(y)$ and $\hat{G}_n(y)$, since this criterion is sometimes used for smoothing, cf. Devroye and Gyrofi (1986). We shall restrict our attention to the case when the randomizing variable X is uniform $(0,1)$, to simplify discussion. In this case, and using Theorem 2.1 (iii), one can write for large n

$$\hat{g}_n(y) - g(y) = \{y^2 h(y) R(k')/na^3\}^{\frac{1}{2}} W + \frac{a^2\sigma^2}{2} y h^{(3)}(y) \tag{2.23}$$

where W is the standard normal variable. Hence we see that, for large n,

$$E|\hat{g}_n(y) - g(y)| = \left\{\frac{y^2 h(y) R(k')}{na^3}\right\}^{\frac{1}{2}} E\left|W - \frac{\sigma^2 h^{(3)}(y)}{2}\left\{\frac{na^7}{h(y)R(k')}\right\}^{\frac{1}{2}}\right| \tag{2.24}$$

But $E|W - \theta| = 2\theta\Phi(\theta) + 2\phi(\theta) - \theta$, for all θ and where Φ is the cdf and ϕ is the pdf of the standard normal. Set

$$\theta = \frac{\sigma^2 h^{(3)}(y)}{2}\left\{\frac{na^7}{h(y)R(k')}\right\}^{\frac{1}{2}}$$

Then

$$a = \left\{\frac{4h(y)R(k')\theta^2}{na^4(h^{(3)}(y))^2}\right\}^{1/7}$$

Therefore, for n large enough,

$$E\left|\hat{g}_n(y) - g(y)\right| = \left\{y^2 h^{\frac{4}{7}}(y)(R(k'))^{\frac{4}{7}}\sigma^{\frac{12}{7}}\left(h^{(3)}(y)\right)^{\frac{6}{7}}/2^{\frac{3}{7}}n^{\frac{3}{7}}\right\}\theta^{-\frac{3}{7}}\delta(\theta) \tag{2.25}$$

where $\delta(\theta) = 2\theta\Phi(\theta) + 2\phi(\theta) - \theta$. Hence, the minimum value of a is $a^{**} = \{4h(y)R(k')\theta_0/na^4(h^{(3)}(y))^2\}^{1/7}$, where θ_0 is the unique minimum of $\theta^{-\frac{3}{7}}\delta(\theta)$. Comparing a^*, the minimizer of amse, with a^{**}, the minimizer of amae, we obtain the ratio $\{\frac{3}{4}\theta_0\}^{1/7}$, which is independent of *all* distributional parameters.

Next, working with $\hat{G}_n(y)$, again for the uniform randomizing mixture, it follows from Theorem 2.3 (iii) that, for large n,

$$\hat{G}_n(y) - G(y) = \{R(k)R(y)y^2/na\}^{\frac{1}{2}} W + \frac{\sigma^2 a^2}{2}\{h'(y) - y h''(y)\} \tag{2.26}$$

where, as above, W is the standard normal variate. Thus the amae of $\hat{G}_n(y)$ is

$$
\mathrm{amae}\left(\hat{G}_n(y)\right) = \left\{\frac{y^2 h(y) R(k)}{na}\right\}^{1/2} E\left|W\right.
$$

$$
- \left\{\frac{na^5}{y^2 h(y) R(k)}\right\}^{\frac{1}{3}} \left\{\frac{\sigma^2 (h'(y) - yh^*(y))}{2}\right\}\right|
$$

Working as above, we see that the value of a that minimizes (2.27) is $\tilde{a} = \{4\gamma_0^2 y^2 h(y) R(k) / n\sigma^4 (h'(y) - yh''(y))^2\}^{\frac{1}{5}}$, where γ_0 is the unique minimum of $\gamma^{-\frac{1}{3}}\delta(\gamma)$. On the other hand, the value of a that minimizes the amse is $a\% = \{y^2 h(y) R(k) / n\sigma^4 (h'(y) - yh''(y))^2\}^{\frac{1}{5}}$. Thus, the ratio of $a\%$ to is $(4\gamma_0)^{-\frac{1}{3}}$, independent of all distributional parameters.

We are now working on data-based choices of the window size a for both $\hat{g}_n(.)$ and $\hat{G}_n(.)$ using a new concept we call "kernel contrasts," and the results will appear, hopefully, before long.

3. ESTIMATING REGRESSION FUNCTIONS

3.1 Case (i). Sensitive Response

As noted in the Introduction, the regression funtion $R(u) = E(Y|U = u)$. is proportional to the regression function $r(u) = E(Z|U = u)$. Since the data is collected on (U,Z), one can employ the readily available literature on $\hat{r}_n(u)$ to deduce *all* properties of $\hat{R}_n(u)$; cf. Scott (1992). For example,

$$
\mathrm{mse}\left(\hat{R}_u(u)\right) = \left[\left(1 + \frac{1}{\beta}\right)/T\right]^2 \mathrm{mse}(\hat{r}_n(u))
$$

$$
\simeq \left[\left(1 + \frac{1}{\beta}\right)/T\right]^2 \frac{R(k)V(Z|u)}{nag_1(u)} \qquad (3.1)
$$

$$
+ \frac{a^4\sigma^4}{4}\left[r''(u) + 2r'(u)\frac{g_1'(u)}{g_1(u)}\right]^2
$$

Also, $(na)^{\frac{1}{2}}(\hat{R})u) - R(u))$ is asymptotically normal with mean 0 and variance $[(1 + 1/\beta)/T]^2 (R(k)\sigma^2 z, u/nag_1(u))$. Hence, for sufficiently large n,

$$
\hat{R}_n(u) - R(u) = \left[\left(1 + \frac{1}{\beta}\right)/T\right]^2 \left\{\left[\frac{R(k)\sigma_{z,u}^2}{nag_1(u)}\right]^{\frac{1}{2}} W - \frac{\sigma^2 a^2}{2}\right.
$$

$$
\left.\left[r''(u) + 2r'(u)\frac{g_1'(u)}{g_1(u)}\right]\right\} \qquad (3.2)
$$

Proceeding as in the previous section, one sees that the value that minimizes the amse is $a^* = \{R(k)\sigma^2_{z,u}/g_1(u)\sigma^4[r''(u) + 2r'(u)(g_1'(u)/g_1(u))]n\}^{1/5}$, whereas that which minimizes amae is $a^{**} = \{4\gamma_0^2 R(k)\sigma^2_{z,u}/g_1(u)\sigma^4[r''(u) + wr'(u)$ $(g_1'(u)/g_1(u))]n\}^{1/5}$, where γ_0 is the unique minimizer of $\gamma^{-1/2}\delta(\gamma)$. Thus, theratio is $\{4\gamma_0^2\}^{1/5}$.

3.2 Case (ii). Sensitive Regressor

Recall the definitions of $S(y)$ and $\hat{S}_n(y)$, given in (1.13) and (1.14), respectively. Let us work with the mse $(\hat{S}_n(y))$. First of all, $E\hat{S}_n(y)$; $E\hat{T}_n(y)/E\hat{g}_{2n}(y)$, where $\hat{T}_n(y)$ is the numerator of $\hat{S}_n(y)$ and $\hat{g}_{2n}(y)$ is as given in (1.4). Now, for large n,

$$E\hat{T}_n(y) = T(y) + \frac{\sigma^2 a^2}{2}\left\{\left[T\left(1 - \frac{1}{\beta}\right)\right]\int u, h^{(0,2)}(u, yT)du\right.$$
$$\left. - (yT^2/\beta)\int u, h^{(0,3)}(u, yT)du\right\}$$

and it follows from (2.7) that, for large n,

$$E\hat{g}_{2n}(y) = g_2(y) + \frac{\sigma^2 a^2}{2}\left\{T\left(1 - \frac{1}{\beta}\right)h_2''(yT) - (yT^2/\beta)h_2^{(3)}(yT)\right\} \tag{3.4}$$

Hence,

$$E\hat{S}_n(y) \simeq S(y) + \frac{\sigma^2 a^2}{2g_2(y)}\left\{T\left(1 - \frac{1}{\beta}\right)\int uh^{(0,2)}(u, yT)du - (yT^2/\beta)\right.$$
$$\left.\int uh^{(0,3)}(y, yT)du\right\} \tag{3.5}$$

Next, let us evaluate the variances:

$$V\hat{T}_n(y) \simeq \frac{1}{na^3}\left(\int u^2 h(u, yT)dy\right)R\left(T\left(1 - \frac{1}{\beta}\right)ak - (yT^2/\beta)k'\right) \tag{3.6}$$

and from (2.7), we have

$$V\hat{g}_{2n}(y) \simeq \frac{g_2(y)}{na^3}R\left(T\left(1 - \frac{1}{\beta}\right)ak - (yT^2/\beta)k'\right) \tag{3.7}$$

Finally, look at

$$\text{cov}\left(\hat{T}_n(y), \hat{g}_{2n}(y)\right) \simeq \frac{g_2(y)}{na^3}R\left(T\left(1 - \frac{1}{\beta}\right)ak - (yT^2/\beta)k'\right)\int uh(y, yT)du \tag{3.8}$$

Collecting terms and using the formula

$$V\left(\frac{U}{V}\right) = \left(\frac{EU}{EV}\right)^2\left\{\frac{V(U)}{(E(U))^2} + \frac{V(V)}{(EV)^3} - \frac{2\text{cov}(UV)}{EUEV}\right\}$$

we get, after careful simplification, that

$$V\hat{S}_n(y) \simeq \frac{R\left(T\left(1 - \frac{1}{\beta}\right)ak - (yT^2/\beta)k'\right)}{na^3 g_2^2(y)}$$

(3.9)

$$\left\{\int u^2 h(y, yT)du + S^2(y)h_2(yT) - 2S(y)\int uh(u, yT)du\right\}$$

From (3.5) and (3.9), the mse of $\hat{S}_n(y)$ is approximated by

$$\text{amse}\left(\hat{S}_n(y)\right) = \frac{R\left(T\left(1 - \frac{1}{\beta}\right)ak - (yT^2/\beta)k'\right)}{na^3 g_2^2(y)}$$

$$\left\{\int u^2 h(u, yT)du + S^2(y)h_2(yT) - 2S(y)\int uh(u, yT)du\right\}$$

$$+ \frac{\sigma^4 a^4}{4g_2^2(y)}\left\{T\left(1 - \frac{1}{\beta}\right)\int uh^{(0,2)}(u, yT)du\right.$$

$$-(yT^2/\beta)\int uh^{(0,3)}(u, yT)du$$

(3.10)

Next, we offer some of the large sample properties of $\hat{S}_n(y)$. Theorem 3.1 below is stated without proof, while the proof of Theorem 3.2 is only briefly sketched.

Theorem 3.1

(i) Let $h^{(0,1)}(., yT)$ be continuous and let $na \to \infty$ as $n \to \infty$ then $\hat{S}_n(y) \to S(y)$ in probability as $n \to \infty$.

(ii) If, in addition, $a \leq U_i \leq b$ and, for any $\varepsilon > 0$, $\sum_{n=1}^{\infty} e^{-\varepsilon na^3} < \infty$, then $\hat{S}_n(y) \to S(y)$ with probability one as $n \to \infty$.

(iii) If, in addition to (i), $na^7 \to 0$, then $\sqrt{na^3}(\hat{S}_n(y) - S(y))$ is asymptotically normal with mean 0 and variance

$$\frac{R((yT^2/\beta)k')}{g_2^2(y)} \left\{ \int u^2 h(u, yT)du + S^2(y)h_2(yT) - 2S(y) \int uh(u, yT)du \right\}$$

Theorem 3.2. Assume that $zh^{(0,1)}(u, z)$ is bounded and uniformly continuous.

(i) Let the condition (i) of Theorem 2.2 hold; then $\sup_{y \in C} |\hat{S}_n(y) - S(y)| \to 0$ in probability for any compact and bounded set C, as n to ∞.

(ii) If, in addition, the condition (ii) of Theorem 2.2 is in force, and if $a \le U_i \le b$, $1 \le i \le n$, then $\sup_{y \in C} |\hat{S}_n(y) - S(y)| \to 0$ with probability one as $n \to \infty$.

Proof. By utilizing the technology of Nadaraya (1970), one can show that

$$P\left[\sup_{y \varepsilon} |\hat{S}_n(y) - S(y)| > \varepsilon\right] \le P\left[\sup_{y \varepsilon} |\hat{T}_n(y) - T(y)| > \varepsilon\right] \tag{3.11}$$

$$P\left[\sup_{y} |\hat{g}_{2n}(y) - g_2(y)| > \varepsilon_2\right]$$

Thus, we prove the results for each of the two terms in (3.11), but the second term follows from Theorem 2.2 while the first term follows similarly by obvious modifications. For sake of brevity, we do not reproduce the details.

If we wish to have the L_1-norm of $\hat{S}_n(y)$, we proceed as in the case of density or distribution. Proceeding as in the density case, we see that the ratio of the optimal choice of a relative to that of amse is for both cases $\{\frac{3}{4}\theta_0\}^{1/7}$, exactly as in the density case.

REFERENCES

Barry, J. and Diggle, P. (1995). Choosing the smoothing parameter in a Fourier approach to nonparametric deconvolution of a density estimate. *J. Nonparam. Statist.*, **4**, 223–232.

Carroll, R.J. and Hall, P. (1988). Optimal rates of convergence for deconvolving a density. *J. Am. Statist. Assoc.*, **83**, 1184–1186.

Chaudhuri, A. and Mukherjee, R. (1987). Randomized response techniques: A review. *Statistica Neerlandica*, **41**, 27–44.

Devroye, L. and Gyrofi, L. (1986). *Nonparametric Density Estimation: The L_1 view*. Wiley, New York.

Diggle, P. and Hall, P. (1993). A Fourier approach to nonparametric deconvolution of a density estimate. *J.R. Statist. Soc., B,* **55**, 523–531.

Duffy, J.C. and Waterton, J.J. (1989). Randomized response models for estimating the distribution function of a qualitative character. *Int. Statist. Rev.,* **52**, 165–171.

Fan, J. (1991a). On the optimal rates of convergence for nonparametric deconvolution problems. *Ann. Statist.,* **19**, 1257–1272.

Fan, J. (1991b). Global behavior of deconvolution kernel estimates. *Statistica Sinica,* **1**, 541–551.

Fan, J. (1992). Deconvolution with supersmooth distributions. *Canad. J. Statist.,* **20**, 155–169.

Liu, M.C. and Taylor, R.L. (1990a). A consistent nonparametric density estimator for the deconvolution problem. *Canad. J. Statist.,* **17**, 427–438.

Liu, M.C. and Taylor, R.L. (1990b). Simulations and computations of nonparametric density estimates of the deconvolution problems. *J. Statist. Comp. Simul.,* **35**, 145–167.

Masry, E. and Rice, J.A. (1992). Gaussian deconvolution via differentiation. *Canad. J. Statist.,* **20**, 9–21.

Nadaraya, E.A. (1965). A nonparametric estimation of density function and regression. *Theor. Prob. Appl.,* **10**, 186–190.

Nadaraya, E.A. (1970). Remarks on nonparametric estimates of density functions and regression curves. *Theor. Prob. Appl.,* **15**, 139–142.

Parzen, E. (1962). On estimation of a probability density function and mode. *Ann. Math. Statist.,* **33**, 1065–1076.

Patil, P. (1996). A note on deconvolutions density estimation. *Statist. Prob. Lett.,* **29**, 79–84.

Poole, W.K. (1974). Estimation of the distribution function of a continuous type random variable through randomized response. *J. Am. Statist. Assoc.,* **69**, 1002–1005.

Scott, D.W. (1992). *Multivariate Density Estimation*. Wiley, New York.

Singh, R.S. (1981). On the exact asymptotic behavior of estimators of a density and its derivatives. *Ann. Statist.,* **6**, 453–456.

Stefanski, L.A. (1990). Rates of convergence of some estimators in a class of deconvolution problems. *Statist. Prob. Lett.*, **9**, 229–235.

Stefanski, L.A. and Carroll, R.J. (1990). Deconvoluting kernel density estimates. *Statistics*, **21**, 169–184.

Warner, S.L. (1965). Randomized response: A survey technique for eliminating evasive answer bias. *J. Am. Statist. Assoc.*, **60**, 63–69.

7
Index-Free, Density-Based Multinomial Choice

JEFFREY S. RACINE University of South Florida, Tampa, Florida

1. INTRODUCTION

Multinomial choice models are characterized by a dependent variable which can assume a limited number of discrete values. Such models are widely used in a number of disciplines and occur often in economics because the decision of an economic unit frequently involves choice, for instance, of whether or not a person joins the labor force, makes an automobile purchase, or chooses the number of offspring in their family unit. The goal of modeling such choices is first to make conditional predictions of whether or not a choice will be made (choice probability) and second to assess the response of the probability of the choice being made to changes in variables believed to influence choice (choice gradient).

Existing approaches to the estimation of multinomial choice models are "index-based" in nature. That is, they postulate models in which choices are governed by the value of an "index function" which aggregates information contained in variables believed to influence choice. For example, the index might represent the "net utility" of an action contemplated by an economic agent. Unfortunately, we cannot observe net utility, we only observe whether an action was undertaken or not; hence such index-based models are known as "latent variable" models in which net utility is a "latent" or unobserved variable.

This index-centric view of nature can be problematic as one must assume that such an index exists, that the form of the index function is known, and typically one must also assume that the distribution of the disturbance term in the latent-variable model is of known form, though a number of existing semiparametric approaches remove the need for this last assumption. It is the norm in applied work to assume that a single index exists ("single-index" model) which aggregates all conditioning information*, however, there is a bewildering variety of possible index combinations including, at one extreme, separate indices for each variable influencing choice. The mere presence of indices can create problems of normalization, identification, and specification which must be addressed by applied researchers. Ahn and Powell (1993, p. 20) note that "parametric estimates are quite sensitive to the particular specification of the index function," and the use of an index can give rise to identification issues which make it impossible to get sensible estimates of the index parameters (Davidson and Mackinnon 1993, pp. 501–521).

For overviews of existing parametric approaches to the estimation of binomial and multinomial choice models the reader is referred to Amemiya (1981), McFadden (1984), and Blundell (1987). Related work on semiparametric binomial and multinomial choice models would include (Coslett (1983), Manski (1985), Ichimura (1986), Rudd (1986), Klein and Spady (1993), Lee (1995), Chen and Randall (1997), Ichimura and Thompson (1998), and Racine (2001).

Rather than beginning from an index-centric view of nature, this paper models multinomial choice via direct estimation of the underlying conditional probability structure. This approach will be seen to complement existing index-based approaches by placing fewer restrictions on the underlying data generating process (DGP). As will be demonstrated, there are a number of benefits that follow from taking a density-based approach. First, by directly modeling the underlying probability structure we can handle a richer set of problem domains than those modeled with standard index-based approaches. Second, the notion of an index is an artificial construct which, as will be seen, is not generally required for the sensible modeling of conditional predictions for discrete variables. Third, nonparametric frameworks for the estimation of joint densities fit naturally into the proposed approach.

---

*Often, in multinational choice models, multiple indices are employed but they presume that all conditioning information enters each index, hence the indices are identical in form apart from the unknown values of the indices' parameters.

The remainder of the paper proceeds as follows: Section 2 presents a brief overview of modeling conditional predictions with emphasis being placed on the role of conditional probabilities, Section 3 outlines a density-based approach towards the modeling of multinomial choice models and sketches potential approaches to the testing of hypotheses, Sections 4 and 5 consider a number of simulations and applications of the proposed technique that highlight the flexibility of the proposed approach relative to standard approaches towards modeling multinomial choice, and Section 6 concludes.

2. BACKGROUND

The statistical underpinnings of models of conditional expectations and conditional probabilities will be briefly reviewed, and the role played by conditional probability density functions (PDFs) will be highlighted since this object forms the basis for the proposed approach.

The estimation of models of conditional expectations for the purpose of conditional prediction of continuous variables constitutes one of the most widespread approaches in applied statistics, and is commonly referred to as "regression analysis." Let $Z = (Y, X) \in \mathbb{R}^{k+1}$ be a random vector for which Y is the dependent variable and $X \in \mathbb{R}^k$ is a set of conditioning variables believed to influence Y, and let $z_i = (y_i, x_i)$ denote a realization of Z. When Y is continuous, interest typically centers on models of the form $m(x_i) = E[Y|x_i]$ where, by definition,

$$
\begin{aligned}
y_i &= E[Y|x_i] + u_i \\
&= \int_{-\infty}^{\infty} y f[y|x_i] dy + u_i
\end{aligned}
\tag{1}
$$

where $f[Y|x_i]$ denotes the conditional density of Y, and where u_i denotes an error process. Interest also typically lies in modeling the response of $E[Y|x_i]$ due to changes in x_i (gradient), which is defined as

$$
\begin{aligned}
\nabla x_i E[Y|x_i] &= \frac{\partial E[Y|x_i]}{\partial x_i} \\
&= \int_{-\infty}^{\infty} y \frac{\partial f[Y|x_i]}{\partial x_i} dy
\end{aligned}
\tag{2}
$$

Another common modeling exercise involves models of choice probabilities for the conditional prediction of discrete variables. In the simplest such situation of binomial choice ($y_i \in \{0, 1\}$), we are interested in models of the form $m(x_i) = f[Y|x_i]$ where we predict the outcome

$$y_i = \begin{cases} 1 & \text{if } f[Y|x_i] > 0.5 \quad (f[Y|x_i] > 1 - f[Y|x_i]) \\ 0 & \text{otherwise} \end{cases} \quad i = 1, \ldots, n \qquad (3)$$

while, for the case of multinomial choice ($y_i \in \{0, 1, \ldots, p\}$), we are again interested in models of the form $m(x_i) = f[Y|x_i]$ where again we predict the outcome

$$y_i = j \text{ if } f[Y = j|x_i] > f[Y = l|x_i] \quad l = 0, \ldots, p, \quad l \neq j, \quad i = 1, \ldots, n \qquad (4)$$

Interest also lies in modeling the response of $f[Y|x_i]$ due to changes in x_i, which is defined as

$$\nabla_{x_i} f[Y|x_i] = \frac{\partial f[Y|x_i]}{\partial x_i}, \qquad (5)$$

and this gradient is of direct interest to applied researchers since it tells us how choice probabilities change due to changes in the variables affecting choice.

It can be seen that modeling conditional predictions involves modeling both conditional probability density functions ("probability functions" for discrete Y) $f[Y|x_i]$ and associated gradients $\partial f[Y|x_i]/\partial x_i$ regardless of the nature of the variable being predicted, as can be seen by examining equations (1), (2), (3), and (5). When Y is discrete $f[Y|x_i]$ is referred to as a "probability function," whereas when Y is continuous $f[Y|x_i]$ is commonly known as a "probability density function." Note that, for the special case of binomial choice where $y_i \in \{0, 1\}$, the conditional expectation $E[Y|x_i]$ and conditional probability $f[Y|x_i]$ are one and the same since

$$E[Y|x_i] = \sum_y y_i f[Y = y_i|x_i]$$
$$= 0 \times (1 - f[Y|x_i]) + 1 \times f[Y|x_i] \qquad (6)$$
$$= f[Y|x_i]$$

however, this does not extend to multinomial choice settings.

When Y is discrete, both the choice probabilities and the gradient of these probabilities with respect to the conditioning variables have been modeled parametrically and semiparametrically using index-based approaches. When modeling choice probabilities via parametric models, it is typically necessary to specify the form of a probability function and an index function which is assumed to influence choice. The probit probability model of binomial choice given by $m(x_i) = \text{CNORM}(-x_i'\beta)$, where CNORM is the cumulative normal distribution function, is perhaps the most common example of a parametric probability model. Note that this model presumes that

$f[Y|x_i] = \text{CNORM}(-x_i'\beta)$ and that $\nabla_{x_i}f[Y|x_i] = \text{CNORM}(-x_i'\beta)(1 - \text{CNORM}(-x_i'\beta))\beta$, where these probabilities arise from the distribution function of an error term from a latent variable model evaluated at the value of the index. That is, $f[Y|x_i]$ is assumed to be given by $F[x_i'\beta]$ where $F[\cdot]$ is the distribution function of a disturbance term from the model $u_i = y_i^* - x_i'\beta$; however, we observe $y_i = 1$ only when $x_i'\beta + u_i > 0$, otherwise we observe $y_i = 0$. Semiparametric approaches permit nonparametric estimation of the probability distribution function (Klein and Spady 1993), but it remains necessary to specify the form of the index function when using such approaches.

We now proceed to modeling multinomial choice probabilities and choice probability gradients using a density-based approach.

3. MODELING CHOICE PROBABILITIES AND CHOICE GRADIENTS

In order to model multinomial choice via density-based techniques, we assume only that the density conditional upon a choice being made $(f[x_i|Y = l], l = 0, \ldots, p$ where $p + 1$ is the number of choices) exists and is bounded away from zero for *at least one* choice*. This assumption is not necessary; however, dispensing with this assumption raises issues of trimming which are not dealt with at this point.

We assume for the time being that the conditioning variables X are continuous† and that only the variable being predicted is discrete. Thus, the object of interest in multinomial choice models is a conditional probability, $f[y_i]$. Letting $p + 1$ denote the number of choices, then if $y \in \{0, \ldots, p\}$, simple application of Bayes' theorem yields

$$
\begin{aligned}
f[Y = j|x_i] &= \frac{f[Y = j \cap x_i]}{f[x_i]} \\
&= \frac{f[x_i|Y = j]f[Y = j]}{\sum_{l=0}^{x} f[x_i|Y = l]f[Y = l]}, \quad j = 0, \ldots, p
\end{aligned}
\tag{7}
$$

*This simply avoids the problem of $f[x_i] = 0$ in equation (7). When nonparametric estimation methods are employed, we may also require continuous differentiability and existence of higher-order moments.

†When the conditioning variables are categorical, we could employ a simple dummy variable approach on the parameters of the distribution of the continuous variables.

The gradient of this function with respect to X will be denoted by $\nabla_X f[Y = j|x_i] = f'[Y = j|x_i] \in R^k$ and is given by

$$\nabla_X f[Y = j|x_i] = \frac{\left(\sum_{l=0}^{p} f[x_i|Y = l]f[Y = l]\right)f[Y = j]f'[x_i|Y = j]}{\left(\sum_{l=0}^{p} f[x_i|Y = l]f[Y = l]\right)^2}$$

$$- \frac{f[Y = j]f[x_i|Y = j]\left(\sum_{l=0}^{p} f'[x_i|Y = l]f[Y = l]\right)}{\left(\sum_{l=0}^{p} f[x_i|Y = l]f[Y = l]\right)^2} \tag{8}$$

Now note that $f[x_i|Y = l]$ is simply the joint density of the conditioning variables for those observations for which choice l is made. Having estimated this object using only those observations for which choice l is made, we then evaluate its value over the entire sample. This then permits evaluation of the choice probability for all observations in the data sample, and permits us to compute choice probabilities for future realizations of the covariates in the same fashion. These densities can be estimated either parametrically assuming a known joint distribution or nonparametrically using, for example, the method of kernels if the dimensionality of X is not too large. We denote such an estimate as $\widehat{f[x_i|Y = l]}$. Note that $f[Y = l]$ can be estimated with the sample relative frequencies $\widehat{f[Y = l]} = n^{-1}\sum_{i=1}^{n} I_l(y_i)$, where $I_l()$ is an indicator function taking on the value 1 if $y_i = l$ and zero otherwise, and where n is the total number of responses*. For prediction purposes, $\widehat{y}_i = j$ if $\widehat{f[Y = j|x_i]} > \widehat{f[Y = l|x_i]} \quad \forall \quad l \neq j, j = 0, 1, \ldots, p$.

Note that this approach has the property that, by definition and construction,

$$\sum_{j=0}^{p} \widehat{f[Y = j|x_i]} = 1 \tag{9}$$

Those familiar with probit models for multinomial choice can appreciate this feature of this approach since there are no normalization or identification issues arising from presumption of an index which must be addressed.

3.1 Properties

Given estimators of $f[x_i|Y = l]$ and $f[Y = l]$ for $l = 0, \ldots, p$, the proposed estimator would be given by

$$\widehat{f[Y = j|x_i]} = \frac{\widehat{f[x_i|Y = j]}\widehat{f[Y = j]}}{\sum_{l=0}^{p} \widehat{f[x_i|Y = l]}\widehat{f[Y = l]}}, \quad j = 0, \ldots, p, \quad i = 1, \ldots, n. \tag{10}$$

---

*This is the maximum likelihood estimator.

Clearly the finite-sample properties of $f[\widehat{Y=j}|x_i]$ will depend on the nature of the estimators used to obtain $f[x_i|\widehat{Y=l}]$ and $f[\widehat{Y=l}]$, $l = 0,\ldots,p$. However, an approximation to the finite-sample distribution based upon asymptotic theory can be readily obtained. Noting that, conditional upon X, the random variable Y equals j with probability $f[Y=j|x_i]$ and is not equal to j with probability $1 - f[Y=j|x_i]$ (i.e. is a Bernoulli random variate), via simple asymptotic theory using standard regularity conditions it can be shown that

$$f[\widehat{Y=j}|x_i] \simeq AN\left(f[Y=j|x_i], \frac{f[Y=j|x_i](1 - f[Y=j|x_i])}{n}\right) \tag{11}$$

Of course, the finite-sample properties are technique-dependent, and this asymptotic approximation may provide a poor approximation to the unknown finite-sample distribution. This remains the subject of future work in this area.

3.2 Estimation

One advantage of placing the proposed method in a likelihood framework lies in the ability to employ a well-developed statistical framework for hypothesis testing purposes; therefore we consider estimation of $f[Y|x_i]$ using the method of maximum likelihood*. If one chooses a parametric route, then one first presumes a joint PDF for X and then estimates the objects constituting $f[Y|x_i]$ in equation (7), thereby yielding $f[\widehat{Y}|x_i]$ given in equation (10). If one chooses a nonparametric route one could, for example, estimate the joint densities using a Parzen (1962) kernel estimator with bandwidths selected via likelihood cross-validation (Stone 1974). The interested reader is referred to Silverman (1986), Härdle (1990), Scott (1992), and Pagan and Ullah (1999) for an overview of such approaches. Unlike parametric models, however, this flexibility typically comes at the cost of a slower rate of convergence which depends on the dimensionality of X. The benefit of employing nonparametric approaches lies in the consistency of the resultant estimator under much less restrictive presumptions than those required for consistency of the parametric approach.

3.3 Hypothesis Testing

There are a number of hypothesis tests that suggest themselves in this setting. We now briefly sketch a number of tests that could be conducted. This

---

*The ML estimator of $f[Y = l]$ is given by $f[\widehat{Y = l}] = n^{-1}\sum_{i=1}^{n} I_l(y_i)$.

section is merely intended to be descriptive as these tests are not the main focus of this paper, though a modest simulation involving the proposed significance test is carried out in Section 4.5.

Significance (orthogonality) tests

Consider hypothesis tests of the form

$$H_0: \quad f[Y = j|x] = f[Y = j] \text{ for all } x \in X$$
$$H_A: \quad f[Y = j|x] \neq f[Y = j] \text{ for some } x \in X$$

(12)

which would be a test of "joint significance" for conditional prediction of a discrete variable, that is, a test of orthogonality of all conditioning variables. A natural way to conduct such a test would be to impose restrictions and then compare the restricted model with the unrestricted one. The restricted model and unrestricted model are easily estimated as the restricted model is simply $f[Y = j]$, which is estimated via sample frequencies.

Alternatively, we could consider similar hypothesis tests of the form

$$H_0: \quad f[Y = j|x] = f[Y = j|x^s] \text{ for all } x \in X$$
$$H_A: \quad f[Y = j|x] \neq f[Y = j|x^s] \text{ for some } x \in X$$

(13)

where $x^s \subset x$, which would be a test of "significance" for a subset of variables believed to influence the conditional prediction of a discrete variable, that is, a test of orthogonality of a subset of the conditioning variables

A parametric approach would be based on a test statistic whose distribution is known under the null, and an obvious candidate would be the likelihood-ratio (LR) test. Clearly the conditional PDF of y_i can always be written as $f[Y = y_i|x_i] = \prod_{l=0}^{p} f[Y = l|x_i^s]^{y_i}$. The restricted log-likelihood would be $\sum_{i=1}^{n} \ln f[Y = y_i|x_i^s]$, whereas the unrestricted log-likelihood would be given by $\sum_{i=1}^{n} \ln f[Y = y_i|x_i]$. Under the null, the LR statistic would be distributed asymptotically as chi-square with j degrees of freedom, where j is the number of restrictions; that is, if $X \in \mathbb{R}^k$ and $X^S \in \mathbb{R}^{k'}$ where $X^s \subset X$, then the number of restrictions is $j = k - k'$. A simple application of this test is given in Section 4.5.

Model specification tests

Consider taking a parametric approach to modeling multinomial choice using the proposed method. The issue arises as to the appropriate joint distribution of the covariates to be used. Given that all estimated distributions are of the form $f[x_i|Y = l]$ and that these are all derived from the joint distribution of X, then it is natural to base model specification tests on existing tests for density functional specification such as Pearson's χ^2 test of goodness of fit. For example, if one assumes an underlying normal distribution for X, then one can conduct a test based upon the difference between observed frequencies and expected sample frequencies if the null of normality is true.

Index specification tests

A test for correct specification of a standard probability model such as the probit can be based upon the gradients generated by the probit versus the density-based approach. If the model is correctly specified, then the gradients should not differ significantly. Such a test could be implemented in this context by first selecting a test statistic involving the integrated difference between the index-based and density-based estimated gradients and then using either asymptotics or resampling to obtain the null distribution of the statistic (Efron 1983, Hall 1992, Efron and Tibshirani 1993).

3.4 Features and Drawbacks

One disadvantage of the proposed approach to modeling multinomial choice is that either we require $f[x_i | Y = l], l = 0, \ldots, p$ to be bounded away from zero for at least one choice or else we would need to impose trimming. This does not arise when using parametric probit models, for example, since the probit CDF is well-defined when $f[x_i | Y = l] = 0$ by construction. However, this does arise for a number of semiparametric approaches, and trimming must be used in the semiparametric approach proposed by Klein and Spady (1993). When dealing with sparse data sets this could be an issue for certain types of DGPs. Of course, prediction in regions where the probability of realizations is zero should be approached with caution in any event (Manski 1999).

Another disadvantage of the proposed approach arises due to the need to estimate the density $f[x_i | Y = l], l = 0, 1 \ldots, p$, for all choices made. If one chooses a parametric route, for example, we would require sufficient observations for which $Y = l$ to enable us to estimate the parameters of the density function. If one chooses a nonparametric route, one would clearly need even more observations for which $Y = l$. For certain datasets this can be problematic if, for instance, there are only a few observations for which we observe a choice being made.

What are the advantages of this approach versus traditional approaches towards discrete choice modeling? First, the need to specify an index is an artificial construct that can create identification and specification problems, and the proposed approach is index-free. Second, estimated probabilities are probabilities naturally obeying rules such as non-negativity and summing to one over available choices for a given realization of the covariates x_i. Third, traditional approaches use a scalar index which reduces the dimensionality of the joint distribution to a univariate one. This restricts the types of situation that can be modeled to those in which there is only one threshold which is given by the scalar index. The proposed approach explicitly models the joint distribution, which admits a richer problem domain than that modeled with standard approaches. Fourth, interest typically centers on the gradient of the choice probability, and scalar-index models restrict the nature of this gradient, whereas the proposed approach does not suffer from such restrictions. Fifth, if a parametric approach is adopted, model specification tests can be based on well-known tests for density functional specification such as the χ^2 test of goodness of fit.

What guidance can be offered to the applied researcher regarding the appropriate choice of method? Experience has shown that data considerations are most likely to guide this choice. When faced with data sets having a large number of explanatory variables one is almost always forced into a parametric framework. When the data permits, however, I advocate randomly splitting the data into two sets, applying *both* approaches on one set, and validating their performance on the remaining set. Experience has again shown that the model with the best performance on the *independent* data will also serve as the most useful one for inference purposes.

3.5   Relation to Bayesian Discriminant Analysis

It turns out that the proposed approach is closely related to Bayesian discriminant analysis which is sometimes used to predict population membership. The goal of discriminant analysis is to determine which population an observation is likely to have come from. We can think of the populations in multinomial choice settings being determined by the choice itself.

By way of example, consider the case of binary choice. Let A and B be two populations with observations X known to have come from either population. The classical approach towards discriminant analysis (Mardia et al. 1979) estimates $f_A(X)$ and $f_B(X)$ via ML and allocates a new observation X to population A if

$$\tilde{f}_A(X) \geq \tilde{f}_B(X) \tag{14}$$

A more general Bayesian approach allocates X to population A if

$$\tilde{f}_A(X) \geq c\tilde{f}_B(X) \tag{15}$$

where c is chosen by considering the prior odds that X comes from population A.

Suppose that we arbitrarily let Population A be characterized by $Y = 1$ and B by $Y = 0$. An examination of equation (10) reveals that $f[\widehat{Y = 1}|x_i] \geq f[\widehat{Y = 0}|x_i]$ if and only if

$$f[\widehat{x_i|Y = 1}]f[\widehat{Y = 1}] \geq f[\widehat{x_i|Y = 0}]\,f[\widehat{Y = 0}] \tag{16}$$

which implies that

$$f[\widehat{x_iY = 1}] \geq \frac{f[\widehat{Y = 0}]}{f[\widehat{Y = 1}]}\,f[\widehat{x_i|Y = 0}] \tag{17}$$

or

$$f[\widehat{x_i|Y = 1}] \geq cf[\widehat{x_i|Y = 0}] \tag{18}$$

Therefore, it can be seen that the proposed approach provides a link between discriminant analysis and multinomial choice analysis through the underpinnings of density estimation. The advantage of the proposed approach relative to discriminant analysis lies in its ability to naturally provide the choice probability gradient with respect to the variables influencing choice which does not exist in the discriminant analysis literature.

4. SIMULATIONS

We now consider a variety of simulation experiments wherein we consider the widely used binomial and multinomial probit models as benchmarks, and performance is gauged by predictive ability. We estimate the model on one data set drawn from a given DGP and then evaluate the predictive ability of the fitted model on an independent data set of the same size drawn from the same DGP. We measure predictive ability via the percentage of correct predictions on the independent data set. For illustrative purposes, the multivariate normal distribution is used throughout for the parametric version of the proposed method.

4.1 Single-Index Latent Variable DGPs

The following two experiments compare the proposed method with correctly specified binomial and multinomial index models when the DGP is in fact generated according to index-based latent variable specifications. We first consider a simple univariate latent DGP for binary choice, and then we consider a multivariate latent DGP for multinomial choice. The parametric index-based models serve as benchmarks for these simulations as we obviously cannot do better than a correctly specified parametric model.

4.1.1 Univariate binomial Specification

For the following experiment, the data were generated according to a latent variable specification $y_i^* = \theta_0 + \theta_1 x_i + u_i$ where we observe only

$$y_i = \begin{cases} 1 & \text{if } \theta_0 + \theta_1 x_1 + u_i \geq 0 \\ 0 & \text{otherwise} \end{cases} \quad i = 1, \dots, n \quad (19)$$

where $x \sim N(0, 1)$ and $U \sim N(0, \sigma_u^2)$.

We consider the traditional probit model and then consider estimating the choice probability and gradient assuming an underlying normal distribution for X. Clearly the probit specification is a correctly specified parametric model hence this serves as our benchmark. The issue is simply to gauge the loss when using the proposed method relative to the probit model.

For illustrative purposes, the parametric version of the proposed method assuming normality yields the model

$$f[\widehat{Y=1}|xi] = \cfrac{\frac{1}{\sqrt{2\pi\tilde{\sigma}_1^2}}\exp\left(-\frac{(x-\tilde{\mu}_1)^2}{2\tilde{\sigma}_1^2}\right)\frac{\sum y_i}{n}}{\frac{1}{\sqrt{2\pi\tilde{\sigma}_0^2}}\exp\left(-\frac{(x-\tilde{\mu}_0)^2}{2\tilde{\sigma}_0^2}\right)\left(1-\frac{\sum y_i}{n}\right)+\frac{1}{\sqrt{2\pi\tilde{\sigma}_1^2}}\exp\left(-\frac{(x-\tilde{\mu}_1)^2}{2\tilde{\sigma}_1^2}\right)\frac{\sum y_i}{n}}$$

$$(20)$$

where $(\tilde{\mu}_0, \tilde{\sigma}_0)$ and $(\tilde{\mu}_1, \tilde{\sigma}_1)$ are the maximum likelihood estimates for those observations on X for which $y_i = 0$ and $y_i = 1$ respectively. The gradient

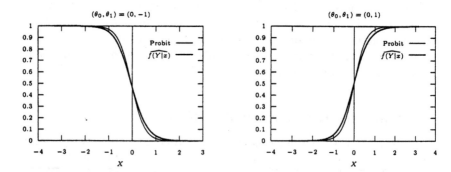

Figure 1. Univariate binomial conditional probabilities via the probit specification and probability specification for one draw with $n = 1000$.

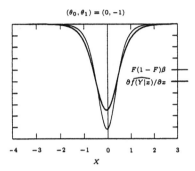
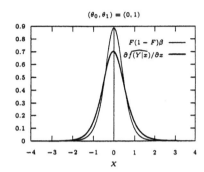

Figure 2. Univariate binomial choice gradient via the probit specifications and probability specifications for one draw with $n = 1000$.

follows in the same manner and is plotted for one draw from this DGP in Figure 2.

Figures 1 and 2 plot the estimated conditional choice probabilities and gradients as a function of X for one draw from this DGP for selected values of (θ_0, θ_1) with $\sigma_u^2 = 0.5$. As can be seen, the density-based and probit approaches are in close agreement where there is only one covariate, and differ slightly due to random sampling since this represents one draw from the DGP.

Next, we continue this example and consider a simple simulation which compares the predictive ability of the proposed method relative to probit regression. The dispersion of U (σ_u), values of the index parameters ($\theta_{(\cdot)}$), and the sample size were varied to determine the performance of the proposed approach in a variety of settings.

Results for the mean correct prediction percentage on independent data are summarized in Table 1. Note that some entries are marked with "–" in the following tables. This arises because, for some of the smaller sample sizes, there was a non-negligible number of Monte Carlo resamples for which either $Y = 0$ or $Y = 1$ for every observation in the resample. When this occurred, the parameters of a probit model were not identified and the proposed approach could not be applied since there did not exist observations for which a choice was made/not made.

As can be seen from Table 1, the proposed approach compares quite favorably to that given by the correctly specified probit model for this simulation as judged by its predictive ability for a range of values for σ_u, (θ_0, θ_1), and n. This suggests that the proposed method can perform as well as a correctly specified index-based model when modeling binomial choice. We now consider a more complex situation of multinomial choice in a

Table 1. Univariate binomial prediction ability of the proposed approach versus probit regression, based on 1000 Monte Carlo replications. Entries marked "—" denote a situation where at least one resample had the property that either $Y = 0$ or $Y = 1$ for each observation in the resample

n	σ_u	(θ_0, θ_1)	Density-based	Probit
100	1.0	(0.0, 0.5)	—	—
		(−1.0, 1.0)	—	—
		(0.0, 1.0)	0.85	0.85
		(1.0, 1.0)	0.94	0.94
	1.5	(0.0, 0.5)	—	—
		(−1.0, 1.0)	—	—
		(0.0, 1.0)	0.81	0.82
		(1.0, 1.0)	0.92	0.93
	2.0	(0.0, 0.5)	—	—
		(−1.0, 1.0)	—	—
		(0.0, 1.0)	—	—
		(1.0, 1.0)	0.91	0.91
500	1.0	(0.0, 0.5)	0.80	0.81
		(−1.0, 1.0)	0.86	0.86
		(0.0, 1.0)	0.86	0.87
		(1.0, 1.0)	0.94	0.95
	1.5	(0.0, 0.5)	0.77	0.78
		(−1.0, 1.0)	0.80	0.80
		(0.0, 1.0)	0.83	0.83
		(1.0, 1.0)	0.93	0.93
	2.0	(0.0, 0.5)	0.75	0.75
		(−1.0, 1.0)	—	—
		(0.0, 1.0)	0.81	0.81
		(1.0, 1.0)	0.92	0.93
1000	1.0	(0.0, 0.5)	0.81	0.82
		(−1.0, 1.0)	0.86	0.86
		(0.0, 1.0)	0.87	0.87
		(1.0, 1.0	0.94	0.95
	1.5	(0.0, 0.5)	0.78	0.78
		(−1.0, 1.0)	0.80	0.80
		(0.0, 1.0)	0.83	0.83
		(1.0, 1.0)	0.93	0.94
	2.0	(0.0, 0.5)	0.76	0.76
		(−1.0, 1.0)	0.75	0.75
		(0.0, 1.0)	0.81	0.81
		(1.0, 1.0)	0.93	0.93

multivariate setting to determine whether this result is more general than would appear from this simple experiment.

Multivariate Multinomial Specification

For the following experiment, the data were generated according to a latent variable multinomial specification $y_i^* = \theta_0 + \theta_1 x_{i1} + \theta_2 x_{i2} + u_i$ where we observe only

$$
y_i = \begin{cases} 2 & \text{if } \theta_0 + \theta_1 x_{i1} + \theta_2 x_{i2} + u_i > 1.5 \\ 1 & \text{if } -0.5 \le \theta_0 + \theta_1 x_{i1} + \theta_2 x_{i2} + u_i \le 1.5 \quad i = 1, \dots, n \\ 0 & \text{otherwise} \end{cases} \quad (21)
$$

Where $X_1 \sim X_2 \sim N(0, \sigma_x^2)$ and $U \sim N(0, \sigma_u^2)$. This is the standard "ordered probit" specification, the key feature being that all the choices depend upon a single index function.

We proceed in the same manner as for the univariate binomial specification, and for comparison with that section we let $\sigma_{x_1} = \sigma_{x_2} = 0.5$. The experimental results mirror those from the previous section over a range of parameters and sample sizes and suggest identical conclusions; hence results in this section are therefore abbreviated and are found in Table 2. The representative results for $\sigma_u = 1$, based on 1000 Monte Carlo replications using $n = 1000$, appear in Table 2.

The proposed method appears to perform as well as a correctly specified multinomial probit model in terms of predictive ability across a wide range of sample sizes and parameter values considered, but we point out that this method does not distract the applied researcher with index-specification issues.

Table 2. Multivariate multinomial prediction ability of the proposed approach versus probit regression with $\sigma_u = 1$, based on 1000 Monte Carlo replications using $n = 1000$.

$(\theta_0, \theta_1, \theta_2)$	Density-based	Probit
$(0, .5, .5)$	0.78	0.78
$(0, 1, 1)$	0.83	0.84
$(0, 1, -1)$	0.83	0.84
$(-1, 1, -1)$	0.82	0.83
$(1, 1, 1)$	0.97	0.97

4.2 Multiple-Index Latent Variable DGPs

Single-index models are universally used in applied settings of binomial choice, whereas multiple-index models are used when modeling multinomial choice. However, the nature of the underlying DGP is restricted when using standard multiple-index models of multinomial choice (such as the multinomial probit) since the index typically presumes the same variables enter each index while only the parameters differ according to the choices made. Also, there are plausible situations in which it is sensible to model binomial choice using multiple indices*. As was seen in Section 4.1, the proposed approach can perform as well as standard index-based models of binomial and multinomial choice. Of interest is how the proposed method performs relative to standard index-based methods for plausible DGPs.

The following two experiments compare the proposed method with standard single-index models when the DGP is in fact generated according to a plausible multiple-index latent variable specification which differs slightly from that presumed by standard binomial and multinomial probit models.

4.3 Multivariate Binomial Choice

For the following experiment, the data were generated according to

$$y_i = \begin{cases} 1 & \text{if } \theta_{01} + \theta_{11}x_{i1} + u_{i1} \geq 0 \text{ and } \theta_{02} + \theta_{12}x_{i2} + u_{i2} \geq 0 \\ 0 & \text{otherwise} \end{cases} \quad i = 1, \ldots, n$$

(22)

where $X_1 \sim N(0,1)$, $X_2 \sim N(0,1)$, $U_1 \sim N(0, \sigma_{u1}^2)$, and $U_2 \sim N(0, \sigma_{u2}^2)$. Again, we consider the traditional probit model and then consider estimating the choice probability and gradient assuming an underlying normal distribution for X.

Is this a realistic situation? Well, consider the situation where the purchase of a minivan is dependent on income and number of children. If after-tax income exceeds a threshold (with error) and the number of children exceeds a threshold (with error), then it is more likely than not that the family will purchase a minivan. This situation is similar to that being modeled here where X_1 could represent after-tax income and X_2 the number of children.

Figures 3 and 4 present the estimated conditional choice probabilities and gradients as a function of X for one draw from this DGP. An examination of Figure 3 reveals that models such as the standard linear index probit

*The semiparametric method of Lee (1995) permits behavior of this sort.

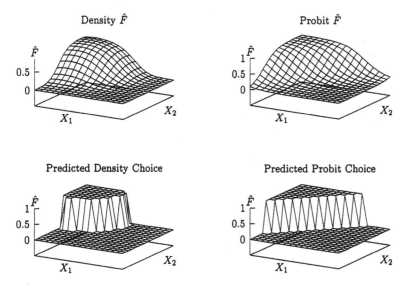

Figure 3. Multivariate binomial choice probability and predicted choices for one draw from simulated data with $X_1, X_2 \simeq N()$, $n = 1000$. The first column contains that for the proposed estimator, the second for the standard probit. The parameter values were $(\theta_{,01}, \theta_{02}, \theta_{11}, \theta_{12}) = (0, 0, -1, 1)$ and $(\sigma_{u1}, \sigma_{u2}) = (0.25, 0.25)$.

cannot capture this simple type of problem domain. The standard probit model can only fit one hyperplane through the input space which falls along the diagonal of the X axes. This occurs due to the nature of the index, and is not rectified by using semiparametric approaches such as that of Klein and Spady (1993). The proposed estimator, however, adequately models this type of problem domain, thereby permitting consistent estimation of the choice probabilities and gradient.

An examination of the choice gradients graphed in Figure 4 reveals both the appeal of the proposed approach and the limitations of the standard probit model and other similar index-based approaches. The true gradient is everywhere negative with respect to X_1 and everywhere positive with respect to X_2. By way of example, note that, when X_2 is at its minimum, the true gradient with respect to X_1 is zero almost everywhere. The proposed estimator picks this up, but the standard probit specification imposes non-zero gradients for this region. The same phenomena can be observed when X_1 is at its maximum whereby the true gradient with respect to X_2 is zero almost everywhere, but again the probit specification imposes a discernibly non-zero gradient.

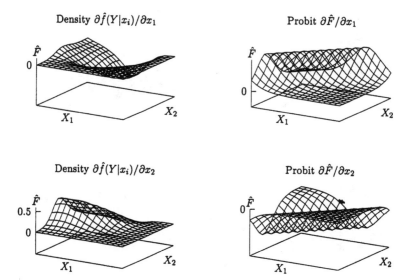

Density $\partial \hat{f}(Y|x_i)/\partial x_1$ Probit $\partial \hat{F}/\partial x_1$

Density $\partial \hat{f}(Y|x_i)/\partial x_2$ Probit $\partial \hat{F}/\partial x_2$

Figure 4. Multivariate binomial choice gradients for simulated data with $X_1, X_2 \simeq N()$, $n = 1000$. The first column contains that for the proposed estimator, the second for the standard probit. The parameter values were $(\theta_{01}, \theta_{02}, \theta_{11}, \theta_{12}) = (0, 0, -1, 1)$.

We now consider a simple simulation to assess the performance of the proposed approach relative to the widely used probit approach by again focusing on the predictive ability of the model on independent data. We consider the traditional probit model, and the proposed approach is implemented assuming an underlying normal distribution for X where the DGP is that from equation (22). A data set was generated from this DGP and the models were estimated, then an independent data set was generated and the percentage of correct predictions based on independent data. The dispersion of U (σ_u) was varied to determine the performance of the proposed approach in a variety of settings, the sample size was varied from $n = 100$ to $n = 1000$, and there were 1000 Monte Carlo replications. Again note that, though we focus only on the percentage of correct predictions, this will not tell the entire story since the standard probit is unable to capture the type of multinomial behavior found in this simple situation*, hence the choice probability gradient will also be misleading. The mean correct prediction percentage is noted in Table 3. It can be seen that the proposed method does better

---

*Note, however, that the proposed approach can model situations for which thestandard probit is appropriate, as demonstrated in Section 4.1.

Table 3. Multivariate binomial prediction ability of the proposed approach versus probit regression, based on 1000 Monte Carlo replications. For all experiments the parameter vector $(\theta_{01}, \theta_{02}, \theta_{11}, \theta_{12},)$ was arbitrarily set to $(0, 0, 1, 1)$. Entries marked "—" denote a situation where at least one resample had the property that either $Y = 0$ or $Y = 1$ for each observation in the resample. Note that $\sigma_u = \sigma_{u1} = \sigma_{u2}$.

n	σ_u	Density-based	Probit
100	0.1	0.86	0.83
	0.5	—	—
	1.0	—	—
250	0.1	0.87	0.83
	0.5	0.83	0.80
	1.0	—	—
500	0.1	0.87	0.83
	0.5	0.83	0.80
	1.0	0.74	0.73
1000	0.1	0.87	0.83
	0.5	0.83	0.80
	1.0	0.74	0.73

than the standard probit model for this DGP based solely upon a prediction criterion, but it is stressed that the choice gradients given by the standard probit specification will be inconsistent and misleading.

4.4 Multivariate Multinomial Choice

For the following experiment, the data were generated according to

$$
y_i = \begin{cases} 2 & \text{if } \theta_{01} + \theta_{11}x_{i1} + u_{i1} \geq 0 \text{ and } \theta_{02} + \theta_{12}x_{i2} + u_{i2} \leq 0 \\ 1 & \text{if } \theta_{01} + \theta_{11}x_{i1} + u_{i1} \leq 0 \text{ and } \theta_{02} + \theta_{02}x_{12} + u_{i2} \geq 0 \quad i = 1, \ldots, n \\ 0 & \text{otherwise} \end{cases}
$$

(23)

where $X_1 \sim N(0, 1)$ and $X_2 \sim N(0, 1)$. Again, we consider the traditional multinomial probit model and then consider estimating the choice probability and gradient assuming an underlying normal distribution for X. The experimental setup mirrors that found in Section 4.3 above.

Table 4 suggests again that the proposed method performs substantially better than the multinomial probit model for this DGP based solely upon a prediction criterion, but again bear in mind that the choice gradients given by the standard probit specification will be inconsistent and misleading for this example.

4.5 Orthogonality Testing

We consider the orthogonality test outlined in Section 3.3 in a simple setting. For the following experiment, the data were generated according to a latent variable binomial specification $y_i^* = \theta_0 + \theta_1 x_{i1} + \theta_2 x_{i2} + u_i$ where we observe only

Table 4. Multivariate multinomial prediction ability of the proposed approach versus probit regression, based on 1000 Monte Carlo replications. For all experiments the parameter vector $(\theta_{01}, \theta_{02}, \theta_{11}, \theta_{12},)$ was arbitrarily set to $(0, 0, 1, 1)$. Entries marked "—" denote a situation where at least one resample had the property that either $Y = 0$, $Y = 1$, or $Y = 2$ for each observation. Note that $\sigma_u = \sigma_{u1} = \sigma_{u2}$

n	σ_u	Density-based	Probit
100	0.1	0.77	0.73
	0.25	—	—
	0.5	—	—
250	0.1	0.78	0.72
	0.25	0.76	0.71
	0.5	0.71	0.67
500	0.1	0.78	0.72
	0.25	0.76	0.71
	0.5	0.72	0.66
1000	0.1	0.78	0.72
	0.25	0.77	0.71
	0.5	0.72	0.66

$$y_i = \begin{cases} 1 & \text{if } \theta_0 + \theta_1 x_{i1} + \theta_2 x_{i2} + u_i \geq 0 \\ 0 & \text{otherwise} \end{cases} \qquad i = 1, \dots, n \qquad (24)$$

where $X_1 \sim X_2 \sim N(0, \sigma_x^2)$ and $U \sim N(0, \sigma_u^2)$.

We arbitrarily set $\sigma_u = 1$, $\sigma_x = 0.5$, $\theta_0 = 0$ and $\theta_1 = 1$. We vary the value of θ_2 from 0 to 1 in increments of 0.1. For a given value of θ_2 we draw 1000 Monte Carlo resamples from the DGP and for each draw compute the value of the proposed likelihood statistic outlined in Section 3.3. The hypothesis to be tested is that X_2 does not influence choices, which is true if and only if $\theta_2 = 0$. We write this hypothesis as

$$H_0 : f[Y = j|x_1, x_2] = f[Y = j|x_1] \text{ for all } x \in X$$

$$H_A : f[Y = j|x_1, x_2] \neq f[Y = j|x_1] \text{ for some } x \in X$$

(25)

We compute the empirical rejection frequency for this test at a nominal size of $\alpha = 0.05$ and sample sizes $n = 25$, $n = 50$, and $n = 100$, and plot the resulting power curves in Figure 5. For comparison purposes, we conduct a t-test of significance for X_2 from a correctly-specified binomial probit model. The proposed test shows a tendency to overreject somewhat for small samples while the probit-based test tends to underreject for small samples, so for comparison purposes each test was size-adjusted to have empirical size equal to nominal size when $\theta_2 = 0$ to facilitate power comparisons.

Figure 5 reveals that these tests behave quite similarly in terms of power. This is reassuring since the probit model is the correct model, hence we have confidence that the proposed test can perform about as well as a correctly specified index model but without the need to specify an index. This modest example is provided simply to highlight the fact that the proposed density-based approach to multinomial choice admits standard tests such as the test of significance, using existing statistical tools.

5. APPLICATIONS

We now consider two applications which highlight the value added by the proposed method in applied settings. For the first application, index models break down. For the second application, the proposed method performs better than the standard linear-index probit model and is in close agreement with a quadratic-index probit model, suggesting that the proposed method frees applied researchers from index-specification issues.

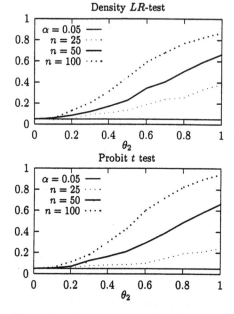

Figure 5. Power curves for the proposed likelihood ratio test and probit-based t-test when testing significance of X_2. The flat lower curve represents the test's nominal size of $\alpha = 0.05$.

5.1 Application to Fisher's Iris Data

Perhaps the best known polychotomous data set is the Iris dataset introduced in Fisher (1936). The data report four characteristics (sepal width, sepal length, petal width and petal length) of three species of Iris flower, and there were $n = 150$ observations.

The goal, given the four measurements, is to predict which one of the three species of Iris flower the measurements are likely to have come from. We consider multinomial probit regression and both parametric and non-parametric versions of the proposed technique for this task.

Interestingly, multinomial probit regression breaks down for this data set and the parameters are not identified. The error message given by TSP© is reproduced below:

Estimation would not converge; coefficients of these
variables would slowly diverge to $+/-$ infinity -- the scale
of the coefficients is not identified in this situation.
See Albert + Anderson, Biometrika 1984 pp.1-10,
or Amemiya, Advanced Econometrics, pp.271-272.

Assuming an underlying normal distribution of the characteristics, the parametric version of the proposed technique correctly predicts 96% of all observations. Using a Parzen (1962) kernel estimator with a gaussian kernel and with bandwidths selected via likelihood cross-validation (Stone 1974), the nonparametric version of the proposed technique correctly predicts 98% of all observations. These prediction values are in the ranges commonly found by various discriminant methods.

As mentioned in Section 1, the notion of an index can lead to problems of identification for some datasets, as is illustrated with this well-known example, but the same is not true for the proposed technique, which works quite well in this situation. The point to be made is simply that applied researchers can avoid issues such as identification problems which arise when using index-based models if they adopt the proposed density-based approach.

5.2   Application—Predicting Voting Behavior

We consider an example in which we model voting behavior given information on various economic characteristics on individuals. For this example we consider the choice of voting "yes" or "no" for a local school tax referendum. Two economic variables used to predict choice outcome in these settings are income and education. This is typically modeled using a probit model in which the covariates are expressed in log( ) form.

The aim of this modest example is simply to gauge the performance of the proposed method relative to standard index-based approaches in a real-world setting. Data was taken from Pindyck and Rubinfeld (1998, pp. 332–333), and there was a total of $n = 95$ observations available. Table 5 summarizes the results from this modest exercise.

We compare a number of approaches: a parametric version of the proposed approach assuming multivariate normality of log(income) and log(education), and probit models employing indices that are linear, quadratic, and cubic in log(income) and log(education). For comparison purposes we consider the percentage of correct predictions given by each estimator, and results are found in Table 5.

As can be seen from Table 5, the proposed method performs better in terms of percentage of correct choices predicted than that obtained using a probit model with a linear index. The probit model incorporating quadratic terms in each variable performs better than the proposed method, but the probit model incorporating quadratic terms and cubic terms in each variable does not perform as well. Of course, the applied researcher using a probit model would need to determine the appropriate functional form for the

Table 5. Comparison of models of voting behavior via percentage of correct predictions. The first entry is that for a parametric version of the proposed approach, whereas the remaining are those for a probit model assuming linear, quadratic, and cubic indices respectively

Method	% correct
Density	67.4%
Probit—linear	65.3%
Probit—quadratic	68.4%
Probit—cubic	66.3%

index, and a typical approach such as the examination of t-stats of the higher-order terms are ineffective in this case since they all fall well below any conventional critical values, hence it is likely that the linear index would be used for this data set. The point to be made is simply that we can indeed model binary choice without the need to specify an index, and we can do better than standard models, such as the probit model employing the widely used linear index, without burdening the applied researcher with index specification issues.

An examination of the choice probability surfaces in Figure 6 is quite revealing. The proposed density-based method and the probit model assuming an index which is quadratic in variables are in close agreement, and both do better than the other probit specifications in terms of predicting choices. Given this similarity, the gradient of choice probabilities with respect to the explanatory variables would also be close. However, both the choice probabilities and gradient would differ dramatically for the alternative probit specifications. Again, we simply point out that we can model binary choice without the need to specify an index, and we note that incorrect index specification can have a marked impact on any conclusions drawn from the estimated model.

6. CONCLUSION

Probit regression remains one of the most popular approaches for the conditional prediction of discrete variables. However, this approach has

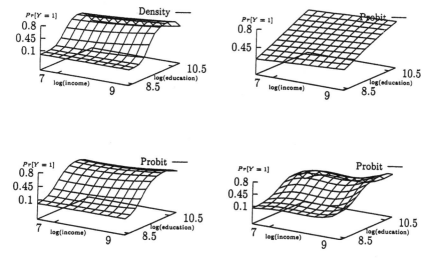

Figure 6. Choice probability surfaces for voting data. The first graph is that for a parametric version of the proposed approach, whereas the remaining graphs are those for a probit model assuming linear, quadratic, and cubic indices respectively

a number of drawbacks arising from the need to specify both a known distribution function and a known index function, which gives rise to specification and identification issues. Recent developments in semiparametric modeling advance this field by removing the need to specify the distribution function, however, these developments remain constrained by their use of an index.

This paper proposes an index-free density-based approach for obtaining choice probabilities and choice probability gradients when the variable being predicted is discrete. This approach is shown to handle problem domains which cannot be properly modeled with standard parametric and semiparametric index-based models. Also, the proposed approach does not suffer from identification and specification problems which can arise when using index-based approaches. The proposed approach assumes that probabilities are bounded away from zero or requires the use of trimming since densities are directly estimated and used to obtain the conditional prediction. Both parametric and nonparametric approaches are considered. In addition, a test of orthogonality is proposed which permits tests of joint significance to be conducted in a natural manner, and simulations suggest that this test has power characteristics similar to correctly specified index models.

Simulations and applications suggest that the technique works well and can reveal data structure that is masked by index-based approaches. This is not to say that such approaches can be replaced by the proposed method, and when one has reason to believe that multinomial choice is determined by an index of known form it is clearly appropriate to use this information in the modeling process. As well, it is noted that the proposed method requires sufficient observations for which choices are made to enable us to estimate a density function of the variables influencing choice either parametrically or nonparametrically. For certain datasets this could be problematic, hence an index-based approach might be preferred in such situations.

There remains much to be done in order to complete this framework. In particular, a fully developed framework for inference based on finite-sample null distributions of the proposed test statistics remains a fruitful area for future research.

ACKNOWLEDGMENT

I would like to thank Eric Geysels, Norm Swanson, Joe Terza, Gabriel Picone, and members of the Penn State Econometrics Workshop for their useful comments and suggestions. All errors remain, of course, my own.

REFERENCES

Ahn, H. and Powell, J. L. (1993), Semiparametric estimation of censored selection models with a nonparametric selection mechanism, *Journal of Econometrics* **58** (1/2), 3–31.

Amemiya, T. (1981), qualitative response models: A survey, *Journal of Economic Literature* **19**,1483–1536.

Blundell, R., ed. (1987), *Journal of Econometrics*, vol. 34, North-Holland, Amsterdam.

Chen, H. Z. and Randall, A. (1997), Semi-nonparametric estimation of binary response models with an application to natural resource valuation, *Journal of Econometrics* **76**, 323–340.

Coslett, S. R. (1983), Distribution-free maximum likelihood estimation of the binary choice model, *Econometrica* **51**, 765–782.

Davidson, R. and MacKinnon, J. G. (1993), *Estimation and Inference in Econometrics*, Oxford University Press, New York.

Efron, B. (1983), *the Jackknife, the Bootstrap, and Other Resampling Plans*, Society for Industrial and Applied Mathematics, Philadelphia, PA.

Efron, B. and Tibshirani, R. (1993), *An Introduction to the Bootstrap*, Chapman and Hall, New York, London.

Fisher, R. A. (1936), The use of multiple measurements in axonomic problems, *Annals of Eugenics* **7**, 179–188.

Hall, P. (1992), *The Bootstrap and Edgeworth Expansion*, Springer Series in Statistics, Springer-Verlag, New York.

Härdle, W. (1990), *Applied Nonparametric Regression*, Cambridge, New Rochelle.

Ichimura, H. and Thompson, T. S. (1998), Maximum likelihood estimation of a binary choice model with random coefficients of unknown distribution, *Journal of Econometrics* **86**(2), 269–295.

Klein, R. W. and Spady, R. H. (1993), An efficient semiparametric estimator for binary response models, *Econometrica* **61**, 387–421.

Lee, L. F. (1995), Semiparametric maximum likelihood estimation of polychotomous and sequential choice models, *Journal of Econometrics* **65**, 381–428.

McFadden, D. (1984), Econometric analysis of qualitative response models, in Z. Griliches and M. Intriligator, eds, *Handbook of Econometrics*, North-Holland, Amsterdam, pp. 1385–1457.

Manski, C. F. (1985), Semiparametric analysis of discrete response: Asymptotic properties of the maximum score estimator, *Journal of Econometrics* **27**, 313–333.

Manski, C. F. (1999), *Identification Problems in the Social Sciences*, Harvard, Cambridge, MA.

Mardia, K. V., Kent, J. T., and Bibby, J. M. (1979), *Multivariate Analysis*, Academic Press, London.

Pagan, A. and Ullah, A. (1999), *Nonparametric Econometrics*, Cambridge University Press.

Parzen, E. (1962), On estimation of a probability density function and mode, *Annals of Mathematical Statistics* **33**, 105–131.

Pindyck, R. S. and Rubinfeld, D. L. (1998), *Econometric Models and Economic Forecasts*, Irwin McGraw-Hill, New York.

Racine, J. S. (2001, forthcoming), Generalized semiparametric binary prediction. Annals of Economics and Finance.

Rudd, P. (1986), Consistent estimation of limited dependent variable models despite misspecification of distribution, *Journal of Econometrics* **32**, 157–187.

Scott, D. W. (1992), *Multivariate Density Estimation: Theory, Practice, and Visualization*, Wiley, New York.

Silverman, B. W. (1986), *Density Estimation for Statistics and Data Analysis*, Chapman and Hall, London.

Stone, C. J. (1974), Cross-validatory choice and assessment of statistical predictions (with discussion), *Journal of the Royal Statistical Society* **36**, 111–147.

8
Censored Additive Regression Models

R. S. SINGH University of Guelph, Guelph, Ontario, Canada

XUEWEN LU Agriculture and Agri-Food Canada, Guelph, Ontario, Canada

1. INTRODUCTION
1.1 The Background

In applied economics, a very popular and widely used model is the transformation model, which has the following form:

$$T(Y) = X\beta + \sigma(X)\epsilon_1 \qquad (1)$$

where T is a strictly increasing function, Y is an observed dependent variable, X is an observed random vector, β is a vector of constant parameters, $\sigma(X)$ is the conditional variance representing the possible heterocedacity, and ϵ_1 is an unobserved random variable that is independent of X. Models of the form (1) are used frequently for the analysis of duration data and estimation of hedonic price functions. Y is censored when one observes not Y but $\min(Y, C)$, where C is a variable that may be either fixed or random. Censoring often arises in the analysis of duration data. For example, if Y is the duration of an event, censoring occurs if data acquisition terminates before all the events under observation have terminated.

Let F denote the cumulative distribution function (CDF) of ϵ_1. In model (1), the regression function is assumed to be parametric and linear. The statistical problem of interest is to estimate β, T, and/or F when they are unknown. A full discussion on these topics is made by Horowitz (1998, Chap. 5). In the regression framework, it is usually not known whether or not the model is linear. The nonparametric regression model provides an appropriate way of fitting the data, or functions of the data, when they depend on one or more covariates, without making assumptions about the functional form of the regression. For example, in model (1), replacing $X\beta$ by an unknown regression function $m(X)$ yields a nonparametric regression model. The nonparametric regression model has some appealing features in allowing to fit a regression model with flexible covariate effect and having the ability to detect underlining relationship between the response variable and the covariates.

1.2 Brief Review of the Literature

When the data are not censored, a tremendous literature on nonparametric regression has appeared in both economic and statistical leading journals. To name a few, Robinson (1988), Linton (1995), Linton and Nielsen (1995), Lavergne and Vuong (1996), Delgado and Mora (1995), Fan et al. (1998), Pagan and Ullah (1999), Racine and Li (2000). However, because of natural difficulties with the censored data, regression problems involving censored data are not fully explored. To our knowledge, the paper by Fan and Gijbels (1994) is a comprehensive one on the censored regression model. The authors have assumed that $E\epsilon_1 = 0$ and $\mathrm{Var}(\epsilon_1) = 1$. Further, $T(\cdot)$ is assumed to be known, so $T(Y)$ is simply Y. They use the local linear approximations to estimate the unknown regression function $m(x) = E(Y| X = x)$ after transforming the observed data in an appropriate simple way.

Fan et al. (1998) introduce the following multivariate regression model in the context of additive regression models:

$$E(Y|X = x) = \mu + f_1(x_1) + f_2(x_2, x_3) \tag{2}$$

where Y is a real-valued dependent variable, $X = (X_1, X_2, X_3)$ is a vector of explanatory variables, and μ is a constant. The variables X_1 and X_2 are continuous with values in $\mathbf{R}^p$ and $\mathbf{R}^q$, respectively, and X_3 is discrete and takes values in $\mathbf{R}^r$. it is assumed that $Ef_1(X_1) = Ef_2(X_2, X_3) = 0$ for identifiability. The novelty of their study is to directly estimate $f_1(x)$ nonparametrically with some good sampling properties. The beauty of model (2) is that it includes both the additive nonparametric regression model with $X = U$

$$E(Y|U = u) = \mu + f_1(u_1) + \cdots + f_p(u_p) \tag{3}$$

and the additive partial linear model with $X = (U, X_3)$

$$E(Y|X = x) = \mu + f_1(u_1) + \cdots + f_p(u_p) + x_3^T \beta \qquad (4)$$

where $U = (U_1, \ldots, U_p)$ is a vector of explanatory variables. These models are much more flexible than a linear model. When data (Y, X) are completely observable, they use a direct method based on "marginal integration," proposed by Linton and Nielsen (1995), to estimate additive components f_1 in model (2) and f_j in models (3) and (4). The resulting estimators achieve optimal rate and other attractive properties. But for censored data, these techniques are not directly applicable; data modification and transformation are needed to model the relationship between the dependent variable Y and the explanatory variables X. There are quite a few data transformations proposed; for example, the KSV transformation (Koul et al. 1981), the Buckley–James transformation (Buckley and James 1979), the Leurgans transformation (Leurgans 1987), and a new class of transformations (Zheng 1987). All these transformations are studied for the censored regression when the regression function is linear. For example, in linear regression models, Zhou (1992) and Srinivasan and Zhou (1994) study the large sample behavior of the censored data least-squares estimator derived from the synthetic data method proposed by Leurgans (1987) and the KSV method. When the regression function is unspecified, Fan and Gijbels (1994) propose two versions of data transformation, called the local average transformation and the NC (new class) transformation, inspired by Buckley and James (1979) and Zheng (1987). They apply the local linear regression method to the transformed data set and use an adaptive (variable) bandwidth that automatically adapts to the design of the data points. The conditional asymptotic normalities are proved. In their article, Fan and Gijbels (1994) devote their attention to the univariate case and indicate that their methodology holds for multivariate case. Singh and Lu (1999) consider the multivariate case with the Leurgans transformation where the asymptotic properties are established using counting process techniques.

1.3 Objectives and Organization of the Paper

In this paper, we consider a censored nonparametric additive regression model which admits continuous and categorical variables in an additive manner. By that, we are able to estimate the lower-dimensional components directly when the data are censored. In particular, we extend the ideas of "marginal integration" and local linear fits to the nonparametric regression analysis with censoring to estimate the low-dimensional components in additive models.

The paper is organized as follows. Section 2 shows the motivation of the data transformation and introduces the local linear regression method. Section 3 presents the main results of this paper and discusses some properties and implications of these results. Section 4 provides some concluding remarks. Procedures for the proofs of the theorems and their conditions are given in the Appendix.

2. DATA TRANSFORMATION AND LOCAL LINEAR REGRESSION ESTIMATOR

2.1 Data Transformation

Consider model (2), and let $m(x) = E(Y|X = x)$. Suppose that $\{Y_i\}$ are randomly censored by $\{C_i\}$, where $\{C_i\}$ are independent identically distributed (i.i.d.) samples of random variable C, independent of $\{(X_i, Y_i)\}$, with distribution $1 - G(t)$ and survival function $G(t) = P(C \geq t)$. We can observe only (X_i, Z_i, δ_i), where $Z_i = \min\{Y_i, C_i\}$, $\delta_i = [Y_i \leq C_i]$, $i = 1, \ldots, n$, $[\cdot]$ stands for the indicator function.

Our model is a nonparametric censored regression model, in which the nonparametric regression function $f_1(x_1)$ is of interest. How does one estimate the relationship between Y and X in the case of censored data? The basic idea is to adjust for the effect by transforming the data in an unbiased way. The following transformation is proposed (see Fan and Gijbels 1996, pp. 160–174):

$$Y^* = \begin{cases} \phi_1(X, Y) & \text{if uncensored} \\ \phi_2(X, C) & \text{if censored} \end{cases}$$
$$= \delta\phi_1(X, Z) + (1 - \delta)\phi_2(X, Z) \tag{5}$$

where a pair of transformations (ϕ_1, ϕ_2), satisfying $E(Y^*|X) = m(X) = E(Y|X)$, is called *censoring unbiased transformation*. For a multivariate setup, assuming that the censoring distribution is independent of the covariates, i.e. $G(c|x) \equiv G(c)$, Fan and Gijbels recommend using a specific subclass of transformation (5) given by

$$\left.\begin{aligned} \phi_1 &= (1 + \alpha)\int_0^y \{G(t)\}^{-1} dt - \alpha y\{G(y)\}^{-1} \\ \phi_2 &= (1 + \alpha)\int_0^y \{G(t)\}^{-1} dt \end{aligned}\right\} \tag{6}$$

where the tuning parameter α is given by

$$\hat{\alpha} = \min_{\{i\,:\,\delta_i = 1\}} \frac{\int_0^{Y_i}\{G(t)\}^{-1} - Y_i}{Y_i\{G(Y_i)\}^{-1} - \int_0^{Y_i}\{G(t)\}^{-1} dt}$$

This choice of α reduces the variability of transformed data. The transformation (6) is a *distribution-based unbiased transformation*, which does not explicitly depend on $F(\cdot|x) = P(Y \le \cdot|X = x)$. The major strength of this kind of transformation is that it can easily be applied to multivariate covariates X when the conditional censoring distribution is independent of the covariates, i.e., $G(c|x) = G(c)$. This fits our needs, since we need to estimate $f_1(x_1)$ through estimating $m(x) = E(Y|X = x)$, which is a multivariate regression function. For a univariate covariate, the *local average unbiased transformation* is recommended by Fan and Gijbels.

2.2 Local Linear Regression Estimation

We treat the transformed data $\{(X_i, Y_i^*) : i = 1, \ldots, n\}$ as uncensored data and apply the standard regression techniques. First, we assume that the censoring distribution $1 - G(\cdot)$ is known; then, under transformation (6), we have an i.i.d. data set $(Y_i^*, X_{1i}, X_{2i}, X_{3i})$ $(i = 1, \ldots, n)$ for model (2). To use the "marginal integration" method, we consider the following local approximation to $f_1(u_1)$:

$$f_1(u_1) \approx a(x_1) + b^T(x_1)(u_1 - x_1)$$

a local linear approximation near a fixed point x_1, where u_1 lies in a neighborhood of x_1; and the following local approximation to $f_2(u_2, x_3)$:

$$f_2(u_2, x_3) \approx c(x_2, x_3)$$

at a fixed point x_2, where u_2 lies in a neighborhood of x_2. Thus, in a neighborhood of (x_1, x_2) and for the given value of x_3, we can approximate the regression function as

$$
\begin{aligned}
m(u_1, u_2, x_3) &\approx \mu + a(x_1) + b^T(x_1)(u_1 - x_1) + c(x_2, x_3) \\
&\equiv \alpha + \beta^T(u_1 - x_1)
\end{aligned}
\tag{7}
$$

The local model (7) leads to the following censored regression problem. Minimize

$$\sum_{i=1}^n (Y_i^* - \alpha - \beta^T(X_{1i} - x_1))^2 K_{h_1}(X_{1i} - x_1) L_{h_2}(X_{2i} - x_2) I\{X_{3i} = x_3\} \tag{8}$$

where

$$k_{h_1}(u) = \frac{1}{h_1^p} K\left(\frac{u}{h_1}\right) \quad \text{and} \quad L_{h_2}(u) = \frac{1}{h_2^q} L\left(\frac{u}{h_2}\right)$$

K and L are kernel functions, h_1 and h_2 are bandwidths. Let $\hat{\alpha}(x)$ and $\hat{\beta}(x)$ be the least-square solutions to (8). Therefore, our partial local linear estimator for $m(\cdot)$ is $\hat{m}(x; \phi_1, \phi_2) = \hat{\alpha}$. We propose the following estimator:

$$\hat{f}_1(x_1) = \hat{g}(x_1; \phi_1, \phi_2) - \bar{g}, \quad \bar{g} = \frac{1}{n}\sum_{i=1}^{n}\hat{g}(X_{1i}; \phi_1, \phi_2)$$

where

$$\hat{g}(x_1; \phi_1, \phi_2) = \frac{1}{n}\sum_{i=1}^{n}\hat{m}(x_1, X_{2i}, X_{3i}; \phi_1, \phi_2)W(X_{2i}, X_{3i}) \tag{9}$$

and $W : \mathbf{R}^{8+r} \to \mathbf{R}$ is a known function with $EW(X_2, X_3) = 1$. Let X be the design matrix and V by the diagonal weight matrix to the least-square problem (8). Then

$$\begin{pmatrix} \hat{\alpha} \\ \hat{\beta} \end{pmatrix} = (X^T V X)^{-1} X^T V Y^*$$

where $Y^* = (Y_1^*, \ldots, Y_n^*)^T$, and it is easy to show that $\hat{m}(x; \phi_1, \phi_2) = \hat{\alpha}$ can be expressed as

$$\hat{m}(x; \phi_1, \phi_2)(x) = \sum_{i=1}^{n} K_n(X_i - x)Y_i^* \tag{10}$$

where, with $S_n(x) = (X^T V X)$ and $e_1^T = (1, 0, \ldots, 0)$,

$$K_n(t_1, t_2, t_3) = e_1^T S_n^{-1}\begin{pmatrix} 1 \\ t_1 \end{pmatrix} K_{h_1}(t_1)L_{h_2}(t_2)I\{t_3 = 0\} \tag{11}$$

3. MAIN RESULTS

3.1 Notations

Let us adopt some notation of Fan et al. (1998). Let $p_1(x_1)$ and $p_{1,2}(x_1, x_2)$ be respectively the density of X_1 and (X_1, X_2), and let $p_{1,2|3}(x_1, x_2|x_3)$, $p_{2|3}(x_2|x_3)$ be respectively the conditional density of (X_1, X_2) given X_3 and of X_2 given X_3. Set $p_3(x_3) = P(X_3 = x_3)$. The conditional variances of $\epsilon = Y - E(Y|X) = \sigma(X)\epsilon_1$ and $\epsilon^* = Y^* - E(Y|X)$ are denoted respectively by

$$\sigma^2(x) = E(\epsilon^2|X = x) = \text{var}(Y|X = x)$$

and

$$\sigma^{*2}(x) = E(\epsilon^{*2}|X = x) = \text{var}(Y^*|X = x)$$

where $X = (X_1, X_2, X_3)$. Let

$$\|K\|^2 = \int K^2 \text{ and } \mu_2(K) = \int uu^T K(u)du$$

3.2 The Theorems, with Remarks

We present our main results in the following theorems.

Theorem 1. Under Condition A given in the Appendix, if the bandwidths are chosen such that $nh_1^p h_2^q / \log n \to \infty$, $h_1 \to 0$, $h_2 \to 0$ in such a way that $h_2^d / h_1^2 \to 0$, then

$$\sqrt{nh_1^p}\left\{\hat{g}(x_1; \phi_1, \phi_2) - f_1(x_1) - \mu_1 - \frac{1}{2}h_1^2 tr(f_1''(x_1)\mu_2(K) + o(h_1^2)\right\} \tag{12}$$
$$\to N(0, v^*(x_1))$$

where

$$v^*(x_1) = \|K\|^2 p_1(x_1)E\left\{\sigma^{*2}(X_1, X_2, X_3)\frac{P_{2|3}^2(X_2|X_3)W^2(X_2, X_3)}{p_{1,2|3}^2(X_1, X_2|X_3)}\bigg|X_1 = x_1\right\}$$

$$\sigma^2(x) = \text{var}(Y^*|X = x)$$

$$= \sigma^2(x) + \int_0^{+\infty}\left[2\int_0^y\int_0^v\frac{1 - G(s)}{G(s)}ds\,dv + \int_0^y\frac{\{(2 - \alpha)y - 2v\}\alpha y}{G^2(y)}dG(v)\right]$$
$$dF(y|x)$$

$$\mu_1 = \mu + Ef_2(X_2, X_3)W(X_2, X_3)$$

It should be pointed out that $\sigma^{*2}(x)$ in Theorem 1 measures the variability of the transformation. It is given by Fan and Gijbels (1994) in a more general data transformation.

We now obtain the optimal weight function $W(\cdot)$. The problem is equivalent to minimizing $v^*(x_1)$ with respect to $W(\cdot)$ subject to $EW(X_2, X_3) = 1$. Applying Lemma 1 of Fan et al. (1998) to this problem, we obtain the optimal solution

$$W(x_2, X_3) = c^{-1}\frac{p(x_1, X_2, X_3)p_1(x_1)}{\sigma^{*2}(x_1, X_2, X_3)p_{2,3}(X_2, X_3)} \tag{13}$$

where $p(x) = p_{1,2|3}(x_1, x_2|x_3)p_3(x_3)$ and $p_{2,3}(x) = p_{2|3}(x_2|x_3)p_3(x_3)$ are respectively the joint "density" of $X = (X_1, X_2, X_3)$ and (X_2, X_3) and where $c = p_1(x_1)^2 E\{\sigma^{*-2}(X)|X_1 = x_1\}$. The minimal variance is

$$\min_W v^*(x_1) = \frac{\|K\|^2}{p_1(x_1)}\left[E\{\sigma^{*-2}(X)|X_1 = x_1\}\right]^{-1} \tag{14}$$

The transformation given by (6) is the "ideal transformation," because it assumes that censoring distribution $1 - G(\cdot)$ is known and therefore the transformation functions $\phi_1(\cdot)$ and $\phi_2(\cdot)$ are known. Usually, the censoring distribution $1 - G(\cdot)$ is unknown and must be estimated. Consistent estimates of G under censoring are available in the literature; for example, take the Kaplan–Meier estimator. Let $\hat{G}$ be a consistent estimator of G and let $\hat{\phi}_1$ and $\hat{\phi}_2$ be the associated transformation functions. We will study the asymptotic properties of $\hat{g}(x; \hat{\phi}_1, \hat{\phi}_2)$. Following the discussion by Fan and Gijbels (1994), a basic requirement in the consistency result for $\hat{g}(x; \hat{\phi}_1, \hat{\phi}_2)$ is that the estimates $\hat{\phi}_1(z)$ and $\hat{\phi}_2(z)$ are uniformly consistent for z in an interval chosen such that the instability of $\hat{G}$ in the tail can be dealt with. Assume that

$$\beta_n(\tau_n) = \max_{j=1,2} \left\{ \sup_{z \in (0, \tau_n)} \left| \hat{\phi}_j(z) - \phi_j(z) \right| \right\} = o_p(1) \tag{15}$$

where $\tau_n > 0$. Redefine $\hat{\phi}_j (j = 1, 2)$ as follows:

$$\left. \begin{array}{ll} \hat{\phi}_j(z) = \hat{\phi}_j(z) & \text{if } z \le \tau_n \\ \quad = z & \text{elswhere} \end{array} \right\} \tag{16}$$

This transformation does not transform the data in the region of instability. This approach is effective when the contribution from the tail is negligible in the following sense: with $\tau > 0$

$$\kappa_n(\tau_n, \tau) = \sup_{x} \max_{j=1,2} \left\{ \sup_{t \in (x \pm \tau)} E\big(I[Z > \tau_n] | Z - \phi_j(Z)|_{X=t}\big) \right\} = o(1)$$

Theorem 2. Assume that the conditions of Theorem 1 hold. Then

$$\hat{g}\left(x; \hat{\phi}_1, \hat{\phi}_2\right) - \hat{g}(x; \phi_1, \phi_2) = O_p(\beta_n(\tau_n) + \kappa_n(\tau_n, \tau)) \tag{17}$$

provided that K is uniformly Lipschitz continuous and has a compact support.

When $\hat{G}$ is the Kaplan–Meier estimator, $\beta_n(\tau_n) = O_p((\log n/n)^{1/2})$. If τ_n and τ are chosen such that $\kappa_n(\tau_n, \tau) = 0$, then the difference in (17) is negligible, implying that the estimator is asymptotically normal.

Theorem 3. Assume that the conditions of Theorem 1 hold, $\kappa_n(\tau_n, \tau) = 0$, and $h_1^p \log n \to 0$. Then

$$\sqrt{nh_1^p}\left\{\hat{g}\left(x_1,\hat{\phi}_1,\hat{\phi}_2\right)-f_1(x_1)-\mu_1-\frac{1}{2}h_1^2tr\left(f_1''(x_1)\mu_2(K)+o(h_1^2)\right)\right\}$$ (18)

$$\to N(0,v^*(x_1))$$

Thus the rate of convergence is the same as that given in Theorem 1.

4. CONCLUDING REMARKS

We remark there that the results developed above can be applied to special models (3) and (4). Some results parallel to Theorems 3–6 of Fan et al. (1998) with censored data formation can also be obtained. For example, for the additive partially linear model (4), one can estimate not only each additive component but also the parameter β with root-n consistency. We will investigate these properties in a separate report.

We also remark that, in contrast to Fan et al. (1998), where the discrete variables enter the model in a linear fashion, Racine and Li (2000) propose a method for nonparametric regression which admits mixed continuous and categorical data in a natural manner without assuming additive structure, using the method of kernels. We conjecture their results hold for censored data as well, after transforming data to account for the censoring.

APPENDIX
Conditions and Proofs

We have used the following conditions for the proofs of Theorems 1–3.

Condition A

(i) $Ef_2^2(X_2,X_3)W^2(X_2,X_3)<\infty$.
(ii) The kernel functions K and L are symmetric and have bounded supports. L is an order d kernel.
(iii) The support of the discrete variable X_3 is finite and

$$\inf_{u_i\in x_1\pm\delta,\,(x_2,x_3)\in S}p_3(x_3)p_{1,2|3}(u_1,x_2|x_3)>0,\quad\text{for some }\delta>0$$

   where S is the support of the function W.
(iv) f_1 has a bounded second derivative in a neighborhood of x_1 and $f(x_2,x_3)$ has a bounded dth-order derivative with respect to x_2. Furthermore, $p_{1,2|3}(u_1,x_2|x_3)$ has a bounded derivative with respect to x_2 up to order d, for u_1 in a neighborhood of x_1.

(v) $E\epsilon^4$ is finite and $\sigma^2(x) = E(\epsilon^2|X = x)$ is continuous, where $\epsilon = Y - E(Y|X)$.

Proof of Theorem 1. Let $x^i = (x_1, X_{2i}, X_{3i})$ and let E_i denote the conditional expectation given by $X_i = (X_{1i}, X_{2i}, X_{3i})$. Let $g(x_1) = Em(x_1, X_2, X_3)W(X_2, X_3) = \mu_1 + f_1(x_1)$; here $\mu_1 = \mu + Ef_2(X_2, X_3) W(X_2, X_3)$. Then, by (9) and condition A(i), we have

$$\hat{g}(x_1; \phi_1, \phi_2) - g(x_1) = n^{-1} \sum_{i=1}^{n} \{\hat{m}(x^i; \phi_1, \phi_2) - m(x^i)\} W(X_{2i}, X_{3i})$$
$$+ O_p(n^{-1/2}) \tag{19}$$

By a similar procedure to that given by Fan et al. (1998) in the proof of their Theorem 1, we obtain

$$\hat{m}(x^i; \phi_1; \phi_2) - m(x^i) = e_1^T S_n^{-1}(x^i) \begin{pmatrix} 1 & \cdots & 1 \\ X_{11} - X_1 & \cdots & X_{1n} - x_1 \end{pmatrix}$$

$$A(x^i)(\hat{r}_i - \bar{\epsilon})$$

$$= \frac{1}{2} h_1^2 tr\{f_1''(x_1)\mu_2(K)\} + o(h_1^2)$$

$$+ n^{-1} \sum_{j=1}^{n} A_j(x^i)\epsilon_j^* \{p_3(X_{3i})p_{1,2|3}(x_1, X_{2i}|X_{3i})\}^{-1}$$

$$+ n^{-1} \sum_{j=1}^{n} \bar{r}_{ij} \{p_3(X_{3i})p_{1,2|3}(x_1, X_{2i}|X_{3i})\}^{-1}$$

where $\hat{r}_i$ is an $n \times 1$ vector with elements $\hat{r}_{ij} = m(X_j) - m(x^i) - f_1'(x_1)^T (X_{1j} - x_1)$, $A(x)$ is a diagonal matrix with diagonal elements $A_j(x) = K_{h1}(X_{1j} - x_1)L_{h2}(X_{2j} - x_2)I\{X_{3j} = x_3\}$, and $\bar{\epsilon}^* = (\epsilon_1^*, \ldots, \epsilon_n^*)^T$ with $\epsilon_i^* = Y_i^* - E(Y_i|X_i)$, $\bar{r}_{ij} = A_j(x^i)\hat{r}_{ij} - E_iA_j(x^i)\hat{r}_{ij}$. Thus, by (19), we have

$$\hat{g}(x_1; \phi_1, \phi_2) - g(x_1) = \frac{1}{2} h_1^2 tr\{f_1''(x_1)\mu_2(K)\} + o(h_1^2) + T_{n,1}$$
$$+ T_{n,2} + O_p(n^{-1/2}) \tag{20}$$

where

$$T_{n,1} = n^{-1} \sum_{j=1}^{n} \epsilon_j^* K_{h_1} (X_{1j} - x_1) n^{-1} \sum_{i=1}^{n} G(X_{2i}, X_{3i})L_{h_2}(X_{2j} - X_{2i})$$
$$I\{X_{3j} = X_{3i}\}$$

and

$$T_{n,2} = n^{-2} \sum_{j=1}^{n} \sum_{i=1}^{n} G(X_{2i}, X_{3i}) \bar{r}_{ij}$$

with

$$G(X_{2i}, X_{3i}) = \frac{W(X_{2i}, X_{3i})}{p_3(X_{3i})p_{1,2|3}(x_1, X_{2i}|X_{3i})}$$

We can show that with $\bar{\epsilon}_j^* = p_3(X_{3j})p_{1,2|3}(x_1, X_{2j}|X_{3j})G(X_{2j}, X_{3j})\epsilon_j^*$

$$T_{n,1} = n^{-1} \sum_{j=1}^{n} K_{h_1}(X_{1j} - x_1)\bar{\epsilon}_j^* + o_p\left((nh_1^p)^{-1/2}\right) \tag{21}$$

and

$$T_{n,2} = o_p\left(n^{-1/2}\right) \tag{22}$$

Combination of (21) and (22) leads to

$$\hat{g}(x_1; \phi_1, \phi_2) - g(x_1) = \frac{1}{2}h_1^2 tr\{f_1''(x_1)\mu_2(K)\} + n^{-1} \sum_{j=1}^{n} K_{h_1}(X_{1j} - x_1)\bar{\epsilon}_j^*$$

$$+ o_p\left((nh_1^p)^{-1/2}\right) \tag{23}$$

Hence, to establish Theorem 1, it suffices to show that

$$\sqrt{nh_1^p}n^{-1} \sum_{j=1}^{n} K_{h1}(X_{1j} - x_1)\bar{\epsilon}_j^* \to N(0, v^*(x_1)) \tag{24}$$

This is easy to verify by checking the Lyapounov condition. For any $\gamma > 0$,

$$\left(\sqrt{nh_1^p}n^{-1}\right)^{2+\gamma} \sum_{j=1}^{n} E(|K_{h_1}(X_{1j} - x_1)\bar{\epsilon}_j^*|)^{2+\gamma} \to 0$$

In fact, it can be shown that

$$E(|K_{h1}(X_{1j} - x_1)\bar{\epsilon}_j^*|)^{2+\gamma} \sim h_1^{p-(2+\gamma)p}$$

Therefore,

$$\left(\sqrt{nh_1^p}n^{-1}\right)^{2+\gamma} \sum_{j=1}^{n} E(|K_{h1}(X_{1j} - x_1)\bar{\epsilon}_j^*|)^{2+\gamma} \sim (nh_1^p)^{-\gamma/2} \to 0$$

since $nh_1^p h_2^q / \log n \to 0$.

Proof of Theorem 2. Denote the estimated transformed data by $\hat{Y}^* = \delta\hat{\phi}_1(Z) + (1 - \delta)\hat{\phi}_2(Z)$, where $\hat{\phi}_1$ and $\hat{\phi}_2$ are given in (16). It follows from (10) that

$$\hat{m}\left(x^i; \hat{\phi}_1, \hat{\phi}_2\right) - \hat{m}(x^i; \phi_1, \phi_2) = e_1^T S_n^{-1}(x^i)\begin{pmatrix} 1 & \cdots & 1 \\ X_{11} - x_1 & \cdots & X_{1n} - x_1 \end{pmatrix}$$

$$A(x^i)\left(\hat{Y}^* - Y^*\right)$$

$$= e^T H\left(HS_n(x^i)H\right)^{-1} H$$

$$\begin{pmatrix} 1 & \cdots & 1 \\ X_{11} - x_1 & \cdots & X_{1n} - x_1 \end{pmatrix} A(x^i)\Delta Y$$

where $\Delta Y = \hat{Y}^* - Y^* = (\Delta Y_1, \ldots, \Delta Y_n)^T$, $H = \text{diag}(1, h_1^{-1}, \ldots, h_1^{-1})$, a $(p+1) \times (p+1)$ diagonal matrix. It can be shown that

$$e^T H\left(n^{-1} HS_n(x)H\right)^{-1} = \{p_3(x_3)p_{1,2|3}(x_1, x_2|x_3)\}^{-1} e^T + o_p(1) \qquad (25)$$

It can be seen that

$$n^{-1}H\begin{pmatrix} 1 & \cdots & 1 \\ X_{11} - x_1 & \cdots & X_{1n} - x_1 \end{pmatrix} A(x^i)\Delta Y$$

$$= n^{-1}\sum_{j=1}^n A_j(x^i)\Delta Y_j\begin{pmatrix} 1 \\ (X_{1j} - x_1)/h_1 \end{pmatrix} \qquad (26)$$

$$= \begin{pmatrix} n^{-1}\sum_{j=1}^n A_j(x^i)\Delta Y_j \\ n^{-1}\sum_{j=1}^n A_j(x^i)\Delta Y_j\{(X_{1j} - x_1)/h_1\} \end{pmatrix}$$

When $Z_j \leq \tau_n$, we have $\Delta Y_j \leq \beta_n(\tau_n)$. When $Z_j > \tau_n$, we have $\Delta Y_j \leq \sum_{k=1}^2 I[Z_j > \tau]|Z_j - \phi_k(Z_j)|$. Hence, we obtain

$$\left|n^{-1}\sum_{j=1}^n A_j(x^i)\Delta_j\right| \leq \beta_n(\tau_n)n^{-1}\sum_{j=1}^n A_j(x^i) + n^{-1}\sum_{k=1}^2\sum_{j=1}^n$$

$$I[Z_j > \tau]|Z_j - \phi_k(Z_j)|A_j(x^i)$$

and

$$\left| n^{-1} \sum_{j=1}^{n} A_j(x^i) \Delta Y_j \{ (X_{1j} - x_1)/h_1 \} \right| \le \beta_n(\tau_n) n^{-1} \sum_{j=1}^{n} A_j(x^i) | (X_{1j} - x_1)/h_1 |$$

$$+ n^{-1} \sum_{k=1}^{2} \sum_{j=1}^{n} I[Z_j > \tau] |Z_j - \phi_k(Z_j)| A_j(x^i) | (X_{1j} - x_1)/h_1 |$$

It can be shown that for any i

$$n^{-1} \sum_{j=1}^{n} A_j(x^i) = E\{p_3(X_{3i}) p_{1,2|3}(x_1, X_{2i}|X_{3i})\} O_p(h_1^2) + O_p(h_2^d) + O_p(n^{-1/2})$$

and

$$n^{-1} \sum_{j=1}^{n} A_j(x^i) | (X_{1j} - x_1)/h_1 | = E\{p_3(X_{3i}) p_{1,2|3}(x_1, X_{2i}|X_{3i})\} \int K(u)|u| du$$

$$+ O_p(h_1) + \left(h_2^d \right) + O_p(n^{-1/2})$$

We see that

$$E\left\{ n^{-1} \sum_{k=1}^{2} \sum_{j=1}^{n} I[Z_j > \tau] |Z_j - \phi_k(Z_j)| A_j(x^i) | X_1, \dots, X_n \right\}$$

$$\le 2\kappa(\tau_n, \tau) n^{-1} \sum_{j=1}^{n} A_j(x^i)$$

and

$$E\left\{ n^{-1} \sum_{k=1}^{2} \sum_{j=1}^{n} I[Z_j > \tau] |Z_j - \phi_k(Z_j)| A_j(x^i) | (X_{1j} - x_1)/h_1 \| X_1, \dots, X_n \right\}$$

$$\le 2\kappa(\tau_n, \tau) n^{-1} \sum_{j=1}^{n} A_j(x^i) | (X_{1j} - x_1)/h_1 |$$

Finally, from the above results, we obtain

$$\left| \hat{m}\left(x^i; \hat{\phi}_1; \hat{\phi}_2\right) - \hat{m}(x^i; \phi_1, \phi_2) \right| \le (\beta_n(\tau_n) + 2\kappa_n(\tau_n, \tau))$$

$$\{p_3(X_{3i}) p_{1,2|3}(x_1, X_{2i}|X_{3i})\}^{-1} \left[E\{p_3(X_3) p_{1,2|3}(x_1, X_2|X_3)\} + o_p(1) \right]$$

By (9) we have

$$\left|\hat{g}\left(x_1; \hat{\phi}_1, \hat{\phi}_2\right) - \hat{g}(x_1; \phi_1, \phi_2)\right| \le n^{-1} \sum_{i=1}^{n} \left|\hat{m}\left(x^i; \hat{\phi}_1, \hat{\phi}_2\right) - \hat{m}(x^i; \phi_1, \phi_2)\right|$$

$$W(X_{2i}, X_{3i}) + O_p\left(n^{-1/2}\right) \le (\beta_n(\tau_n) + 2\kappa_n(\tau_n, \tau)) \sum_{i=1}^{n} W(X_{2i}, X_{3i})$$

$$\{p_3(X_{3i})p_{1,2|3}(x_1, X_{2i}|X_{3i})\}^{-1}\left[E\{p_3(X_3)p_{1,2|3}(x_1, X_2|X_3)\} + o_p(1)\right]$$
$$+ O_p\left(n^{-1/2}\right)$$

$$(27)$$

and the result follows.

REFERENCES

Buckley, J. and James, I. R. (1979). Linear regression with censored data. *Biometrika*, **66**, 429–436.

Delgado, M. A. and Mora, J. (1995). Nonparametric and semiparametric estimation with discrete regressors. *Econometrica*, **63**, 1477–1484.

Fan, J. and Gijbels, I. (1994). Censored regression: local linear approximations and their applications. *Journal of the American Statistical Association*, **89**, 560–570.

Fan, J. and Gijbels, I. (1996). *Local Polynomial Modeling and its Applications*. London, Chapman and Hall.

Fan, J., Härdle, W. and Mammen, E. (1998). Direct estimation of low dimensional components in additive models. *Ann. Statist.*, **26**, 943–971.

Horowitz, J. L. (1998). *Semiparametric Methods in Econometrics*. Lecture Notes in Statistics 131. New York, Springer-Verlag.

Koul, H., Susarla, V. and Van Ryzin, J. (1981). Regression analysis with randomly right-censored data. *Ann. Statist.*, **9**, 1276–1288.

Lavergne, P. and Vuong, Q. (1996). Nonparametric selection of regressors. *Econometrica*, **64**, 207–219.

Leurgans, S. (1987). Linear models, random censoring and synthetic data. *Biometrika*, **74**, 301–309.

Linton, O. (1995). Second order approximation in the partially linear regression model. *Econometrica*, **63**, 1079–1112.

Linton, O. and Nielsen, J. P. (1995). A kernel method of estimating structured nonparametric regression based on marginal integration. *Biometrika,* **82**, 93–100.

Pagan, A. and Ullah, A. (1999). *Nonparametric Econometrics.* Cambridge, Cambridge University Press.

Racine, J. and Li, Q. (2000). Nonparametric estimation of regression functions with both categorical and continuous data. Canadian Econometric Study Group Workshop, University of Guelph, Sep. 29–Oct. 1, 2000.

Robinson, P. (1988). Root–N consistent semiparametric regression. *Econometrica,* **56**, 931–954.

Singh, R. S. and Lu, X. (1999). Nonparametric synthetic data regression estimation for censored survival data. *J. Nonparametric Statist.,* **11**, 13–31.

Srinivasan, C. and Zhou, M. (1994). Linear regression with censoring. *J. Multivariate Ana.,* **49**, 179–201.

Zheng, Z. (1987). A class of estimators of the parameters in linear regression with censored data. *Acta Math. Appl. Sinica,* **3**, 231–241.

Zhou, M. (1992). Asymptotic normality of the 'synthetic data' regression estimator for censored survival data. *Ann. Statist.,* **20**, 1002–1021.

9

Improved Combined Parametric and Nonparametric Regressions: Estimation and Hypothesis Testing

MEZBAHUR RAHMAN   Minnesota State University, Mankato, Minnesota

AMAN ULLAH   University of California, Riverside, Riverside, California

1. INTRODUCTION

Consider the regression model:

$$Y = m(X) + \epsilon \tag{1.1}$$

where $m(x) = E(Y|X = x)$, $x \epsilon R^q$, is the true but unknown regression function. Suppose that n independent and identically distributed observations $\{Y_i, X_i\}_{i=1}^n$ are available from (1.1). If $m(x) = g(\beta, x)$ for almost all x and for some $\beta \epsilon R^p$, then we say that the parametric regression model given by $Y = g(\beta, x) + \epsilon$ is correct. It is well known that, in this case, one can construct a consistent estimate of β, say $\hat{\beta}$, and hence a consistent estimate of $m(x)$ given by $g(\hat{\beta}, x)$. In general, if the parametric regression model is incorrect, then $g(\hat{\beta}, x)$ may not be a consistent estimate of $m(x)$. However, one can still consistently estimate the unknown regression function $m(x)$ by various nonparametric estimation techniques, see Härdle (1990) and Pagan and Ullah (1999) for details. In this paper we will consider

159

the kernel estimator, which is easy to implement and whose asymptotic properties are now well established.

When used individually, both parametric and nonparametric procedures have certain drawbacks. Suppose the econometrician has some knowledge of the parametric form of $m(x)$ but there are regions in the data that do not conform to this specified parametric form. In this case, even though the parametric model is misspecified only over portions of the data, the parametric inferences may be misleading. In particular the parametric fit will be poor (biased) but it will be smooth (low variance). On the other hand the nonparametric techniques which totally depend on the data and have no a priori specified functional form may trace the irregular pattern in the data well (less bias) but may be more variable (high variance). Thus, the problem is that when the functional form of $m(x)$ is not known, a parametric model may not adequately describe the data where it deviates from the specified form, whereas a nonparametric analysis would ignore the important a priori information about the underlying model. A solution is to use a combination of parametric and nonparametric regressions, which can improve upon the drawbacks of each when used individually. Two different combinations of parametric and nonparametric fits, $\hat{m}(x)$, are proposed. In one case we simply add in the parametric start $g(\hat{\beta}, x)$ a nonparametric kernel fit to the parametric residuals. In the other we add the nonparametric fit with weight $\hat{\lambda}$ to the parametric start $g(\hat{\beta}, x)$. Both these combined procedures maintain the smooth fit of parametric regression while adequately tracing the data by the nonparametric component. The net result is that the combined regression controls both the bias and the variance and hence improves the mean squared error (MSE) of the fit. The combined estimator $\hat{m}(x)$ also adapts to the data (or the parametric model) automatically through $\hat{\lambda}$ in the sense that if the parametric model accurately describes the data, then $\hat{\lambda}$ converges to zero, hence $\hat{m}(x)$ puts all the weight on the parametric estimate asymptotically; if the parametric model is incorrect, then $\hat{\lambda}$ converges to one and $\hat{m}(x)$ puts all the weights on the kernel estimate asymptotically. The simulation results suggest that, in small samples, our proposed estimators perform as well as the parametric estimator if the parametric model is correct and perform better than both the parametric and nonparametric estimators if the parametric model is incorrect. Asymptotically, if the parametric model is incorrect, the combined estimates have similar behavior to the kernel estimates. Thus the combined estimators always perform better than the kernel estimator, and are more robust to model misspecification compared to the parametric estimate.

The idea of combining the regression estimators stems from the work of Olkin and Spiegelman (1987), who studied combined parametric and nonparametric density estimators. Following their work, Ullah and Vinod

(1993), Burman and Chaudhri (1994), and Fan and Ullah (1999) proposed combining parametric and nonparametric kernel regression estimators additively, and Glad (1998) multiplicatively. Our proposed estimators here are more general, intuitively more appealing and their MSE performances are better than those of Glad (1998), Fan and Ullah (1999), and Burman and Chaudhri (1994).

Another important objective of this paper is to use $\hat{\lambda}$ as a measure of the degrees of accuracy of the parametric model and hence use it to develop tests for the adequacy of the parametric specification. Our proposed test statistics are then compared with those in Fan and Ullah (1999), Zheng (1996), and Li and Wang (1998), which are special cases of our general class of tests.

The rest of the paper is organized as follows. In Section 2, we present our proposed combined estimators. Then in Section 3 we introduce our test statistics. Finally, in Section 4 we provide Monte Carlo results comparing alternative combined estimators and test statistics.

2. COMBINED ESTIMATORS

Let us start with a parametric regression model which can be written as

$$
\begin{aligned}
Y = m(X) + \epsilon &= g(\beta, X) + \epsilon \\
&= g(\beta, X) + E(\epsilon|X) + \epsilon - E(\epsilon|X) \\
&= g(\beta, X) + \theta(X) + u
\end{aligned}
\tag{2.1}
$$

where $\theta(x) = E(\epsilon|X = x) = E(y|X = x) - E(g(\beta, X)|X = x)$ and $u = \epsilon - E(\epsilon|X)$ such that $E(u|X = x) = 0$. If $m(x) = g(\beta, x)$ is a correctly specified parametric model, then $\theta(x) = 0$, but if $m(x) = g(\beta, x)$ is incorrect, then $\theta(x)$ is not zero.

In the case where $m(x) = g(\beta, x)$ is a correctly specified model, an estimator of $g(\beta, x)$ can be obtained by the least squares (LS) procedure. We represent this as

$$
\hat{m}_1(x) = g(\hat{\beta}, x)
\tag{2.2}
$$

However, if a priori we do not know the functional form of $m(x)$, we can estimate it by the Nadaraya (1964) and Watson (1964) kernel estimator as

$$
\hat{m}_2(x) = \frac{\sum_{i=1}^{n} y_i K_i}{\sum_{i=1}^{n} K_i} = \frac{\hat{r}(x)}{\hat{f}(x)}
\tag{2.3}
$$

where $\hat{r}(x) = (n h^q)^{-1} \sum_{1}^{n} y_i K_i$, $\hat{f}(x) = (n h^q)^{-1} \sum_{1}^{n} K_i$ is the kernel estimator of $f(x)$, the density of X at $X = x$, $K_i = K([x_i - x]/h)$ is a decreasing function of the distance of the regressor vector x_i from the point x, and $h >$

0 is the window width (smoothing parameter) which determines how rapidly the weights decrease as the distance of x_i from x increases.

In practice, the a priori specified parametric form $g(\beta, x)$ may not be correct. Hence $\theta(x)$ in (2.1) may not be zero. Therefore it will be useful to add a smoothed LS residual to the parametric start $g(\hat{\beta}, x)$. This gives the combined estimator

$$\hat{m}_3(x) = g(\hat{\beta}, x) + \hat{\theta}(x) \tag{2.4}$$

where $\hat{\theta}(x)$ is the Nadaraya–Watson kernel estimator of $\theta(x)$,

$$\hat{\theta}(x) = \hat{E}(\hat{\epsilon}|X = x) = \hat{E}(Y|X = x) - \hat{E}(g(\hat{\beta}, x)|X = x)$$

$$= \hat{m}_2(x) - \hat{g}\left(\hat{\beta}, x\right) \tag{2.5}$$

$$= \frac{\sum_{i=1}^{n} \hat{\epsilon}_i K_i}{\sum_{i=1}^{n} K_i}$$

$\hat{g}(\hat{\beta}, x) = \hat{E}\left(g(\hat{\beta}, X)|X = x\right)$ is the smoothed estimator of $g(\beta, x)$ and $\hat{\epsilon}_i = Y_i - g(\hat{\beta}, X_i)$ is the parametric residual. $\hat{m}_3(x)$ is essentially an estimator of $m(x) = g(\beta, x) + \theta(x)$ in (2.1). We note that $\hat{\theta}(x)$ can be interpreted as the nonparametric fit to the parametric residuals.

An alternative way to obtain an estimator of $m(x)$ is to write a compound model

$$Y_i = (1 - \lambda)g(\beta, X_i) + \lambda m(X_i) + v_i$$
$$= g(\beta, X_i) + \lambda[m(X_i) - g(\beta, X_i)] + v_i \tag{2.6}$$

where v_i is the error in the compound model. Note that in (2.6) $\lambda = 0$ if the parametric model is correct; $\lambda = 1$ otherwise. Hence, λ can be regarded as a parameter, the value of which indicates the correctness of the parametric model. It can be estimated consistently by using the following two steps. First, we estimate $g(\beta, X_i)$ and $m(X_i)$ in (2.6) by $g(\hat{\beta}, X_i)$ and $\hat{m}_2^{(i)}(X_i)$, respectively, where $\hat{m}_2^{(i)}(X_i)$ is the leave-one-out version of the kernel estimate $\hat{m}_2(X_i)$ in (2.3). This gives the LS residual $\hat{\epsilon}_i = Y_i - g(\hat{\beta}, X_i)$ and the smoothed estimator $\hat{g}(\hat{\beta}, X_i)$. Second, we obtain the LS estimate of λ from the resulting model

$$\sqrt{w(X_i)}\left(Y_i - g(\hat{\beta}, X_i)\right) = \lambda\sqrt{w(X_i)}\left[\hat{m}_2^{(i)}(X_i) - \hat{g}^{(i)}(\hat{\beta}, X_i)\right] + \hat{v}_i \tag{2.7}$$

or

$$\sqrt{w(X_i)}\,\hat{\epsilon}_i = \lambda\sqrt{w(X_i)}\,\hat{\theta}^{(i)}(X_i) + \hat{v}_i \tag{2.8}$$

where the weight $w(X_i)$ is used to overcome the random denominator problem, the definition of $\hat{v}_i$ is obvious from (2.7), and

$$\hat{\theta}^{(i)}(X_i) = \hat{m}_2^{(i)}(X_i) - \hat{g}^{(i)}\left(\hat{\beta}, X_i\right)$$

$$= \frac{\sum_{j\neq i}^n Y_j K_{ji}}{\sum_{j\neq i}^n K_{ji}} - \frac{\sum_{j\neq i}^n g\left(\hat{\beta}, X_j\right) K_{ji}}{\sum_{j\neq i}^n K_{ji}} \tag{2.9}$$

$$= \frac{\sum_{j\neq i}^n \hat{\epsilon}_j K_{ji}}{\sum_{j\neq i}^n K_{ji}} = \frac{\hat{r}^{(i)}(X_i)}{\hat{f}^{(i)}(X_i)}$$

$K_{ji} = K\left((X_j - X_i)/h\right)$. This gives

$$\hat{\lambda} = \frac{n^{-1}\sum_{i=1}^n w(X_i)\left(\hat{m}_2^{(i)}(X_i) - \hat{g}^{(i)}\left(\hat{\beta}, X_i\right)\right)\left(Y_i - g\hat{\beta}, X_i\right)}{n^{-1}\sum_{i=1}^n w(X_i)\left(\hat{m}_2^{(i)}(X_i) - \hat{g}^{(i)}\left(\hat{\beta}, X_i\right)\right)^2}$$

$$= \frac{n^{-1}\sum_{i=1}^n w(X_i)\hat{\theta}^{(i)}(X_i)\hat{\epsilon}_i}{n^{-1}\sum_{i=1}^n w(X_i)\left(\hat{\theta}^{(i)}(X_i)\right)^2} \tag{2.10}$$

Given $\hat{\lambda}$, we can now propose our general class of combined estimator of $m(x)$ by

$$\hat{m}_4(x) = g\left(\hat{\beta}, x\right) + \hat{\lambda}\left[\hat{m}_2^{(i)}(x) - \hat{g}^{(i)}\left(\hat{\beta}, x\right)\right]$$

$$= g\left(\hat{\beta}, x\right) + \hat{\lambda}\hat{\theta}^{(i)}(x) \tag{2.11}$$

The estimator $\hat{m}_4$, in contrast to $\hat{m}_3$, is obtained by adding a portion of the parametric residual fit back to the original parametric specification. The motivation for this is as follows. If the parametric fit is adequate, then adding the $\hat{\theta}(x)$ would increase the variability of the overall fit. A $\hat{\lambda} \simeq 0$ would control for this. On the other hand, if the parametric model $g(\hat{\beta}, x)$ is misspecified, then the addition of $\hat{\theta}(x)$ should improve upon it. The amount of misspecification, and thus the amount of correction needed from the residual fit, is reflected in the size of $\hat{\lambda}$.

In a special case where $w(X_i) = 1$, $\hat{\theta}^{(i)}(X_i) = \hat{\theta}(X_i) = \hat{m}_2(X_i) - g(\hat{\beta}, X_i)$, and $g(\hat{\beta}, X_i) = X_i\hat{\beta}$ is linear, (2.10) reduces to the Ullah and Vinod (1993) estimator, see also Rahman et al (1997). In Burman and Chaudhri (1994), X_i is fixed, $w(X_i) = 1$, and $\hat{\theta}^{(i)}(X_i) = \hat{m}_2^{(i)}(X_i) - g^{(i)}(\hat{\beta}, X_i)$. The authors provided the rate of convergence of their combined estimator. $\hat{m}_4$ reduces to the Fan and Ullah (1999) estimator when $w(X_i) = \left(\hat{f}^{(i)}(X_i)\right)^2$ and $\hat{\theta}^{(i)}(X_i) = \hat{m}_2^{(i)}(X_i) - g(\hat{\beta}, X_i)$. They provided the asymptotic normality of their combined estimator under the correct parametric specification, incorrect parametric specification, and approximately correct parametric specification.

Note that in Ullah and Vinod, Fan and Ullah, and Burman and Chaudhri $\hat{\theta}(x)$ is not the nonparametric residual fit.

An alternative combined regression is considered by Glad (1998) as

$$m(x) = g(\beta, x)\frac{m(x)}{g(\beta, x)}$$

$$= g(\beta, x)E(Y^*|_{X=x}) \tag{2.12}$$

where $Y^* = Y/g(\beta, X)$. She then proposes the combined estimator of $m(x)$ as

$$\hat{m}_5(x) = g\left(\hat{\beta}, x\right)\frac{\sum_{i=1}^{n}\hat{Y}_i^* K_i}{\sum_{i=1}^{n} K_i} \tag{2.13}$$

where $\hat{Y}_i^* = Y_i/g(\hat{\beta}, X_i)$. At the point $x = X_i$, $\hat{m}_5(X_i) = \sum_{j\neq i}^{n}\hat{Y}_j^* g(\hat{\beta}, X_i)$ $K_{ji}/\sum_{j\neq i} K_{ji}$, where $g(\hat{\beta}, X_j) \neq 0$ for all j. Essentially Glad starts with a parametric estimate, $g(\hat{\beta}, x)$, then multiplies by a nonparametric kernel estimate of the correction function $\hat{m}(x)/g(\hat{\beta}, x)$. We note that this combined estimator is multiplicative, whereas the estimators $\hat{m}_3(x)$ and $\hat{m}_4(x)$ are additive. However the idea is similar, that is to have an estimator that is more precise than a parametric estimator when the parametric model is misspecified. It will be interesting to see how the multiplicative combination performs compared to the additive regressions.

3. MISSPECIFICATION TESTS

In this section we consider the problem of testing a parametric specification against the nonparametric alternative. Thus, the null hypothesis to be tested is that the parametric model is correct:

$$H_0 : P[m(X) = g(\beta_0, X)] = 1 \text{ for some } \beta_0\epsilon\beta \tag{3.1}$$

while, without a specific alternative model, the alternative to be tested is that the null is false:

$$H_1 : P[m(X) = g(\beta, X)] < 1 \text{ for all } \beta \tag{3.2}$$

Alternatively, the hypothesis to be tested is $H_0 : Y = g(\beta, X) + \epsilon$ against $H_0 : Y = m(X) + \epsilon$. A test that has asymptotic power equal to 1 is said to be consistent. Here we propose tests based on our combined regression estimators of Section 2.

First we note from the combined model (2.1) that the null hypothesis of correct parametric specification $g(\beta, X)$ implies the null hypothesis of θ $(X) = E(\epsilon|_X) = 0$ or $E[\epsilon E(\epsilon|_X)f(X)] = E[(E(\epsilon|_X))^2 f(X)] = 0$. Similarly, from the multiplicative model (2.12), the null hypothesis of correct parametric specification implies the null hypothesis of $E(Y^*|X) = 1$ or $E((Y^* - 1)|X) = E(\epsilon|_X) = 0$ provided $g(\beta, X) \neq 0$ for all X. Therefore we can use the sample analog of $E[\epsilon E(\epsilon|_X) f(X)]$ to form a test. This is given by

$$Z_1 = n h^{q/2} V_1 \tag{3.3}$$

where

$$V_1 = \frac{1}{n(n-1)h^q} \sum_{i=1}^{n} \sum_{\substack{j \neq i=1}}^{n} \hat{\epsilon}_i \hat{\epsilon}_j K_{ji} \tag{3.4}$$

is the sample estimate of $E[\epsilon(E\epsilon|_X) f(X)]$. It has been shown by Zheng (1996) and Li and Wang (1998) that under some regularity assumptions and $n \to \infty$

$$Z_1 = n h^{q/2} V_1 \to N(0, \sigma_1^2) \tag{3.5}$$

where $\sigma_1^2 = 2[\int K^2(t) \, dt] \int (\sigma^2(x))^2 f^2(x) \, dx$ is the asymptotic variance of $n h^{q/2} V_1$ and $\sigma^2(x) = E(\epsilon^2|X = x)$ is the conditional variance under H_0. Moreover, σ_1^2 is consistently estimated by

$$\hat{\sigma}_1^2 = \frac{2}{n(n-1)h^q} \sum_{i=1}^{n} \sum_{\substack{j \neq i-1}}^{n} \hat{\epsilon}_i^2 \hat{\epsilon}_j^2 K_{ji}^2 \tag{3.6}$$

where $K_{ji}^2 = K^2((X_j - X_i)/h)$. The standardized version of the test statistic is then given by

$$T_1 = \frac{Z_1}{\hat{\sigma}_1} = \frac{n h^{q/2} V_1}{\hat{\sigma}_1} \to N(0, 1) \tag{3.7}$$

For details on asymptotic distributions, power, and consistency of this test, see Zheng (1996), Fan and Li (1996), and Li and Wang (1998). Also, see Zheng (1996) and Pagan and Ullah (1999, Ch. 3) for the connections of the T_1-test with those of Ullah (1985), Eubank and Spiegelman (1990), Yatchew (1992), Wooldridge (1992), and Härdle and Mammen (1993).

An alternative class of tests for the correct parametric specification can be developed by testing for the null hypothesis $\lambda = 0$ in the combined model (2.6). This is given by

$$Z = \frac{\hat{\lambda}}{\sigma} \tag{3.8}$$

where, from (2.10),

$$\hat{\lambda} = \hat{\lambda}_w = \frac{\hat{\lambda}_N}{\hat{\lambda}_D} = \frac{n^{-1}\sum_{i=1}^{n} w(X_i)\hat{\theta}^{(i)}(X_i)\hat{\epsilon}_i}{n^{-1}\sum_{i=1}^{n} w(X_i)\left(\hat{\theta}^{(i)}(X_i)\right)^2} \tag{3.9}$$

and σ^2 is the asymptotic variance of $\hat{\lambda}$; $\hat{\lambda}_N$ and $\hat{\lambda}_D$, respectively, represent the numerator and denominator of $\hat{\lambda}$ in (2.10) and (3.9). In a special case where $w(X_i) = (\hat{f}^{(i)}(X_i))^2$ and $\hat{\theta}^{(i)}(X_i) = \hat{m}_2^{(i)}(X_i) - g(\hat{\beta}, X_i)$, it follows from Fan and Ullah (1999) that

$$Z_2 = \frac{h^{-q/2}\hat{\lambda}}{\sigma_2} \rightarrow N(0, 1) \tag{3.10}$$

where $\sigma_2^2 = \sigma_u^2/\sigma_D^4$ is the asymptotic variance of $h^{-q/2}\hat{\lambda}$; $\sigma_u^2 = 2\int K^2(t)\,dt \int \sigma^4(x)f^4(x)\,dx$ and $\sigma_D^2 = \int K^2(t)\,dt \int \sigma^2(x)f^2(x)\,dx$. A consistent estimator of σ_2^2 is $\hat{\sigma}_2^2 = \hat{\sigma}_u^2/\hat{\sigma}_D^4$, where

$$\hat{\sigma}_u^2 = \frac{1}{2h^q n(n-1)}\sum_{i=1}^{n}\sum_{j\neq i=1}^{n} \hat{\epsilon}_i^2\hat{\epsilon}_j^2 K_{ji}^2\left(\hat{f}(X_i) + \hat{f}(X_j)\right)^2 \tag{3.11}$$

and

$$\hat{\sigma}_D^2 = \frac{1}{h^q(n-1)^2}\sum_{i=1}^{n}\sum_{j\neq i=1}^{n} \hat{\epsilon}_j^2 K_{ji}^2 \tag{3.12}$$

Using this (Fan and Ullah 1999), the standardized test statistic is given by

$$T_2 = \frac{h^{-q/2}\hat{\lambda}}{\hat{\sigma}_2} \rightarrow N(0, 1) \tag{3.13}$$

Instead of constructing the test statistic based on $\hat{\lambda}$, one can simply construct the test of $\lambda = 0$ based on the numerator of $\hat{\lambda}$, $\hat{\lambda}_N$. This is because the value of $\hat{\lambda}_N$ close to zero implies $\hat{\lambda}$ is close to zero. Also, $\hat{\lambda}_N$ value close to zero implies that the covariance between $\hat{\epsilon}_i$ and $\hat{\theta}^{(i)}(X_i)$ is close to zero, which was the basis of the Zheng–Li–Wang test T_1 in (3.7). Several alternative tests will be considered here by choosing different combinations of the weight $w(X_i)$ and $\hat{\theta}^{(i)}(X_i)$. These are

(i) $w(X_i) = (\hat{f}^{(i)}(X_i))^2$, $\hat{\theta}^{(i)}(X_i) = \hat{m}_2^{(i)}(X_i) - g(\hat{\beta}, X_i)$
(ii) $w(X_i) = (\hat{f}^{(i)}(X_i))^2$, $\hat{\theta}^{(i)}(X_i) = \hat{m}_2^{(i)}(X_i) - \hat{g}(\hat{\beta}, X_i)$
(iii) $w(X_i) = \hat{f}^{(i)}(X_i)$, $\hat{\theta}^{(i)}(X_i) = \hat{m}_2^{(i)}(X_i) - g(\hat{\beta}, X_i)$
(iv) $w(X_i) = \hat{f}^{(i)}(X_i)$, $\hat{\theta}^{(i)}(X_i) = \hat{m}_2^{(i)}(X_i) - \hat{g}(\hat{\beta}, X_i)$

Essentially two choices of $w(X_i)$ are considered, $w(X_i) = \hat{f}^{(i)}(X_i)$ and $w(X_i) = (\hat{f}^{(i)}(X_i))^2$ and with each choice two choices of $\hat{\theta}^{(i)}(X_i)$ are considered; one is with the smoothed estimator $\hat{g}(\hat{\beta}, X)$ and the other with the unsmoothed estimator $g(\hat{\beta}, X)$. The test statistics corresponding to (i) to (iv), respectively, are

$$T_3 = \frac{nh^{q/2}\hat{\lambda}_{N3}}{\hat{\sigma}_u} \to N(0, 1) \tag{3.14}$$

$$T_4 = \frac{nh^{q/2}\hat{\lambda}_{N4}}{\hat{\sigma}_u} \to N(0, 1) \tag{3.15}$$

$$T_5 = \frac{nh^{q/2}\hat{\lambda}_{N5}}{\hat{\sigma}_1} \to N(0, 1) \tag{3.16}$$

$$T_6 = T_1 = \frac{nh^{q/2}\hat{\lambda}_{N6}}{\hat{\sigma}_1} \to N(0, 1) \tag{3.17}$$

where $\hat{\lambda}_{Nr}$, $r = 3$, 4, 5, 6, is $\hat{\lambda}_N$ in (3.9) corresponding to $w(X_i)$ and $\hat{\theta}^{(i)}(X_i)$ given in (i) to (iv), respectively, and $\hat{\sigma}_u^2$ and $\hat{\sigma}_1^2$ are in (3.11) and (3.6), respectively. It can easily be verified that T_6 is the same as T_1. The proofs of asymptotic normality in (3.14) to (3.17) follow from the results of Zheng (1996), Li and Wang (1998), and Fan and Ullah (1999). In Section 4.2 we compare the size and power of test statistics T_1 to T_5.

4. MONTE CARLO RESULTS

In this section we report results from the Monte Carlo simulation study, which examines the finite sample performances of our proposed combined estimators $\hat{m}_3(x)$ and $\hat{m}_4(x)$ with $w(X_i) = \hat{f}^{(i)}(X_i)$, with the parametric estimator $\hat{m}_1(x) = g(\hat{\beta}, x)$, the nonparametric kernel estimator $\hat{m}_2(x)$, the Burman and Chaudhri (1994) estimator $\hat{m}_4(x) = \hat{m}_{bc}(x)$ with $w(X_i) = 1$ and $\hat{\theta}^{(i)}(X_i) = \hat{m}_2(X_i) - g(\hat{\beta}, X_i)$, the Fan and Ullah (1999) estimator $\hat{m}_4(x) = \hat{m}_{fu}(x)$ with $w(X_i) = (\hat{f}^{(i)}(X_i))^2$ and $\hat{\theta}^{(i)}(X_i) = \hat{m}_2^{(i)}(X_i) - g(\hat{\beta}, X_i)$, and the Glad (1998) combined estimator. We note that while $\hat{m}_3$ and $\hat{m}_4$ use the smoothed $\hat{g}(\hat{\beta}, x)$ in $\hat{\theta}^{(i)}(X_i)$, $\hat{m}_{uv}$ and $\hat{m}_{fu}$ use the unsmoothed $g(\hat{\beta}, x)$.

Another Monte Carlo simulation is carried out to study the behavior of test statistics T_1 to T_5 in Section 3.

4.1   Performance of $\hat{m}(x)$

To conduct a Monte Carlo simulation we consider the data generating process

$$Y = \beta_0 + \beta_1 X + \beta_2 X^2 + \delta \left[\gamma_1 \sin \left(\frac{\pi(X-1)}{\nu_2} \right) \right] + \epsilon \qquad (4.1)$$

where $\beta_0 = 60.50$, $\beta_1 = -17$, $\beta_2 = 2$, $\gamma_1 = 10$, $\gamma_2 = 2.25$, $\pi = 3.1416$, and δ is the misspecification parameter which determines the deviation of $m(x) = \gamma_1 \sin(\pi(X-1)/\nu_2)$ from the parametric specification $g(\beta, X) = \beta_0 + \beta_1 X + \beta_2 X^2$. This parameter δ is chosen as 0, 0.3, and 1.0 in order to consider the cases of correct parameter specification ($\delta = 0$), approximately correct parametric specification ($\delta = 0.3$), and incorrect parametric specification ($\delta = 1.0$). In addition to varying δ, the sample size n is varied as $n = 50$, 100, and 500. Both X and ϵ are generated from standard normal populations. Further, the number of replications is 1000 in all cases. Finally, the normal kernel is used in all cases and the optimal window width h is taken as $h = 1.06 n^{-1/5} \hat{\sigma}_X$, where $\hat{\sigma}_X$ is the sample standard deviation of X; see Pagan and Ullah (1999) for details on the choice of kernel and window width.

Several techniques of obtaining the fitted value $\hat{y} = \hat{m}(x)$ are considered and compared. These are the parametric fit $\hat{m}_1(x) = g(\hat{\beta}, x)$, nonparametric kernel fit $\hat{m}_2(x)$, our proposed combined fits $\hat{m}_3(x) = g(\hat{\beta}, x) + \hat{\theta}(x)$ and $\hat{m}_4(x) = g(\hat{\beta}, x) + \hat{\lambda}\hat{\theta}(x)$, Burman and Chaudhri combined estimator $\hat{m}_4(x) = \hat{m}_{bc}(x)$, Fan and Ullah estimator $\hat{m}_4(x) = \hat{m}_{fu}(x)$, and Glad's multiplicative combined estimator $\hat{m}_5(x)$. For this purpose we present in Table 1 the mean (M) and standard deviation (S) of the MSE, that is the mean and standard deviation of $\sum_1^n (y_i - \hat{y}_i)^2/n$ over 1000 replications, in each case and see its closeness to the variance of ϵ which is one. It is seen that when the parametric model is correctly specified our proposed combined fits $\hat{m}_3$ and $\hat{m}_4$ and the other combined fits $\hat{m}_{bc}$, $\hat{m}_{fu}$ and $\hat{m}_5$ perform as well as the parametric fit. The fact that both the additive and multiplicative fits ($\hat{m}_3$ and $\hat{m}_5$) are close to $\hat{m}_4$, $\hat{m}_{bc}$, and $\hat{m}_{fu}$ follows due to $\hat{\lambda} = 1 - \hat{\delta}$ value being close to unity. That also explains why all the combined fits are also close to the parametric fit $\hat{m}_1$. The combined fits however outperform the Nadaraya–Watson kernel fit $\hat{m}_2$ whose mean behavior is quite poor. This continues to hold when $\delta = 0.3$, that is, the parametric model is approximately correct. However, in this case, our proposed combined estimators $\hat{m}_3$ and $\hat{m}_4$ and Glad's multiplicative combined estimator $\hat{m}_5$ perform better than $\hat{m}_{bc}$, $\hat{m}_{fu}$, and the parametric fit $\hat{m}_1$ for all sample sizes. When the parametric model is incorrectly specified ($\delta = 1$) our proposed estimators $\hat{m}_3$, $\hat{m}_4$, and Glad's $\hat{m}_5$ continue to perform much better than the parametric fit $\hat{m}_1$ as well as $\hat{m}_2$, $\hat{m}_{fu}$ and $\hat{m}_{bc}$ for all sample sizes. $\hat{m}_{fu}$ and $\hat{m}_{bc}$ perform better than $\hat{m}_1$ and $\hat{m}_2$ but come close to $\hat{m}_2$ for large samples. Between $\hat{m}_3$, $\hat{m}_4$, and Glad's $\hat{m}_5$ the performance of our proposed estimator $\hat{m}_4$ is the best. In summary, we suggest the use of $\hat{m}_3$, $\hat{m}_4$ and $\hat{m}_5$, especially $\hat{m}_4$, since they perform as

Table 1. Mean (M) and standard deviation (S) of the MSE of fitted values

δ	n		$\hat{m}_1(x)$	$\hat{m}_2(x)$	$\hat{m}_3(x)$	$\hat{m}_4(x)$	$\hat{m}_{bc}(x)$	$\hat{m}_{fu}(x)$	$\hat{m}_5(x)$
0.0	50	M	0.9366	10.7280	0.8901	0.8667	0.9343	0.9656	0.8908
		S	0.1921	2.5773	0.1836	0.1796	0.1915	0.2098	0.1836
	100	M	0.9737	7.5312	0.9403	0.9271	0.9728	0.9895	0.9406
		S	0.1421	1.2578	0.1393	0.1386	0.1420	0.1484	0.1393
	500	M	0.9935	3.2778	0.9794	0.9759	0.9934	0.9968	0.9794
		S	0.0641	0.1970	0.0636	0.0635	0.0641	0.0643	0.0636
0.3	50	M	1.4954	11.7159	1.0323	0.9299	1.4635	1.6320	1.0379
		S	0.5454	3.1516	0.2145	0.1892	0.5162	0.7895	0.2162
	100	M	1.6761	8.2164	1.0587	0.9788	1.6338	1.7597	1.0069
		S	0.4811	1.5510	0.1544	0.1435	0.4427	0.6421	0.1553
	500	M	1.8414	3.5335	1.0361	1.0012	1.7001	1.7356	1.0374
		S	0.2818	0.2485	0.0652	0.0646	0.2067	0.2271	0.0652
1.0	50	M	7.2502	16.9386	2.5007	1.5492	6.3044	8.3949	2.6089
		S	5.2514	5.3785	1.0217	0.5576	4.0529	7.2215	1.0861
	100	M	8.7423	11.8223	2.2527	1.4679	6.6986	8.2644	2.3372
		S	4.8891	2.7239	0.5779	0.3306	2.8179	4.6021	0.6143
	500	M	10.4350	4.8395	1.6101	1.2408	4.5283	5.2413	1.6320
		S	2.9137	0.4689	0.1466	0.1151	0.5352	0.8147	0.1466

$\hat{m}_1(x) = g(\hat{\beta}, x)$ is the parametric fit, $\hat{m}_2(x)$ is the nonparametric kernel fit, $\hat{m}_3(x)$ and $\hat{m}_4(x)$ are proposed combined fits, $\hat{m}_{bc}(x)$ is the Burman and Chaudhri combined fit, $\hat{m}_{fu}(x)$ is the Fan and Ullah combined fit, $\hat{m}_5(x)$ is Glad's multiplicative fit.

well as the parametric fit when the parametric specification is correct and outperform the parametric and other alternative fits when the parametric specification is approximately correct or incorrect.

4.2 Performance of Test Statistics

Here we conduct a Monte Carlo simulation to evaluate the size and power of the T-tests in Section 3. These are the T_1 test due to Zheng (1996) and Li and Wang (1998), T_2 and T_3 tests due to Fan and Ullah (1999) and our proposed tests T_4, T_5, and $T_6 = T_1$.

The null hypothesis we want to test is that the linear regression model is correct:

$$H_0 : m(x) = g(\beta, x) = \beta_0 + \beta_1 x_1 + \beta_2 x_2 \tag{4.2}$$

where βs are parameters and x_1 and x_2 are the regressors. To investigate the size of the test we consider the model in which the dependent variable is generated by

$$Y_i = 1 + X_{1i} + X_{2i} + \epsilon_i \tag{4.3}$$

where the error term ϵ_i is drawn independently from the standard normal distribution, and the regressors X_1 and X_2 are defined as $X_1 = Z_1$, $X_2 = (Z_1 + Z_2)/\sqrt{2}$; Z_1 and Z_2 are vectors of independent standard normal random samples of size n. To investigate the power of the test we consider the following alternative models:

$$Y_i = 1 + X_{1i} + X_{2i} + X_{1i} X_{2i} + \epsilon_i \tag{4.4}$$

and

$$Y_i = (1 + X_{1i} + X_{2i})^{5/3} + \epsilon_i. \tag{4.5}$$

In all experiments, we consider sample sizes of 100 to 600 and we perform 1000 replications. The kernel function K is chosen to be the bivariate standard normal density function and the bandwidth h is chosen to be $c n^{-2/5}$, where c is a constant. To analyze whether the tests are sensitive to the choice of window width we consider c equal to 0.5, 1.0, and 2.0. The critical values for the tests are from the standard normal table. For more details, see Zheng (1996).

Table 2 shows that the size performances of all the tests T_3, T_4, T_5, and $T_6 = T_1$, based on the numerator of $\hat{\lambda}$, $\hat{\lambda}_N$, are similar. This implies that the test sizes are robust to weights and the smoothness of the estimator of $g(\beta, x)$. The size behavior of the test T_2, based on the LS estimator $\hat{\lambda}$, is in general not good. This may be due to the random denominator in $\hat{\lambda}$. Performances of T_5 and $T_6 = T_1$ have a slight edge over T_3 and T_4. This implies that the weighting by $\hat{f}(X_i)$ is better than the weighting by $(\hat{f}(X_i))^2$.

Regarding the power against the model (4.4) we note from Table 3 that, irrespective, of c values, performances of $T_6 = T_1$ and T_5 have a slight edge over T_3 and T_4. The power performance of the T_2 test, as in the case of size performance, is poor throughout. By looking at the findings on size and power performances it is clear that, in practice, either T_5 or $T_6 = T_1$ is preferable to T_2, T_3, and T_4 tests. Though best power results for most of the tests occurred when $c = 0.5$, both size and power of the T_2 test are also sensitive to the choice of window width. The results for the model (4.5) are generally found to be similar, see Table 4.

Table 2. Size of the tests

c	%	Test	n 100	200	300	400	500	600
0.5	1	T_2	0.8	1.2	1.5	1.1	0.5	0.6
		T_3	0.3	0.9	0.8	0.7	0.5	0.8
		T_4	0.2	0.8	0.9	0.7	0.6	0.8
		T_5	0.3	0.8	0.7	0.9	0.5	1.0
		T_6	0.2	0.7	0.6	0.9	0.7	1.1
	5	T_2	6.5	4.5	5.0	4.8	4.0	4.7
		T_3	3.8	4.2	5.4	5.0	4.1	4.9
		T_4	3.9	4.2	5.3	4.8	4.0	4.8
		T_5	4.5	4.5	5.3	5.4	4.4	4.8
		T_6	4.1	4.2	5.5	5.5	4.5	4.8
	10	T_2	11.7	8.4	10.1	10.9	10.1	9.7
		T_3	10.7	8.5	10.2	10.9	10.2	10.2
		T_4	10.1	8.4	10.0	10.9	10.2	10.7
		T_5	12.5	9.3	10.5	11.5	9.2	9.9
		T_6	12.5	9.6	10.1	11.2	9.3	9.8
1.0	1	T_2	4.1	3.1	2.2	2.3	1.5	1.7
		T_3	0.4	0.5	1.0	0.8	1.0	1.5
		T_4	0.3	0.5	1.0	0.8	0.8	1.5
		T_5	0.6	0.2	1.0	0.9	0.8	1.2
		T_6	0.6	0.1	1.1	0.9	0.9	1.1
	5	T_2	9.7	7.2	7.2	8.1	5.5	6.7
		T_3	3.4	3.9	5.5	4.6	3.4	4.9
		T_4	3.5	4.1	5.1	4.4	3.6	4.9
		T_5	3.8	3.9	4.7	4.8	3.9	5.2
		T_6	4.0	3.8	4.5	4.6	4.0	5.0
	10	T_2	15.2	12.2	12.7	14.4	10.3	12.3
		T_3	8.5	8.6	10.5	10.5	8.2	9.9
		T_4	8.3	8.7	10.2	10.1	8.7	9.4
		T_5	8.9	9.8	10.2	10.8	8.5	10.3
		T_6	9.1	9.9	10.7	11.2	8.6	10.2

Table 2 continued

c	%	Test	n					
			100	200	300	400	500	600
2.0	1	T_2	9.1	7.4	5.7	6.4	4.7	4.8
		T_3	0.5	0.5	1.1	0.7	0.6	1.1
		T_4	0.4	0.6	1.1	0.7	0.6	0.9
		T_5	0.4	0.4	0.6	0.6	0.4	0.4
		T_6	0.1	0.3	0.7	0.7	0.4	0.5
	5	T_2	17.3	13.3	11.7	12.1	9.5	11.4
		T_3	2.0	2.0	3.3	3.7	3.3	4.7
		T_4	1.5	1.8	3.0	3.4	3.6	4.0
		T_5	2.6	2.6	3.7	5.2	4.1	4.8
		T_6	1.8	2.9	3.4	4.5	3.8	3.8
	10	T_2	21.6	17.5	16.9	16.4	14.3	16.1
		T_3	5.8	5.9	7.7	8.6	7.2	9.4
		T_4	3.9	5.6	6.3	8.2	6.9	9.5
		T_5	9.0	8.5	9.3	11.1	8.6	10.7
		T_6	6.3	7.2	8.6	11.5	7.8	10.2

Table 3. Power of the test against model (4.4)

c	%	Test	n					
			100	200	300	400	500	600
0.5	1	T_2	0.6	7.0	30.5	47.4	72.0	88.0
		T_3	15.0	44.1	66.2	80.7	91.1	97.7
		T_4	15.3	44.1	65.9	80.8	91.2	97.7
		T_5	25.0	66.6	86.9	95.8	99.2	99.9
		T_6	24.8	66.7	87.1	95.7	99.2	99.9
	5	T_2	10.6	41.0	67.6	81.4	93.0	98.3
		T_3	36.7	68.2	82.9	92.5	96.7	99.3
		T_4	36.5	68.3	82.6	92.5	96.8	99.3
		T_5	48.9	85.4	96.8	99.3	99.7	100.0
		T_6	49.1	85.5	96.8	99.4	99.7	100.0
	10	T_2	24.7	61.4	80.0	91.3	96.4	99.2
		T_3	47.6	77.1	90.3	94.8	98.3	99.5
		T_4	47.8	77.2	90.5	94.8	98.2	99.5
		T_5	61.0	91.6	98.3	99.7	100.0	100.0
		T_6	60.9	91.6	98.3	99.7	100.0	100.0

1.0	1	T_2	0.4	11.4	56.5	83.3	95.9	99.7
		T_3	62.2	93.5	99.5	99.9	100.0	100.0
		T_4	62.0	93.9	99.5	99.9	100.0	100.0
		T_5	76.4	99.4	100.0	100.0	100.0	100.0
		T_6	76.5	99.4	100.0	100.0	100.0	100.0
	5	T_2	14.2	67.8	94.5	98.8	100.0	100.0
		T_3	75.9	97.8	99.7	100.0	100.0	100.0
		T_4	76.2	98.0	99.7	100.0	100.0	100.0
		T_5	89.3	99.9	100.0	100.0	100.0	100.0
		T_6	88.8	99.9	100.0	100.0	100.0	100.0
	10	T_2	37.2	89.2	98.4	99.9	100.0	100.0
		T_3	83.3	98.9	99.9	100.0	100.0	100.0
		T_4	82.5	99.1	99.9	100.0	100.0	100.0
		T_5	92.2	99.9	100.0	100.0	100.0	100.0
		T_6	92.1	99.9	100.0	100.0	100.0	100.0
2.0	1	T_2	0.0	0.0	0.1	4.5	27.9	67.6
		T_3	92.2	100.0	100.0	100.0	100.0	100.0
		T_4	92.6	100.0	100.0	100.0	100.0	100.0
		T_5	97.1	100.0	100.0	100.0	100.0	100.0
		T_6	97.4	100.0	100.0	100.0	100.0	100.0
	5	T_2	0.0	13.2	66.7	93.5	99.4	100.0
		T_3	95.6	100.0	100.0	100.0	100.0	100.0
		T_4	95.5	100.0	100.0	100.0	100.0	100.0
		T_5	98.6	100.0	100.0	100.0	100.0	100.0
		T_6	99.0	100.0	100.0	100.0	100.0	100.0
	10	T_2	7.9	67.6	97.8	100.0	100.0	100.0
		T_3	97.0	100.0	100.0	100.0	100.0	100.0
		T_4	97.3	100.0	100.0	100.0	100.0	100.0
		T_5	99.1	100.0	100.0	100.0	100.0	100.0
		T_6	99.2	100.0	100.0	100.0	100.0	100.0

Table 4. Power against model (4.5)

c	%	Test	\multicolumn{6}{c}{n}
			100	200	300	400	500	600
0.5	1	T_2	2.7	34.9	78.3	93.6	98.7	100.0
		T_3	49.7	89.3	98.4	99.6	99.9	100.0
		T_4	48.6	89.2	98.2	99.6	99.9	100.0
		T_5	67.1	97.6	99.6	100.0	100.0	100.0
		T_6	66.5	97.5	99.6	100.0	100.0	100.0
	5	T_2	28.2	81.1	97.5	99.5	99.5	100.0
		T_3	72.6	96.9	99.3	100.0	100.0	100.0
		T_4	72.7	97.0	99.3	100.0	100.0	100.0
		T_5	86.6	99.8	100.0	100.0	100.0	100.0
		T_6	86.4	99.8	100.0	100.0	100.0	100.0
	10	T_2	54.6	94.1	98.9	99.9	99.9	100.0
		T_3	81.2	98.5	99.6	100.0	100.0	100.0
		T_4	80.6	98.6	99.6	100.0	100.0	100.0
		T_5	92.0	100.0	100.0	100.0	100.0	100.0
		T_6	92.0	100.0	100.0	100.0	100.0	100.0
1.0	1	T_2	0.7	32.6	87.6	98.7	99.6	99.9
		T_3	95.4	100.0	100.0	100.0	100.0	100.0
		T_4	94.9	100.0	100.0	100.0	100.0	100.0
		T_5	98.1	100.0	100.0	100.0	100.0	100.0
		T_6	98.0	100.0	100.0	100.0	100.0	100.0
	5	T_2	32.4	93.9	99.9	100.0	100.0	100.0
		T_3	97.8	100.0	100.0	100.0	100.0	100.0
		T_4	97.9	100.0	100.0	100.0	100.0	100.0
		T_5	99.6	100.0	100.0	100.0	100.0	100.0
		T_6	99.5	100.0	100.0	100.0	100.0	100.0
	10	T_2	65.2	99.5	100.0	100.0	100.0	100.0
		T_3	98.8	100.0	100.0	100.0	100.0	100.0
		T_4	98.6	100.0	100.0	100.0	100.0	100.0
		T_5	99.9	100.0	100.0	100.0	100.0	100.0
		T_6	99.8	100.0	100.0	100.0	100.0	100.0
2.0	1	T_2	0.0	0.0	0.0	1.0	14.6	54.6
		T_3	100.0	100.0	100.0	100.0	100.0	100.0
		T_4	99.9	100.0	100.0	100.0	100.0	100.0
		T_5	100.0	100.0	100.0	100.0	100.0	100.0
		T_6	100.0	100.0	100.0	100.0	100.0	100.0

		0.0	6.3	62.0	92.9	98.9	99.8
5	T_2	0.0	6.3	62.0	92.9	98.9	99.8
	T_3	100.0	100.0	100.0	100.0	100.0	100.0
	T_4	100.0	100.0	100.0	100.0	100.0	100.0
	T_5	100.0	100.0	100.0	100.0	100.0	100.0
	T_6	100.0	100.0	100.0	100.0	100.0	100.0
10	T_2	3.4	68.2	98.6	99.9	100.0	100.0
	T_3	100.0	100.0	100.0	100.0	100.0	100.0
	T_4	100.0	100.0	100.0	100.0	100.0	100.0
	T_5	100.0	100.0	100.0	100.0	100.0	100.0
	T_6	100.0	100.0	100.0	100.0	100.0	100.0

REFERENCES

Burman, P. and P. Chaudhri (1994), A hybrid approach to parametric and nonparametric regression, *Technical Report No. 243*, Division of Statistics, University of California, Davis, CA.

Eubank, R. L. and C. H. Spiegelman (1990), Testing the goodness-of-fit of linear models via nonparametric regression techniques, *Journal of the American Statistical Association*, **85**, 387–392.

Fan, Y. and Q. Li (1996), Consistent model specification tests: Omitted variables and semiparametric functional forms, *Econometrica*, **64**, 865–890.

Fan, Y. and A. Ullah (1999), Asymptotic normality of a combined regression estimator, *Journal of Multivariate Analysis*, **71**, 191–240.

Glad, I. K. (1998), Parametrically guided nonparametric regression, *Scandinavian Journal of Statistics*, **25**(5), 649–668.

Härdle, W. (1990), *Applied Nonparametric Regression*, Cambridge University Press, New York.

Härdle, W. and E. Mammen (1993), Comparing nonparametric versus parametric regression fits, *Annals of Statistics*, **21**, 1926–1947.

Li, Q. and S. Wang (1998), A simple consistent bootstrap test for a parametric regression function, *Journal of Econometrics*, **87**, 145–165.

Nadaraya, E. A. (1964), On estimating regression, *Theory of Probability and its Applications*, **9**, 141–142.

Olkin, I. and C. H. Spiegelman (1987), A semiparametric approach to density estimation, *Journal of the American Statistical Association*, **82**, 858–865.

Pagan, A. and A. Ullah (1999), *Nonparametric Econometrics*, Cambridge University Press, New York.

Rahman, M., D. V. Gokhale, and A. Ullah (1997), A note on combining parametric and nonparametric regression, *Communications in statistics—Simulation and Computation*, **26**, 519–529.

Ullah, A. (1985), Specification analysis of econometric models, *Journal of Quantitative Economics*, **1**, 187–209.

Ullah, A. and H. D. Vinod (1993), General nonparametric regression estimation and testing in econometrics, in G. S. Maddala, C. R. Rao, and H. D. Vinod (eds.), *Handbook of Statistics*, Vol. 11, Elsevier, Amsterdam, pp. 85–116.

Watson, G. S. (1964), Smooth regression analysis, *Sankhya, Series A*, **26**, 359–372.

Wooldridge, J. M. (1992), A test for functional form against nonparametric alternatives, *Econometric Theory*, **8**, 452–475.

Yatchew, A. J. (1992), Nonparametric regression tests based on least squares, *Econometric Theory*, **8**, 435–451.

Zheng, J. X. (1996), A consistent test of functional form via nonparametric estimation techniques, *Journal of Econometrics*, **75**, 263–289.

10

Neyman's Smooth Test and Its Applications in Econometrics

ANIL K. BERA and AUROBINDO GHOSH   University of Illinois at
Urbana–Champaign, Champaign, Illinois

1. INTRODUCTION

Statistical hypothesis testing has a long history. Neyman and Pearson (1933
[80]) traced its origin to Bayes (1763 [8]). However, the systematic use of
hypothesis testing began only after the publication of Pearson's (1900 [86])
goodness-of-fit test. Even after 100 years, this statistic is very much in use in
a variety of applications and is regarded as one of the 20 most important
scientific breakthroughs in the twentieth century. Simply stated, Pearson's
(1900 [86]) test statistic is given by

$$P_{\chi^2} = \sum_{j=1}^{q} \frac{(O_j - E_j)^2}{E_j} \tag{1}$$

where O_j denotes the observed frequency and E_j is the (expected) frequency
that would be obtained under the distribution of the null hypothesis, for the
jth class, $j = 1, 2, \ldots, q$. Although K. Pearson (1900 [86]) was an auspicious
beginning to twentieth century statistics, the basic foundation of the theory
of hypothesis testing was laid more than three decades later by Neyman and

E. S. Pearson (1933 [80]). For the first time the concept of "optimal test" was introduced through the analysis of "power functions." A general solution to the problem of maximizing power subject to a size condition was obtained for the single-parameter case when both the null and the alternative hypotheses were simple [see for example, Bera and Premaratne (2001[15])]. The result was the celebrated Neyman–Pearson (N–P) lemma, which provides a way to construct a uniformly most powerful (UMP) test. A UMP test, however, rarely exists, and therefore it is necessary to restrict optimal tests to a suitable subclass that requires the test to satisfy other criteria such as *local* optimality and *unbiasedness*. Neyman and Pearson (1936 [81]) derived a locally most powerful unbiased (LMPU) test for the one-parameter case and called the corresponding critical region the "type-A region." Neyman and Pearson (1938 [82]) obtained the LMPU test for testing a *multiparameter* hypothesis and termed the resulting critical region the "type-C region."

Neyman's (1937 [76]) smooth test is based on the type-C critical region. Neyman suggested the test to rectify some of the drawbacks of the Pearson goodness-of-fit statistic given in (1). He noted that it is not clear how the class intervals should be determined and that the distributions under the alternative hypothesis were not "smooth." By smooth densities, Neyman meant those that are close to and have few intersections with the null density function. In his effort to find a smooth class of alternative distributions, Neyman (1937 [76]) considered the probability integral transformation of the density, say $f(x)$, under the null hypothesis and showed that the probability integral transform is distributed as uniform in (0, 1) irrespective of the specification of $f(x)$. Therefore, in some sense, "all" testing problems can be converted into testing only *one kind of hypothesis.*

Neyman was not the first to use the idea of probability integral transformation to reformulate the hypothesis testing problem into a problem of testing uniformity. E. Pearson (1938 [84]) discussed how Fisher (1930 [41], 1932 [43]) and K. Pearson (1933 [87], 1934 [88]) also developed the same idea. They did not, however, construct any formal test statistic. What Neyman (1937 [76]) achieved was to integrate the ideas of tests based on the probability integral transforms in a concrete fashion, along with designing "smooth" alternative hypotheses based on normalized Legendre polynomials.

The aim of this paper is modest. We put the Neyman (1937 [76]) smooth test in perspective with the existing methods of testing available at that time; evaluate it on the basis of the current state of the literature; derive the test from the widely used Rao (1948 [93]) score principle of testing; and, finally, we discuss some of the applications of the smooth test in econometrics and statistics.

Section 2 discusses the genesis of probability integral transforms as a criterion for hypothesis testing with Sections 2.1 through 2.3 putting

Neyman's smooth test in perspective in the light of current research in probability integral transforms and related areas. Section 2.4 discusses the main theorem of Neyman's smooth test. Section 3 gives a formulation of the relationship of Neyman's smooth test to Rao's score (RS) and other optimal tests. Here, we also bring up the notion of unbiasedness as a criterion for optimality in tests and put forward the differential geometric interpretation. In Section 4 we look at different applications of Neyman's smooth tests. In particular, we discuss inference using different orthogonal polynomials, density forecast evaluation and calibration in financial time series data, survival analysis and applications in stochastic volatility models. The paper concludes in Section 5.

2. BACKGROUND AND MOTIVATION

2.1 Probability Integral Transform and the Combination of Probabilities from Independent Tests

In statistical work, sometimes, we have a number of *independent* tests of significance for the same hypothesis, giving different probabilities (like p-values). The problem is to combine results from different tests in a single hypothesis test. Let us suppose that we have carried out n independent tests with p-values $y_1, y_2, \ldots, y_n$. Tippett (1931 [112], p. 142) suggested a procedure based on the minimum p-value, i.e., on $y_{(1)} = \min(y_1, y_2, \ldots y_n)$. If all n null hypotheses are valid, then $y_{(1)}$ has a standard beta distribution with parameters $(1, n)$. One can also use any smallest p-value, $y_{(r)}$, the rth smallest p-value in place of $y_{(1)}$, as suggested by Wilkinson (1951[115]). The statistic $y_{(r)}$ will have a beta distribution with parameters $(r, n - r + 1)$. It is apparent that there is some arbitrariness in this approach through the choice of r. Fisher (1932 [43], Section 21.2, pp. 99–100) suggested a simpler and more appealing procedure based on the product of the p-values, $\lambda = \Pi_{i=1}^{n} y_i$. K. Pearson (1933 [87]) also considered the same problem in a more general framework along with his celebrated problem of goodness-of-fit. He came up with the same statistic λ, but suggested a different approach to compute the p-value of the comprehensive test.*

---

*To differentiate his methodology from that of Fisher, K. Pearson added the following note at the end of his paper:

> After this paper had been set up Dr Egon S. Pearson drew my attention to Section 21.1 in the Fourth Edition of Professor R.A. Fisher's *Statistical Methods for Research Workers*, 1932. Professor Fisher is brief, but his method is essentially

(footnote continues)

In the current context, Pearson's goodness-of-fit problem can be stated as follows. Let us suppose that we have a sample of size n, $x_1, x_2, \ldots, x_n$. We want to test whether it comes from a population with probability density function (pdf) $f(x)$. Then the p-values (rather, the $1 - p$-values) y_i ($i = 1, 2, \ldots, n$) can be defined as

$$y_i = \int_{-\infty}^{x_i} f(\omega)d\omega \tag{2}$$

Suppose that we have n tests of significance and the values of our test statistics are T_i, $i = 1, 2, \ldots, n$, then

$$y_i = \int_{-\infty}^{T_i} f_{T_i}(t)dt \tag{3}$$

where $f_T(t)$ is the pdf of T. To find the distribution or the p-value of $\lambda = y_1 y_2 \ldots y_n$ both Fisher and Karl Pearson started in a similar way, though Pearson was more explicit in his derivation. In this exposition, we will follow Pearson's approach.

Let us simply write

$$y = \int_{-\infty}^{x} f(\omega)d\omega \tag{4}$$

and the pdf of y as $g(y)$. Then, from (4), we have

$$dy = f(x)dx \tag{5}$$

and we also have, from change of variables,

$$g(y)dy = f(x)dx \tag{6}$$

Hence, combining (5) and (6),

$$g(y) = 1, 0 < y < 1 \tag{7}$$

i.e., y has a uniform distribution over $(0, 1)$. From this point Pearson's and Fisher's treatments differ. The surface given by the equation

$$\lambda_n = y_1 y_2 \ldots y_n \tag{8}$$

(footnote continued)

what I had thought to be novel. He uses, however a χ^2 method, not my incomplete Γ-function solution; ... As my paper was already set up and illustrates, more amply than Professor Fisher's two pages, some of the advantages and some of the difficulties of the new method, which may be helpful to students, I have allowed it to stand.

is termed "*n-hyperboloid*" by Pearson, and what is needed is the volume of *n*-cuboid (since $0 < y_1 < 1, i = 1, 2, \ldots, n$) cut off by the *n*-hyperboloid. We show the surface λ_n in Figures 1 and 2 for $n = 2$ and $n = 3$, respectively. After considerable algebraic derivation Pearson (1933 [87], p. 382) showed that the *p*-value for λ_n is given by

$$Q_{\lambda_n} = 1 - P_{\lambda_n} = 1 - I(n-1, -\ln\lambda_n) \tag{9}$$

where $I(.)$ is the incomplete gamma function ratio defined by [Johnson and Kotz (1970a [57], p. 167)]

$$I(n-1, u) = \frac{1}{\Gamma(n)} \int_0^{u\sqrt{n}} t^{n-1} e^{-t} dt \tag{10}$$

We can use the test statistic Q_{λ_n} both for combining a number of independent tests of significance and for the goodness-of-fit problem. Pearson (1933 [87], p. 383) stated this very clearly:

If Q_{λ_n} be very small, we have obtained an extremely rare sample, and we have then to settle in our minds whether it is more reasonable to suppose that we have drawn a very rare sample at one trial from the supposed parent population, or that our hypothesis as to the character of the parent population is erroneous, i.e., that the sample $x_1, x_2, \ldots, x_n$ was not drawn from the supposed population.

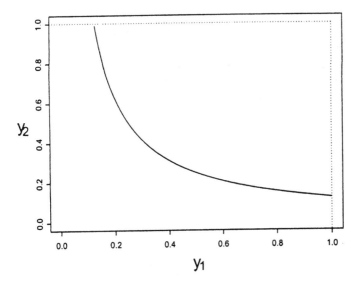

Figure 1. Surface of the equation $y_1 y_2 = \lambda_2$ for $\lambda_2 = 0.125$.

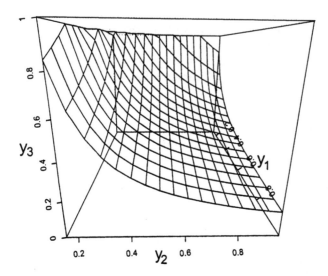

Figure 2. Surface of the equation $y_1 y_2 y_3 = \lambda_3$ for $\lambda_3 = 0.125$.

Pearson (1933 [87], p. 403) even criticized his own celebrated χ^2 statistic, stating that the χ^2 test in equation (1) has the disadvantage of giving the same resulting probability whenever the individuals are in the same class. This criticism has been repeatedly stated in the literature. Bickel and Doksum (1977 [18], p. 378) have put it rather succinctly: "in problems with continuous variables there is a clear loss of information, since the χ^2 test utilizes only the number of observations in intervals rather than the observations themselves." Tests based on P_{λ_n} (or Q_{λ_n}) do not have this problem. Also, when the sample size n is small, grouping the observations in several classes is somewhat hazardous for the inference.

As we mentioned, Fisher's main aim was to combine n p-values from n independent tests to obtain a single probability. By putting $Z = -2 \ln Y \sim U(0, 1)$, we see that the pdf of Z is given by

$$f_z(z) = \frac{1}{2} e^{-z/2} \tag{11}$$

i.e., Z has a χ^2_2 distribution. Then, if we combine n independent z_is by

$$\sum_{i=1}^{n} z_i = -2 \sum_{i=1}^{n} \ln y_i = -2 \ln \lambda_n \tag{12}$$

this statistic will be distributed as χ^2_{2n}. For quite some time this statistic was known as Pearson's P_λ. Rao (1952 [94], p. 44) called it Pearson's P_λ distribution [see also Maddala (1977 [72], pp. 47–48)]. Rao (1952 [94], pp. 217–219) used it to combine several independent tests of the difference between means and on tests for skewness. In the recent statistics literature this is described as Fisher's procedure [for example, see Becker (1977 [9])].

In summary, to combine several independent tests, both Fisher and K. Pearson arrived at the same problem of testing the uniformity of $y_1, y_2, \ldots, y_n$. Undoubtedly, Fisher's approach was much simpler, and it is now used more often in practice. We should, however, add that Pearson had a much broader problem in mind, including testing goodness-of-fit. In that sense, Pearson's (1933 [87]) paper was more in the spirit of Neyman's (1937 [76]) that came four years later.

As we discussed above, the fundamental basis of Neyman's smooth test is the result that when $x_1, x_2, \ldots, x_n$ are independent and identically distributed (IID) with a common density $f(.)$, then the probability integral transforms $y_1, y_2, \ldots, y_n$ defined in equation (2) are IID, $U(0, 1)$ random variables. In econometrics, however, we very often have cases in which x_1, $c_2, \ldots, x_n$ are not IID. In that case we can use Rosenblatt's (1952 [101]) generalization of the above result.

Theorem 1 (Rosenblatt 1952)   Let $(X_1, X_2, \ldots, X_n)$ be a random vector with absolutely continuous density function $f(x_1, x_2, \ldots, x_n)$. Then, the n random variables defined by $Y_1 = P(X_1 \leq x_1)$, $Y_2 = P(X_2 \leq x_2 | X_1 = x_1), \ldots, Y_n = P(X_n \leq x_n | X_1 = x_1, X_2 = x_2, \ldots, X_{n-1} = x_{n-1})$ are IID $U(0, 1)$.

The above result can immediately be seen from the following observation that

$$P(Y_i \leq y_i, i = 1, 2, \ldots, n) = \int_0^{y_1} \int_0^{y_2} \cdots \int_0^{y_n} f(x_1) dx_1 f(x_2 | x_1) dx_2 \cdots$$

$$f(x_n | x_1, \ldots, x_{n-1}) dx_n$$

$$= \int_0^{y_1} \int_0^{y_2} \cdots \int_0^{y_n} dt_1 dt_2 \ldots dt_n$$

$$= y_1 y_2 \cdots y_n$$

(13)

Hence, $Y_1, Y_2, \ldots, Y_n$ are IID $U(0, 1)$ random variables. Quesenberry (1986 [92], pp. 239–240) discussed some applications of this result in goodness-of-fit tests [see also, O'Reilly and Quesenberry (1973[83])]. In Section 4.2, we will discuss its use in density forecast evaluation.

2.2 Summary of Neyman (1937)

As we mentioned earlier, Fisher (1932 [43]) and Karl Pearson (1933 [87], 1934 [88]) suggested tests based on the fact that the probability integral transform is uniformly distributed for an IID sample under the null hypothesis (or the correct specification of the model). What Neyman (1937 [76]) achieved was to integrate the ideas of tests based on probability integral transforms in a concrete fashion along with the method of designing alternative hypotheses using orthonormal polynomials.* Neyman's paper began with a criticism of Pearson's χ^2 test given in (1). First, in Pearson's χ^2 test, it is not clear how the q class intervals should be determined. Second, the expression in (1) does not depend on the order of positive and negative differences $(O_j - E_j)$. Neyman (1980 [78], pp. 20–21) gives an extreme example represented by two cases. In the first, the signs of the consecutive differences $(O_j - E_j)$ are not the same, and in the other there is a run of, say, a number of "negative" differences, followed by a sequence of "positive" differences. These two possibilities might lead to similar values of P_{χ^2}, but Neyman (1937 [76], 1980 [78]) argued that in the second case the goodness-of-fit should be more in doubt, even if the value of χ^2 happens to be small. In the same spirit, the χ^2-test is more suitable for discrete data and the corresponding distributions under the alternative hypotheses are not "smooth." By smooth alternatives Neyman (1937 [76]) meant those densities that have *few* intersections with the null density function and that are *close* to the null.

Suppose we want to test the null hypothesis (H_0) that $f(x)$ is the true density function for the random variable X. The specification of $f(x)$ will be *different* depending on the problem at hand. Neyman (1937 [76], pp. 160–161) first transformed *any* hypothesis testing problem of this type to testing

---

*It appears that Jerzy Neyman was not aware of the above papers by Fisher and Karl Pearson. To link Neyman's test to these papers, and possibly since Neyman's paper appeared in a rather recondite journal, Egon Pearson (Pearson 1938 [84]) published a review article in *Biometrika*. At the end of that article Neyman added the following note to express his regret for overlooking, particularly, the Karl Pearson papers:

"I am grateful to the author of the present paper for giving me the opportunity of expressing my regret for having overlooked the two papers by Karl Pearson quoted above. When writing the paper on the "Smooth test for goodness of fit" and discussing previous work in this direction, I quoted only the results of H. Cramér and R. v. Mises, omitting mention of the papers by K. Pearson. The omission is the more to be regretted since my paper was dedicated to the memory of Karl Pearson."

only *one kind of hypothesis.** Let us state the result formally through the following simple derivation.

Suppose that, under H_0, $x_1, x_2, \ldots, x_n$ are independent and identically distributed with a common density function $f(x|H_0)$. Then, the probability integral transform

$$y \equiv y(x) = \int_{-\infty}^{x} f(u|H_0) du \tag{14}$$

has a pdf given by

$$h(y) = f(x|H_0) \frac{\partial x}{\partial y} \text{ for } 0 < y < 1 \tag{15}$$

Differentiating (14) with respect to y, we have

$$1 = f(x|H_0) \frac{dx}{dy} \tag{16}$$

Substituting this into (15), we get

$$h(y) \equiv h(y|H_0) = 1 \text{ for } 0 < y < 1 \tag{17}$$

Therefore, testing H_0 is equivalent to testing whether the random variable Y has a uniform distribution in the interval $(0, 1)$, irrespective of the specification of the density $f(.)$.

Figure 3, drawn following E. Pearson (1938 [84], Figure 1), illustrates the relationship between x and y when $f(.)$ is taken to be $N(0, 1)$ and $n = 20$. Let us denote $f(x|H_1)$ as the distribution under the alternative hypothesis H_1. Then, Neyman (1937 [76]) pointed out [see also Pearson (1938 [84], p. 138)] that the distribution of Y under H_1 given by

---

*In the context of testing several different hypotheses, Neyman (1937 [76], p. 160) argued this quite eloquently as follows:

If we treat all these hypotheses separately, we should define the set of alternatives for each of them and this would in practice lead to a dissection of a unique problem of a test for goodness of fit into a series of more or less disconnected problems.

However, this difficulty can be easily avoided by substituting for any particular form of the hypotheses H_0, that may be presented for test, another hypotheses, say h_0, which is equivalent to H_0 and which has always the same analytical form. The word equivalent, as used here, means that whenever H_0 is true, h_0 must be true also and inversely, if H_0 is not correct then h_0 must be false.

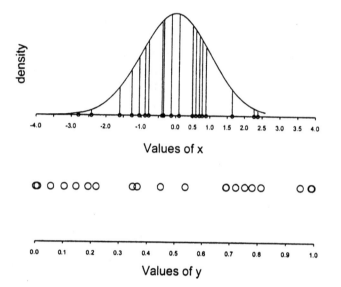

Figure 3. Distribution of the probability integral transform when H_0 is true.

$$f(y|H_1) = f(x|H_1)\frac{dx}{dy} = \frac{f(x|H_1)}{f(x|H_0)}\bigg|_{x=p(y)} \quad \text{for } 0 < y < 1 \tag{18}$$

where $x = p(y)$ means a solution to equation (14). This looks more like a likelihood ratio and will be different from 1 when H_0 is not true. As an illustration, in Figure 4 we plot values of Y when Xs are drawn from $N(2, 1)$ instead of $N(0, 1)$, and we can immediately see that these y values [probability integral transforms of values from $N(2, 1)$ using the $N(0, 1)$ density] are not uniformly distributed.

Neyman (1937 [76], p. 164) considered the following smooth alternative to the uniform density:

$$h(y) = c(\theta)\exp\left[\sum_{j=1}^{k} \theta_j \pi_j(y)\right] \tag{19}$$

where $c(\theta)$ is the constant of integration depending only on $(\theta_1, \ldots, \theta_k)$, and $\pi_j(y)$ are orthonormal polynomials of order j satisfying

$$\int_0^1 \pi_i(y)\pi_j(y)dy = \delta_{ij}, \quad \text{where } \delta_{ij} = 1 \text{ if } i = j$$
$$0 \text{ if } i \neq j \tag{20}$$

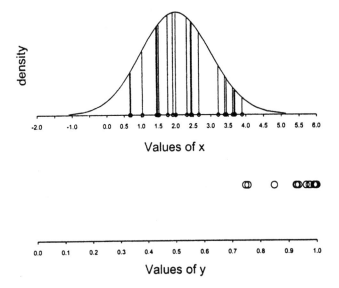

Figure 4. Distribution of the probability integral transform when H_0 is false.

Under $H_0 : \theta_1 = \theta_2 = \ldots = \theta_k = 0$, since $c(\theta) = 1$, $h(y)$ in (19) reduces to the uniform density in (17).

Using the generalized Neyman–Pearson (N–P) lemma, Neyman (1937 [76]) derived the locally most powerful symmetric test for $H_0 : \theta_1 = \theta_2 = \ldots = \theta_k = 0$ against the alternative H_1 : at least one $\theta_i \neq 0$, for small values of θ_i. The test is symmetric in the sense that the asymptotic power of the test depends only on the distance

$$\lambda = (\theta_1^2 + \ldots + \theta_k^2)^{\frac{1}{2}} \tag{21}$$

between H_0 and H_1. The test statistic is

$$\Psi_k^2 = \sum_{j=1}^{k} \frac{1}{n} \left[\sum_{i=1}^{n} \pi_j(y_i) \right]^2 \tag{22}$$

which under H_0 asymptotically follows a central χ_k^2 and under H_1 follows a non-central χ_k^2 with non-centrality parameter λ^2 [for definitions, see Johnson and Kotz (1970a [57], 1970b [58])]. Neyman's approach requires the computation of the probability integral transform (14) in terms of Y. It is, however, easy to recast the testing problem in terms of the original observations on X and pdf, say, $f(x; \gamma)$. Writing (14) as $y = F(x; \gamma)$ and defining

$\pi_i(y) = \pi_i(F(x; \gamma)) = q_i(x; \gamma)$, we can express the orthogonality condition (20) as

$$\int_0^1 \{\pi_i(F(x; \gamma))\}\{\pi_j(F(x; \gamma))\}dF(x; \gamma) = \int_0^1 \{q_i(x; \gamma)\}\{q_j(x; \gamma)\}f(x; \gamma)dx = \delta_{ij}$$

(23)

Then, from (19), the alternative density in terms of X takes the form

$$g(x; \gamma, \theta) = h(F(x; \gamma))\frac{dy}{dx}$$

$$= c(\theta; \gamma)\exp\left[\sum_{j=1}^k \theta_j q_j(x; \gamma)\right]f(x; \gamma)$$

(24)

Under this formulation the test statistic Ψ_k^2 reduces to

$$\Psi_k^2 = \sum_{j=1}^k \frac{1}{n}\left[\sum_{i=1}^n q_j(x_i; \gamma)\right]^2$$

(25)

which has the same asymptotic distribution as before. In order to implement this we need to replace the nuisance parameter γ by an efficient estimate $\hat{\gamma}$, and that will not change the asymptotic distribution of the test statistic [see Thomas and Pierce (1979 [111]), Kopecky and Pierce (1979 [63]), Koziol (1987 [64])], although there could be some possible change in the variance of the test statistic [see, for example, Boulerice and Ducharme (1995 [19])]. Later we will relate this test statistic to a variety of different tests and discuss its properties.

2.3 Interpretation of Neyman's (1937) Results and their Relation to some Later Works

Egon Pearson (1938 [84]) provided an excellent account of Neyman's ideas, and emphasized the need for consideration of the possible alternatives to the hypothesis tested. He discussed both the cases of testing goodness-of-fit and of combining results of independent tests of significance. Another issue that he addressed is whether the upper or the lower tail probabilities (or p-values) should be used for combining different tests. The upper tail probability [see equation (2)]

$$y_i' = \int_{x+i}^\infty f(\omega)d\omega = 1 - y_i$$

(26)

under H_0 is also uniformly distributed in (0, 1), and hence $-2\sum_{i=1}^n \ln y_i'$ is distributed as χ_{2n}^2 following our derivations in equations (11) and (12).

Therefore, the tests based on y_i and y_i' will be the same as far as their size is concerned but will, in general, differ in terms of power. Regarding other aspects of the Neyman's smooth test for goodness-of-fit, as Pearson (1938 [84], pp. 140 and 148) pointed out, the greatest benefit that it has over other tests is that it can detect the direction of the alternative when the null hypothesis of correct specification is rejected. The divergence can come from any combination of location, scale, shape, etc. By selecting the orthogonal polynomials π_j in equation (20) judiciously, we can seek the power of the smooth test in specific directions. We think that is one of the most important advantages of Neyman's smooth test over Fisher and Karl Pearson's suggestion of using only one function of y_i values, namely $\sum_{i=1}^{n}$ ln y_i. Egon Pearson (1938 [84], p. 139) plotted the function $f(y|H_1)$ [see equation (18)] for various specifications of H_1 when $f(x|H_0)$ is $N(0,1)$ and demonstrated that $f(y|H_1)$ can take a variety of nonlinear shapes depending on the nature of the departures, such as the mean being different from zero, the variance being different from 1, and the shape being nonnormal. It is easy to see that a single function like ln y cannot capture all of the nonlinearities. However, as Neyman himself argued, a linear combination of orthogonal polynomials might do the job.

Neyman's use of the density function (19) as an alternative to the uniform distribution is also of fundamental importance. Fisher (1922 [40], p. 356) used this type of exponential distribution to demonstrate the equivalence of the method of moments and the maximum likelihood estimator in special cases. We can also derive (19) analytically by *maximizing* the entropy $-E[\ln h(y)]$ subject to the moment conditions $E[\pi_j(y)] = \eta_j$ (say), $j = 1, 2, \ldots, k$, with parameters $\theta_j, j = 1, 2, \ldots, k$, as the Lagrange multipliers determined by k moment constraints [for more on this see, for example, Bera and Bilias (2001c [13])]. In the information theory literature, such densities are known as *minimum discrimination information* models in the sense that the density $h(y)$ in (19) has the minimum distance from the uniform distributions satisfying the above k moment conditions [see Soofi (1997 [106], 2000 [107])].* We can say that, while testing the density $f(x; \gamma)$, the alternative density function $g(x; \gamma, \theta)$ in equation (24) has a minimum distance from $f(x; \gamma)$, satisfying the moment conditions like $E[q_j(x)] = \eta_j, j = 1, \ldots, k$. From that point of view, $g(x; \gamma, \theta)$ is "truly" a *smooth* alter-

---

*For small values of θ_j $(j = 1, 2, \ldots, k)$, $h(y)$ will be a smooth density close to uniform when k is moderate, say equal to 3 or 4. However, if k is large, then $h(y)$ will present particularities which would not correspond to the intuitive idea of smoothness (Neyman 1937 [76], p. 165). From the maximum entropy point of view, each additional moment condition adds some more roughness and possibly some peculiarities of the data to the density.

native to the density $f(x; \gamma)$. Looking from another perspective, we can see from (19) that $\ln h(y)$ is essentially a *linear* combination of several polynomials in y. Similar densities have been used in the log-spline model literature [see, for instance, Stone and Koo (1986 [109]) and Stone (1990 [108])].

2.4   Formation and Derivation of the Smooth Test

Neyman (1937 [76]) derived a locally most powerful symmetric (regular) unbiased test (critical region) for $H_0 : \theta_1 = \theta_2 = \ldots = \theta_k = 0$ in (19), which he called an unbiased critical region of type C. This type-C critical region is an extension of the locally most powerful unbiased (LMPU) test (type-A region) of Neyman and Pearson (1936 [81]) from a single-parameter case to a multi-parameter situation. We first briefly describe the type-A test for testing $H_0 : \theta = \theta_0$ (where θ is a scalar) for local alternatives of the form $\theta = \theta_0 + \delta/\sqrt{n}, 0 < \delta < \infty$. Let $\beta(\theta)$ be the power function of the test. Then, assuming differentiability at $\theta = \theta_0$ and expanding $\beta(\theta)$ around $\theta = \theta_0$, we have

$$\beta(\theta) = \beta(\theta_0) + (\theta - \theta_0)\beta'(\theta_0) + \frac{1}{2}(\theta - \theta_0)^2 \beta''(\theta_0) + o(n^{-1})$$
$$= \alpha + \frac{1}{2}(\theta - \theta_0)^2 \beta''(\theta_0) + o(n^{-1}) \tag{27}$$

where α is the size of the test, and unbiasedness requires that the "power" should be minimum at $\theta = \theta_0$ and hence $\beta'(\theta_0) = 0$. Therefore, to maximize the local power we need to maximize $\beta''(\theta_0)$. This leads to the well-known LMPU test or the type A critical region. In other words, we can maximize $\beta''(\theta_0)$ subject to two side conditions, namely, $\beta(\theta_0) = \alpha$ and $\beta'(\theta_0) = 0$. These ideas are illustrated in Figure 5. For a locally optimal test, the power curve should have maximum curvature at the point C (where $\theta = \theta_0$), which is equivalent to minimizing distances such as the chord AB.

Using the generalized Neyman–pearson (N–P) lemma, the optimal (type-A) critical region is given by

$$\frac{d^2 L(\theta_0)}{d\theta^2} > k \frac{dL(\theta_0)}{d\theta} + k_2 L(\theta_0) \tag{28}$$

where $L(\theta) = \prod_{i=1}^{n} f(x_i; \theta)$ is the likelihood function, while the constants k_1 and k_2 are determined through the side conditions of size and local unbiasedness. The critical region in (28) can be expressed in terms of the derivatives of the log-likelihood function $l(\theta) = \ln(L(\theta))$ as

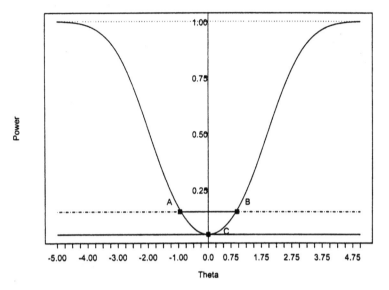

Figure 5. Power curve for one-parameter unbiased test.

$$\frac{d^2 l(\theta_0)}{d\theta^2} + \left[\frac{dl(\theta_0)}{d\theta}\right]^2 > k_1 \frac{dl(\theta_0)}{d\theta} + k_2 \tag{29}$$

If we denote the score function as $s(\theta) = dl(\theta)/d\theta$ and its derivative as $s'(\theta)$, then (29) can be written as

$$s'(\theta_0) + [s(\theta_0)]^2 > k_1 s(\theta_0) + k_2 \tag{30}$$

Neyman (1937 [76]) faced a more difficult problem since his test of $H_0 : \theta_1 = \theta_2 = \ldots = \theta_k = 1$ in (19) involved testing a parameter vector, namely, $\theta = (\theta_1, \theta_2, \ldots, \theta_k)'$. Let us now denote the power function as $\beta(\theta_1, \theta_2, \ldots, \theta_k) = \beta(\theta) \equiv \beta$. Assuming that the power function $\beta(\theta)$ is twice differentiable in the neighborhood of $H_0 : \theta = \theta_0$, Neyman (1937 [76], pp. 166–167) formally required that an unbiased critical region of type C of size α should satisfy the following conditions:

1. $\beta(0, 0, \ldots, 0) = \alpha$ $\tag{31}$

2. $\beta_j = \left.\frac{\partial \beta}{\partial \theta_j}\right|_{\theta=0} = 0, \ j = 1, \ldots, k$ $\tag{32}$

3. $\beta_{ij} = \left.\frac{\partial^2 \beta}{\partial \theta_i \partial \theta_j}\right|_{\theta=0} = 0, \ i, j = 1, \ldots, k, \ i \neq j$ $\tag{33}$

4. $\beta_{jj} = \left.\dfrac{\partial^2 \beta}{\partial \theta_j^2}\right|_{\theta=0} = \left.\dfrac{\partial^2 \beta}{\partial \theta_1^2}\right|_{\theta=0} \quad j = 2, \ldots, k.$ \hfill (34)

And finally, over all such critical regions satisfying the conditions (31)–(34), the common value of $\partial^2 \beta / \partial \theta_j^2|_{\theta=0}$ is the maximum.

To interpret the above conditions it is instructive to look at the $k = 2$ case. Here, we will follow the more accessible exposition of Neyman and Pearson (1938 [82]).[*]

By taking the Taylor series expansion of the power function $\beta(\theta_1, \theta_2)$ around $\theta_1 = \theta_2 = 0$, we have

$$\beta(\theta_1, \theta_2) = \beta(0, 0) + \theta_1 \beta_1 + \theta_2 \beta_2 + \frac{1}{2}(\theta_1^2 \beta_{11} + 2\theta_1 \theta_2 \beta_{12} + \theta_2^2 \beta_{22}) + o(n^{-1})$$

(35)

The type-C *regular* unbiased critical region has the following properties: (i) $\beta_1 = \beta_2 = 0$, which is the condition for any unbiased test; (ii) $\beta_{12} = 0$ to ensure that small positive and small negative deviations in the θs should be controlled *equally* by the test; (iii) $\beta_{11} = \beta_{22}$, so that equal departures from $\theta_1 = \theta_2 = 0$ have the same power in all directions; and (iv) the common value of β_{11} (or β_{22}) is maximized over all critical regions satisfying the conditions (i) and (iii). If a critical region satisfies only (i) and (iv), it is called a *non-regular* unbiased critical region of type C. Therefore, for a type-C regular unbiased critical region, the power function is given by

$$\beta(\theta_1, \theta_2) = \alpha + \frac{1}{2}\beta_{11}(\theta_1^2 + \theta_2^2)$$

(36)

As we can see from Figure 6, maximization of power is equivalent to the minimization of the area of the exposed circle in the figure. In order to find out whether we really have an LMPU test, we need to look at the second-order condition; i.e., the Hessian matrix of the power function $\beta(\theta_1, \theta_2)$ in (35) evaluated at $\theta = 0$,

$$B_2 = \begin{bmatrix} \beta_{11} & \beta_{12} \\ \beta_{12} & \beta_{22} \end{bmatrix}$$

(37)

should be positive definite, i.e., $\beta_{11}\beta_{22} - \beta_{12}^2 > 0$ should be satisfied.

---

[*]After the publication of Neyman (1937 [76]), Neyman in collaboration with Egon Pearson wrote another paper, Neyman and Pearson (1938 [82]), that included a detailed account of the unbiased critical region of type C. This paper belongs to the famous Neyman–Pearson series on the Contribution to the Theory of Testing Statistical Hypotheses. For historical sidelights on their collaboration see Pearson (1966 [85]).

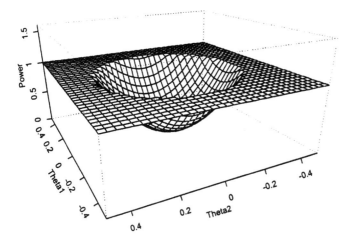

Figure 6. Power surface for two-parameter unbiased test

We should also note from (35) that, for the unbiased test,

$$\theta_1^2\beta_{11} + 2\theta_1\theta_2\beta_{12} + \theta_2^2\beta_2^2 = \text{constant} \tag{38}$$

represents what Neyman and Pearson (1938 [82], p. 39) termed the ellipse of *equidetectability*. Once we impose the further restriction of "regularity," namely, the conditions (ii) and (iii) above, the concentric *ellipses* of equidetectability become concentric circles of the form (see Figure 6)

$$\beta_{11}(\theta_1^2 + \theta_2^2) = \text{constant} \tag{39}$$

Therefore, the resulting power of the test will be a function of the distance measure $(\theta_1^2 + \theta_2^2)$; Neyman (1937 [76]) called this the symmetry property of the test.

Using the generalized N–P lemma, Neyman and Pearson (1938 [82], p. 41) derived the type-C unbiased critical region as

$$L_{11}(\theta_0) \geq k_1[L_{11}(\theta_0) - L_{22}(\theta_0)] + k_2L_{12}(\theta_0) + k_3L_1(\theta_0) + k_4L_2(\theta_0) + k_5L(\theta_0) \tag{40}$$

where $L_i(\theta) = \partial L(\theta)/\partial\theta_i$, $i = 1, 2$, $L_{ij}(\theta) = \partial^2 L(\theta)/\partial\theta_i\partial\theta_j$, $i,j = 1, 2$, and $k_i(1 = 1, 2, \ldots, 5)$ are constants determined from the size and the three side conditions (i)–(iii).

The critical region (40) can also be expressed in terms of the derivatives of the log-likelihood function $l(\theta) = \ln L(\theta)$. Let us denote $s_i(\theta) = \partial l(\theta)/\partial\theta_i$, $i = 1, 2$, and $s_{ij}(\theta) = \partial^2 l(\theta)/\partial\theta_i\partial\theta_j$, $i,j = 1, 2$. Then it is easy to see that

$$L_i(\theta) = s_i(\theta)L(\theta) \tag{41}$$

$$L_{ij}(\theta) = [s_{ij}(\theta) + s_i(\theta)s_j(\theta)]L(\theta) \tag{42}$$

where $i, j = 1, 2$. Using these, (40) can be written as

$$(1 - k_1)s_{11}(\theta_0) + k_1 s_{22}(\theta_0) - k_2 s_{12}(\theta_0) + (1 - k_1)s_1^2(\theta_0) +$$
$$k_1 s_2^2(\theta_0) - k_2 s_1(\theta_0)s_2(\theta_0) - k_3 s_1(\theta_0) - k_4 s_2(\theta_0) + k_5 \geq 0$$
$$\Rightarrow [s_{11}(\theta_0) - s_1^2(\theta_0)] - k_1[s_{11}(\theta_0) - s_{22}(\theta_0) + s_1^2(\theta_0)] -$$
$$k_2[s_1(\theta_0)s_2(\theta_0) + s_{12}(\theta_0)] - k_3 s_1(\theta_0) - k_4 s_2(\theta_0) + k_5 \geq 0 \tag{43}$$

When we move to the general multiple parameter case ($k > 2$), the analysis remains essentially the same. We will then need to satisfy Neyman's conditions (31)–(34). In the general case, the Hessian matrix of the power function evaluated at $\theta = \theta_0$ in equation (37) has the form

$$B_k = \begin{bmatrix} \beta_{11} & \beta_{12} & \cdots & \beta_{1k} \\ \beta_{12} & \beta_{22} & \cdots & \beta_{2k} \\ \cdots & \cdots & \cdots & \cdots \\ \beta_{1k} & \beta_{2k} & \cdots & \beta_{kk} \end{bmatrix} \tag{44}$$

Now for the LMPU test B_k should be positive definite; i.e., all the principal cofactors of this matrix should be positive. For this general case, it is hard to express the type-C critical region in a simple way as in (40) or (43). However, as Neyman (1937 [76]) derived, the resulting test procedure takes a very simple form given in the next theorem.

Theorem 2 (Neyman 1937) For large n, the type-C regular unbiased test (critical region) is given by

$$\Psi_k^2 = \sum_{j=1}^{k} u_j^2 \geq C_\alpha \tag{45}$$

where $u_j = (1\sqrt{n}) \sum_{i=1}^{n} \pi_j(y_i)$ and the critical point C_α is determined from $P[\chi_k^2 \geq C_\alpha] = \alpha$.

Neyman (1937 [76], pp. 186–190) further proved that the limiting form of the power function of this test is given by

$$\left(\frac{1}{\sqrt{2\pi}}\right)^k \int \cdots \int_{\sum u_i^2 \geq C_\alpha} e^{-\frac{1}{2}\sum_{j=1}^{k}(u_j - \theta_j)^2} du_1 du_2 ... du_k \tag{46}$$

In other words, under the alternative hypothesis $H_1 : \theta_j \neq 0$, at least for some $j = 1, 2, \ldots, k$, the test statistic Ψ_k^2 approaches a non-central χ_k^2 distribution with the non-centrality parameter $\lambda = \sum_{j=1}^{k} \theta_j^2$. From (36), we can also see that the power function for this general k case is

$$\beta(\theta_1, \theta_2, \ldots, \theta_k) = \alpha + \frac{1}{2}\beta_{11} \sum_{j=1}^{k} \theta_j^2 = \alpha + \beta_{11}\lambda \tag{47}$$

Since the power depends only on the "distance" $\sum_{j=1}^{k} \theta_j^2$ between H_0 and H_1, Neyman called this test *symmetric*.

Unlike Neyman's earlier work with Egon Pearson on general hypothesis testing, the smooth test went unnoticed in the statistics literature for quite some time. It is quite possible that Neyman's idea of explicitly deriving a test *statistic* from the very first principles under a very general framework was well ahead of its time, and its usefulness in practice was not immediately apparent.* Isaacson (1951 [56]) was the first notable paper that referred to Neyman's work while proposing the type-D unbiased critical region based on Gaussian or total curvature of the power hypersurface. However, D. E. Barton was probably the first to carry out a serious analysis of Neyman's smooth test. In a series of papers (1953a [3], 1955 [5], 1956 [6]), he discussed its small sample distribution, applied the test to discrete data, and generalized the test to some extent to the composite null hypothesis situation [see also Hamdan (1962 [48], 1964 [49]), Barton (1985 [7])]. In the next section we demonstrate that the smooth tests are closely related to some of the other more popular tests. For example, the Pearson χ^2 goodness-of-fit statistic can be derived as a special case of the smooth test. We can also derive Neyman's smooth test statistics Ψ^2 in a simple way using Rao's (1948 [93]) score test principle.

3. THE RELATIONSHIP OF NEYMAN'S SMOOTH TEST WITH RAO'S SCORE AND OTHER LOCALLY OPTIMAL TESTS

Rayner and Best (1989 [99]) provided an excellent review of smooth tests of various categorized and uncategorized data and related procedures. They also elaborated on many interesting, little-known results [see also Bera (2000 [10])]. For example, Pearson's (1900 [86]) P_{χ^2} statistic in (1) can be obtained as a Neyman's smooth test for a categorized hypothesis. To see this result, let us write the probability of the jth class in terms of our density (24) under the alternative hypothesis as

---

*Reid (1982 [100], p. 149) described an amusing anecdote. In 1937, W. E. Deming was preparing publication of Neyman's lectures by the United States Department of Agriculture. In his lecture notes Neyman misspelled *smooth* when referring to the smooth test. "I don't understand the reference to 'Smouth'," Deming wrote to Neyman, "Is that the name of a statistician?".

$$p_j = c(\theta) \exp\left[\sum_{i=1}^{r} \theta_i h_{ij}\right] \theta_{j0} \tag{48}$$

where θ_{j0} is the value p_j under the null hypothesis, $j = 1, 2, \ldots, q$. In (48), h_{ij} are values taken by a random variable H_i with $P(H_i = h_{ij}) = \theta_{j0}$, $j = 1, 2, \ldots, q; i = 1, 2, \ldots, r$. These h_{ij} are also orthonormal with respect to the probabilities θ_{j0}. Rayner and Best (1989 [99], pp. 57–60) showed that the smooth test for testing $H_0 : \theta_1 = \theta_2 = \ldots = \theta_r = 0$ is the same as the Pearson's P_{χ^2} in (1) with $r = q - 1$. Smooth-type tests can be viewed as a compromise between an omnibus test procedure such as Pearson's χ^2, which generally has low power in all directions, and more specific tests with power directed only towards certain alternatives.

Rao and Poti (1946 [98]) suggested a locally most powerful (LMP) one-sided test for the *one-parameter* problem. This test criterion is the precursor to Rao's (1948 [93]) celebrated score test in which the basic idea of Rao and Poti (1946 [98]) is generalized to the *multiparameter* and composite hypothesis cases. Suppose the null hypothesis is composite, like $H_0 : \delta(\theta) = 0$, where $\delta(\theta)$ is an $r \times 1$ vector function of $\theta = (\theta_1, \theta_2, \ldots, \theta_k)$ with $r \leq k$. Then the general form of Rao's score (RS) statistic is given by

$$\text{RS} = s(\tilde{\theta})' \mathcal{I}(\tilde{\theta})^{-1} s(\tilde{\theta}) \tag{49}$$

where $s(\theta)$ is the score function $\partial l(\theta)/\partial \theta$, $\mathcal{I}(\theta)$ is the information matrix $E[-\partial^2 l(\theta)/\partial\theta\partial\theta']$, and $\tilde{\theta}$ is the restricted maximum likelihood estimator (MLE) of θ [see Rao (1973 [95])]. Asymptotically, under H_0, RS is distributed as χ_r^2. Let us derive the RS test statistic for testing $H_0 : \theta_1 = \theta_2 = \cdots = \theta_k = 0$ in (24), so that the number of restrictions are $r = k$ and $\tilde{\theta} = 0$. We can write the log-likelihood function as

$$l(\theta) = \sum_{i=1}^{n} \ln g(x_i; \theta) = \sum_{i=1}^{n} \ln c(\theta) + \sum_{i=1}^{n}\sum_{j=1}^{k} \theta_j q_j(x_i) + \sum_{i=1}^{n} \ln f(x) \tag{50}$$

For the time being we ignore the nuisance parameter γ, and later we will adjust the variance of the RS test when γ is replaced by an efficient estimator $\tilde{\gamma}$.

The score vector and the information matrix under H_0 are given by

$$s(\tilde{\theta}) = n\frac{\partial \ln c(\theta)}{\partial \theta}\bigg|_{\theta=0} + \sum_{j=1}^{k}\sum_{i=1}^{n} q_j(x_i) \tag{51}$$

and

$$\mathcal{I}(\tilde{\theta}) = -n\frac{\partial^2 \ln c(\theta)}{\partial\theta\partial\theta'}\bigg|_{\theta=0} \tag{52}$$

respectively. Following Rayner and Best (1989 [99], pp. 77–80) and differentiating the identity $\int_{-\infty}^{\infty} g(x; \theta)dx = 1$ twice, we see that

$$\frac{\partial \ln c(\theta)}{\partial\theta_j} = -E_g[q_j(x)] \tag{53}$$

$$\frac{\partial^2 \ln c(\theta)}{\partial\theta_j\partial\theta_l} = -\int_{-\infty}^{\infty} q_j(x)\frac{\partial g(x; \theta)}{\partial\theta_l}dx, \quad j, l = 1, 2, \dots, k \tag{54}$$

where $E_g[.]$ is expectation taken with respect to the density under the alternative hypothesis, namely, $g(x; \theta)$. For the RS test we need to evaluate everything at $\theta =$. From (53) it is easy to see that

$$\frac{\partial \ln c(\theta)}{\partial\theta_j}\bigg|_{\theta=0} = 0 \tag{55}$$

and thus the score vector given in equation (51) simplifies to

$$s(\tilde{\theta}) = \sum_{j=1}^{k}\sum_{i=1}^{n} q_j(x_i) \tag{56}$$

From (24) we have

$$\frac{\partial \ln g(x; \theta)}{\partial\theta_l} = \frac{1}{g(x; \theta)}\frac{\partial g(x; \theta)}{\partial\theta_l} = \frac{\partial \ln c(\theta)}{\partial\theta_l} + q_l(x) \tag{57}$$

i.e.,

$$\frac{\partial g(x; \theta)}{\partial\theta_l} = \left[\frac{\partial \ln c(\theta)}{\partial\theta_l} + q_l(x)\right]g(x; \theta) \tag{58}$$

Hence, we can rewrite (54) and evaluate under H_0 as

$$\begin{aligned}
\frac{\partial^2 \ln c(\theta)}{\partial\theta_j\partial\theta_l} &= -\int_{-\infty}^{\infty} q_j(x)\left[\frac{\partial \ln c(\theta)}{\partial\theta_l} + q_l(x)\right]g(x; \theta)\,dx \\
&= -\int_{-\infty}^{\infty} q_j(x)[-E_g[q_l(x)] + q_l(x)]g(x; \theta)\,dx \\
&= \int_{-\infty}^{\infty} q_j(x)q_l(x)g(x; \theta)\,dx \\
&= -\text{Cov}_g(q_j(x), q_l(x)) \\
&= \delta_{jl}
\end{aligned} \tag{59}$$

where $\delta_{jl} = 1$ when $j = l$; $= 0$ otherwise. Then, from (52), $\mathcal{I}(\tilde{\theta}) = nI_k$, where I_k is a k-dimensional identity matrix. This also means that the asymptotic variance–covariance matrix of $(1/\sqrt{n})s(\tilde{\theta})$ will be

$$V\left[\frac{1}{\sqrt{n}}s(\tilde{\theta})\right] = I_k \tag{60}$$

Therefore, using (49) and (56), the RS test can be simply expressed as

$$RS = \sum_{i=1}^{k} \frac{1}{n}\left[\sum_{i=1}^{n} q_j(x_i)\right]^2 \tag{61}$$

which is the "same" as Ψ_k^2 in (25), the test statistic for Neyman's smooth test. To see clearly why this result holds, let us go back to the expression of Neyman's type-C unbiased critical region in equation (40). Consider the case $k = 2$; then, using (56) and (59), we can put $s_j(\theta_0) = \sum_{i=1}^{n} q_j(x_i)$, $s_{jj}(\theta_0) = 1$, $j = 1, 2$, and $s_{12}(\theta_0) = 0$. It is quite evident that the second-order derivatives of the log-likelihood function do not play any role. Therefore, Neyman's test must be based only on score functions $s_1(\theta)$ and $s_2(\theta)$ evaluated at the null hypothesis $\theta = \theta_0 = 0$.

From the above facts, we can possibly assert that Neyman's smooth test is the *first* formally derived RS test. Given this connection between the smooth and the score tests, it is not surprising that Pearson's goodness-of-fit test is nothing but a categorized version of the smooth test as noted earlier. Pearson's test is also a special case of the RS test [see Bera and Bilias (200a[11])]. To see the impact of estimation of the nuisance parameter γ [see equation (24)] on the RS statistic, let us use the result of Pierce (1982 [91]). Pierce established that, for a statistic $U(.)$ depending on parameter vector γ, the asymptotic variances of $U(\gamma)$ and $U(\tilde{\gamma})$, where $\tilde{\gamma}$ is an efficient estimator of γ, are related by

$$\mathrm{Var}\left[\sqrt{n}U(\tilde{\gamma})\right] = \mathrm{Var}\left[\sqrt{n}U(\gamma)\right] - \lim_{n\to\infty} E\left[\frac{\partial U(\gamma)}{\partial \gamma}\right]' \mathrm{Var}\left(\sqrt{n}\tilde{\gamma}\right)\lim_{n\to\infty} E\left[\frac{\partial U(\gamma)}{\partial \gamma}\right] \tag{62}$$

Here, $\sqrt{n}U(\tilde{\gamma}) = (1\sqrt{n})s(\tilde{\theta}, \tilde{\gamma}) = (1/\sqrt{n})\sum_{j=1}^{k}\sum_{i=1}^{n} q_j(x_i; \tilde{\gamma})$, $\mathrm{Var}\left[\sqrt{n}U(\gamma)\right] = I_k$ as in (60), and finally, $\mathrm{Var}\left(\sqrt{n}\tilde{\gamma}\right)$ is obtained from maximum likelihood estimation of γ under the null hypothesis. Furthermore, Neyman (1959 [77]) showed that

$$\lim_{n\to\infty} E\left[\frac{\partial U(\gamma)}{\partial \gamma}\right] = -\mathrm{Cov}\left[\sqrt{n}U(\gamma), \frac{1}{\sqrt{n}}\frac{\partial \ln f(x; \gamma)}{\partial \gamma}\right] = B(\gamma) \text{ (say)} \tag{63}$$

and this can be computed for the given density $f(x; \gamma)$ under the null hypothesis. Therefore, from (62), the adjusted formula for the score function is

$$V\left[\frac{1}{\sqrt{n}}s(\tilde{\theta}, \tilde{\gamma})\right] = I_k - B'(\gamma)\text{var}\,(\sqrt{n}\tilde{\gamma})B(\gamma) = V(\gamma) \text{ say} \qquad (64)$$

which can be evaluated simply by replacing γ by $\tilde{\gamma}$. From (60) and (62), we see that in some sense the variance "decreases" when the nuisance parameter is replaced by its efficient estimator. Hence, the final form of the score or the smooth test will be

$$\begin{aligned} \text{RS} &= \Psi^2 = \frac{1}{n}s(\tilde{\theta}, \tilde{\gamma})'V(\tilde{\gamma})^{-1}s(\tilde{\theta}, \tilde{\gamma}) \\ &= \frac{1}{n}s(\tilde{\gamma})'V(\tilde{\gamma})^{-1}(\tilde{\gamma}) \end{aligned} \qquad (65)$$

since for our case under the null hypothesis $\tilde{\theta} = 0$. In practical applications, $V(\tilde{\gamma})$ may not be of full rank. In that case a generalized inverse of $V(\tilde{\gamma})$ could be used, and then the degree of freedom of the RS statistic will be the rank of $V(\tilde{\gamma})$ instead of k. Rayner and Best (1989 [99], pp. 78–80) also derived the same statistic [see also Boulerice and Ducharme (1995 [19])]; however, our use of Pierce (1982 [91]) makes the derivation of the variance formula much simpler.

Needless to say, since it is based on the score principle, Neyman's smooth test will share the optimal properties of the RS test procedure and will be asymptotically locally most powerful.[*] However, we should keep in mind all the restrictions that conditions (33) and (34) imposed while deriving the test procedure. This result is not as straightforward as testing the *single parameter* case for which we obtained the LMPU test in (28) by maximizing the power function. In the *multiparameter* case, the problem is that, instead of a power function, we have a power *surface* (or a power *hypersurface*). An ideal test would be one that has a power surface with a maximum curvature along *every* cross-section at the point $H_0 : \theta = (0, 0, \ldots, 0)' = \theta_0$, say, subject to the conditions of size and unbiasedness. Such a test, however, rarely exists even for the simple cases. As Isaacson (1951 [56], p. 218) explained, if we maximize the curvature along one cross-section, it will generally cause the curvature to diminish along some other cross-section, and consequently the

---

[*]Recent work in higher-order asymptotics support [see Chandra and Joshi (1983 [21]), Ghosh (1991 [44], and Mukerjee (1993 [75])] the validity of Rao's conjecture about the optimality of the score test over its competitors under local alternatives, particularly in a locally asymptotically unbiased setting [also see, Rao and Mukerjee (1994 [96], 1997 [97])].

curvature cannot be maximized along *all* cross-sections simultaneously. In order to overcome this kind of problem Neyman (1937 [76]) required the type-C critical region to have constant power in the neighborhood of H_0 : $\theta = \theta_0$ along a given family of concentric ellipsoids. Neyman and Pearson (1938 [82]) called these the ellipsoids of equidetectability. However, one can only choose this family of ellipsoids if one knows the relative importance of power in different directions in an infinitesimal neighborhood of θ_0. Isaacson (1951 [56]) overcame this objection to the type-C critical region by developing a natural generalization of the Neyman–Pearson type-A region [see equations (28)–(30)] to the multiparameter case. He maximized the Gaussian (or total) curvature of the power surface at θ_0 subject to the conditions of size and unbiasedness, and called it the type-D region. Gaussian curvature of a function $z = f(x, y)$ at a point (x_0, y_0) is defined as [see Isaacson (1951 [56]), p. 219]

$$
G = \frac{\left.\frac{\partial^2 z}{\partial x^2}\right|_{(x_0,y_0)} \left.\frac{\partial^2 z}{\partial y^2}\right|_{(x_0,y_0)} - \left[\left.\frac{\partial^2 z}{\partial x \partial y}\right|_{(x_0,y_0)}\right]^2}{\left[1 + \left[\left.\frac{\partial z}{\partial x}\right|_{(x_0,y_0)}\right]^2 + \left[\left.\frac{\partial z}{\partial y}\right|_{(x_0,y_0)}\right]^2\right]^2}
\tag{66}
$$

Hence, for the two-parameter case, from (35), we can write the total curvature of the power hypersurface as

$$
G = \frac{\beta_{11}\beta_{22} - \beta_{12}^2}{[1 + 0 + 0]^2} = \det(B_2)
\tag{67}
$$

where B_2 is defined by (37). The Type-D unbiased critical region for testing $H_0 : \theta = 0$ against the two-sided alternative for a level α test is defined by the following conditions [see Isaacson (1951 [56], p. 220)]:

1. $\beta(0, 0) = \alpha$ (68)

2. $\beta_i(0, 0) = 0, \quad i = 1, 2$ (69)

3. B_2 is positive definite (70)

4. And, finally, over all such critical regions satisfying the conditions (68)–(70), det (B_2) is maximized.

Note that for the type-D critical region restrictive conditions like $\beta_{12} = 0$, $\beta_{11} = \beta_{22}$ [see equations (33)–(34)] are not imposed. The type-D critical region maximizes the *total* power

$$
\beta(\theta|\omega) \simeq \alpha + \frac{1}{2}[\theta_1^2\beta_{11} + 2\theta_1\theta_2\beta_{12} + \theta_2^2\beta_{22}]
\tag{71}
$$

among all locally unbiased (LU) tests, whereas the type-C test maximizes power only in "limited" directions. Therefore, for finding the type-D unbiased critical region we minimize the area of the ellipse (for the $k > 2$ case, it will be the volume of an ellipsoid)

$$\theta_1^2 \beta_{11} + 2\theta_1 \theta_2 \beta_{12} + \theta_2^2 \beta_{22} = \delta \tag{72}$$

which is given by

$$\frac{\pi \delta}{\sqrt{\begin{vmatrix} \beta_{11} & \beta_{12} \\ \beta_{12} & \beta_{22} \end{vmatrix}}} = \frac{\pi \delta}{\sqrt{\det(B_2)}} \tag{73}$$

Hence, maximizing the determinant of B_2, as in condition 4 above, is the same as minimizing the volume of the ellipse shown in equation (73). Denoting ω_0 as the type-D unbiased critical region, we can show that inside ω_0 the following is true [see Isaacson (1951 [56])]

$$k_{11}L_{11} + k_{12}L_{12} + k_{21}L_{21} + k_{22}L_{22} \geq k_1 L + k_2 L_1 + k_3 L_2 \tag{74}$$

where $k_{11} = \int_{\omega_0} L_{22}(\theta) dx$, $k_{22} = \int_{\omega_0} L_{11}(\theta) dx$, $k_{12} = k_{21} = -\int_{\omega_0} L_{12}(\theta) dx$, $x = (x_1, x_2, \ldots, x_n)'$ denotes the sample, and k_1, k_2, and k_3 are constants satisfying the conditions for size and unbiasedness (68) and (69).

However, one major problem with this approach, despite its geometric attractiveness, is that one has to *guess* the critical region and then verify it. As Isaacson (1951 [56], p. 223) himself noted, "we must know our region ω_0 in advance so that we can calculate k_{11} and k_{22} and thus verify whether ω_0 has the structure required by the lemma or not." The type-E test suggested by Lehman (1959 [71], p. 342) is the same as the type-D test for testing a composite hypothesis.

Given the difficulties in finding the type-D and type-E tests in actual applications, SenGupta and Vermeire (1986 [102]) suggested a locally most mean powerful unbiased (LMMPU) test that maximizes the *mean* (instead of total) curvature of the power hypersurface at the null hypothesis among all LU level α tests. This average power criterion maximizes the *trace* of the matrix B_2 in (37) [or B_k in (44) for the $k > 2$ case]. If we take an eigenvalue decomposition of the matrix B_k relating to the power function, the eigenvalues, λ_i, give the principal power curvatures while the eigenvectors corresponding to them give the principal power directions. Isaacson (1951 [56]) used the determinant, which is the *product* of the eigenvalues, wheras SenGupta and Vermeire (1986 [102]) used their *sum* as a measure of curvature. Thus, LMMPU critical regions are more easily constructed using just the generalized N–P lemma. For testing $H_0 : \theta = \theta_0$ against $H_1 : \theta \neq \theta_0$, an LMMPU critical region for the $k = 2$ case is given by

$$s_{11}(\theta_0) + s_{22}(\theta_0) + s_1^2(\theta_0) + s_2^2(\theta_0) \geq k + k_1 s_1(\theta_0) + k_2 s_2(\theta_0) \tag{75}$$

where k, k_1 and k_2 are constants satisfying the size and unbiasedness conditions (68) and (69). It is easy to see that (75) is very close to Neyman's type-C region given in (43). It would be interesting to derive the LMMPU test and also the type-D and type-E regions (if possible) for testing $H_0 : \theta = 0$ in (24) and to compare that with Neyman's smooth test [see Choi et al. (1996 [23])]. We leave that topic for future research. After this long discussion of theoretical developments, we now turn to possible applications of Neyman's smooth test.

4. APPLICATIONS

We can probably credit Lawrence Klein (Klein 1991 [62], pp. 325–326) for making the first attempt to introduce Neyman's smooth test to econometrics. He gave a seminar on "Neyman's Smooth Test" at the 1942–43 MIT statistics seminar series.* However, Klein's effort failed, and we do not see any direct use of the smooth test in econometrics. This is particularly astonishing as testing for possible misspecification is central to econometrics. The particular property of Neyman's smooth test that makes it remarkable is the fact that it can be used very effectively both as an omnibus test for detecting departures from the null in several directions and as a more directional test aimed at finding out the exact nature of the departure from H_0 of correct specification of the model.

Neyman (1937 [76], pp. 180–185) himself illustrated a practical application of his test using Mahalanobis's (1934 [73]) data on normal deviates with $n = 100$. When mentioning this application, Rayner and Best (1989 [99], pp. 46–47) stressed that Neyman also reported the *individual* components of the Ψ_k^2 statistic [see equation (45)]. This shows that Neyman (1937 [76]) believed that more specific directional tests identifying the cause and nature of deviation from H_0 can be obtained from these components.

---

*Klein joined the MIT graduate program in September 1942 after studying with Neyman's group in statistics at Berkeley, and he wanted to draw the attention of econometricians to Neyman's paper since it was published in a rather recondite journal. This may not be out of place to mention that Trygve Haavelmo was also very much influenced by Jerzy Neyman, as he mentioned in his Nobel prize lecture (see Haavelmo 1997 [47])

> I was lucky enough to be able to visit the United States in 1939 on a scholarship ... I then had the privilege of studying with the world famous statistician Jerzy Neyman in California for a couple of months.

Haavelmo (1944 [45]) contains a seven-page account of the Neyman–Pearson theory.

4.1   Orthogonal Polynomials and Neyman's Smooth Test

Orthogonal polynomials have been widely used in estimation problems, but their use in hypothesis testing has been very limited at best. Neyman's smooth test, in that sense, pioneered the use of orthogonal polynomials for specifying the density under the alternative hypothesis. However, there are two very important concerns that need to be addressed before we can start a full-fledged application of Neyman's smooth test. First, Neyman used normalized Legendre polynomials to design the "smooth" alternatives; however, he did not justify the use of those over other orthogonal polynomials such as the truncated Hermite polynomials or the Laguerre polynomials (Barton 1953b [4], Kiefer 1985 [60]) or Charlier Type B polynomials (Lee 1986 [70]). Second, he also did not discuss how to choose the number of orthogonal polynomials to be used.* We start by briefly discussing a general model based on orthonormal polynomials and the associated smooth test. This would lead us to the problem of choosing the optimal value of k, and finally we discuss a method of choosing an alternate sequence of orthogonal polynomials.

We can design a smooth-type test in the context of regression model (Hart 1997 [52], Ch. 5)

$$Y_i = r(x_i) + \varepsilon_i, \quad i = 1, 2, \ldots, n \tag{76}$$

where Y_i is the dependent variable and x_i are fixed design points $0 < x_2 < x_2 < \cdots < x_n < 1$, and ε_i are IID $(0, \sigma2)$. We are interested in testing the "constant regression" or "no-effect" hypothesis, i.e., $r(x) = \theta_0$, where θ_0 is an unknown constant. In analogy with Neyman's test, we consider an alternative of the form (Hart 1997 [52], p. 141)

$$r(x) = \theta_0 + \sum_{j=1}^{k} \theta_j \phi_{j,n}(x) \tag{77}$$

where $\phi_{1,n}(x), \ldots, \phi_{k,n}(x)$ are orthonormal over the domain of x,

---

*Neyman (1937 [76], p. 194) did not discuss in detail the choice of the value k and simply suggested:

> My personal feeling is that in most practical cases, there will be no need to go beyond the fourth order test. But this is only an *opinion* and not any mathematical result.

However, from their experience in using the smooth test, Thomas and Pierce (1979 [111], p. 442) thought that for the case of composite hypothesis $k = 2$ would be a better choice.

$$\frac{1}{n}\sum_{q=1}^{n} \phi_{i,n}(x_q)\phi_{j,n}(x_q) = \begin{cases} 1, & \text{if } i=j \\ 0, & \text{if } i \neq j \end{cases} \tag{78}$$

and $\phi_{0,n} \equiv 1$. Hence, a test for $H_0 : \theta_1 = \theta_2 = \cdots = \theta_k = 0$ against $H_1 : \theta_i \neq 0$, for some $i = 1, 2, \ldots, k$ can be done by testing the overall significance of the model given in (76). The least-square estimators of θ_j are given by

$$\hat{\theta}_j = \frac{1}{n}\sum_{i=1}^{n} Y_i \phi_{j,n}(x_i), \ j = 0, \ldots, k \tag{79}$$

A test, which is asymptotically true even if the errors are not exactly Gaussian (so long as they have the same distribution and have a constant variance σ^2), is given by

$$\text{Reject } H_0 \text{ if } T_{N,k} = \frac{n\sum_{j=1}^{k}\hat{\theta}_j^2}{\hat{\sigma}^2} \geq c \tag{80}$$

where $\hat{\sigma}$ is an estimate of σ, the standard deviation of the error terms. We can use any set of orthonormal polynomials in the above estimator including, for example, the normalized Fourier series $\phi_{j,n}(x) = \sqrt{2n}\cos(\pi jx)$ with Fourier coefficients

$$\theta_j = \int_0^1 r(x)\sqrt{2n}\cos(\pi jx)\,dx \tag{81}$$

Observing the obvious similarity in the hypothesis tested, the test procedure in (80) can be termed as a Neyman smooth test for regression (Hart 1997 [52], p. 142).

The natural question that springs to mind at this point is what the value of k should be. Given a sequence of orthogonal polynomials, we can also test for the number of orthogonal polynomials, say k, that would give a desired level of "goodness-of-fit" for the data. Suppose now the sample counterpart of θ_j, defined above, is given by $\hat{\theta}_j = (1/n)\sum_{i=1}^{n} Y_i\sqrt{2n}\cos(\pi jx_i)$. If we have an IID sample of size n, then, given that $E(\hat{\theta}_j) = 0$ and $V(\hat{\theta}_j) = \sigma^2/2n$, let us *normalize* the sample Fourier coefficients using $\hat{\sigma}$, a consistent estimator of σ. Appealing to the central limit theorem for sufficiently large n, we have the test statistic

$$S_k = \sum_{j=1}^{k}\left(\frac{\sqrt{2n}\hat{\theta}_j}{\hat{\sigma}}\right)^2 \overset{a}{\sim} \chi_k^2 \tag{82}$$

for fixed $k \leq n-1$; this is nothing but the Neyman smooth statistic in equation (45) for the Fourier series polynomials.

The optimal choice of k has been studied extensively in the literature of data-driven smooth tests first discussed by Ledwina (1994 [68]) among others. In order to reduce the subjectivity of the choice of k we can use a criterion like the Schwarz information criterion (SIC) or the Bayesian information criterion (BIC). Ledwina (1994 [68]) proposed a test that rejects the null hypothesis that k is equal to 1 for large values of $S_{\bar{k}} = \max_{1 \leq k \leq n-1}\{S_k - k\ln(n)\}$, where S_k is defined in (82). She also showed that the test statistic $S_{\bar{k}}$ asymptotically converges to χ_1^2 random variable [for further insight into data-driven smooth tests see, for example, Hart (1997 [52]), pp. 185–187].

For testing uniformity, Solomon and Stephens (1985 [105]) and D'Agostino and Stephens (1986 [31], p. 352) found that $k = 2$ is optimal in most cases where the location-scale family is used; but $k = 4$ might be a better choice when higher-order moments are required. As mentioned earlier, other papers, including Neyman (1937 [76]) and Thomas and Pierce (1979 [111]), suggested using small values of k. It has been suggested in the literature that, for heavier-tailed alternative distributions, it is better to have more classes for Pearson's P_{χ^2} test in (1) or, equivalently, in the case of Neyman's smooth test, more orthogonal polynomials (see, for example, Kallenberg et al. 1985 [59]). However, they claimed that too many class intervals can be a potential problem for lighter-tailed distributions like normal and some other exponential family distributions (Kallenberg et al. 1985 [59], p. 959). Several studies have discussed cases where increasing the order of the test k slowly to ∞ would have better power for alternative densities having heavier tails (Kallenberg et al. 1985 [59], Inglot et al. 1990 [54], Eubank and LaRiccia 1992 [38], Inglot et al. 1994 [55]).

Some other tests, such as the Cramér-von Mises (CvM) and the Kolmogorov–Smirnov (KS) approaches, are examples of omnibus test procedures that have power against various directions, and hence those tests will be consistent against many alternatives (see Eubank and LaRiccia 1992 [38], p. 2072). The procedure for selecting the truncation point k in the Neyman (1937 [76]) smooth test is similar to the choice of the number of classes in the Pearson χ^2 test and has been discussed in Kallenberg et al. (1985 [59]) and Fan (1996 [39]).

Let us now revisit the problem of choosing an optimal sequence of orthogonal polynomials around the density $f(x; \gamma)$ under H_0. The following discussion closely follows Smith (1989 [103]) and Cameron and Trivedi (1990 [20]). They used the score test after setting up the alternative in terms of orthogonal polynomials with the baseline density $f(x; \gamma)$ under the null hypothesis. Expanding the density $g(x; \gamma, \theta)$ using an orthogonal polynomial sequence with respect to $f(x; \gamma)$, we have

$$g(x; \gamma, \theta) = f(x; \gamma) \sum_{j=0}^{\infty} a_j(\gamma, \theta) p_j(x; \gamma) \tag{83}$$

where

$$a_0(\gamma, \theta) \equiv 1, p_0(x; \gamma) \equiv 1, p_j(x; \gamma) = \sum_{i=0}^{j} \alpha_{ij} x^i \tag{84}$$

The polynomials p_j are orthonormal with respect to density $f(x; \gamma)$.

We can construct orthogonal polynomials through the moments. Suppose we have a sequence of moments $\{\mu_n\}$ of the random variable X, then a necessary and sufficient condition for the existence of a unique orthogonal polynomial sequence is that det $(M_n) > 0$, where $M_n = [M_{ij}] = [\mu_{i+j-2}]$, for $n = 0, 1, \ldots$. We can write $\det(M_n)$ as

$$|M_n| = \begin{vmatrix} M_{n-1} & \mathbf{m} \\ \mathbf{m}' & \mu_{2n} \end{vmatrix} = \mu_{2n}|M_{n-1}| - \mathbf{m}'\mathrm{Adj}(M_{n-1})\mathbf{m} \tag{85}$$

where $\mathbf{m}' = (\mu_n, \mu_{n+1}, \ldots, \mu_{2n-1})$, $|M_{-1}| = |M_0| = 1$, and "Adj" means the adjugate of a matrix. The nth-order orthogonal polynomial can be constructed from

$$P_n(x) = [|M_{n-1}|]^{-1}|D_n(x)|,$$

$$\text{where } |D_n(x)| = \begin{bmatrix} M_{n-1} & \mathbf{m} \\ \mathbf{x}'_{(-n)} & x^n \end{bmatrix} \text{ and } \mathbf{x}'_{(-n)} = (1 \ x \ x^2 \ldots x^{n-1}) \tag{86}$$

This gives us a whole system of orthogonal polynomials $P_n(x)$ [see Cramér (1946 [28]), pp. 131–132, Cameron and Trivedi (1990 [20]), pp. 4–5 and Appendix A].

Smith (1989 [103]) performed a test of $H_0 : g(x; \gamma, \theta) = f(x; \gamma)$ (i.e., $a_j(\gamma, \theta) = 0, j = 1, 2, \ldots, k$ or $\mathbf{a}_k = \{a_j(\gamma, \theta)\}_{j=1}^{k} = 0$) using a truncated version of the expression for the alternative density,

$$g(x; \gamma, \theta) = f(x; \gamma)\left\{1 + \sum_{j=1}^{k} a_j(\gamma, \theta) \sum_{i=1}^{j} \alpha_{ij}(\gamma)[x^i - \mu_{fi}(\gamma)]\right\} \tag{87}$$

where

$$\mu_{fi}(\gamma) = \int x^i f(x; \gamma) \, dx = E_f[x^i] \tag{88}$$

However, the expression $g(x; \gamma, \theta)$ in (87) may not be a proper density function. Because of the truncation, it may not be non-negative for all values of x nor will it integrate to unity. Smith referred to $g(x; \gamma, \theta)$ as a *pseudo-density* function.

If we consider y to be the probability integral transform of the original data in x, then, defining $E_h(y^i|H_0) = \mu_{h_0 i}$, we can rewrite the above density in (87), in the absence of any nuisance parameter γ, as

$$h(y; \theta) = 1 + \sum_{j=1}^{k} \theta_j \sum_{i=1}^{j} \alpha_{ij} [y^i - \mu_{h_0 i}] \tag{89}$$

From this we can get the Neyman smooth test as proposed by Thomas and Pierce (1979 [111]). Here we test $H_0 : \theta = 0$ against $H_1 : \theta \neq 0$, where $\theta = (\theta_1, \theta_2, \dots, \theta_k)'$. From equation (89), we can get the score function as $\partial \ln h(y; \theta)/\partial \mathbf{a} = \mathbf{A}_k \mathbf{m}_k$, where $\mathbf{A}_k = [\alpha_{ij}(\theta)]$ is $k \times k$ lower triangular matrix (with non-zero diagonal elements) and $\mathbf{m}_k$ is a vector of deviations whose ith component is $(y^i - \mu_{h_0 i})$. The score test statistic will have the form

$$S_n = n\mathbf{m}'_{kn}[\mathbf{V}_n]^- \mathbf{m}_{kn} \tag{90}$$

where $\mathbf{m}_{kn}$ is the vector of the sample mean of deviations, $\mathbf{V}_n = \mathcal{I}_{mm} - \mathcal{I}_{m\theta} \mathcal{I}_{\theta\theta}^{-1} \mathcal{I}_{\theta m}$, with $\mathcal{I}_{mm} = E_\theta[\mathbf{m}_k \mathbf{m}'_k]$, $\mathcal{I}_{m\theta} = E_\theta[\mathbf{m}_k \mathbf{s}'_\theta]$, $\mathcal{I}_{m\theta} = E_\theta$ $[\mathbf{s}_\theta \mathbf{s}'_\theta]$, and $\mathbf{s}_\theta = E_\theta[\partial \ln h(y; \theta)/\partial \theta]$, is the conditional variance–covariance matrix and $[\mathbf{V}_n]^-$ is its g-inverse (see Smith 1989 [103], pp. 184–185 for details). Here $E_\theta[.]$ denotes expectation taken with respect to the true distribution of y but eventually evaluated under $H_0 : \theta = 0$. Test statistic (90) can also be computed using an artificial regression of the vector of 1's on the vector of score functions of the nuisance parameters and the deviations from the moments. It can be shown that S_n follows an asymptotic χ^2 distribution with degrees of freedom = rank $(\mathbf{V}_n)$. Possible uses could be in limited dependent variable models like the binary response model and models for duration such as unemployment spells (Smith 1989 [103]). Cameron and Trivedi (1990 [20] derived an analogous test using moment conditions of the exponential family. For testing exponentiality in the context of duration models, Lee (1984 [69]) transformed the "exponentially distributed" random variable X by $z = \Phi^{-1}[F(x)]$, where F is the exponential distribution function and Φ^{-1} is the inverse normal probability integral transform. Lee then proposed testing normality of z using the score test under a Pearson family of distributions as the alternative density for z. If we restrict to the first four moments in Smith (1989 [103]), then the approaches of Lee and Smith are identical.

4.2 Density Forecast Evaluation and Calibration

The importance of density forecast evaluation in economics has been aptly depicted by Crnkovic and Drachman (1997 [30], p. 47) as follows: "At the heart of market risk measurement is the forecast of the probability density functions (PDFs) of the relevant market variables ... a forecast of a PDF is the central input into any decision model for asset allocation and/or hedging ... therefore, the quality of risk management will be considered synonymous with the quality of PDF forecasts." Suppose that we have time series data (say, the daily returns to the S&P 500 Index) given by $\{x_t\}_{t=1}^m$. One of the most important questions that we would like to answer is: what is the sequence of the true density functions $\{g_t(x_t)\}_{t=1}^m$ that generated this particular realization of the data? Since this is time series data, at time t we know all the past values of x_t up to time t or the *information set* at time t, namely, $\Omega_t = \{x_{t-1}, x_{t-2}, \ldots\}$. Let us denote the one-step-ahead forecast of the sequence of densities as $\{f_t(x_t)\}$ conditional on $\{\Omega_t\}$. Our objective is to determine how much the forecast density $\{f_t\}$ depicts the true density $\{g_t\}$. The main problem in performing such a test is that both the actual density $g_t(.)$ and the one-step-ahead predicted density $f_t(.)$ could depend on the time t and thus on the information set Ω_t. This problem is unique since, on one hand, it is a classical goodness-of-fit problem but, on the other, it is also a combination of several different, possibly dependent, goodness-of-fit tests. One approach to handling this particular problem would be to reduce it to a more tractable one in which we have the same, or similar, hypotheses to test, rather than a host of different hypotheses. Following Neyman (1937 [76]), this is achieved using the probability integral transform

$$y_t = \int_{-\infty}^{x_t} f_t(u)du \tag{91}$$

Using equations (3), (6), and (17), the density function of the transformed variable y_t is given by

$$h_t(y_t) = 1, \ 0 < y_t < 1 \tag{92}$$

under the null hypothesis that our forecasted density is the true density for all t, i.e., $H_0 : g_t(.) = f_t(.)$.

If we are only interested in performing a goodness-of-fit test that the variable y_t follows a uniform distribution, we can use a parametric test like Pearson's χ^2 on grouped data or nonparametric tests like the KS or the CvM or a test using the Kuiper statistics (see Crnkovic and Drachman 1997 [30], p. 48). Any of those suggested tests would work as a good *omnibus* test of goodness-of-fit. If we fail to reject the null hypothesis we can conclude that there is not enough evidence that the data is *not* generated from the forecasted density $f_t(.)$; however, a rejection would not throw any light on the possible form of the true density function.

Diebold et al. (1998 [33]) used Theorem 1, discussed in Section 2.1, and tested $H_0 : g_t(.) = f_t(.)$ by checking whether the probability integral transform y_t in (91) follows IID $U(0, 1)$. They employed a graphical (visual) approach to decide on the structure of the alternative density function by a two-step procedure. First, they visually inspected the histogram of y_t to see if it comes from $U(0, 1)$ distribution. Then, they looked at the individual correlograms of each of the first four powers of the variable $z_t = y_t - 0.5$ in order to check for any residual effects of bias, variance, or higher-order moments. In the absence of a more analytical test of goodness-of-fit, this graphical method has also been used in Diebold et al. (1999 [36]) and Diebold et al. (1999 [34]). For reviews on density forecasting and forecast evaluation methods, see Tay and Wallis (2000 [110]), Christoffersen (1998 [24]) and Diebold and Lopez (1996 [35]). The procedure suggested is very attractive due to its simplicity of execution and intuitive justification; however, the resulting size and power of the procedure are unknown. Also, we are not sure about the optimality of such a diagnostic method. Berkowitz (2000 [17], p. 4) commented on the Diebold et al. (1998 [33]) procedure: "Because their interest centers on developing tools for diagnosing *how* models fail, they do not pursue formal testing." Neyman's smooth test (1937 [76]) provides an *analytic* tool to determine the structure of the density under the alternative hypothesis using orthonormal polynomials (normalized Legendre polynomials) $\pi_j(y)$ defined in (20).* While, on one hand, the smooth test provides a basis for a classical goodness-of-fit test (based on the generalized N–P lemma), on the other hand, it can also be used to determine the sensitivity of the power of the test to departures from the null hypothesis in different directions, for example, deviations in variance, skewness, and kurtosis (see Bera and Ghosh 2001 [14]). We can see that the Ψ_k^2 statistic for Neyman's smooth test defined in equation (22) comprises k components of the form $(1/n) (\sum_{i=1}^{n} \pi_j(y_i))^2, j = 1, \ldots k$, which are nothing but the squares of the efficient score functions. Using Rao and Poti (1946 [98]), Rao (1948 [93]), and Neyman (1959 [77]), one can risk the "educated speculation" that an *optimal test* should be based on the *score function* [for more on this, see Bera and Bilias (2001a [11], 2001b [12])]. From that point of view we achieve *optimality* using the smooth test.

---

*Neyman (1937 [76]) used $\pi_j(y)$ as the orthogonal polynomials which can be obtained by using the following conditions,

$$\pi_j(y) = a_{j0} + a_{j1}g + \ldots + a_{jj}y^j, a_{jj} \neq 0$$

given the restrictions of orthogonality given in Section 2.2. Solving these, the first five $\pi_j(y)$ are (Neyman 1937 [76], pp. 163–164)

$$\pi_0(y) = 1$$

$$\pi_1(y) = \sqrt{12}y - \tfrac{1}{2})$$

$$\pi_2(y) = \sqrt{5}(6(y - \tfrac{1}{2})^2 - \tfrac{1}{2})$$

$$\pi_3(y) = \sqrt{7}(20(y - \tfrac{1}{2})^3 - 3(y - \tfrac{1}{2}))$$

$$\pi_4(y) = 210(y - \tfrac{1}{2})^4 - 45(y - \tfrac{1}{2})^2 + \tfrac{9}{8}$$

Neyman's smooth-type test can also be used in other areas of macroeconomics such as evaluating the density forecasts of realized inflation rates. Diebold et al. (1999 [36]) used a graphical technique as did Diebold et al. (1993 [33]) on the density forecasts of inflation from the *Survey of Professional Forecasters*. Neyman's smooth test in its original form was intended mainly to provide an *asymptotic test* of significance for testing goodness-of-fit for "smooth" alternatives. So, one can argue that although we have large enough data in the daily returns of the S&P 500 Index, we would be hard pressed to find similar size data for macroeconomic series such as GNP, inflation. This might make the test susceptible to significant small-sample fluctuations, and the results of the test might not be strictly valid. In order to correct for size or power problems due to small sample size, we can either do a size correction [similar to other score tests, see Harris (1985 [50]), Harris (1987 [51]), Cordeiro and Ferrari (1991 [25]), Cribari-Neto and Ferrari (1995 [29]), and Bera and Ullah (1991 [16]) for applications in econometrics] or use a modified version of the "smooth test" based on Pearson's P_λ test discussed in Section 2.1. This promises to be an interesting direction for future research.

We can easily extend Neyman's smooth test to a multivariate setup of dimension N for m time periods, by taking a combination of Nm sequences of univariate densities as discussed by Diebold et al. (1999 [34]). This could be particularly useful in fields like financial risk management to evaluate densities for high-frequency financial data such as stock or derivative (options) prices and foreign exchange rates. For example, if we have a sequence of the joint density forecasts of more than one, say three, daily foreign exchange rates over a period of 1000 days, we can evaluate its accuracy using the smooth test for 3000 univariate densities. One thing that must be mentioned is that there could be both temporal and contemporaneous dependencies in these observations; we are assuming that taking conditional distribution both with respect to time and across variables is feasible (see, for example, Diebold et al. 1999 [34], p. 662).

Another important area of the literature on the evaluation of density forecasts is the concept of *calibration*. Let us consider this in the light of our formulation of Neyman's smooth test in the area of density forecasts. Suppose that the actual density of the process generating our data, $g_t(x_t)$, is different from the forecasted density, $f_t(x_t)$, say,

$$g_t(x_t) = f_t(x_t)r_t(y_t) \tag{93}$$

where $r_t(y_t)$ is a function depending on the probability integral transforms and can be used to calibrate the forecasted densities, $f_t(x_t)$, recursively. This procedure of calibration might be needed if the forecasts are off in a consistent way, that is to say, if the probability integral transforms $\{y_t\}_{t=1}^m$ are not $U(0, 1)$ but are independent and identically distributed with some other distribution (see, for example, Diebold et al. 1999 [34]).

If we compare equation (93) with the formulation of the smooth test given by equation (24), where $f_t(x)$, the density under H_0, is embedded in $g_t(x)$ (in the absence of nuisance parameter γ), the density under H_1, we can see that

$$
r_t(y_{t+1}) = c(\theta) \exp\left[\sum_{j=1}^{k} \theta_j \pi_j(y_{t+1})\right]
$$

$$
\Leftrightarrow \ln r_t(y_{t+1}) = \ln c(\theta) + \sum_{j=1}^{k} \theta_j \pi_j(y_{t+1})
$$

(94)

Hence, we can actually estimate the calibrating function from (94). It might be worthwhile to compare the method of calibration suggested by Diebold et al. (1999 [34]) using nonparametric (kernel) density estimation with the one suggested here coming from a parametric setup [also see Thomas and Pierce (1979 [111]) and Rayner and Best (1989 [99], p. 77) for a formulation of the alternative hypothesis].

So far, we have discussed only one aspect of the use of Neyman's smooth test, namely, how it can be used for evaluating (and calibrating) density forecast estimation in financial risk management and macroeconomic time-series data such as inflation. Let us now discuss another example that recently has received substantial attention, namely, the Value-at-Risk (VaR) model in finance. VaR is generally defined as an extreme quantile of the value distribution of a financial portfolio. It measures the maximum allowable value the portfolio can lose over a period of time at, say, the 95% level. This is a widely used measure of portfolio risk or exposure to risk for corporate portfolios or asset holdings [for further discussion see Smithson and Minton (1997 [104])]. A common method of calculating VaR is to find the proportion of times the upper limit of interval forecasts has been exceeded. Although this method is very simple to compute, it requires a large sample size (see Kupiec 1995 [65], p. 83). For smaller sample size, which is common in risk models, it is often advisable to look at the entire probability density function or a map of quantiles. Hypothesis tests on the goodness-of-fit of VaRs could be based on the tail probabilities or tail expected loss of risk models in terms of measures of "exceedence" or the number of times that the total loss has exceeded the forecasted VaR. The tail probabilities are often of more concern than the interiors of the density of the distribution of asset returns.

Berkowitz (2000 [17]) argued that in some applications highly specific testing guidelines are necessary, and, in order to give a more formal test for the graphical procedure suggested by Diebold et al. (1998 [33]), he proposed a formal likelihood ratio test on the VaR model. An advantage of his proposed test is that it gives some indication of the nature of the violation when the goodness-of-fit test is rejected. Berkowitz followed Lee's (1984 [69]) approach but used the likelihood ratio test (instead of the score test) based on the inverse standard normal transformation of the probability integral transforms of the data. The main driving forces behind the proposed test are its tractability and the properties of the normal distribution. Let us define the inverse standard normal transform $z_t = \Phi^{-1}(\hat{F}(y_t))$ and consider the following model

$$z_t - \mu = \rho(z_{t-1} - \mu) + \varepsilon_t \tag{95}$$

To test for independence, we can test $H_0 : \rho = 0$ in the presence of nuisance parameters μ and σ^2 (the constant variance of the error term ε_t). We can also perform a joint test for the parameters $\mu = 0$, $\rho = 0$, and $\sigma^2 = 1$, using the likelihood ratio test statistic

$$\text{LR} = -2(l(0, 1, 0) - l(\hat{\mu}, \hat{\sigma}^2, \hat{\rho})) \tag{96}$$

that is distributed as a χ^2 with three degrees of freedom, where $l(\theta) = \ln L(\theta)$ is the log-likelihood function. The above test can be considered a test based on the tail probabilities. Berkowitz (2000 [17]) reported Monte Carlo simulations for the Black–Scholes model and demonstrated superiority of his test with respect to the KS, CvM, and a test based on the Kuiper statistic. It is evident that there is substantial similarity between the test suggested by Berkowitz and the smooth test; the former explicitly puts in the conditions of higher-order moments through the inverse standard Gaussian transform, while the latter looks at a more general exponential family density of the form given by equation (19). Berkowitz exploits the properties of the normal distribution to get a likelihood ratio test, while Neyman's smooth test is a special case of Rao's score test, and therefore, asymptotically, they should give similar results.

To further elaborate, let us point out that finding the distributions of VaR is equivalent to finding the distribution of quantiles of the asset returns. LaRiccia (1991 [67]) proposed a quantile function-based analog of Neyman's smooth test. Suppose we have a sample $(y_1, y_2, \ldots, y_n)$ from a fully specified cumulative distribution function (cdf) of a location-scale family $G(.; \mu, \sigma)$ and define the order statistics as $\{y_{1n}, y_{2n}, \cdots, y_{nn}\}$. We want to test the null hypothesis that $G(.; \mu, \sigma) \equiv F(.)$ is the true data-generating process. Hence, under the null hypothesis, for large sample size n, the expected value of the ith-order statistic, Y_{in}, is given by $E(Y_{in}) = \mu + \sigma Q_0 [i/(n+1)]$, where $Q_0(u) = \inf\{y : F(y) \geq u\}$ for $0 < u < 1$. The covariance matrix under the null hypothesis is approximated by

$$\sigma_{ij} = \text{Cov}(Y_{in}, Y_{jn}) \approx \sigma^{-2} \left[fQ_0\left(\frac{i}{n+1}\right) fQ_0\left(\frac{j}{n+1}\right) \right]$$
$$\times \left[\min\left(\frac{i}{n+1}, \frac{j}{n+1}\right) - \frac{ij}{(n+1)^2} \right] \tag{97}$$

where $fQ_0(.) \equiv f(Q_0(.))$ is the density of the quantile function under H_0. LaRiccia took the alternative model as

$$E(Y_{in}) \simeq \mu + \sigma Q_0\left[\frac{i}{(n+1)}\right] + \sum_{j=1}^{k} \delta_j p_j\left[\frac{i}{(n+1)}\right] \tag{98}$$

with Cov (Y_{in}, Y_{jn}) as given in (97) and where $p_1(.), p_2(.), \ldots, p_k(.)$ are functions for some fixed value of k. LaRiccia (1991 [67]) proposed a likelihood ratio test for $H_0 : \delta = (\delta_1, \delta_2, \ldots, \delta_k)' = 0$, which turns out to be analogous to the Neyman smooth test.

4.3 Smooth Tests in Survival Analysis with Censoring and Truncation

One of the important questions econometricians often face is whether there are one or more unobserved variables that have a significant influence on the outcome of a trial or experiment. Social scientists such as economists have to rely mainly on observational data. Although, in some other disciplines, it is possible to control for unobserved variables to a great extent through experimental design, econometricians are not that fortunate most of the time. This gives rise to misspecification in the model through unobserved heterogeneity (for example, ability, expertise, genetical traits, inherent resistance to diseases), which, in turn, could significantly influence outcomes such as income or survival times. In this subsection we look at the effect of misspecification on distribution of survival times through a random multiplicative heterogeneity in the *hazard* function (Lancaster 1985 [66]) utilizing Neyman's smooth test with generalized residuals.

Suppose now that we observe survival times $t_1, t_2, \ldots, t_n$, which are independently distributed (for the moment, without any censoring) with a density function $g(t; \gamma, \theta)$ and cdf $G(t; \gamma, \theta)$, where γ are parameters. Let us define the hazard function $\lambda(t; \gamma, \theta)$ by

$$P(t < T < t + dt | T > t) = \lambda(t; \gamma, \theta)dt, t > 0 \tag{99}$$

which is the conditional probability of death or failure over the next infinitesimal period dt given that the subject has survived till time t. There could be several different specifications of the hazard function $\lambda(t; \gamma, \theta)$ such as the proportional hazards models. If the survival time distribution is Weibull, then the hazard function is given by

$$\lambda(t; \alpha, \beta) = \alpha t^{\alpha-1} \exp(x'\beta) \tag{100}$$

It can be shown (for example, see Cox and Oakes 1984 [27], p. 14) that if we define the survival function as $\bar{G}(t; \gamma, \theta) = 1 - G(t; \gamma, \theta)$, then we would have

$$\lambda(t; \gamma, \theta) = \frac{g(t; \gamma, \theta)}{\bar{G}(t; \gamma, \theta)} \Rightarrow g(t; \gamma, \theta) = \lambda(t; \gamma, \theta)\bar{G}(t; \gamma, \theta) \tag{101}$$

We can also obtain the survival function as

$$\bar{G}(t; \gamma, \theta) = \exp\left(-\int_0^t \lambda(s; \gamma, \theta)\,ds\right) = \exp(-H(t; \gamma, \theta)) \tag{102}$$

$H(t; \gamma, \theta)$ is known as the integrated hazard function. Suppose we have the function $t_i = T_i(\delta, \varepsilon_i)$, where $\delta = (\gamma', \theta')'$, and also let R_i be uniquely defined so that $\varepsilon_i = R_i(\delta, t_i)$. Then the functional ε_i is called a generalized error, and we can estimate it by $\hat{\varepsilon}_i = R_i(\hat{\delta}, t_i)$. For example, a generalized residual could be the integrated hazard function such as $\hat{\varepsilon}_i = H(t_i; \hat{\gamma}, \hat{\theta}) = \int_0^{t_i} \lambda(s; \hat{\gamma}, \hat{\theta})\,ds$ (Lancaster 1985 [66]), or it could be the distribution function such as $\hat{\varepsilon}_i = G(t_i; \hat{\gamma}, \hat{\theta}) = \int_0^{t_i} g(s; \hat{\gamma}, \hat{\theta})\,ds$ (Gray and Pierce 1985 [45]).

Let us consider a model with hazard function given by $\lambda_z(t) = z\lambda(t)$, where $z = e^u$ is the multiplicative heterogeneity and $\lambda(t)$ is the hazard function with no multiplicative heterogeneity (ignoring the dependence on parameters and covariates, for the sake of simplicity). Hence the survival function, given z, is

$$\bar{G}_z(t|z) = \exp(-z\varepsilon) \tag{103}$$

Let us further define σ_z^2 as the variance of z, $F(t) = E[\exp(-\varepsilon)]$ is the survival function and ε is the integrated hazard function evaluated at t, under the hypothesis of no unobserved heterogeneity. Then, using the integrated hazard function as the generalized residual, the survival function is given by (see Lancaster 1985 [66], pp. 164–166)

$$\bar{G}_z(t) \simeq \bar{F}(t)\left\{1 + \frac{\sigma_z^2}{2}2\right\} \tag{104}$$

Differentiating with respect to t and after some algebraic manipulation of (104), we get for small enough values of σ_z^2

$$g_z(t) \simeq f(t)\left\{1 + \frac{\sigma_z^2}{2}(\varepsilon^2 - 2\varepsilon)\right\} \tag{105}$$

where g_z is the density function with multiplicative heterogeneity z, f is the density with $z = 1$. We can immediately see that, if we used normalized Legendre polynomials to expand g_z, we would get a setup very similar to that of Neyman's (1937 [76]) smooth test with nuisance parameters γ (see also Thomas and Pierce, 1979 [111]). Further, the score test for the existence of heterogeneity ($H_0 : \theta = 0$, i.e., $H_0 : \sigma_z^2 = 0$) is based on the sample counterpart of the score function, $\frac{1}{2}(\varepsilon^2 - 2\varepsilon)$ for $z = 1$. If s^2 is the estimated variance of the generalized residuals $\hat{\varepsilon}$, then the score test, which is also White's (1982 [114]) information matrix (IM) test of specification, is based on the expression $s^2 - 1$, divided by its estimated standard error (Lancaster 1985 [66]). This is a particular case of the result that the IM test is a score test for neglected heterogeneity when the variance of the heterogeneity is small, as pointed out in Cox (1983 [26]) and Chesher (1984 [22]).

Although the procedure outlined by Lancaster (1985 [66]) shows much promise for applying Neyman's smooth test to survival analysis, there are two major drawbacks. First, it is difficult, if not impossible, to obtain real-life survival data without the problem of censoring or truncation; second, Lancaster (1985 [66]) worked within the framework of the Weibull model, and the impact of model misspecification needs to be considered. Gray and Pierce (1985 [45]) focused on the second issue of misspecification in the model for survival times and also tried to answer the first question of censoring in some special cases.

Suppose the observed data is of the form

$$
\left. \begin{array}{l} Y_i = \min\{T_i, V_i\} \\ Z_i = I\{T_i \geq V_i\} \end{array} \right\} \tag{106}
$$

where $I\{A\}$ is an indicator function for event A and V_i are random censoring times generated independently of the data from cdfs $C_i, i = 1, 2, \ldots, n$. Gray and Pierce (1985 [45]) wanted to test the validity of the function $\bar{G}$ rather than the effect of the covariates x_i on T_i. We can look at any survival analysis problem (with or without censoring or truncation) in two parts. First, we want to verify the functional form of the cdf G_i, i.e., to answer the question whether the survival times are generated from a particular distribution like $G_i(t; \beta) = 1 - \exp(-\exp(x_i'\beta)t)$; second, we want to test the effect of the covariates x_i on the survival time T_i. The second problem has been discussed quite extensively in the literature. However, there has been relatively less attention given to the first problem. This is probably because there could be an infinite number of choices of the functional form of the survival function. Techniques based on

Neyman's smooth test provide an opportunity to address the problem of misspecification in a more concrete way.*

The main problem discussed by Gray and Pierce (1985 [45]) is to test H_0, which states that the generalized error $U_i = G_i(T_i; \gamma, \theta = 0) = F_i(T_i; \gamma)$ is IID $U(0, 1)$, against the alternative H_1, which is characterized by the pdf

$$g_i(t; \gamma, \theta) = f_i(t; \gamma) \exp\left\{ \sum_{l=1}^{k} \theta_i \psi_l(F_i(t; \gamma)) \right\} \exp\{-K(\theta, \gamma)\} \tag{107}$$

where $f_i(t; \gamma)$ is the pdf under H_0. Thomas and Pierce (1979 [111]) chose $\psi_l(u) = u^l$, but one could use any system of orthonormal polynomials such as the normalized Legendre polynomials. In order to perform a score test as discussed in Thomas and Pierce (1979 [111]), which is an extension of Neyman's smooth test in presence of nuisance parameters, one must determine the asymptotic distribution of the score statistic. In the case of censored data, the information matrix under the null hypothesis will depend on the covariates, the estimated nuisance parameters, and also on the generally unknown censoring distribution, even in the simplest location-scale setup. In order to solve this problem, Gray and Pierce (1985 [45]) used the distribution conditional on observed values in the same spirit as the EM algorithm (Dempster et al. 1977 [32]). When there is censoring, the true cdf or the survival function can be estimated using a method like the Kaplan–Meier or the Nelson–Aalen estimators (Hollander and Peña 1992 [53], p. 99). Gray and Pierce (1985 [45]) reported limited simulation results where they looked at data generated by exponential distribution with Weibull waiting time. They obtained encouraging results using Neyman's smooth test over the standard likelihood ratio test.

In the survival analysis problem, a natural function to use is the hazard function rather than the density function. Peña (1998a [89]) proposed the

---

*We should mention here that a complete separation of the misspecification problem and the problem of the effect of covariates is not always possible to a satisfactory level. In their introduction, Gray and Pierce (1985 [45]) pointed out:

> Although, it is difficult in practice to separate the issues, our interest is in testing the adequacy of the form of F, rather than in aspects related to the adequacy of the covariables.

This sentiment has also been reflected in Pẽna (1998a, [89]) as he demonstrated that the issue of the effect of covariates is "highly intertwined with the goodness-of-fit problem concerning $\lambda(.)$."

smooth goodness-of-fit test obtained by embedding the baseline hazard function $\lambda(.)$ in a larger family of hazard functions developed through smooth, and possibly random, transformations of $\lambda_0(.)$ using the Cox proportional hazard model $\lambda(t|X(t)) = \lambda(t)\exp(\beta'X(t))$, where $X(t)$ is a vector of covariates. Peña used an approach based on generalized residuals within a counting process framework as described in Anderson et al. (1982 [1], 1991 [2]) and reviewed in Hollander and Peña (1992 [53]).

Suppose, now, we consider the same data as given in (106), (Y_i, Z_i). In order to facilitate our discussion on analyzing for censored data for survival analysis, we define:

1. The number of actual failure times observed without censoring before time t: $N(t) = \sum_{i=1}^{n} I(Y_i \leq t, Z_i = 1)$.
2. The number of individuals who are still surviving at time t: $R(t) = \sum_{i=1}^{n} I(Y_i \geq t)$.
3. The indicator function for *any* survivors at time t: $J(t) = I(R(t) > 0)$.
4. The conditional mean number of survivors at risk at any time $s \in (0, t)$, given that they survived till time s: $A(t) = \int_0^t R(s)\lambda(s)\,ds$.
5. The difference between the observed and the expected (conditional) numbers of failure times at time t: $M(t) = N(t) - A(t)$.*

Let $F = \{\mathcal{F}_t : t \in T\}$ be the history or the information set (filtration) or the predictable process at time t. Then, for the Cox proportional hazards model, the long-run smooth "averages" of N are given by $A = \{A(t) : t \in T\}$, where

$$A(t) = \int_0^t R(s)\lambda(s)\exp\{\beta'X(s)\}\,ds, \; i = 1, \ldots, n \tag{108}$$

and β is a $q \times 1$ vector of regression coefficients and $X(s)$ is a $q \times 1$ vector of predictable (or predetermined) covariate processes.

The test developed by Peña (1998a [89]) is for $H_0 : \lambda(t) = \lambda_0(t)$, where $\lambda_0(t)$ is a completely specified baseline hazard rate function associated with the integrated hazard given by $H_0(t) = \int_0^t \lambda_0(s)\,ds$, which is assumed to be

---

*In some sense, we can interpret $M(t)$ to be the residual or error in the number of deaths or failures over the smooth conditional average of the number of individuals who would die given that they survived till time $s \in (0, t)$. Hence, $M(t)$ would typically be a martingale difference process. The series $A(t)$, also known as the compensator process, is absolutely continuous with respect to the Lebesgue measure and is predetermined at time t, since it is the definite integral up to time t of the predetermined *intensity* process given by $R(s)\lambda(s)$ (for details see Hollander and Peña 1992 [53], pp. 101–102).

strictly increasing. Following Neyman (1937 [76]) and Thomas and Pierce (1979 [111]), the smooth class of alternatives for the hazard function is given by

$$\lambda(t; \theta, \beta) = \lambda_0(t) \exp\{\theta' \psi(t; \beta)\} \tag{109}$$

where $\theta \in \mathbb{R}^k$, $k = 1, 2 \ldots$, and $\psi(t; \beta)$ is a vector of locally bounded predictable (predetermined) processes that are twice continuously differentiable with respect to β. So, as in the case of the traditional smooth test, we can rewrite the null as $H_0 : \theta = 0$. This gives the score statistic process under H_0 as

$$
\begin{aligned}
U_\theta^F(t; \theta, \beta)|_{\theta=0} &= \int_0^t \left[\frac{\partial}{\partial \theta} \log \lambda(s; \theta, \beta) \right] dM(s; \theta, \beta) \Big|_{\theta=0} \\
&= \int_0^t \psi(s; \beta) \, dM(s; 0, \beta)
\end{aligned}
\tag{110}
$$

where $M(t; \theta, \beta) = N(t) - A(t; \theta, \beta)$, $i = 1, \ldots, n$. To obtain a workable score test statistic one has to replace the nuisance parameter β by its MLE under H_0. The efficient score function $(1/\sqrt{n})U_\theta^F(t, 0, \hat{\beta})$ process has an asymptotic normal distribution with 0 mean [see Peña (1998a [89]), p. 676 for the variance–covariance matrix $\Gamma(\cdot, \cdot; \beta)$].

The proposed smooth test statistic is given by

$$s\left(\tau; \hat{\beta}\right) = \frac{1}{n} U_\theta^F\left(\tau; 0, \hat{\beta}\right); \Gamma(\tau, \tau; \hat{\beta})^- U_\theta^F\left(\tau; 0, \hat{\beta}\right) \tag{111}$$

which has an asymptotic $\chi_{k^*}^2$, distribution, $\hat{k}^* = \text{rank}\left[\Gamma\left(\tau, \tau; \hat{\beta}\right)\right]$, where $\Gamma\left(\tau, \tau; \hat{\beta}\right)$ is the asymptotic variance of the score function.[†]

Peña (1998a [89]) also proposed a procedure to combine the different choices of ψ to get an omnibus smooth test that will have power against several possible alternatives. Consistent with the original idea of Neyman (1937 [76]), and as later proposed by Gray and Pierce (1985 [45]) and Thomas and Pierce (1979 [111]), Peña considered the polynomial $\psi(t; \beta) = (1, H_0(t), \ldots, H_0(t)^{k-1})'$, where, $H_0(t)$ is the integrated hazard function under the null [for details of the test see Peña (1998a [89])]. Finally, Peña (1998b [90]), using a similar counting-process approach, suggested a smooth goodness-of-fit test for the composite hypothesis (see Thomas and Pierce 1979 [111], Rayner and Best 1989 [99], and Section 3).

[†]Peña (1998a [89], p. 676) claimed that we cannot get the same asymptotic results in terms of the nominal size of the test if we replace β by any other consistent estimator under H_0. The test statistic might not even be asymptotically χ^2.

4.4 Posterior Predictive *p*-values and Related Tests in Bayesian Statistics and Econometrics

In several areas of research *p*-value might well be the single most reported statistic. However, it has been widely criticized because of its indiscriminate use and relatively unsatisfactory interpretation in the empirical literature. Fisher (1945 [42], pp. 130–131), while criticizing the axiomatic approach to testing, pointed out that setting up fixed probabilities of Type I error *a priori* could yield misleading conclusions about the data or the problem at hand. Recently, this issue gained attention in some fields of medical research. Donahue (1999 [37]) discussed the information content in the *p*-value of a test. If we consider $F(t|H_0) = F(t)$ to be the cdf of a test statistic T under H_0 and $F(t|H_1) = G(t)$ to be the cdf of T under the alternative, the *p*-value, defined as $P(t) = P\{T > t\} = 1 - F(t)$, is a sample statistic. Under H_0, the *p*-value has a cdf given by

$$F_p(p|H_0) = 1 - F[F^{-1}(1-p)|H_0] = p \tag{112}$$

whereas under the alternative H_1 we have

$$F_p(p|H_1) = \Pr\{P \le p|H_1\} = 1 - G((F^{-1}(1-p)|H_0)) \tag{113}$$

Hence, the density function of the *p*-value (if it exists) is given by

$$
\begin{aligned}
f_p(p|H_1) &= \frac{\partial}{\partial p} F_p(p|H_1) \\
&= -g(F^{-1}(1-p)) \cdot \frac{-1}{f(F^{-1}(1-p))} \\
&= \frac{g(F^{-1}(1-p))}{f(F^{-1}(1-p))}
\end{aligned}
\tag{114}
$$

This is nothing but the "likelihood ratio" as discussed by Egon Pearson (1938 [84], p.1 38) and given in equation (18). If we reject H_0 if the same statistic $T > k$, then the probability of Type I error is given by $\alpha = \Pr\{T > k|H_0\} = 1 - F(k)$ while the power of the test is given by

$$\beta = \Pr\{T > k|H_1\} = 1 - G(k) = 1 - G(F^{-1}(1-\alpha)) \tag{115}$$

Hence, the main point of Donahue (1999 [37]) is that, if we have a small *p*-value, we can say that the test is significant, and we can also refer to the strength of the significance of the test. This, however, is usually not the case when we fail to reject the null hypothesis. In that case, we do not have any indication about the probability of Type II error that is being committed. This is reflected by the power and size relationship given in (115).

The p-value and its generalization, however, are firmly embedded in Bayes theory as the tail probability of a predictive density. In order to calculate the p-value, Meng (1994 [74]) also considered having a nuisance parameter in the likelihood function or predictive density. We can see that the classical p-value is given by $p = P\{T(X) \geq T(x)|H_0\}$, where $T(.)$ is a sample statistic and x is a realization of the random sample X that is assumed to follow a density function $f(X|\xi)$, where $\xi = (\delta', \gamma')' \epsilon \Xi$. Suppose, now we have to test $H_o : \delta = \delta_o$ against $H_1 : \delta \neq \delta_o$. In Bayesian terms, we can replace X by a future replication of x, call it x^{rep}, which is like a "future observation." Hence, we define the predictive p-value as $p_B = P\{ T(x^{rep}) \geq T(x)|x, H_0\}$ calculated under the posterior predictive density

$$f(x^{rep}|\Xi x, H_0) = \int_\Theta f(x^{rep}|\xi)\Pi_0(d\xi|x) = \int_{\Gamma_0} f(x^{rep}|\delta_0, \gamma)\pi_0(\xi|x)d\xi \quad (116)$$

where $\Pi_0(\xi|x)$ and $\pi_0(\xi|x)$ are respectively the posterior predictive distribution and density functions of ξ, given x, and under H_0. Simplification in (116) is obtained by assuming $\Gamma_0 = \{\xi : H_0 \text{ is true } \} = \{(\delta_0, \gamma) : \gamma \in A, A \subset \mathbb{R}^d, d \geq 1\}$ and defining $\bar{\pi}(\gamma|\delta_0) = \pi(\delta, \gamma|\delta = \delta_0)$, which gives

$$\pi_0(\xi|x) = \frac{f(x|\xi, H_0)\pi(\xi|H_0)}{\int_{\Gamma_0} f(x|\xi, H_0)\pi(\xi|H_0)d\xi}, \xi \in \Gamma_0$$

$$= \frac{f(x|\delta = \delta_0, \gamma)\pi(\delta, \gamma|\delta = \delta_0)}{\int_{\Gamma_0} f(x|\delta = \delta_0, \gamma)\pi(\delta, \gamma|\delta = \delta_0)d\xi} \quad (117)$$

$$= \frac{f(x|\delta_0, \gamma)\bar{\pi}(\gamma|\delta_0)}{\int_A f(x|\delta_0, \gamma)\bar{\pi}(\gamma|\delta_0)d\gamma}, \gamma \in A$$

This can also be generalized to the case of a composite hypothesis by taking the integral over all possible values of $\delta \in \Delta_0$, the parameter space under H_0. An alternative formulation of the p-value, which makes it clearer that the distribution of the p-value depends on the nuisance parameter γ, is given by $p(\gamma) \equiv P\{D(X, \xi) \geq D(x, \xi)|\delta_0, \gamma\}$, where the probability is taken over the sampling distribution $f(X|\delta_0, \gamma)$, and $D(X, \xi)$ is a test statistic in the classical sense that can be taken as a measure of discrepancy. In order to estimate the p-value $p(\gamma)$ given that γ is unknown, the obvious Bayesian approach is to take the mean of $p(\gamma)$ over the posterior distribution of γ under H_0, i.e., $E[p(\gamma)|x, H_0] = p_B$.

The above procedure of finding the distribution of the p-value can be used in diagnostic procedures in a Markov chain Monte Carlo setting discussed by Kim et al. (1998 [61]). Following Kim et al. (1998 [61], pp. 361–362), let us consider the simple stochastic volatility model

$$y_t = \beta e^{h_t/2} \varepsilon_t, t \geq 1,$$
$$h_{t+1} = \mu + \phi(h_t - \mu) + \sigma_\eta \eta_t,$$
$$h_t \sim N\left(\mu, \frac{\sigma_\eta^2}{1 - \phi^2}\right)$$

(118)

where y_t is the mean corrected return on holding an asset at time t, h_t is the log volatility which is assumed to be stationary (i.e., $|\phi| < 1$) and h_1 is drawn from a stationary distribution and, finally, ε_t and η_t are uncorrelated standard normal white noise terms. Here, β can be interpreted as the modal instantaneous volatility and ϕ is a measure of the persistence of volatility while σ_η is the volatility of log volatility h_t.*

Our main interest is handling of model diagnostics under the Markov chain Monte Carlo method. Defining $\xi = (\mu, \phi, \sigma_\eta^2)'$, the problem is to sample from the distribution of $h_t | Y_t, \xi$, given a sample of draws $h_{t-1}^1, h_{t-1}^2, \ldots, h_{t-1}^M$ from $h_{t-1} | Y_{t-1}, \xi$, where we can assume ξ to be fixed. Using the Bayes rule discussed in equations (116) and (117), the one-step-ahead prediction density is given by

$$f(y_{t+1}|Y_t, \xi) = \int f(y_{t+1}|Y_t, h_{t+1}, \xi) f(h_{t+1}|Y_t, h_t, \xi) f(h_t|Y_t, \xi) \, dh_{t+1} dh_t$$

(119)

and, for each value of $h_t^j (j = 1, 2, \ldots, M)$, we sample h_{t+1}^j from the conditional distribution h_{t+1}^j given h_t^k. Based on M such draws, we can estimate that the probability that y_{t+1}^2 would be less than the observed y_{t+1}^{o2} is given by

$$P(y_{t+1}^2 \leq y_{t+1}^{o2}|Y_t, \theta) \cong u_{t+1}^M = \frac{1}{M} \sum_{j=1}^{M} P(y_{t+1}^2 \leq y_{t+1}^{o2}|h_{t+1}^j, \xi)$$

(120)

which is the sample equivalent of the posterior mean of the probabilities discussed in Meng (1994 74]). Hence, u_{t+1}^M under the correctly specified

---

*As Kim et al. (1998 [61], p. 362) noted, the parameters β and μ are related in the true model by $\beta = \exp(\mu/2)$; however, when estimating the model, they set $\beta = 1$ and left μ unrestricted. Finally, they reported the estimated value of β from the estimated model as $\exp(\mu/2)$.

model will be IID $U(0, 1)$ distribution as $M \to \infty$. This result is an extension of Karl Pearson (1933 [87], 1934 [88]), Egon Pearson (1938 [84]), and Rosenblatt (1952 [101]), discussed earlier, and is very much in the spirit of the goodness-of-fit test suggested by Neyman (1937 [76]). Kim et al. (1998 [61]) also discussed a procedure similar to the one followed by Berkowitz (2000 [17]) where instead of looking at just u_{t+1}^M, they looked at the inverse Gaussian transformation, then carried out tests on normality, autocorrelation, and heteroscedasticity. A more comprehensive test could be performed on the validity of forecasted density based on Neyman's smooth test techniques that we discussed in Section 4.2 in connection with the forecast density evaluation literature (Diebold et al. 1998 [33]). We believe that the smooth test provides a more constructive procedure instead of just checking uniformity of an average empirical distribution function u_{t+1}^M on the square of the observed values y_{t+1}^{o2} given in (120) and other graphical techniques like the Q–Q plots and correlograms as suggested by Kim et al. (1998 [61], pp. 380–382).

5. EPILOGUE

Once in a great while a paper is written that is truly fundamental. Neyman's (1937 [76]) is one that seems impossible to compare with anything but itself, given the statistical scene in the 1930s. Starting from the very first principles of testing, Neyman derived an *optimal* test statistic and discussed its applications along with its possible drawbacks. Earlier tests, such as Karl Pearson's (1900 [86]) goodness-of-fit and Jerzy Neyman and Egon Pearson's (1928 [79]) likelihood ratio tests are also fundamental, but those test statistics were mainly based on intuitive grounds and had no claim for optimality when they were proposed. In terms of its significance in the history of hypothesis testing, Neyman (1937 [76]) is comparable only to the later papers by the likes of Wald (1943 [113]), Rao (1948 [93]), and Neyman (1959 [77]), each of which also proposed fundamental test principles that satisfied certain optimality criteria.

   Although econometrics is a separate discipline, it is safe to say that the main fulcrum of advances in econometrics is, as it always has been, statistical theory. From that point of view, there is much to gain from borrowing suitable statistical techniques and adapting them for econometric applications. Given the fundamental nature of Neyman's (1937 [76]) contribution, we are surprised that the smooth test has not been formally used in econometrics, to the best of our knowledge. This paper is our modest attempt to bring Neyman's smooth test to mainstream econometric research.

ACKNOWLEDGMENTS

We would like to thank Aman Ullah and Alan Wan, without whose encouragement and prodding this paper would not have been completed. We are also grateful to an anonymous referee and to Zhijie Xiao for many helpful suggestions that have considerably improved the paper. However, we retain the responsibility for any remaining errors.

REFERENCES

1. P. K. Anderson, O. Borgan, R. D. Gill, N. Keiding. Linear nonparametric tests for comparison of counting processes, with applications to censored survival data. *International Statistical Review* 50:219–258, 1982.
2. P. K. Anderson, O. Borgan, R. D. Gill, N. Keiding. *Statistical Models Based on Counting Processes*. New York: Springer-Verlag, 1991.
3. D. E. Barton. On Neyman's smooth test of goodness of fit and its power with respect to a particular system of alternatives. *Skandinaviske Aktuarietidskrift* 36:24–63, 1953a.
4. D. E. Barton. The probability distribution function of a sum of squares. *Trabajos de Estadistica* 4:199–207, 1953b.
5. D. E. Barton. A form of Neyman's χ^2 test of goodness of fit applicable to grouped and discrete data. *Skandinaviske Aktuarietidskrift* 38:1–16, 1955.
6] D. E. Barton. Neyman's ψ_k^2 test of goodness of fit when the null hypothesis is composte. *Skandinaviske Aktuarietidskrift* 39:216–246, 1956.
7. D. E. Barton. Neyman's and other smooth goodness-of-fit tests. In: S. Kotz and N. L. Johnson, eds. *Encyclopedia of Statistic Sciences*, Vol. 6. New York: Wiley, 1985, pp. 230–232.
8. T. Bayes, Rev. An essay toward solving a problem in the doctrine of chances. *Philosophical Transactions of the Royal Society* 53:370–418, 1763.
9. B. J. Becker. Combination of p-values. In: S. Kotz, C. B. Read and D. L. Banks, eds. *Encyclopedia of Statistical Sciences*, Update Vol. I. New York: Wiley, 1977, pp. 448–453.
10. A. K. Bera. Hypothesis testing in the 20th century with special reference to testing with misspecified models. In: C. R. Rao and G. Szekely, eds. *Statistics in the 21st Century*. New York: Marcel Dekker, 2000, pp. 33–92.

11. A. K. Bera, Y. Bilias. Rao's score, Neyman's C(α) and Silvey's LM test: An essay on historical developments and some new results. *Journal of Statistical Planning and Inference*, 97:9–44, 2001, (2001a).

12. A. K. Bera, Y. Bilias. On some optimality properties of Fisher–Rao score function in testing and estimation. *Communications in Statistics, Theory and Methods*, 2001, Vol. 30 (2001b).

13. A. K. Bera, Y. Bilias. The MM, ME, MLE, EL, EF, and GMM approaches to estimation: A synthesis. *Journal of Econometrics*, 2001 forthcoming (2001c).

14. A. K. Bera, A. Ghosh. Evaluation of density forecasts using Neyman's smooth test. Paper to be presented at the 2002 North American Winter Meeting of the Econometric Society, 2001.

15. A. K. Bera, G. Premaratne. General hypothesis testing. In: B. Baltagi, B. Blackwell, eds. *Companion in Econometric Theory*. Oxford: Blackwell Publishers, 2001, pp. 38–61.

16. A. K. Bera, A. Ullah. Rao's score test in econometrics. *Journal of Quantitative Economics* 7:189–220, 1991.

17. J. Berkowitz. The accuracy of density forecasts in risk management. Manuscript, 2000.

18. P. J. Bickel, K. A. Doksum. *Mathematical Statistics: Basic Ideas and Selected topics*. Oakland, CA: Holden-Day, 1977.

19. B. Boulerice, G. R. Ducharme. A note on smooth tests of goodness of fit for location-scale families. *Biometrika* 82:437–438, 1995.

20. A. C. Cameron, P. K. Trivedi. Conditional moment tests and orthogonal polynomials. Working paper in Economics, Number 90-051, Indiana University, 1990.

21. T. K. Chandra, S. N. Joshi. Comparison of the likelihood ratio, Rao's and Wald's tests and a conjecture by C. R. Rao. *Sankhyā, Series A* 45:226–246, 1983.

22. A. D. Chesher. Testing for neglected heterogeneity. *Econometrica* 52:865–872, 1984.

23. S. Choi, W. J. Hall, A. Shick. Asymptotically uniformly most powerful tests in parametric and semiparametric models. *Annals of Statistics* 24:841–861, 1996.

24. P. F. Christoffersen. Evaluating interval forecasts. *International Economic Review* 39:841–862, 1998.

25. G. M. Cordeiro, S. L. P. Ferrari. A modified score test statistic having chi-squared distribution to order n^{-1}. *Biometrika* 78:573–582, 1991.

26. D. R. Cox. Some remarks on over-dispersion. *Biometrika* 70;269–274, 1983.

27. D. R. Cox, D. Oakes. *Analysis of Survival Data*. New York: Chapman and Hall, 1984.

28. H. Cramér. *Mathematical Methods of Statistics.* New Jersey: Princeton University Press, 1946.

29. F. Cribari-Neto, S. L. P. Ferrari. An improved Lagrange multiplier test for heteroscedasticity. *Communications in Statistics–Simulation and Computation* 24:31–44, 1995.

30. C. Crnkovic, J. Drachman. Quality control. In; VAR: *Understanding and Applying Value-at-Risk.* London: Risk Publication, 1997.

31. R. B. D'Agostino, M. A. Stephens. *Goodness-of-Fit Techniques.* New York: Marcel Dekker, 1986.

32. A. P. Dempster, N. M. Laird, D. B. Rubin. Maximum likelihood from incomplete data via the EM algorithm. *Journal of the Royal Statistical Society, Series B* 39:1–38, 1977.

33. F. X. Diebold, T. A. Gunther, A. S. Tay. Evaluating density forecasts with applications to financial risk management. *International Economic Review* 39:863–883, 1998.

34. F. X. Diebold, J. Hahn, A. S. Tay. Multivariate density forecast evaluation and calibration in financial risk management: high-frequency returns in foreign exchange. *Review of Economics and Statistics* 81:661–673, 1999.

35. F. X. Diebold, J. A. Lopez. Forecast evaluation and combination. In: G. S. Maddala and C. R. Rao, eds. *Handbook of Statistics*, Vol. 14. Amsterdam: North-Holland, 1996, pp. 241–268.

36. F. X. Diebold, A. S. Tay, K. F. Wallis. Evaluating density forecasts of inflation: the survey of professional forecasters. In: R. F. Engle, H. White, eds. *Cointegration, Causality and Forecasting: Festschrift in Honour of Clive W. Granger.* New York: Oxford University Press, 1999, pp. 76–90.

37. R. M. J. Donahue. A note on information seldom reported via the p value. *American Statistician* 53:303–306, 1999.

38. R. L. Eubank, V. N. LaRiccia. Asymptotic comparison of Cramér–von Mises and nonparametric function estimation techniques for testing goodness-of-fit. *Annals of Statistics* 20:2071–2086, 1992.

39. J. Fan. Test of significance based on wavelet thresholding and Neyman's truncation. *Journal of the American Statistical Association* 91:674–688, 1996.

40. R. A. Fisher. On the mathematical foundations of theoretical statistics. *Philosophical Transactions of the Royal Society* A222:309–368, 1922.

41. R. A. Fisher. Inverse probability. *Proceedings of the Cambridge Philosophical Society* 36:528–535, 1930.

42. R. A. Fisher. The logical inversion of the notion of a random variable. *Sankhyā* 7:129–132, 1945.

43. R. A. Fisher. *Statistical Methods for Research Workers.* 13th ed. New York: Hafner, 1958 (first published in 1932).

44. J. K. Ghosh. Higher order asymptotics for the likelihood ratio, Rao's and Wald's tets. *Statistics & Probability Letters* 12:505–509, 1991.

45. R. J. Gray, D. A. Pierce. Goodness-of-fit tests for censored survival data. *Annals of Statistics* 13:552–563, 1985.

46. T. Haavelmo. The probability approach in econometrics. *Supplements to Econometrica* 12, 1944.

47. T. Haavelmo. Econometrics and the welfare state: Nobel lecture, December 1989. *American Economic Review* 87:13–15, 1997.

48. M. A. Hamdan. The power of certain smooth tests of goodness of fit. *Australian Journal of Statistics* 4:25–40, 1962.

49. M. A. Hamdan. A smooth test of goodness of fit based on Walsh functions. *Australian Journal of Statistics* 6:130–136, 1964.

50. P. Harris. An asymptotic expansion for the null distribution of the efficient score statistic. *Biometrika* 72:653–659, 1985.

51. P. Harris. Correction to "An asymptotic expansion for the null distribution of the efficient score statistic." *Biometrika* 74:667, 1987.

52. J. D. Hart. *Nonparametric Smoothing and Lack of Fit Tests.* New York: Springer-Verlag, 1997.

53. M. Hollander, E. A. Peña. Classes of nonparametric goodness-of-fit tests for censored data. In: A. K. Md.E. Saleh, ed. *A New Approach in Nonparametric Statistics and Related Topics.* Amsterdam: Elsevier, 1992, pp. 97–118.

54. T. Inglot, T. Jurlewicz, T. Ledwina. On Neyman-type smooth tests of fit. *Statistics* 21:549–568, 1990.

55. T. Inglot, W. C. M. Kallenberg, T. Ledwina. Power approximations to and power comparison of smooth goodness-of-fit tests. *Scandinavian Journal of Statistics* 21:131–145, 1994.

56. S. L. Isaacson. On the theory of unbiased tests of simple statistical hypothesis specifying the values of two or more parameters. *Annals of Mathematical Statistics* 22:217–234, 1951.

57. N. L. Johnson, S. Kotz. *Continuous Univariate Distributions—1.* New York: John Wiley, 1970a.

58. N. L. Johnson, S. Kotz. *Continuous Univariate Distributions—2.* New York: John Wiley, 1970b.

59. W. C. M. Kallenberg, J. Oosterhoff, B. F. Schriever. The number of classes in χ^2 goodness of fit test. *Journal of the American Statistical Association* 80:959–968, 1985.

60. N. M. Kiefer. Specification diagnostics based on Laguerre alternatives for econometric models of duration. *Journal of Econometrics* 28:135–154, 1985.

61. S. Kim, N. Shephard, S. Chib. Stochastic volatility: Likelihood inference and comparison with ARCH models. *Review of Economic Studies* 65:361–393, 1998.

62. L. Klein. The Statistics Seminar, MIT, 1942–43. *Statistical Science* 6:320–330, 1991.

63. K. J. Kopecky, D. A. Pierce. Efficiency of smooth goodness-of-fit tests. *Journal of the American Statistical Association* 74:393–397, 1979.

64. J. A. Koziol. An alternative formulation of Neyman's smooth goodness of fit tests under composite alternatives. *Metrika* 34:17–24, 1987.

65. P. H. Kupiec. Techniques for verifying the accuracy of risk measurement models. *Journal of Derivatives*. Winter: 73–84, 1995.

66. T. Lancaster. Generalized residuals and heterogeneous duration models. *Journal of Econometrics* 28:155–169, 1985.

67. V. N. LaRiccia. Smooth goodness-of-fit tests: A quantile function approach. *Journal of the American Statistical Association* 86:427–431, 1991.

68. T. Ledwina. Data-driven version of Neyman's smooth test of fit. *Journal of the American Statistical Association* 89:1000–1005, 1994.

69. L.-F. Lee. Maximum likelihood estimation and a specification test for non-normal distributional assumption for the accelerated failure time models. *Journal of Econometrics* 24:159–179, 1984.

70. L.-F. Lee. Specification test for Poisson regression models. *International Economic Review* 27:689–706, 1986.

71. E. L. Lehmann. *Testing Statistical Hypothesis*. New York: John Wiley, 1959.

72. G. S. Maddala. *Econometrics*. New York: McGraw-Hill, 1977.

73. P. C. Mahalanobis. A revision of Risley's anthropometric data relating to Chitagong hill tribes. *Sankhyā B* 1:267–276, 1934.

74. X.-L. Meng. Posterior predictive *p*-values. *Annals of Statistics* 22:1142–1160, 1994.

75. R. Mukerjee. Rao's score test: recent asymptotic results. In: G. S. Madala, C. R. Rao, H. D. Vinod, eds. *Handbook of Statistics*, Vol. 11. Amsterdam: North-Holland, 1993, pp. 363–379.

76. J. Neyman. "Smooth test" for goodness of fit. *Skandinaviske Aktuarietidskrift* 20:150–199, 1937.

77. J. Neyman. Optimal asymptotic test of composite statistical hypothesis. In: U. Grenander, ed. *Probability and Statistics, the Harold Cramér Volume*. Uppsala: Almqvist and Wiksell, 1959, pp. 213–234.

78. J. Neyman. Some memorable incidents in probabilistic/statistical studies. In: I. M. Chakrabarti, ed. *Asymptotic Theory of Statistical Tests and Estimation*. New York: Academic Press, 1980, pp. 1–32.

79. J. Neyman, E. S. Pearson. On the use and interpretation of certain test criteria for purpose of statistical inference. *Biometrika* 20:175–240, 1928.

80. J. Neyman, E. S. Pearson. On the problem of the most efficient tests of statistical hypothesis. *Philosophical Transactions of the Royal Society, Series A* 231:289–337, 1933.

81. J. Neyman, E. S. Pearson. Contributions to the theory of testing statistical hypothesis I: Unbiased critical regions of Type A and A_1. *Statistical Research Memoirs* 1:1–37, 1936.

82. J. Neyman, E. S. Pearson. Contributions to the theory of testing statistical hypothesis. *Statistical Research Memoirs* 2:25–57, 1938.

83. F. O'Reilly, C. P. Quesenberry. The conditional probability integral transformation and applications to obtain composite chi-square goodness of fit tests. *Annals of Statistics* 1:74–83, 1973.

84. E. S. Pearson. The probability integral transformation for testing goodness of fit and combining independent tests of significance. *Biometrika* 30:134–148, 1938.

85. E. S. Pearson. The Neyman–Pearson story: 1926–34, historical sidelights on an episode in Anglo-Polish collaboration. In: F. N. David, ed. *Research Papers in Statistics, Festscrift for J. Neyman*. New York: John Wiley, 1966, pp. 1–23.

86. K. Pearson. On the criterion that a given system of deviations from the probable in the case of a correlated system of variables is such that it can reasonably be supposed to have arisen from random sampling. *Philosophical Magazine, 5th Series* 50:157–175, 1900.

87. K. Pearson. On a method of determining whether a sample of size n supposed to have been drawn from a parent population having a known probability integral has probably been drawn at random *Biometrika* 25:379–410, 1933.

88. K. Pearson. On a new method of determining "goodness of fit." *Biometrika* 26:425–442, 1934.

89. E. Peña. Smooth goodness-of-fit tests for the baseline hazard in Cox's proportional hazards model. *Journal of the American Statistical Association* 93:673–692, 1998a.

90. E. Peña. Smooth goodness-of-fit tests for composite hypothesis in hazard based models. *Annals of Statistics* 28:1935–1971, 1998b.

91. D. A. Pierce. The asymptotic effect of substituting estimators for parameters in certain types of statistics. *Annals of Statistics* 10:475–478, 1982.

92. C. P. Quesenberry. Some transformation methods in a goodness-of-fit. In: R. B. D'Agostino, M.A. Stephens, eds. *Goodness-of-Fit Techniques*. New York: Marcel Dekker, 1986, pp. 235–277.

93. C. R. Rao. Large sample of tests of statistical hypothesis concerning several parameters with applications to problems of estimation. *Proceedings of the Cambridge Philosophical Society* 44:50–57, 1948.

94. C. R. Rao. *Advanced Statistical Methods in Biometric Research.* New York: John Wiley, 1952.

95. C. R. Rao. *Linear Statistical Inference and its Applications.* New York: John Wiley, 1973.

96. C. R. Rao, R. Mukerjee. Tests based on score statistics: power properties and related results. *Mathematical Methods of Statistics* 3:46–61, 1994.

97. C. R. Rao, R. Mukerjee. Comparison of LR, score and Wald tests in a non-II D setting. *Journal of Multivariate Analysis* 60:99–110, 1997.

98. C. R. Rao, S. J. Poti. On locally most powerful tests when the alternatives are one-sided. *Sankhyā 7:439, 1946.*

99. J. C. W. Rayner, D. J. Best. *Smooth Tests of Goodness of Fit.* New York: Oxford University Press, 1989.

100. C. Reid. *Neyman—From Life.* New York: Springer-Verlag, 1982.

101. M. Rosenblatt. Remarks on a multivariate transformation. *Annals of Mathematical Statistics* 23:470–472, 1952.

102. A. SenGupta, L. Vermeire. Locally optimal tests for multiparameter hypotheses. *Journal of the American Statistical Association* 81:819–825, 1986.

103. R. J. Smith. On the use of distributional misspecification checks in limited dependent variable models. *The Economic Journal* 99 (Supplement Conference Papers):178–192, 1989.

104. C. Smithson, L. Minton. How to calculate VAR. In: *VAR: Understanding and Applying Value-at-Risk.* London: Risk Publications, 1997, pp. 27–30.

105. H. Solomon, M. A. Stephens. Neyman's test for uniformity. In: S. Kotz and N. L. Johnson, eds. *Encyclopedia of Statistical Sciences*, Vol. 6. New York: Wiley, 1985, pp. 232–235.

106. E. S. Soofi. Information theoretic regression methods. In: T. M. Fomby, R. C. Hill, eds. *Advances in Econometrics*, vol. 12. Greenwich: Jai Press,1997, pp. 25–83.

107. E. S. Soofi. Principal information theoretic approaches. *Journal of the American Statistical Association* 95:1349–1353, 2000.

108. C. J. Stone. Large-sample inference for log-spline models. *Annals of Statistics* 18:717–741, 1990.

109. C. J. Stone, C-Y. Koo/ Logspline density estimation. *Contempory Mathematics* 59:1-15, 1986

110. A. S. Tay, K. F. Wallis. Density forecasting: A survey. *Journal of Forecasting* 19:235–254, 2000.

111. D. R. Thomas, D. A. Pierce. Neyman's smooth goodness-of-fit test when the hypothesis is composite. *Journal of the American Statistical Association* 74:441–445, 1979.

112. L. M. C. Tippett. *The Methods of Statistics.* 1st Edition. London: Williams and Norgate, 1931.

113. A. Wald. Tests of statistical hypotheses concerning several parameters when the number of observations is large. *Transactions of the American Mathematical Society* 54:426–482, 1943.

114. H. White. Maximum likelihood estimation of misspecified models. *Econometrica* 50:1–25, 1982.

115. B. Wilkinson. A statistical consideration in psychological research. *Psychology Bulletin* 48:156–158, 1951.

11
Computing the Distribution of a Quadratic Form in Normal Variables

R. W. FAREBROTHER   Victoria University of Manchester, Manchester, England

1. INTRODUCTION

A wide class of statistical problems directly or indirectly involves the evaluation of probabilistic expressions of the form

$$\Pr(u'Ru < x) \tag{1.1}$$

where μ is an $m \times 1$ matrix of random variables that is normally distributed with mean variance Ω,

$$U \sim N(\delta, \Omega) \tag{1.2}$$

and where x is a scalar, δ is an $m \times 1$ matrix, R is an $m \times m$ symmetric matrix, and Ω is an $m \times m$ symmetric positive definite matrix.

In this chapter, we will outline developments concerning the exact evaluation of this expression. In Section 2 we will discuss implementations of standard procedures which involve the diagonalization of the $m \times m$ matrix R, whilst in Section 3 we will discuss procedures which do not require this preliminary diagonalization. Finally, in Section 4 we will apply one of the

231

diagonalization techniques to obtain the 1, 2.5, and 5 percent critical values for the lower and upper bounds of the Durbin–Watson statistics.

2. DIAGONAL QUADRATIC FORMS

Let L be an $m \times m$ lower triangular matrix satisfying $LL' = \Omega$ and lev $v = L^{-1}$ and $Q = L'RL$; then $u'Ru = v'Qv$ and we have to evaluate the expression

$$\Pr(v'Qv < x) \tag{2.1}$$

where $v = L^{-1}u$ is normally distributed with mean $\mu = L^{-1}\delta$ and variance $I_m = L^{-1}\Omega(L')^{-1}$.

Now, let H be an $m \times m$ orthonormal matrix such that $T = H'QH$ is tridiagonal and let $w = H'v$; then $v'Qv = w'Tw$ and we have to evaluate the expression

$$\Pr(w'Tw < x) \tag{2.2}$$

where $w = H'v$ is normally distributed with mean $\kappa = H'\mu$ and variance $I_m = H'H$.

Finally, let G be an $m \times m$ orthonormal matrix such that $D = G'TG$ is diagonal and let $z = G'w$; then $w'Tw = z'Dz$ and we have to evaluate

$$\Pr(z'Dz < x) \tag{2.3}$$

where $z = G'w$ is normally distributed with mean $v = G'_K$ and variance $I_m = G'G$.

Thus expressions (1.1), (2.1), or (2.2) may be evaluated indirectly by determining the value of

$$\Pr\left[\sum_{j=1}^{m} d_{jj}z_j^2 > x\right] \tag{2.4}$$

where, for $j = 1, 2, \cdots, m$, z_j is independently normally distributed with mean v_j and unit variance.

Now, the characteristic function of the weighted sum of noncentral $x^2(1)$ variables $z'Dz$ is given by

$$\Theta(t) = [\Psi(f)]^{-1/2} \tag{2.5}$$

where $I = \sqrt{-1}, f = 2it$, and

$$\Psi(f) = \det(I - fD) \exp[v'v - v'(I - fD)^{-1}v] \tag{2.6}$$

Thus, probabilistic expressions of the form (1.1) may be evaluated by applying the Imhof [1], Ruben [2], Grad and Solomon [3], or Pan [4] procedures to equation (2.4).

(a) The standard Imhof procedure is a general procedure which obtains the desired result by numerical integration. It has been programmed in Fortran by Koerts and Abrahamse [5] and in Pascal by Farebrother [6]. An improved version of this procedure has also been programmed in Algol by Davies [7].

(b) The Ruben procedure expands expression (2.4) as a sum of central $x^2(1)$ distribution functions but its use is restricted to positive definite matrices. It has been programmed in Fortran by Sheil and O'Muircheartaigh [8] and in Algol by Farebrother [9].

(c) The Grad and Solomon (or Pan) procedure uses contour integration to evaluate expression (2.5) but its use is restricted to sums of central $x^2(1)$ variables with distinct weights. It has been programmed in Algol by Farebrother [10, 11].

(d) The Ruben and Grad and Solomon procedures are very much faster than the Imhof procedure but they are not of such general application; the Ruben procedure is restricted to positive linear combinations of noncentral X^2 variables whilst the Grad and Solomon procedure assumes that the X^2 variables are central; see Farebrother [12] for further details.

3. NONDIAGONAL QUADRATIC FORMS

The fundamental problem with the standard procedures outlined in Section 2 is that the matrices R, Q, and T have to be reduced to diagonal form before these methods can be applied. Advances in this area by Palm and Sneek [13], Farebrother [14], Shively, Ansley and Kohn (SAK) [15] and Ansley, Kohn and Shively (AKS) [16] are based on the observation that the transformed characteristic function $\Psi(f)$ may also be written as

$$\Psi(f) = \det(I - fT)\exp\left[\kappa'\kappa - \kappa'(I - fT)^{-1}\kappa\right] \tag{3.1}$$

or as

$$\Psi(f) = \det(I - fQ)\exp\left[\mu'\mu - \mu'(1 - FQ)^{-1}\mu\right] \tag{3.2}$$

or as

$$\Psi(f) = \det(\Omega) \det\left(\Omega^{-1} fR \right) \exp\left[\delta'\Omega^{-1}\delta - \delta'\left(\Omega^{-1} - fR\right)^{-1}\delta \right] \qquad (3.3)$$

so that the numerical integration of the Imhof procedure may be performed using complex arithmetic.

(e) Farebrother's [14] variant of Palm and Sneek's [13] procedure is of general application, but it requires that the matrix Q be constructed and reduced to tridiagonal form. It has been programmed in Pascal by Farebrother [6].

(f) The SAK and AKS procedures are of more restriced application as they assume that R and Ω may be expressed as $R = PAP' - cI_m$ and $\Omega = O\Sigma P'$, where A and Σ (or their inverses) are $n \times n$ symmetric band matrices and where P is an $m \times n$ matrix of rank m and is the orthogonal complement of a known $n \times (n - m)$ matrix, and satisfies $PP' = I_m$.

In this context, with $\delta = 0$ and $x = 0$, and with further restrictions on the structure of the matrix Ω, SAK [15] and AKS [16] respectively used the modified Kalman filter and the Cholesky decomposition to evaluate expression (3.3) and thus (1.1) without actually forming the matrices R, Q, or Ω, and without reducing Q to tridiagonal form.

Both of these procedures are very much faster than the Davies [7], Farebrother [10, 11], and Farebrother [6] procedures for large values of n, but Farebrother [17] has expressed grave reservations regarding their numerical accuracy as the Kalman filter and Cholesky decomposition techniques are known to be numerically unstable in certain circumstances.

Further, the implementation of both procedures is specific to the particular class of A and Σ matrices selected, but Kohn, Shively, and Ansley (KSA) [18] have programmed the AKS procedure in Fortran for the generalized Durbin–Watson statistic.

(g) In passing, we note that Shephard [19] has extended the standard numerical inversion procedure to expressions of the form

$$\Pr\left[\cap_{j=1}^{h} \left(u' R_j u > x_j \right) \right] \qquad (3.4)$$

where $R_1, R_2, \cdots, R_h$ are a set of $m \times m$ symmetric matrices, x_1, $x_2, \cdots, h$ are the associated scalars, and again u is an $m \times 1$ matrix of random variables that is normally distributed with mean δ and variance Ω.

4. DURBIN–WATSON BOUNDS

As an application of this computational technique, we consider the problem of determining the critical values of the lower and upper bounds on the familiar Durbin–Watson statistic when there are n observations and k explanatory variables in the model.

For given values of n, k, and α, we may determine the α-level critical value of the lower bounding distribution by setting $g = 1$ and solving the equation

$$\Pr\left[\sum_{j=1}^{n-k}(\lambda_{j+g} - c)z_j^2 < 0\right] = a \tag{4.1}$$

for c, where

$$\lambda_h = 2 - 2\cos[(\pi(h-1)/n], \qquad h = 1, 2, \cdots, n \tag{4.2}$$

is the hth smallest eigenvalue of a certain $n \times n$ tridiagonal matrix and where, for $j = 1, 2, \cdots, n - k$, the z_j are independently normally distributed random variables with zero means and unit variances. Similarly, we may define the α-level critical value of the upper bounding distribution by setting $g = k$ in equation (4.1).

In a preliminary stage of Farebrother's [20] work on the critical values of the Durbin–Watson minimal bound ($g = 0$), the critical values of the lower and upper bound were determined for a range of values of $n > k$ and for $100\alpha = 1, 2.5, 6$. The values given in the first two diagonals ($n - k = 2$ and $n - k = 3$) of Tables 1 through 6 were obtained using Farebrother's [10] implementation of Grad and Solomon's procedure. For $n - k > 3$ the relevant probabilities were obtained using an earlier Algol version of the Imhof procedure defined by Farebrother [6]. With the exception of a few errors noted by Farebrother [20], these results essentially confirm the figures produced by Savin and White [21].

Table 1. Durbin–Watson one percent lower bound assuming a unit column among the regressors

N	K=1	K=2	K=3	K=4	K=5	K=6	K=7	K=8	K=9	K=10	K=11	K=12	K=13	K=14	K=15	K=16	K=17	K=18	K=19	K=20	K=21
3	1.001																				
4	0.626	0.586																			
5	0.538	0.412	0.382																		
6	0.561	0.390	0.290	0.268																	
7	0.614	0.435	0.294	0.215	0.198																
8	0.665	0.497	0.345	0.229	0.166	0.152															
9	0.709	0.554	0.408	0.279	0.183	0.132	0.120														
10	0.752	0.605	0.466	0.340	0.230	0.150	0.107	0.098													
11	0.792	0.653	0.519	0.396	0.286	0.193	0.124	0.089	0.081												
12	0.828	0.697	0.569	0.449	0.339	0.244	0.164	0.105	0.075	0.068											
13	0.862	0.738	0.616	0.499	0.391	0.294	0.211	0.140	0.090	0.064	0.058										
14	0.893	0.776	0.660	0.547	0.441	0.343	0.257	0.183	0.122	0.078	0.055	0.050									
15	0.922	0.811	0.700	0.591	0.488	0.391	0.303	0.226	0.161	0.107	0.068	0.048	0.043								
16	0.949	0.844	0.737	0.633	0.532	0.437	0.349	0.269	0.200	0.142	0.094	0.060	0.042	0.038							
17	0.974	0.874	0.772	0.672	0.574	0.480	0.393	0.313	0.241	0.179	0.127	0.084	0.053	0.037	0.034						
18	0.998	0.902	0.805	0.708	0.613	0.522	0.435	0.355	0.282	0.217	0.160	0.113	0.075	0.047	0.033	0.030					
19	1.020	0.928	0.835	0.742	0.650	0.561	0.476	0.396	0.322	0.255	0.196	0.145	0.102	0.067	0.043	0.030	0.027				
20	1.041	0.952	0.863	0.773	0.685	0.598	0.515	0.436	0.362	0.294	0.232	0.178	0.131	0.093	0.061	0.038	0.027	0.024			
21	1.060	0.975	0.890	0.803	0.718	0.633	0.552	0.474	0.400	0.331	0.268	0.212	0.162	0.119	0.084	0.055	0.035	0.024	0.022		
22	1.078	0.997	0.914	0.831	0.748	0.667	0.587	0.510	0.437	0.368	0.305	0.246	0.194	0.148	0.109	0.077	0.050	0.032	0.022	0.020	
23	1.096	1.018	0.938	0.858	0.777	0.698	0.620	0.545	0.473	0.404	0.340	0.281	0.227	0.178	0.136	0.100	0.070	0.046	0.029	0.020	0.018
24	1.112	1.037	0.960	0.882	0.805	0.728	0.652	0.578	0.507	0.439	0.375	0.315	0.260	0.209	0.165	0.125	0.092	0.065	0.042	0.027	0.019
25	1.128	1.055	0.981	0.906	0.831	0.756	0.682	0.610	0.540	0.473	0.409	0.348	0.292	0.241	0.194	0.152	0.116	0.085	0.060	0.039	0.025
26	1.143	1.072	1.001	0.928	0.855	0.783	0.711	0.640	0.571	0.505	0.441	0.381	0.324	0.272	0.224	0.180	0.141	0.107	0.079	0.055	0.035
27	1.157	1.089	1.019	0.949	0.878	0.808	0.738	0.669	0.602	0.536	0.473	0.413	0.356	0.303	0.253	0.208	0.167	0.131	0.100	0.073	0.051
28	1.170	1.104	1.037	0.969	0.900	0.832	0.764	0.696	0.630	0.566	0.504	0.444	0.387	0.333	0.283	0.237	0.194	0.156	0.122	0.093	0.068
29	1.183	1.119	1.054	0.988	0.921	0.855	0.788	0.723	0.658	0.595	0.533	0.474	0.417	0.364	0.313	0.265	0.222	0.182	0.146	0.114	0.087
30	1.195	1.133	1.070	1.006	0.941	0.877	0.812	0.748	0.684	0.622	0.562	0.503	0.447	0.393	0.342	0.294	0.249	0.208	0.171	0.137	0.107
31	1.207	1.147	1.085	1.023	0.960	0.897	0.834	0.772	0.710	0.649	0.589	0.531	0.475	0.422	0.370	0.322	0.277	0.234	0.196	0.160	0.129
32	1.218	1.160	1.100	1.040	0.979	0.917	0.856	0.794	0.734	0.674	0.616	0.558	0.503	0.450	0.398	0.350	0.304	0.261	0.221	0.184	0.151
33	1.229	1.172	1.114	1.055	0.996	0.936	0.876	0.816	0.757	0.698	0.641	0.585	0.530	0.477	0.426	0.377	0.331	0.287	0.246	0.208	0.174
34	1.239	1.184	1.128	1.070	1.012	0.954	0.896	0.837	0.779	0.722	0.665	0.610	0.556	0.503	0.452	0.404	0.357	0.313	0.272	0.233	0.197
35	1.249	1.195	1.140	1.085	1.028	0.971	0.914	0.857	0.801	0.744	0.689	0.634	0.581	0.529	0.478	0.430	0.383	0.339	0.297	0.257	0.221
36	1.259	1.206	1.153	1.098	1.043	0.988	0.932	0.877	0.821	0.766	0.711	0.658	0.605	0.554	0.504	0.455	0.409	0.364	0.322	0.282	0.244

236

37	1.268	1.217	1.165	1.112	1.058	1.004	0.950	0.895	0.841	0.787	0.733	0.680	0.628	0.578	0.528	0.480	0.434	0.389	0.347	0.306	0.268
38	1.277	1.227	1.176	1.124	1.072	1.019	0.966	0.913	0.860	0.807	0.754	0.702	0.651	0.601	0.552	0.504	0.458	0.414	0.371	0.330	0.291
39	1.285	1.237	1.187	1.137	1.085	1.034	0.982	0.930	0.878	0.826	0.774	0.723	0.673	0.623	0.575	0.528	0.482	0.438	0.395	0.354	0.315
40	1.293	1.246	1.198	1.148	1.098	1.048	0.997	0.946	0.895	0.844	0.794	0.744	0.694	0.645	0.597	0.551	0.505	0.461	0.418	0.377	0.338
45	1.331	1.288	1.245	1.201	1.156	1.111	1.065	1.020	0.973	0.927	0.881	0.835	0.790	0.744	0.700	0.656	0.612	0.570	0.528	0.488	0.448
50	1.363	1.324	1.285	1.245	1.205	1.164	1.123	1.081	1.039	0.997	0.955	0.913	0.871	0.829	0.787	0.746	0.705	0.665	0.625	0.586	0.548
55	1.391	1.356	1.320	1.284	1.247	1.209	1.172	1.134	1.095	1.057	1.018	0.979	0.940	0.902	0.863	0.825	0.786	0.748	0.711	0.674	0.637
60	1.415	1.383	1.350	1.317	1.283	1.249	1.214	1.179	1.144	1.108	1.072	1.037	1.001	0.965	0.929	0.893	0.857	0.821	0.786	0.751	0.716
65	1.437	1.407	1.377	1.346	1.315	1.283	1.251	1.218	1.186	1.153	1.120	1.087	1.053	1.020	0.986	0.953	0.919	0.886	0.852	0.819	0.786
70	1.457	1.429	1.400	1.372	1.343	1.313	1.283	1.253	1.223	1.192	1.162	1.131	1.099	1.068	1.037	1.005	0.974	0.943	0.911	0.880	0.849
75	1.474	1.448	1.422	1.395	1.368	1.340	1.312	1.284	1.256	1.227	1.199	1.170	1.140	1.111	1.082	1.052	1.023	0.993	0.964	0.934	0.905
80	1.490	1.466	1.441	1.416	1.390	1.364	1.338	1.312	1.286	1.259	1.232	1.205	1.177	1.150	1.122	1.094	1.067	1.039	1.011	0.983	0.955
85	1.505	1.482	1.458	1.435	1.411	1.386	1.362	1.337	1.312	1.287	1.261	1.236	1.210	1.184	1.158	1.132	1.106	1.080	1.053	1.027	1.000
90	1.518	1.496	1.474	1.452	1.429	1.406	1.383	1.360	1.336	1.312	1.288	1.264	1.240	1.215	1.191	1.166	1.141	1.116	1.091	1.066	1.041
95	1.531	1.510	1.489	1.468	1.446	1.425	1.403	1.381	1.358	1.336	1.313	1.290	1.267	1.244	1.221	1.197	1.174	1.150	1.126	1.102	1.079
100	1.542	1.522	1.503	1.482	1.462	1.441	1.421	1.400	1.378	1.357	1.335	1.314	1.292	1.270	1.248	1.225	1.203	1.181	1.158	1.135	1.113
150	1.624	1.611	1.598	1.584	1.571	1.557	1.543	1.529	1.515	1.501	1.487	1.473	1.458	1.444	1.429	1.414	1.400	1.385	1.370	1.355	1.340
200	1.674	1.664	1.654	1.644	1.633	1.623	1.613	1.603	1.592	1.582	1.571	1.561	1.550	1.539	1.528	1.517	1.507	1.496	1.485	1.474	1.462

237

Table 2. Durbin–Watson two and a half percent lower bound assuming a unit column among the regressors

N	K=1	K=2	K=3	K=4	K=5	K=6	K=7	K=8	K=9	K=10	K=11	K=12	K=13	K=14	K=15	K=16	K=17	K=18	K=19	K=20	K=21
3	1.003																				
4	0.684	0.588																			
5	0.664	0.456	0.384																		
6	0.721	0.489	0.323	0.269																	
7	0.779	0.564	0.372	0.241	0.199																
8	0.825	0.633	0.450	0.291	0.186	0.153															
9	0.870	0.690	0.520	0.365	0.234	0.148	0.121														
10	0.911	0.744	0.582	0.433	0.302	0.191	0.120	0.098													
11	0.949	0.793	0.640	0.495	0.366	0.253	0.160	0.100	0.082												
12	0.983	0.838	0.693	0.555	0.426	0.312	0.215	0.135	0.084	0.069											
13	1.014	0.878	0.742	0.609	0.484	0.369	0.269	0.185	0.115	0.072	0.059										
14	1.043	0.915	0.786	0.659	0.538	0.425	0.323	0.234	0.160	0.100	0.062	0.051									
15	1.070	0.949	0.827	0.706	0.589	0.478	0.376	0.284	0.206	0.140	0.087	0.054	0.044								
16	1.094	0.980	0.864	0.749	0.636	0.528	0.427	0.334	0.252	0.185	0.124	0.077	0.048	0.039							
17	1.117	1.009	0.898	0.788	0.680	0.575	0.475	0.383	0.299	0.225	0.162	0.110	0.068	0.042	0.034						
18	1.138	1.035	0.930	0.825	0.720	0.619	0.521	0.430	0.345	0.269	0.202	0.145	0.099	0.061	0.038	0.030					
19	1.158	1.060	0.960	0.859	0.758	0.660	0.565	0.475	0.390	0.313	0.243	0.182	0.131	0.089	0.055	0.034	0.027				
20	1.177	1.083	0.987	0.890	0.794	0.699	0.606	0.517	0.434	0.356	0.285	0.221	0.165	0.118	0.080	0.050	0.030	0.025			
21	1.194	1.104	1.013	0.920	0.827	0.735	0.645	0.558	0.475	0.397	0.325	0.260	0.202	0.150	0.108	0.073	0.045	0.028	0.022		
22	1.210	1.124	1.036	0.947	0.858	0.769	0.681	0.597	0.515	0.438	0.365	0.299	0.238	0.184	0.138	0.098	0.067	0.041	0.025	0.020	
23	1.226	1.143	1.059	0.973	0.887	0.801	0.716	0.633	0.553	0.477	0.404	0.337	0.275	0.219	0.169	0.126	0.090	0.061	0.038	0.023	0.018
24	1.240	1.161	1.080	0.997	0.914	0.831	0.748	0.668	0.589	0.514	0.442	0.375	0.312	0.254	0.202	0.156	0.116	0.083	0.056	0.035	0.021
25	1.254	1.178	1.099	1.020	0.939	0.859	0.779	0.701	0.624	0.550	0.479	0.411	0.348	0.289	0.235	0.187	0.144	0.107	0.077	0.052	0.032
26	1.267	1.193	1.118	1.041	0.963	0.880	0.808	0.732	0.657	0.604	0.514	0.446	0.383	0.324	0.269	0.219	0.173	0.134	0.099	0.071	0.048
27	1.279	1.208	1.135	1.061	0.986	0.911	0.836	0.761	0.688	0.616	0.547	0.481	0.417	0.358	0.302	0.250	0.203	0.161	0.124	0.092	0.066
28	1.291	1.223	1.152	1.080	1.008	0.935	0.862	0.789	0.718	0.648	0.579	0.514	0.451	0.391	0.335	0.282	0.234	0.190	0.150	0.116	0.086

n																					
29	0.108	0.141	0.178	0.219	0.264	0.314	0.367	0.423	0.483	0.546	0.610	0.677	0.745	0.816	0.886	0.957	1.028	1.098	1.168	1.236	1.303
30	0.132	0.166	0.205	0.248	0.295	0.345	0.398	0.455	0.514	0.576	0.640	0.706	0.773	0.841	0.910	0.979	1.047	1.116	1.183	1.249	1.313
31	0.156	0.193	0.233	0.277	0.325	0.375	0.429	0.486	0.544	0.606	0.668	0.733	0.799	0.865	0.932	0.999	1.066	1.132	1.197	1.261	1.323
32	0.181	0.220	0.261	0.306	0.354	0.405	0.459	0.515	0.574	0.634	0.696	0.759	0.823	0.888	0.953	1.018	1.083	1.147	1.211	1.273	1.330
33	0.207	0.246	0.289	0.335	0.383	0.434	0.488	0.544	0.602	0.661	0.722	0.784	0.846	0.910	0.973	1.037	1.100	1.162	1.224	1.284	1.343
34	0.233	0.273	0.317	0.363	0.412	0.463	0.516	0.572	0.629	0.687	0.747	0.807	0.869	0.931	0.992	1.054	1.116	1.176	1.236	1.295	1.352
35	0.259	0.300	0.344	0.391	0.440	0.491	0.544	0.598	0.655	0.712	0.771	0.830	0.890	0.950	1.011	1.071	1.131	1.190	1.248	1.305	1.360
36	0.285	0.327	0.371	0.418	0.467	0.517	0.570	0.624	0.680	0.736	0.794	0.852	0.911	0.969	1.028	1.087	1.145	1.203	1.259	1.315	1.369
37	0.311	0.353	0.398	0.444	0.493	0.544	0.596	0.649	0.704	0.759	0.816	0.873	0.930	0.988	1.045	1.102	1.159	1.215	1.270	1.324	1.377
38	0.336	0.379	0.424	0.470	0.519	0.569	0.620	0.673	0.727	0.782	0.837	0.893	0.949	1.005	1.061	1.117	1.172	1.227	1.280	1.333	1.384
39	0.361	0.404	0.449	0.496	0.544	0.593	0.644	0.696	0.749	0.803	0.857	0.912	0.967	1.022	1.077	1.131	1.185	1.238	1.290	1.342	1.392
40	0.386	0.429	0.474	0.520	0.568	0.617	0.668	0.719	0.771	0.824	0.877	0.930	0.984	1.038	1.091	1.145	1.197	1.249	1.300	1.350	1.399
45	0.503	0.545	0.589	0.634	0.679	0.725	0.772	0.820	0.868	0.916	0.964	1.013	1.061	1.109	1.157	1.205	1.251	1.298	1.343	1.388	1.431
50	0.607	0.648	0.689	0.732	0.774	0.818	0.861	0.905	0.949	0.993	1.037	1.081	1.125	1.169	1.212	1.255	1.297	1.339	1.380	1.420	1.459
55	0.699	0.738	0.777	0.816	0.856	0.897	0.937	0.977	1.018	1.058	1.099	1.139	1.179	1.219	1.258	1.297	1.335	1.373	1.411	1.447	1.483
60	0.780	0.816	0.853	0.890	0.927	0.965	1.002	1.040	1.077	1.115	1.152	1.189	1.225	1.262	1.298	1.334	1.369	1.404	1.438	1.471	1.504
65	0.851	0.885	0.920	0.955	0.989	1.024	1.059	1.094	1.129	1.163	1.198	1.232	1.266	1.299	1.333	1.365	1.398	1.430	1.461	1.492	1.523
70	0.914	0.946	0.979	1.011	1.044	1.076	1.109	1.141	1.174	1.206	1.238	1.269	1.301	1.332	1.363	1.393	1.424	1.453	1.482	1.511	1.540
75	0.970	1.000	1.031	1.061	1.092	1.122	1.153	1.183	1.213	1.243	1.273	1.303	1.332	1.361	1.390	1.418	1.446	1.474	1.501	1.528	1.555
80	1.019	1.048	1.077	1.106	1.135	1.163	1.192	1.220	1.249	1.277	1.305	1.332	1.360	1.387	1.414	1.441	1.467	1.493	1.518	1.544	1.568
85	1.064	1.092	1.119	1.146	1.173	1.200	1.227	1.254	1.280	1.307	1.333	1.359	1.385	1.410	1.436	1.461	1.485	1.510	1.534	1.558	1.581
90	1.105	1.131	1.156	1.182	1.208	1.233	1.259	1.284	1.309	1.334	1.359	1.383	1.407	1.431	1.455	1.479	1.502	1.525	1.548	1.570	1.592
95	1.141	1.166	1.190	1.215	1.239	1.263	1.287	1.311	1.335	1.358	1.382	1.405	1.428	1.451	1.473	1.495	1.518	1.539	1.561	1.582	1.603
100	1.175	1.198	1.221	1.244	1.267	1.290	1.313	1.336	1.358	1.381	1.403	1.425	1.447	1.468	1.490	1.511	1.532	1.552	1.573	1.593	1.613
150	1.395	1.410	1.425	1.440	1.456	1.471	1.485	1.500	1.515	1.529	1.544	1.558	1.573	1.587	1.601	1.615	1.629	1.642	1.656	1.669	1.683
200	1.511	1.523	1.534	1.545	1.556	1.567	1.578	1.589	1.600	1.611	1.621	1.632	1.643	1.653	1.663	1.674	1.684	1.694	1.704	1.715	1.724

Table 3. Durbin–Watson five percent lower bound assuming a unit column among the regressors

N	K=1	K=2	K=3	K=4	K=5	K=6	K=7	K=8	K=9	K=10	K=11	K=12	K=13	K=14	K=15	K=16	K=17	K=18	K=19	K=20	K=21
3	1.012																				
4	0.780	0.594																			
5	0.820	0.527	0.388																		
6	0.890	0.610	0.377	0.272																	
7	0.936	0.700	0.467	0.282	0.201																
8	0.982	0.763	0.559	0.367	0.219	0.155															
9	1.024	0.924	0.629	0.455	0.296	0.174	0.123														
10	1.062	0.879	0.697	0.525	0.376	0.243	0.142	0.100													
11	1.096	0.927	0.758	0.595	0.444	0.316	0.203	0.171	0.083												
12	1.128	0.971	0.812	0.658	0.512	0.380	0.268	0.188	0.099	0.069											
13	1.156	1.010	0.861	0.715	0.574	0.444	0.328	0.230	0.147	0.085	0.059										
14	1.182	1.045	0.905	0.767	0.632	0.505	0.389	0.286	0.200	0.127	0.073	0.051									
15	1.205	1.077	0.946	0.814	0.685	0.562	0.447	0.343	0.251	0.175	0.111	0.064	0.045								
16	1.227	1.106	0.982	0.857	0.734	0.615	0.502	0.398	0.304	0.222	0.155	0.098	0.056	0.039							
17	1.247	1.133	1.015	0.897	0.779	0.664	0.554	0.451	0.356	0.272	0.198	0.138	0.087	0.050	0.035						
18	1.266	1.158	1.046	0.933	0.820	0.710	0.603	0.502	0.407	0.321	0.244	0.177	0.123	0.078	0.045	0.031					
19	1.283	1.180	1.074	0.967	0.859	0.752	0.649	0.549	0.456	0.369	0.290	0.220	0.160	0.111	0.070	0.040	0.028				
20	1.300	1.201	1.100	0.998	0.894	0.792	0.691	0.595	0.502	0.416	0.336	0.263	0.200	0.145	0.100	0.063	0.036	0.025			
21	1.315	1.221	1.125	1.026	0.927	0.829	0.731	0.637	0.546	0.461	0.380	0.307	0.240	0.182	0.132	0.091	0.057	0.033	0.023		
22	1.329	1.239	1.147	1.053	0.958	0.863	0.769	0.677	0.588	0.504	0.424	0.349	0.281	0.220	0.166	0.120	0.083	0.052	0.030	0.021	
23	1.342	1.257	1.168	1.078	0.986	0.895	0.804	0.715	0.628	0.545	0.465	0.391	0.322	0.259	0.202	0.153	0.110	0.076	0.048	0.027	0.019
24	1.355	1.273	1.188	1.101	1.013	0.925	0.837	0.750	0.666	0.584	0.506	0.431	0.362	0.297	0.239	0.186	0.141	0.101	0.070	0.044	0.025
25	1.367	1.288	1.206	1.123	1.038	0.953	0.868	0.784	0.702	0.621	0.544	0.470	0.400	0.335	0.275	0.221	0.172	0.130	0.094	0.065	0.041
26	1.378	1.302	1.224	1.143	1.062	0.979	0.897	0.816	0.735	0.657	0.581	0.508	0.438	0.373	0.312	0.256	0.205	0.160	0.120	0.087	0.060
27	1.389	1.316	1.240	1.162	1.084	1.004	0.925	0.845	0.767	0.691	0.616	0.544	0.475	0.409	0.348	0.291	0.238	0.191	0.149	0.112	0.081
28	1.399	1.328	1.255	1.181	1.104	1.028	0.951	0.874	0.798	0.723	0.649	0.578	0.510	0.445	0.383	0.325	0.271	0.222	0.178	0.138	0.104

29	1.409	1.341	1.270	1.198	1.124	1.050	0.975	0.900	0.826	0.753	0.681	0.612	0.544	0.479	0.418	0.359	0.305	0.254	0.208	0.166	0.129
30	1.418	1.352	1.284	1.214	1.143	1.071	0.998	0.926	0.854	0.782	0.712	0.643	0.577	0.513	0.451	0.393	0.337	0.286	0.238	0.195	0.156
31	1.427	1.363	1.297	1.229	1.160	1.090	1.020	0.950	0.879	0.810	0.741	0.674	0.608	0.545	0.484	0.425	0.370	0.317	0.269	0.224	0.183
32	1.436	1.373	1.309	1.244	1.177	1.109	1.041	0.972	0.904	0.836	0.769	0.703	0.638	0.576	0.515	0.457	0.401	0.349	0.299	0.253	0.211
33	1.444	1.383	1.321	1.258	1.193	1.127	1.061	0.994	0.927	0.861	0.796	0.731	0.667	0.606	0.546	0.488	0.432	0.379	0.329	0.282	0.239
34	1.451	1.393	1.333	1.271	1.208	1.144	1.079	1.015	0.950	0.885	0.821	0.758	0.695	0.634	0.575	0.518	0.462	0.409	0.359	0.312	0.267
35	1.459	1.402	1.343	1.283	1.222	1.160	1.097	1.034	0.971	0.908	0.845	0.783	0.722	0.662	0.603	0.547	0.492	0.439	0.388	0.340	0.295
36	1.466	1.411	1.354	1.295	1.236	1.175	1.114	1.053	0.991	0.930	0.868	0.808	0.748	0.689	0.631	0.575	0.520	0.467	0.417	0.369	0.323
37	1.473	1.419	1.364	1.307	1.249	1.190	1.131	1.071	1.011	0.951	0.891	0.831	0.772	0.714	0.657	0.602	0.548	0.495	0.445	0.397	0.351
38	1.480	1.427	1.373	1.318	1.261	1.204	1.146	1.088	1.029	0.970	0.912	0.854	0.796	0.739	0.683	0.628	0.574	0.523	0.472	0.424	0.378
39	1.486	1.435	1.382	1.328	1.273	1.218	1.161	1.104	1.047	0.990	0.932	0.875	0.819	0.763	0.707	0.653	0.600	0.549	0.499	0.451	0.404
40	1.492	1.442	1.391	1.338	1.285	1.230	1.175	1.120	1.064	1.008	0.952	0.896	0.840	0.785	0.731	0.678	0.626	0.575	0.525	0.477	0.431
45	1.520	1.475	1.430	1.383	1.336	1.287	1.238	1.189	1.139	1.089	1.038	0.988	0.938	0.888	0.837	0.788	0.740	0.691	0.644	0.598	0.553
50	1.543	1.503	1.462	1.421	1.378	1.335	1.291	1.246	1.201	1.156	1.110	1.064	1.019	0.973	0.927	0.882	0.836	0.792	0.747	0.703	0.660
55	1.564	1.528	1.490	1.452	1.414	1.374	1.334	1.294	1.253	1.212	1.170	1.129	1.087	1.045	1.003	0.961	0.919	0.877	0.836	0.795	0.754
60	1.582	1.549	1.514	1.480	1.444	1.408	1.372	1.335	1.298	1.260	1.222	1.184	1.145	1.106	1.068	1.029	0.990	0.951	0.913	0.874	0.836
65	1.598	1.567	1.536	1.503	1.471	1.438	1.404	1.370	1.336	1.301	1.266	1.231	1.195	1.160	1.124	1.088	1.052	1.016	0.980	0.944	0.908
70	1.612	1.583	1.554	1.525	1.494	1.464	1.433	1.401	1.369	1.337	1.305	1.272	1.239	1.206	1.172	1.139	1.105	1.072	1.038	1.005	0.971
75	1.625	1.598	1.571	1.543	1.515	1.487	1.458	1.428	1.399	1.369	1.339	1.308	1.277	1.247	1.215	1.184	1.153	1.121	1.090	1.058	1.027
80	1.636	1.611	1.586	1.560	1.534	1.507	1.480	1.453	1.425	1.397	1.369	1.340	1.312	1.283	1.253	1.224	1.195	1.165	1.136	1.106	1.076
85	1.647	1.624	1.600	1.575	1.550	1.525	1.500	1.474	1.448	1.422	1.396	1.369	1.342	1.315	1.287	1.260	1.232	1.205	1.177	1.149	1.121
90	1.657	1.635	1.612	1.589	1.566	1.542	1.518	1.494	1.469	1.445	1.420	1.395	1.369	1.344	1.318	1.292	1.266	1.240	1.213	1.187	1.161
95	1.666	1.645	1.623	1.602	1.579	1.557	1.535	1.512	1.489	1.465	1.442	1.418	1.394	1.370	1.345	1.321	1.296	1.272	1.247	1.222	1.197
100	1.674	1.654	1.634	1.613	1.592	1.571	1.550	1.528	1.506	1.484	1.462	1.439	1.416	1.394	1.370	1.347	1.324	1.300	1.277	1.253	1.229
150	1.733	1.720	1.706	1.693	1.679	1.665	1.651	1.637	1.622	1.608	1.593	1.579	1.564	1.549	1.534	1.519	1.504	1.489	1.474	1.458	1.443
200	1.768	1.758	1.748	1.738	1.728	1.718	1.707	1.697	1.686	1.675	1.665	1.654	1.643	1.632	1.621	1.610	1.599	1.588	1.577	1.565	1.554

Table 4. Durbin–Watson one percent upper bound assuming a unit column among the regressors

N	K=0	K=1	K=2	K=3	K=4	K=5	K=6	K=7	K=8	K=9	K=10	K=11	K=12	K=13	K=14	K=15	K=16	K=17	K=18	K=19	K=20	K=21
2	0.001																					
3	0.034	1.001																				
4	0.127	0.626	2.001																			
5	0.233	0.538	1.415	2.619																		
6	0.322	0.561	1.142	2.026	3.000																	
7	0.398	0.614	1.036	1.676	2.466	3.247																
8	0.469	0.665	1.003	1.489	2.102	2.782	3.414															
9	0.534	0.709	0.998	1.389	1.875	2.433	3.014	3.532														
10	0.591	0.752	1.001	1.332	1.733	2.193	2.690	3.187	3.618													
11	0.643	0.792	1.010	1.297	1.641	2.029	2.453	2.892	3.319	3.682												
12	0.691	0.828	1.023	1.274	1.575	1.913	2.280	2.665	3.053	3.422	3.732											
13	0.733	0.862	1.038	1.261	1.526	1.826	2.150	2.490	2.838	3.182	3.504	3.771										
14	0.773	0.893	1.054	1.254	1.490	1.757	2.049	2.354	2.667	2.981	6.287	3.570	3.802									
15	0.809	0.922	1.070	1.252	1.464	1.704	1.967	2.244	2.530	2.817	3.101	3.374	3.623	3.827								
16	0.842	0.949	1.086	1.252	1.446	1.663	1.900	2.153	2.416	2.681	2.944	3.201	3.446	3.668	3.847							
17	0.873	0.974	1.102	1.255	1.432	1.630	1.847	2.078	2.320	2.566	2.811	3.053	3.286	3.506	3.705	3.865						
18	0.901	0.998	1.118	1.259	1.422	1.604	1.803	2.015	2.238	2.467	2.697	2.925	3.146	3.358	3.557	3.736	3.879					
19	0.928	1.020	1.132	1.265	1.415	1.584	1.767	1.963	2.169	2.381	2.598	2.913	3.023	3.227	3.420	3.601	3.762	3.891				
20	0.952	1.041	1.147	1.271	1.411	1.567	1.737	1.918	2.110	2.308	5.510	2.714	2.914	3.109	3.297	3.474	3.639	3.785	3.902			
21	0.976	1.060	1.161	1.277	1.408	1.554	1.712	1.881	2.059	2.244	2.434	2.625	2.817	3.004	3.185	3.358	3.521	3.671	3.805	3.911		
22	0.997	1.078	1.174	1.284	1.407	1.543	1.691	1.849	2.015	2.188	2.367	2.548	2.729	2.909	3.084	3.252	3.412	3.562	3.700	3.822	3.919	
23	1.018	1.096	1.187	1.291	1.407	1.535	1.673	1.821	1.977	2.140	2.308	2.479	2.651	2.822	2.991	3.155	3.311	3.459	3.597	3.725	3.837	3.925
24	1.037	1.112	1.199	1.298	1.407	1.528	1.658	1.797	1.944	2.097	2.255	2.417	2.581	2.744	2.906	3.065	3.218	3.363	3.501	3.629	3.747	3.850
25	1.056	1.128	1.211	1.305	1.409	1.523	1.645	1.777	1.915	2.059	2.209	2.362	2.517	2.674	2.829	2.982	3.131	3.274	3.410	3.538	3.657	3.766
26	1.073	1.143	1.222	1.312	1.411	1.518	1.635	1.759	1.889	2.026	2.168	2.313	2.461	2.610	2.758	2.906	3.050	3.191	3.325	3.452	3.572	3.682
27	1.089	1.157	1.233	1.319	1.413	1.515	1.626	1.743	1.867	1.997	2.131	2.269	2.409	2.552	2.694	2.836	2.976	3.113	3.245	3.371	3.490	3.602

28	1.105	1.170	1.244	1.325	1.415	1.513	1.618	1.730	1.847	1.970	2.098	2.229	2.363	2.499	2.635	2.772	2.907	3.040	3.169	3.294	3.412	3.524
29	1.120	1.183	1.254	1.332	1.418	1.512	1.611	1.718	1.830	1.947	2.068	2.193	2.321	2.451	2.582	2.713	2.843	2.972	3.098	3.221	3.338	3.450
30	1.134	1.195	1.263	1.339	1.421	1.511	1.606	1.707	1.814	1.925	2.041	2.160	2.283	2.407	2.533	2.659	2.785	2.909	3.032	3.152	3.268	3.379
31	1.147	1.207	1.273	1.345	1.425	1.510	1.601	1.698	1.800	1.906	2.017	2.131	2.248	2.367	2.487	2.609	2.730	2.851	2.970	3.087	3.201	3.311
32	1.160	1.218	1.282	1.352	1.428	1.510	1.597	1.690	1.787	1.889	1.995	2.104	2.216	2.330	2.446	2.563	2.680	2.797	2.912	3.026	3.137	3.246
33	1.173	1.229	1.291	1.358	1.432	1.510	1.594	1.683	1.776	1.874	1.975	2.080	2.187	2.296	2.408	2.520	2.633	2.746	2.858	2.969	3.078	3.184
34	1.185	1.239	1.299	1.364	1.435	1.511	1.591	1.677	1.766	1.860	1.957	2.057	2.160	2.265	2.373	2.481	2.590	2.699	2.808	2.916	3.022	3.126
35	1.196	1.249	1.307	1.370	1.439	1.512	1.589	1.671	1.757	1.847	1.940	2.037	2.136	2.237	2.340	2.444	2.550	2.655	2.761	2.865	2.969	3.071
36	1.207	1.259	1.315	1.376	1.442	1.513	1.588	1.666	1.749	1.836	1.925	2.018	2.113	2.211	2.310	2.411	2.512	2.614	2.716	2.818	2.919	3.019
37	1.217	1.268	1.323	1.382	1.446	1.514	1.586	1.662	1.742	1.825	1.911	2.001	2.092	2.186	2.282	2.379	2.477	2.576	2.675	2.774	2.872	2.970
38	1.228	1.277	1.330	1.388	1.449	1.515	1.585	1.658	1.735	1.816	1.899	1.985	2.073	2.164	2.256	2.350	2.445	2.540	2.637	2.733	2.828	2.923
39	1.237	1.285	1.337	1.393	1.453	1.517	1.584	1.655	1.729	1.807	1.887	1.970	2.055	2.143	2.232	2.323	2.414	2.507	2.600	2.694	2.787	2.879
40	1.247	1.293	1.344	1.398	1.457	1.518	1.584	1.652	1.724	1.799	1.876	1.956	2.039	2.123	2.209	2.297	2.386	2.476	2.566	2.657	2.747	2.838
45	1.289	1.331	1.376	1.423	1.474	1.528	1.584	1.643	1.704	1.768	1.834	1.902	1.972	2.044	2.118	2.193	2.269	2.346	2.424	2.503	2.582	2.661
50	1.325	1.363	1.403	1.446	1.491	1.538	1.587	1.639	1.692	1.748	1.805	1.864	1.925	1.987	2.051	2.116	2.182	2.250	2.318	2.387	2.456	2.526
55	1.356	1.391	1.427	1.466	1.506	1.548	1.592	1.638	1.685	1.734	1.785	1.837	1.891	1.945	2.001	2.059	2.117	2.176	2.237	2.298	2.359	2.421
60	1.383	1.415	1.449	1.484	1.520	1.558	1.598	1.639	1.682	1.726	1.771	1.817	1.865	1.914	1.964	2.015	2.067	2.119	2.173	2.228	2.283	2.338
65	1.408	1.437	1.468	1.500	1.534	1.568	1.604	1.642	1.680	1.720	1.761	1.803	1.845	1.889	1.934	1.980	2.027	2.074	2.123	2.172	2.221	2.272
70	1.429	1.457	1.485	1.515	1.546	1.578	1.611	1.645	1.680	1.716	1.753	1.792	1.831	1.871	1.911	1.953	1.995	2.038	2.082	2.127	2.172	2.217
75	1.448	1.474	1.501	1.529	1.557	1.587	1.617	1.649	1.681	1.714	1.748	1.783	1.819	1.856	1.893	1.931	1.970	2.009	2.049	2.090	2.131	2.172
80	1.466	1.490	1.515	1.541	1.568	1.595	1.624	1.653	1.683	1.714	1.745	1.777	1.810	1.844	1.878	1.913	1.949	1.985	2.022	2.059	2.097	2.135
85	1.482	1.505	1.528	1.553	1.578	1.603	1.630	1.657	1.685	1.714	1.743	1.773	1.803	1.834	1.866	1.898	1.931	1.965	1.999	2.033	2.068	2.104
90	1.497	1.518	1.540	1.563	1.587	1.611	1.636	1.661	1.687	1.714	1.741	1.769	1.798	1.827	1.856	1.886	1.917	1.948	1.980	2.012	2.044	2.077
95	1.510	1.531	1.552	1.573	1.596	1.618	1.642	1.666	1.690	1.715	1.741	1.767	1.793	1.821	1.848	1.876	1.905	1.934	1.963	1.993	2.023	2.054
100	1.523	1.542	1.562	1.583	1.604	1.625	1.647	1.670	1.693	1.717	1.741	1.765	1.790	1.816	1.842	1.868	1.895	1.922	1.949	1.977	2.006	2.034
150	1.611	1.624	1.637	1.651	1.665	1.679	1.693	1.707	1.722	1.737	1.752	1.767	1.783	1.798	1.814	1.830	1.847	1.863	1.880	1.897	1.914	1.931
200	1.664	1.674	1.684	1.694	1.704	1.714	1.725	1.735	1.746	1.757	1.768	1.779	1.790	1.801	1.813	1.824	1.836	1.848	1.860	1.872	1.884	1.896

Table 5. Durbin–Watson two and a half percent upper bound assuming a unit column among the regressors

N	K=0	K=1	K=2	K=3	K=4	K=5	K=6	K=7	K=8	K=9	K=10	K=11	K=12	K=13	K=14	K=15	K=16	K=17	K=18	K=19	K=20	K=21
2	0.003																					
3	0.085	1.003																				
4	0.230	0.684	2.002																			
5	0.356	0.664	1.464	2.620																		
6	0.448	0.721	1.258	2.065	3.001																	
7	0.536	0.779	1.191	1.775	2.497	3.248																
8	0.612	0.825	1.172	1.628	2.184	2.807	3.415															
9	0.678	0.870	1.163	1.547	1.997	2.502	3.034	3.533														
10	0.737	0.911	1.165	1.493	1.877	2.300	2.749	3.204	3.619													
11	0.789	0.949	1.173	1.456	1.791	2.158	2.545	2.942	3.333	3.683												
12	0.835	0.983	1.183	1.433	1.725	2.051	2.394	2.745	3.096	3.434	3.733											
13	0.877	1.014	1.196	1.418	1.677	1.966	2.276	2.592	2.909	3.219	3.514	3.771										
14	0.915	1.043	1.209	1.409	1.641	1.900	2.178	2.468	2.759	3.044	3.320	3.579	3.802									
15	0.949	1.070	1.222	1.405	1.614	1.848	2.099	2.363	2.633	2.899	3.156	3.403	3.631	3.827								
16	0.981	1.094	1.235	1.403	1.594	1.806	2.035	2.277	2.525	2.775	3.018	3.251	3.471	3.675	3.848							
17	1.010	1.117	1.249	1.403	1.578	1.773	1.983	2.204	2.433	2.666	2.897	3.119	3.330	3.529	3.711	3.865						
18	1.036	1.138	1.261	1.405	1.567	1.746	1.939	2.143	2.356	2.572	2.790	3.003	3.206	3.398	3.578	3.741	3.879					
19	1.061	1.158	1.274	1.407	1.558	1.723	1.902	2.092	2.289	2.491	2.695	2.898	3.095	3.281	3.457	3.620	3.767	3.892				
20	1.084	1.177	1.286	1.411	1.551	1.705	1.871	2.047	2.231	2.421	2.612	2.804	2.993	3.175	3.347	3.507	3.655	3.790	3.902			
21	1.106	1.194	1.297	1.415	1.546	1.690	1.845	2.009	2.181	2.359	2.539	2.721	2.901	3.076	3.245	3.404	3.551	3.687	3.809	3.911		
22	1.126	1.210	1.308	1.419	1.543	1.678	1.823	1.977	2.138	2.305	2.475	2.647	2.818	2.987	3.151	3.307	3.454	3.589	3.714	3.826	3.919	
23	1.145	1.226	1.319	1.424	1.541	1.668	1.804	1.948	2.100	2.257	2.418	2.580	2.744	2.905	3.064	3.217	3.362	3.498	3.623	3.737	3.840	3.926
24	1.162	1.240	1.329	1.429	1.539	1.659	1.788	1.924	2.066	2.214	2.366	2.521	2.676	2.831	2.984	3.132	3.275	3.411	3.537	3.653	3.758	3.853
25	1.179	1.254	1.339	1.434	1.539	1.652	1.773	1.902	2.037	2.177	2.321	2.468	2.616	2.764	2.911	3.055	3.194	3.328	3.454	3.571	3.679	3.777
26	1.195	1.267	1.349	1.439	1.539	1.646	1.761	1.883	2.011	2.143	2.280	2.420	2.561	2.703	2.844	2.983	3.119	3.250	3.375	3.493	3.602	3.702

27	28	29	30	31	32	33	34	35	36	37	38	39	40	45	50	55	60	65	70	75	80	85	90	95	100	150	200
3.630	3.560	3.491	3.425	3.361	3.300	3.242	3.187	3.135	3.085	3.038	2.993	2.951	2.911	2.739	2.606	2.502	2.418	2.351	2.295	2.249	2.210	2.177	2.149	2.125	2.104	1.989	1.947
3.528	3.456	3.387	3.321	3.258	3.198	3.142	3.089	3.038	2.991	2.946	2.903	2.863	2.825	2.663	2.539	2.442	2.364	2.302	2.250	2.208	2.172	2.142	2.117	2.095	2.075	1.972	1.935
3.418	3.346	3.277	3.212	3.151	3.094	3.039	2.988	2.940	2.895	2.852	2.812	2.774	2.738	2.587	2.471	2.382	2.310	2.253	2.206	2.168	2.135	2.108	2.085	2.065	2.047	1.955	1.923
3.300	3.229	3.162	3.100	3.041	2.986	2.934	2.886	2.840	2.798	2.758	2.720	2.684	2.651	2.511	2.404	2.322	2.257	2.205	2.163	2.128	2.098	2.074	2.053	2.035	2.020	1.938	1.911
3.176	3.108	3.044	2.984	2.928	2.876	2.828	2.782	2.740	2.700	2.663	2.628	2.595	2.564	2.435	2.338	2.263	2.205	2.158	2.119	2.088	2.062	2.040	2.022	2.006	1.992	1.922	1.899
3.048	2.983	2.922	2.866	2.814	2.765	2.720	2.678	2.639	2.602	2.568	2.536	2.506	2.478	2.360	2.273	2.205	2.153	2.111	2.077	2.049	2.026	2.007	1.991	1.977	1.965	1.905	1.887
2.916	2.855	2.799	2.747	2.698	2.653	2.612	2.573	2.538	2.504	2.473	2.444	2.417	2.392	2.286	2.208	2.148	2.102	2.065	2.035	2.011	1.991	1.975	1.961	1.949	1.939	1.889	1.875
2.782	2.726	2.674	2.626	2.582	2.542	2.504	2.469	2.437	2.407	2.379	2.353	2.329	2.307	2.213	2.144	2.092	2.051	2.020	1.994	1.973	1.957	1.942	1.931	1.921	1.913	1.873	1.864
2.647	2.596	2.549	2.506	2.467	2.430	2.397	2.366	2.338	2.311	2.287	2.264	2.242	2.223	2.141	2.082	2.037	2.002	1.975	1.954	1.937	1.922	1.911	1.901	1.893	1.887	1.857	1.853
2.511	2.466	2.424	2.387	2.352	2.320	2.291	2.264	2.239	2.217	2.195	2.176	2.157	2.140	2.071	2.020	1.983	1.954	1.932	1.914	1.900	1.889	1.880	1.873	1.866	1.861	1.842	1.841
2.376	2.337	2.301	2.269	2.239	2.212	2.187	2.164	2.143	2.124	2.106	2.089	2.074	2.060	2.002	1.960	1.930	1.907	1.889	1.875	1.865	1.856	1.850	1.844	1.840	1.837	1.826	1.830
2.243	2.210	2.180	2.153	2.128	2.106	2.085	2.066	2.049	2.033	2.019	2.005	1.993	1.981	1.935	1.902	1.878	1.861	1.847	1.838	1.830	1.824	1.820	1.816	1.814	1.812	1.811	1.819
2.113	2.087	2.062	2.040	2.021	2.003	1.986	1.971	1.958	1.945	1.934	1.923	1.914	1.905	1.869	1.845	1.828	1.816	1.807	1.801	1.796	1.793	1.791	1.789	1.788	1.788	1.796	1.808
1.987	1.967	1.948	1.931	1.916	1.903	1.891	1.880	1.869	1.860	1.852	1.844	1.837	1.831	1.806	1.790	1.779	1.772	1.767	1.764	1.763	1.762	1.762	1.763	1.764	1.766	1.781	1.797
1.866	1.851	1.838	1.827	1.816	1.807	1.799	1.791	1.784	1.778	1.773	1.768	1.764	1.760	1.745	1.737	1.732	1.730	1.729	1.729	1.730	1.732	1.734	1.736	1.739	11.739	1.766	1.786
1.751	1.742	1.734	1.727	1.721	1.715	1.711	1.707	1.703	1.700	1.697	1.695	1.693	1.692	1.687	1.685	1.686	1.688	1.692	1.695	1.699	1.703	1.707	1.711	1.715	1.719	1.752	1.776
1.641	1.638	1.634	1.632	1.630	1.628	1.627	1.626	1.626	1.625	1.625	1.626	1.626	1.626	1.630	1.636	1.642	1.649	1.655	1.662	1.668	1.674	1.680	1.686	1.691	1.696	1.737	1.765
1.539	1.540	1.541	1.543	1.544	1.546	1.548	1.550	1.553	1.555	1.557	1.560	1.562	1.564	1.577	1.588	1.600	1.610	1.620	1.630	1.638	1.647	1.654	1.662	1.668	1.675	1.723	1.755
1.445	1.450	1.455	1.460	1.465	1.470	1.474	1.479	1.484	1.488	1.493	1.497	1.502	1.506	1.525	1.543	1.559	1.573	1.587	1.599	1.610	1.620	1.629	1.638	1.646	1.654	1.710	1.745
1.358	1.367	1.375	1.383	1.391	1.399	1.406	1.413	1.420	1.426	1.433	1.439	1.445	1.451	1.477	1.500	1.520	1.538	1.554	1.569	1.582	1.594	1.605	1.615	1.624	1.633	1.696	1.735
1.279	1.291	1.303	1.313	1.323	1.333	1.343	1.352	1.360	1.369	1.377	1.384	1.392	1.399	1.431	1.459	1.483	1.504	1.523	1.540	1.555	1.568	1.581	1.592	1.603	1.613	1.683	1.724
1.210	1.224	1.237	1.250	1.262	1.274	1.285	1.296	1.306	1.316	1.325	1.334	1.343	1.351	1.389	1.420	1.448	1.472	1.493	1.512	1.529	1.544	1.558	1.571	1.582	1.593	1.669	1.715

Table 6. Durbin–Watson five percent upper bound assuming a unit column among the regressors

N	K=0	K=1	K=2	K=3	K=4	K=5	K=6	K=7	K=8	K=9	K=10	K=11	K=12	K=13	K=14	K=15	K=16	K=17	K=18	K=19	K=20	K=21
2	0.012																					
3	0.168	1.012																				
4	0.355	0.780	2.009																			
5	0.478	0.820	1.544	2.624																		
6	0.584	0.890	1.400	2.128	3.005																	
7	0.677	0.936	1.356	1.896	2.547	3.250																
8	0.754	0.982	1.332	1.777	2.287	2.847	3.417															
9	0.820	1.024	1.320	1.699	2.128	2.588	3.067	3.534														
10	0.877	1.062	1.320	1.604	2.016	2.414	2.822	3.231	3.620													
11	0.927	1.096	1.324	1.604	1.928	2.283	2.645	3.004	3.356	3.684												
12	0.972	1.128	1.331	1.579	1.864	2.177	2.506	2.832	3.149	3.454	3.733											
13	1.012	1.156	1.340	1.562	1.816	2.094	2.39	2.692	2.985	3.266	3.531	3.772										
14	1.047	1.182	1.350	1.551	1.779	2.03	2.296	2.572	2.848	3.111	3.36	3.593	3.803									
15	1.079	1.205	1.360	1.543	1.750	1.977	2.22	2.471	2.727	2.979	3.216	3.438	3.644	3.828								
16	1.109	1.227	1.371	1.539	1.728	1.935	2.157	2.388	2.624	2.86	3.090	3.304	3.603	3.686	3.849							
17	1.136	1.247	1.381	1.536	1.710	1.900	2.104	2.318	2.537	2.757	2.975	3.184	3.378	3.557	3.721	3.866						
18	1.160	1.266	1.391	1.535	1.696	1.872	2.060	2.258	2.461	2.668	2.873	3.073	3.265	3.441	3.603	3.75	3.880					
19	1.183	1.283	1.401	1.536	1.685	1.848	2.023	2.206	2.396	2.589	2.783	2.974	3.159	3.335	3.496	3.642	3.775	3.892				
20	1.204	1.300	1.411	1.537	1.676	1.828	1.991	2.162	2.339	2.521	2.704	2.885	3.063	3.234	3.395	3.542	3.676	3.797	3.903			
21	1.224	1.315	1.42	1.538	1.669	1.812	1.964	2.124	2.290	2.461	2.633	2.806	2.976	3.141	3.300	3.448	3.583	3.705	3.815	3.912		
22	1.242	1.329	1.429	1.541	1.664	1.797	1.940	2.09	2.246	2.407	2.571	2.735	2.897	3.057	3.211	3.358	3.495	3.619	3.731	3.832	3.920	
23	1.259	1.342	1.437	1.543	1.66	1.785	1.920	2.061	2.208	2.36	2.514	2.670	2.826	2.979	3.129	3.272	3.409	3.535	3.650	3.753	3.846	3.926
24	1.275	1.355	1.446	1.546	1.656	1.775	1.902	2.035	2.174	2.318	2.464	2.613	2.761	2.908	3.053	3.193	3.327	3.454	3.572	3.678	3.773	3.858
25	1.290	1.367	1.454	1.550	1.654	1.767	1.886	2.013	2.144	2.28	2.419	2.560	2.702	2.844	2.983	3.119	3.251	3.376	3.494	3.604	3.702	3.79
26	1.304	1.378	1.461	1.553	1.652	1.759	1.873	1.992	2.117	2.246	2.379	2.513	2.649	2.784	2.919	3.051	3.179	3.303	3.420	3.531	3.633	3.724
27	1.318	1.389	1.469	1.556	1.651	1.753	1.861	1.974	2.093	2.216	2.342	2.470	2.600	2.730	2.860	2.987	3.112	3.233	3.349	3.46	3.563	3.658
28	1.330	1.399	1.476	1.560	1.65	1.747	1.85	1.959	2.021	2.188	2.309	2.431	2.555	2.68	2.805	2.928	3.05	3.168	3.282	3.392	3.495	3.592
29	1.342	1.409	1.483	1.563	1.65	1.743	1.841	1.944	2.052	2.164	2.278	2.396	2.515	2.634	2.755	2.874	2.992	3.107	3.219	3.327	3.43	3.528
30	1.354	1.418	1.489	1.567	1.65	1.739	1.833	1.931	2.034	2.141	2.251	2.363	2.477	2.592	2.708	2.823	2.937	3.050	3.159	3.266	3.368	3.465
31	1.365	1.427	1.496	1.570	1.65	1.735	1.825	1.92	2.018	2.12	2.226	2.333	2.443	2.553	2.665	2.776	2.887	2.996	3.103	3.208	3.309	3.405
32	1.375	1.436	1.502	1.574	1.65	1.732	1.819	1.909	2.004	2.102	2.203	2.306	2.411	2.518	2.625	2.732	2.840	2.946	3.050	3.153	3.252	3.348

n																						
33	1.385	1.444	1.508	1.577	1.651	1.73	1.813	1.9	1.991	2.085	2.181	2.281	2.382	2.484	2.588	2.692	2.796	2.899	3	3.1	3.198	3.293
34	1.394	1.451	1.514	1.58	1.652	1.728	1.808	1.891	1.978	2.069	2.162	2.257	2.355	2.454	2.553	2.654	2.754	2.854	2.954	3.051	3.147	3.24
35	1.403	1.459	1.519	1.584	1.653	1.726	1.803	1.884	1.967	2.054	2.144	2.236	2.33	2.425	2.521	2.619	2.716	2.813	2.91	3.005	3.099	3.19
36	1.412	1.466	1.525	1.587	1.654	1.724	1.799	1.876	1.957	2.041	2.127	2.216	2.306	2.398	2.492	2.586	2.68	2.774	2.868	2.961	3.053	3.142
37	1.420	1.473	1.53	1.59	1.655	1.723	1.795	1.870	1.948	2.029	2.112	2.197	2.285	2.374	2.464	2.555	2.646	2.738	2.829	2.92	3.009	3.097
38	1.428	1.48	1.535	1.594	1.656	1.722	1.792	1.864	1.939	2.017	2.098	2.180	2.265	2.351	2.438	2.526	2.614	2.703	2.792	2.88	2.968	3.054
39	1.436	1.486	1.54	1.597	1.658	1.722	1.789	1.859	1.932	2.007	2.085	2.164	2.246	2.329	2.413	2.499	2.585	2.671	2.757	2.843	2.929	3.013
40	1.443	1.492	1.544	1.6	1.659	1.721	1.786	1.854	1.924	1.997	2.072	2.149	2.228	2.309	2.391	2.473	2.557	2.641	2.725	2.808	2.892	2.974
45	1.476	1.52	1.566	1.615	1.666	1.72	1.776	1.835	1.895	1.958	2.022	2.088	2.156	2.225	2.296	2.367	2.439	2.512	2.586	2.659	2.733	2.807
50	1.504	1.543	1.585	1.628	1.674	1.721	1.771	1.822	1.875	1.93	1.986	2.044	2.103	2.163	2.225	2.287	2.35	2.414	2.479	2.544	2.61	2.675
55	1.528	1.564	1.601	1.641	1.681	1.724	1.768	1.814	1.861	1.909	1.959	2.010	2.062	2.116	2.17	2.225	2.281	2.338	2.396	2.454	2.512	2.571
60	1.549	1.582	1.616	1.652	1.689	1.727	1.767	1.808	1.85	1.894	1.939	1.984	2.031	2.079	2.127	2.177	2.227	2.278	2.33	2.382	2.434	2.487
65	1.568	1.598	1.629	1.662	1.696	1.731	1.767	1.805	1.843	1.882	1.923	1.964	2.006	2.049	2.093	2.138	2.183	2.229	2.276	2.323	2.371	2.419
70	1.584	1.612	1.641	1.672	1.703	1.735	1.768	1.802	1.838	1.874	1.91	1.948	1.987	2.026	2.066	2.106	2.148	2.19	2.232	2.275	2.318	2.362
75	1.599	1.625	1.652	1.68	1.709	1.739	1.77	1.801	1.834	1.867	1.901	1.935	1.97	2.006	2.043	2.08	2.118	2.156	2.195	2.235	2.275	2.315
80	1.612	1.636	1.662	1.688	1.715	1.743	1.772	1.801	1.831	1.861	1.893	1.925	1.957	1.99	2.024	2.059	2.093	2.129	2.165	2.201	2.238	2.275
85	1.624	1.647	1.671	1.696	1.721	1.747	1.774	1.801	1.829	1.857	1.886	1.916	1.946	1.977	2.008	2.04	2.073	2.105	2.139	2.172	2.206	2.241
90	1.635	1.657	1.679	1.703	1.726	1.751	1.776	1.801	1.827	1.854	1.881	1.909	1.937	1.966	1.995	2.025	2.055	2.085	2.116	2.148	2.179	2.211
95	1.645	1.666	1.687	1.709	1.732	1.755	1.778	1.802	1.827	1.852	1.877	1.903	1.930	1.956	1.984	2.011	2.04	2.068	2.097	2.126	2.156	2.186
100	1.654	1.674	1.694	1.715	1.736	1.758	1.78	1.803	1.826	1.85	1.874	1.898	1.923	1.948	1.974	2.000	2.026	2.053	2.08	2.108	2.135	2.164
150	1.720	1.733	1.747	1.76	1.774	1.788	1.802	1.817	1.832	1.846	1.862	1.877	1.892	1.908	1.924	1.940	1.956	1.972	1.989	2.006	2.023	2.04
200	1.759	1.768	1.779	1.789	1.799	1.809	1.82	1.831	1.841	1.852	1.863	1.874	1.885	1.897	1.908	1.920	1.931	1.943	1.955	1.967	1.979	1.991

REFERENCES

1. J. P. Imhof, Computing the distribution of a quadratic form in normal variables. *Biometrika* 48: 419–426 (1961).
2. H. Ruben, Probability content of regions under spherical normal distribution, IV: The distributions of homogeneous and nonhomogeneous quadratic functions in normal variables. *Annals of Mathematical Statistics* 33: 542–570 (1962).
3. A. Grad and H. Solomon, Distributions of quadratic forms and some applications. *Annals of Mathematical Statistics* 26: 464–477 (1955).
4. J.-J. Pan, Distributions of the noncircular serial correlation coefficients. *Shuxue Jinzhan* 7: 328–337 (1964). Translated by N. N. Chan for *Selected Translations in Mathematical Statistics and Probability* 7: 281–292 (1968).
5. J. Koerts and A. P. J. Abrahamse, *On the Theory and Application of the General Linear Model*: Rotterdam: Rotterdam University Press, 1969.
6. R. W. Farebrother, The distribution of a quadratic form in normal variables. *Applied Statistics* 39: 294–309 (1990).
7. R. B. Davies, The distribution of a linear combination of x^2 random variables. *Applied Statistics* 29:323–333 (1980).
8. J. Sheil and I. O'Muircheartaigh, The distribution of non-negative quadratic forms in normal variables. *Applied Statistics* 26: 92–98 (1977).
9. R. W. Farebrother, The distribution of a positive linear combination of x^2 random variables. *Applied Statistics* 32: 322–339 (1984).
10. R. W. Farebrother, Pan's procedure for the tail probabilities of the Durbin–Watson statistic. *Applied Statistics* 29:224–227 (1980); corrected 30: 189 (1981).
11. R. W. Farebrother, The distribution of a linear combination of central x^2 random variables. *Applied Statistics* 32: 363–366 (1984).
12. R. W. Farebrother, The distribution of a linear combination of x^2 random variables. *Applied Statistics* 33: 366–369 (1984).
13. F. C. Palm and J. N. Sneek, Significance tests and spurious correlation in regression models with autocorrelated errors. *Statistische Hefte* 25: 87–105 (1984).
14. R. W. Farebrother, Eigenvalue-free methods for computing the distribution of a quadratic form in normal variables. *Statistische Hefte,* 26: 287–302 (1985).
15. T. S. Shively, C. F. Ansley, and R. Kohn (1990), Fast evaluation of the Durbin–Watson and other invariant test statistics in time series regression. *Journal of the American Statistical Association* 85: 676–685 (1990).

16. C. F. Ansley, R. Kohn and T. S. Shively, Computing p-values for the Durbin–Watson and other invariant test statistics. *Journal of Econometrics* 54: 277–300 (1992).
17. R. W. Farebrother, A critique of recent methods for computing the distribution of the Durbin–Watson and other invariant test statistics. *Statistische Hefte* 35: 365–369 (1994).
18. R. Kohn, T. S. Shively and C. F. Ansley, Computing p-values for the generalized Durbin–Watson and residual autocorrelations. *Applied Statistics* 42: 249–260 (1993).
19. N. G. Shephard, From characteristic function to distribution function: a simple framework for the theory. *Econometric Theory* 7: 519–529 (1991).
20. R. W. Farebrother, The Durbin–Watson test for serial correlation when there is no intercept in the regression. *Econometrica* 48: 1553–1563 (1980).
21. N. E. Savin and K. J. White, The Durbin–Watson test for serial correlation with extreme sample sizes or many regressors. *Econometrica* 45: 1989–1996 (1977).

12
Improvements to the Wald Test

MAXWELL L. KING   Monash University, Clayton, Victoria, Australia

KIM-LENG GOH   University of Malaya, Kuala Lumpur, Malaysia

1. INTRODUCTION

We typically have two aims in mind when constructing an hypothesis test. These are that the test accurately controls the probability of wrongly rejecting the null hypothesis when it is true, while also having the highest possible probability of correctly rejecting the null hypothesis. In other words, a desirable test is one that has the right size and high power. While the Wald test is a natural test in that it can be performed with relative ease by looking at the ratio of an estimate to its standard error (in its simplest form), it can perform poorly on both accounts.

The Wald test is a member of what is known as the trinity of classical likelihood testing procedures, the other two being the likelihood ratio (LR) and Lagrange multiplier (LM) tests. Because, in the last three decades, Monte Carlo experiments have become relatively easy to conduct, we have seen a large number of studies that have shown the Wald test to have the least accurate asymptotic critical values of these three tests. There are also a number of studies that have questioned the small-sample power properties of the Wald test. (A number of these studies are discussed below.) In particular, in contrast to the LR and LM tests, the Wald test is not invariant to equivalent, but different, formulations of the null hypothesis. By suitably rewriting the null, one can in fact obtain many different versions of the Wald test. Although asymptotically equivalent, they behave differently in small samples, with some of them having rather poor size properties. The test can also suffer from local biasedness and power non-monotonicity, both of which can cause the test's power to be lower than its nominal size in some areas of the alternative hypothesis parameter space.

The aim of this chapter is to review the literature on known properties of the Wald test with emphasis on its small-sample problems such as local biasedness, nonmonotonicity in the power function and noninvariance to reparameterization of the null hypothesis. We review in detail some recent solutions to these problems. Nearly all of the Wald test's problems seem to stem from the use of a particular estimate of the asymptotic covariance matrix of the maximum likelihood estimator (MLE) in the test statistic. An estimate of this covariance matrix is used to determine whether nonzero estimates of function restrictions or parameters under test are significant. A major theme of this review is the use of readily available computer power to better determine the significance of nonzero values.

Briefly, the plan of the remainder of this chapter is as follows. Section 2 introduces the theory of the Wald test and discusses some practical issues in the application of the test. The literature on small-sample problems of the Wald test is reviewed in Section 3 with particular emphasis on local biasedness, possible nonmonotonicity in the power function, and noninvariance to reparameterization of the null hypothesis. The null Wald test, which is designed to overcome nonmonotonicity, is discussed in detail in Section 4, while Section 5 outlines a new method of correcting the Wald test for local biasedness. A Wald-type test, called the squared generalized distance test and which uses simulation rather than asymptotics to estimate the covariance matrix of the parameter estimates, is introduced in Section 6. The chapter concludes with some final remarks in Section 7.

2. THE WALD TEST

Suppose y_t is an observation from the density function $\phi(y_t|x_t, \theta)$ in which x_t is a vector of exogenous variables and θ is an unknown $k \times 1$ parameter vector. The log-likelihood function for n independent observations is

$$l(\theta) = \sum_{t=1}^{n} ln\phi(y_t|x_t, \theta) \tag{1}$$

Let $\theta = (\beta', \gamma')'$ where β and γ are of order $r \times 1$ and $(k-r) \times 1$, respectively. Let $\hat{\theta} = (\hat{\beta}', \hat{\gamma}')'$ denote the unconstrained MLE of θ based on the n observation log-likelihood (1). Under standard regularity conditions (see, for example, Amemiya 1985, Godfrey 1988, and Gouriéroux and Monfort 1995), $\sqrt{n}(\hat{\theta} - \theta)$ is asymptotically distributed as

$$N\left(0, \left[\lim_{n\to\infty} \frac{1}{n} V(\theta)\right]^{-1}\right)$$

where

$$V(\theta) = E\left[-\frac{\partial^2 l(\theta)}{\partial\theta\partial\theta'}\right] \tag{2}$$

is the information matrix.

Consider first the problem of testing exact restrictions on β, namely testing

$$H_0 : \beta = \beta_0 \qquad \text{against} \qquad H_a : \beta \neq \beta_0 \tag{3}$$

where β_0 is a known $r \times 1$ vector. Under H_0,

$$\left(\hat{\beta} - \beta_0\right)'\left(RV(\theta)^{-1}R'\right)^{-1}\left(\hat{\beta} - \beta_0\right) \xrightarrow{d} \chi^2(r) \tag{4}$$

where $R = (I_r, 0)$ is $r \times k$, I_r is an $r \times r$ identity matrix, $\xrightarrow{d}$ denotes convergence in distribution, and $\chi^2(r)$ denotes the chi-squared distribution with r degrees of freedom. The left-hand side (LHS) of (4) forms the basis for the Wald statistic. The asymptotic covariance matrix for the estimated distance from the null value, namely $\hat{\beta} - \beta_0$, is $RV(\theta)^{-1}R'$. Because the true value of θ is not known, it is replaced with $\hat{\theta}$ in $V(\theta)$. Also, in some cases we have difficulty in determining the exact formula for $V(\theta)$, and replace $V(\hat{\theta})$ with $\hat{V}(\hat{\theta})$, a consistent estimator of $V(\hat{\theta})$. The Wald statistic is

$$\left(\hat{\beta} - \beta_0\right)'\left(R\hat{V}(\hat{\theta})^{-1}R'\right)^{-1}\left(\hat{\beta} - \beta_0\right) \tag{5}$$

which is asymptotically distributed as $\chi^2(r)$ under H_0 when appropriate regularity conditions are met. H_0 is rejected for large values of (5).

Now consider the more general problem of testing

$$H_0^h : h(\theta) = 0 \qquad \text{against} \qquad H_a^h : h(\theta) \neq 0 \tag{6}$$

where $h : R^k \to R^r$ is a vector function which is differentiable up to third order. The Wald test (see Stroud 1971 and Silvey 1975) involves rejecting H_0^h for large values of

$$W = h\left(\hat{\theta}\right)' A\left(\hat{\theta}\right)^{-1} h\left(\hat{\theta}\right) \qquad (7)$$

where $A(\theta) = h_1(\theta) V\theta)^{-1} h_1(\theta)'$ and $h_1(\theta)$ is the $r \times k$ matrix of derivatives $\partial h(\theta)/\partial\theta'$. Again, under appropriate regularity conditions, (7) has an asymptotic $\chi^2(r)$ distribution under H_0^h. Observe that when $h(\theta) = \beta - \beta_0$, (7) reduces to (5).

Three estimators commonly used in practice for $V(\theta)$ in the Wald statistic are the information matrix itself, (2), the negative Hessian matrix, $-\partial^2 l(\theta)/\partial\theta\partial\theta'$, and the outer-product of the score vector.

$$\sum_{t=1}^{n} \frac{\partial ln\phi(y_t|x_t, \theta)}{\partial\theta} \frac{\partial ln\phi(y_t|x_t, \theta)}{\partial\theta'}$$

The latter two are consistent estimators of $V(\theta)$. The Hessian is convenient when it is intractable to take expectations of second derivatives. The outer-product estimator was suggested by Berndt et al. (1974) and is useful if it proves difficult to find second derivatives (see also Davidson and MacKinnon 1993, Chapter 13). Although the use of any of these estimators does not affect the first-order asymptotic properties of the Wald test (see, e.g., Gouriéroux and Monfort 1995, Chapters 7 and 17), the tests so constructed can behave differently in small samples.

Using higher-order approximations, Cavanagh (1985) compared the local power functions of size-corrected Wald procedures, to order n^{-1}, based on these different estimators for testing a single restriction. He showed that the Wald tests using the information and Hessian matrix-based estimators have local powers tangent to the local power envelope. This tangency is not found when the outer-product of the score vector is used. The information and Hessian matrix based Wald tests are thus second-order admissible, but the outer-product-based test is not. Cavanagh's simulation study further showed that use of the outer-product-based test results in a considerable loss of power. Parks et al. (1997) proved that the outer-product based test is inferior when applied to the linear regression model. Also, their Monte Carlo evidence for testing a single-parameter restriction in a Box–Cox regression model demonstrated that this test has lower power than that of a Hessian-based Wald test. The available evidence therefore suggests that the outer-product-based Wald test is the least desirable of the three.

There is less of a guide in the literature on whether it is better to use the information matrix or the Hessian matrix in the Wald test. Mantel (1987) remarked that expected values of second derivatives are better than their observed values for use in the Wald test. Davidson and MacKinnon (1993, p. 266) noted that the only stochastic element in the information matrix is $\hat{\theta}$. In addition to $\hat{\theta}$, the Hessian depends on y_t as well, which imports additional noise, possibly leading to a lower accuracy of the test. One would expect that the information matrix, if available, would be best as it represents the limiting covariance matrix and the Hessian should only be used when the information matrix cannot be constructed. Somewhat contrary to this conjecture, empirical evidence provided by Efron and Hinkley (1978) for one-parameter problems showed that the Hessian-based estimator is a better approximation of the covariance matrix than the information matrix-based estimator. The comments on the paper by Efron and Hinkley cautioned against over-emphasis on the importance of the Hessian-based estimator, and some examples illustrated the usefulness of the information matrix approach. More recent work by Orme (1990) has added further evidence in favour of the information matrix being the best choice.

3. SMALL-SAMPLE PROBLEMS OF THE WALD TEST

The Wald procedure is a consistent test that has invariance properties asymptotically. Within the class of asymptotically unbiased tests, the procedure is also asymptotically most powerful against local alternatives (see, e.g., Cox and Hinkley 1974, Chapter 9, Engle 1984, Section 6, and Godfrey 1988, Chapter 1). However, for finite sample sizes, the Wald test suffers from some anomalies not always shared by other large-sample tests. Three major finite-sample problems, namely local biasedness, power non-monotonicity and noninvariance of the Wald test, are reviewed below.

3.1 Local Biasedness

A number of studies have constructed higher-order approximations for power functions of the Wald test, in order to compare this procedure with other first-order asymptotically equivalent tests. One striking finding that emerged from the inclusion of higher-order terms is that the Wald test can be locally biased.

Peers (1971) considered testing a simple null hypothesis of $H_0 : \theta = \theta_0$ against a two-sided alternative. Using an expansion of terms accurate to $n^{-1/2}$, he obtained the power function of the Wald statistic under a sequence of Pitman local alternatives. By examining the local behaviour of the power function at $\theta = \theta_0$, he was able to show that the Wald test is locally biased. Hayakawa (1975) extended this analysis to tests of composite null hypotheses as specified in (3) and also noted that the Wald test is locally biased. Magdalinos (1990) examined the Wald test of linear restrictions on regression coefficients of the linear regression model with stochastic regressors. The first-, second- and third-order power functions of the test associated with k-class estimators were considered, and terms of order $n^{-1/2}$ indicated local biasedness of the Wald test. Magdalinos observed that the two-stage least-squares-based Wald test is biased over a larger interval of local alternatives than its limited information maximum likelihood (ML) counterpart. Oya (1997) derived the local power functions up to order $n^{-1/2}$ of the Wald test of linear hypotheses in a structural equation model. The Wald tests based on statistics computed using the two-stage least squares and limited information MLEs were also found to be locally biased.

Clearly, it would be desirable to have a locally unbiased Wald test. Rothenberg (1984) suggested that such a test can be constructed using the Edgeworth expansion approach but he failed to give details. This approach, however, can be problem specific, for any such expansions are unique to the given testing problem. In addition, Edgeworth expansions can be complicated, especially for nonlinear settings. A general solution to this problem that is relatively easy to implement is discussed in Section 5.

3.2 Nonmonotonicity in the Power Function

The local biasedness problem occurs in the neighborhood of the null hypothesis. As the data generating process (DGP) moves further and further away from the null hypothesis, the Wald statistic and the test's power function can be affected by the problem of nonmonotonicity. A test statistic (power of a test) is said to be nonmonotonic if the statistic (power) first increases, but eventually decreases as the distance between the DGP and the null hypothesis increases. Often the test statistic and power will decrease to zero. This behaviour is anomalous because rejection probabilities are higher for moderate departures from the null than for very large departures when good power is needed most.

Hauck and Donner (1977) first reported the nonmonotonic behavior of the Wald statistic for testing a single parameter in a binomial logit model. They showed that this behavior was caused by the fact that the $R\hat{V}(\hat{\theta})^{-1}R'$ term in (5) has a tendency to decrease to zero faster than $(\hat{\beta} - \beta_0)^2$ increases as $|\beta - \beta_0|$ increases. Vaeth (1985) examined conditions for exponential families under which the Wald statistic is monotonic, using a model consisting of a single canonical parameter with a sufficient statistic incorporated. He demonstrated that the limiting behavior of the Wald statistic hinges heavily on the variance of the sufficient statistic. The conditions (pp. 202–203) under which the Wald statistic behaves well depend on the parameterization of this variance, as well as on the true parameter value. Although these results do not easily generalize to other distribution families, there are two important points for applied econometric research. First, the conditions imply that for discrete probability models, e.g. the logit model, nonmonotonicity of the Wald statistic can be ascribed to the near boundary problem caused by observed values of the sufficient statistic being close to the sample space boundary. Second, the Wald statistic is not always well-behaved for continuous probability models although nonmonotonicity was first reported for a discrete choice model.

An example of the problem in a continuous probability model is given by Nelson and Savin (1988, 1990). They considered the single-regressor exponential model

$$y_t = \exp(\theta x_t) + \epsilon_t, \qquad t = 1, \ldots, n$$

where $\epsilon_t \sim IN(0, 1)$ and $x_t = 1$. The t statistic for testing $H_0 : \theta = \theta_0$ against $H_1 : \theta < \theta_0$ is

$$W^{1/2} = \left(\hat{\theta} - \theta_0\right) \Big/ v\left(\hat{\theta}\right)^{1/2}$$

where $v(\hat{\theta})^{1/2} = n^{-1/2} \exp(-\hat{\theta})$. The statistic $W^{1/2}$ has a minimum value of $-n^{1/2} \exp(\theta_0 - 1)$ which occurs at $\hat{\theta} = \theta_0 - 1$. The statistic increases to zero on the left of this point, and to $+\infty$ on the right, exhibiting nonmonotonicity on the left of the minimum point. To illustrate this further, observe that

$$\frac{d\left(v(\hat{\theta})^{1/2}\right)}{d\hat{\theta}} = -n^{-1/2} \exp\left(-\hat{\theta}\right)$$

so that, as $\hat{\theta} \to -\infty$, $v(\hat{\theta})^{1/2}$ increases at an exponential rate while the rate of change in the estimated distance $\hat{\theta} - \theta_0$ is linear. Thus $W^{1/2} \to 0$.

This anomaly arises from the ill-behaved $V(\hat{\theta})^{-1}$ component in the Wald statistic. As explained by Mantel (1987), this estimator can be guaranteed to provide a good approximation to the limiting covariance matrix only when the null hypothesis is true, but not when $\hat{\theta}$ is grossly different from the null value. For nonlinear models, Nelson and Savin (1988) observed that poor estimates of the limiting covariance matrix can result from near nonidentification of parameters due to a flat likelihood surface in some situations.

Nonmonotonicity can also arise from the problem of nonidentification of the parameter(s) under test. This is different from the near nonidentification problem discussed by Nelson and Savin (1988). Nonidentification occurs when the information in the data does not allow us to distinguish between the true parameter values and the values under the null hypothesis. Examples include the partial adjustment model with autocorrelated disturbances investigated by McManus et al. (1994) and binary response models with positive regressors and an intercept term studied by Savin and Würtz (1996).

Nonmonotonicity of this nature is caused by the fact that, based on the data alone, it is very difficult to tell the difference between a null hypothesis DGP and certain DGPs some distance from the null. This affects not only the Wald test, but also any other procedure such as the LR or LM test. There is little that can be done to improve the tests in the presence of this problem, apart from some suggestions offered by McManus (1992) to relook at the representation of the model as well as the restrictions being imposed.

Studies by Storer et al. (1983) and Fears et al. (1996) provide two examples where the Wald test gives nonsignificant results unexpectedly. In fitting additive regression models for binomial data, Storer et al. discovered that Wald statistics are much smaller than LR test statistics. Vogelsang (1997) showed that Wald-type tests for detecting the presence of structural change in the trend function of a dynamic univariate time series model can also display non-monotonic power behaviour.

Another example of Wald power nonmonotonicity was found by Laskar and King (1997) when testing for MA(1) disturbances in the linear regression model. This is largely a near-boundary problem because the information matrix is not well defined when the moving average coefficient estimate is -1 or 1. To deal with nuisance parameters in the model, Laskar and King employed different modified profile and marginal likelihood functions and their Wald tests are constructed on these functions. Given that only the parameter of interest remains in their likelihood functions, their testing problem exemplifies a single-parameter situation where the test statistic does not depend on any unknown nuisance parameters. Because nonmonotonicity stems from the use of $\hat{\theta}$ in the estimation of the covariance matrix, Mantel (1987) proposed the use of the null value θ_0 instead, for single-parameter testing situations with H_0 as stated in (3). Laskar and King adopted this idea to modify the Wald test and found that it removed the problem of nonmonotonicity. They called this modified test the null Wald test. This test is further discussed and developed in Section 4.

3.3 Noninvariance to Reparameterization of the Null Hypothesis

While the Wald test can behave nonmonotonically for a given null hypothesis, the way the null hypothesis is specified has a direct effect on the numerical results of the test. This is particularly true in testing for nonlinear restrictions such as those under H_0^h in (6). Say an algebraically equivalent form to this null hypothesis can be found, but with a different specification given by

$$H_0^g : g(h(\theta)) = 0$$

where g is monotonic and continuously differentiable. The Wald statistic for this new formulation is different from that given by (7) and therefore can lead to different results. Cox and Hinkley (1974, p. 302) commented that although $h(\hat{\theta})$ and $g(h(\hat{\theta}))$ are asymptotically normal, their individual rates of convergence to the limiting distribution will depend on the parameterization. Although Cox and Hinkley did not relate this observation to the Wald procedure, the statement highlights indirectly the potential noninvariance problem of the test in small samples. This problem was directly noted by Burguete et al. (1982, p. 185).

Gregory and Veall (1985) studied this phenomenon by simulation. For a linear regression model with an intercept term and two slope coefficients, β_1 and β_2, they demonstrated that empirical sizes and powers of the Wald procedure for testing $H_0^A : \beta_1 - 1/\beta_2 = 0$ vary substantially from those obtained for testing $H_0^B : \beta_1 \beta_2 - 1 = 0$. They found that the latter form yields empirical sizes closer to those predicted by asymptotic theory. Lafontaine and White (1986) examined the test for $H_0^C : \beta^q = 1$ where β is a positive slope coefficient of a linear regression model and q is a constant. They illustrated that any value of the Wald statistic can be obtained by suitably changing the value of q without affecting the hypothesis under test. The possibility of engineering any desired value of the Wald statistic was proven analytically by Breusch and Schmidt (1988); see also Dagenais and Dufour (1991). This problem was analyzed by Phillips and Park (1988), using asymptotic expansions of the distributions of the associated Wald statistics up to order n^{-1}. The corrections produced by Edgeworth expansions are substantially larger for the test under H_0^A than those needed for the test under H_0^B, especially when $\beta_2 \to 0$. This indicates that the deviation from the asymptotic distribution is smaller when the choice of restriction is H_0^B. The higher-order correction terms become more appreciable as $|q|$ increases in the case of H_0^C. The main conclusion of their work is that it is preferable to use the form of restriction which is closest to a linear function around the true parameter value.

A way of dealing with this difficulty that can easily be applied to a wide variety of econometric testing problems is to base the Wald test on critical values obtained from a bootstrap scheme. Lafontaine and White (1986) noted that the noninvariance problem leads to different inferences because different Wald statistics are compared to the same asymptotic critical value. They estimated, via Monte Carlo simulation, small-sample critical values (which are essentially bootstrap critical values) for the Wald test of different specifications, and concluded that these values can differ vastly from the asymptotic critical values. Since the asymptotic distribution may not accurately approximate the small-sample distribution of the Wald statistic, they suggested the bootstrap approach as a better alternative.

Horowitz and Savin (1992) showed theoretically that bootstrap estimates of the true null distributions are more accurate than asymptotic approximations. They presented simulation results for the testing problems considered by Gregory and Veall (1985) and Lafontaine and White (1986). Using critical values from bootstrap estimated distributions, they showed that different Wald procedures have empirical sizes close to the nominal level. This reduces greatly the sensitivity of the Wald test to different formulations of the null hypothesis, in the sense that sizes of the different tests can be accurately controlled at approximately the same level (see also Horowitz 1997). In a recent simulation study, Godfrey and Veall (1998) showed that many of the small-sample discrepancies among the different versions of Wald test for common factor restrictions, discovered by Gregory and Veall (1986), are significantly reduced when bootstrap critical values are employed. The use of the bootstrap does not remove the noninvariance nature of the Wald test. Rather, the advantage is that the bootstrap takes into consideration the given formulation of the restriction under test in approximating the null distribution of the Wald test statistic. The results lead to more accurate control of sizes of different, but asymptotically equivalent, Wald tests. This then prevents the possibility of manipulating the result, through using a different specification of the restriction under test, in order to change the underlying null rejection probability.

4. THE NULL WALD TEST

Suppose we wish to test the simple null hypothesis

$$H_0 : \theta = \theta_0 \qquad \text{against} \qquad H_a : \theta \neq \theta_0 \qquad (8)$$

where θ and θ_0 are $k \times 1$ vectors of parameters and known constants, respectively, for the model with log-likelihood (1). If $A(\hat{\theta})$ is a consistent estimator of the limiting covariance matrix of the MLE of θ, then the Wald statistic is

$$W = \left(\hat{\theta} - \theta_0\right)' A\left(\hat{\theta}\right)^{-1} \left(\hat{\theta} - \theta_0\right)$$

which, under the standard regularity conditions, has a $\chi^2(k)$ distribution asymptotically under H_0.

Under H_0, $\hat{\theta}$ is a weakly consistent estimator of θ_0 so, for a reasonably sized sample, $A(\theta)$ would be well approximated by $A(\hat{\theta})$. However, if H_0 is grossly violated, this approximation can be very poor, as noted in Section 3.2, and as $\hat{\theta}$ moves away from θ_0 in some cases $A(\hat{\theta})$ increases at a faster rate, causing $W \to 0$. In the case of testing a single parameter $(k = 1)$, Mantel (1987) suggested the use of θ_0 in place of $\hat{\theta}$ in $A(\hat{\theta})$ to solve this problem. Laskar and King (1997) investigated this in the context of testing for MA(1) regression disturbances using various one-parameter likelihood functions such as the profile, marginal and conditional likelihoods. They called this the null Wald (NW) test. It rejects H_0 for large values of $W_n = (\hat{\theta} - \theta_0)^2 / A(\theta_0)$ which, under standard regularity conditions, has a $\chi^2(1)$ asymptotic distribution under H_0. Although the procedure was originally proposed for testing hypotheses on a single parameter, it can easily be extended for testing k restrictions in the hypotheses (8). The test statistic is

$$W_n = \left(\hat{\theta} - \theta_0 \right)' A(\theta_0)^{-1} \left(\hat{\theta} - \theta_0 \right)$$

which tends to a $\chi^2(k)$ distribution asymptotically under H_0.

The NW concept was introduced in the absence of nuisance parameters. Goh and King (2000) have extended it to multiparameter models in which only a subset of the parameters is under test. When the values of the nuisance parameters are unknown, Goh and King suggest replacing them with the unconstrained MLEs to construct the NW test. In the case of testing (3) in the context of (1), when evaluating $V(\theta)$ in (4), γ is replaced by $\hat{\gamma}$ and β is replaced by β_0. Using $\hat{V}_0(\hat{\gamma})$ to denote this estimate of $V(\theta)$, the test statistic (5) now becomes

$$W_n = \left(\hat{\beta} - \beta_0 \right)' \left(R\hat{V}_0(\hat{\gamma})^{-1} R' \right)^{-1} \left(\hat{\beta} - \beta_0 \right)$$

which asymptotically follows a $\chi^2(r)$ distribution under H_0. This approach keeps the estimated variance stable under H_a, while retaining the practical convenience of the Wald test which requires only estimation of the unrestricted model.

Goh and King (2000) (see also Goh 1998) conducted a number of Monte Carlo studies comparing the small-sample performance of the Wald, NW, and LR tests, using both asymptotic and bootstrap critical values. They investigated a variety of testing problems involving one and two linear restrictions in the binary logit model, the Tobit model, and the exponential model. For these testing problems, they found the NW test was monotonic in its power, a property not always shared by the Wald test. Even in the presence of nuisance parameters, the NW test performs well in terms of power. It sometimes outperforms the LR test and often has comparable power to the LR test. They also found that the asymptotic critical values work best for the LR test, but not always satisfactorily, and work worst for the NW test. They observed that all three tests are best applied with the use of bootstrap critical values, at least for the nonlinear models they considered.

The Wald test should not be discarded altogether for its potential problem of power nonmonotonicity. The test displays nonmonotonic power behavior only in certain regions of the alternative hypothesis parameter space and only for some testing problems. Otherwise it can have better power properties than those of the NW and LR tests. Obviously it would be nice to know whether nonmonotonic power is a possibility before one applies the test to a particular model. Goh (1998) has proposed the following simple diagnostic check for nonmonotonicity as a possible solution to this problem.

For testing (3), let W_1 denote the value of the Wald statistic given by (5) and therefore evaluated at $\hat{\theta} = (\hat{\beta}', \hat{\gamma}')'$ and let W_2 denote the same statistic but evaluated at $\hat{\theta}_\lambda = (\lambda\hat{\beta}', \hat{\gamma}')'$, where λ is a large scalar value such as 5 or 10. The diagnostic check is given by

$$\Delta W = W_2 - W_1$$

A negative ΔW indicates that the sample data falls in a region where the magnitude of the Wald statistic drops with increasing distance from H_0, which is a necessary (but not sufficient) condition for nonmonotonicity. Thus a negative ΔW does not in itself imply power nonmonotonicity in the neighborhood of $\hat{\theta}$ and $\hat{\theta}_\lambda$. At best it provides a signal that nonmonotonic power behavior is a possibility.

Using his diagnostic test, Goh suggested a two-stage Wald test based on performing the standard Wald test if $\Delta W > 0$ and the NW test if $\Delta W < 0$. He investigated the small-sample properties of this procedure for $\lambda = 2.5$, 5, 7.5, and 10, using simulation methods. He found that, particularly for large λ values, the two-stage Wald test tends to inherit the good power properties of either the Wald or the NW test, depending on which is better.

An issue that emerges from all the simulation studies discussed in this section is the possibility of poorly centered power curves of the Wald and NW test procedures. A well centered power curve is one that is roughly symmetric about the null hypothesis, especially around the neighborhood of the null hypothesis. To a lesser extent, this problem also affects the LR test. Its solution involves making these tests less locally biased in terms of power.

5. CORRECTING THE WALD TEST FOR LOCAL BIASEDNESS

As pointed out in Section 3, the Wald test is known to be locally biased in small samples. One possible reason is the small-sample bias of ML estimates—it does seem that biasedness of estimates and local biasedness of tests go hand in hand (see, for example, King 1987).

Consider the case of testing H_0^h against H_a^h, given by (6); where $r = 1$, i.e., $h(\theta)$ is a scalar function. In this case, the test statistic (7) can be written as

$$W = \left(h(\hat{\theta})\right)^2 \Big/ A(\hat{\theta}) \tag{9}$$

which, under appropriate regularity conditions, has an asymptotic $\chi^2(1)$ distribution under H_0^h. A simple approach to constructing a locally unbiased test based on (9) involves working with

$$\sqrt{W} = h(\hat{\theta})A(\hat{\theta})^{-1/2}$$

and two critical values c_1 and c_2. The acceptance region would be $c_1 < \sqrt{W} < c_2$.

If we denote by $\Pi(h(\theta))$ the probability of being in the critical region, given that the restriction function takes the value $h(\theta)$, i.e.,

$$\Pi(h(\theta)) = \Pr\left[\sqrt{W} < c_1 \big| h(\theta)\right] + \Pr\left[\sqrt{W} > c_2 \big| h(\theta)\right]$$

the critical values c_1 and c_2 are determined by solving the size condition

$$\Pi(0) = \alpha \tag{11}$$

and the local unbiasedness condition

$$\left. \frac{\partial \Pi(h)}{\partial h} \right|_{h=0} = 0 \tag{12}$$

where α is the desired significance level of the test. Unfortunately we typically do not have analytical expressions for the LHS of (11) and (12) and have to resort to simulation methods. For sufficiently small $\delta > 0$, the LHS of (12) can be approximated by the numerical derivative

$$D(\delta) = \left\{\Pi(\delta) - \Pi(-\delta)\right\} / 2\delta \tag{13}$$

Thus the problem of finding c_1 and c_2 reduces to repeatedly finding $\Pi(0)$, $\Pi(-\delta)$, and $\Pi(\delta)$ via simulation, using (10) for given c_1 and c_2 and then using a numerical method such as the secant method to find those values of c_1 and c_2 that solve (11) and (12) simultaneously. In the case of testing for autocorrelation in the linear regression model, Grose and King (1991) suggested the use of $\delta = 0.001$.

The disadvantage of this approach is that it cannot be generalized to a vector restriction function, i.e., $r > 1$. A correction for local biasedness that does have the potential to be extended from $r = 1$ to $r > 1$ was suggested by Goh and King (1999). They considered the problem of testing

$$H_0^\beta : h(\beta) = 0 \qquad \text{against} \qquad H_a^\beta : h(\beta) \neq 0$$

where $h(\beta)$ is a continuously differentiable, possibly non-linear scalar function of β, a scalar parameter of interest, with $\theta = (\beta, \gamma')'$. In this case the Wald test statistic is

$$W = \left(h\left(\hat\beta\right)\right)^2 A\left(\hat\theta\right)^{-1}$$

which under H_0 (and standard regularity conditions) asymptotically follows the $\chi^2(1)$ distribution. Given the link between local biasedness of the Wald test and biasedness of MLEs, Goh and King suggested a correction factor for this bias by evaluating W at $\hat\beta_{cw} = \hat\beta - c_w$ rather than at $\hat\beta$. There is an issue of what estimates of γ to use in $A(\theta)$. If $\hat\gamma$ is obtained by maximizing the likelihood function when $\beta = \hat\beta$, then denote by $\hat\gamma\left(\hat\beta_{cw}\right)$ that value of γ that maximizes the likelihood when $\hat\beta = \hat\beta_{cw}$. The suggested test rejects H_0 for

$$CW = \left(h\left(\hat\beta_{cw}\right)\right)^2 A\left(\hat\theta_{cw}\right)^{-1} > c_\alpha \tag{14}$$

where $\hat\theta_{cw} = (\hat\beta_{cw}, \hat\gamma(\hat\beta_{cw})')'$.

The final question is how to determine the correction factor c_w and the critical value c_α. These are found numerically to ensure correct size and the local unbiasedness of the resultant test. This can be done in a similar manner to finding c_1 and c_2 for the test based on $\sqrt{W}$. Both these procedures can be extended in a straightforward manner to the NW test.

Goh and King (1999) (see also Goh 1998) investigated the small-sample properties of the CW test and its NW counterpart which they denoted as the CNW test. They considered two testing problems, namely testing the coefficient of the lagged dependent variable regressor in the dynamic linear regression model and a nonlinear restriction on a coefficient in the static linear regression model. They found the CW test corrects for local biasedness but not for nonmonotonicity whereas the CNW test is locally unbiased, has a well-centered power curve which is monotonic in its behavior, and can have better power properties than the LR test, especially when the latter is locally biased.

It is possible to extend this approach to a $j \times 1$ vector β and an $r \times 1$ vector function $h(\beta)$, with $r \leq j$. The correction to $\hat\beta$ is $\hat\beta - c_w$, where c_w is now a $j \times 1$ vector. Thus we have $j + 1$ effective critical values, namely c_w and c_α. These need to be found by solving the usual size condition simultaneously with j local unbiasedness conditions. This becomes much more difficult, the larger j is.

6. THE SQUARED GENERALIZED DISTANCE TEST

There is little doubt that many of the small-sample problems of the Wald test can be traced to the use of the asymptotic covariance matrix of the MLE in the test statistic. For example, it is the cause of the nonmonotonicity that the NW test corrects for. Furthermore, this matrix is not always readily available because, in some nonlinear models, derivation of the information or Hessian matrix may not be immediately tractable (see, e.g., Amemiya 1981, 1984). In such situations, computer packages often can be relied upon to provide numerical estimates of the second derivatives of the log-likelihood function, for the purposes of computing the Wald test. However, implementation of the NW, CW or CNW tests will be made more difficult because of this. In addition, discontinuities in the derivatives of the log-likelihood can make it impossible to apply the NW or CNW tests by direct substitution of the null parameter values into this covariance matrix. Clearly it would be much better if we did not have to use this covariance matrix in the Wald test.

Consider again the problem of testing $H_0 : \beta = \beta_0$ against $H_a : \beta \neq \beta_0$ as outlined in the first half of Section 2. As discussed, under standard regularity conditions, $\hat{\beta}$ follows a multivariate normal distribution with mean β_0 and a covariance matrix we will denote for the moment by Σ, asymptotically under H_0. Thus, asymptotically, an ellipsoid defined by the random estimator $\hat{\beta}$, and centered at β_0, given by

$$\left(\hat{\beta} - \beta_0\right)' \Sigma^{-1} \left(\hat{\beta} - \beta_0\right) = w_{\alpha;r} \tag{15}$$

where $w_{\alpha;r}$ is the $100(1 - \alpha)$th percentile of the $\chi^2(r)$ distribution, contains $100(1 - \alpha)\%$ of the random observations of $\hat{\beta}$ when H_0 is true. In other words, the probability of an estimate $\hat{\beta}$ falling outside the ellipsoid (15) converges in limit to α, i.e.,

$$\Pr\left[\left(\hat{\beta} - \beta_0\right)' \Sigma^{-1} \left(\hat{\beta} - \beta_0\right) > w_{\alpha;r}\right] \to \alpha \tag{16}$$

under H_0. The quantity

$$\left(\hat{\beta} - \beta_0\right)' \Sigma^{-1} \left(\hat{\beta} - \beta_0\right) \tag{17}$$

is known as the squared generalized distance from the estimate $\hat{\beta}$ to the null value β_0 (see, e.g., Johnson and Wichern, 1988, Chapter 4). For an α–level test, we know that the Wald principle is based on rejecting H_0 if the squared generalized distance (17), computed using some estimator of Σ evaluated at $\hat{\theta}$, is larger than $w_{\alpha;r}$. Alternatively, from a multivariate perspective, a rejection of H_0 takes place if the sample data point summarized by the estimate $\hat{\beta}$ falls outside the ellipsoid (15) for some estimated matrix of Σ. This multivariate approach forms the basis of a new Wald-type test.

Let $\hat{\beta}^0$ be a consistent estimator of β when H_0 is satisfied. Say in small samples, $\hat{\beta}^0$ follows an unknown distribution with mean μ_0 and covariance matrix Σ_0 which is assumed to be positive definite. The new test finds an ellipsoid defined by $\hat{\beta}^0$, which is

$$\left(\hat{\beta}^0 - \mu_0\right)' \Sigma_0^{-1}\left(\hat{\beta}^0 - \mu_0\right) = c \tag{18}$$

where c is a constant such that

$$\Pr\left[\left(\hat{\beta}^0 - \mu_0\right)' \Sigma_0^{-1}\left(\hat{\beta}^0 - \mu_0\right) > c\right] = \alpha \tag{19}$$

for any estimator $\hat{\beta}^0$ when H_0 is true. This implies that the parameter space outside the ellipsoid (18) defines the critical region, so the test involves checking whether the data point, summarized by $\hat{\beta}$, lies inside or outside this ellipsoid. If it is outside, H_0 is rejected. The immediate task is to find consistent estimates of μ_0 and Σ_0 and the constant c.

To accomplish this, a parametric bootstrap scheme is proposed as follows:

(i) Estimate $\hat{\theta} = (\hat{\beta}', \hat{\gamma}')'$ for the given data set, and construct $\hat{\theta}_0 = (\beta_0', \hat{\gamma}')'$.

(ii) Generate a bootstrap sample of size n under H_0. This can be done by drawing n independent observations from $\phi(y_t|x_t, \hat{\theta}_0)$ which approximates the DGP under H_0. Calculate the unconstrained ML estimate of β for this sample. Repeat this process, say, B times. Let $\hat{\beta}_j^0$ denote the estimate for the jth sample, $j = 1, \ldots, B$.

The idea of using this bootstrap procedure is to generate B realizations of the random estimator $\hat{\beta}^0$ under H_0. The bootstrap sample mean

$$\bar{\beta}^0 = \frac{1}{B}\sum_{j=1}^{B} \hat{\beta}_j^0$$

and the sample covariance matrix

$$\hat{\Sigma}_0 = \frac{1}{B-1}\sum_{j=1}^{B}\left(\hat{\beta}_j^0 - \bar{\beta}^0\right)\left(\hat{\beta}_j^0 - \bar{\beta}^0\right)'$$

are consistent estimates for μ_0 and Σ_0, respectively. The critical region must satisfy

$$\Pr\left[\left(\hat{\beta}^0 - \bar{\hat{\beta}}^0\right)' \hat{\Sigma}_0^{-1}\left(\hat{\beta}^0 - \bar{\hat{\beta}}^0\right) > c\right] = \alpha$$

when H_0 is true. Therefore, the critical value c must be found to ensure that the ellipsoid

$$\left(\hat{\beta}^0 - \bar{\hat{\beta}}^0\right)' \hat{\Sigma}_0^{-1}\left(\hat{\beta}^0 - \bar{\hat{\beta}}^0\right) = c \qquad (20)$$

contains $100(1 - \alpha)\%$ of the observations of $\hat{\beta}_j^0$ inside its boundary. This is fulfilled by using the $100(1 - \alpha)\%$ percentile of the array ς_j, where

$$\varsigma_j = \left(\hat{\beta}_j^0 - \bar{\hat{\beta}}_0\right)' \hat{\Sigma}_0^{-1}\left(\hat{\beta}_j^0 - \bar{\hat{\beta}}_0\right), \qquad j = 1, \ldots, B$$

The test rejects H_0 if $\hat{\beta}$ falls outside the boundary of the ellipsoid (20), or, equivalently, if

$$s = \left(\hat{\beta} - \bar{\hat{\beta}}^0\right)' \hat{\Sigma}_0^{-1}\left(\hat{\beta} - \bar{\hat{\beta}}^0\right) > c$$

Our test statistic, s, is based on the squared generalized distance (SGD) from $\hat{\beta}$ to the estimator of μ_0. Because the critical value is determined from a bootstrap procedure, we refer to the test as the bootstrap SGD test.

Under H_0, as $n \to \infty$, $\bar{\hat{\beta}} \overset{p}{\to} \mu_0$ and $\hat{\Sigma}_0 \overset{p}{\to} \Sigma_0$. From (16), this leads to

$$\Pr\left[\left(\hat{\beta} - \bar{\hat{\beta}}^0\right)' \hat{\Sigma}_0^{-1}\left(\hat{\beta} - \bar{\hat{\beta}}^0\right) > w_{\alpha;r}\right] \to \alpha$$

The asymptotic version of the SGD test thus rejects H_0 if $s > w_{\alpha;r}$.

This test procedure applies for testing any number of restrictions but can further be refined in the case of $r = 1$ with the additional aim of centering the test's power function. This has the advantage of resolving the local biasedness problem discussed in Section 5 when testing a single parameter β.

Consider the square root of the SGD statistic in the case of $r = 1$

$$\sqrt{s} = \left(\hat{\beta} - \bar{\hat{\beta}}^0\right) \Big/ \hat{\Sigma}_0^{1/2}$$

and observe that $\bar{\hat{\beta}}^0$ and $\hat{\Sigma}_0$ are bootstrap estimates of unknown parameters, namely μ_0 and Σ_0. If we knew these values, our tests could be based on accepting H_0 for

$$c_1 < \left(\hat{\beta} - \mu_0\right) \Big/ \Sigma_0^{1/2} < c_2$$

where c_1 and c_2 are appropriately chosen critical values or, equivalently, given $\Sigma_0^{1/2} > 0$, accepting H_0 for

$$c_1' < \hat{\beta} < c_2' \qquad (21)$$

The two critical values, c_1' and c_2', in (21) can be found by solving the size condition

$$\pi(\beta)\big|_{H_0} = \alpha \tag{22}$$

and the local unbiasedness condition

$$\frac{\partial \pi(\beta)}{\partial \beta}\bigg|_{H_0} = 0 \tag{23}$$

where

$$\pi(\beta) = \Pr\left[\hat{\beta} < c_1'\middle|\beta\right] + \Pr\left[\hat{\beta} > c_2'\middle|\beta\right]$$

The LHS of (23) does not have an analytical expression, but if δ represents a small but detectable deviation from the null value β_0, then (22) and (23) can be written as

$$\pi(\beta_0) = \alpha \tag{24}$$

and

$$\pi(\beta_0 + \delta) = \pi(\beta_0 - \delta) = 0 \tag{25}$$

respectively.

Let $\hat{\theta}_0^+ = (\beta_0 + \delta, \hat{\gamma}')'$ and $\hat{\theta}_0^- = (\beta_0 - \delta, \hat{\gamma}')'$. The following bootstrap procedure may be employed to find the two critical values.

1. Estimate $\hat{\theta} = (\hat{\beta}, \hat{\gamma}')'$ for the given data set the test is to be conducted on. For a selected δ value, construct $\hat{\theta}_0$, $\hat{\theta}_0^+$, and $\hat{\theta}_0^-$ using $\hat{\gamma}$ and β_0.
2. Generate one bootstrap sample of size n under H_0 by drawing independent observations from $\phi\left(y_t|x_t, \hat{\theta}_0\right)$. Compute the unconstrained ML estimate of β. Repeat this process B times. Let $\hat{\beta}_j^0$, $j = 1, 2, \ldots, B$, denote the estimate for the jth sample. Sort $\hat{\beta}_j^0$ into an array, which we denote by $\psi_{\hat{\beta}_0}$. This array provides the empirical distribution function for $\hat{\beta}^0$.
3. Generate B bootstrap samples of size n under H_0^+ by drawing independent observations from $\phi(y_t|x_t, \hat{\theta}_0^+)$. Compute $\hat{\beta}$ for each of the B samples. Let $\hat{\beta}_j^{0+}$, $j = 1, 2, \ldots, B$, denote the estimate for the jth sample. Repeat this process under H_0^- by using $\phi(y_t|x_t, \hat{\theta}_0^-)$ as the data generating process. Let $\hat{\beta}_j^{0-}$, $j = 1, 2, \ldots, B$, denote the estimate for the jth sample.
4. For a selected ν where $0 < \nu < \alpha$, use the (100ν)th and $100(1 - \alpha + \nu)$th percentiles of $\psi_{\hat{\beta}_0}$ and c_1' and c_2', respectively, in order to satisfy the size condition (24).

5. Using the critical values determined in step 4 and the sequence of $\hat{\beta}_j^{0+}$ from step 3, estimate the power at H_0^+. Repeat using the sequence $\hat{\beta}_j^{0-}$ from step 3 to estimate the power at H_0^-. If the power difference satisfies (25), we have found the required critical values. If not, steps 4 and 5 are repeated using a different v value each time until the solution is found.

6. For the given solution, check whether the local powers at H_0^+ and H_0^- are both greater than α. If not, repeat the whole process by using a larger value for δ.

In the case of testing one restriction, this is the preferred version of the bootstrap SGD test.

We investigated the small-sample properties of the asymptotic and bootstrap SGD tests by repeating the Monte Carlo studies of Goh and King (1999, 2000) discussed in Sections 4 and 5. The testing problems covered included testing one restriction in single- and two-regressor logit and Tobit models, exponential, dynamic, and simple linear models, and testing two restrictions in three-regressor logit, and two-regressor Tobit and exponential models. For full details see Goh (1998, Chapter 6).

For testing one restriction, the simulation results suggest that the asymptotic SGD test has better small-sample sizes than the Wald and LR tests. This is not surprising, given the SGD test is bootstrap based. The power function of the SGD test appears always to be monotonic in contrast to that of the Wald test. The bootstrap SGD test has well centered power curves. In this regard, the bootstrap SGD test performs better than the Wald test for all except the logit model, where both are equally well centered. The bootstrap SGD test shares the good power properties of the LR test in general. There are two situations in which the SGD test is a clear winner compared to the LR test. The first is the higher power of the SGD test relative to that of the LR test at nonlocal alternatives to the left of H_0 for the exponential model. This is due to poor centering of the LR power function. The second is for the dynamic linear model where the LR test is locally biased but the bootstrap SGD test is not. It is also worth noting that, without any effort to recenter its power curve, the asymptotic SGD test has reasonably good power centering properties. The SGD test has similar power behavior to the bootstrap NW test of Sections 4 and 5 except in cases where the NW test is poorly centered. These exceptions occur for the exponential and the dynamic simple linear models.

In the case of testing two restrictions, the simulation results show an SGD test with a slight tendency for higher than nominal test sizes for small-sample sizes. Its sizes are clearly better than those of the Wald test and arguably better than those of the LR test. Overall, the powers of the

SGD and LR tests are comparable for the Tobit model, with both tests sharing the same good power centering properties. These tests perform better than the Wald test in regions of the alternative parameter space where the latter test has nonmonotonic power. The same is also observed for the exponential model. Although better than the Wald test, the SGD test has poorer centering properties than the LR test in the exponential model for $n = 30$. In contrast, the SGD test performs better than the LR test on the same account for $n = 80$. In regions of the alternative parameter space for the exponential model where the power performance of the Wald test is disappointing, the SGD test has power higher than that of the LR test. The reverse is observed in other regions of the parameter space. The SGD test has comparable power to the NW test, particularly for the logit and Tobit models. The SGD test appears to have better power centering properties than the NW test for the exponential model, where the power function of the latter test is ill-centered.

In summary, the new procedure generally has well-behaved sizes in small samples and power comparable to that of the LR test. It can sometimes perform better than the LR test, particularly when the latter is locally biased. For these reasons, it does appear to be a more reliable test than the LR test and particularly the Wald test.

7. SUMMARY AND CONCLUDING REMARKS

As noted in the Introduction, a good test accurately controls the probability of wrongly rejecting the null hypothesis when it is true while also having the highest possible probability of correctly rejecting the null. Unfortunately the Wald test does not always have these desirable properties in small samples. The main thrust of this chapter has been to report on the search for modifications to the Wald test in order to improve its small-sample performance. The main problems of the Wald test are its lack of invariance to different but equivalent formulations of the null hypothesis, local biasedness, and power nonmonotonicity. The first problem means it is difficult to accurately control sizes in small samples, while the latter problems can cause powers to be lower than the nominal size in some areas of the alternative hypothesis parameter space.

The bootstrap-based Wald test can help with the problems caused by the lack of invariance equivalent formulations of the null hypothesis but it does not solve the other power problems. The null Wald test does seem to solve the problem of nonmonotonicity but needs to be applied using bootstrap critical values. There remains the problem of local biasedness of the test in small samples. In the case of testing a single-parameter restriction, this can be overcome by the use of a bootstrap-based procedure with two critical values, one to control the size of the test and the other to ensure local unbiasedness.

It is not difficult to conclude that many of the small-sample problems of the Wald test can be traced to the use of the asymptotic covariance matrix of the MLE in the test statistic. A solution therefore might be to use bootstrap methods to work out the boundary between significance and insignificance of nonzero estimates of parameters or restriction functions under test. These ideas lead to the SGD test, which involves using bootstrap samples to find an ellipsoid for the parameters under test, which in turn defines the acceptance region for the test. A further refinement for testing a single parameter involves two critical values for the estimated parameter, solved simultaneously, to satisfy the size and local unbiasedness conditions. Simulation studies suggest the SGD test is slightly more reliable than the LR test and certainly better than the standard Wald test.

ACKNOWLEDGMENTS

Much of the research reported in this paper was conducted while the second author was a PhD student at Monash University on study leave from the University of Malaya. The research was supported in part by an ARC grant. We are grateful to a referee for very constructive comments which improved the paper.

REFERENCES

T. Amemiya. Qualitative response models: A survey. *Journal of Economic Literature* 19:1483–1536, 1981.

T. Amemiya. Tobit models: A survey. *Journal of Econometrics* 24:3–61, 1984.

T. Amemiya. *Advanced Econometrics*. Cambridge, MA: Harvard University Press, 1985.

E. R. Berndt, B. H. Hall, R. E. Hall, J. A. Hausman. Estimation and inference in non-linear structural models. *Annals of Economic and Social Measurement* 3:653–665, 1974.

T. S. Breusch, P. Schmidt. Alternative forms of the Wald test: How long is a piece of string? *Communications in Statistics (Theory and Methods)* 17:2789–2795, 1988.

J. F. Burguete, A. R. Gallant, G. Souza. On unification of the asymptotic theory of non-linear econometric models. *Econometric Reviews* 1:151–190, 1982.

C. L. Cavanagh, Second-order admissibility of likelihood-based tests. Discussion Paper No. 1148, Institute of Economics Research, Harvard University, 1985.

D. R. Cox, D. V. Hinkley. *Theoretical Statistics*. London: Chapman and Hall, 1974.

M. G. Dagenais, J. M. Dufour. Invariance, non-linear models and asymptotic tests. *Econometrica* 59:1601–1615, 1991.

R. Davidson, J. G. MacKinnon. *Estimation and Inference in Econometrics*. New York: Oxford University Press, 1993.

B. Efron, D. V. Hinkley. Assessing the accuracy of the maximum likelihood estimator: Observed versus expected Fisher information. *Biometrika* 65:457–487, 1978.

R. F. Engle. Wald, likelihood ratio and Lagrange multiplier tests in econometrics. In: Z. Griliches, M. Intriligator, eds., *Handbook of Econometrics*. Amsterdam: North-Holland, 2:776–826, 1984.

T. R. Fears, J. Benichou, M. H. Gail. A reminder of the fallibility of the Wald statistic. *American Statistician* 50:226–227, 1996.

L. G. Godfrey. *Misspecification Tests in Econometrics: The Lagrange Multiplier Principle and Other Approaches*. Cambridge, UK: Cambridge University Press, 1988.

L. G. Godfrey, M. R. Veall. Bootstrap-based critical values for tests of common factor restrictions. *Economics Letters* 59:1–5, 1998.

K. L. Goh. Some solutions to small-sample problems of Wald tests in econometrics. Unpublished PhD thesis, Department of Econometrics and Business Statistics, Monash University, 1998.

K. L. Goh, M. L. King. A correction for local biasedness of the Wald and null Wald tests. *Oxford Bulletin of Economics and Statistics* 61:435–450, 1999.

K. L. Goh, M. L. King. A solution to non-monotonic power of the Wald test in non-linear models. Saleh–Aly Special Edition of *Pakistan Journal of Statistics* 16:195–205, 2000.

C. Gouriéroux, A Monfort. *Statistics and Econometric Models*. Cambridge, UK: Cambridge University Press, 1995.

A. W. Gregory, M. R. Veall. Formulating Wald tests of non-linear restrictions. *Econometrica* 53:1465–1468, 1985.

A. W. Gregory, M. R. Veall. Wald tests of common factor restrictions. *Economics Letters* 22:203–208, 1986.

S. D. Grose, M. L. King. The locally unbiased two-sided Durbin–Watson test. *Economics Letters* 35:401–407, 1991.

W. W. Hauck, Jr, A. Donner. Wald's test as applied to hypotheses in logit analysis. *Journal of the American Statistical Association* 72:851–853, 1977, and Corrigendum, *ibid.,*75:482, 1980.

T. Hayakawa. The likelihood ratio criterion for a composite hypothesis under a local alternative. *Biometrika* 62:451–460, 1975.

J. L. Horowitz. Bootstrap methods in econometrics: Theory and numerical performance. In: D. M. Kreps, K. F. Wallis, eds., *Advances in Economics and Econometrics: Theory and Applications*. New York: Cambridge University Press, 3:188–222, 1997.

J. L. Horowitz, N. E. Savin. Non-invariance of the Wald test: The bootstrap to the rescue. Working Paper Series No. 92-04, Department of Economics, University of Iowa, 1992.

R. A. Johnson, D. W. Wichern. *Applied Multivariate Statistical Analysis*, 2nd edn. Englewood Cliffs: Prentice Hall, 1988.

M. L. King. Testing for autocorrelation in linear regression models: A survey. In: M. L. King, D. E. A. Giles, eds. *Specification Analysis in the Linear Model*. London: Routledge and Kegan Paul, 19–73, 1987.

F. Lafontaine, K. J. White. Obtaining any Wald statistic you want. *Economics Letters* 21:35–40, 1986.

M. R. Laskar, M. L. King. Modified Wald test for regression disturbances. *Economics Letters* 56:5–11, 1997.

D. A. McManus. How common is identification in parametric models? *Journal of Econometrics* 53:5–23, 1992.

D. A. McManus, J. C. Nankervis, N. E. Savin. Multiple optima and asymptotic approximations in the partial adjustment model. *Journal of Econometrics* 62:91–128, 1994.

M. A. Magdalinos. The classical principles of testing using instrumental variables estimates. *Journal of Econometrics* 44:241–279, 1990.

N. Mantel. Understanding Wald's test for exponential families. *American Statistician* 41:147–148, 1987.

F. D. Nelson, N. E. Savin. The non-monotonicity of the power function of the Wald test in non-linear models. Working Paper Series No. 88–7, Department of Economics, University of Iowa, 1988.

F. D. Nelson, N. E. Savin. The danger of extrapolating asymptotic local power. *Econometrica* 58:977–981, 1990.

C. D. Orme. The small sample performance of the information matrix test. *Journal of Econometrics* 46:309–331, 1990.

K. Oya. Wald, LM and LR test statistics of linear hypotheses in a structural equation model. *Econometric Reviews* 16:157–178, 1997.

R. W. Parks, N. E. Savin, A. H. Würtz. The power of Hessian- and outer-product-based Wald and LM tests, Working Paper Series No. 97-02, Department of Economics, University of Iowa, 1997.

H. W. Peers. Likelihood ratio and associated test criteria. *Biometrika* 58:577–587, 1971.

P. C. B. Phillips, J. Y. Park. On the formulation of Wald tests of non-linear restrictions. *Econometrica* 56:1065–1083, 1988.

T. Rothenberg. Approximating the distributions of econometric estimators and test statistics. In: Z. Griliches, M. Intriligator, eds. *Handbook of Econometrics*. Amsterdam: North-Holland, 2:881–935, 1984.

N. E. Savin, A. Würtz. Power of tests in binary response models. Working Paper Series No. 96-06, Department of Economics, University of Iowa, 1996.

S. D. Silvey. *Statistical Inference*. London: Chapman and Hall, 1975.

B. E. Storer, S. Wacholder, N. E. Breslow. Maximum likelihood fitting of general risk models to stratified data. *Applied Statistics* 32:172–181, 1983.

T. W. F. Stroud. On obtaining large-sample tests from asymptotically normal estimators. *Annals of Mathematical Statistics* 42:1412–1424, 1971.

M. Vaeth. On the use of Wald's test in exponential families. *International Statistical Review* 53:199–214, 1985.

T. J. Vogelsang. Wald-type tests for detecting breaks in the trend function of a dynamic time series. *Econometric Theory* 13:818–849, 1997.

13
On the Sensitivity of the *t*-Statistic

JAN R. MAGNUS Tilburg University, Tilburg, The Netherlands

1. INTRODUCTION

There seems to be consensus among statisticians and econometricians that the *t*-statistic (unlike the *F*-statistic) is not sensitive (robust). If we define the *t*-statistic as $\tau = a'x/\sqrt{x'Bx}$, where x is a random vector with mean 0, a is a nonrandom vector, and B a positive semidefinite nonrandom matrix, then τ will in general not follow a Student distribution for three reasons. First, x may not be normally distributed; secondly, even if x is normally distributed, x may not be χ^2-distributed, and thirdly, the numerator and denominator may be dependent. The consensus is that nevertheless the Student distribution can be used to provide a good approximation of the distribution of τ. The purpose of this paper is to analyze some aspects of this situation.

We confine ourselves to the situation where the random vector x is normally distributed; some aspects of the non-normal case are discussed in Ullah and Srivastava (1994). Then all odd moments of τ which exist are 0. This chapter therefore concentrates on the even moments of τ and, in particular, on its variance and kurtosis. The special case where $x'Bx$ follows

277

a χ^2-distribution, but where $a'x$ and $x'Bx$ are dependent, was considered by Smith (1992). He obtains the density for this special case, and, from the density, the moments, but not in an easy-to-use form. He concludes that the t-statistic is robust in most situations. Morimune (1989), in a simultaneous equation context, also concludes that the t-statistic is robust. Maekawa (1980) considered the t-statistic in the seemingly unrelated regression model and derived an Edgeworth expansion up to $\mathcal{O}(n^{-1})$, where n is the sample size.

We shall argue that, although it looks as if the t-statistic is robust because the moments of τ are close to moments of the Student distribution, in fact the conclusion is wrong. The reason for this apparent contradiction is the following. Clearly, the Student distribution is close to the standard normal distribution. Also, the t-statistic τ, properly scaled, is well approximated by the standard normal distribution. For this reason τ is close to a Student distribution since both are close to the standard normal distribution, and in this sense the t-statistic is robust. But in many cases, as we shall see, τ is better approximated by the standard normal distribution than by the appropriate Student distribution, and in this sense the t-statistic is not robust.

The paper is organized as follows. In Section 2 we define the problem and settle the notation. Theorem 1 in Section 3 gives a simple expression for the even moments of τ. This theorem is valid irrespective of whether $x'Bx$ follows a χ^2-distribution or whether $a'x$ and $x'Bx$ are independent. Theorem 2 gives precise conditions when these moments exist. Theorem 1 is a new result and has potential applications in many other situations. In Section 4 we obtain Theorem 3 as a special case of Theorem 1 by assuming that $x'Bx$, properly scaled, follows a χ^2-distribution. The even moments of τ then become extremely simple functions of one "dependence parameter" δ. We analyze the variance and kurtosis for this case and conclude that, if we use the standard normal distribution to approximate the Student distribution, then the approximation will be *better* with dependence than without. All proofs are in the Appendix.

2. SET-UP AND NOTATION

Let x be a normally distributed $n \times 1$ vector with mean 0 and positive definite covariance matrix $\Omega = LL'$. Let a be an $n \times 1$ vector and B a positive semidefinite $n \times n$ matrix with rank $r \geq 1$. We define the "t-type" random variable

$$\tau = \frac{a'x}{\sqrt{x'Bx}}. \tag{1}$$

In order to normalize B we introduce the matrix $B^* = (1/\mathrm{tr}B\Omega)L'BL$, which satisfies $\mathrm{tr}B^* = 1$. We denote by $\lambda_1, \ldots, \lambda_r$ the positive eigenvalues of B^* and by $\sigma_1, \ldots, \sigma_r$ the corresponding normalized eigenvectors, so that $B^*\sigma_i = \lambda_i\sigma_i$, $\sigma_i'\sigma_i = 1$, $\sigma_i'\sigma_j = 0 \, (i \neq j)$.

We normalize the vector a by defining

$$\alpha_i = \frac{\sigma_i'L'a}{\sqrt{a'\Omega a}} \qquad (i = 1, \ldots, r). \tag{2}$$

An important role is played by the scalar

$$\delta = \sum_{i=1}^{r} \alpha_i^2. \tag{3}$$

It is easy to see that $0 \leq \delta \leq 1$. If $\delta = 0$, then $a'x$ and $x'Bx$ are independent and $L'a$ and $L'BL$ are orthogonal; if $\delta > 0$, they are dependent. If $\delta = 1$, then $L'a$ lies in the column space of $L'BL$ (or equivalently, a lies in the column space of B).

3. THE MOMENTS OF τ

All odd moments of τ which exist are 0. As for the even moments that exist, we notice that $\tau^2 = x'Ax/x'Bx$ with $A = aa'$. The exact moments of a ratio of quadratic forms in normal variables was obtained by Magnus (1986), while Smith (1989) obtained moments of a ratio of quadratic forms using zonal polynomials and invariant polynomials with multiple matrix arguments. In the special case where A is positive semidefinite of rank 1 and where the mean of x is 0, drastic simplifications occur and we obtain Theorem 1.

Theorem 1. We have, provided the expectation exists, for $s = 1, 2, \ldots,$

$$E\tau^{2s} = \left(\frac{a'\Omega a}{\mathrm{tr}B\Omega}\right)^s \frac{\kappa_s}{(s-1)!} \int_0^{\infty} t^{s-1} \prod_{i=1}^{r}(1 - \mu_i(t))^{1/2}\left(1 - \sum_{i=1}^{r}\mu_i(t)\alpha_i^2\right)^s dt$$

where

$$\mu_i(t) = \frac{2t\lambda_i}{1 + 2t\lambda_i} \quad \text{and} \quad \kappa_s = 1 \times 3 \times \cdots \times (2s - 1).$$

From Theorem 1 we find the variance $\mathrm{var}(\tau)$ and the kurtosis $\mathrm{kur}(\tau) \equiv E\tau^4/(E\tau^2)^2$ as

$$\text{var}(\tau) = \frac{a'\Omega a}{\text{tr}B\Omega} \int_0^\infty f(t)g(t)dt \qquad \text{kur}(\tau) = 3 \times \frac{\int tf(t)g(t)^2 dt}{(\int f(t)g(t)dt)^2} \qquad (4)$$

where

$$f(t) = \prod_{i=1}^r (1 - \mu_i(t))^{1/2}, \qquad g(t) = 1 - \sum_{i=1}^r \mu_i(t)\alpha_i^2. \qquad (5)$$

Turning now to the existence of the even moments, we notice that for Student's t-distribution Et^{2s} is defined when $r > 2s$. We find the same condition for the moments of τ, except that when $\delta = 1$ all moments of τ exist.

Theorem 2 (existence):

(i) If $r \le n - 1$ and $\delta = 1$, or if $r = n$, then $E\tau^{2s}$ exists for all $s \ge 0$;
(ii) If $r \le n - 1$ and $\delta \ne 1$, then $E\tau^{2s}$ exists for $0 \le s < r/2$ and does not exist for $s \ge r/2$.

The variance and kurtosis given in (4) can be evaluated for any given $\lambda_1, \ldots, \lambda_r$ and $\alpha_1, \ldots, \alpha_r$. To gain insight into the sensitivity of the t-statistic, we consider one important special case, namely the case where $x'Bx$, properly scaled, follows a $\chi^2(r)$ distribution, but where $a'x$ and $x'Bx$ are dependent.

4. SENSITIVITY FOR DEPENDENCE

When $x'Bx$, properly scaled, follows a $\chi^2(r)$ distribution, the only difference between τ and Student's t is that the numerator and denominator in τ are dependent, unless $\delta = 0$. In this section we investigate the impact of this dependence on the moments of τ. We first state Theorem 3, which is a special case of Theorem 1.

Theorem 3. If the nonzero eigenvalues of $L'BL$ are all equal, then we have, provided the expectation exists, for $s = 1, 2, \ldots$,

$$E\tau^{2s} = \left(\frac{a'\Omega a}{\text{tr}B\Omega}\right)^s \frac{\kappa_s}{(s-1)!} \left(\frac{r}{2}\right)^s \sum_{j=0}^s \binom{s}{j} (-\delta)^j B\left(s+j, \frac{r}{2} - s\right),$$

where $B(\cdot, \cdot)$ denotes the Beta function.

It is remarkable that the even moments of τ now depend only on s, r and the "dependence parameter" δ (apart from a scaling factor $a'\Omega a/\text{tr}B\Omega$). Evaluation of $E\tau^{2s}$ is very easy since no integral needs to be calculated. Under the conditions of Theorem 3 we obtain

$$\text{var}(\tau) = \frac{r}{r-2}\left(1 - \frac{2\delta}{r}\right) \tag{6}$$

and

$$\text{kur}(\tau) = 3 \times \frac{r-2}{r-4} \times \frac{r^2(r+2) - 8\delta r(r+2) + 24\delta^2 r}{r^2(r+2) - 4\delta r(r+2) + 4\delta^2(r+2)}. \tag{7}$$

For fixed r, both $\text{var}(\tau)$ and $\text{kur}(\tau)$ are monotonically decreasing functions on the [0, 1] interval, and we find

$$1 \le \text{var}(\tau) \le \frac{r}{r-2} \tag{8}$$

and

$$3\frac{r}{r+2} \le \text{kur}(\tau) \le 3\frac{r-2}{r-4}. \tag{9}$$

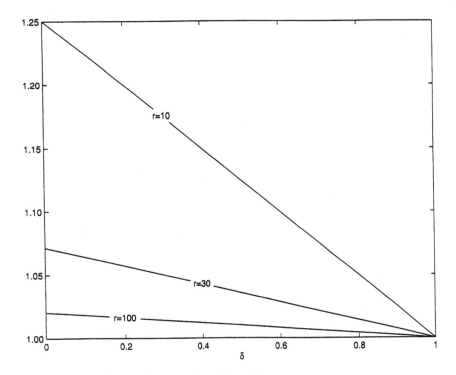

Figure 1. Variance of τ as a function of δ.

Figures 1 and 2 illustrate the behaviour of var(τ) and kur(τ) for $r = 10$, $r = 30$, and $r = 100$ (and, of course, $r = \infty$).

The variance of τ is linear in δ. For $\delta = 0$ we find var(τ) $= r/(r-2)$, the variance of the $t(r)$-distribution. When $\delta > 0$, var(τ) decreases monotonically to 1. Hence, the more dependence there is (the higher δ is), the *better* is var(τ) approximated by the variance of the standard normal distribution.

The kurtosis of τ is *not* linear in δ, see (7), but in fact is almost linear, slightly curved, and convex on $[0, 1]$; see Figure 2. When $\delta = 0$ (independence) we find kur(τ) $= 3(r-2)/(r-4)$, the kurtosis of the $t(r)$-distribution. When $\delta > 0$, the kurtosis decreases, becomes 3 (at some δ between 0.500 and 0.815), and decreases further to $3r/(r+2)$. The deviation of the kurtosis from normality is *largest* in the case of independence ($\delta = 0$).

In conclusion, if we use the standard normal distribution to approximate the t-distribution, then the approximation will be *better* with dependence than without. The t-statistic τ is thus better approximated by the standard normal distribution than by the appropriate Student distribution. In this sense the t-statistic is *not* robust.

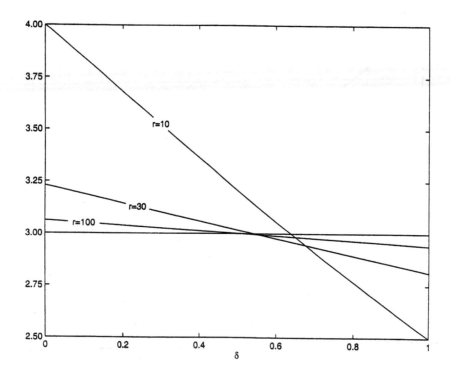

Figure 2. Kurtosis of τ as a function of δ.

APPENDIX: PROOFS

Proof of Theorem 1. Since $\tau^2 = x'Ax/x'Bx$ with $A = aa'$, the even moments of τ are the moments of a ratio of two quadratic forms. Using Theorem 6 of Magnus (1986), let P be an orthogonal $n \times n$ matrix and Λ^* a diagonal $n \times n$ matrix such that

$$P'L'BLP = \Lambda^*, \qquad P'P = I_n.$$

Then,

$$E\tau^{2s} = \frac{1}{(s-1)!} \sum_\nu \gamma_s(\nu) \int_0^\infty t^{s-1} |\Delta| \prod_{j=1}^s (\mathrm{tr}R^j)^{n_j} dt$$

where the summation is over all $1 \times s$ vectors $\nu = (n_1, n_2, \ldots, n_s)$ whose elements n_j are non-negative integers satisfying $\sum_{j=1}^s jn_j = s$,

$$\gamma_s(\nu) = s! \, 2^s \prod_{j=1}^s \left(n_j!(2j)^{n_j}\right)^{-1},$$

and

$$\Delta = (I_n + 2t\Lambda^*)^{-1/2}, \qquad R = \Delta P'L'aa'LP\Delta.$$

Now, R has rank 1, so that

$$\left(\mathrm{tr}R^j\right)^{n_j} = \left(a'LP\Delta^2 P'L'a\right)^{jn_j}$$

and hence, since $\sum_j jn_j = s$,

$$\prod_{j=1}^s (\mathrm{tr}R^j)^{n_j} = \left(a'LP\Delta^2 P'L'a\right)^s.$$

Also, using Lemma 3 of Magnus (1986) for the special case $n = 1$, we see that $\sum_\nu \gamma_s(\nu) = \kappa_s$. Hence,

$$E\tau^{2s} = \frac{\kappa_s}{(s-1)!} \int_0^\infty t^{s-1} |\Delta| \left(a'LP\Delta^2 P'L'a\right)^s dt.$$

Letting $\lambda_1^*, \ldots, \lambda_r^*$ denote the nonzero diagonal elements of Λ^*, we see that $\lambda_i^* = (\mathrm{tr}B\Omega)\lambda_i$. Letting $t^* = (\mathrm{tr}B\Omega)t$, we thus obtain

$$|\Delta| = \prod_{i=1}^r (1 + 2t\lambda_i^*)^{-1/2} = \prod_{i=1}^r (1 - \mu_i(t^*))^{1/2}$$

and

$$a'LP\Delta^2P'L'a = (a'\Omega a)\left(1 - \sum_{i=1}^{r}\mu_i(t^*)\alpha_i^2\right).$$

Proof of Theorem 2. Let Q be an $n \times (n - r)$ matrix of full column rank $n - r$ such that $BQ = 0$. Then, using Magnus (1986, Theorem 7) or Magnus (1990, Theorem 1) and noticing that $Q'a = 0$ if and only if $\delta = 1$, the result follows.

Proof of Theorem 3. Let $\lambda_i = \lambda = 1/r$ and $\mu = 2t\lambda/(1 + 2t\lambda)$. Theorem 1 implies that

$$E\tau^{2s} = \left(\frac{a'\Omega a}{\text{tr} B\Omega}\right)^s \frac{\kappa_s}{(s-1)!}\int_0^\infty t^{s-1}(1-\mu)^{r/2}\left(1 - \mu\sum_{i=1}^{r}\alpha_i^2\right)^s dt.$$

By making the change of variable

$$t = \frac{1}{2\lambda}\cdot\frac{\mu}{1-\mu}$$

we obtain

$$E\tau^{2s} = \left(\frac{a'\Omega a}{\text{tr} B\Omega}\right)^s \frac{\kappa_s}{(s-1)!}\left(\frac{r}{2}\right)^s\int_0^1 \mu^{s-1}(1-\mu)^{r/2-s-1}(1-\mu\delta)^s d\mu.$$

Now,

$$\int_0^1 \mu^{s-1}(1-\mu)^{r/2-s-1}(1-\mu\delta)^s d\mu$$

$$= \sum_{j=0}^{s}\binom{s}{j}(-\delta)^j\int_0^1 \mu^{s+j-1}(1-\mu)^{r/2-s-1}d\mu$$

$$= \sum_{j=0}^{s}\binom{s}{j}(-\delta)^j B(s+j, r/2-s)$$

and the result follows.

REFERENCES

K. Maekawa. An asymptotic expansion of the distribution of the test statistics for linear restrictions in Zellner's SUR model. *Hiroshima Economic Review* 4:81–97, 1980.

J. R. Magnus. The exact moments of a ratio of quadratic forms in normal variables. *Annales d'Economie et de Statistique* 4:95–109, 1986.

J. R. Magnus. On certain moments relating to ratios of quadratic forms in normal variables: further results. *Sankhya, Series B* 52:1–13, 1990.

K. Morimune. *t* test in a structural equation. *Econometrica* 57:1341–1360, 1989.

M. D. Smith. On the expectation of a ratio of quadratic forms in normal variables. *Journal of Multivariate Analysis* 31:244–257, 1989.

M. D. Smith. Comparing the exact distribution of the *t*-statistic to Student's distribution when its constituent normal and chi-squared variables are dependent. *Communications in Statistics, Theory and Methods* 21(12):3589–3600, 1992.

A. Ullah, V. K. Srivastava. Moments of the ratio of quadratic forms in non-normal variables with econometric examples. *Journal of Econometrics* 62:129–141, 1994.

14

Preliminary-Test and Bayes Estimation of a Location Parameter under "Reflected Normal" Loss

DAVID E. A. GILES   University of Victoria, Victoria, British Columbia, Canada

1. INTRODUCTION

In statistics and econometrics, the expression "preliminary-test estimation" refers to a situation where the choice of estimator for some parameter (vector) is essentially randomized through the prior application of an hypothesis test. The test need not relate to the parameters of interest—for example, it generally relates to a set of nuisance parameters. Neither is it necessary that the same set of data be used for the prior test as for the primary estimation problem. This randomization of the choice of estimator complicates its sampling properties significantly, as was first recognized by Bancroft (1944). Extensive surveys of the subsequent literature on preliminary-test estimation are given by Bancroft and Han (1977) and Giles and Giles (1993).

Preliminary-test estimators generally have quite adequate (if not optimal) large-sample properties. For example, if the "component" estimators (between which a choice is made) and the prior test are each consistent, then the pre-test estimator will also be consistent. On the other hand, as pre-test estimators are discontinuous functions of the data, it is well known (e.g., Cohen 1965) that they are inadmissible under conventional loss functions. Despite their inadmissibility, and the fact that generally they are not minimax, pre-test estimators are of considerable practical interest for at least two reasons. First, they represent estimation strategies of precisely the type that is frequently encountered in practice. Typically, in many areas of applied statistics, a model will be estimated and then the model will be subjected to various specification tests. Subsequently, the model in question may be simplified (or otherwise amended) and then re-estimated. Second, in all of the pre-test problems that have been considered in the literature, pre-test estimators dominate each of their component estimators over different parts of the parameter space in terms of risk. Indeed, in some cases (e.g., Giles 1991) they may even dominate their components simultaneously over the *same* part of the parameter space.

Although the sampling properties of various preliminary-test estimators have been studied by a range of authors, little is known about their complete sampling distributions. The only exceptions appear to be the results of Giles (1992), Giles and Srivastava (1993), Ohtani and Giles (1996b), and Wan (1997). Generally the finite-sample properties of pre-test estimators have been evaluated in terms of risk, and usually on the assumption that the loss function is quadratic. Some exceptions (using absolute-error loss, the asymmetric "LINEX" loss function, or "balanced" loss) include the contributions of Giles (1993), Ohtani et al. (1997), Giles *et al.* (1996), Ohtani and Giles (1996a), Giles and Giles (1996), and Geng and Wan (2000), among others.

Despite its tractability and historical interest, two obvious practical shortcomings of the quadratic loss function are its unboundedness and its symmetry. Recently, Spiring (1993) has addressed the first of these weaknesses (and to a lesser degree the second) by analyzing the "reflected normal" loss function. He motivates this loss with reference to problems in quality assurance (e.g., Taguchi 1986). The reflected normal loss function has the particular merit that it is bounded. It can readily be made asymmetric if this is desired for practical reasons. An alternative loss structure is the "bounded LINEX" (or "BLINEX") loss discussed by Levy and Wen (1997a,b) and Wen and Levy (1999a,b). The BLINEX loss is both bounded and asymmetric.

In this paper, we consider a simple preliminary-test estimation problem where the analyst's loss structure is "reflected normal." Specifically, we consider the estimation of the location parameter in a normal sampling problem, where a preliminary test is conducted for the validity of a simple restriction on this parameter. The exact finite-sample risk of this pre-test estimator is derived under reflected normal loss, and this risk is compared with those of both the unrestricted and restricted maximum likelihood estimators. This appears to be the first study of a pre-test estimator when the loss structure is bounded, and comparisons are drawn between these results and those obtained under conventional (unbounded) quadratic loss. Our results extend naturally to the case of estimating the coefficients in a normal linear multiple regression model. Although we consider only a symmetric loss function in this paper, the extension to the asymmetric case is also straightforward.

In the next section we formulate the problem and the notation. Exact expressions for the risk functions are derived in Section 3, and these are evaluated, illustrated, and discussed in Section 4. Some related Bayesian analysis is provided in Section 5; and Section 6 offers some concluding remarks and suggests some directions for further research.

2. FORMULATION OF THE PROBLEM

The problem that we consider here is cast in simple terms to facilitate the exposition. However, the reader will recognize that it generalizes trivially to more interesting situations, such as the estimation of the coefficient vector in a standard linear multiple regression model when potentially there are exact linear restrictions on the coefficients. In that sense, our analysis here extends that of Judge and Bock (1978), and of others,[*] through the consideration of a different loss structure. We will be concerned with the estimation of the location parameter, μ, in a normal population with unknown scale parameter. We have a simple random sample of values:

$$x_i \sim \text{i.i.d. } N[\mu, \sigma^2]; \qquad i = 1, 2, 3, \ldots, n$$

and we will be concerned with a prior "t-test" of

$$H_0 : \mu = \mu_0 \qquad \text{vs.} \qquad H_A : \mu \neq \mu_0$$

The choice of estimator for μ will depend on the outcome of this preliminary test. If H_0 were false we would use the unrestricted maximum likelihood estimator (UMLE), $\mu_1 = (\sum_{i=1}^{n} x_i/n) = \bar{x}$. On the other hand, if H_0 were true we would use the restricted maximum likelihood estimator (RMLE), which here is just μ_0 itself. So, the preliminary-test estimator (PTE) of μ in this situation is

$$\mu_p = [I_R(t) \times \mu_1] + [I_A(t) \times \mu_0]$$

---

[*]See Giles and Giles (1993) for detailed references.

where t is the usual t-statistic for testing H_0, defined as

$$t = [\mu_1 - \mu_0]/[s^2/n]^{1/2}$$

and

$$s^2 = \left[\sum_{i=1}^{n}(x_i - \bar{x})^2\right]/(n-1)$$

It will be more convenient to use $F = t^2$ as the test statistic, recalling that $F \sim F(1, n-1; \lambda)$ where the non-centrality parameter is $\lambda = n\delta^2/(2\sigma^2)$, and $\delta = (\mu - \mu_0)$. So, the PTE of μ may be written as

$$\mu_p = [I_R(F) \times \mu_1] + [I_A(F) \times \mu_0] \tag{1}$$

where $I_\Omega(.)$ is an indicator function taking the value unity if its argument is in the subscripted interval, and zero otherwise. In our case, the rejection region is the set of values $R = \{F : F > c_\alpha\}$ and the "acceptance" (strictly, "non-rejection") region for the "t-test" is $A = \{F : F \leq c_\alpha\}$, where c_α is the critical value for a chosen significance level α. It should be noted that

$$I_R(F) = [1 - I_A(F)] \tag{2}$$

$$[I_R(F) \times I_A(F)] = 0 \tag{3}$$

and

$$[I_\Omega(F)]^p = I_\Omega(F); \quad \Omega = \{A, R\}; \quad p \geq 1 \tag{4}$$

If we let θ be a scalar parameter to be estimated, and let τ be a statistic used as an estimator of θ, then the "reflected normal" loss function is defined as

$$\mathcal{L}(\tau, \theta) = K\{1 - \exp[-(\tau - \theta)^2/(2\gamma^2)]\} \tag{5}$$

where K is the maximum loss, and γ is a pre-assigned shape parameter that controls the rate at which the loss approaches its upper bound. For example, if we set $\gamma = (\Delta/4)$, for some Δ, then $\mathcal{L} \geq (0.9997K)$ for all values $\tau > \theta \pm \Delta$. The "reflected normal" loss structure arises in the context of "M-estimation" (e.g., Huber 1977), and in the context of robust estimation its influence function is known to have rather good properties. Figure 1 compares the reflected normal and conventional quadratic loss functions.

3. DERIVATION OF THE RISK FUNCTIONS

The risk function of τ as an estimator of θ is $\mathcal{R}(\tau) = E[\mathcal{L}(\tau, \theta)]$, where expectation is taken over the sample space. Given that the reflected normal

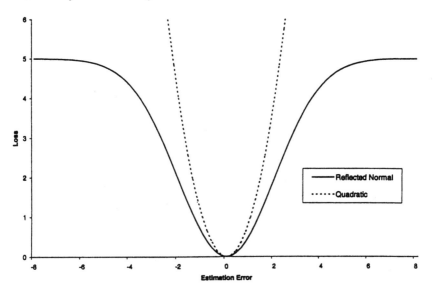

Figure 1. Reflected normal and quadratic loss functions ($n = 10$, $\sigma = 1$, $K = 5$, $\gamma = 1$).

loss function is bounded, the existence of the associated risk function is assured. As Wen and Levy (1999b) note, this is not always the case for *unbounded* loss functions in conjunction with certain densities—e.g., the LINEX loss function when the likelihood is Student-t. Let us consider the risks of the RMLE, UMLE, and PTE estimators of μ in turn.

3.1 RMLE

In our case the risk of the RMLE is trivial, as μ_0 is a constant, and is simply

$$\mathcal{R}(\mu_0) = K\{1 - \exp[-\delta^2/(2\gamma^2)]\} \tag{6}$$

3.2 UMLE

$$\mathcal{R}(\mu_1) = K \int_{-\infty}^{\infty} \{1 - \exp[-(\mu - \mu_1)^2/(2\gamma^2)]\}p(\mu_1)d\mu_1 \tag{7}$$

where

$$p(\mu_1) = (n/2\pi\sigma^2)^{1/2}\exp[-(\mu_1 - \mu)^2/(2\sigma^2/n)] \tag{8}$$

Substituting (8) in (7), completing the square, and using the result that a normal density integrates to unity, we obtain

$$\mathcal{R}(\mu_1) = K\{1 - \gamma/[(\sigma^2/n) + \gamma^2]\} \tag{9}$$

Before proceeding to the derivation of the risk function for the PTE of μ, some comments on the risks of these "component estimators" are in order. First, as is the case with a quadratic loss function, $\mathcal{R}(\mu_1)$ does not depend on the (squared) "estimation error" δ^2, and hence is also constant with respect to λ. Second, as is also the case with a quadratic loss function for this problem, $\mathcal{R}(\mu_0)$ is an increasing function of δ^2 (or λ). Under quadratic loss this risk increases linearly with λ, and so it is unbounded. Here, however, it increases from zero at a decreasing rate, and *limit* $[\mathcal{R}(\mu_0)] = K$ when $\delta^2 \to \infty$. That is, the risk of the RMLE is *bounded*. Finally, equating $\mathcal{R}(\mu_0)$ and $\mathcal{R}(\mu_1)$, we see that the risk functions intersect at

$$\delta^2 = -2\gamma^2 \ln\{\gamma/[(\sigma^2/n) + \gamma^2]\}$$

These results are reflected in the figures in the next section.

3.3 PTE

The derivation of the risk of the preliminary-test estimator of μ is somewhat more complex, and we will use the following result from Clarke (1986, Appendix 1).

Lemma. Let w be a non-central chi-square variate with g degrees of freedom and non-centrality parameter θ, let $\phi(\cdot)$ be any real-valued function, and let n be any real value such that $n > (-g/2)$. Then

$$E[w^n\phi(w)] = 2^n \sum_{m=0}^{\infty} (e^{-\theta}\theta^m/m!)\left[\Gamma\left(\frac{1}{2}(g+2n+2m)\right)\Big/\Gamma\left(\frac{1}{2}(g+2m)\right)\right]$$
$$E[\phi(\chi^2(g+1+2m))]$$

Now, to evaluate $\mathcal{R}(\mu_p)$, we note that it can be written as

$$\mathcal{R}(\mu_p) = K\{1 - E(\exp[-(\mu - \mu_p)^2/(2\gamma^2)])\}$$
$$= K\sum_{r=1}^{\infty}\{[E(\mu_p - \mu)^{2r}]/[(-1)^r(2\gamma^2)^r(r!)]\}$$

From (1), we have

$$(\mu_p - \mu) = [I_R(F) \times \mu_1] + [I_A(F) \times \mu_0] - \mu$$
$$= I_R(F)(\mu_1 - \mu) + \delta I_A(F)$$

and so, using (2)–(4)

$$(\mu_p - \mu)^{2r} = (\mu_1 - \mu)^{2r} + [\delta^{2r} - (\mu_1 - \mu)^{2r}]I_A(F); \qquad r = 1, 2, 3, \ldots$$

Then

$$\mathcal{R}(\mu_p) = K \sum_{r=1}^{\infty} \{ [E(\mu_1 - \mu)^{2r} + E\{I_A(F)[\delta^{2r} - (\mu_1 - \mu)^{2r}]\}] /$$
$$[(-1)^r (2\gamma^2)^r (r!)] \} \tag{10}$$

Now, from the moments of the normal distribution (e.g., Zellner 1971, pp. 364–365),

$$E(\mu_1 - \mu)^{2r} = [2r(\sigma^2/n)^r/(2\pi)]\Gamma(r + 1/2); \qquad r = 1, 2, 3, \ldots \tag{11}$$

Also,

$$E\{I_A(F)[\delta^{2r} - (\mu_1 - \mu)^{2r}]\} = E\left\{ I_A(F)\left[\delta^{2r} - \sum_{j=0}^{2r} \delta^j (\mu_1 - \mu_0)^{2r-j} (^{2r}C_j) \right] \right\}$$
$$= E\{I_A(F)$$
$$\left[\delta^{2r} - \sum_{j=0}^{2r} \delta^j \left(n\chi_{(1;\lambda)}^2/\sigma^2 \right)^{r-j/2} (^{2r}C_j) \right] \} \tag{12}$$

where $\chi_{(q;\lambda)}^2$ denotes a non-central chi-square variate with q degrees of freedom and non-centrality parameter $\lambda(= n\delta^2/(2\sigma^2))$.

Recalling that $F = [(n - 1)(\chi_{(1;\lambda)}^2/\chi_{(n-1;0)}^2)]$, where the two chi-square variates are independent, we can re-express (12) as

$$E\{I_A(F)[\delta^{2r} - (\mu_1\mu)^{2r}]\} = \delta^{2r} \mathrm{Pr}[(\chi_{(1;\lambda)}^2/\chi_{(n-1;0)}^2) < c_\alpha^*]$$
$$- \sum_{j=0}^{2} \{\delta^j (^{2r}C_j)(n/\sigma^2)^{r-j/2} E[I_A((n-1)\chi_{(1;\lambda)}^2/$$
$$\chi_{(n-1;0)}^2)(\chi_{(1;\lambda)}^2)^{r-j/2}]\} \tag{13}$$

where $c_\alpha^* = [c_\alpha/(n - 1)]$.

The expectation in (13) can be evaluated by repeatedly using the result of Clarke (1986), stated in the above lemma, and the independence of the associated chi-square variates[*]

$$
E\left[I_A\left((n-1)\chi^2_{(1;\lambda)}/\chi^2_{(n-1;0)}\right)\left(\chi^2_{(1;\lambda)}\right)^{r-j/2}\right]
$$

$$
= 2^{r-j/2}\sum_{i=0}^{\infty}(e^{-\lambda}\lambda^i/i!)\left[\Gamma\left(\frac{1}{2}(1+2r-j+2i)\right)/\Gamma\left(\frac{1}{2}(1+2i)\right)\right] \qquad (14)
$$

$$
\times \Pr\left[\left(\chi^2_{(1+2r-j+2i;\lambda)}/\chi^2_{(n-1)}\right) < c^*_\alpha\right]
$$

Finally, using the results in (11)–(14), we can write the risk of the pre-test estimator, (10), as

$$
\mathcal{R}(\mu_p) = K\sum_{r=1}^{\infty}\left[1/\{(-1)^r(2\gamma^2)^r(r!)\}\right]\{[2r(\sigma^2/n)^r/(2\pi)]\Gamma(r+1/2)
$$

$$
+\delta^{2r}\Pr\left[\left(\chi^2_{(1;\lambda)}/\chi^2_{(n-1;0)}\right) < c^*_\alpha\right] - \sum_{j=0}^{2r}\{\delta^j\left(^{2r}C_j\right)(n/\sigma^2)^{r-j/2}2^{r-j/2}
$$

$$
\times \sum_{i=0}^{\infty}(e^{-\lambda}\lambda^i/i!)[\Gamma(1/2(1+2r-j+2i))/\Gamma(1/2(1+2i))]
$$

$$
\Pr\left[\left(\chi^2_{1+2r-j+2i;\lambda)}/\chi^2_{(n-1)}\right) < c^*_\alpha\right]\}\}
$$

4. SOME ILLUSTRATIVE EVALUATIONS

The risk functions for our various estimators under a conventional quadratic loss function are well known,[†] and for comparative purposes they are illustrated in Figure 2. In this case the pre-test estimator has bounded risk, although its inadmissibility was noted above. The risk functions for the restricted and unrestricted maximum likelihood estimators under reflected normal loss, as in (6) and (9), are easily evaluated for particular choices of the parameters and sample size, and these are illustrated in Figure 3. In particular, we see there that $\mathcal{R}(\mu_0)$ is bounded above by K. The evaluation of the risk of the preliminary-test estimator is rather more tedious, but it can readily be verified by simulation. Some examples of this appear in Figures 4–6. There, the range of values for δ^2 is such that the boundedness of $\mathcal{R}(\mu_0)$ is not visually apparent.

---

[*]Further details of the proof of this result are available from the author on request.
[†]For example, see Judge and Bock (1978, Chapter 3), and Giles and Giles (1993).

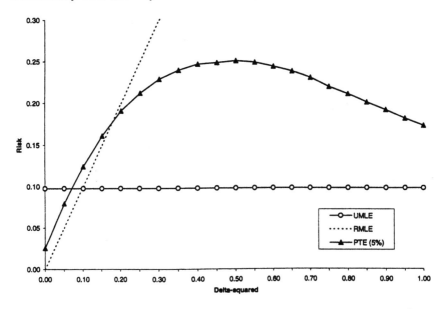

Figure 2. Risks under quadratic loss ($n = 10$, $\sigma = 1$, $K = 1$, $\gamma = 1$).

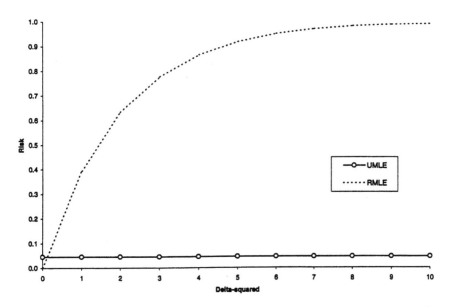

Figure 3. Risk under reflected normal loss ($n = 10$, $\sigma = 1$, $K = 1$, $\gamma = 1$).

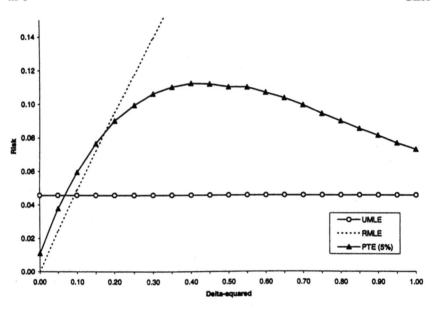

Figure 4. Risks under reflected normal loss ($n = 10$, $\sigma = 1$, $K = 1$, $\gamma = 1$).

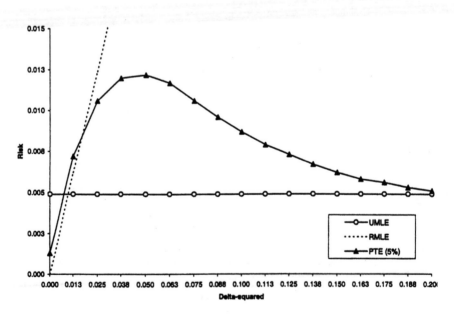

Figure 5. Risks under reflected normal loss ($n = 100$, $\sigma = 1$, $K = 1$, $\gamma = 1$).

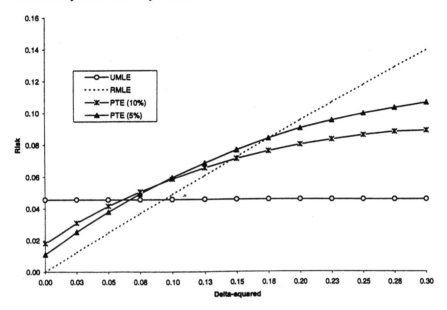

Figure 6.   Risks under reflected normal loss ($n = 10$, $\sigma = 1$, $K = 1$, $\gamma = 1$).

In particular, Figure 4 compares $\mathcal{R}(\mu_p)$ with $\mathcal{R}(\mu_0)$ and $\mathcal{R}(\mu_1)$ for a small size ($n = 10$) and illustrative parameter values. The general similarity between these results and their counterparts under quadratic loss (as in Figure 2) is striking. In particular, there is a region of the parameter space where μ_p is least preferred among the three estimators under consideration. Similarly, there are regions where each of μ_0 and μ_1 are least preferred. There are regions where either μ_0 or μ_1 is most preferred among the three estimators, but there is no region of the parameter space where the pretest estimator is preferred over both μ_0 and μ_1 simultaneously. The effect of increasing the sample size from $n = 10$ to $n = 100$ can be seen by comparing Figures 4 and 5. In each of these figures the convergence of $\mathcal{R}(\mu_p)$ to $\mathcal{R}(\mu_0)$ as $\delta^2 \to \infty$ is as expected. The preliminary test has a power function that approaches unity in this case, so in the limit the PTE and the UMLE of μ coincide

The similarity of the reflected normal results to those under quadratic loss extends to the fact that in the latter case $\mathcal{R}(\mu_p)$ is again bounded. Cohen's (1965) results imply that this estimator is inadmissible. Parenthetically, this can be related in turn to Brown's (1971) necessary and sufficient conditions for the admissibility of an estimator that has bounded risk. Specifically, Brown proves that for every potentially admissible estimator of an m-element parameter vector there exists a diffusion in m-dimensional space. The estimator is admissible if and only if this diffusion is recurrent (rather than transient), provided that its risk is bounded. Taken in conjunction with Cohen's result this implies, of course, that our pre-test estimator is associated with a transient diffusion, whether the loss function is quadratic or "reflected normal."

Figure 6 depicts the effects of increasing the significance level for the preliminary test from 5% to 10%. A larger significance level implies that greater weight is given to the UMLE when $\delta^2 = 0$ (when H_0 is true). An increase in the size of the test also increases the power, so μ_p also gives greater weight to μ_1 when $\delta^2 > 0$. It is clear that in principle it should be possible to bound the region in which $\mathcal{R}(\mu_p)$ and $\mathcal{R}(\mu_0)$ intersect, as is done in the quadratic loss case by Judge and Bock (1978, p. 73). However, the extremely complex nature of the expression for the former risk function (in (15)) makes this rather impractical from an analytical standpoint.

5. BAYESIAN ESTIMATION

In this section we briefly consider the Bayes estimator of μ under reflected normal loss, as this estimator has the desirable attribute of being admissible if the prior p.d.f. is "proper." We will take the Bayes estimator, τ_B, of a parameter θ to be the "minimum (posterior) expected loss" (MELO) estimator. That is, τ_B minimizes

$$\text{EL} = \int_{-\infty}^{\infty} \mathcal{L}(\tau_B, \theta)p(\theta|x)d\theta \tag{16}$$

where $p(\theta|x)$ is the posterior p.d.f. for θ, given $x = (x_1, \ldots, x_n)$. If $p(\theta)$ is the prior p.d.f. for θ, it is well known[*] that the τ_B that minimizes (15) will also minimize the Bayes risk

$$\text{Er}(\theta) = \int_{-\infty}^{\infty} \int_{-\infty}^{\infty} p(\theta)\mathcal{L}(\tau_B, \theta)p(x|\theta)dxd\theta \tag{17}$$

as long as the Bayes risk is finite. As is also well known, if $\mathcal{L}(\tau_B, \theta)$ is quadratic, then τ_B is the mean of $p(\theta|x)$.

However, when $\mathcal{L}(\tau, \theta)$ is reflected normal, there is no simple closed-form expression for the τ_B that minimizes (15) for an arbitrary posterior p.d.f. Of course, more progress can be made for specific posterior cases. Let us consider some particular choices of prior (and hence posterior) p.d.f. for μ in our problem, assuming that σ^2 is known. The posterior analysis for μ in the

---

[*]For instance, see Zellner (1971, pp. 24–26).

case of unknown σ^2 does not lend itself to a simple analysis, and is not considered further below.

5.1 Case (i): Known Variance, Conjugate Prior

Recalling the normality of our data, the natural-conjugate prior for μ is $N[\mu', \sigma']$ and the posterior for μ is $N[\mu'', \sigma'']$. It is well known[*] that

$$\mu'' = \left[\mu_1 \sigma'^2 + \mu'(\sigma^2/n)\right]/\left[\sigma'^2 + (\sigma^2/n)\right] \tag{18}$$

and

$$\sigma''^2 = \left[\sigma'^2 \sigma^2/n\right]/\left[\sigma'^2 + (\sigma^2/n)\right] \tag{19}$$

So, under quadratic loss, the Bayes estimator of μ is $\mu_B = \mu''$. The Bayes estimator under the reflected normal loss is the μ_B that minimizes

$$\text{EL} = \int_{-\infty}^{\infty} \mathcal{L}(\mu_B, \mu) p(\mu|x) d\mu \tag{20}$$

where

$$\mathcal{L}(\mu_B, \mu) = K\{1 - \exp[-(\mu_B - \mu)^2/(2\gamma^2)]\} \tag{21}$$

and

$$p(\mu|x) = \left(\sigma''\sqrt{2\pi}\right)^{-1} \exp\left[-(\mu - \mu'')^2/(2\sigma''^2)\right]; \qquad -\infty < \mu < \infty \tag{22}$$

Substituting (21) and (22) in (20), setting the derivative of EL with respect to μ_B equal to zero, completing the square on μ, and solving for μ_B, it emerges after a little manipulation that $\mu_B = \mu''$. So, for this case it is particularly interesting to note that the Bayes estimator of μ is the same under either quadratic or reflected normal loss functions.

5.2 Case (ii): Known Variance, Diffuse Prior

In this case the (improper) prior p.d.f. for μ is

$$p(\mu) \propto \text{constant}; \qquad -\infty < \mu < \infty$$

and the corresponding posterior p.d.f. is well known[†] to be

$$p(\mu|x) = \sqrt{n}\left(\sigma\sqrt{2n}\right)^{-1} \exp[-n(\mu - \mu_1)^2/(2\sigma^2)] \tag{23}$$

---

[*]See Raiffa and Schlaifer (1961, p. 55) and Zellner (1971, pp. 14–15).
[†]For example, see Zellner (1971, p. 20).

That is, the posterior is $N[\mu_1, \sigma^2/n]$, and so the Bayes estimator of μ under quadratic loss is $\mu_B = \mu_1$. The corresponding estimator under reflected normal loss is obtained by substituting (23) and (22) in (20), setting the derivative of EL with respect to μ_B equal to zero, completing the square on μ, and solving for μ_B. It transpires after some manipulation that $\mu_B = \mu_1$, so for this case as well the Bayes estimator of μ is the same under either quadratic or reflected normal loss functions. This is simply a reflection of the algebra of the problem, rather than a consequence of any deep statistical result, and would not be expected to hold in more general estimation situations.

6. CONCLUDING REMARKS

Preliminary-test estimation is commonplace, but often little attention is paid to the implications that such prior testing has for the sampling properties of estimators. When these implications have been studied, generally the analysis has been in terms of the risk function of the pre-test estimator and its 'component" estimators. The majority of this risk analysis has been based on very restrictive loss functions, such as quadratic loss. One aspect of such loss structures is that they are symmetric with respect to the "direction" of the estimation error, and this may be unrealistic in practice. This condition has been relaxed by several authors, as is discussed in Section 1. Another feature of conventional loss functions (and the asymmetric ones that have been considered in a pre-test context) is that they are unbounded as the estimation error grows. This may also be unrealistic in practice. The (bounded) "reflected normal" loss function is considered in this paper, in the context of estimating a normal mean after a pre-test of a simple restriction. With this loss structure the risk of the restricted maximum likelihood "estimator" is also bounded, in contrast to the situation under quadratic loss. In other respects, however, the *qualitative* risk properties of the pre-test estimator under reflected normal loss are the same as under quadratic loss. Interestingly, the Bayes estimator of the mean is the same under both loss structures, with either a conjugate or diffuse prior, at least in the case where the precision of the process is known.

ACKNOWLEDGMENT

An earlier version of this paper was presented at the University of Victoria Econometrics Colloquium, and benefited accordingly. I am especially grateful to Judith Giles for sharing her wealth of knowledge regarding the properties of preliminary-test strategies.

REFERENCES

T. A. Bancroft. On the biases in estimation due to the use of preliminary tests of significance. *Annals of Mathematical Statistics* 15:190–204, 1994.

T. A. Bancroft, C.-P. Han. Inference based on conditional specification: a note and bibliography. *International Statistical Review* 45:117–127, 1977.

L. D. Brown. Admissible estimators, recurrent diffusions and insoluble boundary-value problems. *Annals of Mathematical Statistics* 42:855–903, 1971

J. A. Clarke. Some implications of estimating a regression scale parameter after a preliminary test of restrictions. MEc minor thesis, Department of Econometrics and Operations Research, Monash University, 1986.

A. Cohen. Estimates of the linear combinations of parameters in the mean vector of a multivariate distribution. *Annals of Mathematical Statistics* 36:299–304, 1965.

W. J. Geng, A. T. K. Wan. On the sampling performance of an inequality pre-test estimator of the regression error variance under LINEX loss. *Statistical Papers,* 41:453–472, 2000.

D. E. A. Giles. The exact distribution of a simple pre-test estimator. In: W. E. Griffiths, H. Lutkepohl, M. E. Bock, eds. *Readings in Econometric Theory and Practice: In Honor of George Judge.* Amsterdam: North-Holland, 1992, pp. 57–74.

D. E. A. Giles. Pre-test estimation in regression under absolute error loss. *Economics Letters* 41:339–343, 1993.

D. E. A. Giles, V. K. Srivastava. The exact distribution of a least squares regression coefficient estimator after a preliminary *t*-test. *Statistics and Probability Letters* 16:59–64, 1993.

J. A. Giles. Pre-testing for linear restrictions in a regression model with spherically symmetric disturbances. *Journal of Econometrics* 50:377–398, 1991.

J. A. Giles, D. E. A. Giles. Pre-test estimation in econometrics: Recent developments. *Journal of Economic Surveys* 7:145–197, 1993.

J. A. Giles, D. E. A. Giles. Estimation of the regression scale after a pre-test for homoscedasticity under LINEX loss. *Journal of Statistical Planning and Inference* 50:21–35, 1996.

J. A. Giles, D. E. A. Giles , K. Ohtani. The exact risks of some pre-test and Stein-type estimators under balanced loss. Communications in Statistics, A 25:2901–2924, 1996.

P. J. Huber. *Robust Statistical Procedures*. Philadelphia: SIAM, 1977, pp. 1–56.

G. G. Judge, M. E. Bock. *The Statistical Implications of Pre-Test and Stein-Rule Estimators in Econometrics*. New York: Wiley, 1978, pp.1–340.

M. S. Levy, D. Wen. Bayesian estimation under the BLINEX loss. Unpublished mimeo, Department of Quantitative Analysis and Operations Management, University of Cincinnati, 1997a.

M. S. Levy, D. Wen. BLINEX: a bounded asymmetric loss function. Unpublished mimeo, Department of Quantitative Analysis and Operations Management, University of Cincinnati, 1997b.

K. Ohtani, D. E. A. Giles. On the estimation of regression 'goodness of fit' under absolute error loss. Journal of Quantitative Economics 12:17–26, 1996a.

K. Ohtani, J. A. Giles. The density function and MSE dominance of the pre-test estimator in a heteroscedastic linear regression model with omitted variables. *Statistical Papers* 37:323–342, 1996b.

K. Ohtani, D. E. A. Giles, J. A. Giles. The exact risk performance of a pre-test estimator in a heteroscedastic linear regression model under the balanced loss function. *Econometric Reviews* 16:119–130, 1997.

H. Raiffa, R. Schlaifer. *Applied Statistical Decision Theory*. Cambridge, MA: MIT Press, 1961, pp. 1–356.

F. A. Spiring. The reflected normal loss function. *Canadian Journal of Statistics* 21:321–330, 1993.

G. Taguchi. *Introduction to Quality Engineering: Designing Quality into Products and Processes*. White Plains, NY: Krauss, 1986.

A. T. K. Wan. The exact density and distribution functions of the inequality constrained and pre-test estimators. *Statistical Papers* 38:329–341, 1997.

D. Wen, M. S. Levy. Admissibility of Bayes estimates under BLINEX loss for the normal mean problem. Unpublished mimeo, Department of Quantitative Analysis and Operations Management, University of Cincinnati, 1999a.

D. Wen, M. S. Levy. BLINEX: a bounded asymmetric loss function with application to Bayesian estimation. Unpublished mimeo, Department of Quantitative Analysis and Operations Management, University of Cincinnati, 1999b.

A. Zellner. *An Introduction to Bayesian Inference in Econometrics*. New York: Wiley, 1971.

15

MSE Performance of the Double *k*-Class Estimator of Each Individual Regression Coefficient under Multivariate *t*-Errors

AKIO NAMBA and KAZUHIRO OHTANI   Kobe University, Kobe, Japan

1. INTRODUCTION

In the context of the linear regression model, the Stein-rule (SR) estimator proposed by Stein (1956) and James and Stein (1961) dominates the ordinary least squares (OLS) estimator in terms of predictive mean squared error (PMSE) if the number of the regression coefficients is more than or equal to three and the normal error terms are assumed. Further, Baranchik (1970) established that the positive-part Stein-rule (PSR) estimator dominates the SR estimator under normality.

As an improved estimator, Theil (1971) proposed the minimum mean squared error (MMSE) estimator. However, since Theil's (1971) MMSE estimator includes unknown parameters, Farebrother (1975) suggested the operational variant of the MMSE estimator, which can be obtained by replacing the unknown parameters by the OLS estimators. Hereafter, we call the operational variant of the MMSE estimator the MMSE estimator simply. As an extension of the MMSE estimator, Ohtani (1996) considered the adjusted minimum mean squared error (AMMSE) estimator, which is obtained by adjusting the degrees of freedom of the component of the MMSE estimator.

The SR estimator, the MMSE estimator, and the AMMSE estimator can be regarded as special cases of the double k-class (KK) estimator proposed by Ullah and Ullah (1978). There are several studies of the sampling properties of the KK estimators, such as Vinod (1980), Carter (1981), Menjoge (1984), Carter et al. (1993), and Vinod and Srivastava (1995). Ohtani (2000) examined the MSE performance of a pre-test KK estimator.

Though most of the studies on small sample properties of shrinkage estimators have assumed that all the regression coefficients are estimated simultaneously, there are several researches which examine the sampling properties of estimators for each individual regression coefficient. In particular, Ohtani and Kozumi (1996) derived explicit formulae for the moments of the SR estimator and the PSR estimator for each individual regression coefficient, when the error terms are normally distributed. Furthermore, Ohtani (1997) derived the moments of the MMSE estimator for each individual regression coefficient under normality.

In most of the theoretical and applied works, it is assumed that error terms follow a normal distribution. However, there exist many economic data (specifically, financial data) which may be generated by distributions with fatter tails than normal distributions (see, for example, Fama 1965, Blattberg and Gonedes 1974). One example of such a distribution is a multivariate t distribution. The multivariate t distribution has often been used to examine the effects of departure from normality of error terms on the sampling performance of estimators and test statistics in a linear regression model. Some examples are Zellner (1976), Prucha and Kelejian (1984), Ullah and Zinde-Walsh (1984), Judge et al. (1985), Sutradhar and Ali (1986), Sutradhar (1988), Singh (1988, 1991), Giles (1991, 1992), Ohtani (1991, 1993), Ohtani and Giles (1993), and Ohtani and Hasegawa (1993). However, sampling properties of the SR, PSR, MMSE, and AMMSE estimators for each individual regression coefficient under multivariate t errors have not been examined so far.

In this paper, for estimating the coefficients vector of a linear regression model with multivariate t error terms, a family of pre-test double k-class (PTKK) estimators has been considered which includes the SR, PSR, MMSE, and AMMSE estimators as special cases. In Section 2, we present the model and the estimators. We derive the explicit formulae for the moments of the PTKK estimator in Section 3. It is shown analytically that the PSR estimator for each individual regression coefficient dominates the SR estimator in terms of MSE even when the error terms have a multivariate t distribution. In Section 4 we compare the MSEs of the estimators by numerical evaluations. It is shown numerically that the AMMSE estimator has the smallest MSE over a wide region of parameter space when the number of regression coefficients is small. It is also shown that the MMSE estimator has a smaller MSE than the OLS estimator over a wide region of parameter space when the error terms have a multivariate *t* distribution with three degrees of freedom. Finally, some concluding remarks are given in Section 5.

2. THE MODEL AND THE ESTIMATORS

Consider a linear regression model

$$y = X\beta + u, \tag{1}$$

where y is an $n \times 1$ vector of observations on a dependent variable, X is an $n \times k$ matrix of full column rank of observations on nonstochastic independent variables, and β is a $k \times 1$ vector of regression coefficients.

It is assumed that the error term u has multivariate t distribution with the probability density given as

$$p(u|v, \sigma_u) = \frac{v^{v/2}\Gamma((v+n)/2)}{\pi^{n/2}\Gamma(v/2)\sigma_u^n} \frac{1}{\{v + u'u/\sigma_u^2\}^{(n+v)/2}} \tag{2}$$

where σ_u is the scale parameter and v is the degrees of freedom parameter. For $v > 2$, it is easy to show that $E[u] = 0$ and $E[uu'] = [v\sigma_u^2/(v-2)]I_n$. When $v \to \infty$, u approaches the normal distribution with mean 0 and covariance matrix $\sigma_u^2 I_n$.

As shown in Zellner (1976), the multivariate t distribution can be viewed as a mixture of multivariate normal and inverted gamma distributions:

$$p(u|v, \sigma_u) = \int_0^\infty p_N(u|\sigma)p_{IG}(\sigma|v, \sigma_u)d\sigma \tag{3}$$

where

$$p_N(u|\sigma) = (2\pi\sigma^2)^{-n/2} \exp[-u'u/(2\sigma^2)] \tag{4}$$

$$p_{IG}(\sigma|v,\sigma_u) = \frac{2(v\sigma_u^2/2)^{v/2}}{\Gamma(v/2)} \sigma^{-(v+1)} \exp[-v\sigma_u^2/(2\sigma^2)] \tag{5}.$$

Following Judge and Yancey (1986, p. 11), we reparameterize the model (1) and work with the following orthonormal counterpart:

$$y = Z\gamma + u \tag{6}$$

where $Z = XS^{-1/2}$, $\gamma = S^{1/2}\beta$, and $S^{1/2}$ is the symmetric matrix such that $S^{-1/2}SS^{-1/2} = Z'Z = I_k$, where $S = X'X$. Then, the ordinary least squares (OLS) estimator of γ is

$$c = Z'y \tag{7}$$

In the context of the reparameterized model, the Stein-rule (SR) estimator proposed by Stein (1956) is defined as

$$c_{SR} = \left(1 - \frac{ae'e}{c'c}\right)c \tag{8}$$

where $e = y - Zc$ and a is a constant such that $0 \le a \le 2(k-2)/(n-k+2)$. Although the Gauss–Markov theorem ensures that the OLS estimator is the best linear unbiased estimator for $v > 2$, the SR estimator dominates the OLS estimator under predictive mean squared error (PMSE) for $k \ge 3$, if normal error terms are assumed. As shown in James and Stein (1961), the PMSE of the SR estimator is minimized when $a = (k-2)/(n-k+2)$. Thus, we have used this value of a hereafter.

Although the SR estimator dominates the OLS estimator, Baranchik (1970) showed that under the normality assumption the SR estimator is further dominated by the positive-part Stein-rule (PSR) estimator, defined as

$$c_{PSR} = \max\left[0, 1 - \frac{ae'e}{c'c}\right]c \tag{9}$$

The minimum mean squared error (MMSE) estimator, proposed by Farebrother (1975), is

$$c_M = \left(\frac{c'c}{c'c + e'e/(n-k)}\right)c \tag{10}$$

As an extension of the MMSE estimator, Ohtani (1996) considered the following estimator, which is obtained by adjusting the degrees of freedom of $c'c$ (i.e., k):

$$c_{AM} = \left(\frac{c'c/k}{c'c/k + e'e/(n-k)}\right)c \tag{11}$$

The double k-class estimator proposed by Ullah and Ullah (1978) is

$$c_{KK} = \left(1 - \frac{k_1 e'e}{y'y - k_2 e'e}\right)c \tag{12}$$

where k_1 and k_2 are constants chosen appropriately.

Further, we consider the following pre-test double k-class (PTKK) estimator:

$$\hat{\gamma}_\tau = I(F \geq \tau)\left(1 - \frac{k_1 e'e}{y'y - k_2 e'e}\right)c \qquad (13)$$

where $I(A)$ is an indicator function such that $I(A) = 1$ if an event A occurs and $I(A) = 0$ otherwise, $F = (c'c/k)/(e'e/(n-k))$ is the test statistic for the null hypothesis $H_0 : \gamma = 0$ against the alternative $H_1 : \gamma \neq 0$, and τ is a critical value of the pre-test. When $k_1 = a$, $k_2 = 1$, and $\tau = 0$, the PTKK estimator reduces to the SR estimator. When $k_1 = a$, $k_2 = 1$, and $\tau = a(n-k)/k$, the PTKK estimator reduces to the PSR estimator. When $k_1 = 1/(n-k)$, $k_2 = 1 - 1/(n-k)$, and $\tau = 0$, the PTKK estimator reduces to the MMSE estimator. Also, when $k_1 = k/(n-k)$, $k_2 = 1 - k/(n-k)$, and $\tau = 0$, the PTKK estimator reduces to the AMMSE estimator.

Let h be a $k \times 1$ vector with known elements. If h' is the ith row vector of $S^{-1/2}$, the estimator $h'\hat{\gamma}_\tau$ is the ith element of the PTKK estimator for β. Since the elements of h are known, we assume that $h'h = 1$ without loss of generality.

Ohtani and Kozumi (1996) and Ohtani (1997) derived the MSE of the SR, PSR, and MMSE estimators for each individual regression coefficient when the normal error terms are assumed. However, the MSE performance of these estimators under multivariate t error terms has not been examined so far. Thus, we derive the explicit formulae for the moments of $h'\hat{\gamma}_\tau$ and examine MSE performances in the next section.

3. MOMENTS

Substituting $y'y = c'c + e'e$ in (13), we have

$$\begin{aligned}
\hat{\gamma}_\tau &= I(F \geq \tau)\left(1 - \frac{k_1 e'e}{y'y - k_2 e'e}\right)c \\
&= I(F \geq \tau)\left(\frac{c'c + \alpha_1 e'e}{c'c + \alpha_2 e'e}\right)
\end{aligned} \qquad (14)$$

where $\alpha_1 = 1 - k_1 - k_2$ and $\alpha_2 = 1 - k_2$. Then the MSE of $h'\hat{\gamma}_\tau$ is

$$\begin{aligned}
\mathrm{MSE}[h'\hat{\gamma}_\tau] &= E\left[(h'\hat{\gamma}_\tau - h'\gamma)^2\right] \\
&= E\left[I(F \geq \tau)\left(\frac{c'c + \alpha_1 e'e}{c'c + \alpha_2 e'e}\right)^2 (h'c)^2\right] \\
&\quad - 2h'\gamma E\left[I(F \geq \tau)\left(\frac{c'c + \alpha_1 e'e}{c'c + \alpha_2 e'e}\right)(h'c)\right] + (h'\gamma)^2
\end{aligned} \qquad (15)$$

We define the functions $H(p, q; \alpha_1, \alpha_2; \tau)$ and $J(p, q; \alpha_1, \alpha_2; \tau)$ as

$$H(p, q; \alpha_1, \alpha_2; \tau) = E\left[I(F \geq \tau)\left(\frac{c'c + \alpha_1 e'e}{c'c + \alpha_2 e'e}\right)^p (h'c)^{2q}\right] \tag{16}$$

$$J(p, q; \alpha_1, \alpha_2; \tau) = E\left[I(F \geq \tau)\left(\frac{c'c + \alpha_1 e'e}{c'c + \alpha_2 e'e}\right)^p (h'c)^{2q}(h'c)\right] \tag{17}$$

Then, the MSE of $h'\hat{\gamma}_\tau$ is written as

$$\text{MSE}[h'\hat{\gamma}_\tau] = H(2, 1; \alpha_1, \alpha_2; \tau) - 2h'\gamma J(1, 0; \alpha_1, \alpha_2; \tau) + (h'\gamma)^2 \tag{18}$$

As shown in the Appendix, the explicit formulae for $H(p, q, ; \alpha_1, \alpha_2; \tau)$ and $J(p, q, ; \alpha_1, \alpha_2; \tau)$ are

$$H(p, q, ; \alpha_1, \alpha_2; \tau) = (v\sigma_u^2)^{\nu/2} \sum_{i=0}^{\infty} \sum_{j=0}^{\infty} \left(\frac{\theta_1^i}{i!}\right)\left(\frac{\theta_2^j}{j!}\right)(\theta_1 + \theta_2 + v\sigma_u^2)^{q-i-j-\nu/2}$$

$$\times \frac{\Gamma(\nu/2 + i + j - q)}{\Gamma(\nu/2)} G_{ij}(p, q; \alpha_1, \alpha_2; \tau) \tag{19}$$

$$J(p, q, ; \alpha_1, \alpha_2; \tau) = h'\gamma(v\sigma_u^2)^{\nu/2} \sum_{i=0}^{\infty} \sum_{j=0}^{\infty} \left(\frac{\theta_1^i}{i!}\right)\left(\frac{\theta_2^j}{j!}\right)(\theta_1 + \theta_2 + v\sigma_u^2)^{q-i-j-\nu/2}$$

$$\times \frac{\Gamma(\nu/2 + i + j - q)}{\Gamma(\nu/2)} G_{i+1,j}(p, q; \alpha_1, \alpha_2; \tau) \tag{20}$$

where

$$G_{ij}(p, q; \alpha_1, \alpha_2; \tau) = \frac{\Gamma(1/2 + q + i)\Gamma(n/2 + q + i + j)}{\Gamma((n-k)/2)\Gamma(1/2 + i)\Gamma(k/2 + q + i + j)}$$

$$\times \int_{\tau^*}^{1} \left(\frac{\alpha_1 + (1 - \alpha_1)t}{\alpha_2 + (1 - \alpha_2)t}\right)^p t^{k/2+q+i+j-1}(1 - t)^{(n-k)/2-1} dt \tag{21}$$

$\theta_1 = (h'\gamma)^2$, $\theta_2 = \gamma'(I_k - hh')\gamma$ and $\tau^* = k\tau/(k\tau + n - k)$. Substituting (19) and (20) in (18), we obtain the explicit formula for the MSE of $h'\hat{\gamma}_\tau$.

Using the following formula

$$\frac{\partial G_{ij}(p, q; \alpha_1, \alpha_2; \tau)}{\partial \tau} = -\frac{\Gamma(1/2 + q + i)\Gamma(n/2 + q + i + j)}{\Gamma((n-k)/2)\Gamma(1/2 + i)\Gamma(k/2 + q + i + j)}$$

$$\times \left(\frac{\alpha_1(n-k) + k\tau}{\alpha_2(n-k) + k\tau}\right)^p \tag{22}$$

$$\times \frac{k^{k/2 + q + i + j}(n-k)^{(n-k)/2}\tau^{k/2 + q + i + j - 1}}{(k\tau + n - k)^{n/2 + q + i + j}}$$

and differentiating $\text{MSE}[h'\hat{\gamma}_\tau]$ with respect to τ, we obtain:

$$\frac{\partial \text{MSE}[h'\hat{\gamma}_\tau]}{\partial \tau} = -(v\sigma_u^2)^{v/2}D(\tau)\sum_{i=0}^{\infty}\sum_{j=0}^{\infty}\left(\frac{\theta_1^i}{i!}\right)\left(\frac{\theta_2^j}{j!}\right)(\theta_1 + \theta_2 + v\sigma_u^2)^{-i-j-v/2}$$

$$\times \frac{\Gamma(v/2 + i + j - 1)\Gamma(1/2 + i + 1)\Gamma(n/2 + i + j + 1)}{\Gamma(v/2)\Gamma((n-k/2))\Gamma(1/2 + i)\Gamma(k/2 + i + j + 1)}$$

$$\times \frac{k^{k/2 + i + j + 1}(n-k)^{(n-k)/2}\tau^{k/2 + i + j}}{(k\tau + n - k)^{n/2 + i + j + 1}}$$

$$\times \left[(\theta_1 + \theta_2 + v\sigma_u^2)D(\tau) - 2\theta_1\frac{v/2 + i + j - 1}{1/2 + i}\right] \tag{23}$$

where

$$D(\tau) = \frac{\alpha_1(n-k) + k\tau}{\alpha_2(n-k) + k\tau} \tag{24}$$

From (23), $\partial \text{MSE}[h'\hat{\gamma}_\tau]/\partial \tau < 0$ if $D(\tau) < 0$. Thus, $\text{MSE}[h'\hat{\gamma}_\tau]$ is monotonically decreasing on $\tau \in [\min\{0, -\alpha_1(n-k)/k, -\alpha_2(n-k)/k\}, \max\{0, -\alpha_1 (n-k)/k, -\alpha_2(n-k)/k\}]$. For $k_1 = a$ and $k_2 = 1$, $\partial \text{MSE}[h'\hat{\gamma}_\tau]/\partial \tau \leq 0$ for $0 \leq \tau \leq a(n-k)/k$. Since the PTKK estimator reduces to the SR estimator when $k_1 = a, k_2 = 1$ and $\tau = 0$, and to the PSR estimator when $k_1 = a, k_2 = 1$ and $\tau = a(n-k)/k$, the following theorem is obtained.

Theorem 1. The PSR estimator dominates the SR estimator for each individual regression coefficient in terms of MSE when the multivariate *t* error terms are assumed.

This is an extension of Theorem 2 in Ohtani and Kozumi (1996). Although they showed that the PSR estimator dominates the SR estimator under normality assumption, the same result can be obtained even when the error terms are extended to a multivariate *t* distribution.

Also, since $h'\gamma = E[h'c] = J(0, 0; \alpha_1, \alpha_2; 0)$, the bias of $h'\hat{\gamma}_\tau$ is

$$\text{Bias}[h'\hat{\gamma}_\tau] = E[h'\hat{\gamma}_\tau] - h'\gamma$$

$$= J(1, 0; \alpha_1, \alpha_2; \tau) - J(0, 0; \alpha_1, \alpha_2; \tau)$$

$$= h'\gamma(v\sigma_u^2)^{\nu/2} \sum_{i=0}^{\infty} \sum_{j=0}^{\infty} \left(\frac{\theta_1^i}{i!}\right)\left(\frac{\theta_2^j}{j!}\right)(\theta_1 + \theta_2 + v\sigma_u^2)^{-i-j-\nu/2}$$

$$\times \frac{\Gamma(\nu/2 + i + j)\Gamma(n/2 + i + j + 1)}{\Gamma(\nu/2)\Gamma((n-k)/2)\Gamma(k/2 + i + j + 1)} \tag{25}$$

$$\times \left\{ \int_{\tau^*}^{1} \left(\frac{\alpha_1 + (1-\alpha_1)t}{\alpha_2 + (1-\alpha_2)t}\right) t^{k/2+i+j}(1-t)^{(n-k)/2-1} dt \right.$$

$$\left. - \int_0^1 t^{k/2+i+j}(1-t)^{(n-k)/2-1} dt \right\}$$

Since

$$0 < \frac{\alpha_1 + (1-\alpha_1)t}{\alpha_2 + (1-\alpha_2)t} < 1 \tag{26}$$

when $0 < k_1 < 1$, $0 < k_1 + k_2 < 1$, and $0 \le t \le 1$, we obtain

$$\int_{\tau^*}^{1} \left(\frac{\alpha_1 + (1-\alpha_1)t}{\alpha_2 + (1-\alpha_2)t}\right) t^{k/2+i+j}(1-t)^{(n-k)/2-1} dt < \int_0^1 t^{k/2+i+j}(1-t)^{(n-k)/2-1} dt \tag{27}$$

Thus we obtain the following theorem.

Theorem 2. When $0 < k_1 < 1$ and $0 < k_1 + k_2 < 1$, $h'\hat{\gamma}_\tau$ has negative bias if $h'\gamma > 0$ and has positive bias if $h'\gamma < 0$.

Since theoretical analysis is difficult because of the intricate nature of exact expressions, we compare the MSEs of the SR, PSR, MMSE, and AMMSE estimators numerically in the next section.

4. NUMERICAL ANALYSIS

Using (18), we compared the MSEs of the SR, PSR, MMSE, and AMMSE estimators by numerical evaluations. To compare the MSEs of the estimators, we have evaluated the values of relative MSE, defined as $\text{MSE}[h'\bar{\gamma}]/\text{MSE}[h'c]$, where $\bar{\gamma}$ is any estimator of γ. Thus, the estimator $h'\bar{\gamma}$ has a smaller MSE than the OLS estimator when the value of the relative MSE is smaller than unity. It is seen from (18) that the relative MSE of $h'\hat{\gamma}_\tau$

depends on the values of k, n, v, θ_1/σ_u^2, and θ_2/σ_u^2. Thus, it can be assumed without loss of generality that $\sigma_u^2 = 1$ in the numerical evaluations.

The numerical evaluations were executed on a personal computer, using the FORTRAN code. In evaluating the integral in $G_{ij}(p, q; \alpha_1, \alpha_2; \tau)$ given in (21), Simpson's 3/8 rule with 200 equal subdivisions has been used. The double infinite series in $H(p, q; \alpha_1, \alpha_2; \tau)$ and $J(p, q; \alpha_1, \alpha_2; \tau)$ were judged to converge when the increment of the series got smaller than 10^{-12}.

The relative MSEs for $k = 3$, $n = 20$, and $v = 3$ are shown in Table 1. The AMMSE estimator has the smallest MSE over a wide region of the parameter space among the estimators considered here. In particular, it is evident from Table 1 that the AMMSE estimator has the smallest MSE when $\theta_1 + \theta_2 \leq 10$. However, when $\theta_1 + \theta_2 = 50$, the MSE of the AMMSE estimator gets larger than those of the other estimators as θ_1 increases from zero to $\theta_1 + \theta_2$. Although the MSE of the AMMSE estimator is larger than that of the OLS estimator when the value of $\theta_1 + \theta_2$ is large and θ_1 is close to $\theta_1 + \theta_2$ (e.g., $\theta_1 + \theta_2 = 50$ and $\theta_1 \geq 40$), the difference between the MSEs of the AMMSE and OLS estimators is small relative to the difference when the value of θ_1 is close to zero. Also, the MMSE estimator has smaller MSE than the OLS estimator even when $\theta_1 + \theta_2 = 50$, though the MSE improvement around $\theta_1 = 0$ is small relative to those of the other estimators.

The relative MSEs for $k = 3$, $n = 20$, and $v = 20$ are shown in Table 2. It can be seen from Tables 1 and 2, that as the degrees of freedom of error term (i.e., v) increases from 3 to 20, the region of $\theta_1 + \theta_2$ and θ_1 where the estimators have smaller MSE than the OLS estimator gets narrow. We see from Table 2 that the MSE of the AMMSE estimator is smallest when $\theta_1 + \theta_2 = 0.5$. Although the AMMSE estimator has smaller MSE than the SR and PSR estimators when $\theta_1 + \theta_2 = 3$, the MSE of the AMMSE estimator is larger than that of the MMSE estimator when θ_1 is close to $\theta_1 + \theta_2$. Also, when $\theta_1 + \theta_2 \geq 10$ and θ_1 is close to $\theta_1 + \theta_2$, the MSE of the AMMSE estimator is considerably larger than unity. Although the MSE of the MMSE estimator is smaller than unity when $\theta_1 + \theta_2 \geq 10$ and θ_1 is close to zero, it gets larger than unity as θ_1 exceeds about half of $\theta_1 + \theta_2$.

The relative MSEs for $k = 8$, $n = 20$, and $v = 3$ are shown in Table 3. From Table 3 it is clear that the PSR estimator has the smallest MSE when $\theta_1 + \theta_2 = 0.5$. This indicates that, as k increases from 3 to 8, the MSE dominance of the AMMSE estimator at $\theta_1 + \theta_2 = 0.5$ is violated. Although the MSE of the PSR estimator is smaller than that of the AMMSE estimator when $\theta_1 + \theta_2 = 3$, 10, and θ_1 is close to zero, the former gets larger than the latter as θ_1 approaches $\theta_1 + \theta_2$. Also, when $\theta_1 + \theta_2 = 50$, the MSE of the AMMSE estimator is smaller than that of the PSR estimator. Similar to the case of $k = 3$, $n = 20$, and $v = 3$, the MMSE estimator has smaller MSE than the OLS estimator even when $\theta_1 + \theta_2 = 50$.

Table 1. Relative MSEs for $k = 3$, $n = 20$, and $v = 3$

$\theta_1 + \theta_2$	θ_1	SR	PSR	MMSE	AMMSE
0.5	.00	.6586	.5892	.6421	.3742
	.05	.6630	.5928	.6444	.3788
	.10	.6673	.5964	.6467	.3835
	.15	.6717	.6000	.6490	.3882
	.20	.6760	.6036	.6513	.3928
	.25	.6803	.6072	.6536	.3975
	.30	.6847	.6108	.6559	.4022
	.35	.6890	.6144	.6582	.4069
	.40	.6933	.6180	.6605	.4115
	.45	.6977	.6216	.6628	.4162
	.50	.7020	.6252	.6651	.4209
3.0	.00	.6796	.6261	.6706	.4081
	.30	.6954	.6398	.6799	.4286
	.60	.7112	.6535	.6892	.4490
	.90	.7270	.6673	.6986.	.4695
	1.20	.7428	.6810	.7079	.4900
	1.50	.7586	.6947	.7172	.5105
	1.80	.7744	.7085	.7265	.5310
	2.10	.7903	.7222	.7359	.5515
	2.40	.8061	.7359	.7452	.5720
	2.70	.8219	.7497	.7545	.5924
	3.00	.8377	.7634	.7638	.6129
10.0	.00	.7231	.6865	.7189	.4733
	1.00	.7469	.7081	.7347	.5120
	2.00	.7707	.7296	.7506	.5506
	3.00	.7945	.7512	.7664	.5893
	4.00	.8184	.7727	.7823	.6279
	5.00	.8422	.7943	.7981	.6666
	6.00	.8660	.8159	.8140	.7052
	7.00	.8898	.8374	.8298	.7439
	8.00	.9136	.8590	.8457	.7825
	9.00	.9375	.8806	.8615	.8212
	10.00	.9613	.9021	.8774	.8598

Table 1.   Continued

$\theta_1 + \theta_2$	θ_1	SR	PSR	MMSE	AMMSE
50.0	.00	.8153	.7971	.8126	.6235
	5.00	.8382	.8185	.8302	.6730
	10.00	.8610	.8400	.8478	.7224
	15.00	.8839	.8615	.8653	.7719
	20.00	.9068	.8830	.8829	.8214
	25.00	.9296	.9044	.9004	.8708
	30.00	.9525	.9258	.9180	.9203
	35.00	.9753	.9473	.9355	.9697
	40.00	.9982	.9688	.9531	1.0192
	45.00	1.0211	.9902	.9706	1.0687
	50.00	1.0439	1.0117	.9882	1.1181

We see from Tables 3 and 4 that, as v increases from 3 to 20, the region of $\theta_1 + \theta_2$ and θ_1 where the estimators have smaller MSE than the OLS estimator gets narrow. It has been observed from Table 4 that the MSE of the PSR estimator is smaller than that of the AMMSE estimator when $\theta_1 + \theta_2 \leq 3$ and θ_1 is close to zero, though the former gets larger than the latter as θ_1 approaches $\theta_1 + \theta_2$. When $\theta_1 + \theta_2 = 10$, the MSE of the AMMSE estimator is smaller than that of the PSR estimator. Although the MSE of the AMMSE estimator is smaller than that of the PSR estimator when $\theta_1 + \theta_2 = 50$ and θ_1 is close to zero, the former gets larger than the latter as θ_1 approaches $\theta_1 + \theta_2$. Although the MSE improvement of the MMSE estimator is not large when $\theta_1 + \theta_2 \geq 10$ and θ_1 is close to zero, the MSE of the MMSE estimator is much smaller than those of the other estimators when θ_1 is close to $\theta_1 + \theta_2$.

5.   CONCLUDING REMARKS

In this paper, we have derived the explicit formulae for the moments of the PTKK estimator under multivariate t error terms. In Section 3, it was shown analytically that the PSR estimator for each individual regression coefficient dominates the SR estimator in terms of MSE even when the multivariate t error terms are assumed. In Section 4, the MSEs of the SR, PSR, MMSE, and AMMSE estimators are compared by numerical evaluations. It is shown numerically that the MSE of the AMMSE estimator for each individual coefficient is smallest over a wide region of the parameter space when the number of the regression coefficients is three. Also, the MMSE estimator

Table 2.   Relative MSEs for $k = 3$, $n = 20$, and $\nu = 20$

$\theta_1 + \theta_2$	θ_1	SR	PSR	MMSE	AMMSE
0.5	.00	.6656	.6051	.6543	.3876
	.05	.6773	.6148	.6604	.4002
	.10	.6889	.6245	.6666	.4128
	.15	.7006	.6342	.6728	.4254
	.20	.7122	.6438	.6790	.4380
	.25	.7239	.6535	.6852	.4505
	.30	.7355	.6632	.6914	.4631
	.35	.7471	.6728	.6976	.4757
	.40	.7588	.6825	.7038	.4883
	.45	.7704	.6922	.7100	.5009
	.50	.7821	.7018	.7162	.5134
3.0	.00	.7243	.7018	.7290	.4780
	.30	.7617	.7351	.7523	.5305
	.60	.7992	.7685	.7755	.5830
	.90	.8367	.8019	.7987	.6356
	1.20	.8742	.8352	.8219	.6881
	1.50	.9117	.8686	.8451	.7406
	1.80	.9491	.9019	.8684	.7931
	2.10	.9866	.9353	.8916	.8456
	2.40	1.0241	.9687	.9148	.8982
	2.70	1.0616	1.0020	.9380	.9507
	3.00	1.0990	1.0354	.9612	1.0032
10.0	.00	.8355	.8333	.8371	.6365
	1.00	.8710	.8676	.8655	.7152
	2.00	.9064	.9019	.8939	.7939
	3.00	.9419	.9362	.9224	.8725
	4.00	.9774	.9705	.9508	.9512
	5.00	1.0129	1.0048	.9792	1.0299
	6.00	1.0483	1.0391	1.0077	1.1086
	7.00	1.0838	1.0735	1.0361	1.1873
	8.00	1.1193	1.1078	1.0645	1.2660
	9.00	1.1548	1.1421	1.0930	1.3446
	10.00	1.1902	1.1764	1.1214	1.4233

Table 2. Continued

$\theta_1 + \theta_2$	θ_1	SR	PSR	MMSE	AMMSE
50.0	.00	.9573	.9573	.9539	.8736
	5.00	.9678	.9678	.9650	.9157
	.10.00	.9783	.9783	.9762	.9578
	15.00	.9888	.9888	.9874	.9999
	20.00	.9993	.9993	.9985	1.0419
	25.00	1.0098	1.0098	1.0097	1.0840
	30.00	1.0203	1.0203	1.0209	1.1261
	35.00	1.0308	1.0308	1.0320	1.1682
	40.00	1.0413	1.0413	1.0432	1.2103
	45.00	1.0518	1.0518	1.0543	1.2523
	50.00	1.0623	1.0623	1.0655	1.2944

for each individual regression coefficient has smaller MSE than the OLS estimator over a wide region of parameter space even if the error terms have a multivariate *t* distribution with three degrees of freedom. Further, our numerical results show that the MSE performance of all the estimators is much improved when the error terms depart from a normal distribution.

ACKNOWLEDGMENTS

The authors are grateful to Alan T. K. Wan and the referees for their many useful comments and suggestions which have greatly improved the presentation of the paper.

APPENDIX

In the appendix, the formulae for $H(p, q; \alpha_1, \alpha_2; \tau)$ and $J(p, q; \alpha_1, \alpha_2; \tau)$ have been derived. First, we derive the formula for

$$E\left[I(F \geq \tau)\left(\frac{c'c + \alpha_1 e'e}{c'c + \alpha_2 e'e}\right)^p (h'c)^{2q}\bigg|\sigma\right] \tag{28}$$

Let $u_1 = (h'c)^2/\sigma_2$, $u_2 = c'[I_k - hh']c/\sigma^2$, and $u_3 = e'e/\sigma^2$. Then, for given σ, $u_1 \sim \chi_1'^2(\lambda_1)$ and $u_2 \sim \chi_{k-1}'^2(\lambda_2)$, where $\lambda_1 = (h'\gamma)^2/\sigma^2$ and $\lambda_2 = \gamma'(I_k - hh')\gamma/\sigma^2$, and $\chi_f'^2(\lambda)$ is the noncentral chi-square distribution with f degrees of freedom and noncentrality parameter λ. Further, u_3 is

Table 3. Relative MSEs for $k = 8$, $n = 20$, and $\nu = 3$

$\theta_1 + \theta_2$	θ_1	SR	PSR	MMSE	AMMSE
0.5	.00	.3575	.2662	.8055	.3241
	.05	.3674	.2736	.8062	.3285
	.10	.3773	.2809	.8070	.3329
	.15	.3872	.2882	.8077	.3373
	.20	.3971	.2956	.8084	.3417
	.25	.4070	.3029	.8091	.3461
	.30	.4169	.3102	.8099	.3505
	.35	.4268	.3176	.8106	.3549
	.40	.4367	.3249	.8113	.3593
	.45	.4466	.3322	.8120	.3637
	.50	.4565	.3396	.8128	.3681
3.0	.00	.3663	.2924	.8162	.3427
	.30	.4094	.3269	.8195	.3646
	.60	.4525	.3613	.8228	.3866
	.90	.4957	.3958	.8261	.4085
	1.20	.5388	.4302	.8293	.4305
	1.50	.5819	.4647	.8326	.4524
	1.80	.6250	.4991	.8359	.4743
	2.10	.6682	.5336	.8392	.4963
	2.40	.7113	.5680	.8424	.5182
	2.70	.7544	.6025	.8457	.5402
	3.00	.7976	.6369	.8490	.5621
10.0	.00	.4029	.3502	.8365	.3842
	1.00	.4841	.4203	.8429	.4345
	2.00	.5653	.4903	.8494	.4849
	3.00	.6465	.5604	.8559	.5352
	4.00	.7277	.6304	.8623	.5855
	5.00	.8089	.7005	.8688	.6358
	6.00	.8901	.7705	.8752	.6861
	7.00	.9713	.8406	.8817	.7364
	8.00	1.0525	.9106	.8882	.7868
	9.00	1.1337	.9807	.8946	.8371
	10.00	1.2149	1.0507	.9011	.8874

Table 3. Continued

$\theta_1 + \theta_2$	θ_1	SR	PSR	MMSE	AMMSE
50.0	.00	.5364	.5093	.8826	.5089
	5.00	.6413	.6063	.8915	.5974
	.10.00	.7462	.7034	.9005	.6859
	15.00	.8511	.8004	.9094	.7744
	20.00	.9560	.8975	.9183	.8629
	25.00	1.0608	.9945	.9272	.9514
	30.00	1.1657	1.0915	.9361	1.0399
	35.00	1.2706	1.1885	.9450	1.1284
	40.00	1.3755	1.2855	.9539	1.2169
	45.00	1.4804	1.3825	.9628	1.3054
	50.00	1.5853	1.4796	.9717	1.3939

distributed as the chi-square distribution with $n - k$ degrees of freedom and u_1, u_2, and u_3 are mutually independent.

Using u_1, u_2, and u_3, (28) can be expressed as

$$E\left[I(F \geq \tau) \left(\frac{c'c + \alpha_1 e'e}{c'c + \alpha_2 e'e} \right)^p (h'c)^{2q} \Big| \sigma \right]$$

$$= (\sigma^2)^q \sum_{i=0}^{\infty} \sum_{j=0}^{\infty} K_{ij} \int\int\int_R \left(\frac{u_1 + u_2 + \alpha_1 u_3}{u_1 + u_2 + \alpha_2 u_3} \right)^p \qquad (29)$$

$$u_1^{1/2+q+i-1} u_2^{(k-1)/2+j-1} u_3^{(n-k)/2-1}$$

$$\times \exp[-(u_1 + u_2 + u_3)/2] du_1 \, du_2 \, du_3$$

where

$$K_{ij} = \frac{w_i(\lambda_1) w_j(\lambda_2)}{2^{n/2+i+j} \Gamma(1/2 + i) \Gamma((k-1)/2) \Gamma((n-k)/2)} \qquad (30)$$

$w_i(\lambda) = \exp(-\lambda/2)(\lambda/2)^i/i!$, and R is the region such that $(u_1 + u_2)/u_3 \geq k\tau/(n-k) = \tau^{**}$.

Let $v_1 = (u_1 + u_2)/u_3$, $v_2 = u_1 u_3/(u_1 + u_2)$, and $v_3 = u_3$. Using v_1, v_2, and v_3 in (29), we obtain

$$(\sigma^2)^q \sum_{i=0}^{\infty} \sum_{j=0}^{\infty} K_{ij} \int_0^{\infty} \int_0^{v_3} \int_{\tau^{**}}^{\infty} \left(\frac{\alpha_1 + v_1}{\alpha_2 + v_1} \right)^p v_1^{k/2+q+i+j-1} v_2^{1/2+q+i-1} v_3^{(n-k)/2}$$

$$\times (v_3 - v_2)^{(k-1)/2+j-1} \exp[-(1 + v_1)v_3/2] dv_1 \, dv_2 \, dv_3 \qquad (31)$$

Table 4.   Relative MSEs for $k = 8$, $n = 20$, and $v = 20$

$\theta_1 + \theta_2$	θ_1	SR	PSR	MMSE	AMMSE
0.5	.00	.3583	.2758	.8098	.3310
	.05	.3849	.2956	.8118	.3428
	.10	.4116	.3154	.8137	.3547
	.15	.4382	.3352	.8157	.3665
	.20	.4649	.3550	.8176	.3784
	.25	.4915	.3747	.8196	.3903
	.30	.5182	.3945	.8216	.4021
	.35	.5448	.4143	.8235	.4140
	.40	.5715	.4341	.8255	.4258
	.45	.5981	.4539	.8274	.4377
	.50	.6248	.4736	.8294	.4495
3.0	.00	.3854	.3464	.8383	.3809
	.30	.4957	.4368	.8467	.4390
	.60	.6059	.5271	.8552	.4972
	.90	.7162	.6175	.8637	.5553
	1.20	.8264	.7079	.8722	.6135
	1.50	.9367	.7982	.8806	.6717
	1.80	1.0469	.8886	.8891	.7298
	2.10	1.1572	.9789	.8976	.7880
	2.40	1.2674	1.0693	.9060	.8461
	2.70	1.3777	1.1596	.9145	.9043
	3.00	1.4879	1.2500	.9230	.9624
10.0	.00	.5037	.4972	.8869	.4889
	1.00	.6684	.6509	.9006	.6096
	2.00	.8332	.8047	.9144	.7302
	3.00	.9979	.9584	.9282	.8509
	4.00	1.1627	1.1122	.9419	.9715
	5.00	1.3274	1.2659	.9557	1.0922
	6.00	1.4922	1.4197	.9694	1.2128
	7.00	1.6569	1.5734	.9832	1.3335
	8.00	1.8217	1.7272	.9970	1.4541
	9.00	1.9864	1.8809	1.0107	1.5748
	10.00	2.1512	2.0346	1.0245	1.6954

Table 4. Continued

$\theta_1 + \theta_2$	θ_1	SR	PSR	MMSE	AMMSE
50.0	.00	.7984	.7984	.9593	.7462
	5.00	.8908	.8907	.9681	.8705
	.10.00	.9831	.9830	.9768	.9947
	15.00	1.0755	1.0753	.9855	1.1189
	20.00	1.1679	1.1676	.9942	1.2431
	25.00	1.2602	1.2599	1.0030	1.3673
	30.00	1.3526	1.3522	1.0117	1.4915
	35.00	1.4450	1.4445	1.0204	1.6157
	40.00	1.5373	1.5368	1.0292	1.7399
	45.00	1.6297	1.6291	1.0379	1.8641
	50.00	1.7221	1.7214	1.0466	1.9883

Again, substituting $z_1 = v_2/v_3$, (31) reduces to

$$(\sigma^2)^q \sum_{i=0}^{\infty} \sum_{j=0}^{\infty} K_{ij} \frac{\Gamma(1/2 + q + i)\Gamma((k-1)/2 + j)}{\Gamma(k/2 + q + i + j)}$$

$$\times \int_0^{\infty} \int_{\tau^{**}}^{\infty} \left(\frac{\alpha_1 + v_1}{\alpha_2 + v_1}\right)^p v_1^{k/2+q+i+j-1} v_3^{n/2+q+i+j-1} \exp[-(1 + v_1)v_3/2] dv_1 \, dv_3$$

$$(32)$$

Further, substituting $z_2 = (1 + v_1)v_3/2$, (32) reduces to

$$(\sigma^2)^q \sum_{i=0}^{\infty} \sum_{j=0}^{\infty} K_{ij} 2^{n/2+q+i+j} \frac{\Gamma(1/2 + q + i)\Gamma((k-1)/2 + j)\Gamma(n/2 + q + i + j)}{\Gamma(k/2 + q + i + j)}$$

$$\times \int_{\tau^{**}}^{\infty} \left(\frac{\alpha_1 + v_1}{\alpha_2 + v_1}\right)^p \frac{v_1^{k/2+q+i+j-1}}{(1 + v_1)^{n/2+q+i+j}} dv_1$$

$$(33)$$

Finally, making use of the change of variable $t = v_1/(1 + v_1)$, we obtain

$$E\left[I(F \geq \tau)\left(\frac{c'c + \alpha_1 e'e}{c'c + \alpha_2 e'e}\right)^p (h'c)^{2q} \Big| \sigma\right]$$

$$= (2\sigma^2)^q \sum_{i=0}^{\infty} \sum_{j=0}^{\infty} w_i(\lambda_1) w_j(\lambda_2) G_{ij}(p, q; \alpha_1, \alpha_2; \tau)$$

$$(34)$$

where

$$G_{ij}(p, q; \alpha_1, \alpha_2; \tau) = \frac{\Gamma(1/2 + q + i)\Gamma(n/2 + q + i + j)}{\Gamma((n-k)/2)\Gamma(1/2 + i)\Gamma(k/2 + q + i + j)}$$

$$\times \int_{\tau^*}^1 \left(\frac{\alpha_1 + (1-\alpha_1)t}{\alpha_2 + (1-\alpha_2)t}\right)^p t^{k/2+q+i+j-1}(1-t)^{(n-k)/2-1} dt \qquad (35)$$

and $\tau^* = k\tau/(k\tau + n - k)$.

Next, we derive the formula for

$$E\left[I(F \geq \tau)\left(\frac{c'c + \alpha_1 e'e}{c'c + \alpha_2 e'e}\right)^p (h'c)^{2q}(h'c)\Big|\sigma\right] \qquad (36)$$

Differentiating (34) with respect to γ, we have

$$\frac{\partial E\left[I(F \geq \tau)\left(\frac{c'c + \alpha_1 e'e}{c'c + \alpha_2 e'e}\right)^p (h'c)^{2q}\Big|\sigma\right]}{\partial\gamma}$$

$$= (2\sigma^2)^q \sum_{i=0}^{\infty}\sum_{j=0}^{\infty}\left[\frac{\partial w_i(\lambda_1)}{\partial\gamma}w_j(\lambda_2) + w_i(\lambda_1)\frac{\partial w_j(\lambda_2)}{\partial\gamma}\right]G_{ij}(p, q; \alpha_1, \alpha_2; \tau)$$

$$= -\frac{hh'\gamma}{\sigma^2}(2\sigma^2)^q \sum_{i=0}^{\infty}\sum_{j=0}^{\infty}w_i(\lambda_1)w_j(\lambda_2)G_{ij}(p, q; \alpha_1, \alpha_2; \tau)$$

$$+ \frac{hh'\gamma}{\sigma^2}(2\sigma^2)^q \sum_{i=0}^{\infty}\sum_{j=0}^{\infty}w_i(\lambda_1)w_j(\lambda_2)G_{i+1,j}(p, q; \alpha_1, \alpha_2; \tau)$$

$$- \frac{(I_k - hh')\gamma}{\sigma^2}(2\sigma^2)^q \sum_{i=0}^{\infty}\sum_{j=0}^{\infty}w_i(\lambda_1)w_j(\lambda_2)G_{ij}(p, q; \alpha_1, \alpha_2; \tau)$$

$$+ \frac{(I_k - hh')\gamma}{\sigma^2}(2\sigma^2)^q \sum_{i=0}^{\infty}\sum_{j=0}^{\infty}w_i(\lambda_1)w_j(\lambda_2)G_{i,j+1}(p, q; \alpha_1, \alpha_2; \tau) \qquad (37)$$

where we define $w_{-1}(\lambda_1) = w_{-1}(\lambda_2) = 0$. Since $h'h = 1$ and $h'(I_k - hh') = 0$, we obtain

$$h' \frac{\partial E\left[I(F \geq \tau)\left(\frac{c'c+\alpha_1 e'e}{c'c+\alpha_2 e'e}\right)^p (h'c)^{2q}\Big|\sigma\right]}{\partial \gamma}$$

$$= -\frac{h'\gamma}{\sigma^2} E\left[I(F \geq \tau)\left(\frac{c'c+\alpha_1 e'e}{c'c+\alpha_2 e'e}\right)^p (h'c)^{2q}\Big|\sigma\right] \tag{38}$$

$$+ \frac{h'\gamma}{\sigma^2}(2\sigma^2)^q \sum_{i=0}^{\infty}\sum_{j=0}^{\infty} w_i(\lambda_1)w_j(\lambda_2)G_{i+1,j}(p, q; \alpha_1, \alpha_2; \tau)$$

Expressing (36) by c and $e'e$, we have

$$E\left[I(F \geq \tau)\left(\frac{c'c+\alpha_1 e'e}{c'c+\alpha_2 e'e}\right)^p (h'c)^{2q}\Big|\sigma\right]$$

$$= \iint_{F \geq \tau}\left(\frac{c'c+\alpha_1 e'e}{c'c+\alpha_2 e'e}\right)^p (h'c)^{2q}p_N(c|\sigma)p_e(e'e|\sigma)dc\,d(e'e), \tag{39}$$

where $p_e(e'e|\sigma)$ is the conditional density function of $e'e$ given σ, and

$$p_N(c|\sigma) = \frac{1}{(2\pi)^{k/2}\sigma^k}\exp\left[-\frac{(c-\gamma)'(c-\gamma)}{2\sigma^2}\right] \tag{40}$$

is the conditional density function of c given σ.

Differentiating (39) with respect to γ and multiplying h' from the left, we obtain

$$h' \frac{\partial E\left[I(F \geq \tau)\left(\frac{c'c+\alpha_1 e'e}{c'c+\alpha_2 e'e}\right)^p (h'c)^{2q}\Big|\sigma\right]}{\partial \gamma}$$

$$= -\frac{h'\gamma}{\sigma^2} E\left[I(F \geq \tau)\left(\frac{c'c+\alpha_1 e'e}{c'c+\alpha_2 e'e}\right)^p (h'c)^{2q}\Big|\sigma\right] \tag{41}$$

$$+ \frac{1}{\sigma^2} E\left[I(F \geq \tau)\left(\frac{c'c+\alpha_1 e'e}{c'c+\alpha_2 e'e}\right)^p (h'c)^{2q}(h'c)\Big|\sigma\right]$$

Equating (38) and (41), we obtain

$$E\left[I(F \geq \tau)\left(\frac{c'c+\alpha_1 e'e}{c'c+\alpha_2 e'e}\right)^p (h'c)^{2q}(h'c)\Big|\sigma\right]$$

$$= h'\gamma(2\sigma^2)^q \sum_{i=0}^{\infty}\sum_{j=0}^{\infty} w_i(\lambda_1)w_j(\lambda_2)G_{i+1,j}(p, q; \alpha_1, \alpha_2; \tau) \tag{42}$$

Finally, we derive the formulae for $H(p, q; \alpha_1, \alpha_2; \tau)$ and $J(p, q; \alpha_1, \alpha_2; \tau)$. From (3), we have

$$H(p, q; \alpha_1, \alpha_2; \tau) = \int_0^\infty E\left[I(F \geq \tau)\left(\frac{c'c + \alpha_1 e'e}{c'c + \alpha_2 e'e}\right)^p (h'c)^{2q}\Big|\sigma\right] p_{IG}(\sigma|v, \sigma_u)d\sigma$$

(43)

$$J(p, q; \alpha_1, \alpha_2; \tau) = \int_0^\infty E\left[I(F \geq \tau)\left(\frac{c'c + \alpha_1 e'e}{c'c + \alpha_2 e'e}\right)^p (h'c)^{2q}(h'c)\Big|\sigma\right]$$

$$\times \; p_{IG}(\sigma|v, \sigma_u)d\sigma$$

(44)

Substituting (34) and (42) in (43) and (44), and making use of the change of variable $z = (\theta_1 + \theta_2 + v\sigma_u^2)/2\sigma^2$, we obtain (16) and (17) in the text.

REFERENCES

Baranchik, A.J. (1970), A family of minimax estimators of the mean of a multivariate normal distribution, *Annals of Mathematical Statistics*, 41, 642–645.

Blattberg, R.C., and N.J. Gonedes (1974), A comparison of the stable and Student distributions as statistical models for stock prices, *Journal of Business*, 47, 244–280.

Carter, R.A.L. (1981), Improved Stein-rule estimator for regression problem, *Journal of Econometrics*, 17, 113–123.

Carter, R.A.L., V.K. Srivastava, and A. Chaturvedi (1993), Selecting a double k-class estimator for regression coefficients, *Statistics and Probability Letters*, 18, 363–371.

Fama, E.F. (1965), The behavior of stock market prices, *Journal of Business*, 38, 34–105.

Farebrother, R.W. (1975), The minimum mean square error linear estimator and ridge regression. *Technometrics*, 17, 127–128.

Giles, J.A. (1991), Pre-testing for linear restrictions in a regression model with spherically symmetric disturbances, *Journal of Econometrics*, 50, 377–398.

Giles, J.A. (1992), Estimation of the error variance after a preliminary test of homogeneity in a regression model with spherically symmetric disturbances, *Journal of Econometrics*, 53, 345–361.

James, W. and C. Stein (1961), Estimation with quadratic loss, *Proceedings of the Fourth Berkeley Symposium on Mathematical Statistics and Probability 1* (Berkeley, University of California Press), 361–379.

Judge, G., S. Miyazaki. and T. Yancey (1985), Minimax estimators for the location vectors of spherically symmetric densities, *Econometric Theory*, 1, 409–417.

Judge, G., and T. Yancey (1986), *Improved Methods of Inference in Econometrics* (North-Holland, Amsterdam).

Menjoge, S.S. (1984), On double *k*-class estimators of coefficients in linear regression, *Economics Letters*, 15, 295–300.

Ohtani, K. (1991), Small sample properties of the interval constrained least squares estimator when error terms have a multivariate *t* distribution, *Journal of the Japan Statistical Society*, 21, 197–204.

Ohtani, K. (1993), Testing for equality of error variances between two linear regressions with independent multivariate *t* errors, *Journal of Quantitative Economics*, 9, 85–97.

Ohtani, K. (1996), On an adjustment of degrees of freedom in the minimum mean squared error estimator, *Communications in Statistics—Theory and Methods*, 25, 3049–3058.

Ohtani, K. (1997), Minimum mean squared error estimation of each individual coefficient in a linear regression model, *Journal of Statistical Planning and Inference*, 62, 301–316.

Ohtani, K. (2000), Pre-test double *k*-class estimators in linear regression, *Journal of Statistical Planning and Inference*, 87, 287–299.

Ohtani, K., and J. Giles (1993), Testing linear restrictions on coefficients in a linear regression model with proxy variables and spherically symmetric disturbances, *Journal of Econometrics*, 57, 393–406.

Ohtani, K., and H. Hasegawa (1993), On small sample properties of R^2 in a linear regression model with multivariate *t* errors and proxy variables, *Econometric Theory*, 9, 504–515.

Ohtani, K., and H. Kozumi (1996), The exact general formulae for the moments and the MSE dominance of the Stein-rule and positive-part Stein-rule estimators, *Journal of Econometrics*, 74, 273–287.

Prucha, I.R., and H. Kelejian (1984), The structure of simultaneous equation estimators: a generalization towards nonnormal distributions, *Econometrica*, 52, 721–736.

Singh, R.S. (1988), Estimation of error variance in linear regression models with error having multivariate Student-*t* distribution with unknown degrees of freedom, *Economics Letters*, 27, 47–53.

Singh, R.S. (1991), James–Stein rule estimators in linear regression models with multivariate-t distributed error, *Australian Journal of Statistics*, 33, 145–158.

Stein, C. (1956), Inadmissibility of the usual estimator for the mean of a multivariate normal distribution, *Proceedings of the Third Berkeley Symposium on Mathematical Statistics and Probability* (Berkeley, University of California Press), 197–206.

Sutradhar, B.C. (1988), Testing linear hypothesis with t error variable, *Sankhȳa-B*, 50, 175–180.

Sutradhar, B.C., and M.M. Ali (1986), Estimation of the parameters of a regression model with a multivariate t error variable, *Communications in Statistics—Theory and Methods*, 15, 429–450.

Theil, H. (1971), *Principles of Econometrics* (John Wiley, New York).

Ullah, A., and S. Ullah (1978), Double k-class estimators of coefficients in linear regression, *Econometrica*, 46, 705–722.

Ullah, A. and V. Zinde-Walsh (1984), On the robustness of LM, LR, and Wald tests in regression models, *Econometrica*, 52, 1055–1066.

Vinod, H.D. (1980), Improved Stein-rule estimator for regression problems, *Journal of Econometrics*, 12, 143–150.

Vinod, H.D., and V.K. Srivastava (1995), Large sample properties of the double k-class estimators in linear regression models, *Econometric Reviews*, 14, 75–100.

Zellner, A. (1976), Bayesian and non-Bayesian analysis of the regression model with multivariate Student-t error terms, *Journal of the American Statistical Association*, 71, 400–405.

16

Effects of a Trended Regressor on the Efficiency Properties of the Least-Squares and Stein-Rule Estimation of Regression Coefficients

SHALABH   Panjab University, Chandigarh, India

1. INTRODUCTION

The least-squares method possesses the celebrated property of providing the optimal estimator for the coefficient vector in a linear regression model in the class of linear and unbiased estimators. If we take the performance criterion as risk under a quadratic loss function, James and Stein [1] have demonstrated that it is possible to find nonlinear and biased estimators with smaller risk in comparison with the least-squares estimator. This pioneering result has led to the development of several families of estimators having superior performance under the risk criterion. Among them, the Stein-rule family characterized by a single scalar has acquired considerable popularity and importance, see, e.g., [2] and [3].

The properties of Stein-rule estimators have been extensively studied in the literature but most of the investigations, particularly dealing with large sample properties, have been conducted under the specification that the regressors in the model are asymptotically cooperative, i.e., the limiting form of the variance–covariance matrix of the explanatory variables as

327

the number of observations tends to infinity is a finite and nonsingular matrix. Such an assumption may be violated in many practical situations, for instance, where some regressors are trended. In particular, when one of the explanatory variables has a linear or, more generally, a polynomial trend, its variance tends to infinity and consequently the limiting form of the variance–covariance matrix of the explanatory variables is no longer finite. Similarly, if the variable follows an exponential trend, its variance tends to zero and thus the limiting form of the variance–covariance matrix of the explanatory variables becomes a singular matrix.

The purpose of this article is to study the effect of trended variables on the performance properties of least = squares and Stein-rule estimators in large samples. In Section 2, we describe a linear regression model with simply one trended regressor and state the least-squares and Stein-rule estimators for the regression coefficients. Section 3 presents their large sample properties when all the regressors are asymptotically cooperative. Relaxing this specification and assuming that the values of an explanatory variable follow a linear trend, we analyze the performance properties of the least-squares and Stein-rule estimators in Section 4. A similar exercise is reported in Section 5 for the nonlinear trend case, using two simple formulations, viz., quadratic trend and exponential trend. Some concluding remarks are then offered in Section 6. Finally, the Appendix gives the derivation of the main results.

2. MODEL SPECIFICATION AND THE ESTIMATORS

Let us postulate the following linear regression model:

$$y = \alpha e + X\beta + \delta Z + \epsilon \tag{2.1}$$

where y is an $n \times 1$ vector of n observations on the study variable, α is a scalar representing the intercept term in the regression relationship, e is an $n \times 1$ vector with all elements unity, X is an $n \times p$ matrix of n observations on p explanatory variables, β is a $p \times 1$ vector of p regression coefficients, Z is an $n \times 1$ vector of n observations on another explanatory variable, δ is the regression coefficient associated with it, and ϵ is an $n \times 1$ vector of disturbances.

It is assumed that the elements of vector ϵ are independently and identically distributed following a normal distribution with mean 0 and finite but unknown variance σ^2

If we define

$$A = I_n - \frac{1}{n}ee' \tag{2.2}$$

we can write the model as

$$Ay = AX\beta + \delta AZ + A\epsilon$$
$$= AQ\gamma + A\epsilon \tag{2.3}$$

were $Q = (X \ Z)$ and $\gamma = (\beta' \ \delta)'$.
The least-squares estimator of γ is given by

$$C = (Q'AQ)^{-1}Q'Ay \tag{2.4}$$

whence the least-squares estimator b and d of β and δ respectively are

$$d = \frac{Z'Ay - Z'AX(X'AX)^{-1}X'AY}{Z'AZ - Z'AX(X'AX)^{-1}X'AZ} \tag{2.5}$$

$$b = (X'AX)^{-1}X'A(y - dZ) \tag{2.6}$$

The Stein-rule family of estimators of γ is defined by

$$\hat{\gamma} = \left[1 - \left(\frac{k}{n}\right)\frac{(y - QC)'A(y - QC)}{C'Q'AQC}\right]C \tag{2.7}$$

where k is the nonstochastic scalar, independent of n, characterizing the estimator. Thus the Stein-rule estimators of β and δ are

$$\hat{\beta} = \left[1 - \left(\frac{k}{n}\right)\frac{(y - Xb - dZ)'A(y - Xb - dZ)}{(Xb + dZ)'A(Xb + dZ)}\right]b \tag{2.6}$$

$$\hat{\delta} = \left[1 - \left(\frac{k}{n}\right)\frac{(y - Xb - dZ)'A(y - Xb - dZ)}{(Xb + dZ)'A(Xb + dZ)}\right]d \tag{2.7}$$

In the next three sections, we analyze the performance properties of estimators employing the large sample asymptotic theory.

3. EFFICIENCY PROPERTIES: THE ASYMPTOTICALLY COOPERATIVE CASE

When all the $(p + 1)$ explanatory variables are asymptotically cooperative, it means that, as n tends to infinity, $n^{-1}X'AX$ tends to a finite and nonsingular matrix, $n^{-1}X'AZ$ tends to a finite vector, and $n^{-1}Z'AZ$ tends to a finite positive scalar.

Under the above specification, the least-squares estimators d and b are unbiased whereas the Stein-rule estimators $\hat{\delta}$ and $\hat{\beta}$ are biased with the bias expressions to order $O(n^{-1})$ as

$$B(\hat{\delta}) = E\left(\hat{\delta} - \delta\right)$$

$$= -\frac{\sigma^2 k}{(X\beta + \delta Z)'A(X\beta + \delta Z)}\delta \qquad (3.1)$$

$$B(\hat{\beta}) = E\left(\hat{\beta} - \beta\right)$$

$$= -\frac{\sigma^2 k}{(X\beta + \delta Z)'A(X\beta + \delta Z)}\beta \qquad (3.2)$$

Further, the variance of d is larger than the mean squared error of $\hat{\delta}$ to order $O(n^{-2})$ when

$$0 < k < 2(p - 1); \qquad p > 1 \qquad (3.3)$$

If we define the predictive risks associated with the estimators of β as

$$R(b) = \frac{1}{n}E(Xb - X\beta)'A(Xb - X\beta) \qquad (3.4)$$

$$R(\hat{\beta}) = \frac{1}{n}E\left(X\hat{\beta} - X\beta\right)'A\left(X\hat{\beta} - X\beta\right) \qquad (3.5)$$

and consider their approximations to order $O(n^{-2})$, the estimator $\hat{\beta}$ is found to have smaller risk in comparison with b under the condition (3.3); see [4] for the case of disturbances that are not necessarily normally distributed.

In the next two sections, we assume that the elements of Z are trended so that the limiting value of $n^{-1}Z'AZ$ as n goes to infinity is either zero or infinite. However, the remaining p regressors continue to be asymptotically cooperative so that $n^{-1}X'AX$ tends to a finite and nonsingular matrix. Accordingly, we define

$$S = \frac{1}{n}X'AX \qquad (3.6)$$

$$h = \frac{1}{(nZ'AZ)^{1/2}}X'AZ \qquad (3.7)$$

and assume that the elements of S and h are of order $O(1)$, following [5] (p. 24). We consider both the cases of linear and nonlinear trends in the last explanatory variable.

4. EFFICIENCY PROPERTIES: THE LINEAR TREND CASE

Let us assume that the elements of Z follow a linear trend specified by

$$Z_t = \theta_0 + \theta_1 t \quad (t = 1, 2, \ldots, n) \tag{4.1}$$

where θ_0 and θ_1 are constants characterizing the nature of the trend. It is easy to see that

$$Z'AZ = \frac{n^3\theta_1^2}{12}\left(1 - \frac{1}{n^2}\right) \tag{4.2}$$

so that the limiting value of $n^{-1}Z'AZ$ is infinite as n tends to infinity.

Using (4.2), it is easy to see that

$$V(d) = E(d - \delta)^2$$

$$= \frac{12\sigma^2}{n^3\theta_1^2(1 - h'S^{-1}h)} + O(n^{-5}) \tag{4.3}$$

$$V(b) = E(b - \beta)(b - \beta)'$$

$$= \frac{\sigma^2}{n}\left[S^{-1} + \left(\frac{1}{1 - h'S^{-1}h}\right)S^{-1}hh'S^{-1}\right] \tag{4.4}$$

Similar results for the estimators $\hat{\delta}$ and $\hat{\beta}$ are derived in the Appendix and are stated below.

Theorem 1. The bias of estimator $\hat{\delta}$ to order $O(n^{-3})$ is given by

$$B(\hat{\delta}) = E(\hat{\delta} - \delta)$$

$$= -\frac{12\sigma^2 k}{n^3\theta_1^2\delta} \tag{4.5}$$

while the difference between the variance of d and the mean squared error of $\hat{\delta}$ to order $O(n^{-6})$ is

$$D(\hat{\delta}; d) = E(\hat{\delta} - \delta)^2 - E(d - \delta)^2$$

$$= \frac{144\sigma^4 k^2}{n^6\theta_1^4\delta^2} \tag{4.6}$$

Similarly, the bias vector of $\hat{\beta}$ to order $O(n^{-3})$ is

$$B(\hat{\beta}) = E(\hat{\beta} - \beta)$$

$$= -\frac{12\sigma^2 k}{n^3\theta_1^2\delta^2}\beta \tag{4.7}$$

and the difference between the variance–covariance matrix of b and the mean squared error matrix of $\hat{\beta}$ to order $O(n^{-4})$ is

$$\Delta(\hat{\beta}) = E\left(\hat{\beta} - \beta\right)\left(\hat{\beta} - \beta\right)' - E(b - \beta)(b - \beta)'$$

$$= -\frac{24\sigma^4 k}{n^4 \theta_1^2 \delta^2}\left[S^{-1} + \frac{S^{-1}hh'S^{-1}}{1 - h'S^{-1}h}\right] \qquad (4.8)$$

We thus observe that the estimator d is not only unbiased but has smaller variance than the mean squared error of $\hat{\delta}$, at least to the order of our approximation. On the other hand, the estimator $\hat{\beta}$ is biased but is uniformly superior to the unbiased estimator b for all positive values of k under the strong criterion of the mean squared error matrix. Thus the presence of a trended regressor leads to a substantial change in the performance properties of Stein-rule estimators.

When all the variables are asymptotically cooperative, the leading term in the bias of Stein-rule estimators is of order $O(n^{-1})$. This order is $O(n^{-3})$ when one regressor is trended. This implies that Stein-rule estimators become nearly unbiased in large samples in the presence of linear trend in a regressor.

Similarly, we observe from (4.3) that the mean squared error of d and $\hat{\delta}$ is of order $O(n^{-1})$ in the asymptotically cooperative case but becomes of order $O(n^{-3})$ in the presence of linear trend. However, no such change occurs in the case of estimation of β and the mean squared error matrix of b, as well as $\hat{\beta}$, remains of order $O(n^{-1})$, see (4.4).

It is interesting to note that the magnitude of the slope of the trend line, besides other parameters of the model, plays an important role in the distributional properties of estimators whereas the intercept term in the trend line exhibits no influence.

5. EFFICIENCY PROPERTIES: THE NONLINEAR TREND CASE

There are numerous specifications for the nonlinear trend, but we restrict our attention to two popular formulations. One is the quadratic trend and the other is the exponential trend.

5.1 Quadratic Trend

Let us suppose that the elements of Z follow a quadratic trend given by

$$Z_t = \theta_0 + \theta_1 t + \theta_2 t^2 \qquad (t = 1, 2, \ldots, n) \tag{5.1}$$

where θ_0, θ_1, and θ_2 are the constants determining the nature of the trend.

The trend equation (5.1) describes an upwardly or downwardly concave curve with a bend and continually changing slope. Using the results

$$\frac{1}{n}\sum t = \frac{(n+1)}{2}, \quad \frac{1}{n}\sum t^2 = \frac{(n+1)(2n+1)}{6},$$

$$\frac{1}{n}\sum t^3 = \frac{n(n+1)^2}{4}, \quad \frac{1}{n}\sum t^4 = \frac{(n+1)(2n+1)(3n^2 + 3n - 1)}{30}$$

it is easy to see that

$$
\begin{aligned}
Z'AZ = \frac{4n^5 \theta_2^2}{45} & \left[1 + \frac{15}{8n}\left(1 + \frac{\theta_1}{\theta_2}\right) \right. \\
& - \frac{5}{16n^2}\left(1 - 6\frac{\theta_1}{\theta_2} - 3\frac{\theta_1^2}{\theta_2^2}\right) \\
& \left. - \frac{15}{8n^3}\left(1 + \frac{\theta_1}{\theta_2}\right) - \frac{1}{16n^4}\left(11 + 30\frac{\theta_1}{\theta_2} + 15\frac{\theta_1^2}{\theta_2^2}\right) \right]
\end{aligned}
\tag{5.2}
$$

Clearly, the limiting value of $n^{-1}Z'AZ$ as n tends to infinity is infinite. Further, the quantity $n^{-1}Z'AZ$ approaches infinity at a faster rate when compared with the case of linear trend; see (4.2).

Using (5.2), we observe that the variance of d is

$$
\begin{aligned}
V(d) &= E(d - \delta)^2 \\
&= \frac{45\sigma^2}{4n^3\theta_2^2(1 - h'S^{-1}h)} + O(n^{-6})
\end{aligned}
\tag{5.3}
$$

while the variance–covariance matrix of b remains the same as that given by (4.4).

In the Appendix, the following results are obtained.

Theorem 2. The bias of $\hat{\delta}$ to order $O(n^{-5})$ is given by

$$
\begin{aligned}
B(\hat{\delta}) &= E(\hat{\delta} - \delta) \\
&= -\frac{45\sigma^2 k}{4n^3\theta_2^2\delta}
\end{aligned}
\tag{5.4}
$$

and the difference between the mean squared error of $\hat{\delta}$ and the variance of d to order $O(n^{-10})$ is

$$D\left(\hat{\delta}; d\right) = E\left(\hat{\delta} - \delta\right)^2 - E(d - \delta)^2$$

$$= \frac{2025\sigma^4 k^2}{16n^{10}\theta_2^4\delta^2} \tag{5.5}$$

Similarly, the bias vector of $\hat{\beta}$ to order $O(n^{-5})$ is

$$B(\hat{\beta}) = E\left(\hat{\beta} - \beta\right)$$

$$= -\frac{45\sigma^2 k}{4n^5\theta_2^2\delta^2}\beta \tag{5.6}$$

while the mean squared error matrix of $\hat{\beta}$ to order $O(n^{-6})$ exceeds the variance–covariance matrix of b by a positive semi-definite matrix specified by

$$\Delta\left(\hat{\beta}; b\right) = E\left(\hat{\beta} - \beta\right)\left(\hat{\beta} - \beta\right)' - E(b - \beta)(b - \beta)'$$

$$= -\frac{45\sigma^4 k}{2n^6\theta_2^2\delta^2}\left[S^{-1} + \frac{S^{-1}hh'S^{-1}}{1 - h'S^{-1}h}\right] \tag{5.7}$$

As is observed in the case of linear trend, the Stein-rule estimator of δ is not only biased but also inefficient in comparison with the least-squares estimator d. However, when estimation of β is considered, it is seen that the estimator b is exactly unbiased whereas the Stein-rule estimator $\hat{\beta}$ is nearly unbiased in the sense that the bias to order $O(n^{-4})$ is zero. Further, from (5.7), the variance–covariance matrix of b exceeds the mean squared error matrix of $\hat{\beta}$ by a non-negative definite matrix, implying the strong superiority of $\hat{\beta}$ over b.

The difference between the linear trend case and the quadratic trend case lies only at the level of approximation. For instance, the leading term in the bias of $\hat{\delta}$ and $\hat{\beta}$ is of order $O(n^{-3})$ in the case of linear trend and $O(n^{-5})$ in the case of quadratic trend.

5.2 Exponential Trend

Let us now assume that the elements of Z follow an exponential trend specified by

$$Z_t = \theta_0 + \theta_1\theta_2^{t-1} \qquad (t = 1, 2, \ldots, n) \tag{5.8}$$

where θ_0, θ_1, and θ_2 are constants. Using (5.8), we observe that

$$Z'AZ = \frac{\theta_1^2(1-\theta_2^{2n})}{(1-\theta_2^2)} - \frac{\theta_1^2(1-\theta_2^n)^2}{n(1-\theta_2)^2} \tag{5.9}$$

When $-1 < \theta_2 < 1$, i.e., the trend is non-explosive, it can be easily seen that the quantity $n^{-1}Z'AZ$ approaches 0 as n tends to infinity.

Under the specification (5.10), the variance of d is

$$V(d) = E(d-\delta)^2$$

$$= \frac{\sigma^2(1-\theta_2^2)}{\theta_1^2(1-\theta_2^{2n})(1-h'S^{-1}h)}\left[1 + \frac{(1-\theta_2^n)^2(1+\theta_2)}{n(1-\theta_2^{2n})(1-\theta_2)}\right] + O(n^{-2}) \tag{5.10}$$

while the variance–covariance matrix of b is the same as that specified by (4.4).

From the Appendix, we have the following results.

Theorem 3.   The bias of estimator d of δ to order $O(n^{-1})$ is

$$B\left(\hat{\delta}\right) = E\left(\hat{\delta}-\delta\right)$$

$$= -\frac{\sigma^2 k}{n\beta'S\beta}\delta \tag{5.11}$$

and the difference between the mean squared error of $\hat{\delta}$ and the variance of d to order $O(n^{-1})$ is

$$D\left(\hat{\delta}; d\right) = E\left(\hat{\delta}-\delta\right)^2 - E(d-\delta)^2$$

$$= -\frac{2\sigma^2 K(1-\theta_2^2)}{n\theta_1^2(1-\theta_2^{2n})\beta'S\beta} \tag{5.12}$$

Similarly, the bias vector of $\hat{\beta}$ to order $O(n^{-1})$ is

$$B\left(\hat{\beta}\right) = E\left(\hat{\beta}-\beta\right)$$

$$= -\frac{\sigma^2 k}{n\beta'S\beta}\beta \tag{5.13}$$

while the difference between the mean squared error matrix of $\hat{\beta}$ and the variance-covariance matrix of b to order $O(n^{-2})$ is

$$\Delta\left(\hat{\beta}; b\right) = E\left(\hat{\beta} - \beta\right)\left(\hat{\beta} - \beta\right)' - E(b - \beta)(b - \beta)'$$

$$= \frac{\sigma^4 k}{n^2 \beta' S \beta}\left[\left(\frac{k+4}{\beta' S \beta}\right)\beta\beta'\right.$$

$$\left. -2\left(S^{-1} + \frac{1}{1 - h'S^{-1}h}S^{-1}hh'S^{-1}\right)\right] \tag{5.14}$$

From (5.11) and (5.12), we see that the Stein-rule estimator of δ is biased but more efficient than the least squares estimator for all positive values of k. Similarly, it is observed from (5.13) that the Stein-rule estimator of β is biased, with every element of the bias vector possessing a sign opposite to that of the corresponding regression coefficient.

Now, for the comparison of b and $\hat{\beta}$ under the criterion of mean squared error matrix, we state two results for a positive definite matrix G and a vector g; see, e.g., [6] (p. 370).

Lemma 1. The matrix $(gg' - G^{-1})$ can never be non-negative definite except when g is a scalar.

Lemma 2. The matrix $(G^{-1} - gg')$ is non-negative definite if and only if $g'Gg$ does not exceed 1.

Employing Lemma 1, it follows from (5.14) that b cannot be superior to $\hat{\beta}$ except in the trivial case $p = 1$. On the other hand, applying Lemma 2, it is seen that $\hat{\beta}$ is superior to b if and only if

$$\left(\frac{k+4}{2\beta' S \beta}\right)\beta'\left(S^{-1} + \frac{1}{1 - h'S^{-1}h}S^{-1}hh'S^{-1}\right)^{-1}\beta < 1$$

or

$$\left(\frac{k+4}{2\beta' S \beta}\right)\beta'(S - hh')\beta < 1$$

or

$$\frac{(\beta'h)^2}{\beta' S \beta} > \left(\frac{k+2}{k+4}\right) \tag{5.15}$$

Next, let us compare b and $\hat{\beta}$ under a weak criterion, say, the predictive risk. From (5.14), we observe that

$$\mathrm{tr} S\Delta\left(\hat{\beta}; b\right) = E\left(\hat{\beta} - \beta\right)' S\left(\hat{\beta} - \beta\right) - E(b - \beta)' S(b - \beta)$$

$$= \frac{\sigma^4 k}{n^2 \beta' S\beta}\left[k - 2\left(p - 2 + \frac{h'S^{-1}h}{1 - h'S^{-1}h}\right)\right] \tag{5.16}$$

from which it is interesting to see that $\hat{\beta}$ has smaller predictive risk to order $O(n^{-2})$ in comparison with b when k satisfies the constraint

$$0 < k < 2\left(p - 2 + \frac{h'S^{-1}h}{1 - h'S^{-1}h}\right); \quad p > 2 - \frac{h'S^{-1}h}{1 - h'S^{-1}h} \tag{5.17}$$

As $h'S^{-1}h$ lies between 0 and 1, this condition is satisfied as long as

$$0 < k < 2(p - 2); \quad p > 2 \tag{5.18}$$

which is slightly different from (3.3).

6. SOME REMARKS

We have considered a linear regression model containing a trended regressor and have analyzed the performance properties of the least-squares (LS) and Stein-rule (SR) estimators employing the large sample asymptotic theory.

It is well known that the SR estimators of regression coefficients are generally biased and that the leading term in the bias is of order $O(n^{-1})$ with a sign opposite to that of the corresponding regression coefficient whereas the LS estimators are always unbiased. Further, the SR estimators have smaller predictive risk, to order $O(n^{-2})$, in comparison with the LS estimators when the positive characterizing scalar k is less than $2(p - 1)$ where there are $(p + 2)$ unknown coefficients in the regression relationship. These inferences are deduced under the assumption of the asymptotic cooperativeness of all the variables, i.e., the variance–covariance matrix of regressors tends to a finite and nonsingular matrix as n goes to infinity. Such an assumption is violated, for instance, when trend is present in one or more regressors.

Assuming that simply one regressor is trended while the remaining regressors are asymptotically cooperative, we have considered two specific situations. In the first situation, the limiting form of the variance–covariance matrix of regressors is singular, as is the case with nonexplosive exponential trend, while in the second situation, it is infinite, as is the case with linear and quadratic trends.

Comparing the two situations under the criterion of bias, we observe that the bias of SR estimators of δ and β is of order $O(n^{-1})$ in the first situation.

In the second situation, it is of order $O(n^{-3})$ in the linear trend case and $O(n^{-5})$ in the quadratic trend case. In other words, the SR estimators are nearly unbiased in the second situation when compared with the first situation. Further, the variances and mean squared errors of LS and SR estimators of δ are of order $O(1)$ in the first situation but they are of order $O(n^{-3})$ in the linear trend case and $O(n^{-5})$ in the quadratic trend case. This, however, does not remain true when we consider the estimation of β, consisting of the coefficients associated with asymptotically cooperative regressors. The variance–covariance and mean squared error matrices of LS and SR estimators of β continue to remain of order $O(n^{-1})$ in every case and the trend does not exert its influence, at least asymptotically.

Analyzing the superiority of the SR estimator over the LS estimator, it is interesting to note that the mean squared error of the biased SR estimator of δ is smaller than the variance of the unbiased LS estimator for all positive values of k in the first situation. A dramatic change is seen in the second situation where the LS estimator of δ remains unbeaten by the SR estimator on both the fronts of bias and mean squared error. When we consider the estimation of β, we observe that the SR estimator may dominate the LS estimator, under a certain condition, with respect to the strong criterion of mean squared error matrix in the first situation. However, if we take the weak criterion of predictive risk, the SR estimator performs better than the LS estimator at least as long as k is less than $2(p-2)$, with the rider that there are three or more regressors in the model besides the intercept term. In the second situation, the SR estimator of β is found to perform better than the LS estimator, even under the strong criterion of the mean squared error matrix, to the given order of approximation.

Next, let us compare the performance properties of estimators with respect to the nature of the trend, i.e., linear versus nonlinear as specified by quadratic and exponential equations. When linear trend is present, the leading term in the bias of SR estimators is of order $O(n^{-3})$. This order is $O(n^{-5})$ in the case of quadratic trend, whence it follows that the SR estimators tend to be unbiased at a slower rate in the case of linear trend when compared with quadratic trend or, more generally, polynomial trend. However, the order of bias in the case of exponential trend is $O(n^{-1})$, so that the SR estimators may have substantial bias in comparison to situations like linear trend where the bias is nearly negligible in large samples.

Looking at the variances and mean squared errors of the estimators of δ, it is observed that the difference between the LS and SR estimators appears at the level of order $O(n^{-6})$ in the case of linear trend while this level varies in nonlinear cases. For instance, it is $O(n^{-10})$ in the case of quadratic trend. Thus both the LS and SR estimators are equally efficient up to order $O(n^{-9})$ in the case of quadratic trend whereas the poor performance of the SR estimator starts precipitating at the level of $O(n^{-6})$ in the case of linear trend. The situation takes an interesting turn when we compare linear and exponential trends. For all positive values of k, the SR estimator of δ is invariably inferior to the LS estimator in the case of linear trend but it is invariably superior in the case of exponential trend.

So far as the estimation of regression coefficients associated with asymptotically cooperative regressors is concerned, the LS and SR estimators of β have identical mean squared error matrices up to order $O(n^{-3})$ in the linear trend case and $O(n^{-5})$ in the quadratic trend case. The superior performance of the SR estimator under the strong criterion of mean squared error matrix appears at the level of $O(n^{-4})$ and $O(n^{-6})$ in the cases of linear and quadratic trends respectively. In the case of exponential trend, neither estimator is found to be uniformly superior to the other under the mean squared error matrix criterion. However, under the criterion of predictive risk, the SR estimator is found to be better than the LS estimator, with a rider on the values of k.

Similarly, our investigations have clearly revealed that the asymptotic properties of LS and SR estimators deduced under the specification of asymptotic cooperativeness of regressors lose their validity and relevance when one of the regressors is trended. The consequential changes in the asymptotic properties are largely governed by the nature of the trend, such as its functional form and the coefficients in the trend equation. Interestingly enough, the presence of trend in a regressor not only influences the efficiency properties of the estimator of its coefficient but also jeopardizes the properties of the estimators of the coefficients associated with the regressors that are asymptotically cooperative.

We have envisaged three simple formulations of trend for analyzing the asymptotic performance properties of LS and SR estimators in a linear regression model. It will be interesting to extend our investigations to other types of trend and other kinds of models.

APPENDIX

Besides (3.6), let us first introduce the following notation:

$$u = \frac{1}{n^{1/2}} X'A\epsilon$$

$$w = \frac{1}{(Z'AZ)^{1/2}} Z'A\epsilon$$

$$v = n^{1/2} \left(\frac{\epsilon'A\epsilon}{n} - \sigma^2 \right)$$

Now, it is easy to see from (2.5) and (2.6) that

$$(d - \delta) = \frac{w - h'S^{-1}u}{(Z'AZ)^{1/2}(1 - h'S^{-1}h)} \tag{A.1}$$

$$(b - \beta) = (X'AX)^{-1}X'A[\epsilon - (d - \delta)Z]$$

$$= \frac{1}{n^{1/2}}S^{-1}\left[u - \left(\frac{w - h'S^{-1}u}{1 - h'S^{-1}h}\right)h\right] \tag{A.2}$$

Similarly, we have

$$(Xb + dz)'A(Xb + dZ)$$

$$= y'AX(X'AX)^{-1}X'AY + \frac{[Z'Ay - Z'AX(X'AX)^{-1}X'Ay]^2}{[Z'AZ - Z'AX(X'AX)^{-1}X'AZ]}$$

$$= n\beta'S\beta + 2n^{1/2}\beta'u + u'S^{-1}u + (Z'AZ)\delta^2$$

$$+ 2(Z'AZ)^{1/2}(w + n^{1/2}h'\beta)\delta + \frac{(w - h'S^{-1}u)^2}{(1 - h'S^{-1}h)} \tag{A.3}$$

$$(y - Xb - dZ)'A(y - Xb - dZ)$$

$$= [\epsilon - (d - \delta)Z]'[A - AX(X'AX)^{-1}X'A][\epsilon - (d - \delta)Z]$$

$$= \epsilon'[A - AX(X'AX)^{-1}X'A]\epsilon$$

$$- \frac{[Z'A\epsilon - Z'AX(X'AX)^{-1}X'A\epsilon]^2}{[Z'AZ - Z'AX(X'AX)^{-1}X'AZ]}$$

$$= n\sigma^2 + n^{1/2}v - u'S^{-1}u - (w - h'S^{-1}u)^2(1 - h'S^{-1}h)^{-1} \tag{A.4}$$

Proof of Theorem 1

Substituting (4.2) in (A.1) and (A.3), we find

$$(d - \delta) = \frac{\sqrt{12}}{n^{3/2}n} \left(\frac{w - h'S^{-1}u}{1 - h'S^{-1}h} \right) \left(1 - \frac{1}{n^2} \right)^{-7/2}$$

$$= \frac{\sqrt{12}}{n^{3/2}\theta_1} \left(\frac{w - h'S^{-1}u}{1 - h'S^{-1}h} \right) + O_p\left(n^{-7/2}\right) \tag{A.5}$$

$$(Xb + dZ)'A(Xb + dZ) = \frac{n^3\theta_1^2\delta^2}{12} + O_p\left(n^{3/2}\right) \tag{A.6}$$

whence it follows that

$$\left(\frac{1}{n}\right) \frac{(y - Xb - dZ)'A(y - Xb - dZ)}{(Xb + dZ)'A(Xb + dZ)}$$

$$= \frac{12}{n^3\theta_1^2\delta^2} \left(\sigma^2 + \frac{v}{n^{1/2}} \right) + O_p\left(n^{-4}\right) \tag{A.7}$$

Using (A.7) and (A.5) in (2.9), we can express

$$\left(\hat{\delta} - \delta\right) = \frac{\sqrt{12}}{n^{3/2}\theta_1} \left(\frac{w - h'S^{-1}u}{1 - h'S^{-1}h} \right)$$

$$- \frac{12k}{n^3\theta_1^2\delta} \left(\sigma^2 + \frac{v}{n^{1/2}} \right) + O_p\left(n^{-4}\right) \tag{A.8}$$

Thus the bias to order $O(n^{-3})$ is

$$B\left(\hat{\delta}\right) = E\left(\hat{\delta} - \delta\right)$$

$$= -\frac{12\sigma^2 k}{n^3\theta_1^2\delta} \tag{A.9}$$

which is the result (4.5) of Theorem 1.

Similarly, the difference between the variance of d and the mean squared error of $\hat{\delta}$ to order $O(n^{-6})$ is

$$D\left(\hat{\delta}; d\right) = E\left(\hat{\delta} - \delta\right)^2 - E(d - \delta)^2$$

$$= E\left[\left(\hat{\delta} - \delta\right) - (d - \delta) \right]\left[\left(\hat{\delta} - \delta\right) + (d - \delta) \right] \tag{A.10}$$

$$= \frac{144\sigma^4 k^2}{n^6\theta_1^4\delta^2}$$

which is the result (4.6).

Similarly, using (A.2) and (A.7), we observe that

$$\left(\hat{\beta} - \beta\right) = \frac{1}{n^{1/2}} S^{-1}\left[u - \left(\frac{w - h'S^{-1}u}{1 - h'S^{-1}h}\right)h\right] - \frac{12k\sigma^2}{n^3\theta_1^2\delta^2}\beta$$

$$- \frac{12k}{n^{7/2}\theta_1^2\delta^2}\left[v\beta + \sigma^2 S^{-1}u - \sigma^2\left(\frac{w - h'S^{-1}u}{1 - h'S^{-1}h}\right)S^{-1}h\right] \qquad \text{(A.11)}$$

$$+ O_p(n^{-4})$$

from which the bias vector to order $O(n^{-3})$ is

$$B\left(\hat{\beta}\right) = E\left(\hat{\beta} - \beta\right)$$

$$= -\frac{12k\sigma^2}{n^3\theta_1^2\delta^2}\beta \qquad \text{(A.12)}$$

providing the result (4.7).

Further, it is easy to see from (A.2) and (A.11) that the difference to order $O(n^{-4})$ is

$$\Delta\left(\hat{\beta}; b\right) = E\left(\hat{\beta} - \beta\right)\left(\hat{\beta} - \beta\right)' - E(b - \beta)(b - \beta)'$$

$$= -\frac{12k}{n^4\theta_1^2\delta^2}\left[E\left(vS^{-1}u\beta' + v\beta u'S^{-1}\right)\right.$$

$$+ 2\sigma^2 S^{-1}E\left\{u - \left(\frac{w - h'S^{-1}u}{1 - h'S^{-1}}\right)h\right\} \qquad \text{(A.13)}$$

$$\left.\left\{u' - \left(\frac{w - h'S^{-1}u}{1 - h'S^{-1}h}\right)h'\right\}S^{-1}\right]$$

$$= -\frac{24\sigma^4 k}{n^4\theta_1^2\delta^2}\left[S^{-1} + \frac{1}{1 - h'S^{-1}h}S^{-1}hh'S^{-1}\right]$$

which is the result (4.8).

Proof of Theorem 2

Under the specification (5.1), we observe from (A.1) and (A.2) that

$$(d - \delta) = \frac{3\sqrt{5}}{2n^{5/2}\theta_2}\left[1 + \frac{15}{8n}\left(1 + \frac{\theta_1}{\theta_2}\right) - \frac{5}{16n^2}\left(1 - 6\frac{\theta_1}{\theta_2} + 3\frac{\theta_1^2}{\theta_2^2}\right)\right.$$

$$\left. + O(n^{-3})\right]^{-1/2}\left(\frac{w - h'S^{-1}u}{1 - h'S^{-1}h}\right) \tag{A.14}$$

$$= \frac{3\sqrt{5}}{2n^{5/2}\theta_2}\left(\frac{w - h'S^{-1}u}{1 - h'S^{-1}h}\right) + O_p(n^{-7/2})$$

Proceeding in the same manner as indicated in the case of linear trend, we can express

$$\left(\frac{1}{n}\right)\frac{(y - Xb - dZ)'A(y - Xb - dZ)}{(Xb + dZ)'A(Xb + dZ)}$$

$$= \frac{45}{4n^5\theta_2^2\delta^2}\left(\sigma^2 + \frac{v}{n^{1/2}}\right) + O_p(n^{-6}) \tag{A.15}$$

whence the bias of $\hat{\delta}$ to order $O(n^{-5})$ is

$$B\left(\hat{\delta}\right) = E(d - \delta) - \frac{45\sigma^2 k}{4n^5\theta_2^2\delta}$$

$$= -\frac{45\sigma^2 k}{4n^5\theta_2^2\delta} \tag{A.16}$$

and the difference between the mean squared error of $\hat{\delta}$ and the variance of d to order $O(n^{-10})$ is

$$D\left(\hat{\delta}; d\right) = E\left(\hat{\delta} - \delta\right)^2 - E(d - \delta)^2$$

$$= E\left[\left(\hat{\delta} - \delta\right) - (d - \delta)\right]\left[\left(\hat{\delta} - \delta\right) + (d - \delta)\right] \tag{A.17}$$

$$= \frac{2025\sigma^4 k^2}{16n^{10}\theta_2^4\delta^2}$$

These are the results (5.4) and (5.5) of Theorem 2. The other two results (5.6) and (5.7) can be obtained in a similar manner.

Proof of Theorem 3

Writing

$$\lambda = \frac{1}{\theta_1}\left(\frac{1-\theta_2^2}{1-\theta_2^{2n}}\right)^{1/2}$$

$$\lambda^* = \frac{(1+\theta_2)(1-\theta_2^n)^2}{(1-\theta_2)(1-\theta_2^{2n})}$$

(A.18)

and using (5.9) in (A.1), we get

$$(d-\delta) = \lambda\left(\frac{w-h'S^{-1}u}{1-h'S^{-1}h}\right)\left(1-\frac{\lambda^*}{n}\right)^{-1/2}$$

$$= \lambda\left(\frac{w-h'S^{-1}u}{1-h'S^{-1}h}\right)\left(1+\frac{\lambda^*}{2n}\right) + O_p(n^{-2})$$

(A.19)

Similarly, from (A.3) and (A.4), we have

$$\frac{(y-Xb-dZ)'A(y-Xb-dZ)}{(Xb+dZ)'A(Xb+dZ)}$$

$$= \frac{1}{\beta'S\beta}\left[\sigma^2 + \frac{1}{n^{1/2}}\left(v - \frac{2\sigma^2\beta'u}{\beta'S\beta}\right)\right] + O_p(n^{-1})$$

(A.20)

whence we can express

$$\left(\hat{\delta}-\delta\right) = \lambda\left(\frac{w-h'S^{-1}u}{1-h'S^{-1}h}\right)$$

$$+ \frac{1}{n}\left[\lambda\left(\frac{\lambda^*}{2} - \frac{\sigma^3 k}{\beta'S\beta}\right)\left(\frac{w-h'S^{-1}u}{1-h'S^{-1}h}\right) - \frac{\sigma^2 k\delta}{\beta'S\beta}\right] + O_p(n^{-3/2})$$

(A.21)

Thus the bias to order $O(n^{-1})$ is

$$B\left(\hat{\delta}\right) = E\left(\hat{\delta}-\delta\right)$$

$$= -\frac{\sigma^2 k\delta}{n\beta'S\beta}$$

(A.22)

and the difference between the mean squared error of $\hat{\delta}$ and the variance of d to order $O(n^{-1})$ is

$$D\left(\hat{\delta}; d\right) = E\left(\hat{\delta} - \delta\right)^2 - E(d - \delta)^2$$

$$= -\frac{2\sigma^2\lambda^2 k}{n\beta'S\beta} \tag{A.23}$$

These are the results (5.11) and (5.12) stated in Theorem 3.
Similarly, from (A.2) and (A.20), we can express

$$\left(\hat{\beta} - \beta\right) = \frac{1}{n^{1/2}}S^{-1}\left[u - \left(\frac{w - h'S^{-1}u}{1 - h'S^{-1}h}\right)h\right] - \frac{\sigma^2 k}{n\beta'S\beta}\beta$$

$$- \frac{k}{n^{3/2}\beta'S\beta}\left[v\beta + \sigma^2 S^{-1}u - \sigma^2\left(\frac{w - h'S^{-1}u}{1 - h'S^{-1}h}\right)S^{-1}h \right. \tag{A.24}$$

$$\left. - \frac{2\sigma^2\beta'u}{\beta'S\beta}\beta\right] + O_p\left(n^{-2}\right)$$

Thus the bias vector to order $O(n^{-1})$ is

$$B\left(\hat{\beta}\right) = E\left(\hat{\beta} - \beta\right)$$

$$= -\frac{\sigma^2 k}{n\beta'S\beta}\beta \tag{A.25}$$

Further, to order $O(n^{-2})$, we have

$$\Delta\left(\hat{\beta}; b\right) = E\left(\hat{\beta} - \beta\right)\left(\hat{\beta} - \beta\right)' - E(b - \beta)(b - \beta)'$$

$$= \frac{\sigma^4 k}{n^2\beta'S\beta}\left[\left(\frac{k+4}{\beta'S\beta}\right)\beta\beta' - 2S^{-1} - \frac{2}{1 - h'S^{-1}h}S^{-1}hh'S^{-1}\right] \tag{A.26}$$

which is the last result (5.14) of Theorem 3.

REFERENCES

1. James, W., and C. Stein (1961): Estimation with quadratic loss. *Proceedings of the Fourth Berkeley Symposium on Mathematical Statistics and Probability*, University of California Press, 361–379.
2. Judge, G.G., and M.E. Bock (1978): *The Statistical Implications of Pre-Test and Stein-Rule Estimators In Econometrics*, North-Holland, Amsterdam.

3. Judge, G.G., W.E. Griffiths, R.C. Hill, H. Lütkepohl,and T. Lee (1985): *The Theory and Practice of Econometrics,* 2nd edition, John Wiley, New York.
4. Vinod, H.D., and V.K. Srivastava (1995): Large sample asymptotic properties of the double *k*-class estimators in linear regression models, *Econometric Reviews,* **14,** 75–100.
5. Krämer, W. (1984): On the consequences of trend for simultaneous equation estimation *Economics Letters,* **14,** 23–30.
6. Rao, C.R., and H. Toutenburg (1999): *Linear Models: Least Squares and Alternatives,* 2nd edition, Springer, New York.

17

Endogeneity and Biased Estimation under Squared Error Loss

RON C. MITTELHAMMER   Washington State University, Pullman, Washington

GEORGE G. JUDGE   University of California, Berkeley, Berkeley, California

1. INTRODUCTION

The appearance in 1950 of the Cowles Commission monograph 10, *Statistical Inference in Economics*, alerted econometricians to the possible simultaneous–endogenous nature of economic data sampling processes and the consequent least-squares bias in infinite samples (Hurwicz 1950, and Bronfenbrenner 1953). In the context of a single equation in a system of equations, consider the linear statistical model $y = X\beta + \epsilon$, where we observe a vector of sample observations $y = (y_1, y_2, \ldots, y_n)'$, X is an $(n \times k)$ matrix of stochastic variables, $\epsilon \sim (0, \sigma^2 I_n)$ and $\beta \in B$ is a $(k \times 1)$ vector of unknown parameters. If $E[n^{-1}X'\epsilon] \neq 0$ or $\text{plim}[n^{-1}X'\epsilon] \neq 0$, traditional maximum likelihood (ML) or least squares (LS) estimators, or equivalently the method of moments (MOM)-extremum estimator $\hat{\beta}_{\text{mom}} = \arg_{\beta \in B}[n^{-1}X'(y - X\beta) = 0]$, are biased and inconsistent, with unconditional expectation $E[\hat{\beta}] \neq \beta$ and plim $[\hat{\beta}] \neq \beta$.

Given a sampling process characterized by nonorthogonality of X and ϵ , early standard econometric practice made use of strong underlying distributional assumptions and developed estimation and inference procedures with known desirable asymptotic sampling properties based on maximum likelihood principles. Since in many cases strong underlying distributional assumptions are unrealistic, there was a search for at least a consistent estimation rule that avoided the specification of a likelihood function. In search of such a rule it became conventional to introduce additional information in the form of an $(n \times h)$, $h \geq k$ random matrix $\mathbf{Z}$ of instrumental variables, whose elements are correlated with $\mathbf{X}$ but uncorrelated with ϵ. This information was introduced in the statistical model in sample analog form as

$$\mathbf{h}(\mathbf{y}, \mathbf{X}, \mathbf{Z}; \beta) = n^{-1}[\mathbf{Z}'(\mathbf{y} - \mathbf{X}\beta)] \xrightarrow{P} \mathbf{0} \tag{1.1}$$

and if $h = k$ the sample moments can be solved for the instrumental variable (IV) estimator $\hat{\beta}_{iv} = (\mathbf{Z}'\mathbf{X})^{-1}\mathbf{Z}'\mathbf{y}$. When the usual regularity conditions are fulfilled, this IV estimator is consistent, asymptotically normally distributed, and is an optimal estimating function (EF) estimator (Godambe 1960, Heyde 1997, Mittelhammer et al. 2000).

When $h \geq k$, other estimation procedures remain available, such as the EF approach (Godambe 1960, Heyde and Morton 1998) and the empirical likelihood (EL) approach (Owen 1988, 1991, 2001, DiCiccio et al. 1991, Qin and Lawless 1994), where the latter identifies an extremum-type estimator of β as

$$\hat{\beta} = \arg\max_\beta \left[\ell_E(\beta) = \max_\mathbf{W} \left\{ \phi(\mathbf{w}) \middle| \sum_{i=1}^n w_i \mathbf{z}_i'(y_i - \mathbf{x}_i\beta) = \mathbf{0}, \right. \right.$$

$$\left. \left. \sum_{i=1}^n \mathbf{w}_i = 1, \ w_i \geq 0\, \forall i, \ \beta \in \mathbf{B} \right\} \right] \tag{1.2}$$

Note that $\ell_E(\beta)$ in (1.2) can be interpreted as a profile empirical likelihood for β (Murphy and Van Der Vaart 2000). The estimation objective function in (1.2), $\phi(\mathbf{w})$, may be either the traditional empirical log-likelihood objective function, $\sum_{i=1}^n \log(w_i)$, leading to the maximum empirical likelihood (MEL) estimate of β, or the alternative empirical likelihood function, $-\sum_{i=1}^n w_i \log(w_i)$, representing the Kullback–Leibler information or entropy estimation objective and leading to the maximum entropy empirical likelihood (MEEL) estimate of β. The MEL and MEEL criteria are connected through the Cressie–Read statistics (Cressie and Read 1984, Read and Cressie 1988, Baggerly 1998). Given the estimating equations under consideration, both EL type estimators are consistent, asymptotically normally distributed, and asymptotically efficient relative to the optimal estimating function (OptEF) estimator. Discussions regarding the solutions and the asymptotic properties for these types of problems can be found in Imbens et al. (1998) and Mittelhammer et al. (2000), as well as the references therein.

Among other estimators in the over-identified case, the GMM estimators (Hansen 1982), which minimize a quadratic form in the sample moment information

$$\hat{\beta}(\mathbf{W}) = \arg \min_{\beta \in \mathbf{B}}[Q_n(\beta)] = \arg \min_{\beta \in \mathbf{B}} {}^{[n^{-2}(\mathbf{Y}-\mathbf{X}\beta)'\mathbf{Z}\mathbf{W}\mathbf{Z}'(\mathbf{Y}-\mathbf{X}\beta)]} \qquad (1.3)$$

can be shown to have optimal asymptotic properties for an appropriate choice of the weighting matrix $\mathbf{W}$. Under the usual regularity conditions, the optimal GMM estimator, as well as the other extremum type estimators noted above, are first order asymptotically equivalent. However, when $h > k$ their finite sample behavior in terms of bias, precision, and inference properties may differ, and questions regarding estimator choice remain. Furthermore, even if the $\mathbf{Z}'\mathbf{X}$ matrix has full column rank, the correlations between economic variables, normally found in practice, may be such that these matrices exhibit ill-conditioned characteristics that result in estimates with low precision for small samples. In addition, sampling processes for economic data may not be well behaved in the sense that extreme outliers from heavy-tailed sampling distributions may be present. Moreover, the number of observations in a sample of data may be considerably smaller than is needed for any of the estimators to achieve even a modicum of their desirable asymptotic properties. In these situations, variants of the estimators noted above that utilize the structural constraint (1.1) may lead, in finite samples, to estimates that are highly unstable and serve as an unsatisfactory basis for estimation and inference.

In evaluating estimation performance in situations such as these, a researcher may be willing to trade off some level of bias for gains in precision. If so, the popular squared error loss (SEL) measure may be an appropriate choice of estimation metric to use in judging both the quality of parameter estimates, measured by $\mathbf{L}(\beta, \hat{\beta}) = \|(\beta - \hat{\beta})\|^2$, as well as dependent variable prediction accuracy, measured by $\mathbf{L}(\mathbf{y}, \hat{\mathbf{y}}) = \|(\mathbf{y} - \hat{\mathbf{y}})\|^2$. Furthermore, one usually seeks an estimator that provides a good finite sample basis for inference as it relates to interval estimation and hypothesis testing. With these objectives in mind, in this paper we introduce a semiparametric estimator that, relative to traditional estimators for the overdetermined non-orthogonal case, exhibits the potential for significant finite sample performance gains relating to precision of parameter estimation and prediction, while offering hypothesis testing and confidence interval procedures with comparable size and coverage probabilities, and with potentially improved test power and reduced confidence interval length.

In Section 2 we specify a formulation to address these small sample estimation and inference objectives in the form of a nonlinear inverse problem, provide a solution, and discuss the solution characteristics. In Section 3 estimation and inference finite sample results are presented and evaluated. In Section 4 we discuss the statistical implications of our results and comment on how the results may be extended and refined.

2. THE ELDBIT CONCEPT

In an attempt to address the estimation and inference objectives noted in Section 1, and in particular to achieve improved finite sampling performance relative to traditional estimators, we propose an estimator that combines the EL estimator with a variant of a data-based information theoretic (DBIT) estimator proposed by van Akkeren and Judge (1999). The fundamental idea underlying the definition of the ELDBIT estimator is to combine a consistent estimator that has questionable finite sample properties due to high small sample variability with an estimator that is inconsistent but that has small finite sample variability. The objective of combining these estimations is to achieve a reduction in the SEL measure with respect to both overall parameter estimation precision, $L(\beta, \hat{\beta}) = \|\beta - \hat{\beta}\|^2$, and accuracy of predictions, $L(\mathbf{y}, \hat{\mathbf{y}}) = \|\mathbf{y} - \hat{\mathbf{y}}\|^2$.

2.1   The EL and DBIT Components

When $h = k$, the solutions to either the MEL or the MEEL extremum problems in (1.2) degenerate to the standard IV estimator with $w_i = n^{-1} \forall i$. When $h > k$, the estimating equations–moment conditions over-determine the unknown parameter values and a nontrivial EL solution results that effectively weights the individual observations contained in the sample data set in formulating moments that underlie the resulting estimating equations. The result is a consistent estimator derived from weighted observations of the form identified in (1.2) that represents the empirical moment equations from which the parameter estimates are solved.

As a first step, consider an alternative estimator, which is a variation of the DBIT estimator concept initially proposed by van Akkeren and Judge (1999) and coincides closely with the revised DBIT estimator introduced by van Akkeren et al. (2000) for use in instrumental variable contexts characterized by unbalanced design matrices. In this case, sample moment information for estimating β is used in the form

$$n^{-1}\left[\mathbf{Z}'\mathbf{y} - \mathbf{Z}'\mathbf{X}(n^{-1}\mathbf{X}'\mathbf{X})^{-1}(\mathbf{X} \odot \mathbf{p})^{-}{}'\mathbf{y}\right] = \mathbf{0} \tag{2.1}$$

and

$$\hat{\beta}(\mathbf{p}) = \left(n^{-1}\mathbf{X}'\mathbf{X}\right)^{-1}(\mathbf{X} \odot \mathbf{p})'\mathbf{y} \qquad (2.2)$$

is an estimate of the unknown β vector, $\mathbf{p}$ is an unknown $(n \times 1)$ weighting vector, and $\odot$ is the Hadamard (element-wise) product. The observables $\mathbf{y}$, $\mathbf{Z}$, and $\mathbf{X}$ are expressed in terms of deviations about their means in this formulation (without loss of generality).

Consistent with the moment information in the form of (2.1), and using the K–L information measure, the extremum problem for the DBIT estimator is

$$\min_{\mathbf{p}}\left\{I(\mathbf{p}, \mathbf{q})|n^{-1}\mathbf{Z}'y = n^{-1}\mathbf{Z}'\mathbf{X}\left(n^{-1}\mathbf{X}'\mathbf{X}\right)^{-1}(\mathbf{X} \odot \mathbf{p})'y, \ \mathbf{1}'\mathbf{p} = 1, \ \mathbf{p} \geq \mathbf{0}\right\}$$
$$(2.3)$$

where $I(\mathbf{p}, \mathbf{q}) \equiv \mathbf{p}'\log(\mathbf{p}/\mathbf{q})$ and $\mathbf{q}$ represents the reference distribution for the K–L information measure. In this extremum problem, the objective is to recover the unknown $\mathbf{p}$ and thereby derive the optimal estimate of β, $\hat{\beta}(\hat{\mathbf{p}}) = (n^{-1}\mathbf{X}'\mathbf{X})^{-1} (\mathbf{X} \odot \hat{\mathbf{p}})'y$. Note that, relative to the β parameter space B, the feasible values of the estimator are defined by $(n^{-1}\mathbf{X}'\mathbf{X})^{-1}(\mathbf{X} \odot \mathbf{p})'\mathbf{y} = (n^{-1}\mathbf{X}' \mathbf{X})^{-1} \sum_{i=1}^{n} p_i\mathbf{X}_i'y_i$, for all non-negative choices of the vector $\mathbf{p}$ that satisfy $\mathbf{1}'\mathbf{p} = 1$. The *unconstrained* optimum is achieved when all of the convexity weights, $p_i = 1, \ldots, n$, are drawn to n^{-1}, which represents an empirical probability distribution applied to the sample observations that would be maximally uninformative, and coincides with the standard uniformly distributed empirical distribution function (EDF) weights on the sample observations.

It is evident that there is a tendency to draw the DBIT estimate of β towards the least-squares estimator $(\mathbf{X}'\mathbf{X})^{-1}\mathbf{X}'\mathbf{y}$ which, in the context of nonorthogonality, is a biased and inconsistent estimator. However, the instrument-based moment conditions represent constraining sample information relating to the true value of β, which generally prevents the unconstrained solution, and thus the LS estimator, from being achieved. In effect, the estimator is drawn to the LS estimator only as closely as the sample IV-moment equations will allow, and, under general regularity conditions relating to IV estimation, the limiting solution for $\hat{\beta}$ satisfying the moment conditions $n^{-1}\mathbf{Z}'(\mathbf{y} - \mathbf{X}\hat{\beta}) = \mathbf{0}$ will be the true value of β with probability converging to 1. The principal regularity condition for the consistency of the DBIT estimator, over and above the standard regularity conditions for consistency of the IV estimator, is that the true value of β is contained in the feasible space of DBIT estimator outcomes as $n \to \infty$. See vanAkkeren et al. (2000) for proof details.

2.2   The ELDBIT Estimation Problem

In search of an estimator that has improved finite sampling properties relative to traditional estimators in the nonorthogonal case, and given the concepts and formulations of both the EL and DBIT estimators discussed
above, we define the ELDBIT estimator as the solution to the following
extremum problem

$$\min_{p,w} \mathbf{p}' \ln(\mathbf{p}) + \mathbf{w}' \ln(\mathbf{w}) \tag{2.4}$$

subject to

$$(\mathbf{Z} \odot \mathbf{w})' \mathbf{y} = (\mathbf{Z} \odot \mathbf{w})' \mathbf{X} \beta(\mathbf{p}) \tag{2.5}$$

$$\mathbf{1}'\mathbf{p} = 1, \ \mathbf{p} \geq 0 \quad \text{and} \quad \mathbf{1}'\mathbf{w} = 1, \ \mathbf{w} \geq 0 \tag{2.6}$$

In the above formulation, $\beta(\mathbf{p})$ is a DBIT-type estimator as defined in (2.2),
with $\mathbf{p}$ being an $(n \times 1)$ vector of probability or convexity weights on the
sample outcomes that define the convex weighted combination of support
points defining the parameter values, and $\mathbf{w}$ is an $(n \times 1)$ vector of EL-type
sample weights that are used in defining the empirical instrumental variable-
based moment conditions. The specification of any cross entropy–KL reference distribution is a consideration in practice, and, in the absence of a
priori knowledge to the contrary, a uniform distribution, leading to an
equivalent maximum entropy estimation objective, can be utilized, as is
implicit in the formulation above.

In the solution of the ELDBIT estimation problem, an $(n \times 1)$ vector of
EL-type sample weights, $\mathbf{w}$ will be chosen to effectively transform, with
respect to solutions for the β vector, the over-determined system of empirical moment equations into a just-determined one. Evaluated at the optimal
(or any feasible) solution value of $\mathbf{w}$, the moment equations will then admit a
solution for the value of $\beta(\mathbf{p})$. Choosing support points for the beta values
based on the DBIT estimator, as

$$\beta(\mathbf{p}) = \left(n^{-1} \mathbf{X}' \mathbf{X}\right)^{-1} (\mathbf{X} \odot \mathbf{p})' \mathbf{y} \tag{2.7}$$

allow the LS estimate to be one of the myriad of points in the support space,
and indeed the LS estimate would be defined when each element of $\mathbf{p}$ equals
n^{-1}, corresponding to the unconstrained solution of the estimation objective. Note that if $\mathbf{Z}$ were chosen to equal $\mathbf{X}$, then in fact the LS solution
would be obtained when solving (2.4)–(2.7). The entire space of potential
values of beta is defined by conceptually allowing $\mathbf{p}$ to assume all of its
possible non-negative values, subject to the adding-up constraint that the
$\mathbf{p}$ elements sum to 1. Any value of β that does not reside within the convex
hull defined by vertices consisting of the n component points underlying the

LS estimator, $\beta_{(i)} = (n^{-1}\mathbf{X}'\mathbf{X})\mathbf{X}[i,.]'y[i]$, $i = 1, \ldots, n$, is not in the feasible space of ELDBIT estimates.

2.3 Solutions to The ELDBIT Problem

The estimation problem defined by (2.4)–(2.7) is a nonlinear programming problem having a nonlinear, strictly convex, and differentiable objective function and a compact feasible space defined via a set of nonlinear equality constraints and a set of non-negativity constraints on the elements of $\mathbf{p}$ and $\mathbf{w}$. Therefore, so long as the feasible space is nonempty, a global optimal solution to the estimation problem will exist.

The estimation problem can be expressed in Lagrangian form as

$$L = \mathbf{p}'\ln(\mathbf{p}) + \mathbf{w}'\ln(\mathbf{w}) - \lambda'\left[((\mathbf{Z} \odot \mathbf{w})'\mathbf{y} - (\mathbf{Z} \odot \mathbf{w})'\mathbf{X}\beta(\mathbf{p}))\right]$$
$$- \eta[\mathbf{1}'\mathbf{p} - 1] - \xi[\mathbf{1}'\mathbf{w} - 1] \tag{2.8}$$

where $\beta(\mathbf{p})$ is as defined in (2.7) and the non-negativity conditions are taken as implicit. The first-order necessary conditions relating to $\mathbf{w}$ and $\mathbf{p}$ are given by

$$\frac{\partial L}{\partial \mathbf{p}} = \mathbf{1} = \ln(\mathbf{p}) + (\mathbf{X} \odot \mathbf{y})(n^{-1}\mathbf{X}'\mathbf{X})^{-1}\mathbf{X}'(\mathbf{Z} \odot \mathbf{w})\lambda - \mathbf{1}\eta = \mathbf{0} \tag{2.9}$$

and

$$\frac{\partial L}{\partial \mathbf{w}} = \mathbf{1} + \ln(\mathbf{w}) - \left(\mathbf{Z} \odot \left(\mathbf{y} - \mathbf{X}(n^{-1}\mathbf{X}'\mathbf{X})^{-1}(\mathbf{X} \odot \mathbf{p})'\mathbf{y}\right)\right)\lambda - \mathbf{1}\xi = \mathbf{0} \tag{2.10}$$

The first-order conditions relating to the Lagrange multipliers in (2.8) reproduce the equality constraints of the estimation problem, as usual. The conditions (2.9)–(2.10) imply that the solutions for $\mathbf{p}$ and $\mathbf{w}$ can be expressed as

$$\mathbf{p} = \frac{\exp\left(-(\mathbf{X} \odot \mathbf{y})(n^{-1}\mathbf{X}'\mathbf{X})^{-1}\mathbf{X}'(\mathbf{Z} \odot \mathbf{w})\lambda\right)}{\Omega_\mathbf{p}} \tag{2.11}$$

and

$$\mathbf{w} = \frac{\exp\left(\left(\mathbf{Z} \odot e\left(\mathbf{y} - \mathbf{X}(n^{-1}\mathbf{X}'\mathbf{X})^{-1}(\mathbf{X} \odot \mathbf{p})'\mathbf{y}\right)\right)\lambda\right)}{\Omega_\mathbf{w}} \tag{2.12}$$

where the denominators in (2.11) and (2.12) are defined by

$$\Omega_\mathbf{p} \equiv \mathbf{1}'\exp\left(-(\mathbf{X} \odot \mathbf{y})(n^{-1}\mathbf{X}'\mathbf{X})^{-1}\mathbf{X}'(\mathbf{Z} \odot \mathbf{w})\lambda\right)$$

and

$$\Omega_\mathbf{w} \equiv \mathbf{1}'\exp\left(\left(\mathbf{Z} \odot \left(\mathbf{y} - \mathbf{X}(n^{-1}\mathbf{X}'\mathbf{X})^{-1}(\mathbf{X} \odot \mathbf{p})'\mathbf{y}\right)\right)\lambda\right)$$

As in the similar case of the DBIT estimator (see VanAkkeren and Judge 1999 and Van Akkeren et al. 2000) the first-order conditions characterizing the solution to the constrained minimization problem do not allow a closed-form solution for the elements of either the **p** or the **w** vector, or the Lagrangian multipliers associated with the constraints of the problem. Thus, solutions must be obtained through the use of numerical optimization techniques implemented on a computer.

We emphasize that the optimization problem characterizing the definition of the ELDBIT estimator is well defined in the sense that the strict convexity of the objective function eliminates the need to consider local optima or saddlepoint complications. Our experience in solving thousands of these types of problems using sequential quadratic programming methods based on various quasi-Newton algorithms (CO-constrained optimization application module, Aptech Systems, Maple Valley, Washington) suggests that the solutions are both relatively quick and not difficult to find, despite their seemingly complicated nonlinear structure.

3.   SAMPLING PROPERTIES IN FINITE SAMPLES

In this section we focus on finite sample properties and identify, in the context of a Monte Carlo sampling experiment, estimation and inference properties of the ELDBIT approach. We comment on the asymptotic properties of the ELDBIT approach in the concluding section of this paper.

Since the solution for the optimal weights $\hat{\mathbf{p}}$ and $\hat{\mathbf{w}}$ in the estimation problem framed by equations (2.4)–(2.7) cannot be expressed in closed form, the finite sample properties of the ELDBIT estimator cannot be derived from direct evaluation of the estimator's functional form. In this section the results of Monte Carlo sampling experiments are presented that compare the finite sample performance of competing estimators in terms of the expected squared error loss associated with estimating β and predicting y. Information is also provided relating to (i) average bias and variances and (ii) inference performance as it relates to confidence interval coverage (and test size via duality) and power. While these results are, of course, specific to the collection of particular Monte Carlo experiments analyzed, the computational evidence does provide an indication of relative performance gains that are possible over a range of correlation scenarios characterizing varying degrees of nonorthogonality in the model, as well as for alternative interrelationships among instrumental variables and between instrumental variables and the explanatory variable responsible for nonorthogonality.

3.1 Experimental Design

Consider a sampling process of the following form:

$$y_{i1} = z_{i1}\beta_1 + y_{i2}\beta_2 + e_i = \mathbf{x}_i' \boldsymbol{\beta} + \varepsilon_i \tag{3.1}$$

$$y_{i2} = \pi_0 + \sum_{j=1}^{4} \pi_j z_{ij} + v_i \tag{3.2}$$

where $\mathbf{x}_i = (z_{i1}, y_{i2})'$ and $i = 1, 2, \ldots, n$. The two-dimensional vector of unknown parameters, $\boldsymbol{\beta}$, in (3.1) is arbitrarily set equal to the vector $[-1, 2]'$. The outcomes of the (6×1) random vector $[y_{i2}, \varepsilon_i, z_{i1}, z_{i2}, z_{i3}, z_{i4}]$ are generated *iid* from a multivariate normal distribution having a zero mean vector, unit variances, and under various conditions relating to the correlations existent among the five scalar random variables. The values of the π_js in (3.2) are clearly determined by the regression function between y_{i2} and $[z_{i1}, z_{i2}, z_{i3}, z_{i4}]$, which is itself a function of the covariance specification relating to the marginal normal distribution associated with the (5×1) random vector $[y_{i2}, z_{i1}, z_{i2}, z_{i3}, z_{i4}]$. Thus the π_js generally change as the scenario postulated for the correlation matrix of the sampling process changes. In this sampling design, the outcomes of $[y_{i1}, v_i]$ are then calculated by applying equations (3.1)–(3.2) to the outcomes of $[y_{i2}, \varepsilon_i, z_{i1}, z_{i2}, z_{i3}, z_{i4}]$.

Regarding the details of the sampling scenarios simulated for this set of Monte Carlo experiments, the focus was on sample sizes of $n = 20$ and $n = 50$. Thus the experiments were deliberately designed to investigate the behavior of estimation methods in small samples. The outcomes of ε_i were generated independently of the vector $[z_{i1}, z_{i2}, z_{i3}, z_{i4}]$ so that the correlations between ε_i and the z_{ij}s were zeros, thus fulfilling one condition for $[z_{i1}, z_{i2}, z_{i3}, z_{i4}]$ to be considered a set of valid instrumental variables for estimating the unknown parameters in (3.1). Regarding the degree of nonorthogonality existing in (3.1), correlations of .25, .50, and .75 between the random variables y_{i2} and ε_i were utilized to simulate weak, moderate, and relatively strong correlation/nonorthogonality relationships between the explanatory variable y_{i2} and the model disturbance ε_i .

For each sample size, and for each level of correlation between y_{i2} and ε_i, four scenarios were examined relating to both the degree of correlation existent between the instruments and the y_{i2} variable, and the levels of collinearity existent among the instrumental variables. Specifically, the pairwise correlation between y_{i2} and each instrument in the collection $[z_{i1}, z_{i2}, z_{i3}, z_{i4}]$ was set equal to .3, .4, .5, and .6, respectively, while the corresponding pairwise correlations among the four instruments were each set equal to 0, .3, .6, and .9. Thus, the scenarios range from relatively weak but independent instruments (.3, 0) to stronger but highly collinear instruments (.6, .9). All told, there were 24 combinations of sample sizes, y_{i2} and ε_i correlations, and instrument collinearity examined in the experiments.

The reported sampling properties relating to estimation objectives are based on 2000 Monte Carlo repetitions, and include estimates of the empirical risks, based on a SEL measure, associated with estimating β with $\hat{\beta}$ (parameter estimation risk) and estimating y with $\hat{y}$ (predictive risk). We also report the average estimated bias in the estimates, Bias $(\beta) = E[\hat{\beta}] - \beta$, and the average estimated variances of the estimates, Var $(\hat{\beta}_i)$.

The reported sampling properties relating to inference objectives are based on 1000 Monte Carlo repetitions, the lower number of repetitions being due to the substantially increased computational burden of jackknifing the inference procedures (discussed further in the next subsection). We report the actual estimated coverage probabilities of confidence intervals intended to provide .99 coverage of the true value of β_2 (and, by duality, intended to provide a size .01 test of the true null hypothesis $H_0 : \beta_2 = 2$), and also include the estimated power in rejecting the false null hypothesis $H_0 : \beta_2 = 1$. The statistic used in defining the confidence intervals and tests is the standard T-ratio, $T = (\hat{\beta}_2 - \beta_2^0)/\text{std}(\hat{\beta}_2)$. As discussed ahead, the estimator $\hat{\beta}$ in the numerator of the test statistic is subject to a jackknife bias correction, and the denominator standard deviation is the jackknife estimate of the standard deviation. Critical values for the confidence intervals and tests are set conservatively based on the T–distribution, rather than on the asymptotically valid standard normal distribution.

Three estimators were examined, specifically, the GMM estimator based on the asymptotically optimal GMM weighting matrix (GMM-2SLS), GMM based on an identity matrix weighting (GMM-I), and the ELDBIT estimator. Estimator outcomes were obtained using GAUSS software, and the ELDBIT solutions were calculated using the GAUSS constrained optimization application module provided by Aptech Systems, Maple Valley, Washington. The estimator sampling results are displayed in Appendix Tables A.1 and A.2 for sample sizes $n = 20$ and $n = 50$, respectively. A summary of the characteristics of each of the 12 different correlation scenarios defining the Monte Carlo experiments is provided in Table 3.1.

We note that other MC experiments were conducted to observe sampling behavior from non-normal, thicker-tailed sampling distributions (in particular, sampling was based on a multivariate T distribution with four degrees of freedom), but the numerical details are not reported here. These estimation and inference results will be summarized briefly ahead.

Table 3.1. Monte Carlo experiment definitions, with $\beta = [-1, 2]'$ and $\sigma_{e_i}^2 = \sigma_{Y_{2i}}^2 = \sigma_{Z_{ij}}^2 = 1$, $\forall i$ and j

Experiment number	ρ_{y_{2i}, e_i}	$\rho_{y_{2i}, x_{ij}}$	$\rho_{z_{ij}, z_{ik}}$	$R^2_{\mathbf{Y}_1, \hat{\mathbf{Y}}_1}$	Condition number of Cov(**Z**)
1	.75	.3	0	.95	1.0
2	.75	.4	.3	.95	2.7
3	.75	.5	.6	.96	7.0
4	.75	.6	.9	.98	37.0
5	.50	.3	0	.89	1.0
6	.50	.4	.3	.89	2.7
7	.50	.5	.6	.89	7.0
8	.50	.6	.9	.89	37.0
9	.25	.3	0	.84	1.0
10	.25	.4	.3	.83	2.7
11	.25	.5	.6	.82	7.0
12	.25	.6	.9	.80	37.0

Note: ρ_{y_{2i}, e_i} denotes the correlation between Y_{2i} and e_i and measures the degree of nonorthogonality; $\rho_{y_{2i}, z_{ij}}$ denotes the common correlation between Y_{2i} and each of the four instrumental variables, the Z_{ij}s; $\rho_{z_{ij}, z_{ik}}$ denotes the common correlation between the four instrumental variables; and $R^2_{\mathbf{Y}_1, \hat{\mathbf{Y}}_1}$ denotes the population squared correlation between $\mathbf{Y}_1$ and $\hat{\mathbf{Y}}_1$.

3.2 Jackknifing to Mitigate Bias in Inference and Estimate ELDBIT Variance

A jackknife resampling procedure was incorporated into the ELDBIT approach when pursuing inference objectives (but not for estimation) in order to mitigate bias, estimate variance, and thereby attempt to correct confidence interval coverage (and test size via duality). In order for the GMM-2SLS and GMM-I estimators to be placed on an equal footing for purposes of comparison with the ELDBIT results, jackknife bias and standard deviation estimates were also applied when inference was conducted in the GMM-2SLS and GMM-I contexts. We also computed MC inference results (confidence intervals and hypothesis tests) based on the standard traditional uncorrected method; while the details are not reported here, we discuss these briefly ahead.

Regarding the specifics of the jackknifing procedure, n "leave one observation out" solutions were calculated for each of the estimators and for each simulated Monte Carlo sample outcome. Letting $\hat{\theta}_{(i)}$ denote the ith jacknifed estimator outcome and $\hat{\theta}$ denote the estimate of the parameter vector based on the full sample of n observations, the jackknife estimate of the bias vector, $E[\hat{\theta}] - \theta$, was calculated as $\text{bias}_{\text{jack}} = (n-1)(\hat{\theta}_{(\cdot)} - \hat{\theta})$, where $\hat{\theta}_{(\cdot)} = n^{-1} \sum_{i=1}^{n} \hat{\theta}_{(i)}$. The corresponding jackknife estimate of the standard deviations of the estimator vector elements was calculated in the usual way based on $\text{std}_{\text{jack}} = [(\{n-1\}/n) \sum_{i=1}^{n} (\hat{\theta}_{(i)} - \hat{\theta}_{(\cdot)})^2]^{1/2}$. Then when forming the standard pivotal quantity, $T = (\hat{\beta}_2 - \beta_2^0)/\text{std}(\hat{\beta}_2)$, underlying the confidence interval estimators for β_2 (and test statistics, via duality), the estimators in the numerator were adjusted (via subtraction) by the bias estimate, and the jackknifed standard deviation was used in the denominator in place of the usual asymptotically valid expressions used to approximate the finite sample standard deviations.

We note that one could contemplate using a "delete-d" jackknifing procedure, $d > 1$, in place of the "delete-1" jackknife that was used here. The intent of the delete-d jackknife would be to improve the accuracy of the bias and variance estimates, and thereby improve the precision of inferences in the correction mechanism delineated above. However, the computational burden quickly escalates in the sense that $(n!/\{d!(n-d)!\})$ recalculations of the ELDBIT, 2SLS, or GMM estimators are required. One could also consider bootstrapping to obtain improved estimates of bias and variance, although it would be expected that notably more than n resamples would be required to achieve improved accuracy. This issue remains an interesting one for future work. Additional useful references on the use of the jackknife in the context of IV estimation include Angrist et al.(1999) and Blomquist and Dahlberg (1999).

3.3   MC Sampling Results

The sampling results for the 24 scenarios, based on sampling from the multivariate normal distribution and reported in Appendix Tables A.1–A.2, provide the basis for Figures 3.1–3.4 and the corresponding discussion.

Estimator bias, variance, and parameter estimation risk

Regarding the estimation objective, the following general conclusions can be derived from the simulated estimator outcomes presented in the appendix tables and Figure 3.1:

(1)   ELDBIT dominates both GMM-2SLS and GMM-I in terms of parameter estimation risk for all but one case (Figure 3.1 and Tables A.1–A.2).

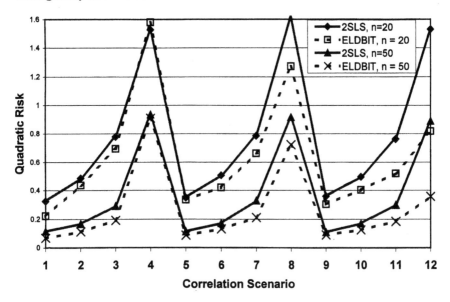

Figure 3.1. ELDBIT vs GMM-2SLS parameter estimation risk.

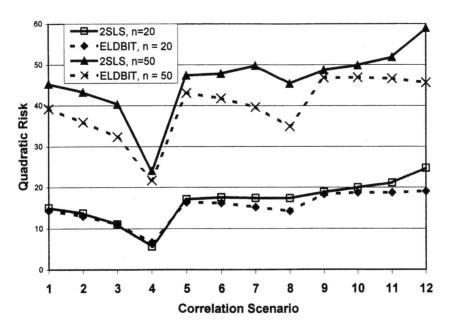

Figure 3.2. ELDBIT vs GMM-2SLS predictive risk.

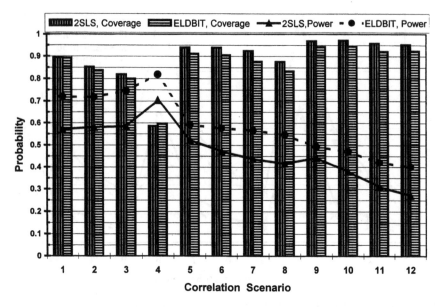

Figure 3.3.   Coverage (target = .99) and power probabilities, $n = 20$.

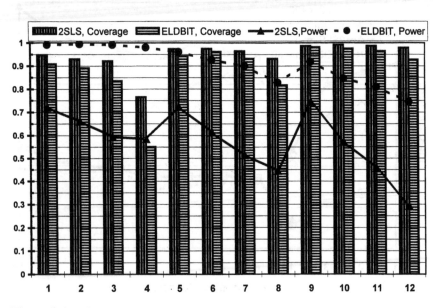

Figure 3.4.   Coverage (target = .99) and power probabilities, $n = 50$.

(2) ELDBIT is most often more biased than either GMM-2SLS or GMM-I, although the magnitudes of the differences in bias are in most cases not large relative to the magnitudes of the parameters being estimated.

(3) ELDBIT has lower variance than either GMM-2SLS or GMM-I in all but a very few number of cases, and the magnitude of the decrease in variance is often relatively large.

Comparing across all of the parameter simulation results, ELDBIT is strongly favored relative to more traditional competitors. The use of ELDBIT involves trading generally a small amount of bias for relatively large reductions in variance. This is, of course, precisely the sampling results desired for the ELDBIT at the outset.

Empirical predictive risk

The empirical predictive risks for the GMM-2SLS and ELDBIT estimators for samples of size 20 and 50 are summarized in Figure 3.2; detailed results, along with the corresponding empirical predictive risks for the GMM-I estimator, are given in Tables A.1–A.2. The ELDBIT estimator is superior in predictive risk to both the GMM-2SLS and GMM-I estimators in all but two of the possible 24 comparisons, and in one case where dominance fails, the difference is very slight (ELDBIT is uniformly superior in prediction risk to the unrestricted reduced form estimates as well, which are not examined in detail here). The ELDBIT estimator achieved as much as a 25% reduction in prediction risk relative to the 2SLS estimator.

Coverage and power results

The sampling results for the jackknifed ELDBIT and GMM-2SLS estimates, relating to coverage probability/test size and power, are summarized in Figures 3.3 and 3.4. Results for the GMM-I estimator are not shown, but were similar to the GMM-2SLS results. The inference comparison was in terms of the proximity of actual confidence interval coverage to the nominal target coverage and, via duality, the proximity of the actual test size for the hypothesis $H_0 : \beta_2 = 2$ to the nominal target test size of .01. In this comparison the GMM-2SLS estimator was closer to the .99 and .01 target levels of coverage and test size, respectively. However, in terms of coverage/test size, in most cases the performance of the ELDBIT estimator was a close competitor and also exhibited a decreased confidence interval width.

It is apparent from the figures that all of the estimators experienced difficulty in achieving coverage/test size that was close to the target level when nonorthogonality was high (scenarios 1–4). This was especially evident when sample size was at the small level of $n = 20$ and there was high multicollinearity among the instrumental variables (scenario 4). In terms of the test power observations, the ELDBIT estimator was clearly superior to GMM-2SLS, and usually by a wide margin.

We note that the remarks regarding the superiority of GMM-2SLS and GMM-I relative to ELDBIT in terms of coverage probabilities and test size would be substantially altered, and generally reversed, if comparisons had been made between the ELDBIT results, with jackknife bias correction and standard error computation, and the traditional GMM-2SLS and GMM-I procedure based on uncorrected estimates and asymptotic variance expressions. That is, upon comparing ELDBIT with the *traditional* competing estimators in this empirical experiment, ELDBIT is the superior approach in terms of coverage probabilities and test sizes. Thus, in this type of contest, ELDBIT is generally superior in terms of the dimensions of performance considered here, including parameter estimation, predictive risk, confidence interval coverage, and test performance. This of course only reinforces the advisability of pursuing some form of bias correction and alternative standard error calculation method when basing inferences on the GMM-2SLS or GMM-I approach in small samples and/or in contexts of ill-conditioning.

Sampling results—Multivariate $T_{(4)}$

To investigate the impact on sampling performance of observations obtained from distributions with thicker tails than the normal, sampling results for a multivariate $T_{(4)}$ sampling distribution were developed. As is seen from Figure 3.5, over the full range of correlation scenarios the parameter estimation risk comparisons mirrored those for the normal distribution case, and in many cases the advantage of ELDBIT over GMM-2SLS (and GMM-I) was equal to or greater than those under normal sampling. These sampling results, along with the corresponding prediction risk performance (not shown here), illustrate the robust nature of the ELDBIT estimator when sampling from heavier-tailed distributions.

Using the jackknifing procedure to mitigate finite sample bias and improve finite sample estimates of standard errors, under $T_{(4)}$ sampling, each of the estimators performed reasonably well in terms of interval coverage and implied test size. As before, the power probabilities of the ELDBIT estimator were uniformly superior to those of the GMM-2SLS estimator. All estimators experienced difficulty in achieving coverage and test size for correlation scenarios 1–4 corresponding to a high level of nonorthogonality. This result was particularly evident in the case of small sample size 20 and high multicollinearity among the instrumental variables.

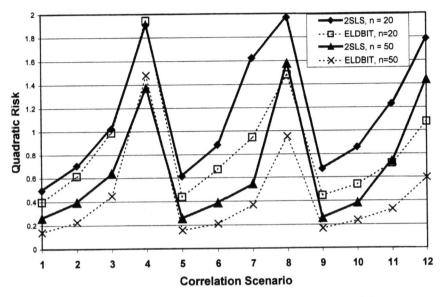

Figure 3.5.   Multivariate T sampling: parameter estimation risk.

4. CONCLUDING REMARKS

In the context of a statistical model involving nonorthogonality of explanatory variables and the noise component, we have demonstrated a semiparametric estimator that is completely data based and that has excellent finite
sample properties relative to traditional competing estimation methods. In
particular, using empirical likelihood-type weighting of sample observations
and a squared error loss measure, the new estimator has attractive estimator
precision and predictive fit properties in small samples. Furthermore, inference performance is comparable in test size and confidence interval coverage
to traditional approaches while exhibiting the potential for increased test
power and decreased confidence interval width. In addition, the ELDBIT
estimator exhibits robustness with respect to both ill-conditioning implied
by highly correlated covariates and instruments, and sample outcomes from
non-normal, thicker-tailed sampling processes. Moreover, it is not difficult
to implement computationally. In terms of estimation and inference, if one
is willing to trade off limited finite sample bias for notable reductions in
variance, then the benefits noted above are achievable.

There are a number of ways the ELDBIT formulation can be extended
that may result in an even more effective estimator. For example, higher-
order moment information could be introduced into the constraint set and,
when valid, could contribute to greater estimation and predictive efficiency.

Along these lines, one might consider incorporating information relating to heteroskedastic or autocorrelated covariance structures in the specification of the moment equations if the parameterization of these structures were sufficiently well known (see Heyde (1997) for discussion of moment equations incorporating heteroskedastic or autocorrelated components). If non-sample information is available relating to the relative plausibility of the parameter coordinates used to identify the ELDBIT estimator feasible space, this information could be introduced into the estimation objective function through the reference distribution of the Kullback–Leibler information, or cross entropy, measure.

One of the most intriguing possible extensions of the ELDBIT approach, given the sampling results observed in this study, is the introduction of differential weighting of the IV-moment equations and the DBIT estimator components of the ELDBIT estimation objective function. It is conceivable that, by tilting the estimation objective toward or away from one or the other of these components, and thereby placing more or less emphasis on one or the other of the two sources of data information that contribute to the ELDBIT estimate, the sampling properties of the estimator could be improved over the balanced (equally weighted) criterion used here. For example, in cases where violations in the orthogonality of explanatory variables and disturbances are not substantial, it may be beneficial to weight the ELDBIT estimation objective more towards the LS information elements inherent in the DBIT component and rely less heavily on the IV component of information. In fact, we are now conducting research on such an alternative in terms of a weighted ELDBIT (WELDBIT) approach, where the weighting is adaptively generated from data information, with very promising SEL performance.

To achieve post-data *marginal*, as opposed to *joint*, weighting densities on the parameter coordinates that we have used to define the ELDBIT estimator feasible space, an $(n \times k)$ matrix of probability or convexity weights, P, could be implemented. For example, one $(n \times 1)$ vector could be used for each element of the parameter vector, as opposed to using only one $(n \times 1)$ vector of convexity weights as we have done. In addition, alternative data-based definitions of the vertices underlying the convex hull definition of $\beta(\mathbf{p})$ could also be pursued.

More computationally intensive bootstrapping procedures could be employed that might provide more effective bias correction and standard deviation estimates. The bootstrap should improve the accuracy of hypothesis testing and confidence interval procedures beyond the improvements achieved by the delete-1 jackknifing procedures we have employed.

Each of the above extensions could have a significant effect on the solved values of the $\mathbf{p}$ and $\mathbf{w}$ distributions and ultimately on the parameter estimate, $\beta(\hat{\mathbf{p}})$, produced by the ELDBIT procedure. All of these possibilities provide an interesting basis for future research.

We note a potential solution complication that, although not hampering the analysis of the Monte Carlo experiments in this study, is at least a conceptual possibility in practice. Specifically, the solution space for β in the current ELDBIT specification is not R^k, but rather a convex subset of R^k defined by the convex hull of the vertex points $\beta_{(i)} = (n^{-1}X'X)^{-1}X[i, .]'y[i]$, $i = 1, \ldots, n$. Consequently, it is possible that there does not exist a choice of β within this convex hull that also solves the empirical moment constraints $(Z \odot w)'y = (Z \odot w)'X\beta$ for some choice of feasible w within the simplex $1'w = 1$, $w \geq 0$. A remedy for this potential difficulty will generally be available by simply altering the definition of the feasible space of ELDBIT estimates to $\beta(p, \tau) = \tau(n^{-1}X'X)^{-1}(X \odot p)'y$, where $\tau \geq 1$ is a scaling factor (recall that data in the current formulation is measured in terms of deviations about sample means). Expanding the scaling, and thus the convex hull, appropriately will generally lead to a non-null intersection of the two constraint sets on β. The choice of τ could be automated by including it as a choice variable in the optimization problem.

The focus of this paper has been on small sample behavior of estimators with the specific objective of seeking improvements over traditional approaches in this regard. However, we should note that, based on limited sampling results for samples of size 100 and 250, the superiority of ELDBIT relative to GMM-2SLS and GMM-I, although somewhat abated in magnitude, continues to hold in general. Because all empirical work is based on sample sizes an infinite distance to the left of infinity, we would expect that ELDBIT would be a useful alternative estimator to consider in many empirical applications.

Regarding the asymptotics of ELDBIT, there is nothing inherent in the definition of the estimator that guarantees its consistency, even if the moment conditions underlying its specification satisfy the usual IV regularity conditions. For large samples the ELDBIT estimator emulates the sampling behavior of a large sum of random vectors, as $\sum_{i=1}^{n}(n^{-1}X'X)^{-1}p[i]X[i,]'y[i] \stackrel{a}{=} Q^{-1}\sum_{i=1}^{n}p[i]X[i, .]'y[i] = Q^{-1}\sum_{i=1}^{n}v_i$, where $\text{plim}(n^{-1}X'X) = Q$. Consequently, sufficient regularity conditions relating to the sampling process could be imagined that would result in estimator convergence to an asymptotic mean vector other than β, with a scaled version of the estimator exhibiting an asymptotic normal distribution.

Considerations of optimal asymptotic performance have held the high ground in general empirical practice over the last half century whenever estimation and inference procedures were considered for addressing the nonorthogonality of regressors and noise components in econometric analyses. All applied econometric work involves finite samples. Perhaps it is time to reconsider the opportunity cost of our past emphasis on infinity.

APPENDIX

Table A.1. ELDBIT, GMM-2SLS, and GMM-I sampling performance with $\beta = [-1, 2]'$, $n = 20$, and 2000 Monte Carlo repetitions

$\rho y_{2i}, e_i$	$\rho y_{2i}, z_{ij}$	$\rho z_{ij}, z_{ik}$	Sampling property	Estimators		
				GMM-2SLS	GMM-I	ELDBIT
.75	.3	0	SEL($\hat{\beta}, \beta$), SEL($\hat{y}, y$)	.33, 14.97	.39, 16.21	.22, 14.32
			bias($\hat{\beta}_1$), bias($\hat{\beta}_2$)	−.05, .17	−.04, .14	−.05, .18
			var($\hat{\beta}_1$), var($\hat{\beta}_2$)	.07, .22	.09, .28	.07, .12
	.4	.3	SEL($\hat{\beta}, \beta$), SEL($\hat{y}, y$)	.49, 13.62	.59, 16.55	.44, 12.97
			bias($\hat{\beta}_1$), bias($\hat{\beta}_2$)	−.12, .30	−.13, .20	−.13, .32
			var($\hat{\beta}_1$), var($\hat{\beta}_2$)	.09, .29	.13, .40	.09, .23
	.5	.6	SEL($\hat{\beta}, \beta$), SEL($\hat{y}, y$)	.78, 11.00	.90, 14.25	.69, 11.04
			bias($\hat{\beta}_1$), bias($\hat{\beta}_2$)	−.24, .49	−.25, .39	−.23, .47
			var($\hat{\beta}_1$), var($\hat{\beta}_2$)	.12, .36	.17, .51	.12, .30
	.6	.9	SEL($\hat{\beta}, \beta$), SEL($\hat{y}, y$)	1.53, 5.59	1.61, 6.80	1.58, 6.65
			bias($\hat{\beta}_1$), bias($\hat{\beta}_2$)	−.53, .90	−.54, .86	−.53, .88
			var($\hat{\beta}_1$), var($\hat{\beta}_2$)	.13, .31	.17, .41	.16, .37
.50	.3	0	SEL($\hat{\beta}, \beta$), SEL($\hat{y}, y$)	.35, 17.08	.38, 17.60	.33, 16.40
			bias($\hat{\beta}_1$), bias($\hat{\beta}_2$)	−0.5, .15	−.04, .14	−.06, .21
			var($\hat{\beta}_1$), var($\hat{\beta}_2$)	.09, .24	.10, .26	.09, .20
	.4	.3	SEL($\hat{\beta}, \beta$), SEL($\hat{y}, y$)	.51, 17.54	.55, 19.13	.42, 16.17
			bias($\hat{\beta}_1$), bias($\hat{\beta}_2$)	−.09, .21	−.09, .13	−.11, .28
			var($\hat{\beta}_1$), var($\hat{\beta}_2$)	.12, .33	.14, .38	.10, .23
	.5	.6	SEL($\hat{\beta}, \beta$), SEL($\hat{y}, y$)	.79, 17.34	.86, 18.95	.56, 15.16
			bias($\hat{\beta}_1$), bias($\hat{\beta}_2$)	−.17, .34	−.18, .27	−.20, .42
			var($\hat{\beta}_1$), var($\hat{\beta}_2$)	.18, .46	.22, .54	.14, .30
	.6	.9	SEL($\hat{\beta}, \beta$), SEL($\hat{y}, y$)	1.62, 17.32	1.75, 18.65	1.27, 14.21
			bias($\hat{\beta}_1$), bias($\hat{\beta}_2$)	−.37, .63	−.38, .60	−.41, .69
			var($\hat{\beta}_1$), var($\hat{\beta}_2$)	.34, .74	.39, .86	.21, .41
.25	.3	0	SEL($\hat{\beta}, \beta$), SEL($\hat{y}, y$)	.36, 18.79	.40, 19.26	.30, 18.30
			bias($\hat{\beta}_1$), bias($\hat{\beta}_2$)	−.03, .09	−.04, .09	−.04, .14
			var($\hat{\beta}_1$), var($\hat{\beta}_2$)	.09, .25	.10, .28	.09, .19
	.4	.3	SEL($\hat{\beta}, \beta$), SEL($\hat{y}, y$)	.50, 19.92	.58, 21.09	.40, 18.70
			bias($\hat{\beta}_1$), bias($\hat{\beta}_2$)	−.05, .11	−.05, .08	−.07, .18
			var($\hat{\beta}_1$), var($\hat{\beta}_2$)	.13, .35	.15, .42	.12, .25
	.5	.6	SEL($\hat{\beta}, \beta$), SEL($\hat{y}, y$)	.76, 21.10	.92, 22.4	.52, 18.73
			bias($\hat{\beta}_1$), bias($\hat{\beta}_2$)	−.06, .13	−.07, .09	−.11, .21
			var($\hat{\beta}_1$), var($\hat{\beta}_2$)	.20, .54	.24, .67	.16, .30
	.6	.9	SEL($\hat{\beta}, \beta$), SEL($\hat{y}, y$)	1.53, 24.75	1.65, 25.84	.82, 19.03
			bias($\hat{\beta}_1$), bias($\hat{\beta}_2$)	−.19, .32	−.19, .31	−.22, .36
			var($\hat{\beta}_1$), var($\hat{\beta}_2$)	.42, .97	.46, 1.06	.23, .40

Table A.2. ELDBIT, GMM-2SLS, and GMM-I sampling performance, with $\beta = [-1, 2]'$, $n = 50$, and 2000 Monte Carlo repetitions

$\rho y_{2i}, e_i$	$\rho y_{2i}, z_{ij}$	$\rho z_{ij}, z_{ik}$	Sampling property	Estimators		
				GMM-2SLS	GMM-I	ELDBIT
.75	.3	0	SEL($\hat{\beta}, \beta$), SEL($\hat{y}, y$)	.11, 45.33	.12, 45.70	.07, 39.18
			bias($\hat{\beta}_1$), bias($\hat{\beta}_2$)	−.02, .06	−.02, .06	−.03, .13
			var($\hat{\beta}_1$), var($\hat{\beta}_2$)	.03, .08	.03, .09	.02, .03
	.4	.3	SEL($\hat{\beta}, \beta$), SEL($\hat{y}, y$)	.17, 43.31	.18, 47.48	.11, 35.91
			bias($\hat{\beta}_1$), bias($\hat{\beta}_2$)	−.04, .11	−.04, .06	−.07, .20
			var($\hat{\beta}_1$), var($\hat{\beta}_2$)	.04, .11	.05, .13	.02, .04
	.5	.6	SEL($\hat{\beta}, \beta$), SEL($\hat{y}, y$)	.29, 40.29	.32, 44.91	.19, 32.30
			bias($\hat{\beta}_1$), bias($\hat{\beta}_2$)	−.10, .19	−.10, .13	−.14, .28
			var($\hat{\beta}_1$), var($\hat{\beta}_2$)	.06, .18	.07, .21	.03, .06
	.6	.9	SEL($\hat{\beta}, \beta$), SEL($\hat{y}, y$)	.94, 24.05	.95, 26.01	.91, 21.69
			bias($\hat{\beta}_1$), bias($\hat{\beta}_2$)	−.36, .61	−.37, .57	−.38, .64
			var($\hat{\beta}_1$), var($\hat{\beta}_2$)	.13, .31	.14, .34	.10, .26
.50	.3	0	SEL($\hat{\beta}, \beta$), SEL($\hat{y}, y$)	.11, 47.42	.12, 47.56	.09, 43.15
			bias($\hat{\beta}_1$), bias($\hat{\beta}_2$)	−.02, .05	−.02, .04	−.04, .13
			var($\hat{\beta}_1$), var($\hat{\beta}_2$)	.03, .08	.03, .08	.02, .05
	.4	.3	SEL($\hat{\beta}, \beta$), SEL($\hat{y}, y$)	.17, 47.82	.19, 50.11	.13, 41.80
			bias($\hat{\beta}_1$), bias($\hat{\beta}_2$)	−.03, .07	−.03, .03	−.07, .18
			var($\hat{\beta}_1$), var($\hat{\beta}_2$)	.04, .13	.04, .14	.03, .06
	.5	.6	SEL($\hat{\beta}, \beta$), SEL($\hat{y}, y$)	.33, 49.76	.35, 52.14	.21, 39.67
			bias($\hat{\beta}_1$), bias($\hat{\beta}_2$)	−.05, .10	−.05, .05	−.13, .26
			var($\hat{\beta}_1$), var($\hat{\beta}_2$)	.08, .24	.08, .26	.04, .09
	.6	.9	SEL($\hat{\beta}, \beta$), SEL($\hat{y}, y$)	.92, 45.40	.93, 46.55	.72, 34.91
			bias($\hat{\beta}_1$), bias($\hat{\beta}_2$)	−.26, .43	−.26, .41	−.35, .58
			var($\hat{\beta}_1$), var($\hat{\beta}_2$)	.19, .47	.20, .49	.08, .18
.25	.3	0	SEL($\hat{\beta}, \beta$), SEL($\hat{y}, y$)	.11, 48.71	.12, 48.95	.09, 46.88
			bias($\hat{\beta}_1$), bias($\hat{\beta}_2$)	−.01, .03	−.01, .03	−.03, .09
			var($\hat{\beta}_1$), var($\hat{\beta}_2$)	.03, .08	.03, .08	.03, .05
	.4	.3	SEL($\hat{\beta}, \beta$), SEL($\hat{y}, y$)	.17, 49.85	.18, 50.61	.13, 46.90
			bias($\hat{\beta}_1$), bias($\hat{\beta}_2$)	−.02, .04	−.02, .02	−.05, .12
			var($\hat{\beta}_1$), var($\hat{\beta}_2$)	.04, .13	.04, .14	.04, .08
	.5	.6	SEL($\hat{\beta}, \beta$), SEL($\hat{y}, y$)	.30, 51.81	.32, 52.95	.18, 46.65
			bias($\hat{\beta}_1$), bias($\hat{\beta}_2$)	−.04, .07	−.03, .04	−.09, .18
			var($\hat{\beta}_1$), var($\hat{\beta}_2$)	.08, .21	.09, .23	.05, .09
	.6	.9	SEL($\hat{\beta}, \beta$), SEL($\hat{y}, y$)	.89, 58.96	.96, 60.65	.36, 45.63
			bias($\hat{\beta}_1$), bias($\hat{\beta}_2$)	−.13, .21	−.13, .21	−.20, .33
			var($\hat{\beta}_1$), var($\hat{\beta}_2$)	.23, .60	.25, .65	.07, .14

REFERENCES

Angrist, J. D., Imbens, G. W., and Krueger, A. B. Jackknifing instrumental variables estimation. *Journal of Applied Econometrics*, 14:57–67, 1999.

Baggerly, K. A. Empirical likelihood as a goodness of fit measure. *Biometrika*, 85: 535–547, 1998.

Blomquist, S., and Dahlberg, M. Small sample properties of LIML and jackknife IV estimators: Experiments with weak instruments. *Journal of Applied Econometrics, 14:69–88, 1999.*

Bronfenbrenner, J. Sources and size of least-squares bias in a two-equation model. In: W.C. Hood and T.C. Koopmans, eds. *Studies in Econometric Method.* New York: John Wiley, 1953, pp. 221–235.

Cressie, N., and Read, T. Multinomial goodness of fit tests. *Journal of the Royal Statistical Society, Series B* 46:440–464, 1984.

DiCiccio, T., Hall, P., and Romano, J. Empirical likelihood is Bartlett-correctable. *Annals of Statistics* 19: 1053–1061, 1991.

Godambe, V. An optimum property of regular maximum likelihood estimation. *Annals of Mathematical Statistics* 31: 1208–1212, 1960.

Hansen, L. P. Large sample properties of generalized method of moment estimates. *Econometrica* 50:1029–1054, 1982.

Heyde, C. *Quasi-Likelihood and its Application.* New York: Springer-Verlag, 1997.

Heyde, C., and Morton, R. Multiple roots in general estimating equations. *Biometrika* 85(4):954–959, 1998.

Hurwicz, L. Least squares bias in time series. In: T. Koopmans, ed. *Statistical Inference in Dynamic Economic Models.* New York: John Wiley, 1950.

Imbens, G. W., Spady, R. H. and Johnson, P. Information theoretic approaches to inference in moment condition models. *Econometrica* 66:333–357, 1998.

Mittelhammer, R., Judge, G., and Miller, D. *Econometric Foundations.* New York: Cambridge University Press, 2000.

Murphy, A., and Van Der Vaart, A. On profile likelihood. *Journal of the American Statistical Association* 95:449–485, 2000.

Owen, A. Empirical likelihood ratio confidence intervals for a single functional. *Biometrika* 75:237–249, 1988.

Owen, A. Empirical likelihood for linear models. *Annals of Statistics* 19(4): 1725–1747, 1991.

Owen, A. *Empirical Likelihood.* New York: Chapman and Hall, 2001.

Qin, J., and Lawless, J. Empirical likelihood and general estimating equations. *Annals of Statistics* 22(1): 300–325, 1994.

Read, T. R., and Cressie, N. A. *Goodness of Fit Statistics for Discrete Multivariate Data.* New York: Springer-Verlag, 1988.

van Akkeren, M., and Judge, G. G. Extended empirical likelihood estimation and inference. Working paper, University of California, Berkeley, 1999, pp. 1–49.

van Akkeren, M., Judge, G. G., and Mittelhammer, R.C. Generalized moment based estimation and inference. *Journal of Econometrics* (in press), 2001.

18

Testing for Two-Step Granger Noncausality in Trivariate VAR Models

JUDITH A. GILES   University of Victoria, Victoria, British Columbia, Canada

I. INTRODUCTION

Granger's (1969) popular concept of causality, based on work by Weiner (1956), is typically defined in terms of predictability for one period ahead. Recently, Dufour and Renault (1998) generalized the concept to causality at a given horizon h, and causality up to horizon h, where h is a positive integer that can be infinite ($1 \leq h < \infty$); see also Sims (1980), Hsiao (1982), and Lütkepohl (1993a) for related work. They show that the horizon h is important when auxiliary variables are available in the information set that are not directly involved in the noncausality test, as causality may arise more than one period ahead indirectly via these auxiliary variables, even when there is one period ahead noncausality in the traditional sense. For instance, suppose we wish to test for Granger noncausality (GNC) from Y to X with an information set consisting of three variables, X, Y and Z, and suppose that Y does not Granger-cause X, in the traditional one-step sense. This does not preclude two-step Granger causality, which will arise when Y Granger-causes Z and Z Granger-causes X; the auxiliary variable Z enables predictability to result two periods ahead. Consequently, it is important to examine for causality at horizons beyond one period when the information set contains variables that are not directly involved in the GNC test.

Dufour and Renault (1998) do not provide information on testing for GNC when $h > 1$; our aim is to contribute in this direction. In this chapter we provide an initial investigation of testing for two-step GNC by suggesting two sequential testing strategies to examine this issue. We also provide information on the sampling properties of these testing procedures through simulation experiments. We limit our study to the case when the information set contains only three variables, for reasons that we explain in Section 2.

The layout of this chapter is as follows. In the next section we present our modeling framework and discuss the relevant noncausality results. Section 3 introduces our proposed sequential two-step noncausality tests and provides details of our experimental design for the Monte Carlo study. Section 4 reports the simulation results. The proposed sequential testing strategies are then applied to a trivariate data set concerned with money–income causality in the presence of an interest rate variable in Section 5. Some concluding remarks are given in Section 6.

2. DISCUSSION OF NONCAUSALITY TESTS

2.1 Model Framework

We consider an n-dimensional vector time series $\{y_t: t = 1, 2, \ldots, T\}$, which we assume is generated from a vector autoregression (VAR) of finite order p^*:

$$y_t = \sum_{i=1}^{p} \Pi_i y_{t-i} + \epsilon_t \tag{1}$$

where Π_i is an $(n \times n)$ matrix of parameters, ϵ_t is an $(n \times 1)$ vector distributed as $IN(0, \Sigma)$, and (1) is initialized at $t = -p + 1, \ldots, 0$; the initial values can be any random vectors as well as nonstochastic ones. Let y_t be partitioned as $y_t = (X_t^T, Y_t^T, Z_t^T)^T$, where, for $Q = X, Y, Z, Q_t$ is an $(n_Q \times 1)$ vector, and $n_X + n_Y + n_Z = n$. Also, let Π_i be conformably partitioned as

---

*We limit our attention to testing GNC within a VAR framework to remain in line with the vast applied literature; the definitions proposed by Dufour and Renault (1998) are more general.

$$\Pi_i = \begin{bmatrix} \pi_{i,XX} & \pi_{i,XY} & \pi_{i,XZ} \\ \pi_{i,YX} & \pi_{i,YY} & \pi_{i,YZ} \\ \pi_{i,ZX} & \pi_{i,ZY} & \pi_{i,ZZ} \end{bmatrix}$$

where, for $Q, R = X, Y, Z$, $\pi_{i,QR}$ is an $(n_Q \times n_R)$ matrix of coefficients.

2.2 h-Step Noncausality

Suppose we wish to determine whether or not Y Granger-noncauses X one period ahead, denoted as $Y \underset{1}{\not\rightarrow} X$, in the presence of the auxiliary variables contained in Z. Traditionally, within the framework we consider, this is examined via a test of the null hypothesis H_{01}: $P_{XY} = 0$ where $P_{XY} = [\pi_{1,XY}, \pi_{2,XY}, \ldots, \pi_{p,XY}]$, using a Wald or likelihood ratio (LR) statistic. What does the result of this hypothesis test imply for GNC from Y to X at horizon $h(> 1)$, which we denote as $Y \underset{h}{\not\rightarrow} X$, and for GNC from Y to X up to horizon h that includes one period ahead, denoted by $Y \underset{(h)}{\not\rightarrow} X$?

There are three cases of interest.

(1) $n_Z = 0$; i.e., there are no auxiliary variables in Z so that all variables in the information set are involved in the GNC test under study. Then, from Dufour and Renault (1998, Proposition 2.2), the four following properties are equivalent:

(i) $Y \underset{1}{\not\rightarrow} X$; (ii) $Y \underset{h}{\not\rightarrow} X$; (iii) $Y \underset{h}{\not\rightarrow} X$; (iv) $Y \underset{\infty}{\not\rightarrow} X$

That is, when all variables are involved in the GNC test, support for the null hypothesis H_0: $P_{XY} = 0$ is also support for GNC at, or up to, any horizon h. This is intuitive as there are no auxiliary variables available through which indirect causality can occur. For example, the bivariate model satisfies these conditions.

(2) $n_Z = 1$; i.e., there is one auxiliary variable in the information set that is not directly involved in the GNC test of interest. Then, from Dufour and Renault (1998, Proposition 2.4 and Corollary 3.6), we have that $Y \underset{(2)}{\not\rightarrow} X$ (which implies $Y \underset{(\infty)}{\not\rightarrow} X$) if and only if at least one of the two following conditions is satisfied:

C1. $Y \underset{1}{\not\rightarrow} (X^T, Z)^T$;

C2. $(Y^T, Z)^T \underset{1}{\not\rightarrow} X$.

The null hypothesis corresponding to condition C1 is H_{02}: $P_{XY} = 0$ and $P_{ZY} = 0$ where P_{XY} is defined previously and $P_{ZY} = [\pi_{1,ZY}, \pi_{2,ZY}, \ldots, \pi_{p,ZY}]$, while that corresponding to condition C2 is H_{03}: P_{XY}

$= 0$ and $P_{XZ} = 0$, where $P_{XZ} = [\pi_{1,XZ}, \pi_{2,XZ}, \ldots, \pi_{p,XZ}]$; note that the restrictions under test are linear. When one or both of these conditions holds there cannot be indirect causality from Y to X via Z, while failure of both conditions implies that we have either one-step Granger causality, denoted as $Y \xrightarrow{1} X$, or two-step Granger causality, denoted by $Y \xrightarrow{2} X$ via Z; that is, $Y \xrightarrow{(2)} X$. As Z contains only one variable it directly follows that two-step GNC implies GNC for all horizons, as there are no additional causal patterns, internal to Z, which can result in indirect causality from Y to X. For example, a trivariate model falls into this case.

(3) $n_Z > 1$. This case is more complicated as causality patterns internal to Z must also be taken into account. If Z can be appropriately partitioned as $Z = (Z_1^T, Z_2^T)^T$ such that $(Y^T, Z_2^T)^T \xrightarrow{1} \!\!\!\!\!\!/ \; (X^T, Z_1^T)^T$, then this is sufficient for $Y \xrightarrow{(\infty)} \!\!\!\!\!\!/ \; X$. Intuitively, this result follows because the components of Z that can be caused by Y (those in Z_2) do not cause X, so that indirect causality cannot arise at longer horizons. Typically, the zero restrictions necessary to test this sufficient condition in the VAR representation are nonlinear functions of the coefficients. Dufour and Renault (1998, p. 1117) note that such "restrictions can lead to Jacobian matrices of the restrictions having less than full rank under the null hypothesis," which may lead to nonstandard asymptotic null distributions for test statistics.

For these reasons, and the preliminary nature of our study, we limit our attention to univariate Z and, as the applied literature is dominated by cases with $n_X = n_Y = 1$, consequently to a trivariate VAR model; recent empirical examples of the use of such models to examine for GNC include Friedman and Kuttner (1992), Kholdy (1995), Henriques and Sadorsky (1996), Lee et al. (1996), Riezman et al. (1996), Thornton (1997), Cheng (1999), Black et al. (2000), and Krishna et al. (2000), among many others.

2.3   Null Hypotheses, Test Statistics, and Limiting Distributions

The testing strategies we propose to examine for two-step GNC within a trivariate VAR framework involve the hypotheses H_{01}, H_{02}, and H_{03} detailed in the previous sub-section and the following conditional null hypotheses:

H_{04}: $P_{ZY} = 0 | P_{XY} = 0$;

H_{05}: $P_{XZ} = 0 | P_{XY} = 0$.

The null hypotheses involve linear restrictions on the coefficients of the VAR model and their validity can be examined using various methods, including Wald statistics, LR statistics, and model selection criteria. We limit our attention to the use of Wald statistics, though we recognize that other approaches may be preferable; this remains for future exploration.

Each of the Wald statistics that we consider in Section 3 is obtained from an appropriate model, for which the lag length p must be determined prior to testing; the selection of p is considered in that section. In general, consider a model where θ is an ($m \times 1$) vector of parameters and let R be a known nonstochastic ($q \times m$) matrix with rank q. To test H_0: $R\theta = 0$, a Wald statistic is

$$W = T\hat{\theta}^T R^T \left\{ R\hat{V}\left[\hat{\theta}\right] R^T \right\}^{-1} R\hat{\theta} \tag{2}$$

where $\hat{\theta}$ is a consistent estimator of θ and $\hat{V}[\hat{\theta}]$ is a consistent estimator of the asymptotic variance–covariance matrix of $\sqrt{T}(\hat{\theta} - \theta)$. Given appropriate conditions, W is asymptotically distributed as a ($\chi^2(q)$ variate under H_0.

The conditions needed for W's null limiting distribution are not assured here, as y_t may be nonstationary with possible cointegration. Sims et al. (1990) and Toda and Phillips (1993, 1994) show that W is asymptotically distributed as a χ^2 variate under H_0 when y_t is stationary or is nonstationary with "sufficient" cointegration; otherwise W has a nonstandard limiting null distribution that may involve nuisance parameters. The basic problem with nonstationarity is that a singularity may arise in the asymptotic distribution of the least-squares (LS) estimators, as some of the coefficients or linear combinations of them may be estimated more efficiently with a faster convergence rate than $\sqrt{T}$. Unfortunately, there seems no basis for testing for "sufficient" cointegration within the VAR model, so that Toda and Phillips (1993, 1994) recommend that, in general, GNC tests should not be undertaken using a VAR model when y_t is nonstationary.

One possible solution is to map the VAR model to its equivalent vector error correction model (VECM) form and to undertake the GNC tests within this framework; see Toda and Phillips (1993, 1994). The problem then is accurate determination of the cointegrating rank, which is known to be difficult with the currently available cointegration tests due to their low power properties and sensitivity to the specification of other terms in the model, including lag order and deterministic trends. Giles and Mirza (1999) use simulation experiments to illustrate the impact of this inaccuracy on the finite sample performance of one-step GNC tests; they find that the practice of pretesting for cointegration can often result in severe over-rejections of the noncausal null.

An alternative approach is the "augmented lags" method suggested by Toda and Yamamoto (1995) and Dolado and Lütkepohl (1996), which results in asymptotic χ^2 null distributions for Wald statistics of the form we are examining, irrespective of the system's integration or cointegration properties. These authors show that overfitting the VAR model by the highest order of integration in the system eliminates the covariance matrix singularity problem. Consider the augmented VAR model

$$y_t = \sum_{i=1}^{p} \Pi_i y_{t-i} + \sum_{i=1}^{d} \Pi_{p+i} Y_{t-p-i} + \epsilon_t \tag{3}$$

where we assume that y_t is at most integrated of order $d(I(d))$. Then, Wald test statistics based on testing restrictions involving the coefficients contained in $\Pi_1, \ldots, \Pi_p$ have standard asymptotic χ^2 null distributions; see Theorem 1 of Dolado and Lütkepohl (1996) and Theorem 1 of Toda and Yamamoto (1995). This approach will result in a loss in power, as the augmented model contains superfluous lags and we are ignoring that some of the VAR coefficients, or at least linear combinations of them, can be estimated more efficiently with a higher than usual rate of convergence. However, the simulation experiments of Dolado and Lütkepohl (1996), Zapata and Rambaldi (1997), and Giles and Mirza (1999) suggest that this loss is frequently minimal and the approach often results in more accurate GNC outcomes than the VECM method, which conditions on the outcome of preliminary cointegration tests. Accordingly, we limit our attention to undertaking Wald tests using the augmented model (3).

Specifically, assuming a trivariate augmented model with $n_X = n_Y = n_Z = 1$, let θ be the $9(p+d)$ vector given by $\theta = \text{vec}[\Pi_1, \Pi_2, \ldots, \Pi_{p+d}]$, where vec denotes the vectorization operator that stacks the columns of the argument matrix. The LS estimator of θ is $\hat{\theta}$. Then, the null hypothesis $H_{01}: P_{XY} = 0$ can be examined using the Wald statistic given by (2), with R being a selector matrix such that $R\theta = \text{vec}[P_{XY}]$. We denote the resulting Wald statistic as W_1. Under our assumptions, W_1 is asymptotically distributed as a $\chi^2(p)$ variate under H_{01}.

The Wald statistic for examining $H_{02}: P_{XY} = 0$ and $P_{ZY} = 0$, denoted W_2, is given by (2), where $\hat{\theta}$ is the estimator of $\theta = \text{vec}[\Pi_1, \Pi_2, \ldots, \Pi_{p+d}]$ and R is a selector matrix such that $R\theta = \text{vec}[P_{XY}, P_{ZY}]$. Under our assumptions, the statistic W_2 has a limiting $\chi^2(2p)$ null distribution. Similarly, let W_3 be the Wald statistic for examining $H_{03}: P_{XY} = 0$ and $P_{XZ} = 0$. The statistic W_3 is then given by (2) with the selector matrix chosen to ensure that $R\theta = \text{vec}[P_{XY}, P_{XZ}]$; under our assumptions W_3 is an asymptotic $\chi^2(2p)$ variate under H_{03}.

We denote the Wald statistics for testing the null hypotheses H_{04}: $P_{ZY} = 0|P_{XY} = 0$ and H_{05}: $P_{XZ} = 0|P_{XY} = 0$ as W_4 and W_5 respectively. These statistics are obtained from the restricted model that imposes that $P_{XY} = 0$. Let θ^* be the vector of remaining unconstrained parameters and $\hat{\theta}^*$ be the LS estimator of θ^*, so that, to test the conditional null hypothesis H_0: $R^*\theta^* = 0$, a Wald statistic is

$$W^* = T\hat{\theta}^{*T} R^{*T} \left\{ R^* \hat{V}\left[\hat{\theta}^*\right] R^{*T} \right\}^{-1} R^*\hat{\theta}^*$$

where $\hat{V}[\hat{\theta}^*]$ is a consistent estimator of the asymptotic variance–covariance matrix of $\sqrt{T}(\hat{\theta}^* - \theta^*)$. Under our assumptions, W^* is asymptotically distributed as a χ^2 variate under H_0 with the rank of R^* determining the degrees of freedom; Theorem 1 of Dolado and Lütkepohl (1996) continues to hold in this restricted case as the elements of $\Pi_{p+1}, \ldots, \Pi_{p+d}$ are not constrained under the conditioning or by the restrictions under test. Our statistics W_4 and W_5 are then given by W^* with, in turn, the vector $R^*\theta^*$ equal to $\text{vec}(P_{ZY})$ and $\text{vec}(P_{XZ})$; in each case p is the degrees of freedom. We now turn, in the next section, to detail our proposed testing strategies.

3. PROPOSED SEQUENTIAL TESTING STRATEGIES AND MONTE CARLO DESIGN

3.1 Sequential Testing Strategies

Within the augmented model framework, the sequential testing procedures we consider are:

(M1)

$$\text{Test } H_{02} \text{ and test } H_{03} \begin{cases} \text{If } H_{02} \text{ and } H_{03} \text{ are rejected, reject the} \\ \text{hypothesis of } Y \xrightarrow{\nrightarrow}_{(2)} X. \\ \text{Otherwise, support the null hypothesis} \\ \text{of } Y \xrightarrow{\nrightarrow}_{(2)} X. \end{cases}$$

(M2)

$$\text{Test } H_{01} \begin{cases} \text{If } H_{01} \text{ is rejected, reject the null hypothesis of } Y \xrightarrow{\nrightarrow}_{1} X. \\ \text{Otherwise, test } H_{04} \text{ and } H_{05} \begin{cases} \text{If } H_{04} \text{ and } H_{05} \text{ are} \\ \text{rejected, reject } Y \xrightarrow{\nrightarrow}_{2} X \\ \text{Otherwise, support the} \\ \text{null hypothesis of } Y \xrightarrow{\nrightarrow}_{2} X \end{cases} \end{cases}$$

The strategy (M1) provides information on the null hypothesis H_{0A}: $Y \xrightarrow{\nrightarrow}_{(2)} X$ as it directly tests whether the conditions C1 and C2, outlined in the previous section, are satisfied. Each hypothesis test tests $2p$ exact linear restrictions. The approach (M1) does not distinguish between one-step and two-step GNC; that is, rejection here does not inform the researcher whether the causality arises directly from Y to X one step ahead or whether there is GNC one step ahead with the causality then arising indirectly via the auxiliary variable at horizon two. This is not a relevant concern when interest is in only answering the question of the presence of Granger causality at any horizon.

The strategy (M2), on the other hand, provides information on the horizon at which the causality, if any, arises. The first hypothesis test, H_{01}, undertakes the usual one-step test for direct GNC from Y to X, which, when rejected, implies that no further testing is required as causality is detected. However, the possibility of indirect causality at horizon two via the auxiliary variable Z is still feasible when H_{01} is supported, and hence the second layer of tests that examine the null hypothesis H_{0B}: $Y \underset{2}{\nrightarrow} X | Y \underset{1}{\nrightarrow} X$. We require both of H_{04} and H_{05} to be rejected for a conclusion of causality at horizon two, while we accept H_{0B} when we support one or both of the hypotheses H_{04} and H_{05}. Each test requires us to examine the validity of p exact linear restrictions.

That the sub-tests in the two strategies have different degrees of freedom may result in power differences that may lead us to prefer (M2) over (M1). In our simulation experiments described below, we consider various choices of nominal significance level for each sub-test, though we limit each to be identical at, say, $100\alpha\%$. We can say the following about the asymptotic level of each of the strategies. In the case of (M1) we are examining nonnested hypotheses using statistics that are not statistically independent. When both H_{02} and H_{03} are true we know from the laws of probability that the level is smaller than $200\alpha\%$, though we expect to see asymptotic levels less than this upper bound, dependent on the magnitude of the probability of the union of the events. When one of the hypotheses is false and the other is true, so that $Y \underset{(2)}{\nrightarrow} X$ still holds, the asymptotic level for strategy (M1) is smaller than or equal to $100\alpha\%$.

An asymptotic level of $100\alpha\%$ applies for strategy (M2) for testing for $Y \underset{1}{\nrightarrow} X$, while it is at most $200\alpha\%$ for testing for H_{0B} when H_{04} and H_{05} are both true, and the level is at most $100\alpha\%$ when only one is true. As noted with strategy (M1), when both hypotheses are true we would expect levels that are much smaller than $200\alpha\%$. Note also that W_4 and W_5 are not (typically) asymptotically independent under their respective null hypotheses, though the statistics W_4 and W_1 are asymptotically independent under H_{04} and H_{01}, given the nesting structure.

Finally, we note that the testing strategies are consistent, as each component test is consistent, which follows directly from the results presented by, e.g., Dolado and Lütkepohl (1996, p. 375).

3.2 Monte Carlo Experiment

In this subsection, we provide information on our small-scale simulation design that we use to examine the finite sample performance of our proposed sequential test procedures. We consider two basic data generating processes (DGPs), which we denote as DGPA and DGPB; for each we examine three cases: DGPA1, DGPA2, DGPA3, DGPB1, DGPB2, and DGPB3. To avoid potential confusion, we now write y_t as $y_t = [y_{1t}, y_{2t}, y_{3t}]^T$ to describe the VAR system and the GNC hypotheses we examine; we provide the mappings to the variables X, Y, and Z from the last subsection in a table. For all cases the series are I(1). The first DGP, denoted as DGPA, is

$$\begin{bmatrix} y_{1t} \\ y_{2t} \\ y_{3t} \end{bmatrix} = \begin{bmatrix} 1 & 0 & 0 \\ 1 & 0.5 & 0 \\ a & -0.5 & 1 \end{bmatrix} \begin{bmatrix} y_{1t-1} \\ y_{2t-1} \\ y_{3t-1} \end{bmatrix} + \begin{bmatrix} \epsilon_{1t} \\ \epsilon_{2t} \\ \epsilon_{3t} \end{bmatrix}$$

We consider $a = 1$ for DGPA1, which implies one cointegrating vector. The parameter a is set to zero for DGPA2 and DGPA3, which results in two cointegrating vectors among the I(1) variables. Toda and Phillips (1994) and Giles and Mirza (1999) also use this basic DGP. The null hypotheses H_{01} to H_{05} are true for the GNC effects we examine for DGPA1 and DGPA3. Using Corollary 1 of Toda and Phillips (1993, 1994), it is clear that there is insufficient cointegration with respect to the variables whose causal effects are being studied for all hypotheses except H_{05}. That is, Wald statistics for H_{01} to H_{04} applied in the non-augmented model (1) do not have their usual χ^2 asymptotic null distributions, though the Wald statistics we use in the augmented model (3) have standard limiting null distributions for all the null hypotheses.

Our second basic DGP, denoted as DGPB, is

$$\begin{bmatrix} y_{1t} \\ y_{2t} \\ y_{3t} \end{bmatrix} = \begin{bmatrix} 0.32 & b & 0.44 \\ 0 & 1.2325 & 0.31 \\ 0 & -0.285 & 0.62 \end{bmatrix} \begin{bmatrix} y_{1t-1} \\ y_{2t-1} \\ y_{3t-1} \end{bmatrix} + \begin{bmatrix} \epsilon_{1t} \\ \epsilon_{2t} \\ \epsilon_{3t} \end{bmatrix}$$

We set the parameter $b = -0.265$ for DGPB1 and $b = 0$ for DGPB2 and DGPB3; for each DGP there are two cointegrating vectors. Zapata and Rambaldi (1997) and Giles and Mirza (1999) also use a variant of this DGP in their experiments. The null hypotheses H_{01}, H_{02} and H_{04} are each true for DGPB1 and DGPB2, while H_{03} and H_{05} are false. The cointegration for this DGP is sufficient with respect to the variables whose causal effects are being examined, so that standard Wald statistics in the non-augmented model for H_{01}, H_{02}, and H_{04} are asymptotic χ^2 variates under their appropriate null hypotheses.

We provide summary information on the DGPs in Table 1. We include the mappings to the variables X, Y, and Z used in our discussion of the sequential testing strategies in the last subsection, the validity of the null hypotheses H_{01} to H_{05}, and the GNC outcomes of interest. Though our range of DGPs is limited, they enable us to study the impact of various causal patterns on the finite sample performance of our strategies (M1) and (M2).

The last row of the table provides the "1-step GC map" for each DGP, which details the pair-wise 1-step Granger causal (GC) patterns. For example, the 1-step GC map for DGPB1 in terms of X, Y, and Z is

This causal map is an example of a "directed graph," because the arrows lead from one variable to another; they indicate the presence of 1-step GC, e.g., $X \underset{1}{\rightarrow} Y$. In addition to being a useful way to summarize the 1-step GC relationships, the 1-step GC map allows us to visualize the possibilities for 2-step GC. To illustrate, we consider the GC relations from Y to X. The map indicates that $Y \underset{1}{\nrightarrow} X$, so that for $Y \underset{2}{\rightarrow} X$ we require $Y \underset{1}{\rightarrow} Z$ and $Z \underset{1}{\rightarrow} X$; directly from the map we see that the latter causal relationship holds but not the former: Z is not operating as an auxiliary variable through which 2-step GC from Y to X can occur.

As our aim is to examine the finite-sample performance of (M1) and (M2) at detecting 2-step GNC, and at distinguishing between GC at the different horizons, each of our DGPs imposes that $Y \underset{1}{\nrightarrow} X$, but for two DGPs—DGPA2 and DGPB3—$Y \underset{2}{\rightarrow} X$, while for the other DGPs we have $Y \underset{2}{\nrightarrow} X$. This allows us to present rejection frequencies when the null hypothesis $Y \underset{2}{\nrightarrow} X$ (or $Y \underset{(2)}{\nrightarrow} X$) is true and false.

When the null is true we denote the rejection frequencies as $FI(\alpha)$, because they estimate the probability that the testing strategy makes a Type I error when the nominal significance level is α, that is, the probability that the testing strategy rejects a correct null hypothesis. The strategy (M1) provides information on the null hypothesis H_{0A}: $Y \underset{(2)}{\nrightarrow} X$, which is true when we support both H_{02} and H_{03} or when only one is accepted. Let $PIA(\alpha)$ be the probability of a Type I error associated with testing H_{0A}, at nominal significance level α, using (M1), so $PIA(\alpha) = \mathrm{Pr}_{H_{0A}}$ (reject H_{03} and reject $H_{02} | Y \underset{(2)}{\nrightarrow} X$). We estimate $PIA(\alpha)$ by $FIA(\alpha) = N^{-1} \sum_{i=1}^{N} I(P_{3i} \leq \alpha$ and

Table 1. DGP descriptions

	DGPA1	DGPA2	DGPA3	DGPB1	DGPB2	DGPB3
X	y_1	y_3	y_1	y_2	y_2	y_1
Y	y_3	y_1	y_3	y_1	y_1	y_2
Z	y_2	y_2	y_2	y_3	y_3	y_3
$H_{01}: P_{XY} = 0$	True	True	True	True	True	True
$H_{02}: P_{XY} = 0 \ \& \ P_{ZY} = 0$	True	False	True	True	True	False
$H_{03}: P_{XY} = 0 \ \& \ P_{XZ} = 0$	True	False	True	False	False	False
$H_{04}: P_{ZY} = 0 \vert P_{XY} = 0$	True	False	True	True	True	False
$H_{05}: P_{XZ} = 0 \vert P_{XY} = 0$	True	False	True	False	False	False
$Y \underset{1}{\nrightarrow} X$	Yes	Yes	Yes	Yes	Yes	Yes
$Y \underset{2}{\nrightarrow} X$	Yes	No	Yes	Yes	Yes	No
1-step GC map						

$P_{2i} \leq \alpha$), where N denotes the number of Monte Carlo samples, P_{ji} is the ith Monte Carlo sample's P value associated with testing H_{0j} that has been calculated from a $\chi^2(2p)$ distribution using the statistic W_j for $j = 2, 3$, and $I(\cdot)$ is the indicator function.

We desire $FIA(\alpha)$ to be "close" to the asymptotic nominal level for the strategy, which is bounded by $100\alpha\%$ when only one of H_{02} or H_{03} is true and by $200\alpha\%$ when both are true, though $FIA(\alpha)$ will differ from the upper bound because of our use of a finite number of simulation experiments and the asymptotic null distribution to calculate the P values; the latter is used because our statistics have unknown finite sample distributions. In particular, our statistics, though asymptotically pivotal, are not pivotal in finite samples, so $FIA(\alpha)$ (and also $PIA(\alpha)$) depends on where the true DGP is in the set specified by H_{0A}. One way of solving the latter problem is to report the size of the testing strategy, which is the supremum of the $PIA(\alpha)$ values over all DGPs contained in H_{0A}. In principle, we could estimate the size as the supremum of the $FIA(\alpha)$ values, though this task is in reality infeasible here because of the multitude of DGPs that can satisfy H_{0A}. Accordingly, we report $FIA(\alpha)$ values for a range of DGPs.

The testing strategy (M2) provides information on two null hypotheses: $H_{01}: Y \not\underset{1}{\rightarrow} X$ and $H_{0B}: Y \not\underset{2}{\rightarrow} X | Y \not\underset{1}{\rightarrow} X$. The statistic W_1, used to test H_{01}, has asymptotic level $100\alpha\%$. We estimate the associated probability of a Type I error, denoted as $PI1(\alpha) = \Pr_{H_{01}}(\text{reject } H_{01})$, by $FI1(\alpha) = N^{-1} \times \sum_{i=1}^{n} I$ $(P_{1i} \leq \alpha)$, where P_{1i} is the P value associated with testing H_{01} for the ith Monte Carlo sample and is generated from a $\chi^2(p)$ distribution, α is the assigned nominal level, and $I(\cdot)$ is the indicator function. Further, let $PIB(\alpha)$ denote the probability of a Type I error for using (M2) to test H_{0B}, and let $FIB(\alpha) = N_{1*}^{-1} \sum_{i=1}^{N_{1*}} I(P_{5i} \leq \alpha \text{ and } P_{4i} \leq \alpha)$ be an estimator of $PIB(\alpha)$, where N_{1*} is the number of Monte Carlo samples that accepted H_{01}, and P_{ji} is the P value for testing H_{0j} for the ith Monte Carlo sample calculated from a $\chi^2(p)$ distribution, $j = 4, 5$. We report $FI1(\alpha)$ and $FIB(\alpha)$ values for our DGPs that satisfy H_{01} and H_{0B}. Recall that the upper bound on $PIB(\alpha)$ is $100\alpha\%$ when only one of H_{04} or H_{05} is true and it is $200\alpha\%$ when both are true.

We also report rejection frequencies for the strategies when $Y \underset{2}{\rightarrow} X$; these numbers aim to indicate "powers" of our testing strategies. In econometrics, there is much debate on how powers should be estimated, with many studies advocating that only so-called "size-corrected" power estimates be provided. Given the breadth of the set under the null hypotheses of interest here, it is computationally infeasible to contemplate obtaining "size-corrected" power estimates for our study. Some researchers approach this problem by obtaining a critical value for their particular DGP that ensures that the nominal and true probabilities of a Type I error are equal; they then claim, incorrectly, that the reported powers are "size-corrected." A further complication in attempting to provide estimates of size-corrected power is that the size-corrected critical value may often be infinite, which implies zero power for the test (e.g, Dufour 1997).

Given these points, when dealing with a composite null hypothesis, Horowitz and Savin (2000) suggest that it may be preferable to form power estimates from an estimate of the Type I critical value that would be obtained if the exact finite sample distribution of the test statistic under the true DGP were known. One such estimator is the asymptotic critical value, though this estimator may not be accurate in finite-samples. Horowitz and Savin (2000) advocate bootstrap procedures to estimate the pseudo-true value of the parameters from which an estimate of the finite-sample Type I critical value can be obtained; this approach may result in higher accuracy in finite samples. In our study, given its preliminary nature, we use the asymptotic critical value to estimate the powers of our strategies; we leave the potential application of bootstrap procedures for future research. We denote the estimated power for strategy (M1) that is associated with testing H_{0A} by $FIIA(\alpha) = N^{-1}\sum_{i=1}^{N} I(P_{3i} \leq \alpha$ and $P_{2i} \leq \alpha)$, and the estimated powers for strategy (M2) for testing H_{0B} by $FIIB(\alpha) = N_{I^*}^{-1} I(P_{5i} \leq \alpha$ and $P_{4i} \leq \alpha)$ respectively.

For each of the cases outlined in Table 1 we examine four net sample sizes: $T = 50, 100, 200, 400$. The number of Monte Carlo simulation repetitions is fixed to be 5000, and for each experiment we generate $(T + 100 + 6)$ observations from which we discard the first 100 to remove the effect of the zero starting values; the other six observations are needed for lagging. We limit attention to an identity innovation covariance matrix, though we recognize that this choice of covariance matrix is potentially restrictive and requires further attention in future research.* We generate FI and FII values for twelve values of the nominal significance level: $\alpha = 0.0001, 0.0005, 0.001, 0.005, 0.01, 0.025, 0.05, 0.1, 0.15, 0.2, 0.25$, and 0.30.

The conventional way to report the results is by tables, though this approach has two main drawbacks. First, there would be many tables, and second, it is often difficult to see how changes in the sample size, DGPs, and α values affect the rejection frequencies. In this paper, as recommended by Davidson and MacKinnon (1998), we use graphs to provide the information on the performance of our testing strategies. We use P-value plots to report the results on FIA, $FI1$ and FIB; these

---

*Note that Yamada and Toda (1998) show that the finite sample distribution of Wald statistics, such as those considered here, is invariant to the form of the innovation covariance matrix when the lag length order of the VAR is known. However, this result no loner holds once we allow for estimation of the lag order.

plot the *FI* values against a nominal level. In the ideal case, each of our *P*-values would be distributed as uniform $(0, 1)$, so that the resulting graph should be close to the $45°$ line. Consequently, we can easily see when a strategy is over-rejecting, or under-rejecting, or rejecting the right proportion reasonably often. The asymptotic nominal level for DGPB1 and DGPB2 is α for each of H_{01}, H_{0A}, and H_{0B}, it is α for H_{01} for DGPA1 and DGPA3, and the upper bound is 2α for H_{0A} and H_{0B}. However, we anticipate levels less than 2α, as this upper bound ignores the probability of the union event, so we use α as the level to provide the $45°$ line for the plots.

We provide so-called "size–power" curves to report the information on the power of our testing strategies; see Wilk and Gnanadesikan (1968) and Davidson and MacKinnon (1998). In our case we use *FI* values rather than size estimates and our power estimates are based on the *FII* values; accordingly, we call these graphs *FI–FII* curves. The horizontal axis gives the *FI* values, computed when the DGP satisfies the null hypothesis, and the vertical axis gives the *FII* values, generated when the DGP does not satisfy $Y \not\rightarrow_2 X$ in a particular way. The lower left-hand corner of the curve arises when the strategy always supports the null hypothesis, while the upper right hand corner results when the test always rejects the null hypothesis. When the power of a testing strategy exceeds its associated probability of a Type I error, the *FI–FII* curve lies above the $45°$ line, which represents the points of equal probability.

To undertake the hypothesis tests of interest, we need to choose the lag order for the VAR, which is well known to impact on the performance of tests; see Lütkepohl (1993b), Dolado and Lütkepohl (1996), Giles and Mirza (1999), among others. We examine four approaches to specifying the lag order: p is correctly specified at 1, which we denote as the "True" case; p is always over-estimated by 1, that is, $p = 2$, which we denote as "Over"; and p is selected by two common goodness-of-fit criteria—Schwarz's (1978) Bayesian criterion (SC) and Akaike's (1973) information criterion (AIC). The AIC does not consistently estimate the lag order (e.g., Nishi 1988, Lütkepohl 1993b), as there is a positive probability of over-estimation of p, which does not, nevertheless, affect consistent estimation of the coefficients, though over-estimation may result in efficiency and power losses. The SC is a consistent estimator of p, though evidence suggests that this estimator may be overly parsimonious in finite samples, which may have a detrimental impact on the performance of subsequent hypothesis tests. In our experiments that use the AIC and SC we allow p to be at most 6. Though the "True" case is somewhat artificial, we include it as it can be regarded as a best-case scenario. The over-specified case illustrates results on using a pre-specified, though incorrect, lag order.

4. SIMULATION RESULTS

4.1 *P*-Valve Plots

Figure 1 shows P-value plots for the testing strategy (M1) for DGPA1 when $T = 50, 100, 200$, and 400 and for the four lag selection methods we outlined in the previous section. It is clear that the strategy systematically results in levels that are well below α, irrespective of the sample size. For instance, the estimated probability of a Type I error is typically close to $\alpha/2$ when $T = 200$, irrespective of the specified nominal level. In contrast, the strategy seems to work better for the smaller sample of 50 observations.

As expected from our discussion, these observed features for DGPA1 differ from DGPB1 and DGPB2 when only one of the hypotheses is true. In Figure 2 we provide the P-value plots for DGPB1, from which we see that the strategy (M1) systematically over-rejects, with the degree of over-rejection becoming more pronounced as T falls. Qualitatively, the P-value plots for DGPA3 and DGPB2 match those for DGPA1 and DGPB1 respectively, so we do not include them here.

Over-specifying or estimating the lag order does little to alter the P-value plots from those for the correctly specified model when $T \geq 200$, though there are some observable differences when T is smaller, in particular for $T = 50$. The performance of the AIC and Over cases is always somewhat worse than for the SC, although we expect that this is a feature of the low-order VAR models that we examine for our DGPs. Figure 1 and Figure 2 show that the AIC rejects more often than does Over for small α, but this is reversed for higher levels, say greater than 10%.

We now turn to the P-value plots for the procedure (M2). We provide the P-value plots for testing H_{01}: $Y \nrightarrow_1 X$ for DGPA3 in Figure 3. This hypothesis forms the first part of strategy (M2) and is undertaken using the Wald statistic W_1. The plots for the other DGPs are qualitatively similar. It is clear that the test systematically over-rejects H_{01}, especially when $T \leq 100$ and irrespective of the lag selection approach adopted.

Figure 4 shows P-value plots for the second part of strategy (M2) that examines $Y \nrightarrow_2 X | Y \nrightarrow_1 X$ for DGPA1. It is clear that there is a pattern of the FI values being less than α, with the difference increasing with T, irrespective of the lag selection approach. This systematic pattern is similar to that observed for strategy (M1) with this DGP, and also with DGPA3. As in that case, we do not observe this feature with DGPB1 and DGPB2, as we see from Figure 5 that shows P-value plots for DGPB1, which are representative for both DGPB1 and DGPB2. Here the testing procedure works well, though there is a small tendency to over-reject for $T \geq 100$, while only the Over approach leads to over-rejection when $T = 50$ with the other lag selection cases slightly under-rejecting. These over- and under-rejections (see-

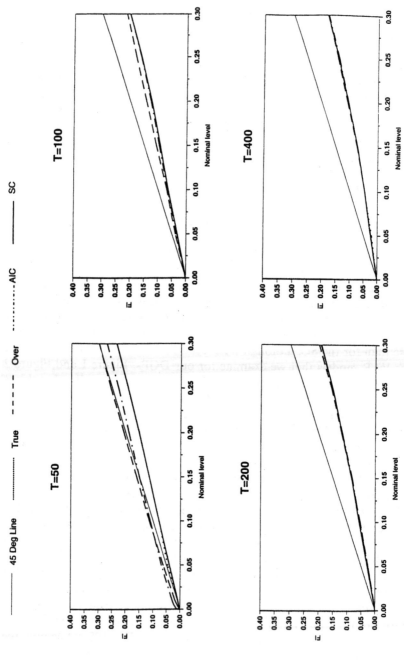

Figure 1. *P*-value plots for strategy (M1), *FIA*, DGPA1.

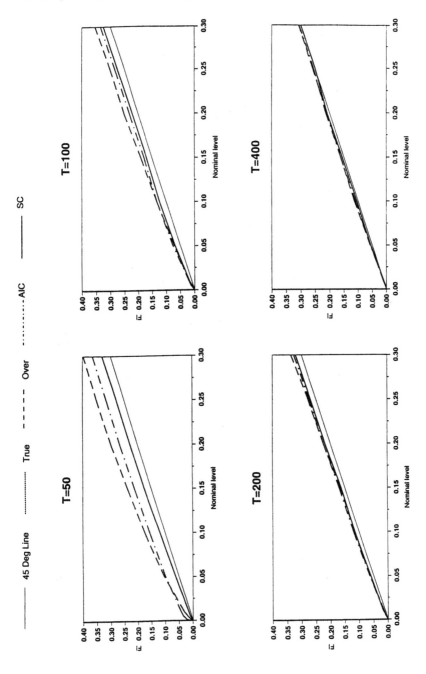

Figure 2. *P*-value plots for strategy (M1), *FIA*, DGPB1.

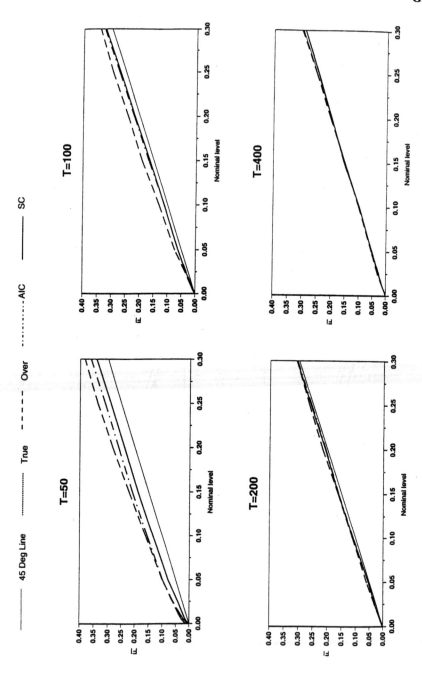

Figure 3. *P*-value plots for strategy (M2), *FII, DGPA3*.

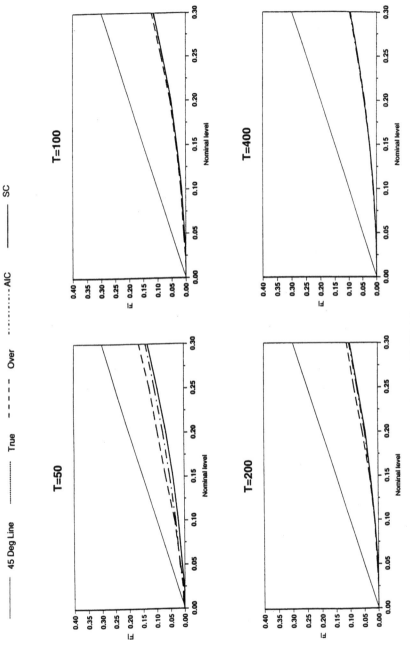

Figure 4. *P*-value plots for strategy (M2), *FIB*, DGPA1.

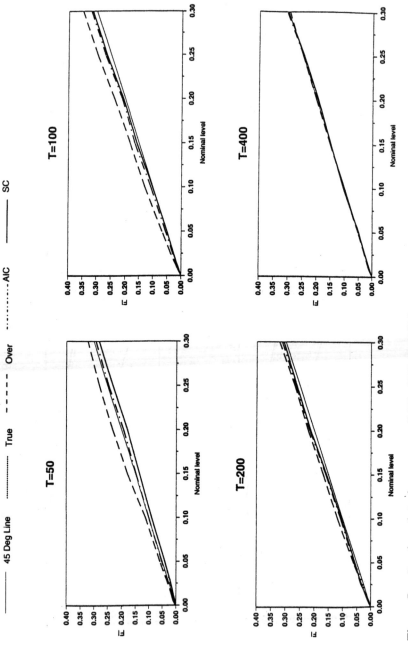

Figure 5. *P*-value plots for strategy (M2), *FIB*, DGPB1.

mingly) become more pronounced as the nominal level rises, though, as a proportion of the nominal level, the degree of over- or under-rejection is in fact declining as the nominal level rises.

4.2 *FI–FII* Plots

Figure 6 presents *FIA–FIIA* curves associated with strategy (M1) generated from DGPA1 (for which the null $Y \underset{(2)}{\not\to} X$ is true) and DGPA3 (for which the null $Y \underset{2}{\not\to} X$ is false in a particular way) when $T = 50$ and $T = 100$; the consistency property of the testing procedure is well established for this DGP and degree of falseness of the null hypothesis by $T = 100$, so we omit the graphs for $T = 200$ and $T = 400$. Several results are evident from this figure. We see good power properties irrespective of the sample size, though this is a feature of the chosen value for the parameter *a* for DGPA3. Over-specifying the lag order by a fixed amount does not result in a loss in power, while there is a loss associated with use of the AIC.

The *FIA–FIIA* curves for strategy (M1) generated from DGPB2 (for which the null $Y \underset{(2)}{\not\to} X$ is true) and DGPB3 (for which the null $Y \underset{2}{\not\to} X$ is false in a particular way) when $T = 50$ and $T = 100$ are given in Figure 7; we again omit the curves for $T = 200$ and $T = 400$ as they merely illustrate the consistency feature of the strategy. In this case, compared to that presented in Figure 6, the *FII* levels are lower for a given *FI* level, which reflects the degree to which the null hypothesis is false. Nevertheless, the results suggest that the strategy does well at rejecting the false null hypothesis, especially when $T \geq 100$.

The *FIB–FIIB* curves for strategy (M2) for testing H_{0B} display qualitatively similar features to those just discussed for strategy (M1). To illustrate, Figure 8 provides the curves generated from DGPB2 and DGPB3; figures for the other cases are available on request. It is of interest to compare Figure 7 and Figure 8. As expected, and irrespective of sample size, there are gains in power from using strategy (M2) over (M1) when $Y \underset{1}{\not\to} X$ but $Y \underset{2}{\to} X$, as the degrees of freedom for the former strategy are half those of the latter.

Overall, we conclude that the strategies perform well at detecting a false null hypothesis, even in relatively small samples. Our results illustrate the potential gains in power with strategy (M2) over strategy (M1) when $Y \underset{1}{\not\to} X$ and $Y \underset{2}{\to} X$. In practice, this is likely useful as we anticipate that most researchers will first test for $Y \underset{1}{\not\to} X$, and only proceed to a second stage test for $Y \underset{2}{\not\to} X$ when the first test is not rejected. There is some loss in power in using the AIC to select the lag order compared with the other

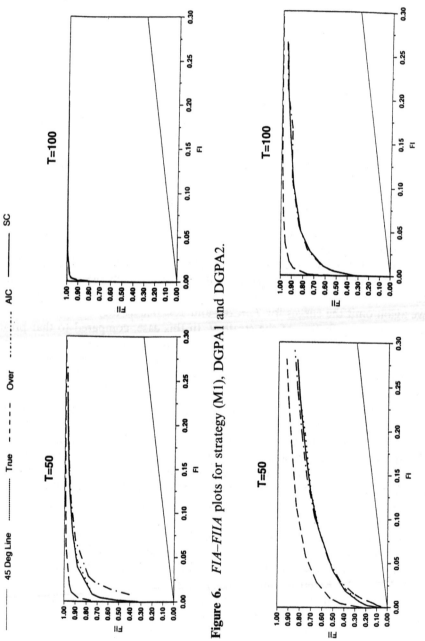

Figure 6. *FIA–FIIA* plots for strategy (M1), DGPA1 and DGPA2.

Figure 7. *FIA – FIIA* plots for strategy (M1), DGPB2 and DGPB3.

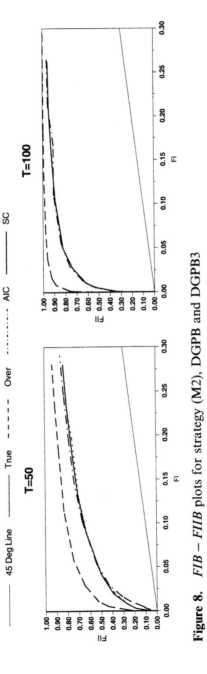

Figure 8. *FIB – FIIB* plots for strategy (M2), DGPB and DGPB3

approaches we examine, though a study of DGPs with longer lag orders may alter this outcome. Our results support the finding from other studies that lag order selection is important.

5. EMPIRICAL EXAMPLE

To illustrate the application of the testing strategies on the outcome of GNC tests in trivariate systems, we have re-examined the data set used by Hoffman and Rasche (1996), which enables us to consider the well-studied issue of the causal relationships between money and income. We downloaded the quarterly, seasonally adjusted US time series data from the *Journal of Applied Econometrics* Data Archive; the data were originally obtained from Citibase. Let X be real money balances, which is calculated by deflating the nominal series by the GDP deflator; let Y be real income, represented by real GDP; and let Z be nominal interest rates, the auxiliary variable, which is represented by the Treasury bill rate. Both real balances and real GDP are expressed in natural logarithms. Allowing for lagging, we use observations from 1950:3 to 1995:2 (164 observations). Our Monte Carlo study ignored the realistic possibility that there may be deterministic trends, which we incorporate here by extending model (1) as

$$(y_t - \mu - \delta t) = \sum_{i=1}^{p} \Pi_i (y_{t-i} - \mu - \delta(t - i)) + \epsilon_t \tag{4}$$

where μ and δ are vectors of unknown coefficients. We can write (4) equivalently as

$$y_t = \mu^* + \delta^* t + \sum_{i=1}^{p} \Pi_i y_{t-i} + \epsilon_t \tag{5}$$

where

$$\mu^* = (-\Pi\mu + \Pi^*\delta), \quad \delta^* = -\Pi\delta, \quad \Pi^* = \sum_{i=1}^{p} i\Pi_i$$

and

$$\Pi = -\left(I_3 - \sum_{i=1}^{p} \Pi_i \right)$$

The matrix Π is the usual potentially reduced rank matrix that indicates the number of cointegrating relationships. Often applied researchers impose $\delta = 0$ a priori, which implies that μ^* is forced to zero when there is no cointegration; this seems a limiting restriction, so we use (5) as stated.

We assume that each of our time series is integrated of order one, as do also Hoffman and Rasche (1996); this implies that we augment our model with one extra lag of y. We use the AIC to choose the lag order p allowing for up to ten possible lags; the results support six lags. Table 2 reports asymptotic P-values for examining for the six possible 1-step GNC relationships. Using a (nominal) 10% significance level, the outcomes imply that the 1-step GC map is

from which it is clear that there is support for the common finding that real money balances Granger-causes real income without feedback. However, the map illustrates the potential for 2-step GC from real income to real money balances to arise indirectly via the interest rate variable. To explore this possibility we apply our testing strategies (M1) and (M2); the

Table 2. Money–income example
P-values

Null Hypothesis	P-value
$Y \not\rightarrow_1 X$	0.170
$Z \not\rightarrow_1 X$	< 0.001
$X \not\rightarrow_1 Y$	0.001
$Z \not\rightarrow_1 Y$	< 0.001
$X \not\rightarrow_1 Z$	0.090
$Y \not\rightarrow_1 Z$	0.086
Strategy (M1)	
H_{02}	0.070
H_{03}	< 0.001
Strategy (M2)	
H_{04}	< 0.001
H_{05}	0.095

asymptotic P-values are reported in Table 2. Both strategies support 2-step Granger causality from real income to real money balances via the interest rate variable. The example serves to illustrate the changes in causality conclusions that may occur once an allowance is made for indirect causality via an auxiliary variable.

6. CONCLUDING REMARKS

This paper has provided two testing strategies for examining for 2-step Granger noncausality in a trivariate VAR system that may arise via the auxiliary variable that is included for explanatory power but is not directly involved in the noncausality test under examination. The testing strategy (M1) is advantageous in that it can be applied without a prior test for the traditional 1-step noncausality as it tests for noncausality up to horizon two. This feature is appealing when the interest is in knowing whether or not causality exists, irrespective at which horizon. However, strategy (M2) should be used when information on whether detected causality arises directly one step ahead or indirectly two steps ahead is also desired. As a result, our expectation is that most researchers will more likely use M2 rather than M1.

We investigated the finite sample properties of our sequential strategies through Monte Carlo simulations. Though the data generating processes that we employed were relatively simple and should be expanded on in a more elaborative study, our findings may be summarized as follows:

(i)   The testing procedures perform reasonably well, irrespective of sample size and lag selection method.

(ii) The form of the underlying DGP can impact substantially on the probability of the Type I error associated with that DGP. The actual level is closer to a notion of an assigned level when the auxiliary variable is sufficiently causal with one of the variables under test. That is, when testing for $Y \underset{2}{\nrightarrow} X$ via Z, there are two causal relationships of interest: $Y \underset{1}{\nrightarrow} Z$ and $Z \underset{1}{\nrightarrow} X$. If only one of these is true, so that either $Y \underset{1}{\rightarrow} Z$ or $Z \underset{1}{\rightarrow} Y$, then our results suggest that we can have some confidence about the probability of the Type I error that we are likely to observe. However, when both are true, so that Z is not sufficiently involved with X or Y, then the probability of the Type I error can be small and quite different from any nominal notion of a level that we may have.

(iii) The testing strategies do well at detecting when $Y \underset{2}{\not\to} X$ is false, irrespective of the form of the DGP. Our results suggest that a sample size of at least 100 is preferable, though this depends, naturally, on the degree to which the null is false.

(iv) The choice of the lag length in the performance of the test is important. This issue requires further attention, with DGPs of higher lag order.

An obvious extension of this work is to the development of tests that examine for multi-step noncausality in higher-dimensional systems. This will typically involve testing for zero restrictions on multilinear functions of the VAR coefficients, which may result in Jacobian matrices of the restrictions having less than full rank under the null hypothesis and so lead to nonstandard asymptotic null distributions.

ACKNOWLEDGMENTS

I am grateful to David Giles, Nilanjana Roy, seminar participants at the University of Victoria, and the referee for helpful discussions and/or comments and suggestions on earlier drafts. This research was supported by a grant from the Social Sciences and Humanities Council of Canada.

REFERENCES

H. Akaike. Information theory and an extension of the maximum likelihood principle. In: B. N. Petrov and F. Csáki, eds. *2nd International Symposium on Information Theory*. Budapest: Académiai Kiadó, pp. 267–281, 1973.

D. C. Black, P. R. Corrigan, and M. R. Dowd. New dogs and old tricks: do money and interest rates still provide information content for forecasts of output and prices? *International Journal of Forecasting* 16: 191–205, 2000.

B. S. Cheng. Beyond the purchasing power parity: testing for cointegration and causality between exchange rates, prices and interest rates. *Journal of International Money and Finance* 18: 911–924, 1999.

R. Davidson and J. G. MacKinnon. Graphical methods for investigating the size and power of hypothesis tests. *The Manchester School* 66: 1–26, 1998.

J. J. Dolado and H. Lütkepohl. Making Wald tests work for cointegrated VAR systems. *Econometric Reviews* 15: 369–385, 1996.

J.-M. Dufour. Some impossibility theorems in econometrics with applications to structural and dynamic models. *Econometrica* 65: 103–132, 1997.

J.-M. Dufour and E. Renault. Short run and long run causality in time series: theory. *Econometrica* 66: 1099–1125, 1998.

B. M. Friedman and K. N. Kuttner. Money, income, prices and interest rates. *American Economic Review* 82: 472–492, 1992.

J. A. Giles and S. Mirza. Some pretesting issues on testing for Granger noncausality. Econometrics Working Paper EWP9914, Department of Economics, University of Victoria, 1999.

C. W. J. Granger. Investigating causal relations by econometric models and cross-spectral methods. *Econometrica* 37: 424–459, 1969.

I. Henriques and P. Sadorsky. Export-led growth or growth-driven exports? The Canadian case. *Canadian Journal of Economics* 96: 540–555, 1996.

D. L. Hoffman and R. H. Rasche. Assessing forecast performance in a cointegrated system. *Journal of Applied Econometrics* 11, 495–517, 1996.

J. L. Horowitz and N. E. Savin. Empirically relevant critical values for hypothesis tests: a bootstrap approach. *Journal of Econometrics* 95: 375–389, 2000.

C. Hsiao. Autoregressive modeling and causal ordering of economic variables. *Journal of Economic Dynamics and Control* 4: 243–259, 1982.

S. Kholdy. Causality between foreign investment and spillover efficiency. *Applied Economics* 27: 745–749, 1995.

K. Krishna, A. Ozyildirim, and N. R. Swanson. Trade, investment and growth: nexus, analysis and prognosis. Mimeograph, Department of Economics, Texas A&M University, 2000.

J. Lee, B. S. Shin, and I. Chung. Causality between advertising and sales: new evidence from cointegration. *Applied Economics Letters* 3: 299–301, 1996.

H. Lütkepohl. Testing for causation between two variables in higher dimensional VAR models. In: H. Schneeweiss and K. Zimmerman, eds. *Studies in Applied Econometrics*. Heidelberg: Springer-Verlag, 1993a.

H. Lütkepohl. *Introduction to Multiple Time Series Analysis*, 2nd Edition. Berlin: Springer-Verlag, 1993b.

R. Nishi. Maximum likelihood principle and model selection when the true model is unspecified. *Journal of Multivariate Analysis* 27: 392–403, 1988.

R. G. Riezman, P. M. Summers, and C. H. Whiteman. The engine of growth or its handmaiden? A time series assessment of export-led growth. *Empirical Economics* 21: 77–113, 1996.

G. Schwarz. Estimating the dimension of a model. *Annals of Statistics* 6: 461–464, 1978.

C. A. Sims. Macroeconomics and reality. *Econometrica* 48: 10–48, 1980.

C. A. Sims, J. H. Stock, and M. J. Watson. Inference in linear time series models with some unit roots. *Econometrica* 58: 113–144, 1990.

J. Thornton. Exports and economic growth: evidence from nineteenth century Europe. *Economics Letters* 55: 235–240, 1997.

H. Y. Toda and P. C. B. Phillips. Vector autoregression and causality. *Econometrica* 61: 1367–1393, 1993.

H. Y. Toda and P. C. B. Phillips. Vector autoregression and causality: a theoretical overview and simulation study. *Econometric Reviews* 13: 259–285, 1994.

H. Y. Toda and T. Yamamoto. Statistical inference in vector autoregressions with possibly integrated processes. *Journal of Econometrics* 66: 225–250, 1995.

N. Weiner. The theory of prediction. In: E. F. Beckenback, ed.*The Theory of Prediction*. New York: McGraw-Hill, Chapter 8, 1956.

M. B. Wilk and R. Gnanadesikan. Probability plotting methods for the analysis of data. *Biometrika* 33: 1–17, 1968.

H. Yamada and H. Y. Toda. Inference in possibly integrated vector autoregressive models: some finite sample evidence. *Journal of Econometrics* 86: 55–95, 1998.

H. O. Zapata and A. N. Rambaldi. Monte Carlo evidence on cointegration and causation. *Oxford Bulletin of Economics and Statistics* 59: 285–298, 1997.

19

Measurement of the Quality of Autoregressive Approximation, with Econometric Applications

JOHN W. GALBRAITH and VICTORIA ZINDE-WALSH   McGill University, Montreal, Quebec, Canada

1. INTRODUCTION

There are many circumstances in which one stochastic process is taken as a model of another, either inadvertently through mis-specification or deliberately as an approximation. In the present paper we are concerned with cases in which an autoregressive-moving average (ARMA) or moving-average process is, explicitly or implicitly, approximated by a pure autoregressive process. It is well known (see for example [1]) that an ARMA process with all latent roots of the moving-average polynomial inside the unit circle can be approximated arbitrarily well by an autoregressive process of order ℓ, as $\ell \to \infty$. The technique has been used for the estimation of moving-average or ARMA models by, among others, [2]–[8]. References [9] and [10] address the estimation of the spectral density through autoregression; [11] uses autoregression to approximate an ARMA error process in the residuals of a regression model; and [12] addresses the impact of that approximation on the asymptotic distribution of the ADF statistic.

   In problems such as these, the quality of the approximation affects some statistic of interest, and an ideal measure of the quality of the approximation would be monotonically related to the deviations caused by replacing the true process by the approximate one. As an example of the use of such a measure, consider a forecast based on a mis-specified model. If the accuracy of the forecast is monotonically related to some measure of the divergence between the true and mis-specified models, one can make an immediate use of the divergence measure in designing Monte Carlo experiments to evaluate forecast performance for different models and types of mis-specification; the measure allows us to identify cases where the approximation will do relatively well or badly, and to be sure of examining both.

The present study presents an approach to problems of this type. Our primary emphasis is on understanding the factors affecting quality of autoregressive approximation, and the impact of approximation as used in treating data, not on the development of new techniques for carrying out approximation or modeling. We treat autoregressive approximation and mis-specification in a common framework, implying replacement of the true model with one from another class, and use distance in the space of stationary stochastic processes as a measure of the severity of mis-specification, or quality of approximation. A measure of the distance from a process to a class of processes is defined, and may be minimized to find the closest member of that class.* We are able to indicate the order of AR process necessary to approximate particular MA(1) or MA(2) processes well, and are also able to give some general results on the value of the distance between processes as an indicator of the adequacy of an approximation in particular circumstances. For MA(1) processes the magnitude of the root is often mentioned as the factor determining the degree to which autoregressive approximation will be successful; here we are able to give a more general result.

It is important to distinguish these results about the appropriate order of approximating process from the use of sample-dependent criteria such as the Akaike or Schwarz information criteria to choose the order. While the two approaches may to some extent be complementary, the present study offers a priori information about the ability of an AR(ℓ) process, for given ℓ, to approximate a particular ARMA. In empirical applications, this information may be combined with information about the process being approximated, and a loss function, to generate a specific choice of order. Distance measures may also be used to evaluate information criteria in particular contexts, as in the example of Section 4.4.

---

*There are various possible measures of distance or divergence, including the well-known Kullback–Leibler and Hilbert distances; we concentrate here on the latter.

We offer several other econometric applications in Section 4. The distance measure is defined and described in Section 2, while Section 3 discusses its use in examining AR approximations.

2. DEFINITIONS AND PROPERTIES OF THE DISTANCE MEASURES

This section concentrates on the distance measures, particularly the Hilbert distance, that will be used for the problem of autoregressive approximation. For general reviews of information theory and distance measures, see [13] and [14].

We consider a discrete-time stochastic process $\{X_t\}$. The space of zero-mean, finite-variance stochastic processes can be represented as a real Hilbert space H with the scalar product (X, Y) defined by $E(XY)$; the Hilbert norm $\|X\|$ is given by $[E(X^2)]^{1/2}$. The values of the stochastic process $\{X_t\}$, $t \in Z$ (where the index set Z is the set of integers), span a subspace $H_x \subset H$ of the Hilbert space, which is itself a separable Hilbert space and thus has a countable basis. The lag operator L is defined such that $LX_t = X_{t-1}$; [15] and [16], for example, describe the relevant definitions and properties of the stationary stochastic processes and the Hilbert spaces used here. For the purpose of examining mis-specification, we restrict ourselves to the space H_x.

2.1 The Hilbert Distance

The Hilbert distance is the primary measure that we will use.

Since the space of second-order stationary stochastic processes is a Hilbert space, the *distance* between two processes X and Y is given by the norm of the difference, $d_H(X, Y) = \|X - Y\| = [E(X - Y)^2]^{1/2}$. In [17], this distance is used to examine mis-specification in first-order processes. In a Hilbert space, we can easily define the distance from a process to a *class* of processes (or the distance between classes), obtained by minimizing the distance over all processes in the class: for example, for the distance to the AR(ℓ) class, $d_H(X, AR(\ell)) = \inf_{Y \in AR(\ell)} d_H(X, Y)$.

The distance can also be expressed using the innovations representation of the processes in terms of the stationary uncorrelated process: i.e., the orthogonal basis $\{e_t\}_{-\infty}^{\infty}$. If $X_t = \sum_{i=0}^{\infty} \beta_i e_{t-i}$ and $Y_t = \sum_{i=0}^{\infty} \xi_i \varepsilon_{t-i}$ with $\varepsilon_t = \sigma e_t$, then

$$d_H(X, Y) = \|X - Y\| = \left[\sum_{i=0}^{\infty} (\beta_i - \sigma \xi_i)^2 \right]^{1/2} \sigma_e \qquad (2.1)$$

where $\|e_t\| = \sigma_e$, the standard error of the $\{e_t\}$. Without loss of generality, we will consider $\sigma_e = 1$ below unless otherwise specified.

We will consider processes that can be represented as $X_t = f(L)e_t$, where e_t is a white noise process and $f(L)$ is a rational polynomial, so that $f(L) = Q(L)/P(L)$ where Q and P are polynomials; we will express this as $P(L) = I - \alpha_1 L - \cdots - \alpha_p L^p$, and $Q(L) = I + \theta_1 L + \cdots + \theta_q L^q$. An ARMA$(p, q)$ process is described by $P(L)X_t = Q(L)e_t$, and is stationary if and only if the latent (i.e., inverse) roots of the polynomial $Q(L)$ are within the unit circle. If the process is invertible, then the inverse process $\{X_t^-\}$ defined by $Q(L)X_t^- = P(L)e_t$ is stationary. It is normally assumed that $P(L)$ and $Q(L)$ have no common factors. If $P(L) \equiv I$, then $\{X_t\}$ is an MA process; if $Q(L) \equiv I$, it is an AR process.

A stationary, zero-mean ARMA(p, q) process $\{X_t\}$ can be approximated arbitrarily well by an MA(k) process for some k: for an arbitrary bound δ on the approximation error, fix k such that $\sum_{i=k+1}^{\infty} \beta_i^2 < \delta$, and set the parameters θ_i of the approximating MA(k) process $\{Y_t\}$ such that $\theta_i = \beta_i$ for $i = 1, \ldots, k$. It follows that $\|X - Y\| < \delta^{1/2}$. If $\{X_t\}$ is an invertible process, then, for sufficiently large k, $\{Y_t\}$ will also be invertible.

Moreover, if $\{X_t\}$ is invertible, then it is also possible to express X_t as a convergent weighted sum of past values X_{t-i}, so that we can also find an AR(ℓ) process which approximates $\{X_t\}$ arbitrarily well. Consider an invertible kth-order moving-average lag polynomial represented by $Q_k(L)$, corresponding to $\{Y_t\}$ above. It has an infinite AR representation with autoregressive polynomial $P_k(L) \equiv [Q_k(L)]^{-1}$. If $Q_k(L) = I + \theta_1 L + \cdots + \theta_k L^k$, then $P_k(L) = (I + \theta_1 L + \cdots + \theta_k L^k)^{-1} = I - \theta_1 L + (\theta_1^2 - \theta_2)L^2 + \cdots = \sum_{i=0}^{\infty} \gamma_i L^i$. Denoting by ν_i the latent (i.e., inverse) roots of $Q_k(L)$, note that $\gamma_i \approx O(\bar{\nu}^i)$, where $\bar{\nu} = \max_{1 \le i \le k} |\nu_i|$, and $|\,.\,|$ represents the modulus of the root. Thus $\sum_{\ell+1}^{\infty} \gamma_i^2 \approx O(\bar{\nu}^{2\ell+2})$, and, for suitable order ℓ of the approximating process, this can be made less than any chosen δ. Denoting by $\{Z_t\}$ the AR(ℓ) process with coefficients $\alpha_i = \gamma_i$, $i = 1, \ldots, \ell$, we have $\|X - Z\| = \|X - Y + Y - Z\| \le \|X - Y\| + \|Y - Z\| = (\sum_{k+1}^{\infty} \beta_i^2)^{1/2} + (\sum_{\ell+1}^{\infty} \gamma_i^2)^{1/2}$. Hence an AR$(\ell)$ process can be found which is arbitrarily close to $\{X_t\}$ in the Hilbert metric. Also, convergence in the Hilbert metric implies convergence of the Fourier coefficients of the representation in the orthogonal basis of the processes.

As an example, consider an invertible MA(q) process $X_t = e_t + \sum_{j=1}^{q} \theta_j e_{t-j}$ with var$(e_t) = 1$, which is approximated by the AR(p) process $Z_t = \sum_{j=1}^{p} \alpha_j Z_{t-j} + \varepsilon_t$ with var$(\varepsilon_t) = \sigma^2$, that minimizes the Hilbert distance. As $p \to \infty$, the Hilbert distance between $\{X_t\}$ and $\{Z_t\}$ approaches zero, $\sigma \to 1$, and the first q coefficients $\{\alpha_j\}$ approach the values $\alpha_1 = \theta_1$, $\alpha_2 = -\theta_1 \alpha_1 + \theta_2$, $\alpha_i = -\theta_1 \alpha_{i-1} - \theta_2 \alpha_{i-2} - \cdots - \theta_{i-1} \alpha_1 + \theta_i$ for $i \le q$, and $\alpha_j = \sum_{i=1}^{q} -\theta_i \alpha_{j-i}$ for $j \ge q + 1$. These relations are used for parameter estimation in [7] and [8].

The Hilbert distance between second-order stationary processes in H corresponds to convergence in probability in that class. In fact, since it is defined through the mean square, convergence in this metric implies convergence in probability. On the other hand, convergence in probability to a process in H implies that the processes converge in mean square. Of course, if the processes in H converge in probability to a non-stationary process, they do not converge in this metric. The correspondence to convergence in probability makes the Hilbert metric a valuable measure of "closeness" in the space H, which can be used to evaluate the quality of various approximations. Unlike measures in finite parameter spaces, this measure can be used to compare processes of different types and orders.

2.2 The Kullback–Leibler and Kullback–Leibler–Jeffreys Divergence Measures

These two divergence measures are based on information functionals; see, for example, [18] or the review in [14]. For the Shannon entropy functional the Kullback–Leibler (K–L) divergence from a distribution with a density function $\phi_1(y)$ to a distribution with density $\phi_2(y)$ is given by

$$I(\phi_1 : \phi_2) = \int [\log(\phi_1/\phi_2) - 1]\phi_1 \, dy$$

This measure of divergence is not symmetric; it is sometimes called directional. The Kullback–Leibler–Jeffreys (K–L–J) divergence measure is non-directional (symmetric) and is defined as

$$d_{\mathrm{KLJ}}(\phi_1, \phi_2) = \frac{1}{2}[I(\phi_1 : \phi_2) + I(\phi_2 : \phi_1)]$$

Note that, although symmetric, d_{KLJ} is not a distance since it does not satisfy the triangle inequality. For Gaussian processes $X_t = f_1(L)e_t$ and $Y_t = f_2(L)e_t$, these divergence measures can be calculated as

$$I(X : Y) = (2\pi)^{-1} \int_0^{2\pi} [f_1(e^{iw})f_1(e^{-iw})f_2^{-1}(e^{iw})f_2^{-1}(e^{-iw}) - 1] \, dw$$

and we can compute $I(Y : X)$ similarly; then

$$d_{\mathrm{KLJ}}(X, Y) = \frac{1}{2}[I(X : Y) + I(Y : X)] \tag{2.2}$$

The Hilbert distance can be represented through f_1, f_2 as

$$\|X - Y\|^2 = (2\pi)^{-1} \int_0^{2\pi} [f_1(e^{iw})f_1(e^{-iw}) + f_2(e^{iw})f_2(e^{-iw}) - f_1(e^{iw})f_2(e^{-iw})$$
$$- f_2(e^{iw})f_1(e^{-iw})]dw$$

We can also represent $d_{\text{KLJ}}(X, Y)$ via the Hilbert norm. If we define a process $\{Z_t\}$ via $Z = f_1(L)/f_2(L)e = \sum_i \omega_i e_{t-i}$, and define $\overline{Z} = f_2(L)/f_1(L)e = \sum_i \overline{\omega}_i e_{t-i}$, then

$$d_{\text{KLJ}}(X, Y) = \frac{1}{2}\left[\|Z\|^2 + \|\overline{Z}\|^2\right] - 1 \qquad (2.3)$$

where $\|Z\|^2 = \sum \omega_i^2$ and $\|\overline{Z}\|^2 = \sum \overline{\omega}_i^2$. The formula (2.3) can be used to compute the Kullback–Leibler–Jeffreys divergence from one process to another and can be minimized over a particular class to find the minimum divergence from a given process to a class of processes. While our primary focus in this paper is on the use of the Hilbert distance, we will incorporate K–L–J distance measures into several examples below for purposes of comparison.

Before addressing some applications of these concepts, we note that it may be useful to restrict somewhat the class of mis-specified models considered in the applications. We may assume that some characteristics will be shared between the true and mis-specified models; in particular, if we know that some population moments exist, we may wish to consider a mis-specified process with the same population moments. Indeed, if we were to use sample moments in the estimation they would come from the same time series data regardless of which model was specified. Since covariance stationary stochastic processes possess at least two moments, here we consider as the approximation the closest process in the approximating class, subject to the restriction that the first two moments are the same as those of the process being approximated. That is, in cases for which we compute the theoretical best approximation within a class, this restriction on population moments is imposed in using both Hilbert and K–L–J distances. The K–L–J distance then becomes

$$d_{\text{KLJ}}(X, Y) = \frac{1}{2}\left[\sum \omega_i^2(v_2/v_1) + \sum \overline{\omega}_i^2(v_1/v_2)\right] - 1$$

where v_1, v_2 are the variances of the processes X_t and Y_t defined above. In the case of the Hilbert distance, we normalize one of the sets of squared projection coefficients by the ratio of variances.

3. EVALUATION OF APPROXIMATIONS USING DISTANCE MEASURES

When we use techniques that approximate one process by a process from another class, we can identify some member or members of the approximating class that are closest to the original process by the Hilbert (or other) distance. We will refer to the distance between the original process and an approximating process in a given class as the *approximation distance,* and will be interested in calculating the minimum approximation distance achievable.* As discussed in Section 2.2, the approximate process is restricted to have the same mean and variance as the original process.

In order to evaluate this minimum (Hilbert) approximation distance, we express the original process and a candidate approximating process in terms of the projections onto past innovations. The function describing the distance between them, (2.1), is the sum of squared differences between the coefficients of these innovations' representations. Truncating this expression at a large value, the distance may be calculated, and with subsequent iterations the function can be minimized numerically over the parameters of the approximating process. In the calculations below we use a Powell algorithm (see [20], p. 299) to minimize the distance function.

Tables 1 and 2 give these examples of the approximation distances from specific invertible MA(1) and MA(2) processes to the closest members of the AR(p) class, $p = 1, 2, 4, 8, 12$; the approximating process is constrained to have the same variance as the original process. Table 2b gives the parameter values and roots of the processes appearing in Table 2a. These distances cannot be guaranteed to be global minima, but appear to be very close to them, at least for distances on the order of 10^{-8} or greater. The tables also report the distances from the original processes to the uncorrelated, or white noise, process having the same variance. For MA(1) processes, the distances are unaffected by the sign of the parameter. While for MA(1) processes the distance is a monotonic function of the modulus of the root, note that this is not the case with respect to the largest root of MA(2) processes.

These examples suggest at least two conclusions. First, through most of the MA(1) or MA(2) parameter spaces, the approximation distance can be made quite small with moderate orders of approximating process. For MA(1) processes, order 8 is sufficient in all cases to make the approximation distance less than 1% of the distance of the original process to the uncorre-

Parzen in [19] discusses a related concept, the approximation bias arising from the use of a finite-order AR(p) in place of the AR(∞) representation of a process. Parzen introduces a particular penalty function with which to estimate the approximating order, yielding the *criterion of autoregressive transfer function* for order selection.

Table 1. Approximation distances:* Distance from MA(1) process to nearest AR(p)

Root θ	Order, p, of approximating AR process					
	0	1	2	4	8	12
.999	1.081	0.570	0.366	0.199	9.37×10^{-2}	5.70×10^{-2}
.99	1.071	0.563	0.360	0.195	9.04×10^{-2}	5.43×10^{-2}
.95	1.023	0.530	0.335	0.176	7.64×10^{-2}	4.29×10^{-2}
.90	0.964	0.490	0.303	0.152	6.02×10^{-2}	3.04×10^{-2}
.70	0.734	0.335	0.185	7.16×10^{-2}	1.51×10^{-2}	3.54×10^{-3}
.50	0.514	0.196	8.75×10^{-2}	2.05×10^{-2}	1.27×10^{-3}	8.09×10^{-5}
.30	0.303	8.05×10^{-2}	2.36×10^{-2}	2.11×10^{-3}	1.72×10^{-5}	1.46×10^{-7}
.10	0.100	9.85×10^{-3}	9.85×10^{-4}	9.86×10^{-6}	9.86×10^{-10}	1.00×10^{-13}
.05	0.050	2.49×10^{-3}	1.25×10^{-4}	3.11×10^{-7}	1.96×10^{-12}	1.22×10^{-18}
.01	0.010	1.00×10^{-4}	1.00×10^{-6}	1.00×10^{-10}	1.00×10^{-18}	1.00×10^{-26}

*In Tables 1 and 2a, the column headed "0" gives the distance to the white noise process having the same variance. Results in Table 1 are unaffected by multiplying the moving-average parameter by -1.

Table 2a. Approximation distances:[†] Distance from MA(2) process to nearest AR(p)

Case	Order, p, of approximating AR process					
	0	1	2	4	8	12
1	2.605	1.326	0.792	0.376	0.137	6.57×10^{-2}
2	2.368	1.178	0.683	0.299	8.34×10^{-2}	2.78×10^{-2}
3	1.095	0.569	0.362	0.194	8.95×10^{-2}	5.36×10^{-2}
4	1.785	0.818	0.421	0.128	5.58×10^{-2}	2.03×10^{-2}
5	1.225	0.477	0.189	8.47×10^{-2}	2.05×10^{-2}	5.08×10^{-3}
6	0.990	0.404	0.188	4.85×10^{-2}	3.55×10^{-3}	2.41×10^{-4}
7	0.604	0.446	0.259	0.139	5.28×10^{-2}	2.41×10^{-2}
8	1.680	0.792	0.436	0.171	4.01×10^{-2}	1.15×10^{-2}
9	0.142	0.108	2.19×10^{-2}	3.39×10^{-3}	7.03×10^{-5}	1.36×10^{-6}
10	0.457	0.305	0.158	6.60×10^{-2}	1.38×10^{-2}	3.09×10^{-3}
11	0.766	0.245	6.87×10^{-2}	1.14×10^{-2}	2.52×10^{-4}	1.29×10^{-5}
12	0.0283	1.96×10^{-2}	8.89×10^{-4}	2.47×10^{-5}	1.50×10^{-8}	8.08×10^{-12}

[†]The case numbers refer to Table 2b, where the processes are described

Table 2b. Features of MA(2) processes used in Table 2a

Case	MA parameters		Real parts		Imaginary parts		Moduli	
	θ_1	θ_2						
1	−1.96	0.98	0.980	0.980	0.140	−0.140	0.990	0.990
2	−1.80	0.90	0.900	0.900	0.300	−0.300	0.949	0.949
3	−1.01	0.0198	0.990	0.020	0.00	0.00	0.990	0.020
4	−1.40	0.70	0.700	0.700	0.458	−0.458	0.837	0.837
5	1.00	0.50	−0.500	−0.500	0.500	−0.500	0.707	0.707
6	0.90	0.20	−0.500	−0.400	0.00	0.00	0.500	0.400
7	−0.50	−0.30	0.852	−0.352	0.00	0.00	0.852	0.352
8	−1.40	0.49	0.700	0.700	0.00	0.00	0.700	0.700
9	0.10	−0.10	−0.370	0.270	0.00	0.00	0.370	0.270
10	0.40	−0.20	−0.690	0.290	0.00	0.00	0.690	0.290
11	−0.70	0.20	0.350	0.350	0.278	−0.278	0.447	0.447
12	0.020	0.02	−0.010	−0.010	0.141	−0.141	0.141	0.141

lated process (that is, the approximation has picked up 99% of the original process, by our distance measure). For the MA(2) processes used in these examples, order 12 is sufficient in most cases to meet the same condition, but is not sufficient in cases 1, 2, 3, 4, and 7, where there is one or more root with modulus greater than 0.85 in absolute value. Nonetheless, in most cases it is clearly possible to make the approximation distances very small with orders of AR process that are well within the range estimable with typical samples of data.

Second, these results give an a priori indication of the appropriate order of approximating AR process. For moving average processes with the largest root near zero, there is little gain in increasing the order, p, beyond fairly small values. For processes with a root near the boundary of the invertibility region, there are still substantial gains in increasing p beyond 12, and the order of AR process necessary to make the approximation distance negligible may be large. This requirement imposes a lower bound on the sample size necessary to provide a good approximation with an autoregressive process.* Note, however, that these results do not embody the effect of increased model order on efficiency of parameter estimation; results bearing on this question are presented in Section 4.1.

---

*As well, small reductions in approximation distance become more important with increasing sample size, since overall distance from the estimated representation to the true process is itself declining in expectation.

The magnitude of approximation distance that is tolerable will depend upon the application. Nonetheless, it is worth emphasizing that this information about the order of the approximating process is not sample dependent. It is well known that widely-used sample-based criteria for order selection, such as the Akaike information criterion, may systematically suggest over- or under-parameterization; see [21] and the examples in Section 4.4 below. A criterion such as the distance in the space of population models, by contrast, provides a guide to order selection prior to estimation.

4. ECONOMETRIC APPLICATIONS

There are two types of problem that we can distinguish as being of interest in the context of mis-specified or approximate models. In the first type, the statistic is directly related to the mis-specification; and an example is given in Section 4.2, where we examine a test for the null of uncorrelated residuals in a model where MA errors are modeled by autoregression. In the second type, a statistic may estimate or test some property not directly related to the mis-specification; the mis-specification is nonetheless relevant because the distribution of the statistic will differ from the distribution that it would have with a correctly specified model. Examples are given in Section 4.3, where we consider the forecast error arising when MA processes are forecast using AR models, and in 4.4, where we examine the performance of information criteria in selecting the order of model which is the best approximation to an unknown process of more general form.

In each of these cases, we expect that the more severe the mis-specification, or the poorer the approximation, the more substantial will be the effect on the statistic of interest. Ideally, we would like to have a measure of the extent of mis-specification that has predictive power in a wide variety of circumstances. In the examples just mentioned, this would allow us to predict which of various MA processes will show the higher mean squared forecast error when forecasting is done via an AR, or which MA process in the errors of a regression model will lead to the largest average test statistic in an autocorrelation test when modeled as AR. In these examples, we show that the Hilbert distance performs well as such a measure, and, in particular, that it is a much better indicator than is the largest of the moduli of MA roots. While the Hilbert distance is the primary focus of our interest, we will also refer for comparison to the Kullback–Leibler–Jeffreys distance in two of the applications.

Before exploring these examples, in which the Hilbert distance measure is used to predict the values of sample-based criteria and is thereby evaluated, we apply this distance measure directly to the general problem of choice of AR order, which forms an element of the examples in Sections 4.2–4.4.

4.1 Choice of AR Order

Mis-specification or approximation can be thought of as yielding two sources of error: one caused by the mismatch between the mis-specified process (e.g., x) and the true process (e.g. y), and the other resulting from estimation of the mis-specified model (yielding $\hat{x}$ rather than x). Each of these, the approximation error and the estimation error, plays a role in determining the best approximating process, as the following application illustrates.

Consider the estimation of an AR model of a pure MA process. In choosing the best order for the AR model, there are two offsetting effects: first, as Section 3 showed, the best available approximation within the $AR(k)$ class will be closer to the true process as k increases; second, as k increases the efficiency of parameter estimation will be reduced, leading to a higher mean distance to the true process. We will use the Hilbert distance to investigate the optimal model order, $k^* = \text{argmin}_{k:\hat{x} \in AR(k)} E(\|y - \hat{x}\|)$, given these two effects.

For a given process y, and an approximating $AR(k)$ model, there is a closest process x within the $AR(k)$ class, and an estimated model $\hat{x}$. As k increases, x becomes a better approximation by the Hilbert distance ($\|y - x\|$ decreases monotonically). Parameter estimation becomes less efficient, however, and the mean distance of the estimated model to the best approximating model, $\|\hat{x} - x\|$, increases. The overall distance between true and estimated processes, $\|y - \hat{x}\|$, will have a minimum at some finite value of k.

Figures 1 to 3 present the results of simulations designed to estimate the relation between $\|y - \hat{x}\|$ and k for several examples of MA processes. There are 10,000 replications on sample sizes of $T = \{200, 1000\}$, and $k = \{1, 2, \ldots, 10\}$. Values on the vertical axis are the average values of $\|y - \hat{x}\|$ for the given MA process, y, across the 10,000 samples.

Note first that the optimal order increases in T, reflecting diminished relative importance of parameter estimation error, at a given k, as T increases. Optimal order also increases, subject to the integer constraint on k, as the distance between the true process and the closest process in the $AR(k)$ class increases (see again Tables 1 and 2a). For $\theta = 0.90$, optimal orders are 5 and 9 at $T = 200$ and 1000, for $\theta = 0.50$, optimal orders are 3 and 4, while for $\theta = 0.10$ there is no gain in approximating with an order greater than 1 at either sample size.

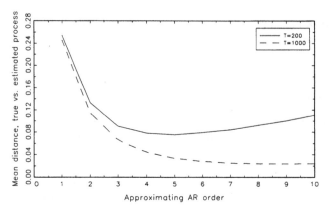

Figure 19.1.   MA(1) parameter = 0.90.

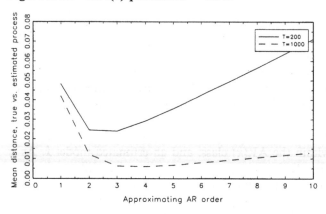

Figure 19.2.   MA(1) parameter = 0.50.

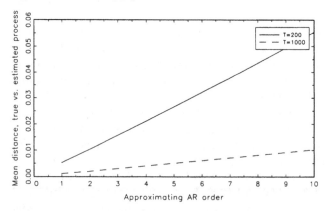

Figure 19.3.   MA(1) parameter = 0.10.

These results are purely illustrative. However, we can summarize the results of a larger number of such experiments by estimating a response surface for the optimal order as a function of the parameter of an MA(1) model and sample size, with $\theta = \{0.05, 0.1, 0.3, 0.5, 0.7, 0.8, 0.9, 0.95, 0.99\}$ and $T = \{25, 50, 75, 100, 200, 300, 400, 500, 1000\}$, yielding 81 cases. The response surface (standard errors in brackets)

$$k^* = -2.82 \quad +4.67(1-\theta)^2 \quad -0.23T^{1/4} \quad +2.56\theta T^{1/4} + u$$
$$\quad\;\;(0.17) \quad\;\;\; (0.32) \qquad\qquad (0.06) \qquad\;\; (0.08)$$

was found to provide a reasonable fit ($R^2 = 0.98$) to the points. For example, using the processes examined in Figures 1–3, we have for each combination (θ, T) the following estimated optimal orders: $(0.1, 200)$, $\hat{k}^* = 1.07$; $(0.1, 1000)$, $\hat{k}^* = 1.12$; $(0.5, 200)$, $\hat{k}^* = 2.30$; $(0.5, 1000)$, $\hat{k}^* = 4.26$; $(0.9, 200)$, $\hat{k}^* = 5.02$; $(0.9, 1000)$, $\hat{k}^* = 8.89$. Each of these is quite close to the actual optimum for the given (θ, T).

To summarize: the distance measure allows us to indicate the best approximating model for a given process and sample size, taking into account the estimation method. In these cases of MA processes approximated by autoregressions, the optimal orders are fairly modest.

4.2 Dynamic Specification

Appropriate specification of dynamics is an important problem in time series regression; see Hendry [22] for a thorough review of this literature. One of the most commonly applied techniques is the imposition of a low-order autoregressive structure on the errors of a regression model (which may be a static regression apart from the error dynamics). It is well known that this implies a common-factor restriction on the coefficients of a corresponding autoregressive-distributed lag model with white noise errors: that is,

$$y_t = \beta x_t + u_t, \qquad \rho(L)u_t = \varepsilon_t \tag{4.1}$$

is equivalent to

$$\rho(L)y_t = \rho(L)\beta x_t + \varepsilon_t \tag{4.2}$$

where $\{\varepsilon_t\}$ is a white-noise process, implying a set of restrictions on the coefficients of the regression model (4.2) arising from the common lag polynomial $\rho(L)$. If $\rho(L)$ is of degree k there are k such restrictions; for example, for $k = 2$, $\rho(L) = 1 - \rho_1 L - \rho_2 L^2$ and

$$y_t = \rho_1 y_{t-1} + \rho_2 y_{t-2} + \beta x_t + \delta_1 x_{t-1} + \delta_2 x_{t-2} + \varepsilon_t \tag{4.3}$$

with $\delta_1 = \rho_1 \beta$ and $\delta_2 = \rho_2 \beta$.

Consider now the effect of using an AR model of error dynamics in this way when the true process contains a moving-average component: that is, the true error process in (4.1) is instead $\gamma(L)u_t = \theta(L)\varepsilon_t$. The autoregressive-distributed lag (ADL) representation of the model now embodies sets of coefficients on both lagged Y and lagged X, from the approximating AR polynomial $\rho(L)$, which decline geometrically but are nonzero at any finite lag. There is a corresponding (infinite) set of common-factor restrictions. Truncating the representation at any finite lag length k might be expected to perform relatively well as the Hilbert distance to this approximating AR(k) model is smaller. If the distance measure is useful in indicating the order of AR polynomial necessary to model a relation with ARMA errors via an ADL model, there must be a close correspondence between the distance from the ARMA to a given AR, and sample-based measures of the degree of *inadequacy* of the dynamic specification. The indicator that we use is a standard LM statistic for the null of no autocorrelation from lags 1 to s. The mis-specification considered is the use of an AR(2) error process instead of the true MA(2). Note that the distance measure is not an *alternative* to the sample-based (LM) statistic; instead, it is intended to help understand the process of approximation by describing the degree to which particular known processes may be well approximated by particular models. If successful, the distance measure should in some degree predict the results actually observed in sample-based indicators of adequacy of approximation, such as the LM statistic for residual autocorrelation.

Table 3 reports the results of a simulation experiment designed to check this performance. Using the MA(1) and MA(2) models of Tables 1 and 2b, 5000 replications on samples of size $T = 200$ were generated from the DGP $y_t = \alpha + \beta x_t + u_t$, $\gamma(L)u_t = \theta(L)\varepsilon_t$, with $\alpha = \beta = 1$, $\gamma(L) = I$, and $\theta(L)$ as given in Tables 1, 2b.* The innovations $\{\varepsilon_t\}$ have unit variance. The process is modelled with the ADL model corresponding to an AR(2) model of the errors,

$$y_t = \alpha_0 + \sum_{i=1}^{2} \alpha_i y_{t-i} + \sum_{i=1}^{2} \gamma_i x_{t-i} + e_t \tag{4.4}$$

On each sample, the residuals are tested for autocorrelation up to order ℓ, $\ell = \{1, 2, 12\}$ via an LM test which is asymptotically χ_ℓ^2 under the null of no autocorrelation. If the approximation is adequate, then there should be little evidence of residual autocorrelation in these tests. Table 3 gives the mean values of the LM statistics, and ranks both these and the corresponding Hilbert and K–L–J measures of the distance between the true process and

---

*Results on samples of size 1000 are very similar and are therefore not reported.

Table 3a. LM tests for residual autocorrelation: MA errors modelled by AR approximation; $T = 200$

Case	θ_1	θ_2	$\ell = 1$	$\ell = 2$	$\ell = 12$
1	−1.96	0.98	31.64	51.36	95.95
2	−1.80	0.90	34.39	55.86	92.77
3	−1.01	0.0198	12.73	21.12	46.11
4	−1.40	0.70	28.58	39.66	53.69
5	1.00	0.50	3.882	5.555	20.03
6	0.90	0.20	8.507	12.55	22.85
7	−0.50	−0.30	7.529	13.57	31.30
8	−1.40	0.49	25.22	39.76	63.77
9	0.10	−0.10	1.055	2.138	12.31
10	0.40	−0.20	4.308	7.556	18.95
11	−0.70	0.20	1.999	3.106	13.14
12	0.020	0.02	1.026	2.093	12.21

approximating model; since the approximating model is in every case AR(2), the ranks by distance are in all cases based on the distance to the nearest AR(2).

Both distance measures provide very good a priori indicators of the degree to which residual autocorrelation will be detected; that is, they explain the variation in mean LM test statistics very well. The Hilbert distance is especially good; as the order of test increases to measure

Table 3b. Cases ranked by approximation distance and LM test (rank of given case by: K–L–J distance, Hilbert distance, LM statistic); $T = 200$

Case	θ_1	θ_2	$\ell = 1$	$\ell = 2$	$\ell = 12$
1	−1.96	0.98	(1,1,2)	(1,1,2)	(1,1,1)
2	−1.80	0.90	(2,2,1)	(2,2,1)	(2,2,2)
3	−1.01	0.0198	(3,5,5)	(3,5,5)	(3,5,5)
4	−1.40	0.70	(5,4,3)	(5,4,4)	(5,4,4)
5	1.00	0.50	(8,7,9)	(8,7,9)	(8,7,8)
6	0.90	0.20	(7,8,6)	(7,8,7)	(7,8,7)
7	−0.50	−0.30	(6,6,7)	(6,6,6)	(6,6,6)
8	−1.40	0.49	(4,3,4)	(4,3,3)	(4,3,3)
9	0.10	−0.10	(11,11,11)	(11,11,11)	(11,11,11)
10	0.40	−0.20	(9,9,8)	(9,9,8)	(9,9,9)
11	−0.70	0.20	(10,10,10)	(10,10,10)	(10,10,10)
12	0.020	0.02	(12,12,12)	(12,12,12)	(12,12,12)

autocorrelations up to 12, the match by ranks becomes virtually perfect for the Hilbert measure, differing only in the ranking of cases 5 and 6 (ranked 7th and 8th by the Hilbert measure, but 8th and 7th by K–L–J and mean LM). These cases are extremely close, having distances to the nearest AR(2) of 0.189 and 0.188 respectively. The first twelve lagged innovations capture a smaller part of the total variation for process 5 than for 6; k higher than twelve is necessary in the LM test in order to reproduce exactly the Hilbert distance rankings.

The use of ADL models to capture dynamics easily through LS regression is commonplace, and is a successful strategy in cases where the error dynamics can be well modeled by a low-order AR. However, where there are MA components with substantial roots, or other components for which the PACF does not approach zero quickly, the Hilbert distance from the DGP of the errors to the AR approximation implicitly used in the ADL specification is a reliable measure of the adequacy of the implicit approximation. These distances are not sample-based measures, but aids to understanding a priori the features of a process that make it relatively easy or difficult to approximate with a model of a given order.

4.3 Forecasting

Consider next the problem of forecasting a time series process, which may have a moving average component, using a pure autoregression. In this case, a measure of the distance between a given process and the nearest AR(p) will be useful insofar as it gives an a priori indication of the degree to which mean squared error of the forecast is increased by the use of the AR approximation in the place of a model containing MA parts. The process to be forecast is a stationary process $\{y_t\}$, with a Wold representation which we can write as

$$y_{t+1} = f(L)e_{t+1} = f_1(L)e_t + f_0\varepsilon_{t+1} \tag{4.5}$$

where $e_t = \{\varepsilon_t, \varepsilon_{t-1}, \ldots, \varepsilon_1\}'$, and the $\{\varepsilon_t\}$ are white noise. Given a sample of data, we obtain implicitly an estimated lag polynomial $\hat{f}_1(L)$.[*] The one-step-ahead forecast is generated by

$$\hat{y}_{t+1|t} = \hat{f}_1(L)\hat{e}_t \tag{4.6}$$

where $\hat{y}_{t+1|t}$ indicates a forecast made at time t of the $t+1$ value of Y. The one-step-ahead forecast error is then

[*]For example, if we fit an AR model to the data, $\hat{f}(L)$ represents the projection of the estimated AR polynomial onto past innovations.

$$(\hat{y}_{t+1|t} - y_{t+1}) = \hat{f}_1(L)\hat{e}_t - f_1(L)e_t - f_0\varepsilon_{t+1} \tag{4.7}$$

Table 4 gives the mean squared errors of one-step-ahead forecasts made from AR(1), AR(2), and AR(4) models of the MA(2) processes listed in Table 2b, again for $T = 200$ and 5000 replications. Once again, the ordering given by distances of the example processes to the relevant AR approxima-

Table 4a. MSEs of one-step-ahead forecasts: MA processes modelled by AR approximation; $T = 200$

Case	θ_1	θ_2	AR(1)	AR(2)	AR(4)
1	−1.96	0.98	3.278	2.460	1.821
2	−1.80	0.90	2.792	2.058	1.476
3	−1.01	0.0198	1.514	1.340	1.219
4	−1.40	0.70	1839	1.357	1.098
5	1.00	0.50	1.268	1.077	1.065
6	0.90	0.20	1.236	1.088	1.039
7	−0.50	−0.30	1.261	1.141	1.081
8	−1.40	0.49	1.872	1.457	1.177
9	0.10	−0.10	1.022	1.017	1.032
10	0.40	−0.20	1.125	1.55	1.041
11	−0.70	0.20	1.081	1.023	1.032

Table 4b. Cases ranked by approximation distance and one-step MSE (rank of given case by: K–L–J distance, Hilbert distance, MSE); $T = 200$

Case	θ_1	θ_2	AR(1)	AR(2)	AR(4)
1	−1.96	0.98	(1,1,1)	(1,1,1)	(1,1,1)
2	−1.80	0.90	(2,2,2)	(2,2,2)	(3,2,2)
3	−1.01	0.0198	(3,5,5)	(3,5,5)	(2,3,3)
4	−1.40	0.70	(5,3,4)	(5,4,4)	(6,6,5)
5	1.00	0.50	(8,6,6)	(8,7,8)	(7,7,7)
6	0.90	0.20	(7,8,8)	(7,8,7)	(9,9,9)
7	−0.50	−0.30	(6,7,7)	(6,6,6)	(5,5,6)
8	−1.40	0.49	(4,4,3)	(4,3,3)	(4,4,4)
9	0.10	−0.10	(11,11,11)	(11,11,11)	(11,11,11)
10	0.40	−0.20	(9,9,9)	(9,9,9)	(8,8,8)
11	−0.70	0.20	(10,10,10)	(10,10,10)	(10,10,10)
12	0.020	0.02	(12,12,12)	(12,12,12)	(12,12,12)

tion matches very well the ordering of the estimated MSEs. In the AR(4) case, the distance and MSE rankings differ only by interchanging cases 4 and 7, which have distances to the nearest AR(4) of 0.128 and 0.139 respectively. Mean squared errors tend to be very close to unity, the correct value for a properly specified model, for approximation distances of less than 0.1.

Again, both distance measures explain the results well, providing an a priori understanding of the MA or ARMA parameter values that allow a good approximation to be made with an AR of given order. The Hilbert distance seems again to have some advantage. For the AR(4) case, the Hilbert ranking differs from that of the forecast errors only for cases 4 and 7 (ranked 6th and 5th, respectively, rather than 5th and 6th). The K–L–J ranking is similar to that of the Hilbert distance, but makes an additional interchange relative to the ranking of cases by forecast error, in cases 2 and 3.

4.4 Evaluation of Information Criteria

Sample-based selection of appropriate lag length (or, more generally, model order) is often based on information criteria such as those of Akaike, Schwarz, and others; see [21] and [23] for recent reviews. In the context of problems for which the DGP is a special case of more general estimated models, we can investigate these criteria by simulation, preferring those which tend to yield lag lengths close to the optimal values. Where the model is an approximation, however, it may be unclear what the best lag length is even in a constructed example, so that evaluation of the criteria in cases such as that of AR models, which are being used to approximate more general processes, cannot proceed.

However, using a distance measure of the difference between DGP and AR approximation, we can proceed as in Section 4.1 to an answer to the question of what the optimal approximating model order is, given a DGP and sample size. From this it is possible to evaluate the information criteria by examining the degree to which the typical selected lag length differs from the optimum. This section provides a brief example of such an exercise, using the AIC, BIC, Schwarz, and FPE criteria.* The exercise may be viewed as an extension of the application of Section 4.1, in that we now investigate the ability of a posteriori, sample-based criteria to reproduce the optimal orders obtained in that section.

---

*For this linear regression problem the criteria can be reduced to the following expressions in the sample size, T, number of autoregressive terms, k, and sum of squared residuals, $\hat{\epsilon}'\hat{\epsilon}$: AIC: $\ln(\hat{\epsilon}'\hat{\epsilon}/T) + 2k/T$; BIC: $\ln(\hat{\epsilon}'\hat{\epsilon}/T) + k\ln(T)/T$; Schwarz: $\ln(\hat{\epsilon}'\hat{\epsilon}/(T-k)) + k\ln(T)/T$; FPE : $((T+k)/(T-k))(\hat{\epsilon}'\hat{\epsilon}/(T-k))$.

For the data generation processes and sample sizes in Section 4.1, we compute the average lag lengths selected by each of these criteria, in 2500 simulated samples. The results are recorded in Table 5, along with the optimal approximating lag lengths from Figures 1–3. Where the objective function is nearly flat near the optimum lag length, we report a range of optimal values (e.g., 8–10 for $T = 1000$ and $\theta = 0.9$). The set of lag lengths considered ranged from 1 to 20; with even larger values included, averages for the AIC would rise slightly.

The BIC and Schwarz criteria, which are very similar and closely related, produce very good results. The AIC, as has been observed in contexts where approximation and mis-specification play no role, over-parameterizes dramatically. The FPE falls in between, over-parameterizing consistently, but less substantially than the AIC.

5. CONCLUDING REMARKS

There are many circumstances in which it is convenient to approximate an ARMA process by a pure AR(p) process. But while the technique is widely used, often implicitly, there are relatively few results concerning the order of autoregression necessary to provide a good approximation. This paper addresses the question of the quality of an approximation using measures of the distance between processes, primarily the Hilbert distance. By minimizing this distance from a process to a class of processes, we are able to find the closest process of given order in the target class, and by incorporating information about estimation of differing approximate models, we can find by simulation the best approximating model at a particular sample size. The results offer a general contribution to understanding of the relations

Table 5. Estimated optimal AR order vs. mean selected order, various criteria

Case		Optimal order	Averaged selected order			
θ	T		AIC	BIC	Schwarz	FPE
0.1	200	1	12.1	1.14	1.06	3.47
0.1	1000	1	10.1	1.05	1.02	2.89
0.5	200	2–3	12.7	2.20	1.94	4.91
0.5	1000	3–4	11.4	2.96	2.80	5.49
0.9	200	4–5	16.2	5.98	5.09	11.1
0.9	1000	8–10	17.8	9.91	9.16	15.8

between ARMA processes, of the gains available from more elaborate modeling, and of the use of autoregressive approximations in various applied problems including the traditional problem of choice of order.

ACKNOWLEDGMENTS

The authors thank Alfred Haug, David Hendry, Aman Ullah, and seminar participants at Amsterdam, Erasmus, and Oxford universities, CORE, the Canadian Econometric Study Group, and Société canadienne des sciences économiques for valuable comments. The Fonds pour la Formation de chercheurs et l'aide à la recherche (Quebec) and the Social Sciences and Humanities Research Council of Canada provided financial support for this research.

REFERENCES

1. W.A. Fuller. *Introduction to Statistical Time Series*. New York: Wiley, 1976.
2. J. Durbin. Efficient estimation of parameters in moving-average models. *Biometrika* 46:306–316, 1959.
3. J. Durbin. The fitting of time series models. *Review of the International Statistical Institute* 28:233–243, 1960.
4. E. J. Hannan, J. Rissanen. Recursive estimation of mixed autoregressive-moving average order. *Biometrika* 69:81–94, 1982.
5. P. Saikkonen. Asymptotic properties of some preliminary estimators for autoregressive moving average time series models. *Journal of Time Series Analysis* 7:133–155, 1986.
6. S. Koreisha, T. Pukkila. A generalized least squares approach for estimation of autoregressive moving average models. *Journal of Time Series Analysis* 11:139–151, 1990.
7. J. W. Galbraith, V. Zinde-Walsh. A simple, non-iterative estimator for moving-average models. *Biometrika* 81:143–155, 1994.
8. J. W. Galbraith, V. Zinde-Walsh. Simple estimation and identification techniques for general ARMA models. *Biometrika* 84:685–696, 1997.
9. H. Akaike. Power spectrum estimation through autoregressive model fitting. *Annals of the Institute of Statistical Mathematics* 21:407–419, 1969.
10. K. N. Berk. Consistent autoregressive spectral estimates. *Annals of Statistics* 2:489–502, 1974.

11. S. E. Said, D.A. Dickey. Testing for unit roots in autoregressive-moving average models of unknown order. *Biometrika* 71:599–607, 1984.
12. J. W. Galbraith, V. Zinde-Walsh. On the distributions of Augmented Dickey–Fuller statistics in processes with moving average components. *Journal of Econometrics* 93:25–47, 1999.
13. E. Maasoumi. A compendium to information theory in economics and econometrics. *Econometric Reviews* 12:137–181, 1993.
14. A. Ullah. Entropy, divergence and distance measures with econometric applications. *Journal of Statistical Planning and Inference* 49:137–162, 1996.
15. Y. A. Rozanov. *Stationary Random Process*. San Francisco: Holden-Day, 1967.
16. M. B. Priestley. *Spectral Analysis and Time Series*. London: Academic Press, 1981.
17. V. Zinde-Walsh. The consequences of mis-specification in time series processes. *Economics Letters* 32:237–241, 1990.
18. J. Burbea, C.R. Rao. Entropy differential metric, distance and divergence measures in probability spaces: a unified approach. *Journal of Multivariate Analysis* 12:575–596, 1982.
19. E. Parzen. Autoregressive spectral estimation. In: D.R. Brillinger, P. R. Krishnaiah, eds., *Handbook of Statistics,* vol. 3. Amsterdam: North-Holland, 1983, pp 221–247.
20. W. H. Press, B.P. Flannery, S.A. Teukolsky, W.T. Vetterling. *Numerical Recipes: the Art of Scientific Computing*. Cambridge: Cambridge University Press, 1986.
21. B.S. Choi. *ARMA Model Identification*. New York: Springer-Verlag, 1992.
22. D. F. Hendry. *Dynamic Econometrics*. Oxford: Oxford University Press, 1995.
23. J. A. Mills, K. Prasad. A comparison of model selection criteria. *Econometric Reviews* 11:201–233, 1992.

20

Bayesian Inference of a Dynamic Linear Model with Edgeworth Series Disturbances

ANOOP CHATURVEDI   University of Allahabad, India

ALAN T. K. WAN   City University of Hong Kong, Hong Kong S.A.R., China

GUOHUA ZOU   Chinese Academy of Sciences, Beijing, China

1. INTRODUCTION

Linear dynamic models containing a one-period lagged-dependent variable and an arbitrary number of fixed regressors, termed as ARX(1) models, are used frequently in econometrics. Often these models arise in the form of partial adjustment or first-order autoregressive distributed lag models, or an AR(1) model with an intercept, linear trend or seasonal dummy variables. Various results have been published on the moments and the limiting and exact finite-sample distributional properties of the least squares estimator of the autoregressive parameter in AR(1) models with or without exogenous information. Some contributions to this area are by Mann and Wald (1943), Anderson (1959), Phillips (1977), Tanaka (1983), Grubb and Symons (1987), Peters (1989), Kiviet and Phillips (1993), Kiviet et al. (1995), Dufour and Kiviet (1998), among others. Zellner (1971, Ch.7) put forward a Bayesian approach to analyzing dynamic regression models using a vague prior distribution assumption for the model's parameters. Building on Zellner's work, Broemeling (1985, Ch. 5), assumed a normal-gamma prior for the regression coefficients and precision of disturbances derived the posterior density functions of the model's coefficients.

The bulk of the research reported so far has been conducted on the premise that the distribution of the model's disturbances is normal, though it is well recognized that such an assumption is often questionable and may lead to varying effects in a variety of situations. See, e.g., Kendall (1953), Mandelbrot (1963, 1967), Fama (1965), Press (1968), and Praetz (1972). As a model for non-normality, the Gram–Charlier or the Edgeworth series distribution (ESD) has received a great deal of attention over the years. See the work of Barton and Dennis (1952), Davis (1976), Kocherlakota and Chinganda (1978), Balakrishnan and Kocherlakota (1985), Knight (1985, 1986), Peters (1989), Chaturvedi et al. (1997a,b), and Hasegawa et al. (2000). Of particular relevance here are the results of Peters (1989), who studied the sensitivity of the least-squares estimator of the autoregressive coefficient in an ARX(1) model with disturbances drawn from an ESD, and Chaturvedi et al. (1997a), who derived the posterior distributions and Bayes estimators of the regression coefficients in a linear model with a non-normal error process characterized by an ESD.

This paper considers Bayesian analysis of the ARX(1) model with Edgeworth series errors. Under a diffuse prior, the posterior distribution of the model's autoregressive coefficient is derived. The results are obtained along the lines suggested by Davis (1976) and Knight (1985) in some earlier work. Numerical results are then presented to demonstrate the effects of the departure from normality of disturbances on the posterior distribution of the model's autoregressive coefficient. It is found that the posterior distribution is sensitive to both the skewness and kurtosis of the distribution and the increase in posterior risk of the Bayes estimator from erroneously ignoring non-normality in the model's error process is non-negligible.

2. MODEL AND POSTERIOR ANALYSIS

Consider the stable ARX(1) model,

$$y = \rho y_{-1} + X\beta + u \tag{2.1}$$

where $y = (y_1, y_2, \ldots, y_T)'$ is a $T \times 1$ vector of observations on a dependent variable, $y_{-1} = (y_0, y_1, \ldots, y_{T-1})'$ is the y vector lagged one period, X is a full column-rank $T \times p$ matrix on p fixed regressors, ρ is a scalar constant, β is a $p \times 1$ vector of fixed coefficients, and $u = (u_1, u_2, \ldots, u_T)'$ is the $T \times 1$ vector of i.i.d. disturbances assumed to follow an ESD, i.e., $u_t \sim \text{ESD}$ $(0, \tau^{-1})$, $t = 1, \ldots, T$. Denote the rth cumulant of the distribution by k_r, $(r = 1, 2, \ldots)$ so that k_3 and k_4 measure the distribution's skewness and the kurtosis, respectively. Throughout the analysis it is assumed that the cumulants of order greater than four are negligible.

Now, write the distribution of u as

$$p(u) = E_z[\phi_T(u|z, \tau I_T)] \tag{2.2}$$

where $z = (z_1, z_2, \ldots, z_T)'$, the z_t, $t = 1, \ldots, T$, are i.i.d. pseudo variates whose mean and variance are zero and higher cumulants are the same as u_t and $\phi_T(u|z, \tau I_T)$ is the p.d.f. of a normal distribution with mean vector z and precision matrix τI_T (see Davis 1976). For the specification of the prior of the parameters it is assumed that $|\rho| < 1$ holds with certainty, but within the region $-1 < \rho < 1$ the prior for the parameters is noninformative. Thus it seems reasonable to write the prior p.d.f. as

$$p(\beta, \rho, \tau) \propto \frac{1}{\tau(1 - \rho^2)^{1/2}}, \qquad -1 < \rho < 1; \ 0 < \tau < \infty; \ \beta \in R^p \tag{2.3}$$

Further, conditioning on the initial value y_0, it follows that,

$$L(y|X, y_0, \rho, \beta, \tau) = E_z\left[\left(\frac{\tau}{2\pi}\right)^{T/2}\right.$$
$$\left. \exp\left(-\frac{\tau}{2}(y - \rho y_{-1} - X\beta - z)'(y_{-1} - X\beta - z)\right)\right] \tag{2.4}$$

is the likelihood function for the vector of (ρ, β, τ).

Theorem 2.1. Given the stated assumptions, the marginal posterior p.d.f. of ρ is given by

$$p^*(\rho) = \frac{1}{(1 - \rho^2)^{1/2} m_{00}[\eta + \vartheta(\rho - \bar{\rho})^2]^{v/2}} \left[1 + \sum_{r=2}^{6} \frac{\gamma_r 2^r (v/2)_r}{r![\eta + \vartheta(\rho - \bar{\rho})^2]^r}\right]$$
$$\times \left[1 + \sum_{r=2}^{6} \frac{\xi_r 2^r (v/2)_r}{r! m_{oo}}\right]^{-1} \tag{2.5}$$

where

$$\eta = y'My - \frac{(y'My_{-1})^2}{y'_{-1}My_{-1}}, \qquad M = I - X(X'X)^{-1}X', \qquad v = T - p$$

$$\vartheta = y'_{-1} M y_{-1}, \quad m_{ij} = \int_{-1}^{1} \frac{(\rho - \bar{\rho})^i}{(1 - \rho^2)^{1/2} \{\eta + \vartheta(\rho - \bar{\rho})^2\}^{v/2+j}} \, d\rho,$$

$$\bar{\rho} = \frac{y' M y_{-1}}{y'_{-1} M y_{-1}},$$

$$\xi_2 = \left[-k_3 \sum_{t=1}^{T} \bar{u}_t + \frac{1}{4} k_4 \sum_{t=1}^{T} (1 - q_{tt})^2 \right] m_{02} + k_3 \sum_{t=1}^{T} \delta_t m_{12}$$

$$+ k_3 \sum_{t=1}^{T} (\bar{u}_t m_{02} - \delta_t m_{12}) q_{tt},$$

$$\xi_3 = k_3 \sum_{t=1}^{T} (\bar{u}_t^3 m_{03} - 3\bar{u}_t^2 \delta_t m_{13} + 3\bar{u}_t \delta_t^2 m_{23} - \delta_t^3 m_{33}) - \frac{3}{2} k_4 \sum_{t=1}^{T} (\bar{u}_t^2 m_{03}$$

$$- 2\bar{u}_t \delta_t m_{13} + \delta_t^2 m_{23}) \times (1 - q_{tt}) + \frac{3}{4} k_3^3 m_{03} \sum_{t \neq s}^{T} q_{ts} (1 - 2q_{tt})$$

$$+ \frac{1}{4} k_3^3 m_{03} \sum_{t \neq s}^{T} (3 q_{tt} q_{ts} q_{ss} + 2 q_{ts}^3)$$

$$\xi_4 = k_4 \sum_{t=1}^{T} (\bar{u}_t^4 m_{04} - 4\bar{u}_t^3 \delta_t m_{14} + 6\bar{u}_t^2 \delta_t^2 m_{24} - 4\bar{u}_t \delta_t^3 m_{34} + \delta_t^4 m_{44})$$

$$+ 3k_3^2 \sum_{t \neq S}^{T} (\bar{u}_t \bar{u}_s m_{04} - 2\bar{u}_t \delta_s m_{14} + \delta_t \delta_s m_{24}) - 6k_3^2 \sum_{t \neq S}^{T} \{ (\bar{u}_t^2 m_{04}$$

$$- 2\bar{u}_t \delta_t m_{14} + \delta_t^2 m_{24}) q_{ts} + (\bar{u}_t \bar{u}_s m_{04} - \bar{u}_s \delta_t m_{14} - \bar{u}_t \delta_s m_{14}$$

$$+ \delta_t \delta_s m_{24}) q_{tt} \} + 3k_3^2 \sum_{t \neq s}^{T} \{ 2(\bar{u}_t^2 m_{04} - 2\bar{u}_t \delta_t m_{14} + \delta_t^2 m_{24}) q_{ts} q_{ss}$$

$$+ (\bar{u}_t \bar{u}_s m_{04} - 2\bar{u}_t \delta_s m_{14} + \delta_t \delta_s m_{24})(q_{tt} q_{ss} + 2q_{ts}^2) \}$$

$$\xi_5 = -10 k_3^2 \sum_{t \neq s}^{T} (\bar{u}_t^3 \bar{u}_s m_{05} - \bar{u}_t^3 \delta_s m_{15} - 3\bar{u}_t^2 \bar{u}_s \delta_t m_{15} + 3\bar{u}_t^2 \delta_t \delta_s m_{25}$$

$$+ 3\bar{u}_t \bar{u}_s \delta_t^2 m_{25} - 3\bar{u}_t \delta_t^2 \delta_s m_{35} - \bar{u}_s \delta_t^3 m_{35} + \delta_t^3 \delta_s m_{45})(1 - q_{ss})$$

$$+ 15 k_3^2 \sum_{t \neq s}^{T} (\bar{u}_t^2 \bar{u}_s^2 m_{05} - 4\bar{u}_t^2 \bar{u}_s \delta_s m_{15} + 2\bar{u}_t^2 \delta_s^2 m_{25} + 4\bar{u}_t \bar{u}_s \delta_t \delta_s m_{25}$$

$$- 4\bar{u}_t \delta_t \delta_s \delta_s^2 m_{35} + \delta_t^2 \delta_s^2 m_{45}) q_{ts}$$

$$\xi_6 = 10k_3^2 \sum_{t \neq S}^{T} (\bar{u}_t^3 \bar{u}_s^3 m_{06} - 6\bar{u}_t^3 \bar{u}_s^2 \delta_s m_{16} + 6\bar{u}_t^3 \bar{u}_s \delta_s^2 m_{26} + 9\bar{u}_t^2 \bar{u}_s^2 \delta_t \delta_s m_{26}$$

$$-2\bar{u}_t^3 \delta_s^3 m_{36} - 18\bar{u}_s^2 \bar{u}_s \delta_t \delta_s^2 m_{36} + 6\bar{u}_t^2 \delta_t \delta_s^3 m_{46} + 9\bar{u}_t \bar{u}_s \delta_t^2 \delta_s^2 m_{46}$$

$$-6\bar{u}_t \delta_t^2 \delta_s^3 m_{56} + \delta_t^3 \delta_s^3 m_{66})$$

$$\gamma_2 = -k_3 \sum_{t=1}^{T} \bar{u}_t + \frac{1}{4}k_4 \sum_{t=1}^{T} (1 - q_{tt})^2 + k_3(\rho - \bar{\rho}) \sum_{t=1}^{T} \delta_t + k_3 \sum_{t=1}^{T}$$

$$[\bar{u}_t - (\rho - \bar{\rho})\delta_t]q_{tt}$$

$$\gamma_3 = k_3 \sum_{t=1}^{T} [\bar{u}_t - \delta_t(\rho - \bar{\rho})]^3 - \frac{3}{2}k_4 \sum_{t=1}^{T} [\bar{u}_t - \delta_t(\rho - \bar{\rho})]^2 (1 - q_{tt})$$

$$+ \frac{3}{4}k_3^2 \sum_{t \neq s}^{T} (1 - 2q_{tt})q_{ts} + \frac{1}{4}k_3^2 \sum_{t \neq s}^{T} (3q_{tt}q_{ts}q_{ss} + 2q_{ts}^3)$$

$$\gamma_4 = k_4 \sum_{t=1}^{T} [\bar{u}_t - \delta_t(\rho - \bar{\rho})]^4 + 3k_3^2 \sum_{t \neq s}^{T} [\bar{u}_t - \delta_t(\rho - \bar{\rho})][\bar{u}_s - \delta_s(\rho - \bar{\rho})]$$

$$- 6k_3^2 \sum_{t \neq s}^{T} [\bar{u}_t - \delta_t(\rho - \bar{\rho})]^2 q_{ts} - 6k_3^2 \sum_{t \neq s}^{T} [\bar{u}_t - \delta_t(\rho - \bar{\rho})]$$

$$[\bar{u}_s - \delta_s(\rho - \bar{\rho})]q_{tt} + 6k_3^2 \sum_{t \neq s}^{T} [\bar{u}_t - \delta_t(\rho - \bar{\rho})]^2 q_{ts}q_{ss} + 3k_3^2$$

$$\times \sum_{t \neq s}^{T} [\bar{u}_t - \delta_t(\rho - \bar{\rho})][\bar{u}_s - \delta_s(\rho - \bar{\rho})](q_{tt}q_{ss} + 2q_{ts}^2)$$

$$\gamma_5 = -10k_3^2 \sum_{t \neq s}^{T} [\bar{u}_t - \delta_t(\rho - \bar{\rho})]^3 [\bar{u}_s - \delta_s(\rho - \bar{\rho})](1 - q_{ss})$$

$$+ 15k_3^2 \sum_{t \neq s}^{T} [\bar{u}_t - \delta_t(\rho - \bar{\rho})]^2 [\bar{u}_s - \delta_s(\rho - \bar{\rho})]^2 \times q_{ts}$$

$$\gamma_6 = 10k_3^2 \sum_{t \neq s}^{T} [\bar{u}_t - \delta_t(\rho - \bar{\rho})]^3 [\bar{u}_s - \delta_s(\rho - \bar{\rho})]^3$$

$$q_{ts} = x_t'(X'X)^{-1}x_s, \quad \delta_t = y_{t-1} - x_t'(X'X)^{-1}X'y_{-1},$$

$$\bar{u}_t = y_t - \bar{\rho}y_{t-1} - x_t'\bar{\beta}(\bar{\rho})$$

x_t' is the tth row of X' and

$$\bar{\beta}(\bar{\rho}) = (X'X)^{-1}X'(y - \bar{\rho}y_{-1})$$

Proof. see Appendix.

Obviously, under normal errors (i.e., $k_3 = k_4 = 0$), equation (2.5) reduces to

$$p^*(\rho) = \frac{1}{(1 - \rho^2)^{1/2} m_{oo}[\eta + \vartheta(\rho - \bar{\rho})^2]^{v/2}} \tag{2.6}$$

The following corollary gives the expression, under a quadratic loss structure, for the Bayes estimator, which is the mean of the posterior distribution of ρ.

Corollary 2.1. Under the stated assumptions, the Bayes estimator of ρ is given by

$$\hat{\rho} = \bar{\rho} + \frac{m_{10}}{m_{00}}\left[1 + \sum_{r=2}^{6}\frac{\xi_r^* 2^r(v/2)_r}{r! m_{10}}\right]\left[1 + \sum_{r=2}^{6}\frac{\xi_r 2^r(v/2)_r}{r! m_{00}}\right]^{-1} \tag{2.7}$$

where the expressions for ξ_r^* are the same as those of $\xi_{r,,}$ except that the m_{ij} are replaced by $m_{i+1,j}$ $\forall$ $i,j = 0, 1, 2, \ldots$

Proof. see Appendix.

Substituting $k_3 = k_4 = 0$ in (2.7) yields the following expression for the Bayes estimator of ρ when the distribution of the disturbances is normal:

$$\tilde{\rho} = \bar{\rho} + \frac{m_{10}}{m_{00}} \tag{2.8}$$

3. NUMERICAL RESULTS

To gain further insights of the effects of non-normality on the posterior p.d.f of ρ, we numerically evaluate (2.5) and (2.6) for a range of chosen values of parameters. The model on which the numerical exercises are based is

$$y_t = -0.5y_{t-1} + 1.0 + 2.2x_{1t} - 1.3x_{2t} + u_t; \qquad t = 1, \ldots, 20 \tag{3.1}$$

where x_1 and x_2 are $N(0, 1)$ random variables. The vector of non-normal disturbances u, is obtained through the transformation

$$\varphi = \sinh\left(\frac{z - f_1}{f_2}\right) \tag{3.2}$$

where $z \sim N(0, 1)$ and f_1 and f_2 are combined to determine the skewness and the spread of the distribution (see Johnson 1949 and Srivastava and Maekawa 1995). Now, it can be shown that

$$E(\varphi) = -\omega^{1/2} \sinh(\Omega) \tag{3.3}$$

where $\omega = \exp(f_2^{-2})$ and $\Omega = f_1/f_2$. In face of (3.3), it seems reasonable to generate u using the formula

$$u = \varphi + \omega^{1/2} \sinh(\Omega) = \varphi + \exp(1/(2f_2^2)) \sinh(\Omega) \tag{3.4}$$

Note also that

$$E(u) = 0 = m_1 \tag{3.5}$$

$$E(u^2) = \frac{1}{2}(\omega - 1)[\omega \cosh(2\Omega) + 1] = m_2 \tag{3.6}$$

$$E(u^3) = -\frac{1}{4}\omega^{1/2}(\omega - 1)^2[\omega(\omega + 2) \sinh(3\Omega) + 3 \sinh(\Omega)] = m_3 \tag{3.7}$$

and

$$E(u^4) = \frac{1}{8}(\omega - 1)^2[\omega^2(\omega^4 + 2\omega^3 + 3\omega^2 - 3) \cosh(4\Omega)$$
$$+ 4\omega^2(\omega + 2) \cosh(2\Omega) + 3(2\omega + 1)]. \tag{3.8}$$
$$= m_4$$

Hence we have,

$$k_3 = m_3 \tag{3.9}$$

and

$$k_4 = m_4 - 3m_2^2 \tag{3.10}$$

Also, the ranges of k_3 and k_4 are chosen such that $k_3^2 < 0.5$ and $0 < k_4 < 2.4$ to ensure that the ESD is a well-behaved density function. Table 1 gives a glance of the relationship between (f_1, f_2) and $(k_3, .k_4)$.

All of the computations have been undertaken using MATHEMATICA, version 4. The posterior densities corresponding to some selected values of k_3 and k_4 are graphed in Figures 1–3, alongside their normal counterpart when $k_3 = k_4 = 0$ is erroneously assumed. That is, the posterior p.d.f.s corresponding to $k_3 = k_4 = 0$ are calculated on the basis of the equation described by (2.6), when the disturbances are in fact generated by an ESD with the values of k_3 and k_4 as shown in the figures. The diagrams reflect

Table 1. Relationship between (f_1, f_2) and (k_3, k_4)

k_3	k_4		
	0.2	1.2	2.2
−0.7	$f_1 = 198.47804$	$f_1 = 4.46274$	$f_1 = 0.95396$
	$f_2 = 38.98192$	$f_2 = 2.93507$	$f_2 = 1.70850$
−0.25	$f_1 = 10.24461$	$f_1 = 0.32645$	$f_1 = 0.19130$
	$f_2 = 5.12042$	$f_2 = 1.62127$	$f_2 = 1.50024$
0	$f_1 = 0.00000$	$f_1 = 0.00000$	$f_1 = 0.00000$
	$f_2 = 1.94322$	$f_2 = 1.58147$	$f_2 = 1.48531$
0.25	$f_1 = -10.24461$	$f_1 = -0.32645$	$f_1 = -0.19130$
	$f_2 = 5.12042$	$f_2 = 1.621277$	$f_2 = 1.50024$
0.7	$f_1 = -198.47080$	$f_1 = -4.46274$	$f_1 = -0.95396$
	$f_2 = 38.98192$	$f_2 = 2.93507$	$f_2 = 1.70850$

that ignoring non-normality when the latter is present has the effect of increasing the posterior risk of the Bayes estimator of ρ, and it becomes more pronounced as $|k_3|$ and k_4 increase. The ESD based posterior p.d.f.s follow basically the same shape as their normal counterparts, but are more centered around the posterior mean and thus have lower variance. All of the

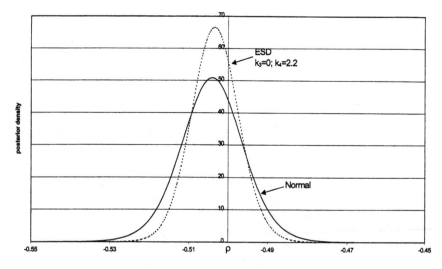

Figure 1. Posterior density for $k_3 = 0.0$ and $k_4 = 2.2$.

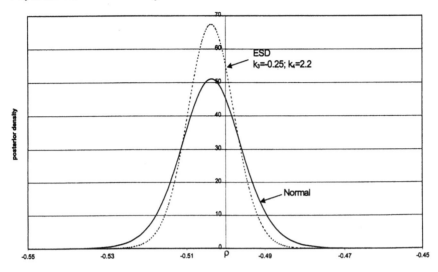

Figure 2. Posterior density function for $k_3 = -0.25$ and $k_4 = 2.2$.

posterior p.d.f.s have the characteristic of a slight asymmetry around the mean. Both the normal and ESD-based posterior p.d.f.s tend to have a mild increase in values when $\rho \to \pm 1$, reflecting the effect of the prior p.d.f. described in (2.3) on the posterior p.d.f. See Figure 4, which depicts the behaviour of the ESD-based posterior p.d.f. when $\rho \to -1$.

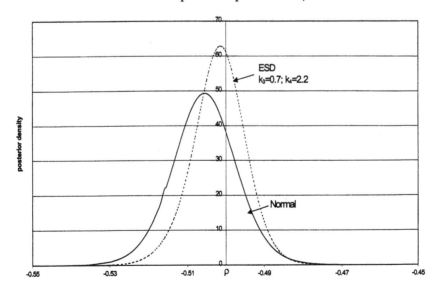

Figure 3. Posterior density functions for $k_3 = 0.7$ and $k_4 = 2.2$.

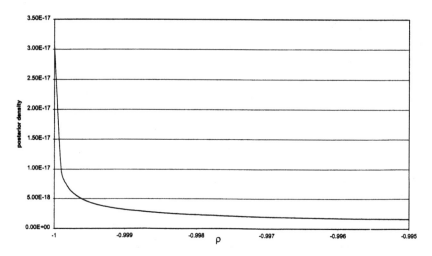

Figure 4.   Posterior density functions for $k_3 = 0.0$ and $k_4 = 2.2$.

For an exact assessment of the effect of ignoring non-normality on the posterior risk, consider the following:

Corollary 3.1.   Under the stated assumptions, the increase in the posterior risk of $\hat{\rho}$ when non-normality is erroneously ignored is given by

$$E\left[(\rho - \tilde{\rho})^2\right] - E\left[(\rho - \hat{\rho})^2\right] = \left(\hat{\rho} - \bar{\rho} - \frac{m_{10}}{m_{00}}\right)^2 \tag{3.11}$$

Proof.   see Appendix.

Equation (3.11) confirms that ignoring non-normality when it is present always results in an increase in the posterior risk of the Bayes estimator of ρ. In the case that $k_3 = k_4 = 0$, $\hat{\rho} = \bar{\rho} + m_{10}/m_{00} = \tilde{\rho}$ and (3.11) reduces to zero. In Table 2, in order to gain further insights, we calculate, on the basis of model (3.1), the percentage increase in posterior risk should normality of the

Table 2.   Percentage increase in posterior risk

k_3	k_4	% increase
0	2.2	1.4742
−0.25	2.2	0.0158
0.25	2.2	5.6104
−0.7	2.2	17.3373

model's disturbances be mistakenly assumed when the disturbances are generated by an ESD with the given values of k_3 and k_4.

It is interesting to note that the percentage increase in posterior risk for wrongly assuming normality is greater when the underlying ESD is characterized by $k_3 = 0$ and $k_4 = 2.2$ than when $k_3 = -0.25$ and $k_4 = 2.2$. This can be explained by observing that, for model (3.1), the posterior mean of ρ for $k_3 = 0$ and $k_4 = 2.2$ is located to the right of that corresponding to the assumption of normality; but as k_3 decreases from zero, the posterior mean shifts to the left. Note also that equation (3.11) is equivalent to $(\hat{\rho} - \tilde{\rho})^2$, which decreases as $\hat{\rho}$ lies closer to $\tilde{\rho}$. However, as k_3 decreases further from zero, the difference between the two posterior means increases and hence the deviation in risk also increases.

4. CONCLUDING REMARKS

The literature on dynamic models involving lagged dependent variables has paid only scant attention to models with non-normal error terms. Even scarcer is the analysis of the problem from a Bayesian perspective. This paper derives and evaluates the posterior density function of the autoregressive coefficient of a stationary ARX(1) model based on a diffuse prior on the model's parameters and under the assumption of an Edgeworth series distribution on the model's disturbances. It is demonstrated that unwittingly ignoring non-normality can lead to an increase in the posterior risk and hence a widening of the highest posterior density interval. A difficulty does arise with the numerical computation of the posterior p.d.f., which is prohibitively complicated in the present context. It is rightly remarked that a Markov chain Monte Carlo algorithm such as Gibbs sampling or the Hastings–Metropolis algorithm as a means for posterior simulation would be more preferable (see, for example, Geweke 1999 for a description of the techniques). In practice k_3 and k_4 are unknown, but Bayes estimators of these quantities can be obtained along the lines of Chaturvedi et al. (1997a).

ACKNOWLEDGMENT

The bulk of this work was undertaken while the first and third authors were visiting the City University of Hong Kong during the summer of 2000. Thanks are due to the City University of Hong Kong for financial support, and to Hikaru Hasegawa, Koichi Maekawa, Viren Srivastava and a referee for helpful comments. The first author also acknowledges support from C.S.I.R.O., India. The usual disclaimer applies.

APPENDIX

Proof of Theorem 2.1

Combining the likelihood function (2.4) and the prior distribution (2.3), and ignoring cumulants of order greater than four, we obtain the posterior distribution for (β, ρ, τ) as

$$p^*(\beta, \rho, \tau) \propto E_Z \left[\frac{1}{(1-\rho^2)^{1/2}} \tau^{T/2-1} \exp\left(-\frac{\tau}{2}(y - \rho y_{-1} - X\beta - Z)' \right. \right.$$

$$\left. \left. (y - \rho y_{-1} - X\beta - Z) \right) \right]$$

$$\propto \frac{1}{(1-\rho^2)^{1/2}} \tau^{T/2-1} \exp\left[-\frac{\tau}{2}(y - \rho y_{-1} - X\beta)'(y - \rho y_{-1} - X\beta)\right]$$

$$E_Z \left[\exp\left\{ \tau\left(Z'u - \frac{Z'Z}{2}\right) \right\} \right]$$

$$\propto \frac{1}{(1-\rho^2)^{1/2}} \tau^{T/2-1} \exp\left[-\frac{\tau}{2}(y - \rho y_{-1} - X\beta)'(y - \rho y_{-1} - X\beta)\right] \qquad \text{(A.1)}$$

$$\left[1 + \sum_{r=2}^{6} \frac{\sigma_r \tau^r}{r!} \right]$$

where

$$\sigma_2 = -k_3 \sum_{t=1}^{T} u_t + \frac{T}{4}k_4, \qquad \sigma_3 = k_3 \sum_{t=1}^{T} u_t^3 - \frac{3}{2}k_4 \sum_{t=1}^{T} u_t^2,$$

$$\sigma_4 = k_4 \sum_{t=1}^{T} u_t^4 + 3k_3^2 \sum_{t \neq S} u_t u_s, \qquad \sigma_5 = -10k_3^2 \sum_{t \neq s} u_t^3 u_s, \quad \text{and}$$

$$\sigma_6 = 10k_3^2 \sum_{t \neq s} u_t^3 u_s^3$$

Observing that

$$(y - \rho y_{-1} - X\beta)'(y - \rho y_{-1} - X\beta) = \beta'X'X\beta - 2\beta'X'(y - \rho y_{-1})$$

$$+ (y - \rho y_{-1})'(y - \rho y_{-1})$$

$$= [\beta - \bar{\beta}(\rho)]'X'X[\beta - \bar{\beta}(\rho)] + (y - \rho y_{-1})'M(y - \rho y_{-1})$$

$$= [\beta - \bar{\beta}(\rho)]'X'X[\beta - \bar{\beta}(\rho)] + \rho^2 y'_{-1}My_{-1} - 2\rho y'My_{-1} + y'My$$

$$= \eta + [\beta - \bar{\beta}(\rho)]'X'X[\beta - \bar{\beta}(\rho)] + \vartheta(\rho - \bar{\rho})^2$$

$$\text{(A.2)}$$

where

$$\bar{\beta}(\bar{\rho}) = (X'X)^{-1}X'(y - \rho y_{-1})$$

Hence the posterior p.d.f. of (β, ρ) can be written as

$$p^*(\beta, \rho) \propto \int_0^\infty \frac{1}{(1 - \rho^2)^{1/2}} \tau^{T/2-1} \exp\left\{-\frac{\tau}{2}[\eta + (\beta - \bar{\beta}(\rho)))'X(\beta - \bar{\beta}(\rho))\right.$$

$$\left. + \vartheta(\rho - \bar{\rho})^2]\right\} \times \left[1 + \sum_{r=2}^6 \frac{\sigma_r \tau^r}{r!}\right] d\tau$$

Thus,

$$p^*(\beta, \rho) = C \frac{2^{T/2}\Gamma(T/2)}{(1 - \rho^2)^{1/2}\left\{\eta + [\beta - \bar{\beta}(\rho)]'X'X[\beta - \bar{\beta}(\rho)] + \vartheta(\rho - \bar{\rho})^2\right\}^{T/2}}$$

$$\times \left[1 + \sum_{r=2}^6 \frac{2^r \sigma_r(T/2)_r}{r!\left\{\eta + [\beta - \bar{\beta}(\rho)]'X'X[\beta - \bar{\beta}(\rho)] + \vartheta(\rho - \bar{\rho})^2\right\}^r}\right]$$

$$(A.3)$$

Now, to work out the normalizing constant C, note that

$$C^{-1} = \int_{-1}^{+1} \int_0^\infty \int_{R^p} \frac{1}{(1 - \rho^2)^{1/2}} \tau^{T/2-1} \exp\left\{-\frac{\tau}{2}[\eta + (\beta - \bar{\beta}(\rho))'X'X\right.$$

$$\left.(\beta - \bar{\beta}(\rho)) + \vartheta(\rho - \bar{\rho})^2]\right\} \times \left[1 + \sum_{r=2}^6 \frac{\sigma_r \tau^r}{r!}\right] d\beta d\tau d\rho$$

$$= 2^{p/2}\pi^{p/2}|X'X|^{-1/2} \int_{-1}^1 \frac{1}{(1 - \rho^2)^{1/2}} \left\{\int_0^\infty \tau^{(T-p)/2-1}\right.$$

$$\exp\left[-\frac{\tau}{2}(\eta + \vartheta(\rho - \bar{\rho})^2)\right]\left[\int_{R^p}\left[1 + \sum_{r=2}^6 \frac{\sigma_r \tau^r}{r!}\right] \times g(\beta)d\beta\right]d\tau\right\}d\rho$$

$$(A.4)$$

where

$$g(\beta) = \left(\frac{\tau}{2\pi}\right)^{p/2}|X'X|^{1/2} \exp\left\{-\frac{\tau}{2}[\beta - \bar{\beta}(\rho)]'X'X[\beta - \bar{\beta}(\rho)]\right\}$$

denotes the density function of a normal distribution with mean vector $\bar{\beta}(\rho)$ and precision matrix $\tau X'X$. Now, it can be verified that

$$\int_{R^p} g(\beta)d\beta = 1$$

$$\int_{R^p} \sigma_2 g(\beta)d\beta = -k_3 \sum_{t=1}^{T}[\bar{u}_t - \delta_t(\rho - \bar{\rho})] + \frac{T}{4}k_4$$

$$\int_{R^p} \sigma_3 g(\beta)d\beta = k_3 \sum_{t=1}^{T}\left\{[\bar{u}_t - \delta_t(\rho - \bar{\rho})]^3 + \frac{3}{\tau}[\bar{u}_t - \delta_t(\rho - \bar{\rho})]q_{tt}\right\}$$

$$- \frac{3}{2}k_4 \sum_{t=1}^{T}\left\{[\bar{u}_t - \delta_t(\rho - \bar{\rho})]^2 + \frac{q_{tt}}{\tau}\right\}$$

$$\int_{R^p} \sigma_4 g(\beta)d\beta = k_4 \sum_{t=1}^{T}\left\{[\bar{u}_t - \delta_t(\rho - \bar{\rho})]^4 + \frac{6}{\tau}[\bar{u}_t - \delta_t(\rho - \bar{\rho})]^2 q_{tt} + \frac{3}{\tau^2}q_{tt}^2\right\}$$

$$+ 3k_3^2 \sum_{t \neq s}^{T}\left\{[\bar{u}_t - \delta_t(\rho - \bar{\rho})][\bar{u}_s - \delta_s(\rho - \bar{\rho})] + \frac{q_{ts}}{\tau}\right\}$$

$$\int_{R^p} \sigma_5 g(\beta)d\beta = -10k_3^2 \sum_{t \neq s}^{T}\left\{[\bar{u}_t - \delta_t(\rho - \bar{\rho})]^3[\bar{u}_s - \delta_s(\rho - \bar{\rho})]\right.$$

$$+ \frac{3}{\tau}[\bar{u}_t - \delta_t(\rho - \bar{\rho})]^2 q_{ts} + \frac{3}{\tau}[\bar{u}_t - \delta_t(\rho - \bar{\rho})]$$

$$\left. [\bar{u}_s - \delta_s(\rho - \bar{\rho})]q_{tt} + \frac{3}{\tau^2}q_{tt}q_{ts}\right\}$$

and

$$\int_{R^p} \sigma_6 g(\beta)d\beta = 10k_3^2 \sum_{t \neq s}^{T}\left\{[\bar{u}_t - \delta_t(\rho - \bar{\rho})]^3[\bar{u}_s - \delta_s(\rho - \bar{\rho})]^3\right.$$

$$+ \frac{6}{\tau}[\bar{u}_t - \delta_t(\rho - \bar{\rho})]^3[\bar{u}_s - \delta_s(\rho - \bar{\rho})]q_{ss}$$

$$+ \frac{9}{\tau}[\bar{u}_t - \delta_t(\rho - \bar{\rho})]^2[\bar{u}_s - \delta_s(\rho - \bar{\rho})]^2 q_{ts}$$

$$+ \frac{18}{\tau^2}[\bar{u}_t - \delta_t(\rho - \bar{\rho})]^2 q_{ts}q_{ss}$$

$$+ \frac{9}{\tau^2}[\bar{u}_t - \delta_t(\rho - \bar{\rho})][\bar{u}_s - \delta_s(\rho - \bar{\rho})](q_{tt}q_{ss} + 2q_{ts}^2)$$

$$\left. + \frac{3}{\tau^3}(3q_{tt}q_{ts}q_{ss} + 2q_{ts}^3)\right\}$$

Utilizing these integrals, we have,

$$\int_{R^p} \left[1 + \sum_{r=2}^{6} \frac{\sigma_r \tau^r}{r!}\right] g(\beta) d\beta = 1 + \sum_{r=2}^{6} \frac{\gamma_r \tau^r}{r!} \tag{A.5}$$

Substituting (A.5) in (A.4), and after some manipulations, we obtain

$$C^{-1} = 2^{T/2} \pi^{p/2} |X'X|^{-1/2} \Gamma(v/2) m_{00} \left[1 + \sum_{r=2}^{6} \frac{\xi_r 2^r (v/2)_r}{r! m_{00}}\right] \tag{A.6}$$

Substituting the value of C in (A.3) leads to the expression for the joint posterior p.d.f. of (β, ρ).

Now, to derive the marginal posterior p.d.f. of ρ, note that

$$p^*(\rho) = C \int_0^\infty \int_{R^p} \frac{1}{(1 - \rho^2)^{1/2}} \tau^{T/2-1}$$

$$\exp\left\{-\frac{\tau}{2}\left[\eta + (\beta - \bar{\beta}(\rho))' X'X(\beta - \bar{\beta}(\rho)) + \vartheta(\rho - \bar{\rho})^2\right]\right\}$$

$$\times \left[1 + \sum_{r=2}^{6} \frac{\sigma_r \tau^r}{r!}\right] d\beta d\tau \tag{A.7}$$

$$= C \frac{2^{p/2} \pi^{p/2} |X'X|^{-1/2}}{(1 - \rho^2)^{1/2}} \int_0^\infty \tau^{v/2-1} \exp\left\{-\frac{\tau}{2}\left[\eta + \vartheta(\rho - \bar{\rho})^2\right]\right\}$$

$$\times \left\{\int_{R^p} \left[1 + \sum_{r=2}^{6} \frac{\sigma_r \tau^r}{r!}\right] g(\beta) d\beta\right\} d\tau$$

Using (A.5), we obtain

$$p^*(\rho) = C \frac{2^{p/2} \pi^{p/2} |X'X|^{-1/2}}{(1 - \rho^2)^{1/2}} \int_0^\infty \tau^{v/2-1} \exp\left\{-\frac{\tau}{2}\left[\eta + \vartheta(\rho - \bar{\rho})^2\right]\right\}$$

$$\times \left[1 + \sum_{r=2}^{6} \frac{\gamma_r \tau^r}{r!}\right] d\tau$$

$$= C \frac{2^{T/2} \pi^{p/2} |X'X|^{-1/2} \Gamma(v/2)}{(1 - \rho^2)^{1/2} [\eta + \varphi(\rho - \bar{\rho})^2]^{v/2}} \left[1 + \sum_{r=2}^{6} \frac{\gamma_r 2^r (v/2)_r}{r! [\eta + \vartheta(\rho - \bar{\rho})^2]^r}\right]$$

$$\tag{A.8}$$

which, after substituting the value of the normalizing constant C from (A.6), leads to the required result (2.5).

Proof of Corollary 2.1

To obtain the Bayes estimator $\hat{\rho}$ under a quadratic loss structure, observe that

$$\hat{\rho} = E(\rho) = \bar{\rho} + E(\rho - \bar{\rho})$$

$$= \bar{\rho} + C \int_{-1}^{1} \int_{0}^{\infty} \int_{R^p} (\rho - \bar{\rho}) \frac{1}{(1 - \rho^2)^{1/2}} \tau^{T/2-1}$$

$$\times \exp\left\{-\frac{\tau}{2}\left[\eta + (\beta - \bar{\beta}(\rho))' X'X(\beta - \bar{\beta}(\rho))\right.\right. \tag{A.9}$$

$$\left.\left. + \vartheta(\rho - \bar{\rho})^2\right]\right\}\left[1 + \sum_{r=2}^{6} \frac{\sigma_r \tau^r}{r!}\right] d\beta d\tau d\rho$$

Proceeding along the lines for the derivation of C in (A.6), it can be shown that,

$$\int_{-1}^{1} \int_{0}^{\infty} \int_{R^p} (\rho - \bar{\rho}) \frac{1}{(1 - \rho^2)^{1/2}} \tau^{T/2-1}$$

$$\times \exp\left\{-\frac{\tau}{2}\left[\eta + (\beta - \bar{\beta}(\rho))' X'X(\beta - \bar{\beta}(\rho)) + \vartheta(\rho - \bar{\rho})^2\right]\right\}$$

$$\times \left[1 + \sum_{r=2}^{6} \frac{\sigma_r \tau^r}{r!}\right] d\beta d\tau d\rho$$

$$= 2^{T/2} \pi^{p/2} |X'X|^{-1/2} \Gamma(v/2) m_{10} \left[1 + \sum_{r=2}^{6} \frac{\xi_r^* 2^r (v/2)_r}{r! m_{10}}\right]$$

$$\tag{A.10}$$

Substituting (A.10) and the expression of C in (A.9), the required expression of $\hat{\rho}$ is obtained.

Proof of Corollary 3.1

Note that,

$$E(\rho - \hat{\rho})^2 = E\left[(\rho - \bar{\rho} + \bar{\rho} - \hat{\rho})^2\right]$$

$$= E(\rho - \bar{\rho})^2 + (\bar{\rho} - \hat{\rho})^2 + 2(\bar{\rho} - \hat{\rho})E(\rho - \bar{\rho}) \tag{A.11}$$

$$= E(\rho - \bar{\rho})^2 - (\bar{\rho} - \hat{\rho})^2$$

and expectation is taken with respect to the posterior distribution of ρ. Also,

$$E(\rho - \bar{\rho})^2 = \frac{m_{20}}{m_{00}}\left[1 + \sum_{r=2}^{6}\frac{\xi_r^{**}2^r(v/2)_r}{r!m_{20}}\right]\left[1 + \sum_{r=2}^{6}\frac{\xi_r 2^r(v/2)_r}{r!m_{00}}\right]^{-1} \tag{A.12}$$

where ξ_r^{**} are obtained by replacing m_{ij} by $m_{i+2,j}$ in the expression for ξ_r, and

$$(\bar{\rho} - \hat{\rho})^2 = \frac{m_{10}^2}{m_{00}^2}\frac{\left[1 + \sum_{r=2}^{6}\dfrac{\xi_r^* 2^r(v/2)_r}{r!m_{10}}\right]^2}{\left[1 + \sum_{r=2}^{6}\dfrac{\xi_r 2^r(v/2)_r}{r!m_{00}}\right]} \tag{A.13}$$

Note also that

$$E(\rho - \tilde{\rho})^2 = E(\rho - \bar{\rho})^2 + \frac{m_{10}^2}{m_{00}^2} - 2\frac{m_{10}}{m_{00}}E(\rho - \bar{\rho}) \tag{A.14}$$

Therefore,

$$\begin{aligned}
E(\rho - \tilde{\rho})^2 - E(\rho - \hat{\rho})^2 &= (\bar{\rho} - \hat{\rho})^2 - 2\frac{m_{10}}{m_{00}}E(\rho - \bar{\rho}) + \frac{m_{10}^2}{m_{00}^2} \\
&= (\hat{\rho} - \bar{\rho})^2 - 2\frac{m_{10}}{m_{00}}(\hat{\rho} - \bar{\rho}) + \frac{m_{10}^2}{m_{00}^2} \\
&= \left(\hat{\rho} - \bar{\rho} - \frac{m_{10}}{m_{00}}\right)^2
\end{aligned} \tag{A.15}$$

which is the required expression.

REFERENCES

T. W. Anderson. On asymptotic distributions of estimates of parameters of stochastic difference equations. *Annals of Mathematical Statistics* 30: 676–687, 1959.

N. Balakrishnan, S. Kocherlakota. Robustness to non-normality of the linear discriminant function: mixtures of normal distribution. *Communications in Statistics: Theory and Methods* 14: 465–478, 1985.

D. E. Barton, K. E. Dennis. The conditions under which Gram–Charlier and Edgeworth curves are positive definite and unimodal. *Biometrika* 39: 425–427, 1952.

L. D. Broemeling. *Bayesian Analysis of Linear Models*. New York: Marcel Dekker, 1985.

A. Chaturvedi, H. Hasegawa, S. Asthana. Bayesian analysis of the linear regression model with non-normal disturbances. *Australian Journal of Statistics* 39: 277–293, 1997a.

A. Chaturvedi, H. Hasegawa, S. Asthana. Bayesian predictive analysis of the linear regression model with an Edgeworth series prior distribution. *Communications in Statistics: Theory and Methods* 24: 2469–2484, 1997b.

A. W. Davis. Statistical distributions in univariate and multivariate Edgeworth populations. *Biometrika* 63: 661–670, 1976.

J. M. Dufour, J. F. Kiviet. Exact inference methods for first-order autoregressive distributed lag models. *Econometrica* 66: 79–104, 1998.

E. F. Fama. The behaviour of stock market prices. *Journal of Business* 38: 34–105, 1965.

J. Geweke. Using simulation methods for Bayesian econometric models: inference, development and communication. *Econometric Reviews* 18: 1–73, 1999.

D. Grubb, J. Symons. Bias in regressions with a lagged dependent variable. *Econometric Theory* 3: 371–386, 1987.

H. Hasegawa, A. Chaturvedi, Tran Van Hoa. Bayesian unit root test in non-normal ARX(1) model. *Journal of Time Series Analysis* 21: 261–280, 2000.

N. L. Johnson. System of frequency curves generated by methods of translation. *Biometrika* 36: 149–176, 1949.

M. G. Kendall. The analysis of economic time series, Part 1: prices. *Journal of the Royal Statistical Society, Series A* 96: 11–25, 1953.

J. F. Kiviet, G. D. A. Phillips. Alternative bias approximations in regressions with a lagged-dependent variable. *Econometric Theory* 9: 62–80, 1993.

J. F. Kiviet, G. D. A. Phillips, B. Schipp. The bias of OLS, GLS and ZEF estimators in dynamic seemingly unrelated regression models. *Journal of Econometrics* 69: 241–266, 1995.

J. L. Knight. The moments of OLS and 2SLS when the disturbances are non-normal. *Journal of Econometrics* 27: 39–60, 1985.

J. L. Knight. The distribution of the Stein-rule estimator in a model with non-normal disturbances. *Econometric Theory* 2: 202–219, 1986.

S. Kocherlakota, E. F. Chinganda. Robustness of the linear discriminant function to non-normality: Edgeworth series distribution. *Journal of Statistical Planning and Inference* 2: 79–91, 1978.

B. B. Mandelbrot. The variation of certain speculative prices. *Journal of Business* 36: 394–419, 1963.

B. B. Mandelbrot. The variation of some other speculative prices. *Journal of Business* 40: 393–413, 1967.

H. B. Mann, A. Wald. On the statistical treatment of linear stochastic difference equations. *Econometrica* 11: 223–239, 1943.

T. A. Peters. The exact moments of OLS in dynamic regression models with non-normal errors. *Journal of Econometrics* 40: 279–305, 1989.

P. C. B. Phillips. Approximation to some finite sample distributions associated with a first order stochastic difference equation. *Econometrica* 45: 1517–1534, 1977.

P. D. Praetz. The distribution of share price changes. *Journal of Business* 45: 49–55, 1972.

S. J. Press. A compound events model for security prices. *Journal of Business* 45: 317–335, 1968.

V. K. Srivastava, K Maekawa. Efficiency properties of feasible generalized least squares estimators in SURE models under non-normal disturbances. *Journal of Econometrics* 66: 99–121, 1995.

K. Tanaka. Asymptotic expansions associated with the AR(1) model with unknown mean. *Econometrica* 51: 1221–1231, 1983.

A. Zellner. *An Introduction to Bayesian Inference in Econometrics.* New York: Wiley, 1971.

21

Determining an Optimal Window Size for Modeling Volatility

XAVIER CHEE HOONG YEW   University of Oxford, Oxford, England

MICHAEL McALEER   University of Western Australia, Perth, Western Australia

SHIQING LING   Hong Kong University of Science and Technology, Hong Kong, China

1. INTRODUCTION

The keen interest in modeling volatility over an extended period has largely been motivated by the importance of risk considerations in economic and financial markets. This area of research has resulted in the development of several types of volatility models. Without doubt, the Generalized Autoregressive Conditional Heteroskedasticity (GARCH) model, initiated by Engle (1982) and generalized by Bollerslev (1986), is the most popular and successful model for analyzing time-varying volatility, due to its relative ease of interpretation and estimation. Recently, extensions to the GARCH model have been developed to accommodate additional features of high-frequency financial data, namely the Glosten, Jagannathan, and Runkle (1993) (GJR) model and the Exponential GARCH (EGARCH) model of Nelson (1990a). These volatility models explicitly accommodate observed asymmetric effects in financial time series, which were first noted by Black (1976). In essence, large negative shocks or innovations result in higher observed volatility than positive shocks of equal magnitude. Both the

443

GJR and EGARCH models are regarded as being able to accommodate such empirical regularities.

However, Rabemanjara and Zakoian (1993) reported that, while large negative shocks generate higher volatility than large positive shocks, the reverse holds for small shocks. As a result, the resultant asymmetric behavior of volatility depends heavily on the size of the news impact. This has led to the development of Fornari and Mele's (1997) Voltility-Switching GARCH (VS-GARCH) model, which captures the reversion of asymmetric reactions to news. Fornari and Mele observed that VS-GARCH outperformed GJR for the stock indices of six countries, on the basis of in-sample likelihood tests and Engle and Ng's (1993) sign-bias tests. These results notwithstanding, the conclusions derived from such single estimation windows, albeit for six stock indices, need to be interpreted with caution as the performance of volatility models can be highly sensitive to the choice of estimation period. Moreover, the failure to satisfy appropriate regularity conditions can yield inferences which are invalid. Although inferences are crucial in reaching Fornari and Mele's (1997) conclusions, no regularity conditions seem yet to have been established for the VS-GARCH model.

The purpose of this paper is to analyze recursive estimates of the GARCH and GJR models over the period January 1994 to December 1997, beginning with a sample period of one year. It is clear that the sample period includes extreme observations generated during the Asian financial crisis.

The GJR model is preferred to EGARCH for the purpose of accommodating asymmetric conditional variance because the regularity conditions have been derived recently for GJR in Ling and McAleer (1999a). These conditions for the strict stationarity of the model and the existence of its moments are simple to check and should prove useful in practice.

We systematically add one day to the initial window sample until it extends to December 1997, thereby obtaining estimates of the following:

(1) GARCH parameters and their corresponding t-ratios;
(2) GJR parameters and their corresponding t-ratios;
(3) GARCH second and fourth moment regularity conditions;
(4) GJR second and fourth moment regularity conditions.

The derivation of the regularity conditions for the GJR model is given in Ling and McAleer(1999a), where simple and practical conditions for the existence of the second and fourth moments are presented. The empirical results obtained here have not previously been analyzed. Henceforth, the validity of inferences can be interpreted in the GJR model.

GARCH-type models generally require large sample sizes to maximize the likelihood function efficiently which, in turn, can yield unstable parameter estimates. Through the use of recursive estimation, we can derive the smallest range of robust window samples, which are subsequently used for a recursive analysis of volatility. Pagan and Schwert (1990) underscored the implications of the role of window sizes in modeling volatility. Consequently, the recursive plots of the parameter estimates and their corresponding t-ratios should provide some useful insights into the determination of an optimal window size for modeling volatility.

The plan of the paper is as follows. In Section 2 the symmetric and asymmetric GARCH (1,1) models are presented, and their structural and statistical properties are discussed. The empirical results are presented in Section 3. Some concluding remarks are given in Section 4.

2. SYMMETRIC AND ASYMMETRIC GARCH MODELS

Consider a first-order autoregressive mean returns equation, where R_1 is the return on the stock index:

$$R_t = \phi_0 + \phi_0 R_{t-1} + \epsilon_t \tag{1}$$

$$\epsilon_t = \eta_t \sqrt{h_t}, \qquad \eta \sim \text{IID}(0, 1) \tag{2}$$

in which ϵ_t is assumed to follow a GARCH (1,1) process; that is, the conditional distribution of ϵ_t, given the information set at time t, is specified as

$$h_t = \omega + \alpha \epsilon_{t-1}^2 + \beta h_{t-1} \tag{3}$$

where $\omega > 0$, $\alpha \geq 0$, and $\beta \geq 0$. When $\beta = 0$ in (3), the GARCH (1,1) model reduces to the first-order ARCH model, ARCH (1). Bollerslev (1986) showed that the necessary and sufficient condition for the second-order stationarity of the GARCH (1,1) model is

$$\alpha + \beta < 1 \tag{4}$$

Nelson (1990b) obtained the necessary and sufficient condition for strict stationarity and ergodicity as

$$E\big(\ln(\alpha \eta_t^2 + \beta)\big) < 0 \tag{5}$$

which allows $\alpha + \beta$ to exceed 1, in which case $E\epsilon_t^2 = \infty$. Assuming that the GARCH process starts infinitely far in the past with finite $2m$th moment, as in Engle (1982), Bollerslev (1986) provided the necessary and sufficient condition for the existence of the $2m$th moment of the GARCH (1,1) model. Without making such a restrictive assumption, Ling (1999) showed that a sufficient condition for the existence of the $2m$th moment of the GARCH (1,1) model is

$$\rho[E(A_t^{\otimes m})] < 1 \tag{6}$$

where the matrix A is a function of α, β, and η_t, $\rho(A) = \max$ {eigenvalues of A}, $A^{\otimes m} = A \otimes A \otimes K \otimes A$ (m factors), and $\otimes$ is the Kronecker product. (Note that condition (4) is a special case of (6) when $m = 1$.) Ling and McAleer (1999b) showed that condition (6) is also necessary for the existence of the $2m$th moment.

Defining $\delta = (\omega, \alpha, \beta)'$, maximum likelihood estimation can be used to estimate δ. Given observations ϵ_t, $t = 1, \ldots, n$, the conditional log-likelihood can be written as

$$L(\delta) = \sum_{t-1}^{n} l_t \tag{7}$$

$$l_t = \tfrac{1}{2} \ln h_t - \tfrac{1}{2} \epsilon_t^2 / \sqrt{h_t}$$

in which $h + t$ is treated as a function of ϵ_t. Let $\delta \in \Delta$, a compact subset of R^3, and define $\hat{\delta} = \arg\max_{\delta \in \Delta} L(\delta)$. As the conditional error η_t is not assumed to be normal, $\hat{\delta}$ is called the quasi-maximum likelihood estimator (QMLE). Lee and Hansen (1994) and Lumsdaine (1996) proved that the QMLE is consistent and asymptotically normal under the condition given in (5). However, Lee and Hansen (1994) required all the conditional expectations of $\eta_t^{2+\kappa}$ to be finite for $\kappa > 0$, while Lumsdaine (1996) required that $E\eta_t^{32} < 0$ (that is, for $m = 16$ in (5)), both of which are strong conditions. Ling and Li (1998) proved that the local QMLE is consistent and asymptotically normal under fourth-order stationarity. Ling and McAleer (1996c) proved the consistency of the global QMLE under only the second moment condition, and derived the asymptotic normality of the global QMLE under the 6th moment condition.

The asymmetric GJR (1,1) process of Glosten et al. (1993) specifies the conditional variance h_t as

$$h_t = \omega + \alpha \epsilon_{t-1}^2 + \gamma D_{t-1} \epsilon_{t-1}^2 + \beta h_{t-1} \tag{8}$$

where $\gamma \geq 0$, and $D_{t-1} = 1$ when $\epsilon_{t-1} < 0$ and $D_{t-1} = 0$ when $\epsilon_{t-1} \geq 0$. Thus, good news (or shocks) with $\epsilon_t > 0$, and bad news with $\epsilon_t < 0$, have different impacts on the conditional variance: good news has an impact of α, while bad news has an impact of $\alpha + \gamma$. Consequently, the GJR model in (8) seeks to accommodate the stylized fact of observed asymmetric/leverage effects in financial markets.

Both the GARCH and GJR models are estimated by maximizing the log-likelihood function, assuming that η_t is conditionally normally distributed. However, financial time series often display non-normality. When η_t is not normal, the QMLE is not efficient; that is, its asymptotic covariance matrix is not minimal in the class of asymptotically normal estimators. In order to obtain an efficient estimator it is necessary to know, or to be able to estimate, the density function of η_t an to use an adaptive estimation procedure. Drost and Klaasen (1997) investigated adaptive estimation of the GARCH (1,1) model. It is possible that non-normality is caused by the presence of extreme observations and/or outliers. Thus, the Bollerslev and Wooldridge (1992) procedure is used to calculate robust t-ratios for purposes of valid inferences. Although regularity conditions for the GJR model have recently been established by Ling and McAleer (1999a), it has not yet been shown whether they are necessary and/or sufficient for consistency and asymptotic normality.

3. EMPIRICAL RESULTS

The data sets used are obtained from Datastream, and relate to the Hong Kong Seng index, Nikkei 225 index, and Standard and Poor's composite 500 (SP500) index. Figures 1(a–c) plot the Hang Seng, Nikkei 225, and SP500 close-to-close, daily index returns, respectively. The results are presented for GARCH, followed by those for GJR. The Jarque and Bera (1980) Lagrange multiplier statistics rejects normality for all three index returns, which may subsequently yield non-normality for the conditional errors. Such non-normality may be caused, among other factors, by the presence of extreme observations.

The Asian financial crisis in 1997 had a greater impact on the Hang Seng index (HSI) and Standard and Poor's 500 (SP500) than on the Nikkei index (NI). However, the reversions to the mean of volatility seem to be immediate in all three stock markets, which implies relatively short-lived, or non-persistent, shocks. The persistence of shocks to volatility can be approximated by $\alpha + \beta$ for the GARCH model. Nelson (1990a) calculated the *half-life* persistence of shocks to volatility, which essentially converts the parameter β into the corresponding number of days. The persistence of shocks to volatility after a large negative shock is usually short lived (see, for example, Engle an Mustafa 1992, Schwert 1990, and Pagan an Schwert 1990). Such findings of persistence highlight the importance of conditional variance for the financial management of long-term risk issues. However, measures of persistence from conditional variance models have typically been unable to identify the origin of the shock. Consequently, the persistence of a shock may be masked, or measured spuriously from GARCH-type models.

Figures 2 and 3 reveal some interesting patterns from the estimated α (ARCH) and β (GARCH) parameters for the three stock indices. Estimates of α and β tend to move in opposite directions, so that when the estimate of

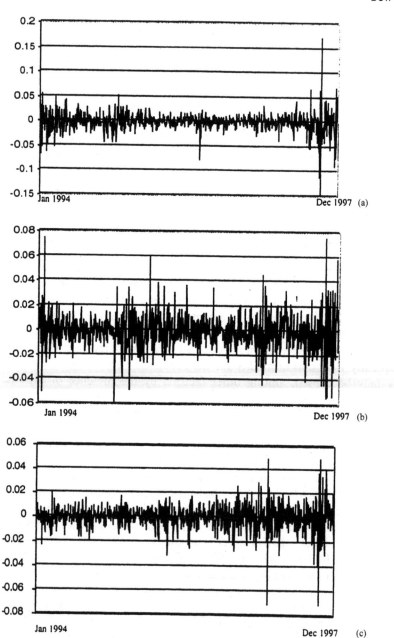

Figure 1. (a) Hang Seng index returns for January 1994 to December 1997, (b) Nikkei 225 index returns for January 1994 to December 1997, (c) SP500 index returns for January 1994 to December 1997.

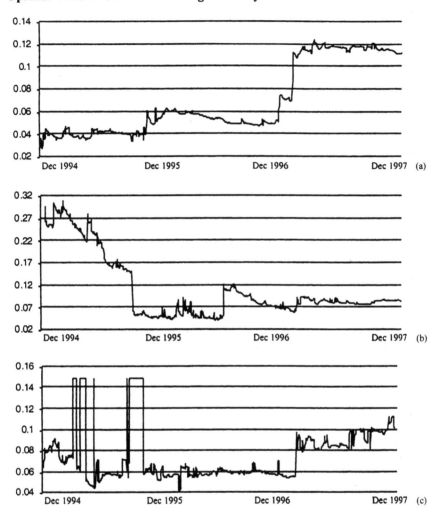

Figure 2. (a) Recursive plot of GARCH-$\hat{\alpha}$ for the Hang Seng index, (b) Recursive plot of GARCH-$\hat{\alpha}$ for the Nikkei 225 index, (c) Recursive plot of GARCH-$\hat{\alpha}$ for the SP500 index.

α is high the corresponding estimate of β is low, and vice-versa. The main differences for the three stock indices reside in the first and third quartiles of the plots, where the estimates of α for HSI and SP500 are much smaller than those for the NI. Moreover, the estimated α for the SP500 exhibit volatile movements in the first quartile. The estimated α for GARCH (1,1) reflect the level of conditional kurtosis, in the sense that larger (smaller) values of α

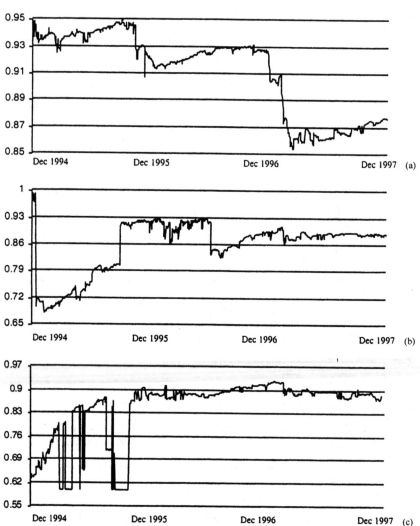

Figure 3. (a) Recursive plot of GARCH-$\hat{\beta}$ for the Hang Seng index, (b) Recursive plot of GARCH-$\hat{\beta}$ for the Nikkei 25 index, (c) Recursive plot of GARCH-$\hat{\beta}$ for the SP500 index.

reflect higher (lower) conditional kurtosis. Figures 3(a–c) imply that the unconditional returns distribution for NI exhibits significantly higher kurtosis than those for the HSI and SP500. This outcome could be due to the presence of extreme observations, which increases the value of the estimated α, and the conditional kurtosis. Note that all three stock indices display leptokurtic unconditional distributions, which render inappropriate the assumption of conditional normal errors.

The estimated β for NI and SP500 are initially much lower than those for the HSI. Interestingly, when we compare Figures 2(a–c) with 3(a–c), the estimated α and β for the stock indices tend toward a similar range when the estimation window is extended from three to four years, that is, the third to fourth quartile. This strongly suggests that the optimum window size is from three to four years, as the recursive plots reveal significant stability in the estimated parameters for these periods.

Figures 4(a–c) reveal that the robust t-ratios for the three indices increase as the estimation window size is increased. In addition, the robust t-ratios for the estimated α are highly significant in the fourth quartile, which indicates significant ARCH effects when the estimation window size is from three to four years. Figures 5(a–c) display similar movements for the robust t-ratios of the estimated β for the three indices. Moreover, the estimated β are highly significant in all periods. Consequently, the fourth quartile strongly suggests that the GARCH model adequately describes the in-sample volatility. Together with the earlier result of stable ARCH parameters present in the fourth quartile, the optimum estimation window size is from three to four years for daily data.

In order to establish a valid argument for the optimal window size, it is imperative to investigate the regularity conditions for the GARCH model. Figures 6(a–c) and 7(a–c) present the second and fourth moment regularity conditions, respectively. The necessary and sufficient condition for second-order stationarity is given in equation (4). From Figures 6(a) and 6(c), the GARCH processes for the HSI and SP500 display satisfactory recursive second moment conditions throughout the entire period. On the other hand, the second moment condition is violated once in the first quartile for NI, and the conditional variance seems to be integrated in one period during the first quartile. This class of models is referred to as IGARCH, which implies that information at a particular point in time remains important for forecasts of conditional variance for all horizons. Figure 6(b) for the NI, also exhibits satisfactory second moment regularity conditions for the fourth quartile.

The fourth moment regularity condition for normal errors is given by

$$(\alpha + \beta)^2 + 2\alpha^2 < 1 \qquad (9)$$

Surprisingly, Figure 7(a) reveals the fourth moment condition being violated only in the fourth quartile for the HSI. On the other hand, in Figure 7(b) the fourth moment condition is violated only in the first quartile for the NI. Figure 7(c) indicates a satisfactory fourth moment condition for SP500, with the exception of some observations in the fourth quartile. This result is due

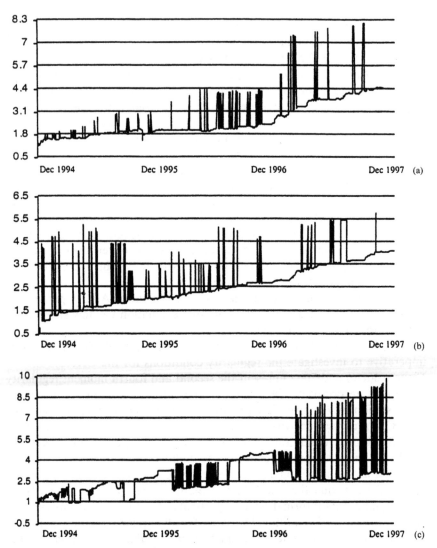

Figure 4. (a) Recursive GARCH-$\hat{\alpha}$ robust t-ratios for the Hang Seng index, (b) Recursive GARCH-$\hat{\alpha}$ robust t-ratios for the Nikkei 225 index, (c) Recursive GARCH-$\hat{\alpha}$ robust t-ratios for the SP500 index.

largely to the presence of extreme observations during the Asian financial crisis in October 1997. Moreover, with the exception of the HSI, there seems to be stability in the fourth moment condition in the fourth quartile, which points to a robust window size.

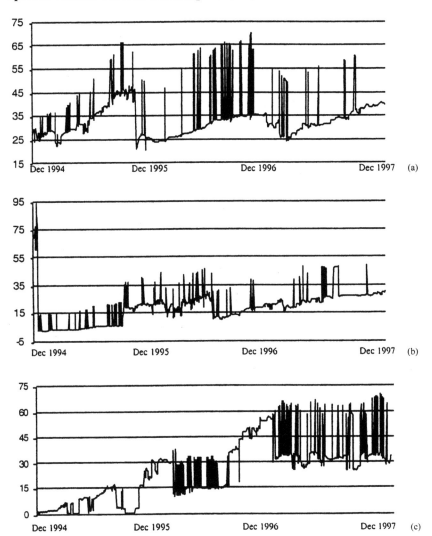

Figure 5. (a) Recursive GARCH-$\hat{\beta}$ robust t-ratios for the Hang Seng index, (b) Recursive GARCH-$\hat{\beta}$ robust t-ratios for the Nikkei 225 index, (c) Recursive GARCH-$\hat{\beta}$ robust t-ratios for the SP500 index.

GARCH (1,1) seems to detect in-sample volatility well, especially when the estimation window size is from three to four years for daily observations. This raises the interesting question of asymmetry and the appropriate volatility model, in particular, a choice between GARCH and GJR. Note that

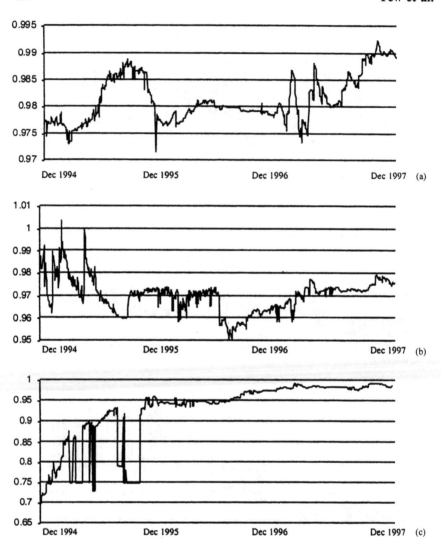

Figure 6. (a) Recursive GARCH second moments for the Hang Seng index, (b) Recursive GARCH second moments for the Nikkei 225 index, (c) Recursive GARCH second moments for the SP500 index.

the symmetric GARCH model is nested within the asymmetric GJR model, which is preferred when the QMLE of γ is significant in (8). Moreover, a significant estimate of γ implies that asymmetric effects are present within the sample period. Figures 8(a–c), 9(a–c), and 10(a–c) present the estimates

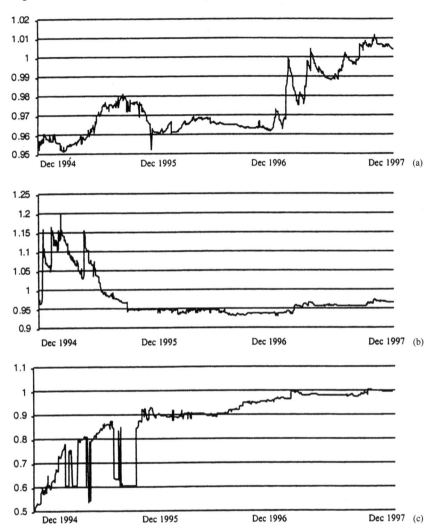

Figure 7. (a) Recursive GARCH fourth moments for the Hang Seng index, (b) Recursive GARCH fourth moments for the Nikkei 225 index, (c) Recursive GARCH fourth moments for the SP500 index.

of the α, β, and γ parameters of GJR for the HSI, NI, and SP500, respectively.

Within the GJR formulation, asymmetry is detected by the estimated γ parameter in Figures 10(a–c), which typically yields smaller estimates of α. In fact, the estimates of α for the HSI are all negative in the first and last

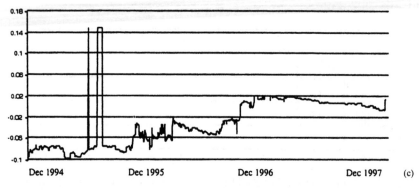

Figure 8. (a) Recursive GJR-$\hat{\alpha}$ for the Hang Seng index, (b) Recursive GJR-$\hat{\alpha}$ for the Nikkei 225 index, (c) Recursive GJR-$\hat{\alpha}$ for the SP500 index.

quartiles, while the NI has negative estimates of α only in the fourth quartile. Importantly, the estimates of α tend to move in parallel when the estimation window is extended to the fourth quartile for all three indices. The estimates of β in GJR in Figures 9(1–c) display similar recursive patterns to those of β in GARCH: (i) the estimates of β tend to move in opposite directions to their

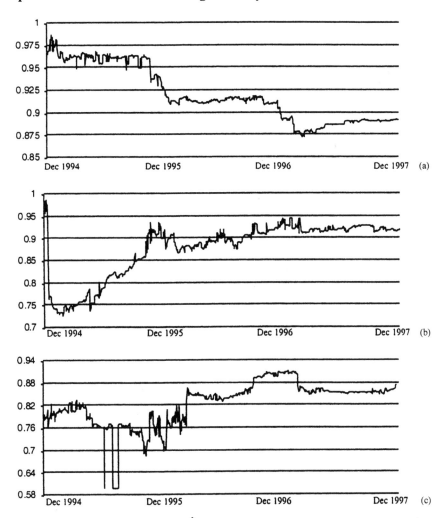

Figure 9. (a) Recursive GJR-$\hat{\beta}$ for the Hang Seng index, (b) Recursive GJR-$\hat{\beta}$ for the Nikkei 225 index, (c) Recursive GJR-$\hat{\beta}$ for the SP500 index.

estimated α counterparts; (ii) the movements in the estimates of β in GJR are stable throughout the fourth quartile; (iii) the estimates of β exhibit less volatility as compared with their GARCH counterparts, and move almost in parallel for all three indices in the fourth quartile; and (iv) the financial crisis in October 1997 does not influence parameter stability during the fourth quartile. Hence, there exists a robust estimation window size from three to four years, with daily data in the case of GJR.

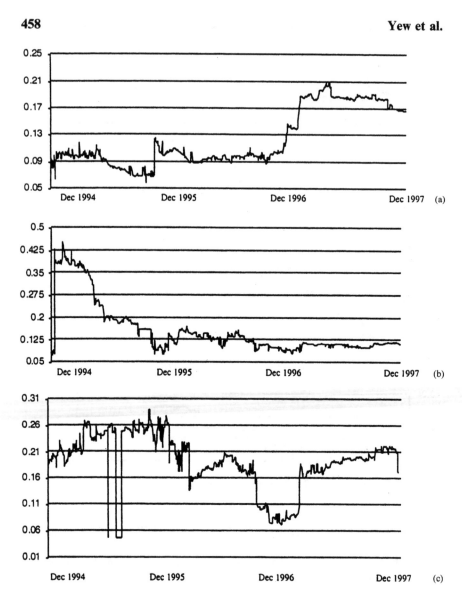

Figure 10. (a) Recursive GJR-$\hat{\gamma}$ for the Hang Seng index, (b) Recursive GJR-$\hat{\gamma}$ for the Nikkei 225 index, (c) Recursive GJR-$\hat{\gamma}$ for the SP500 index.

The estimated γ of GJR move closely with respect to their α counterparts, with the former having a greater magnitude due to the presence of extreme negative observations. This is more apparent for the NI than for the HSI. As in the previous analysis of estimated coefficients, the estimated γ exhibit stable and parallel movements in all three stock indices when the estimation window is extended to the fourth quartile.

Figures 11(a–c) reveal the robust t-ratios for the estimated α to be negative and significant for the first quartile for the HSI. Interestingly, the t-

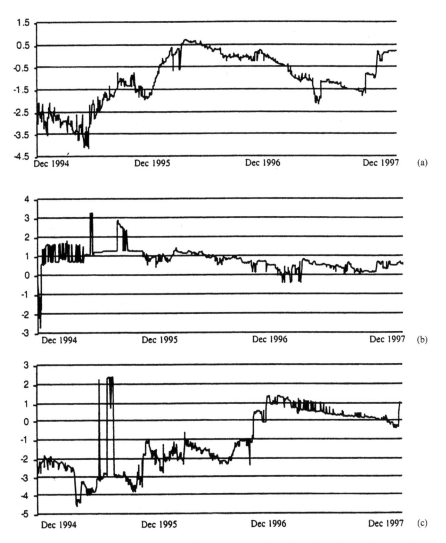

Figure 11. (a) Recursive GJR-$\hat{\alpha}$ robust t-ratios for the Hang Seng index, (b) Recursive GJR-$\hat{\alpha}$ robust t-ratios for the Nikkei 225 index, (c) Recursive GJR-$\hat{\alpha}$ robust t-ratios for the SP500 index.

ratios for all three indices are generally not significant in the fourth quartile, which suggests the presence of parameter stability as well as asymmetry.

The recursive plots of the robust t-ratios for the estimated β in Figures 12(a–c) are similar to those from the GARCH model. Movements of the t-ratios for GJR exhibit a significant positive trend for the NI and SP500, while the HSI has a similar trend, but with a sharp drop in December 1995. In particular, the t-ratios have similar trends in the fourth quartile.

In Figures 13(a–c), there exists a strong presence of asymmetry in the HSI, apart from the second quartile. On the other hand, the NI displays considerably greater volatility in its asymmetric effect, as shown by its fluctuating t-ratios. However, the t-ratios for the estimated γ seem to be positively tended as the estimation window size is increased. In particular, the fourth quartile t-ratios are highly significant for both the HSI and the NI. The t-ratios are significant virtually throughout the sample for SP500. Consequently, asymmetry and leverage effects are detected when the estimation window size is from three to four years of daily data. Moreover, the significant estimates of γ suggest that the in-sample volatility for the fourth quartile is described by GJR rather than GARCH.

It is important to note that even the robust t-ratios can be quite volatile throughout the estimation period, so that the significance of the GJR or the GARCH model can hinge on the inclusion or exclusion of a particular observation. However, as noted above, such *oversensitive* characteristics of GARCH-type models can be dampened by extending the estimation window from three to four years of daily observations, so that a significant model of in-sample volatility can be determined. It is also important to check whether the regularity conditions of the model are satisfied. Figures 14(a–c) and 15(a–c) present the second and fourth moment regularity conditions for GJR, respectively.

The second moment regularity condition for GJR is given in Ling and McAleer (1999a) as

$$\alpha + \beta + \tfrac{1}{2}\gamma < 1 \tag{10}$$

which reduces to condition (4) in the absence of asymmetry ($\gamma = 0$). A comparison of (4) and (10) makes clear that the admissible region for (α, β) for second-order stationarity of GJR is smaller than for GARCH because the asymmetry increases the uncertainty in GJR. It is apparent from Figures 14(a) and 14(c) that the second moment condition of GJR for the HSI and SP500 are satisfied for all periods. However, the second moment conditions for the NI are not satisfied in the first quartile, so that valid inferences cannot be drawn using GJR for the NI when the estimation window size is small. By the fourth quartile, the second moment conditions

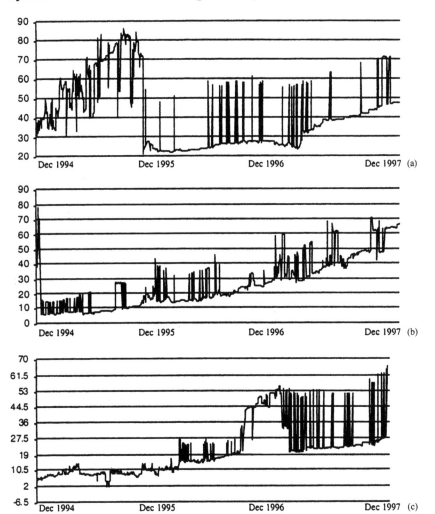

Figure 12. (a) Recursive GJR-$\hat{\beta}$ robust t-ratios for the Hang Seng index, (b) Recursive GJR-$\hat{\beta}$ robust t-ratios for the Nikkei 225 index, (c) Recursive GJR-$\hat{\beta}$ robust t-ratios for the SP500 index.

for all three stock indices are satisfied, and generally indicate stable estimated parameters of the GJR model.

The fourth moment regularity condition for the GJR model is given in
Ling and McAleer (199a) as

$$(\alpha + \beta)^2 + 2\alpha^2 + \beta\gamma + 3\alpha\gamma + \tfrac{3}{2}\gamma^2 < 1 \tag{11}$$

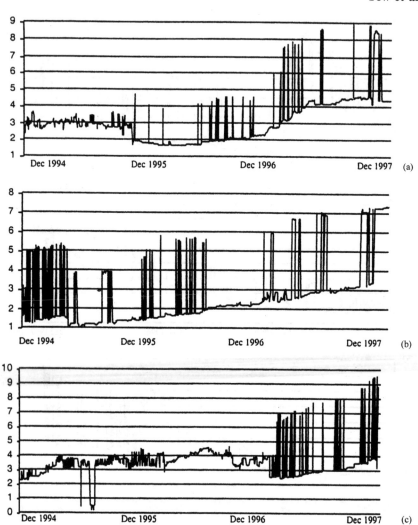

Figure 13. (a) Recursive GJR-$\hat{\gamma}$ robust t-ratios for the Hang Seng index, (b) Recursive GJR-$\hat{\gamma}$ robust t-ratios for the Nikkei 225 index, (c) Recursive GJR-$\hat{\gamma}$ robust t-ratios for the SP500 index.

which reduces to condition (9) in the absence of asymmetry ($\gamma = 0$). A comparison of (9) and (11) also makes clear that the admissible region for (α, β) for fourth-order stationarity of GJR is smaller than for GARCH because of the asymmetry ($\gamma 0$). As observed in Figures 15(a–c), GJR for the NI clearly violates the fourth moment condition when the estimation

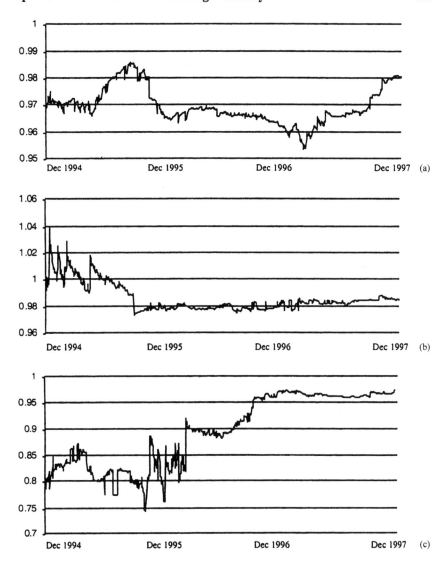

Figure 14. (a) Recursive GJR second moments for the Hang Seng Index, (b) Recursive GJR second moments for the Nikkei 225 index, (c) Recursive GJR second moments for the SP500 index.

window size has less than three years of daily data. However, the fourth moment condition is also violated in several periods during he last quartile for SP500. It is satisfied throughout for the HSI, apart from a few observations at the end of the sample. The presence of extreme observations is

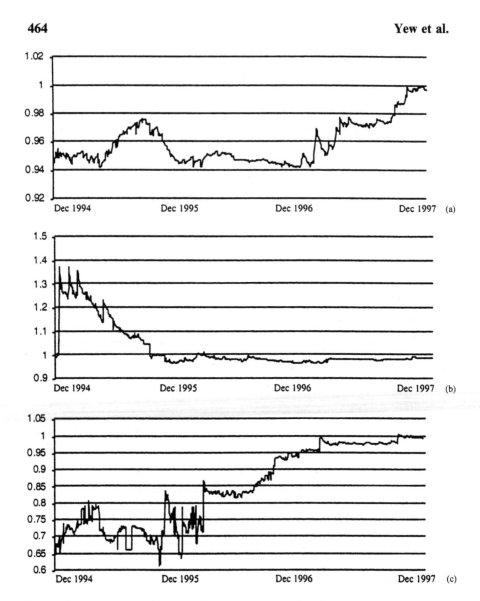

Figure 15. (a) Recursive GJR fourth moments for the Hang Seng index, (b) Recursive GJR fourth moments for the Nikkei 225 index, (c) Recursive GJR fourth moments for the SP500 index.

largely responsible or the non-normality in th unconditional distribution, which increases the conditional kurtosis and leads to the violation of the fourth-order moment regularity condition.

Notwithstanding the presence of extreme observations, the recursive second moment conditions and some forth moment conditions for GJR are generally satisfied in the third and fourth quartiles. However, there is still a need to examine the influence of extreme observations on the fourth-order moment conditions for both GARCH and GJR.

4. CONCLUDING REMARKS

In this paper, we have investigated the sensitivity of the estimated parameters of the GARCH and GJR models and concluded that there exists a robust estimation window size from three to four years of daily observations. The conclusions were reached by estimating the GARCH and GJR models recursively until the minimum number of observations within the window exhibited significant parameter stability. These findings provide an important insight into robust in-sample volatility analysis, as the choice of the estimation window size and data frequency can often yield alternative empirical models. Such conclusions are based on estimating the GARCH and GJR models using the Hang Seng, Nikkei 225, and SP500 index returns.

In addition, the empirical properties of both GARCH and GJR were also evaluated by checking the second and fourth moment regularity conditions, which have often been ignored in the existing empirical literature. Based on the derivations in Ling and McAleer (1999a), who obtained the second and fourth moment regularity conditions for the GJR model, it is now possible to analyse the empirical validity of GJR. Inferences based on the empirical results strongly suggest that the asymmetric GJR model outperforms the symmetric GARCH in-sample, especially when the estimation window size is extended to a robust range from three to four years. The presence of extreme observations was shown to have adverse effects on the regularity conditions for both models. Finally, the empirical analysis has provided a foundation for examining the relationship between optimal window sizes and volatility forecasts.

ACKNOWLEDGMENTS

The second and third authors wish to acknowledge the financial support of the Australian Research Council. This paper was written while the second author was a Visiting Scholar from Abroad at the Institute of Monetary and Economic Studies, Bank of Japan.

REFERENCES

Black, F. (1976), Studies in stock price volatility changes. *Proceedings of the 1976 Meeting of the Business and Economic Statistics Section, American Statistical Association* pp. 177–181.

Bollerslev, T. (1986), Generalized autoregressive conditional heteroskedasticity. *Journal of Econometrics*, 31, 307–327.

Bollerslev, T., and J. M. Wooldridge (1992), Quasi-maximum likelihood estimation and inference in dynamic models with time varying covariances. *Econometric Reviews*, 11, 143–172.

Drost, F. C., and C. A. J. Klaassen (1997), Efficient estimation in semiparametric GARCH models. *Journal of Econometrics*, 81, 193–221.

Engle, R. F. (1982), Autoregressive conditional heteroskedasticity with estimates of the variance of United Kingdom inflation. *Econometrica*, 50, 987–1007.

Engle, R. F., and C. Mustafa (1992), Implied ARCH models from options prices. *Journal of Econometrics,* 52, 289–311.

Engle, R. F., and V. K. Ng (1993), Measuring and testing the impact of news on volatility. *Journal of Finance*, 48, 1749–1778.

Fornari, F., and A. Mele (1997), Sign- and volatility-switching ARCH models: Theory and applications to international stock markets. *Journal of Applied Econometrics*, 12, 49–65.

Glosten, L., R, Jagannathan, and D. Runkle (1993), On the relations between the expected value and the volatility on the nominal excess returns on stocks. *Journal of Finance*, 48, 1779–1801.

Jarque. C. M., and A. K. Bera (1980), Efficient tests for normality, heteroskedasticity, and serial independence of regression residuals. *Economics Letters*, 6, 225–229.

Lee, S.-W., and B. E. Hansen (1994), Asymptotic theory for the GARCH (1,1) quasi-maximum likelihood estimator. *Econometric Theory*, 10, 29–52.

Ling, S. (1999), On the probabilistic properties of a double threshold ARMA conditional heteroskedasticity model. *Journal of Applied Probability*, 36, 1–18.

Ling, S., and W. K. Li (1998), Limiting distributions of maximum likelihood estimators for unstable ARMA models with GARCH errors. *Annals of Statistics*, 26, 84–125.

Ling ,S., and M. McAleer (1999a), Stationarity and the existence of moments of a family of GARCH processes. Submitted.

Ling, S., and M. McAleer (1999b), Necessary and sufficient moment conditions for the ARCH (r, s) and asymmetric power GARCH (r, s) models. Submitted.

Ling, S., and M. McAleer (1999c), Asymptotic theory for a new vector ARMA–GARCH model. Submitted.

Lumsdaine, R. L. (1996), Consistency and asymptotic normality of the quasi-maximum likelihood estimator in IGARCH (1,1) and covariance stationary GARCH (1,1) models. *Econometrica*, 64, 575–596.

Nelson, D. B. (1990a), Conditional heteroskedasticity in asset returns: a new approach. *Econometrica*, 59, 347–370.

Nelson, D. B. (1990b), Stationarity and persistence in the GARCH (1,1) model. *Econometric Theory*, 6, 318–334,

Pagan, A. R., and G. W. Schwert (1990), Alternative models of conditional stock volatility. *Journal of Econometrics*, 45, 267–290.

Rabemanjara, R., and J. M. Zakoian (1993), Threshold ARCH models and asymmetries in volatility. *Journal of Applied Econometrics*, 8, 31–49.

Schwert, G. W. (1990). Stock volatility and the crash of 87, *Review of Financial Studies* 3, 71–102.

22
SUR Models with Integrated Regressors

KOICHI MAEKAWA   Hiroshima University, Hiroshima, Japan

HIROYUKI HISAMATSU   Kagawa University, Takamatsu, Japan

1. EXOGENOUS I(1) REGRESSORS
1.1 n-Equation SUR System

We consider the following n-equation system

Model 1

$$y_{it} = \beta_i x_{it} + u_{it}, \quad i = 1, 2, \ldots, n \tag{1}$$

where the explanatory variables x_{it} are integrated processes of order one denoted by I(1) for each i,

$$x_{it} = x_{it-1} + v_{it}$$

and the suffixes $i = 1, 2, \ldots, n$ and $t = 1, 2, \ldots, T$ denote the equation numbers and time respectively. If the disturbances between equations are correlated in this system, then this is Zellner's SUR system (see [5],[6]). This system is also a special case of Park and Phillips [2].

In this paper we confine ourselves to the SUR system and investigate the statistical properties of the SUR estimator. We assume that u_{it} are normally distributed $N(0, \sigma_{ii})$ and have the covariance $\mathrm{Cov}(u_{it}, u_{jt}) = \sigma_{ij}$ for all t, and that u_{it} and v_{it} are independent.

To write the system compactly we introduce the following vectors and matrices:

$$x_i' = (x_{i1}, x_{i2}, \ldots, x_{iT})$$
$$y_i' = (y_{i1}, y_{i2}, \ldots, y_{iT})$$
$$u_i' = (u_{i1}, u_{i2}, \ldots, u_{iT})$$
$$\beta = (\beta_1, \beta_2, \ldots, \beta_n)$$

then we can write the system as

$$
\begin{pmatrix} y_1 \\ y_2 \\ \cdot \\ \cdot \\ \cdot \\ y_n \end{pmatrix}
=
\begin{pmatrix}
x_1 & 0 & \cdot & \cdot & 0 \\
0 & x_2 & \cdot & \cdot & 0 \\
\cdot & \cdot & \cdot & \cdot & \cdot \\
\cdot & \cdot & \cdot & \cdot & \cdot \\
0 & \cdot & \cdot & \cdot & x_n
\end{pmatrix}
\beta +
\begin{pmatrix} u_1 \\ u_2 \\ \cdot \\ \cdot \\ \cdot \\ u_n \end{pmatrix}
$$

or more compactly

$$Y = X\beta + U$$

where the definitions of Y, X, and U are self-evident. The distribution of U can be written as

$$U \sim N(0, \Omega)$$

where

$$\Omega = (\Omega_{ij}) = (\sigma_{ij}I) = \Sigma_U \otimes I$$

where $\Sigma_U = (\sigma_{ij})$ is assumed positive definite and $\otimes$ signifies the Kronecker product. The disturbance vector $v_i' = (v_{i1}, v_{i2}, \ldots, v_{iT})$ is normally distributed as

$$v_i \sim N(0, \sigma_{vi}^2 I)$$

for all i, where I is the $T \times T$ identity matrix. The OLS and SUR estimators of β are respectively written as

$$\text{OLS: } \hat{\beta} = (X'X)^{-1}X'Y \tag{2}$$

and

$$\text{SUR: } \tilde{\beta} = \left(X'\tilde{\Omega}^{-1}X\right)^{-1}X'\tilde{\Omega}^{-1}Y \tag{3}$$

where $\tilde{\Omega}$ is the estimated covariance matrix:

$$\tilde{\Omega} = (s_{ij}) \otimes I$$

where s_{ij} is the estimator of σ_{ij}. The Zellner's restricted SUR (abbreviated RSUR) estimator, denoted by $\tilde{\beta}_R$, uses the estimator

$$s_{ij} = \frac{\hat{u}_i'\hat{u}_j}{T-1}$$

where $\hat{u}_i = y_i - x_i\hat{\beta}_i$, $\hat{\beta}_i = (x_i'x_i)^{-1}x_i'y_i$, $i,j = 1, 2, \ldots, n$, and the Zellner's unrestricted SUR (abbreviated USUR) estimator, denoted by $\tilde{\beta}_U$, uses the estimator

$$s_{ij}^* = \frac{e_i'e_j}{T-n}$$

where $e_i = y_i - Z\hat{\gamma}$, $Z = (x_1 \; x_2 \; \cdots \; x_n)$, $\hat{\gamma} = (Z'Z)^{-1}y_i$, $i,j = 1, 2, \ldots, n$. Since both s_{ij} and s_{ij}^* are written as

$$s_{ij}, s_{ij}^* = \sigma_{ij} + O_p\left(\frac{1}{\sqrt{T}}\right), \qquad i,j = 1, 2, \ldots, n$$

they are consistent estimators of σ_{ij}, $i,j = 1, 2, \ldots, n$ (see Maekawa and Hisamatsu [1]). It is noted that the USUR is a special case of the GLS estimators dealt with in Park and Phillips [2], but the RSUR is not included in [2].

1.2 Asymptotic Distributions of the SUR and OLS Estimators

Following Park and Phillips [2] we derive the asymptotic distributions of the RSUR, USUR, and OLS estimators. We define n-dimensional vectors U_t, V_t, and W_t:

$$U_t' = (u_{1t}, u_{2t}, \cdots, u_{nt}), \; V_t' = (v_{1t}, v_{2t}, \cdots, v_{nt}), \; W_t = \begin{pmatrix} U_t \\ V_t \end{pmatrix}$$

where V_t has a covariance matrix Σ_V:

$$\Sigma_V = \text{diag}\left(\sigma_{v1}^2, \sigma_{v2}^2, \cdots, \sigma_{vn}^2\right)$$

The standardized partial sum of $\mathbf{W}_t$ converges in distribution as

$$\frac{1}{\sqrt{T}} \sum_{j=1}^{[Tr]} \mathbf{W}_j \Rightarrow B(r) = \begin{pmatrix} B_1(r) \\ B_2(r) \end{pmatrix}, \quad (j-1)/T \le r < j/T, \ j = 1, \ldots, T$$

where [ ] denotes the integer part of its argument and "$\Rightarrow$" signifies convergence in distribution, and $B(r)$ is a $2n$-dimensional Brownian motion with a $2n \times 2n$ covariance matrix

$$\Theta = \begin{pmatrix} \Sigma_U & 0 \\ 0 & \Sigma_V \end{pmatrix}$$

and $B_1(r)$ and $B_2(r)$ have elements such as

$$B_1(r)' = (B_{11}(r), B_{12}(r), \cdots, B_{1n}(r)), \ B_2(r)' = (B_{21}(r), B_{22}(r), \cdots, B_{2n}(r))$$

It is easy to see that the asymptotic distribution of the OLS estimator is given by

$$T(\hat{\beta}_i - \beta_i) \Rightarrow \frac{\int_0^1 B_{2i}(r)dB_{1i}(r)}{\int_0^1 B_{2i}(r)^2 dr}, \quad i = 1, 2, \ldots, n. \tag{4}$$

The asymptotic distribution of the RSUR estimator $\tilde{\beta}_R$ is written as

$$T(\tilde{\beta}_R - \beta) = T(X'\tilde{\Omega}^{-1}X)^{-1}X'\tilde{\Omega}^{-1}U$$

$$= T\left[s^{ij}\mathbf{x}_i'\mathbf{x}_j\right]^{-1}\left[s^{ij}\mathbf{x}_i'\mathbf{u}_j\right]\mathbf{i} \tag{5}$$

where $[a_{ij}]$ denotes a matrix A with the (i, j) element a_{ij} and a^{ij} is the (i, j) element of the inverse matrix $(a_{ij})^{-1}$, $i, j = 1, \ldots, n$ and $\mathbf{i}' = (1, 1, \ldots, 1)$ with n elements of 1. By standard calculation we have

$$\frac{1}{T^2}\mathbf{x}_i'\mathbf{x}_i \Rightarrow \int_0^1 B_{2i}(r)^2 dr$$

$$\frac{1}{T\mathbf{x}_i'\mathbf{u}_j} \Rightarrow \int_0^1 B_{2i}(r)dB_{1j}(r)$$

$$\frac{1}{T^2}\mathbf{x}_i'\mathbf{x}_j \Rightarrow \int_0^1 B_{2i}(r)B_{2j}(r)dr$$

$$s_{ij}, s_{ij}^* \to \sigma_{ij}, \quad i, j = 1, 2, \ldots, n$$

$$s^{ij}, s^{ij*} \to \sigma^{ij}, \quad i, j = 1, 2, \ldots, n$$

where "$\rightarrow$" signifies convergence in probability. By substitution we have

$$T(\tilde{\beta}_R - \beta) \Rightarrow \left[c\sigma^{ij} \int_0^1 B_{2i}(r)B_{2j}(r)dr \right]^{-1} \left[\sigma^{ij} \int_0^1 B_{2i}(r)dB_{1j}(r) \right] \mathbf{i} \tag{6}$$

Since both s_{ij} and s_{ij}^* are consistent, the RSUR and USUR estimators are asymptotically equivalent. It is noted that if $\sigma_{ij} = \sigma_{ji} = 0$ $(i \neq j)$, then the RSUR estimator is asymptotically identical to the OLS estimator

$$T(\tilde{\beta}_R - \beta) = T\left(\hat{\beta} - \beta\right) \tag{7}$$

When $n = 2$, the asymptotic distribution of $T(\tilde{\beta}_R - \beta)$ is reduced to a simple formula as

$$T(\tilde{\beta}_R - \beta) \Rightarrow \frac{1}{D} \left[\begin{array}{c} \sigma_{11} \int_0^1 B_{22}(r)^2 dr \cdot P + \sigma_{12} \int_0^1 B_{21}(r)B_{22}(r)dr \cdot Q \\ \sigma_{21} \int_0^1 B_{21}(r)B_{22}(r)dr \cdot P + \sigma_{22} \int_0^1 B_{11}(r)^2 dr \cdot Q \end{array} \right] \tag{8}$$

where

$$D = \sigma_{11}\sigma_{22} \int_0^1 B_{21}(r)^2 dr \int_0^1 B_{22}(r)^2 dr - \sigma_{12}\sigma_{21} \left(\int_0^1 B_{21}(r)B_{22}(r)dr \right)^2$$

$$P = \sigma_{22} \int_0^1 B_{21}(r)dB_{11}(r) - \sigma_{12} \int_0^1 B_{21}(r)dB_{12}(r)$$

$$Q = -\sigma_{21} \int_0^1 B_{22}(r)dB_{11}(r) + \sigma_{11} \int_0^1 B_{22}(r)dB_{12}(r)$$

It is straightforward to extend the above model so as to include a constant term in each equation. We can show that the SUR estimators of β and a constant term are T and $\sqrt{T}$ consistent respectively and that the asymptotic distributions are also nonstandard (see [1] for details).

1.3 Monte Carlo Experiment

Small sample distributions of $\tilde{\beta}_R$, $\tilde{\beta}_U$, and $\hat{\beta}$ in the 2-equation SUR system are examined by a Monte Carlo experiment with 5000 iterations for sample sizes $T = 30$. We controlled the two variance ratios $\eta = \sigma_{22}/\sigma_{11}$ (the variance ratio between u_{1t} and u_{2t}), $\kappa = \sigma_{ii}/\sigma_{vi}^2$, $i = 1, 2$ (the variance ratio between u_{it} and v_{it}), and the correlation between u_{1t} and u_{2t}, $\rho = \sigma_{12}/\sqrt{\sigma_{11}\sigma_{22}}$. The results are presented in Figure 1. This shows the empirical cumulative distribution functions (CDFs) of the RSUR, USUR, and OLS estimators. It is seen that the two lines for the RSUR and USUR estimators cannot be distinguished but those distributions are more concen-

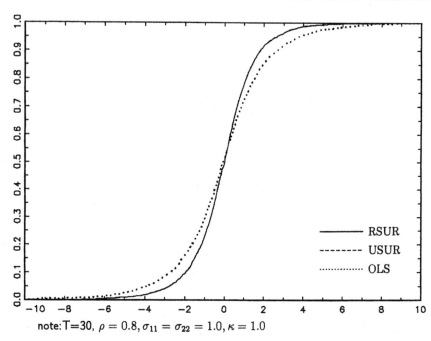

note: T=30, $\rho = 0.8$, $\sigma_{11} = \sigma_{22} = 1.0$, $\kappa = 1.0$

Figure 1.1. Distributions of the RSUR, USUR, and OLS in Model 1.

trated around the origin than the distribution of the OLS estimator. The three distributions are symmetric around the origin. From this observation we can say that the RSUR and USUR estimators are more efficient than the OLS estimator. To see the effect of κ and ρ, we conducted further experiments for $\rho = 0.2$ and for $\rho = 0.8$. Although figures are omitted here, we observed that the distributions are slightly concentrated for $\kappa = 0.5$ and that the empirical CDFs of the three distributions are much closer for $\rho = 0.2$ than for $\rho = 0.8$, and hence the SUR estimation has little gain over the OLS estimation when ρ is small.

2. SUR RANDOM WALKS

2.1 *n*-Equation SUR Random Walks

This section deals with the *n*-equation SUR system: y_{it} are seemingly unrelated random walks but are actually related through the correlated disturbances. Such a model is written as

Model 2

$$y_{it} = \beta_i y_{it-1} + u_{it}, \quad i = 1, 2, \ldots, n \\ \beta_i = 1 \text{ for all } i \Bigg\} \tag{9}$$

We assume that $y_{i0} = 0$ for all i, $E(u_{it}u_{jt}) = \sigma_{ij} \neq 0$, and, for any i, $u_{it} \sim$ iid $N(0, \sigma_{ii})$ for any i, j. Using vectors and matrices:

$$y_i' = (y_{i1}, y_{i2}, \ldots, y_{iT})$$
$$y_{i,-1}' = (y_{i0}, y_{i,1}, \ldots, y_{i,T-1})$$
$$u_i' = (u_{i1}, u_{i2}, \ldots, u_{iT})$$

(9) is expressed by

$$\begin{pmatrix} y_1 \\ y_2 \\ \cdot \\ \cdot \\ \cdot \\ y_n \end{pmatrix} = \begin{pmatrix} y_{1,-1} & 0 & \cdot & \cdot & 0 \\ 0 & y_{2,-1} & \cdot & \cdot & 0 \\ & & \cdot & \cdot & \cdot \\ 0 & & \cdot & \cdot & y_{n,-1} \end{pmatrix} \begin{pmatrix} \beta_1 \\ \beta_2 \\ \cdot \\ \cdot \\ \beta_n \end{pmatrix} + \begin{pmatrix} u_1 \\ u_2 \\ \cdot \\ \cdot \\ u_n \end{pmatrix}$$

or more compactly by

$$Y = X\beta + U$$

where the definitions of $Y, X, \beta,$ and U are self-evident. U is normally distributed as

$$U \sim N(0, \Omega)$$

where

$$\Omega = (c\Omega_{ij}) = \Sigma_U \otimes I$$

The OLS and SUR estimators are defined by

$$\text{OLS: } \hat{\beta} = (X'X)^{-1}X'Y \tag{10}$$

$$\text{SUR: } \tilde{\beta} = (X'\tilde{\Omega}^{-1}X)^{-1}X'\tilde{\Omega}^{-1}Y \tag{11}$$

where

$$\tilde{\Omega} = (s_{ij}) \otimes I$$

The RSUR estimator $\tilde{\beta}_R$ applies a covariance estimator s_{ij}, defined by

$$s_{ij} = \frac{\hat{u}_i'\hat{u}_j}{T-1}$$

where $\hat{u}_i = y_i - y_{i,-1}\hat{\beta}_i$, $\hat{\beta}_i = (y'_{i,-1}y_{i,-1})^{-1}y'_{i,-1}y_i$, $i, j = 1, 2, \ldots, n$, and the USUR estimator β_U applies s^*_{ij}, defined by

$$s^*_{ij} = \frac{e'_i e_j}{T - n}$$

where $e_i = y_i - Z\hat{\gamma}$, $Z = (y_{1,-1}\ y_{2,-1}\ \cdots\ y_{n,-1})$, $\hat{\gamma} = (X'Z)^{-1}Z'y_i$, $i, j = 1, 2, \ldots, n$.

As s_{ij} and s^*_{ij} are consistent estimators of σ_{ij}, $i, j = 1, 2, \ldots, n$ (see [1]), $\tilde{\beta}_R$ and $\tilde{\beta}_U$ are asymptotically equivalent.

2.2 Asymptotic Distributions of the SUR and OLS Estimators

The asymptotic distributions of the OLS and SUR estimators can be obtained following Phillips and Durlauf [4]. There exists a triangular $n \times n$ matrix $\Sigma_U^{1/2} = (\omega^{ij})$ by which Σ_U can be decomposed as $\Sigma_U = (\Sigma_U^{1/2})'(\Sigma_U^{1/2})$. We define vectors $\mathbf{U}_t$ and $\mathbf{Y}_t$ as

$$\mathbf{U}'_t = (u_{1t}, u_{2t}, \ldots, u_{nt}), \quad \mathbf{Y}'_t = (y_{1t}, y_{2t}, \ldots, y_{nt})$$

Let S_t denote the partial sum of $\mathbf{U}_t$, i.e.,

$$S_t = \sum_{j=1}^{t} \mathbf{U}_j$$

then $(1/\sqrt{T})S_{[Tr]}$ converges in distribution to a Brownian motion by the central limit theorem,

$$\frac{1}{\sqrt{T}}S'_{[Tr]} \Rightarrow B(r)' = (B_1(r), B_2(r), \ldots, B_n(r))$$

$$\frac{1}{\sqrt{T}}S'_{[T]} \Rightarrow B(1)' = (B_1(1), B_2(1), \ldots, B_n(1))$$

where $B(r)$ is an n-dimensional Brownian motion. The standardized partial sum converges in distribution to the standardized Wiener process $W(r)$, i.e.,

$$\frac{1}{\sqrt{T}}\left(\Sigma_U^{-1/2}S_{[Tr]}\right)' \Rightarrow W(r)' = (W_1(r), W_2(r), \ldots, W_n(r))$$

and

$$\frac{1}{\sqrt{T}}\left(\Sigma_U^{-1/2}S_{[T]}\right)' \Rightarrow W(1)' = (W_1(1), W_2(1), \ldots, W_n(1))$$

where the relation $\Sigma_U^{-1/2}B(r) = W(r)$ is used.

Then the asymptotic distribution of the OLS estimator is written as

$$T(\hat{\beta}_i - \beta_i) \Rightarrow \frac{\frac{1}{2}\{W_i^2(1) - 1\}}{\int_0^1 W_i^2(r)dr}, \qquad i = 1, 2, \ldots, n \tag{12}$$

as is well-known.

The Zellner's restricted estimator is given by

$$T(\tilde{\beta}_R - \beta) = T(X'\tilde{\Omega}^{-1}X)^{-1}X\tilde{\Omega}^{-1}U$$

$$= \left[s^{ij}\Sigma_{t=1}^T y_{it-1}y_{jt-1}/T^2\right]^{-1}\left[s^{ij}\Sigma_{t=1}^T y_{it-1}u_{jt-1}/T\right]\mathbf{i} \tag{13}$$

$$= T\left[s^{ij}y_{i,-1}'y_{j,-1}\right]^{-1} \times \left[s^{ij}y_{i,-1}', u_{j,-1}\right]\mathbf{i}$$

From Lemma 3.1 in Phillips and Durlauf [4] we have

$$T^{-3/2}\sum_{t=1}^T Y_t \Rightarrow \int_0^1 B(r)dr = \Sigma_U^{1/2}\int_0^1 W(r)dr$$

$$T^{-2}\sum_{t=1}^T Y_tY_t' = T^{-2}\left[\sum_{t=1}^T y_{it}y_{jt}\right] \Rightarrow \Sigma_U^{1/2}\int_0^1 W(r)W(r)'dr\left(\Sigma_U^{1/2}\right)'$$

$$T^{-1}\sum_{t=1}^T Y_{t-1}U_t' = T^{-1}\left[\sum_{t=1}^T y_{i,t-1}u_{jt}\right] \Rightarrow \Sigma_U^{1/2}\int_0^1 W(r)dW(r)'\left(\Sigma_U^{1/2}\right)'$$

Therefore we have

$$T(\tilde{\beta}_R - \beta) \Rightarrow \left[(\sigma^{ij}) \odot \left(\Sigma_U^{1/2}\int_0^1 W(r)W(r)'dr\left(\Sigma_U^{1/2}\right)'\right)\right]^{-1}$$

$$\times \left[(\sigma^{ij}) \odot \left(\Sigma_U^{1/2}\int_0^1 W(r)dW(r)'\left(\Sigma_U^{1/2}\right)'\right)\right]\mathbf{i}$$

where $\odot$ is the Hadamard product. As $s_{ij}^* \to \sigma_{ij}$ the two estimators $\tilde{\beta}_U$ and $\tilde{\beta}_R$ are asymptotically equivalent. If $\sigma_{ij} = 0$ for all $i \neq j$, $i, j = 1, 2, \ldots, n$, we have

$$T(\tilde{\beta}_R - \beta) = T(\hat{\beta} - \beta) \tag{14}$$

When $n = 2$ we can choose the triangular matrix as

$$\Sigma_U^{1/2} = \begin{pmatrix} \sqrt{\sigma_{11}} & \rho\sqrt{\sigma_{22}} \\ 0 & \sqrt{(1-\rho^2)\sigma_{22}} \end{pmatrix} \equiv \begin{pmatrix} a & b \\ 0 & c \end{pmatrix}$$

to standardize $B(r)$ so that we have $\Sigma_U^{-1/2}B(r) = W(r) = (W_1(r), W_2(r))'$. Using this, we can rewrite the asymptotic distribution of $T(\tilde{\beta}_R - \beta)$ as

$$
\begin{bmatrix} T(\tilde{\beta}_{R1} - \beta_1) \\ T(\tilde{\beta}_{R2} - \beta_2) \end{bmatrix} \Rightarrow \frac{1}{D} \begin{bmatrix} \sigma_{11} \cdot c^2 \int_0^1 W_2^2(r)dr \cdot P \\ +\sigma_{12}\left\{ac \int_0^1 W_1(r)W_2(r)dr + bc \int_0^1 W_2^2(r)dr\right\} \cdot Q \\ \sigma_{21}\left\{ac \int_0^1 W_1(r)W_2(r)dr + bc \int_0^1 W_2^2(r)dr\right\} \cdot P \\ +\sigma_{22}\left\{\int_0^1 (aW_1(r) + bW_2(r))^2 dr\right\} \cdot Q \end{bmatrix}
$$

$$(15)$$

where

$$
D = \sigma_{11}\sigma_{22}\left\{\int_0^1 (aW_1(r) + bW_2(r))^2 dr\right\}\left\{c^2 \int_0^1 W_2^2(r)dr\right\}
$$

$$
- \sigma_{12}\sigma_{21}\left\{ac \int_0^1 W_1(r)W_2(r)dr + bc \int_0^1 W_2^2(r)dr\right\}^2
$$

$$
P = \sigma_{22}\left\{ \begin{array}{l} a\left(a\int_0^1 W_1(r)dW_1(r) + b\int_0^1 W_2(r)dW_1(r)\right) \\ +b\left(a\int_0^1 W_1(r)dW_2(r) + b\int_0^1 W_2(r)dW_2(r)\right) \end{array}\right\}
$$

$$
- \sigma_{12}\left\{ac \int_0^1 W_1(r)dW_2(r) + bc \int_0^1 W_2(r)dW_2(r)\right\}
$$

$$
Q = \sigma_{21}\left\{ac \int_0^1 W_2(r)dW_1(r) + bc \int_0^1 W_2(r)dW_2(r)\right\}
$$

$$
+ \sigma_{11}\left\{c^2 \int_0^1 W_2(r)dW_2(r)\right\}
$$

When $\sigma_{12} = \sigma_{21} = 0$ or $\rho = 0$ we have

$$
T(\tilde{\beta}_{Ri} - \beta_i) \Rightarrow \frac{\int_0^1 W_i(r)dW_i(r)}{\int_0^1 W_i^2(r)dr} = \frac{\frac{1}{2}\{W_i^2(1) - 1\}}{\int_0^1 W_i^2(r)dr}, \qquad i = 1, 2,
$$

and hence we have

$$
T(\tilde{\beta}_R - \beta) = T(\hat{\beta} - \beta). \tag{16}
$$

If a constant term is included in each equation the resulting asymptotic distributions are normal (see [1] for details).

2.3 Monte Carlo Experiment

Small sample distributions of $\tilde{\beta}_R$, $\tilde{\beta}_U$, and $\hat{\beta}$ in the 2-equation SUR random walks are examined by a Monte Carlo experiment with 5000 iterations for the sample sizes $T = 30$ and 100. We controlled the variance ratio $\eta = \sigma_{22}/\sigma_{11}$ (the variance ratio between u_{1t} and u_{2t}) and the correlation between u_{1t} and u_{2t}, $\rho = \sigma_{12}/\sqrt{\sigma_{11}\sigma_{22}}$. The results are shown in Figures 2.1 and 2.2.

Figure 2.1 shows empirical cumulative distributions of the RSUR, USUR, and OLS estimators when $T = 30$ and $\rho = 0.8$. It is seen that the distributions are very skewed to the left but the distributions of the RSUR and USUR estimators are almost indistinguishable. As in the previous case (Section 1), Figure 2.1 indicates that the distributions of the RSUR and USUR estimates are more concentrated around the true value, 0, than the OLS estimator. Observing these figures, we can say that the SUR estimators are more efficient than the OLS estimator in the sense that the distributions of the two SUR estimators are more concentrated around the origin than the OLS estimator.

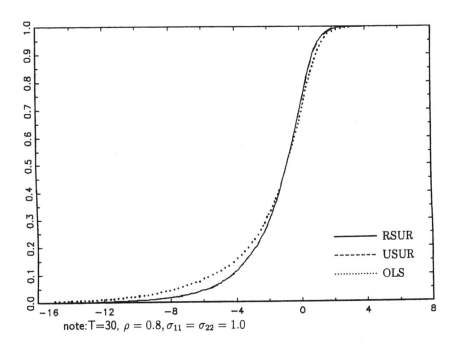

Figure 2.1. Distribution of the RSUR, USUR and OLS in Model 2.

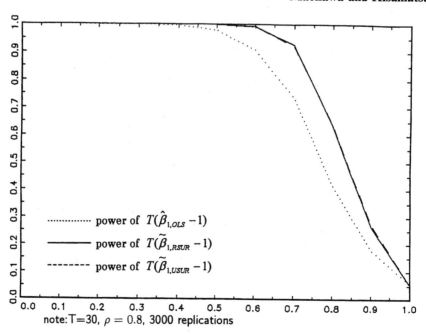

note: T=30, $\rho = 0.8$, 3000 replications

Figure 2.2. Power of Unit Root Tests

Next we consider the unit root test for $H_0 : \beta_1 = 1$ against $H_a : \beta_1 < 1$. In the above simulation, we calculated 5% critical values for $T = 30$ and 100 as follows:

	OLS	RSUR	USUR
$T = 30$, $\rho = 0.8$	−7.394	−5.657	−5.631
$T = 100$, $\rho = 0.8$	−7.933	−5.945	−5.883

The empirical power of the three tests are compared for $T = 30$ in Figure 2.2. This figure shows that the unit root tests based on both RSUR and USUR are more powerful than the test based on the OLS estimator for $\rho = 0.8$. When ρ is small, such as 0.2, the three empirical power curves are much closer, although the related figures are omitted here.

2.4   The SUR Mixed Model

2.4.1 The asymptotic distribution of the SUR estimator

This section deals with the 2-equation SUR system: y_{1t} is a random walk and y_{2t} is stationary AR(1) . Such a model is written as

Model 3

$$\left.\begin{array}{l} y_{1t} = \beta_1 y_{1t-1} + u_{1t} \\ y_{2t} = \beta_2 y_{2t-1} + u_{2t} \\ \beta_1 = 1 \quad \text{and} \quad |\beta_2| < 1 \end{array}\right\} \tag{17}$$

We assume that $y_{i0} = 0$ for all i $E(u_{1t}u_{2t}) = \sigma_{12} \neq 0$, and $u_{it} \sim$ iid $N(0, \sigma_{ii}, i = 1, 2$.

After some tedious calculation, we have the asymptotic distribution of the three estimators in the first equation as follows:

$$T(\tilde{\beta}_{R1} - \beta_1),\ T(\tilde{\beta}_{U1} - \beta_1),\ \text{and}\ T(\hat{\beta}_1 - \beta_1) \Rightarrow \frac{(\frac{1}{2})\{W_1(1)^2 - 1\}}{\int_0^1 W_1(r)^2 dr} \tag{18}$$

This shows that the three estimators are T-consistent and have the same nonstandard asymptotic distribution. Thus the three estimators are equivalent in a large sample in estimating the first equation.

In the second equation the asymptotic distributions are as follows:

$$\sqrt{T}(\tilde{\beta}_{R2} - \beta_2)\ \text{and}\ \sqrt{T}(\tilde{\beta}_{U2} - \beta_2) \Rightarrow N(0, (1 - \rho^2)(1 - \beta_2^2))$$

$$\sqrt{T}(\hat{\beta}_2 - \beta_2) \Rightarrow N(0, (1 - \beta_2^2))$$

These estimators are both $\sqrt{T}$-consistent and the normalized estimators are asymptotically normally distributed with different variances. Comparing the variances, the SUR estimators always have smaller variance than the OLS estimator and hence the former are more efficient than the latter. Furthermore, in finite samples, we find that the SUR is more efficient than the OLS by Monte Carlo simulation.

It is easy to include constant terms in model 3:

$$y_{1t} = \mu_1 + \beta_1 y_{1t-1} + u_{1t}$$

$$y_{2t} = \mu_2 + \beta_2 y_{2t-1} + u_{2t}$$

$$\beta_1 = 1 \quad \text{and} \quad |\beta_2| < 1$$

$$y_{i0} = 0 \quad \text{for all } i$$

where the first equation is a random walk with a drift and the second one is a stationary AR(1) process. After some calculation, we can show that the asymptotic distributions of the standardized OLS estimators, $T^{1/2}(\hat{\mu}_1 - \mu_1)$ and $T^{3/2}(\hat{\beta}_1 - \beta_1)$, and the standardized SUR estimators, $T^{1/2}(\tilde{\mu}_2 - \mu_2)$ and $T^{1/2}(\tilde{\beta}_2 - \beta_2)$, are normal.

Note that if a constant term exists the asymptotic distributions for both OLS and SUR estimators of β_1 are normal; if not, they are nonstandard distribution.

2.4.2 Monte Carlo experiment

Small sample distributions of $\tilde{\beta}_R$, $\tilde{\beta}_U$, and $\hat{\beta}$ in the 2-equation SUR mixed model are examined by a Monte Carlo experiment with 5000 iterations for the sample size $T = 30$. We controlled the correlation between u_{1t} and u_{2t}, $\rho = \sigma_{12}/\sqrt{\sigma_{11}\sigma_{22}}$. The results are shown in Figures 2.3 and 2.4.

Figure 2.3 shows empirical cumulative distributions of three standardized estimators, RSUR $T(\tilde{\beta}_{1R} - \beta_1)$, USUR $T(\tilde{\beta}_{1U} - \beta_1)$, and OLS $T(\hat{\beta}_1 - \beta_1)$, for the first nonstationary equation when $T = 30$ and $\rho = 0.8$. It is seen that the distributions are very skewed to the left and the distributions of the RSUR and USUR estimators are almost indistinguishable. They are more concentrated around the true value, 0, than the empirical distribution of the OLS estimator. Observing these figures, we can say that the SUR estimators are more efficient than the OLS estimator in the sense that the distributions

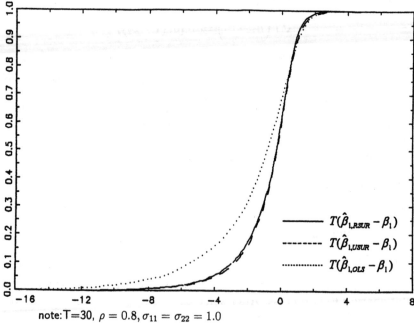

note: T=30, $\rho = 0.8$, $\sigma_{11} = \sigma_{22} = 1.0$

Figure 2.3. Distributions of the RSUR, USUR, and OLS in the first equation of Model 3.

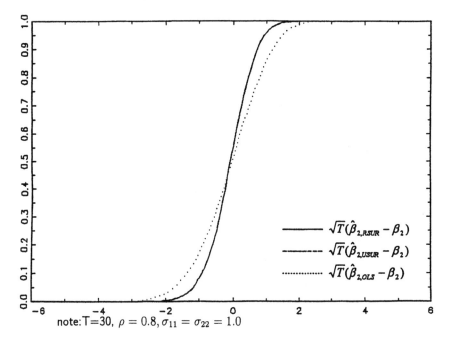

note: T=30, $\rho = 0.8, \sigma_{11} = \sigma_{22} = 1.0$

Figure 2.4. Distribution of the RSUR, USUR, and OLS in the second equation of Model 3.

of the two SUR estimators are more concentrated around the origin than the distribution of the OLS estimator.

Figure 2.4 is the empirical cumulative distribution of three estimators for the second stationary equation. We also calculated the empirical pdf of the standardized estimators $\sqrt{T}(\tilde{\beta}_{2R} - \beta_2)$, $\sqrt{T}(\tilde{\beta}_{2U} - \beta_2)$, and $\sqrt{T}(\hat{\beta}_2 - \beta_2)$ (graphs are omitted here), and observed that the three distributions in the stationary case look like the normal distribution, even for $T = 30$. In this case the SUR estimators are more efficient than the OLS estimator. The other experiments, which are not shown here, indicate that the distributions of the SUR estimators and the OLS estimator get closer as ρ becomes smaller, and hence the SUR estimation has little gain over the OLS estimation when ρ is small.

3 COINTEGRATION TEST IN THE SUR MODEL

3.1 Cointegration Test

In this section we consider the 2-equation SUR model

Model 4

$$\left.\begin{array}{l} y_{1t} = \beta_1 x_{1t} + u_{1t} \\ y_{2t} = \beta_2 x_{2t} + u_{2t} \\ t = 1, 2, \ldots, T \end{array}\right\} \tag{19}$$

where

$$\begin{pmatrix} u_1 \\ u_2 \end{pmatrix} \sim \left(0, \begin{pmatrix} \sigma_{11}I & \sigma_{12}I \\ \sigma_{21}I & \sigma_{22}I \end{pmatrix} \right)$$

Note that y and x are cointegrated if x_{it} are I(1) and u_{it} are stationary.

We assume that the explanatory variables x_{it} are I(1) for each $i = 1, 2$,

$$cx_{it} = x_{it-1} + v_{it}$$

where $v_{it} \sim$ iid $N(0, \sigma_{vi}^2)$ and v_{1t} and v_{2t} are independent for $t = 1, 2, \ldots, T$.

We may wonder whether there exists a cointegration relation between y_{1t} and y_{2t} or not. To see this, we can rearrange the 2-equation SUR model as

$$y_{1t} = y_{1t-1} + \beta_1 v_{1t} + u_{1t} u_{1t-1} = y_{1t-1} + v_t$$

$$y_{2t} = y_{2t-1} + \beta_2 v_{2t} + u_{2t} - u_{2t-1} = {}_{2t-1} + w_t$$

and calculate the covariance matrix as follows:

$$\Sigma = E \begin{pmatrix} v_t \\ w_t \end{pmatrix} (v_t \quad w_t)$$

$$\begin{pmatrix} \beta_1^2 \sigma_{v1}^2 & 2\sigma_{12} \\ 2\sigma_{12} & \beta_2^2 \sigma_{v2}^2 \end{pmatrix}$$

By analogous reasoning in Phillips [3] it is easy to prove that y_{1t} and y_{2t} are cointegrated if

$$\left| \Sigma \right| = \beta_1^2 \beta_2^2 \sigma_{v1}^2 \sigma_{v2}^2 - 4\sigma_{12}^2 = 0$$

To deal with the cointegration test we further assume that u_{it} are generated by

$$u_{it} = \rho_i u_{it-1} + \epsilon_{it}$$

If $|\rho_i| < 1$, x and y are cointegrated. Therefore the null hypothesis for no cointegration is $H_0 : \rho_i = 1$.

This model is the SUR system if the covariance $\text{Cov}(\epsilon_{it}, \epsilon_{jt}) = \sigma_{ij} \neq 0$ for all t. We assume that ϵ_{it} are normally distributed $N(0, \sigma_{ii})$ and that ϵ_{it} and v_{it} are independent.

We consider a cointegration test statistic for testing the hypothesis

$$H_0 : \rho_i = 1, \quad i = 1, 2$$

as follows

$$T(\hat{\rho}_i - 1) = T \left\{ \frac{\sum_{t=1}^{T} \hat{u}_{it}\hat{u}_{it-1}}{\sum_{t=1}^{T} \hat{u}_{it-1}^2} - 1 \right\}, \quad i = 1, 2 \tag{20}$$

where $\hat{u}_{it}$, $i = 1, 2$, are the residuals calculated from OLS, RSUR, and USUR estimation. The asymptotic null distribution of the test statistic for $\rho_i = 1$, $i = 1, 2$ can be obtained as follows:

$$T(\hat{\rho}_i - 1) \Rightarrow \frac{\mathrm{Asy}\left(\hat{\beta}_i^{(k)}\right)^2 \left(\frac{1}{2}\right)\{B_{2i}(1)^2 - \sigma_{vi}^2\} + \left(\frac{1}{2}\right)\{B_{1i}(1)^2 - \sigma_{ii}\}}{\mathrm{Asy}\left(\hat{\beta}_i^{(k)}\right)^2 \int_0^1 B_{2i}(r)^2 dr - 2\mathrm{Asy}\left(\hat{\beta}_i^{(k)}\right) \int_0^1 B_{2i}(r)B_{1i}(r)dr + \int_0^1 B_{1i}(r)^2 dr}$$

where $\mathrm{Asy}(\hat{\beta}_i^{(k)})$ are the asymptotic distributions of $\hat{\beta}_i - \beta_i$, for $\rho_i = 1$, $i = 1, 2$, where $k = 1$ for the OLS estimator and $k = 2$ for the RSUR (or USUR) estimator. These asymptotic distributions are given in [1].

We have also derived the asymptotic distributions of the $T(\hat{\rho}_i - 1)$ for the following SUR model with a constant.

$$y_{1t} = \mu_1 + \beta_1 x_{1t} + u_{1t}$$

$$y_{2t} = \mu_2 + \beta_2 x_{2t} + u_{2t}$$

$$t = 1, 2, \ldots, T$$

Although the results are omitted, the test statistics have nonstandard asymptotic distribution. The detailed derivations are given in the earlier version [1].

3.2 Monte Carlo Experiment

Small sample distributions of $\tilde{\beta}_R$, $\tilde{\beta}_U$, and $\hat{\beta}$ in Model 4 are examined by a Monte Carlo experiment with 5000 iterations for sample sizes $T = 30$ and 100. We controlled the three parameters $\eta = \sigma_{22}/\sigma_{11}$ (the variance ratio between u_{1t} and u_{2t}), $\kappa = \sigma_{ii}/\sigma_{vi}^2$, $i = 1, 2$ (the variance ratio between ϵ_{it} and v_{it}), and the correlation between u_{1t} and u_{2t}, $\rho = \sigma_{12}/\sqrt{\sigma_{11}\sigma_{22}}$. The results are presented in Figure 3.1.

Figure 3.1 shows the empirical cumulative distribution functions (CDF) of the null distribution of the RSUR, USUR, and OLS estimators for $\rho_i = 1$, $i = 1, 2$. It is seen that the three distributions are symmetric around the origin and that the distribution of the USUR estimator is the most concentrated around the origin. From this observation we can say that the USUR

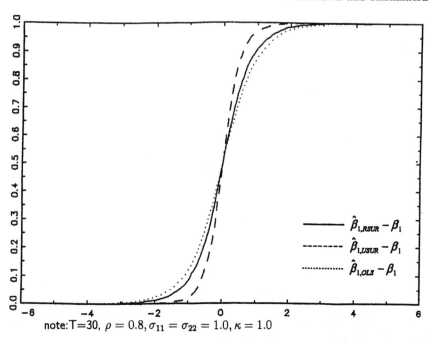

note: T=30, $\rho = 0.8, \sigma_{11} = \sigma_{22} = 1.0, \kappa = 1.0$

Figure 3.1. Distributions of the RSUR, USUR, and OLS in Model 4.

estimator is the most efficient, followed by the RSUR and OLS estimators in that order.

Figure 3.2 shows the cumulative distribution functions of the test statistics $T(\widehat{\rho}_{1,\text{RSUR}} - 1)$, $T(\widehat{\rho}_{1,\text{USUR}} - 1)$, $T(\widehat{\rho}_{1,\text{OLS}} - 1)$, where suffixes RSUR, USUR, OLS denote that $\widehat{\rho}_{1,*}$ is calculated from RSUR-, USUR-, and OLS-residuals respectively. We see that the distribution of test statistic $T(\widehat{\rho}_{1,\text{USUR}} - 1)$ is the most concentrated around the origin, followed by the distributions of $T(\widehat{\rho}_{1,\text{RSUR}} - 1)$ and of $T(\widehat{\rho}_{1,\text{OLS}} - 1)$ in that order.

Figure 3.3 shows the empirical power curves of the test statistics $T(\widehat{\rho}_{1,\text{RSUR}} - 1)$, $T(\widehat{\rho}_{1,\text{USUR}} - 1)$, $T(\widehat{\rho}_{1,\text{OLS}} - 1)$ for $T = 30$. We calculate the 5% critical values of these tests for $T = 30, 100$ by a Monte Carlo experiment with 5000 iterations. The critical values are -13.217, -11.313, -14.128 for $T = 30$ and -14.110, -11.625, -15.100 for $T = 100$, respectively. Using these critical points, we calculated the empirical power curves of these tests with 3000 replications. It is seen that the test statistic $T(\widehat{\rho}_{1,\text{USUR}} - 1)$ is the most powerful among the three tests and in the other two tests $T(\widehat{\rho}_{1,\text{RSUR}} - 1)$ is slightly more powerful than $T(\widehat{\rho}_{1,\text{OLS}} - 1)$.

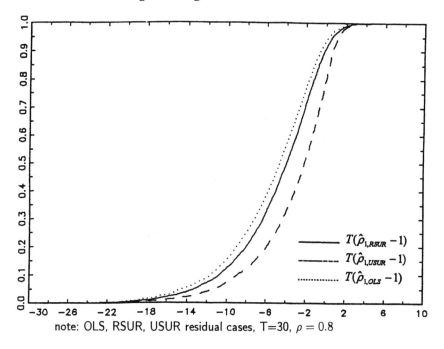

note: OLS, RSUR, USUR residual cases, T=30, $\rho = 0.8$

Figure 3.2. Null distributions of the $T(\hat{\rho}_1 - 1)$.

4. CONCLUDING REMARKS

We have considered the following Zellner's SUR model with contempora-
neously correlated and iid disturbances and with and without a constant
term:

Model 1: regressors are I(1) processes.
Model 2: all equations in the system are random walk.
Model 3: some equations in the system are random walk.
Model 4: errors are replaced with AR(1) process in Model 1.

In these models we derived the asymptotic distributions of the OLS,
unrestricted SUR (USUR), and restricted SUR (RSUR) estimators and
analyzed small sample properties of them by Monte Carlo experiments.

In Model 1 the asymptotic distributions of the three estimators have
nonstandard asymptotic distributions. Our Monte Carlo experiment for
the 2-equation system showed the following distributional properties:

(1) The distributions of the three estimators are symmetric and the dis-
     tributions of the RSUR and USUR estimators are much the same.

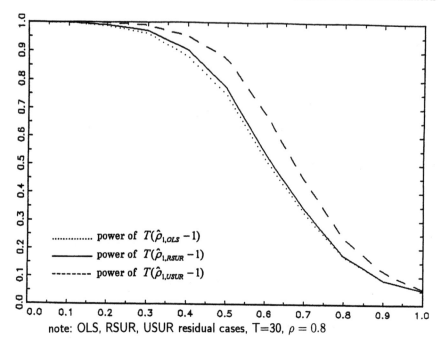

note: OLS, RSUR, USUR residual cases, T=30, $\rho = 0.8$

Figure 3.3. Power of the $T(\hat{\rho}_1 - 1)$.

(2) The distributions of two standardized SUR estimators are more concentrated around the origin than that of the OLS estimator for the 2-equation system. This means that the SUR estimators are more efficient than the OLS estimator.

(3) As ρ (the correlation of the disturbances across the equations) approaches 1, the SUR estimators become more efficient. When $\rho = 0$ the three asymptotic distributions are the same.

In Model 2, our Monte Carlo experiment showed similar results to Model 1. In addition we observe that

(1) The three distributions are very skewed to the left.
(2) The unit root test based on the SUR is more powerful than that based on the OLS estimator.

In Model 3, our Monte Carlo experiment was performed for the 2-equation system where the first equation is a random walk and the second equation is a stationary AR(1). The experiment showed that for the first equation (random walk process) the distributions of the three estimators are very skewed to the left and the distributions of the RSUR and USUR estimators are almost indistinguishable. We can say that the SUR estimators are more efficient than the OLS estimator in the sense that the distributions of the two standardized SUR estimators are more concentrated around the origin than the OLS estimator. For the second equation (stationary AR(1) process) the normalized three estimators are asymptotically normally distributed with different variances and the empirical distributions of these three estimators are symmetric, and the SUR estimators are more concentrated around the origin than the OLS.

In Model 4, the asymptotic distributions of the three estimators have nonstandard asymptotic distributions. Our Monte Carlo experiment for the 2-equation system showed the following distributional properties:

(1)   The distributions of the three estimators are symmetric.
(2)   The distributions of the two SUR estimators are more concentrated around the origin than that of the OLS estimator for the 2-equation system. This means that the SUR estimators are more efficient than the OLS estimator.
(3)   As ρ approaches 1 the SUR estimators become more efficient. When $\rho = 0$ the three asymptotic distributions are the same.
(4)   The distributions of the cointegration test statistics based on the residuals of OLS, RSUR, and OLS estimation are asymmetric.
(5)   The cointegration test based on USUR residual is more powerful than the OLS or RSUR residual cases.

From the above analysis we can conclude that in the nonstationary SUR system the SUR method is superior to the single equation OLS method, as in the stationary cases.

ACKNOWLEDGMENTS

This paper is based on [1] which was presented at the Far Eastern Meeting of the Econometric Society held in Singapore in July 1999. We are grateful to Peter C. B. Phillips for his comments at the meeting and to Jun-ichiro Fukuchi for his pointing out an error in an earlier version of this paper. This research was supported by Grant-in-Aid for Scientific Research 08209114 of the Ministry of Education of Japan, for which we are most grateful.

REFERENCES

1. K. Maekawa, H. Hisamatsu. Seemingly unrelated regression models with integrated regressors. Paper presented at the Far Eastern Meeting of the Econometric Society, 1999.
2. J. Y. Park, P. C. B. Phillips. Statistical inferences in regressions with integrated processes: Part 1. *Econometric Theory* 4: 468–497, 1988.
3. P. C. B. Phillips. Understanding spurious regressions in econometrics. *Journal of Econometrics* 33: 311–340, 1986.
4. P. C. B. Phillips, S. N. Durlauf. Multiple time series regression with integrated processes. *Review of Economic Studies* 53: 473–496, 1986.
5. A. Zellner. An efficient method of estimating seemingly unrelated regressions and tests for aggregation bias. *Journal of American Statistical Association 57: 348–368, 1962.*
6. A. Zellner. Estimators for seemingly unrelated regressions: some finite sample results. *Journal of American Statictical Association* 58: 977–992, 1963.

23

Estimating Systems of Stochastic Coefficients Regressions When Some of the Observations Are Missing

GORDON FISHER and MARCEL-CHRISTIAN VOIA   Concordia University, Montreal, Quebec, Canada

1. INTRODUCTION

This chapter considers systems of regressions with stochastic coefficients when some of the observations are missing. Estimation in this context applies: (i) to the unknown stochastic coefficients corresponding to relations whose variables are observed, and (ii) to the coefficients of their corresponding mean effect, as well as (iii) to the prediction of the stochastic coefficients corresponding to the regressions whose observations are missing. The problems of estimating models of this kind have their origins in several branches of the literature: random coefficients regressions (e.g. [1–3]); missing observations in regression analysis (e.g. [4,5]; see also [6] and articles quoted therein); predictions in time-series regressions (e.g. [7–9]); and cross-section regressions with coefficients varying across different sub-groups (e.g. [10]) or over clusters of units when some of the clusters are missing [11] . It is convenient to refer to this literature as classical in view of its dependence on traditional least-squares and maximum-likelihood methods.

　Pfefferman [12] critically reviews the classical statistical literature on stochastic coefficients regression (SCR) models, with complete and with missing data, and succinctly consolidates a broad range of results. More recently, Swamy, Conway, and LeBlanc [13–15] argue forcefully against the application of the traditional fixed coefficients regression (FCR) model in econometrics and reveal a broad range of ten econometric specifications that are encompassed by the SCR framework; their analysis is then illustrated with practical applications to forecasting and stability analysis. Despite the wide range of literature cited in [13–15], there is no mention of Pfefferman's work. The same holds true of recent econometric literature on panel data and Kalman filter models (e.g., [16–18]), both of which may be regarded as SCR models; but there is a citation in the survey of SCR models in [19, p. 552].

The general aim of this chapter is to clarify, extend and illustrate the classical results surveyed in [12–15] and thereby, in particular, to introduce to econometricians the results of Pfefferman [12] in a modern setting.

Following an introduction to the basic framework of interest in Section 2, there is a brief synthesis of the basis for applications of SCR models in economics [13] in Section 3. In Section 4 various results noted in [12] are amplified and extended. A new geometric proof of an extended Gauss–Markov theorem is presented in Section 5. This applies to minimum-mean-square, unbiased linear estimation and prediction of the coefficients in an SCR model that has some of its observations missing [12, Section 6 and Appendix].

There are two interesting features to the theorem: the first concerns an apparent asymmetry in its conclusions; the second concerns the method of proof. First, on the one hand, the direct generalized least squares (GLS) estimator of the random coefficients corresponding to the observed relations is not generally the efficient unbiased linear estimator. On the other hand, the prediction of the random coefficients corresponding to the unobserved relations bears a strong resemblance to standard optimal predictions based on normal theory, even though normality is not presumed. A simple rank condition lies behind these conclusions.

Second, Pfefferman's proof of the theorem is based on a procedure of Chipman [5] and Duncan and Horn [8]. This uses an orthogonality condition which is masked behind a series of matrix manipulations. In the new proof, the vector of observations is resolved into component vectors lying in specific subspaces which are chosen, a priori, to be orthogonal relative to a conveniently selected scalar product. In this setting, minimization of sampling variation is achieved by straightforward application of Pythagoras' theorem. Thus the new procedure is placed in a familiar modern econometric setting which is both straightforward and intuitively appealing.

The whole commentary of the chapter is then illustrated with an empirical example in Section 6. This uses American financial data applied to a Kalman filter model of stochastic volatility.

2. NOTATION AND BASIC FRAMEWORK

R^n will denote Euclidean n-space on which the natural scalar product is $(\cdot, \cdot)$. If A is an $n \times r$ matrix, $\mathcal{R}[A]$ denotes the range of A, that is, $\{x \in R^n : x = A\lambda,\ \lambda \in R^r\}$, and $\mathcal{N}[A^T]$ denotes the null of A^T, $\{z \in R^n : A^T z = 0\}$; of course, $\mathcal{N}[A^T] = \mathcal{R}[A]^\perp$. The notation $q \sim [R^n; \mu, \Xi]$ denotes a random vector q distributed on R^n with vector mean μ and dispersion Ξ.

The column vector $y \in R^n$ comprises n observations; the matrix X, of rank k, comprises n observations on each of k explanatory variables; the $k \times p$ matrix B, of known coefficients, has rank p; and $\theta \in R^p$ is an unknown fixed vector. These matrices and vectors are related together in the following framework:

$$y = X\beta + \varepsilon \tag{1}$$

$$\varepsilon \sim [R^n;\ 0,\ \Sigma] \tag{2}$$

$$\beta \sim [R^k;\ B\theta,\ \Delta] \tag{3}$$

$$E[\varepsilon\beta^\top] = 0 \tag{4}$$

In (2), Σ is positive definite on R^n while, in (3), Δ is positive definite on R^k. Equation (1) comprises, inter alia, k unknown random coefficients, β, distributed according to (3), and n disturbances, ε, distributed as in (2); equation (4) ensures that ε and β are uncorrelated. By (3), the vector β may also be written in freedom equation form as

$$\beta = B\theta + v, \qquad v \sim [R^k;\ 0,\ \Delta] \tag{5}$$

Estimation of the unknown coefficients θ and β in (1)–(4) may be represented as seeking an unbiased linear function of the observations y, say $\alpha(y)$, where

$$\alpha(y) = \alpha_0 + \alpha_1^T y \qquad \alpha_0 \in R,\ \alpha_1 \in R^n \tag{6}$$

to estimate the parametric function $\psi(\theta, \beta)$, defined by

$$\psi(\theta, \beta) = c_1^T \theta + c_2^T \beta \qquad c_1 \in R^p,\ c_2 \in R^k \tag{7}$$

which has minimum-mean-square error (MSE). Unbiased in this context is defined under ξ, the joint distribution of y and β; thus $\alpha(y)$ is unbiased if

$$E_\xi[\alpha(y) - \psi(\theta, \beta)] = 0 \tag{8}$$

The MSE of $\alpha(y)$ is defined by

$$\mathrm{MSE}[\alpha(y)] = E_\xi[\alpha(y) - \psi(\theta, \beta)]^2 \tag{9}$$

(c.f. [20]). An estimator of (7) in the form of (6), which obeys (8) and minimizes (9), is said to be a minimum-mean-square unbiased linear estimator or MIMSULE.

3. APPLICATIONS

The SCR model was originally introduced as a practical device to characterize situations in which the coefficients of an FCR model could vary over certain regimes or domains: for example, over different time regimes; or across different families, or corporations, or other economic micro-units; or across strata; and so on.

In macroeconomics, variation of coefficients over different policy regimes is an empirical consequence of the Lucas critique ([21], see also [22]) that an FCR model must ultimately involve a contradiction of dynamic optimizing behavior. This is because each change in policy will cause a change in the economic environment and hence, in general, a change in the existing set of optimal coefficients. Given this position, an SCR model is clearly preferable, on fundamental grounds, to an FCR model.

Secondly, whenever the observations are formed of aggregates of micro-units, then, on an argument used in [23], Swamy et al. [13] develop technical conditions to demonstrate that an SCR model requires less stringent conditions to ensure its logical existence than is the case for the logical existence of an FCR model. This is not surprising. Given the SCR model (1)–(4), it is always possible to impose additional restrictions on it to recover the corresponding FCR model. This implies that an FCR model is never less restricted than the corresponding SCR model, and hence it is not surprising that an SCR model in aggregates is less restrictive than the corresponding FCR model in the same aggregates.

A third argument, implying greater applicability of an SCR model than the corresponding FCR model, is that the former is more adaptable in the sense that it provides a closer approximation to nonlinear specifications than does the latter (see, e.g., [24]).

Two subsidiary justifications for SCR models rely on arguments involving omitted variables and the use of proxy variables, both of which can induce coefficient variation (see [25] for further details).

When the basic framework (1)–(4) varies across certain regimes, there are M distinct relationships of the kind

$$y_i = X_i \beta_i + \varepsilon_i \qquad i = 1, 2, \ldots, M \tag{10}$$

of n_i observations and p explanatory variables such that $n = \sum_i n_i$ and $Mp = k$. The vectors y and ε of (1) are now stacked vectors of vector components y_i and ε_i respectively, X is a block-diagonal matrix with X_i in (10) forming the ith block; β is also a stacked vector of M components β_i, making β an $Mp \times 1$, that is a $k \times 1$, vector as in (1) above. The β_i are presumed to be drawn independently from $[R^p; \theta, \Lambda]$, whereupon

$$\beta \sim [R^k; \ (e \otimes I_p)\theta, \ I_M \otimes \Lambda] \tag{11}$$

In (11) : e is the equiangular vector in R^M (the $M \times 1$ vector of ones) yielding, in the notation of (5), $B = (e \otimes I_p)$, a $k \times p$ known matrix; θ is an unknown, fixed, $p \times 1$ vector; and $\Delta = (I_M \otimes \Lambda)$. In this case, then, p and k are related through $Mp = k$ and θ is the vector mean of the p-dimensional distribution from which the vectors $\beta_1, \beta_2, \ldots, \beta_M$ are drawn and hence retains the same number of elements (p) as each of them. The regime model (10) and (11) applies to time series or to cross sections or to combinations of the two (c.f. [3, 26]).

The empirical model considered later in the paper is a pure time-series application of the model (10) and (11). Here the β_i in (10) are generated by the Markovian scheme $\beta_{i+1} = T\beta_i + \eta_{i+1}$, $i = 0, 1, 2, \ldots, (M-1)$, in which T is a $p \times p$ transition matrix and the $\eta_{i+1} \sim [R^p; 0, \Lambda]$ independently, with $\beta_0 = \theta$. This is the familiar Kalman filter model in which θ is the starting point from which the β_{i+1} are generated. Let

$$A = \begin{bmatrix} I_p & 0 & 0 & \cdots & 0 \\ T & I_p & 0 & \cdots & 0 \\ T^2 & T & I_p & \cdots & 0 \\ \vdots & \vdots & \vdots & \ddots & \vdots \\ T^{M-1} & T^{M-2} & T^{M-3} & \cdots & I_p \end{bmatrix}$$

and set $v = A\eta$, in which $\eta = [\eta_1^T, \eta_2^T, \ldots, \eta_M^T]^T$; the matrix B is given by

$$B = \begin{bmatrix} T \\ T^2 \\ \vdots \\ T^M \end{bmatrix}$$

Then β is the $Mp \times 1$ (or $k \times 1$ with $Mp = k$) stacked vector of components β_i and

$$\beta = B\theta + v \tag{12}$$

$$v \sim [R^{Mp}; \ 0, \ A(I_M \otimes \Lambda)A^T] \tag{13}$$

hence $\Delta = A(I_M \otimes \Lambda)A^T$ according to (5).

Extensions of the models outlined in this section, especially the Kalman filter model, are discussed below in Sections 5 and 6.

4. EFFICIENT ESTIMATION OF $\psi(\theta, \beta)$

There are three separate problems considered in the literature concerning MIMSULE of the model (1)–(4) according to the definitions in equations (6)–(9): (i) estimation of $c_1^T \theta$; (ii) estimation of $\psi(\theta, \beta)$ when θ is known; and (iii) estimation of $\psi(\theta, \beta)$ when θ is unknown. Solutions to these problems will be shown to depend on the structure of Σ and Δ and the rank of B.

4.1 Optimal Estimation of $c_1^T \theta$

When (5) is substituted into (1) there results

$$y = XB\theta + (Xv + \varepsilon) \tag{14}$$

$$(Xv + \varepsilon) \sim [R^n;\ 0,\ \Omega], \qquad \Omega = (X\Delta X + \Sigma) \tag{15}$$

a positive definite matrix. Let $XB = X_0$, an $n \times p$ matrix of rank $p \leq k$. Clearly $\mathcal{R}[X_0] \equiv \mathcal{X}_0 \subseteq \mathcal{R}[X] \equiv \mathcal{X}$. Direct application of generalized least-squares GLS to (14) then implies [2, 3] that the MIMSULE of $c_1^T \theta$ is $c_1^T \hat{\theta}$, where

$$\hat{\theta} = (X_0^T \Omega^{-1} X_0)^{-1} X_0^T \Omega^{-1} y \tag{16}$$

Pfefferman [12, p. 141] notes an interesting property of (16), namely that it is equivalent to the estimator obtained from substituting $\hat{\beta} = (X^T \Sigma^{-1} X)^{-1} X^T \Sigma^{-1} y$ for β and applying GLS to estimate θ in (5), rewritten as

$$\hat{\beta} = B\theta + \{(\hat{\beta} - \beta) + v\} \tag{17}$$

$$\{(\hat{\beta} - \beta) + v\} \sim [R^k;\ 0,\ \Gamma], \quad \Gamma = \{(X^T \Sigma^{-1} X)^{-1} + \Delta\} \tag{18}$$

This leads to the estimator

$$\theta^* = (B^T \Gamma^{-1} B)^{-1} B^T \Gamma^{-1} \hat{\beta} \tag{19}$$

That $\hat{\theta}$ in (16) is equal to θ^* in (19) is far from obvious. In fact, the result is easily obtained by manipulation using the identity $X^T = \Gamma^{-1} \Gamma X^T = \Gamma^{-1} (X^T \Sigma^{-1} X)^{-1} X^T \Sigma^{-1} \Omega$; when X^T and the last expression are each post-multiplied by $\Omega^{-1} X$, there results

$$X^T \Omega^{-1} X = \Gamma^{-1} \tag{20}$$

The reason why (16) and (19) are equal essentially depends on an invariance condition. The subspace $\mathcal{X}$ is invariant under $X\Delta X^T \Sigma^{-1} = Q$ [27] . This is clearly the case because, if P is any projection matrix on $\mathcal{X}$, $PQP = QP$ [28, p. 76, Theorem 1]. The implication of such invariance is that the projection matrix on $\mathcal{X}$ orthogonal relative to the scalar product $(\cdot, \Omega^{-1}\cdot)$ is precisely the same as the corresponding projection matrix orthogonal relative to the scalar product $(\cdot, \Sigma^{-1}\cdot)$; in symbols, $P_{\mathcal{X}:\Omega^{-1}} = X(X^T\Omega^{-1}X)^{-1}X^T\Omega^{-1}$ $= P_{\mathcal{X}:\Sigma^{-1}} = X(X^T\Sigma^{-1}X)^{-1}X^T\Sigma^{-1}$. Another way of putting the same point is that the direction of projection on $\mathcal{X}$ is the same in each case because $\mathcal{N}[X^T\Omega^{-1}] = \mathcal{N}[X^T\Sigma^{-1}]$. This is because $X\Delta X^T$, whose columns lie in $\mathcal{X}$, plays no role in the projection matrix $P_{\mathcal{X}:\Omega^{-1}}$, leaving Σ^{-1} alone to have influence on the direction of projection. Moreover, since $\mathcal{X}_0 \subseteq \mathcal{X}$, from (16)

$$X_0\hat{\theta} = P_{\mathcal{X}_0:\Omega^{-1}}y = P_{\mathcal{X}_0:\Omega^{-1}}P_{\mathcal{X}:\Omega^{-1}}y = P_{\mathcal{X}_0:\Omega^{-1}}P_{\mathcal{X}:\Sigma^{-1}}y = P_{\mathcal{X}_0:\Omega^{-1}}X\hat{\beta} \qquad (21)$$

where $P_{\mathcal{X}_0:\Omega^{-1}} = X_0(X_0^T\Omega^{-1}X_0)^{-1}X_0^T\Omega^{-1}$. Now

$$P_{\mathcal{X}_0:\Omega^{-1}}X\hat{\beta} = X\{B(B^TX^T\Omega^{-1}XB)^{-1}B^TX^T\Omega^{-1}X\}\hat{\beta} = XG_{\mathcal{B}:X^T\Omega^{-1}X}\hat{\beta} \qquad (22)$$

where $G_{\mathcal{B}:X^T\Omega^{-1}X}$ is the expression in braces in (22), the projection matrix from R^k on $\mathcal{B} = \mathcal{R}[B]$ orthogonal relative to the scalar product $(\cdot, X^T\Omega^{-1}X\cdot)$. But, by (20), $G_{\mathcal{B}:X^T\Omega^{-1}X} = G_{\mathcal{B}:\Gamma^{-1}}$ and so, from (21) and (22),

$$X_0\hat{\theta} = XG_{\mathcal{B}:\Gamma^{-1}}\hat{\beta} = X_0(B^T\Gamma^{-1}B)^{-1}B^T\Gamma^{-1}\hat{\beta} = X_0\theta^* \qquad (23)$$

whereupon, identically in X_0, $\hat{\theta} = \theta^*$.

When B is non-singular, $\mathcal{X}_0 = \mathcal{X}$ and $G_{\mathcal{B}:\Gamma^{-1}} = I_k$; then, by (23), $XB\hat{\theta} = X\hat{\beta}$. Hence, identically in X, $B\hat{\theta} = \hat{\beta}$. However, in this case $\hat{\theta} = (B^TX^T\Omega^{-1}XB)^{-1}B^TX^T\Omega^{-1}y = B^{-1}(X^T\Sigma^{-1}X)^{-1}X^T\Sigma^{-1}y = B^{-1}\hat{\beta}$ since $\mathcal{X}$ is invariant under $X\Delta X^T\Sigma^{-1} = Q$. Thus, when B is nonsingular, Δ play no role in the MIMSULE of $c_1^T\theta$. This result extends [1] as quoted by Pfefferman [12, p. 141], who reports only the special case when $B = I_k$. Finally, if $\mathcal{X}$ is invariant under Σ, then $P_{\mathcal{X}\Omega^{-1}} = P_{\mathcal{X}\Sigma^{-1}} = P_{\mathcal{X}I_n}$ $= X(X^TX)^{-1}X^T$ and neither Δ nor Σ will have any influence on the MIMSULE of $c_1^T\theta$ (c.f. [12, pp. 141–142]).

4.2 MIMSULE of $\psi(\theta, \beta)$ for θ Known and for θ Unknown

When θ is known, the MIMSULE of $\psi(\theta, \beta)$ is $\tilde{\psi}(\theta, \beta)$, which may be expressed as

$$\tilde{\psi}(\theta, \beta) = c_1^T\theta + c_2^T\{B\theta + \Delta\Gamma^{-1}(\hat{\beta} - B\theta)\} \qquad (24)$$

This is essentially the formula quoted in [12, equation (4.1), p. 142] from [5] and [29, p. 234]. The implicit MIMSULE of β in (24) is the expression in braces, $\tilde{\beta} = \{B\theta + \Delta\Gamma^{-1}(\hat{\beta} - B\theta)\}$ or

$$\tilde{\beta} = \hat{\beta} - (I_k - \Delta\Gamma^{-1})(\hat{\beta} - B\theta)$$

$$= \hat{\beta} - (X^T\Sigma^{-1}X)^{-1}X^T\Omega^{-1}X(\hat{\beta} - B\theta) \tag{25}$$

Although the invariance of $\mathcal{X}$ under $X\Delta X^T\Sigma^{-1}$ ensures that $P_{\mathcal{X}\Omega^{-1}} = P_{\mathcal{X}_{-\Sigma^{-1}}}$, $X^T\Sigma^{-1}X$ is not equal to $X^T\Omega^{-1}X$; however, the same invariance does ensure that there exists a nonsingular matrix M to reflect the divergence of $(X^T\Sigma^{-1}X)^{-1}X^T\Omega^{-1}X$ from I_k; that is $X^T\Omega^{-1}X = X^T\Sigma^{-1}XM$ and

$$\tilde{\beta} = \hat{\beta} - M(\hat{\beta} - B\theta) \tag{26}$$

When θ is unknown and $p \le k$, Harville [30] derives the MIMSULE of $\psi(\theta, \beta)$ as $\hat{\psi}(\theta, \beta)$, which is precisely (24) with θ replaced by $\hat{\theta}$ of (16):

$$\hat{\psi}(\theta, \beta) = c_1^T\hat{\theta} + c_2^T\{B\hat{\theta} + \Delta\Gamma^{-1}(\hat{\beta} - B\hat{\theta})\} \tag{27}$$

When $p = k$ and B is invertible, it has been established, via invariance, that $\hat{\beta} = B\hat{\theta}$. Then, from (27),

$$\hat{\psi}(\theta, \beta) = c_1^T\hat{\theta} + c_2^T B\hat{\theta} = c_1^T\hat{\theta} + c_2^T\hat{\beta} \tag{28}$$

The distinction between (27) and (28) is important. When B is non-singular the information that β is stochastic, with mean $B\theta = \mu \in R^k$ and dispersion Δ, is not informative and hence is of no use in predicting β: thus estimating θ and β essentially comes down to estimating the one or the other, since B is given. When the rank of B is $p < k$, on the other hand, the knowledge that $E(\beta) \in \mathcal{R}[B]$ is informative, and hence it is possible to find, as in (27), a more efficient estimator than the GLS estimator; that is, by (26),

$$\beta^* = \hat{\beta} - M(\hat{\beta} - B\hat{\theta})$$

Equation (26) represents a novel form of (24) resulting from the invariance of $\mathcal{X}$ under $X\Delta X^T\Sigma^{-1}$; by the same token, $\hat{\psi}(\theta, \beta)$ in (27) reduces to $c_1^T\hat{\theta} + c_2^T\beta^*$. These formulae augment and simplify corresponding results in [12, p. 143].

5. AN EXTENSION
5.1 A New Framework

Interest will now focus on equation (10) with $n_1 = n_2 = \cdots = n_M = n$. The blocks $i = 1, 2, \ldots, m$ will be regarded as group 1 and blocks $i = (m+1)$,

$(m + 2), \ldots, M$ will form group 2. With a natural change in notation, (10) may be rewritten in terms of groups 1 and 2:

$$\begin{bmatrix} y_1 \\ y_2 \end{bmatrix} = \begin{bmatrix} X_1 & 0 \\ 0 & X_2 \end{bmatrix} \begin{bmatrix} \beta_1 \\ \beta_2 \end{bmatrix} + \begin{bmatrix} \varepsilon_1 \\ \varepsilon_2 \end{bmatrix} \tag{29}$$

The vectors y_1 and ε_1 are each $nm \times 1$ while y_2 and ε_2 are each $n(M - m) \times 1$, β_1 and β_2 are of order $mk \times 1$ and $(M - m)k \times 1$ respectively. Let ε be the vector of components ε_1 and ε_2; then

$$\varepsilon \sim [R^{nM}; \; 0, \; \Sigma] \tag{30}$$

and Σ may be partitioned, according to the partitioning of ε, into blocks Σ_{rs} where r and s each take on values 1, 2. Corresponding to (5),

$$\begin{bmatrix} \beta_1 \\ \beta_2 \end{bmatrix} = \begin{bmatrix} B_1 \\ B_2 \end{bmatrix} + \begin{bmatrix} \nu_1 \\ \nu_2 \end{bmatrix} \tag{31}$$

B_1 and B_2 have mk and $(M - m)k$ rows and p columns, θ is still $p \times 1$. In a notation corresponding to (30)

$$\nu \sim [R^{Mk}; \; 0, \; \Delta] \tag{32}$$

and, like Σ, Δ may be partitioned in blocks Δ_{rs} corresponding to ν_1 and ν_2. Finally, $E[\varepsilon \nu^T] = 0$, as in (4).

The model defined by (29)–(32) and (4) represents the complete model to be estimated, but the estimation to be undertaken presumes that the observations corresponding to y_2 and X_2 are not available. What is available may be consolidated as

$$\begin{bmatrix} y_1 \\ 0 \\ 0 \end{bmatrix} = \begin{bmatrix} X_1 & 0 & 0 \\ -I_{mk} & 0 & B_1 \\ 0 & -I_{(M-m)k} & B_2 \end{bmatrix} \begin{bmatrix} \beta_1 \\ \beta_2 \\ \theta \end{bmatrix} + \begin{bmatrix} \varepsilon_1 \\ \nu_1 \\ \nu_2 \end{bmatrix} \tag{33}$$

where X_1 is an $nm \times mk$ matrix, θ is a $p \times 1$ vector, β_1 is an $mk \times 1$ vector, and β_2 is an $(M - m)k \times 1$ vector. Equation (33) will be written compactly as

$$y = W\lambda + \delta \tag{34}$$

y being the $(nm + Mk) \times 1$ vector on the left-hand side of (33), W the $(nm + Mk) \times (Mk + p)$ matrix, of full column rank, which has leading block X_1, λ the $(Mk + p) \times 1$ vector comprising stochastic coefficients β_1 and β_2 and the fixed vector θ, and δ the last disturbance vector in (33). The disturbance vector

$$\delta \sim [R^{nm+Mk}; \; 0, \; \Omega] \tag{35}$$

in which the redefined Ω is given by

$$\Omega = \begin{bmatrix} \Sigma_{11} & 0 & 0 \\ 0 & \Delta_{11} & \Delta_{12} \\ 0 & \Delta_{21} & \Delta_{22} \end{bmatrix} \tag{36}$$

a positive-definite matrix on R^{nm+Mk}, Σ_{11} is $nm \times nm$, Δ_{11} is $mk \times mk$, Δ_{12} is $mk \times (M-m)k$, Δ_{21} is $(M-m)k \times mk$, and Δ_{22} is $(M-m)k \times (M-m)k$. It will be convenient to define the row dimension of (33) and (34) as $(nm+Mk) = N$ and the column dimension of W as $(Mk+p) = K$. From (34),

$$E[\lambda] = \begin{bmatrix} B_1 \\ B_2 \\ I_p \end{bmatrix} \theta = B\theta \tag{37}$$

where B is the $\{mk + (M-m)k + p\} \times p$ or the $K \times p$ matrix of blocks B_1, B_2 and I_p; the rank of B is assumed to be $p \le mk$. If $n = 1$, that is, there is only one observation for each block, then the row dimension of X_1 is m, and $N = m + Mk$. Also, W must have rank $K \le N$, implying that $nm + Mk \ge Mk + p \Leftrightarrow nm \ge p$. Therefore, when $n = 1$, $p \le m$ and a fortiori $p \le mk$; otherwise $nm \ge p$; if $k \ge n$ then clearly $mk \ge p$ a fortiori. It follows that $\mathcal{N}[B] = \{x \in R^p : Bx = 0\} = \emptyset$; thus there are no nonzero vectors in R^p such that $Bx = 0$; the same holds for $B_1 x = 0$. This becomes important later on.

In a time-series application, the order in which the observations appear is obviously important and, if the application is a time series of cross-section equations, then the order of equations will be important. It is straightforward to augment model (33)–(37) to handle situations of this kind. As an illustration, consider cross-sections of n firms for each of m years. The m years are divided into four groups: the first m_1 years form group 1; the next m_2 years form group 2; the following m_3 years form group 3, and the last m_4 years form group 4. Thus $i = 1, 2, 3, 4$ and $\sum_i m_i = m$. Extending the notation of (29) to four groups, $y_i = X_i \beta_i + \varepsilon_i$ wherein y_i and ε_i now have nm_i rows while X_i is block diagonal of $n \times k$ blocks, thus forming a matrix of order $nm_i \times km_i$; β_i has km_i rows and follows the rule [c.f. (31) above] $\beta_i = B_i \theta + v_i$, where B_i is a $km_i \times p$ known matrix, θ is $p \times 1$, and v_i is $km_i \times 1$. Groups 2 and 4 will now represent the missing observations and so, corresponding to (33), the model of four groups, with groups 2 and 4 missing, may be written

$$
\begin{bmatrix} y_1 \\ y_3 \\ 0 \\ 0 \\ 0 \\ 0 \end{bmatrix} = \begin{bmatrix} X_1 & 0 & 0 & 0 & 0 \\ 0 & 0 & X_3 & 0 & 0 \\ -I_1 & 0 & 0 & 0 & B_1 \\ 0 & -I_2 & 0 & 0 & B_2 \\ 0 & 0 & -I_3 & 0 & B_3 \\ 0 & 0 & 0 & -I_4 & B_4 \end{bmatrix} \begin{bmatrix} \beta_1 \\ \beta_2 \\ \beta_3 \\ \beta_4 \\ \theta \end{bmatrix} + \begin{bmatrix} \varepsilon_1 \\ \varepsilon_2 \\ \nu_1 \\ \nu_2 \\ \nu_3 \\ \nu_4 \end{bmatrix}
$$

in which I_i is the identity of order km_i. This equation may also be written $y = W\lambda + \delta$, corresponding to (34) with $\delta \sim [R^{n(m_1+m_3)+mk}; 0, \Omega]$ in which Ω is a two-block-diagonal, square matrix of order $n(m_1 + m_3) + mk$ having Σ as the leading block for the dispersion of the stacked vector of ε_1 and ε_3, and $\varDelta$ in the other position as the dispersion of the stacked vector ν_1, ν_2, ν_3, and ν_4. Thus the new situation corresponds closely to the model (33)–(37); moreover, corresponding to (37),

$$
E[\lambda] = \begin{bmatrix} B_1 \\ B_2 \\ B_3 \\ B_4 \\ I_p \end{bmatrix} \theta = B\theta
$$

and the ranks of B, B_1, and B_3 are each taken to be p; hence the null spaces of these matrices are empty. Thus, in considering the optimal estimation of (33)–(37), a rather wide class of time-series–cross-section models is naturally included. Having made this point, it is appropriate to return to the model (33)–(37) and introduce some additional notation.

5.2   Further Notation

Given that the natural scalar product on R^N is $(\cdot, \cdot)$, then $\langle \cdot, \cdot \rangle = (\cdot, \Omega^{-1}\cdot)$. The length of a vector in the metric $\langle \cdot, \cdot \rangle$ will be denoted $\| \cdot \|_{\Omega^{-1}}$; that is, $\|x\|_{\Omega^{-1}}^2 = (x, \Omega^{-1}x) = x^T\Omega^{-1}x$. Let $\mathcal{L} = \mathcal{R}[W] = \{w \in R^N : w = W\gamma, \; \gamma \in R^K\}$. Then

$$
\mathcal{L}^0 = \{z \in R^N : \langle z, w \rangle = 0 \; \forall \; w \in \mathcal{L}\}
$$

Dim $\mathcal{L} = K$, whereupon dim $\mathcal{L}^0 = (N - K)$ because $\mathcal{L}^0$ is the orthocomplement of $\mathcal{L}$ in R^N, orthogonality being relative to $\langle \cdot, \cdot \rangle$. Thus $\mathcal{L} \cap \mathcal{L}^0 = \emptyset$ and $\mathcal{L} \oplus \mathcal{L}^0 = R^N$. The row space of W is $\mathcal{R}[W^T] = R^K$. For the purpose of defining linear estimators, it will be convenient to work in the metric $\langle \cdot, \cdot \rangle$; there is nothing restrictive about this because, for any $q \in R^N$, there will always exist an $a \in R^N$ such that $q = \Omega^{-1}a$. Therefore $(q, y) = (\Omega^{-1}a, \ y) = (a, \ \Omega^{-1}y) = \langle a, y \rangle$, Ω^{-1} being positive definite and therefore symmetric.

As before, a parametric function of the coefficients λ is denoted $\psi(\lambda) = c^T \lambda$, $c \in R^K$. A parametric function is said to be estimable if there exists an $\alpha_0 \in R$ and an $a \in R^N$ such that $(\alpha_0 + \langle a, y \rangle)$ is ξ-unbiased in the sense of (8).

5.3 Results

Lemma 1. The parametric function $\psi(\lambda)$ is estimable if and only if $\alpha_0 = 0$ and $c \in \mathcal{R}[W^T]$, identically in θ.

Proof. Let $\psi(\lambda)$ be estimable. Then, identically in θ,

$$E_\xi[\alpha_0 + \langle a, y \rangle - c^T \lambda] = \alpha_0 + a^T \Omega^{-1} WB\theta - c^T B\theta = 0$$

Clearly, for any θ, $\alpha_0 = 0$. Moreover, since there are no nonzero vectors θ such that $B\theta = 0$, $a^T \Omega^{-1} W - c^T = 0$ identically in θ; that is, $c \in \mathcal{R}[W^T]$.

Let $c = W^T \gamma$ for some $\gamma \in R^n$ and $\alpha_0 = 0$. Then

$$E_\xi[\langle a, y \rangle - \gamma^T W \lambda] = a^T \Omega^{-1} WB\theta - \gamma^T WB\theta$$

Now $a^T \Omega^{-1} WB\theta = a_1^T \Sigma_{11}^{-1} X_1 B_1 \theta$ and $\gamma^T WB\theta = \gamma_1^T X_1 B_1 \theta$, where a_1 and γ_1 are the first nm elements of a and γ respectively. Since $\mathcal{N}[B_1] = \emptyset$, $(a^T \Omega^{-1} - \gamma^T)W = 0$. Setting $\gamma = \Omega^{-1}a$, the lemma is established. □

Notice that in a and γ the last $\{(M - m)n + p\}$ elements are arbitrary, as expected in view of the structure (33).

Lemma 2. Given that $\psi(\lambda)$ is estimable, there exists a unique ξ-unbiased linear estimator $\langle a^*, y \rangle$ with $a^* \in \mathcal{L}$. If $\langle a, y \rangle$ is any ξ-unbiased estimator of $\psi(\lambda)$, then a^* is the orthogonal projection of a on $\mathcal{L}$ relative to $\langle \cdot, \cdot \rangle$.

Proof. Since $\psi(\lambda)$ is estimable, there exists an $a \in R^N$ such that $\langle a, y \rangle$ is ξ-unbiased for $\psi(\lambda)$. Let $a = a^* + (a - a^*)$ with $a^* \in L$ and $(a - a^*)$ in $\mathcal{L}^0$. Then

$$\langle a, y \rangle = \langle a^*, y \rangle + \langle a - a^*, y \rangle$$

But $E_\xi[\langle a - a^*, y \rangle] = (a - a^*)^T \Omega^{-1} WB\theta = 0$ since $(a - a^*) \in \mathcal{L}^0$. Hence $\langle a^*, y \rangle$ is ξ-unbiased for $\psi(\lambda)$ because this holds true for $\langle a, y \rangle$. Now suppose that the same holds true for $\langle b, y \rangle$, $b \in \mathcal{L}$. Then

$$E_\xi[\langle a^*, y \rangle - c^T \lambda] = E_\xi[\langle b, y \rangle - c^T \lambda] = 0$$

and hence $E_\xi[\langle a^*, y \rangle - \langle b, y \rangle] = 0$, implying that $(a^* - b)^T \Omega^{-1} WB\theta = 0$ identically in θ. Since $\mathcal{N}[B] = \emptyset$, identically in θ $(a^* - b)$ must lie in $\mathcal{L}^0$. But a^* and b both lie in $\mathcal{L}$ by construction and $\mathcal{L} \cap \mathcal{L}^0 = \emptyset$. Thus $(a^* - b) = 0$ or

$a^* = b$. It follows that a^* is unique in $\mathcal{L}$ for any a in R^N such that $\langle a, y \rangle$ is ξ-unbiased for $\psi(\lambda)$. For any such a,

$$a^* = P_{\mathcal{L}:\Omega^{-1}}\Omega^{-1}a = W(W^T\Omega^{-1}W)^{-1}W^T\Omega^{-1}a$$

where $P_{\mathcal{L}:\Omega^{-1}}$ is the unique orthogonal projection on $\mathcal{L}$ relative to $\langle \cdot, \cdot \rangle$, or on $\mathcal{L}$ along $\mathcal{L}^0$.   □

In the theorem now to be introduced, the following notation is used. For any column vector function $f = f(y)$ of q components, $D_\xi[f(y)]$ will denote the $q \times q$ matrix

$$D_\xi[f(y)] = E_\xi[f - E_\xi(f)][f - E_\xi(f)]^T$$

Theorem 1. Under the assumptions of the model (33)–(37) every estimable function $\psi(\lambda) = c^T\lambda$, $c \in R^K$, has a unique ξ-unbiased estimator $\langle a^*, y \rangle = c^T\hat{\lambda}$, with $a^* \in \mathcal{L}$ and $\hat{\lambda}$ the GLS estimator of λ from (34) using the inverse of (36), and

$$D_\xi[\langle a, y \rangle] \geq D_\xi[\langle a^*, y \rangle]$$

for every $\langle a, y \rangle$ which is ξ-unbiased for $\psi(\lambda)$.

Proof. Let $\langle a, y \rangle$ be any ξ-unbiased estimator of $\psi(\lambda)$ and let $a = a^* + (a - a^*)$ where $a^* = P_{\mathcal{L}:\Omega^{-1}}a$ is unique according to Lemma 2. Then $\{\langle a, y \rangle - c^T\lambda\} = \{\langle a^*, y \rangle - c^T\lambda\} + \langle a - a^*, y \rangle$, each component having ξ-expectation equal to zero. Moreover,

$$E_\xi\{\langle a^*, y \rangle - c^T\lambda\}\{\langle a - a^*, y \rangle\}^T = 0$$

and $D_\xi[\langle a, y \rangle] = \|a\|_{\Omega^{-1}}^2$, $D_\xi[\langle a^*, y \rangle] = \|a^*\|_{\Omega^{-1}}^2$, and $D_\xi[\langle a - a^*, y \rangle] \geq 0$, implying that

$$D_\xi[\langle a, y \rangle] \geq D_\xi[\langle a^*, y \rangle]$$

Finally,

$$\langle a^*, y \rangle = (P_{\mathcal{L}:\Omega^{-1}}a, \Omega^{-1}y) = (a, \Omega^{-1}P_{\mathcal{L}:\Omega^{-1}}y) = (W^T\Omega^{-1}a, \hat{\lambda})$$

where $\hat{\lambda} = (W\Omega^{-1}W)^{-1}W\Omega^{-1}y$. But, from Lemma 1, $c = W^T\gamma$ for $\gamma = \Omega^{-1}a$. Hence $\langle a^*, y \rangle = c^T\hat{\lambda} = \psi(\hat{\lambda})$.   □

The theorem may be regarded as a vector version of Pfefferman's matrix theorem [12, pp. 145 and 147–148]. The disadvantage of the matrix theorem is that the required orthogonality condition is not recognized explicitly at the outset and must be established by matrix manipulations, rather than as a consequence of the geometric setting of GLS estimation. When the orthogonality condition embedded in the scalar product $\langle \cdot, \cdot \rangle$ is recognized explicitly, it is easy to establish a corresponding matrix result as a corollary to the theorem.

Let C be a $K \times r$ matrix of rank r whose columns may be used to form r linearly independent parametric functions $c_1^T \lambda, c_2^T \lambda, \ldots, c_r^T \lambda$. In summary, $C^T \lambda = \Psi(\lambda)$ is a set of $r \leq K$ parametric functions of λ. Let A be an $N \times r$ matrix of fixed numbers whose columns $a_1, a_2, \ldots, a_r$ may be used to form the r ξ-unbiased linear estimators $\langle a_1, y \rangle, \langle a_2, y \rangle, \ldots, \langle a_r, y \rangle$ corresponding to $c_1^T \lambda, c_2^T \lambda, \ldots, c_r^T \lambda$; that is, such that $E_\xi[\langle a_i, y \rangle - c_i^T \lambda] = 0$, $i = 1, 2, \ldots, r$, or collectively $E_\xi[A^T \Omega^{-1} y - C^T \lambda] = 0$. Notice that the $D_\xi[\hat{\lambda}]$ $= (W^T \Omega^{-1} W)^{-1} \equiv Q_{\hat{\lambda}}^{-1}$. By Lemma 2, ξ-unbiasedness requires that $C = W^T \Omega^{-1} A$ so that $c_i \in R[W^T]$ for all i. The matrix A may be decomposed into two orthogonal parts, relative to $\langle \cdot, \cdot \rangle$, $A^* = P_{\mathcal{L}:\Omega^{-1}} A$ and $A^{**} = (I_n - P_{\mathcal{L}:\Omega^{-1}})A$, and hence $A^{*T} \Omega^{-1} A^{**} = 0$. Then

$$A^T \Omega^{-1} y - C^T \lambda = \{A^{*T} \Omega^{-1} y - C^T \lambda\} + A^{**} \Omega^{-1} y$$

which, using $C = W^T \Omega^{-1} A$, implies

$$D_\xi[A^T \Omega^{-1} y] = C^T Q_{\hat{\lambda}}^{-1} C + A^T [\Omega^{-1} - \Omega^{-1} W (W^T \Omega^{-1} W)^{-1} W^T \Omega^{-1}] A$$

$$= C^T Q_{\hat{\lambda}}^{-1} C + R$$

where R is a non-negative definite matrix. This result is a direct consequence of the theorem and may therefore be summarized as

Corollary 1. Let $\Psi(\lambda) = C^T \lambda$ be a set of $r \leq K$ linearly independent parametric functions of λ and let $A^T \Omega^{-1} y$ be a corresponding set of r ξ-unbiased linear estimators of $\Psi(\lambda)$. Then

$$D_\xi[A^T \Omega^{-1} y] = C^T Q_{\hat{\lambda}}^{-1} C + R$$

in which $\hat{\lambda}$ is the GLS estimator of λ from (34), $Q_{\hat{\lambda}}^{-1}$ is $D_\xi(\hat{\lambda})$, and R is a non-negative definite matrix.

6. DISCUSSION OF THE EMPIRICAL RESULTS
6.1 GLS Solutions and Data

The component vectors of $\hat{\lambda}$ are $\hat{\beta}_1$, $\hat{\beta}_2$, and $\hat{\theta}$, while the component vectors of $\hat{\beta}$ are obviously $\hat{\beta}_1$ and $\hat{\beta}_2$. The solutions $\hat{\beta}_1$, $\hat{\beta}_2$, and $\hat{\theta}$ are readily obtained by expanding $\hat{\lambda} = (W^T \Omega^{-1} W)^{-1} W^T \Omega^{-1} y$, yielding the following relations:

$$\hat{\theta} = (B^T \Delta^{-1} B)^{-1} B^T \Delta^{-1} \hat{\beta} \tag{39}$$

$$\hat{\beta}_1 = b_1 - [\Delta_{11} X_1^T \Sigma_{11}^{-1} X_1]^{-1} [\hat{\beta}_1 - B_1 \hat{\theta}] \tag{40}$$

$$\hat{\beta}_2 = B_2 \hat{\theta} + \Delta_{21} \Delta_{11}^{-1} (\hat{\beta}_1 - B_1 \hat{\theta}) \tag{41}$$

in which $b_1 = (X_1^T \Sigma_{11}^{-1} X_1)^{-1} X_1^T \Sigma_{11}^{-1} y_1$, the direct GLS estimator of β_1 for the SCR corresponding to the observations in group 1.

In general, $\hat{\beta}_1 \neq b_1$ and b_1 must be modified by a factor which depends on its own estimated dispersion $(X_1^T \Sigma_{11}^{-1} X_1)^{-1}$, the inverse of the natural dispersion inherent in the stochastic coefficient β_1, namely Δ_{11}^{-1}, and the difference between $\hat{\beta}_1$ and its estimated mean effect $B_1 \hat{\theta}$. If $\hat{\beta}_1 = B_1 \hat{\theta}$, then $\hat{\beta}_1 = b_1$ regardless of Δ_{11}; if B_1 is the identity, then $\beta_1 = \theta$ and $\hat{\beta}_1 = \hat{\theta} = b_1$, regardless of Δ_{11}.

The formula for $\hat{\beta}_2$ in equation (41) is essentially the formula quoted in [7, Chapter 4.3, equations (10) and (11)], except that β_2 is replaced by $\hat{\beta}_2$ because it is not observed directly. When $\hat{\beta}_1 = B_1 \hat{\theta}$, the optimal predictor of β_2 is the optimal estimator of its expectation $(B_2 \hat{\theta})$. When B_1 is the identity matrix, $\hat{\beta}_2 = B_2 \hat{\theta}$ regardless of the covariance term Δ_{21}. The equation for $\hat{\theta}$ is the GLS solution to equation (39) when β is replaced by $\hat{\beta}$ without adjusting the error term as in (17).

The time series analyzed in [31] is the long (1000 observations in the complete set of observations and 900 observations in the missing-observations case) daily return series of the S&P500 stock index originally compiled by William Schwert (see [32]). Observations are first differences of natural logarithms. Thus the first observation is dated 30 June 1987 and the last observation is dated 30 April 1991 in the complete set of observations. In the missing-observations case the last 100 observations are excluded. In order to analyze these data the mean of the series is first subtracted, then the natural logarithm of the square of daily returns is determined (see Section 6.2 below). These data are in the files *sp*500.inf and *spm*500.inf, in the form of a long column vector, which is used for running GAUSSX programs.

6.2 The Model with No Missing Observations

A stochastic volatility (SV) model can be represented as

$$R_t = \mu + \sigma e^{\frac{1}{2} h_t} u_t, \quad \text{where } u_t \sim N[0, 1]$$

$$h_t = \gamma h_{t-1} + v_t, \quad v_t \sim N[0, \sigma_v^2]; \qquad t = 1, 2, \ldots, 1000$$

The measurement equation is obtained by linearizing the equation for R_t. Taking logarithms of the square of the mean adjusted return $r_t = R_t - \mu$, and setting $y_t = \ln(r_t^2)$, there results

$$y_t = \alpha + h_t + \varepsilon_t \tag{42}$$

in which $\alpha = \ln(\sigma^2)$, $\varepsilon_t = \ln(u_t^2)$, and ε_t is approximated by a normal distribution with mean $\mu_\varepsilon = -1.27$ and variance $\sigma_\varepsilon^2 = \pi^2/2$. The state vector is defined as $\beta_t = [h_t \ \alpha]^T$ and, setting $x = [1 \ 1]$, the measurement equation is defined as

$$y_t = x^T \beta_t + \varepsilon_t \tag{43}$$

In (43), β_t is generated by the Markovian scheme

$$\beta_t = T\beta_{t-1} + Rv_t \tag{44}$$

in which

$$T = \begin{bmatrix} \gamma & 0 \\ 0 & 1 \end{bmatrix}$$

is the transition matrix,

$$R = \begin{bmatrix} 1 \\ 0 \end{bmatrix},$$

and $E(\varepsilon_t, v_t) = 0$ for all t.

This model may be estimated using quasi-maximum likelihood through a Kalman filter, because β_t is unobserved. From the measurement equation, the conditional distribution of y_t is normal with mean

$$E_{t-1}(y_t) = \bar{y}_{t|t-1} = x^T \beta_{t|t-1} + \mu_\varepsilon \tag{45}$$

In (45),

$$\beta_{t|t-1} = \bar{T}\beta_{t-1|t-2} + \bar{K}y_{t-1} \tag{46}$$

is the recursion for the state equation with the transition and gain matrices; these are defined by $\bar{T} = T - \bar{K}x$, where $\bar{K}$ is given by $T\Delta x(x^T\Delta x + \Sigma)^{-1}$. The covariance matrix is $\Omega = x^T\Delta x + \Sigma$, where

$$\Delta = \begin{bmatrix} \dfrac{\sigma_v^2}{1 - \gamma^2} & 0 \\ 0 & 0 \end{bmatrix}$$

is the covariance matrix of the unobserved error v_t, and $\Sigma = \sigma_\varepsilon^2 = \pi^2/2$. Thus

$$\Omega = \frac{\sigma_v^2}{1 - \gamma^2} + \sigma_\varepsilon^2 = \frac{\sigma_v^2 + \sigma_\varepsilon^2(1 - \gamma^2)}{1 - \gamma^2}$$

For a Gaussian model, the likelihood function can be written immediately as

$$\ln L = -\frac{NT}{2}\ln(2\pi) - \frac{1}{2}\sum_{t=1}^{T}\ln|\Omega| - \frac{1}{2}\sum_{t=1}^{T}v_t{'}\Omega^{-1}v_t \tag{47}$$

where $v_t = y_t - \bar{y}_{t|t-1}$, $t = 1, 2, \ldots, 1000$, and $\bar{y}_{t|t-1}$ is defined according to (45). Also θ, the coefficient of the mean effect, is reflected in (46). The resulting estimates for α, γ, and θ are GLS estimates. These are presented in Table 1 along with estimates for σ_v.

6.3 The Model with Missing Observations

The measurement equation in this case is defined as

$$\begin{bmatrix} y_{1,t} \\ -B_2\theta \end{bmatrix} = \begin{bmatrix} x_1 & 0 \\ 0 & -I_1 \end{bmatrix}\begin{bmatrix} \beta_{1,t} \\ \beta_{2,t} \end{bmatrix} + \begin{bmatrix} \varepsilon_{1,t} \\ v_{2,t} \end{bmatrix}$$

where $y_{1,t} = y_{mt} - \alpha$, y_{mt} is the vector of 900 observations, $x_1 = 1$, $B_2 = \gamma$, and $I_1 = 1$, or, in the obvious notation,

$$y_t = W\beta_t + \delta_t \tag{48}$$

The transition equation is defined as

$$\beta_t = T\beta_{t-1} + v_t \tag{49}$$

where $T = \gamma I_2$, $v_t = [v_{1,t} \quad v_{2,t}]^T$ and $E(\varepsilon_{1,t}, v_{1,t}) = 0$ for all t. The transition covariance matrix is

$$\Delta = \frac{1}{1-\gamma^2}\begin{bmatrix} \sigma_{v_1}^2 & \sigma_{v_{12}} \\ \sigma_{v_{12}} & \sigma_{v_2}^2 \end{bmatrix}$$

Table 1. Estimates and associated statistics: no missing data

Parameter	Estimate	Standard error	t-statistic	p-value
α	−9.090	0.0267	−339.629	[.000]
γ	0.989	0.0154	63.923	[.000]
θ	−10.650	0.0269	−394.954	[.000]
σ_v	0.080	0.0326	2.452	[.014]

The covariance matrix is defined as

$$\Omega = W\Delta W^T + \Sigma = \frac{2}{1-\gamma^2}\begin{bmatrix} 2\sigma_{v_1}^2 + \pi^2(1-\gamma^2) & -2\sigma_{v_{12}} \\ -2\sigma_{v_{12}} & 2\sigma_{v_2}^2(2-\gamma^2) \end{bmatrix}$$

$$\Sigma = \begin{bmatrix} \frac{\pi^2}{2} & 0 \\ 0 & \sigma_{v_2}^2 \end{bmatrix}$$

The model is solved, as in the previous case, using quasi-maximum likelihood through a Kalman filter, because, as before, β_t is unobserved. The resulting estimates of this model are presented in Table 2.

In order to compute the estimates of β_1 and β_2, formulae (40) and (41) are used. Knowing that $\hat{\theta} = -9.966$, $\Delta_{11} = 0.027 \Rightarrow \Delta_{11}^{-1} = 36.916$, $X_1 = 1$, $\Sigma_{11}^{-1} = 2/\pi^2 = 0.2026$, $\Delta_{21} = -0.4005$, and using the formula for the GLS estimator of β_1, $b_1 = (X_1^T\Sigma_{11}^{-1}X_1)^{-1}X_1^T\Sigma_{11}^{-1}y_1$, $b_1 = -9.54$. Then, $\hat{\beta}_1 = -9.852$ and $\hat{\beta}_1 - B_1\hat{\theta} = 0.007$. This difference confirms that B_1 is close to the identity matrix (I_1), but not equal to it. In fact $B_1 = 0.993$.

For the second group of observations, $\hat{\beta}_2 = -10.002$ and $\hat{\beta}_2 - B_2\hat{\theta} = 0.103$. The increase of the deviation of $\hat{\beta}_2$ from its estimated mean effect $(B_2\hat{\theta})$ in the second group of observations is due to the product $\Delta_{21}\Delta_{11}^{-1}$, which is equal to -14.78.

A comparison between the estimates from the complete set of observations and those when some observations are missing reveals that the estimated transition coefficient (γ) is hardly affected, moving from 0.989 to 0.993. This movement depends in part on the estimated covariance $\sigma_{v_{12}}$. The estimated coefficient of α, however, decreases from -9.090 to -10.995, a substantial change caused by the estimated difference between α and its mean effect modified by $(\sigma_\varepsilon^2/\Delta_{11})$, Δ_{11} being a function of the transition coefficient.

Table 2. Estimates and associated statistics: with missing data

Parameter	Estimate	Standard error	t-statistic	p-value
α	-10.995	0.0099	-1105.427	[.000]
γ	0.993	0.0002	4437.801	[.000]
θ	-9.966	0.0098	-1008.822	[.000]
σ_{v_1}	0.164	0.0038	43.245	[.000]
σ_{v_2}	1.293	0.0095	135.311	[.000]
$\sigma_{v_{12}}$	-0.400	0.0075	-53.069	[.000]

The empirical results illustrate the ability of the Kalman filter model to predict well, even though the assumption of normality is made as an approximation. In the case considered, the empirical model does not fully satisfy the assumptions of the theoretical model; for example, the matrices B, Σ, and Δ are assumed known in theory but in practice must be estimated. This being the case, the Kalman filter is revealed to be an efficient algorithm for producing feasible GLS estimates of the coefficients of interest, even though these are not the ideal of the theory presented in the paper. This is a common feature of GLS estimation.

In an FCR model, the matrix B would essentially represent a set of fixed restrictions on β, expressed in freedom equation form as in $\beta = B\theta$, implying a set of linearly independent restrictions $A^T\beta = 0$ for some matrix A, B having columns in $N[A^T]$. If, then, B is nonsingular, no essential estimation problem remains and so it is crucial that B be singular in a reasonable model. In the SCR model, a singular B implies an informative model and a nonsingular B ensures no information gain over a corresponding FCR model. In the case of the empirical model outlined in the paper, the crucial matrix is B, and the gain in an SCR model over an FCR model may be measured by the deviation of B_1 from 1; evidently the gain is small but significant.

ACKNOWLEDGMENT

The authors are grateful to an anonymous referee for helpful comments on an earlier draft of this chapter.

REFERENCES

1. Rao, C. R., The theory of least squares when the parameters are stochastic and its application to the analysis of growth curves. *Biometrika*, 52, 1965, 447–458.
2. Fisk, P. R., Models of the second kind in regression analysis. *Journal of the Royal Statistical Society B*, 29, 1967, 266–281.
3. Swamy, P. A. V. B., *Statistical Inference in Random Coefficient Regression Models*. New York: Springler-Verlag, 1971.
4. Anderson, T. W., Maximum likelihood estimates for a multivariate normal distribution when some observations are missing. *Journal of the American Statistical Association*, 52, 1957, 200–203.
5. Chipman, J. S., On least squares with insufficient observations. *Journal of the American Statistical Association*, 59, 1964, 1078–1111.

6.   Kmenta, J., On the problem of missing measurements in the estimation of economic relationships. Chapter 10 of E. G. Charatsis (editor), *Proceedings of the Econometric Society European Meeting 1979.* Amsterdam: North Holland, 1981.

7.   Whittle, P., *Prediction and Regulation by Linear Least-Squares Methods*, 2nd edition, revised. Oxford: Basil Blackwell, 1984.

8.   Duncan, D. B., and S. D. Horn, Linear dynamic recursive estimation from the viewpoint of regression analysis. *Journal of the American Statistical Association*, 67, 1972, 815–822.

9.   Cooley, T. F., and E. Ç. Prescott, Varying parameter regression: a theory and some applications. *Annals of Economic and Social Measurement*, 2, 1973, 463–473.

10. Rubin, D. B., Using empirical Bayes' techniques in the Law School validity studies (with Discussion). *Journal of the American Statistical Association*, 75, 1980, 801–827.

11. Pfefferman, D., and G. Nathan, Regression analysis of data from a cluster sample. *Journal of the American Statistical Association*, 76, 1981, 681–689.

12. Pfefferman, D, On extensions of the Gauss–Markov theorem to the case of stochastic regression coefficients. *Journal of the Royal Statistical Society B*, 46, 1984, 139–148.

13. Swamy, P. A. V. B., R. K. Conway, and M. R. LeBlanc., The stochastic coefficients approach to econometric modeling, part I: a critique of fixed coefficients models. *The Journal of Agricultural Economics Research*, 40, 1988, 2–10.

14. Swamy, P. A. V. B., R. K. Conway, and M. R. LeBlanc, The stochastic coefficients approach to econometric modeling, part II: description and motivation. *The Journal of Agricultural Economics Research*, 40, 1988, 21–30.

15. Swamy, P. A. V. B., R. K. Conway, and M. R. LeBlanc, The stochastic coefficients approach to econometric modeling, part III: estimation, stability testing, and prediction. *The Journal of Agricultural Economics Research*, 41, 1989, 4–20.

16. Hsiao, C., *Analysis of Panel Data*. Cambridge: Cambridge University Press, 1986.

17. Baltagi, B. H., *Econometric Analysis of Panel Data*. Chichester: Wiley, 1995.

18. Harvey, A. C., *Forecasting Structural Time Series Models and the Kalman Filter*. Cambridge: Cambridge University Press, 1989.

19. Stuart, A., K. Ord, and S. Arnold, *Kendall's Advanced Theory of Statistics, Vol. 2A, Classical Inference and the Linear Model*. London: Arnold, 1999.

20. Durbin, J., Estimation of parameters in time-series regression models. *Journal of the Royal Statistical Society B*, 22, 1960, 139–153.
21. Lucas, R. E., Econometric policy evaluation: a critique. In: K. Brunner and A. Meltzer (editors), *The Phillips Curve and Labour Markets*. Carnegie–Rochester Conference Series, Vol. 1, supplement to *Journal of Monetary Economics*, 1975.
22. Lucas, R. E., and T. J. Sargent, After Keynesian macroeconomics. In: *After the Phillips Curve: Persistence of High Inflation and High Unemployment*. Boston: FRB Boston Conference Series No. 19, 1978.
23. Zellner, A., On the aggregation problem: a new approach to a troublesome problem. In: K. A. Fox, J. K. Sengupta, and G. V. L Narasimham (editors), *Economic Models, Estimation and Risk Programming: Essays in Honor of Gerhard Tintner*. New York: Springer-Verlag, 1969.
24. Rausser, G. C., Y. Mundlak, and S. R. Johnson, Structural change, updating, and forecasting. In: G. C. Rausser (editor), *New Directions in Econometric Modeling and Forecasting in U.S. Agriculture*. Amsterdam: North-Holland, 1983.
25. Duffy, W. J., Parameter variation in a quarterly model of the post-war U.S. economy, PhD thesis, University of Pittsburgh, 1969.
26. Rosenberg, B., The estimation of stationary stochastic regression parameters re-examined. *Journal of the American Statistical Association*, 67, 1972, 650–654.
27. Fisher, G., An invariance condition for the random coefficients regression model. Working Paper, Concordia University, 1999.
28. Halmos, P. R., *Finite-Dimensional Vector Spaces*, 2nd edition. Princeton: Van Nostrand, 1958.
29. Rao, C. R., *Linear Statistical Inference and its Applications*, 2nd edition. New York: Wiley, 1973.
30. Harville, D., Extensions of the Gauss–Markov theorem to include the estimation of random effects. *Annals of Statistics*, 4, 1976, 384–396.
31. Voia, M. C., ARCH and stochastic volatility models for the SP500. MA paper, Concordia University, 1999.
32. French, K. R., G. W. Schwert, and R. F. Stambaugh, Expected stock return and volatility. *Journal of Financial Economics*, 19, 1987, 3–30.

24

Efficiency Considerations in the Negative Exponential Failure Time Model

JOHN L. KNIGHT University of Western Ontario, London, Ontario, Canada

STEPHEN E. SATCHELL Cambridge University, Cambridge, England

1. INTRODUCTION

In this paper we derive some exact properties for log-linear least squares (LLSE) and maximum likelihood estimators (MLE) for the negative exponentially distributed regression model in the case when there are two exogenous variables, one a constant and the other a dummy variable. The exponential regression model has been frequently used by applied economists to estimate hazard functions, especially for unemployment duration. Choices of exogenous variables in these models typically include several dummy (binary) variables such as gender, ethnicity, and marital status. For this reason an exact study should shed light on questions of estimation and inference. In economics, all previous analysis of this model has, to our knowledge, relied on large-sample properties of maximum likelihood. This position has been justified on the basis of data sets of size 400 or more but there has been little attempt to see how confident one can be about asymptotic results.

Generalizations of our results are possible and, for the general two-variable case including a constant, one can derive the joint distribution function, see Knight and Satchell (1996). In the case of K-variable regression, although it is not possible to calculate explicit formulae for the estimators, their exact properties can be calculated numerically (see Knight and Satchell 1993). The dummy and constant model, examined here, seems a suitable trade-off between generality and tractability and should lead to insights about the general case.

In Section 2 we present the model, formulae for the estimators, and their exact distributions and moments, where derivable. We stress that many of these results have been known to statisticians for some time under various guises; see Lehmann (1983), for example, and Kalbfleisch and Prentice (1980). Section 3 contains the large-sample properties while in Section 4 we compare the estimators and develop minimum variance unbiased estimators (MVUE) for the dummy exogenous variable model. Section 5 extends the MVUE analysis to cases where the exogenous variable takes on r distinct values.

2. THE MODEL AND EXACT RESULTS

We assume that observations of failure times, t_i, $i = 1, \ldots, N$, come from a negative exponential distribution with hazard function $\lambda_i = \exp(\alpha_0 + \beta_0 X_i)$, where $X_i = 1$ or 0 with $X_i = 1$ N_1 times, $N = N_1 + N_0$, and α_0 and β_0 are the true parameter values.

The LLSEs of α_0 and β_0 are calculated from running a regression of log $t_i + c$ on -1 and $-X_i$, where c is Euler's constant. Amemiya (1985, p. 439) shows that these estimators are unbiased with variance–covariance matrix equal to

$$\frac{\pi^2}{6} \begin{bmatrix} 1/N_0 & -1/N_0 \\ -1/N_0 & 1/N_1 + 1/N_0 \end{bmatrix}$$

If we write the regression as

$$\log t_i + c = -\alpha_0 - X_i\beta_0 + w_i \tag{1}$$

then, by direct calculation, the LLSEs $\hat{\alpha}$ and $\hat{\beta}$ are*

$$\hat{\alpha} = -T_0 - c$$

$$\hat{\beta} = T_0 - T_1$$

where $T_1 = \sum_{j=1}^{N_1} \ln t_j / N_1$ and $T_0 = \sum_{j=N_1+1}^{N} \ln t_j / N_0$. If we transform from t_j to w_j from (1) we can readily derive the moment-generating functions associated with $\hat{\alpha}$ and $\hat{\beta}$. The results are given in the following theorem.

*We have assumed that the N_1 observations have $X_i = 1$, without loss of generality.

Theorem 1. Using (1) and transforming from $t_j \rightarrow w_j$ we have

$$\hat{\alpha} = \alpha_0 - W_0$$

$$\hat{\beta} = \beta_0 + W_0 - W_1 \tag{2}$$

where $W_1 = \sum_{j=1}^{N_1} w_j/N_1$ and $W_0 = \sum_{j=N_1+1}^{N} w_j/N_0$ and their respective moment-generating functions are given by

$$\psi_1(s) = \mathrm{mgf}(\hat{\alpha} - \alpha_0) = \exp(-cs)(\Gamma(1 - s/N_0))^{N_0} \tag{3}$$

and

$$\psi_2(s) = \mathrm{mgf}(\hat{\beta} - \beta_0) = (\Gamma(1 + s/N_0))^{N_0}(\Gamma(1 - s/N_1))^{N_1}$$

Proof. The mgf results follow immediately upon noting that the mgf of w_i is

$$\phi(s) = E[e^{w_i s}] = \exp(cs)\Gamma(1 + s)$$

It should be possible to invert $\psi_1(is)$ and $\psi_2(is)$ to derive their distributions, although we have all the moments as they are. It is worth noting that the marginal distribution of $(\hat{\beta} - \beta_0)$ does not depend on c.

Turning to the maximum likelihood problem, where L is the likelihood, we have

$$\ln L = N\alpha + N_1\beta - \sum_{j=1}^{N_1} t_j \exp(\alpha + \beta) - \sum_{j=N_1+1}^{N} t_j \exp(\alpha)$$

$$\frac{\partial \ln L}{\partial \alpha} = N - \sum_{j=1}^{N_1} \exp(\tilde{\alpha} + \tilde{\beta})t_j - \sum_{j=N_1+1}^{N} \exp(\tilde{\alpha})t_j = 0 \tag{4}$$

$$\frac{\partial \ln L}{\partial \beta} = N_1 - \sum_{j=1}^{N_1} \exp(\tilde{\alpha} + \tilde{\beta})t_j = 0$$

where $\tilde{\alpha}$ and $\tilde{\beta}$ are the MLE of α_0 and β_0. From (4), we see that

$$\tilde{\alpha} = -\ln\left(\sum_{j=N_1+1}^{N} t_j/N_0\right)$$

$$\tilde{\beta} = \ln\left(\sum_{j=N_1+1}^{N} t_j/N_0\right) - \ln\left(\sum_{j=1}^{N_1} t_j/N_1\right) \tag{5}$$

We now define $\tilde{t}_j = \exp(\alpha_0 + X_i\beta_0)t_j$, which will be negative exponential with parameter 1, (NE(1)). If we transform t_j to $\tilde{t}_j$ in (5) and simplify, we see that

$$\tilde{\alpha} - \alpha_0 = -\ln \tilde{T}_0$$

$$\tilde{\beta} - \beta_0 = \ln \tilde{T}_0 - \ln \tilde{T}_1 \tag{6}$$

where $\tilde{T}_1 = \sum_{j=1}^{N_1} \tilde{t}_j / N_1$, $\tilde{T}_0 = \sum_{j=N_1+1}^{N} \tilde{t}_j / N_0$.

The results in (5) should be compared with those in (2); note that the MLEs are the logarithm of the arithmetic mean whilst the LLSEs are the logarithm of the geometric mean. Since the geometric mean is always less than or equal to the arithmetic mean, it follows that $(\prod_{j=1}^{N^*} t_j)^{1/N^*} \le \sum_{j=1}^{N^*} t_j / N^*$ for either N_1 or $N_0 = N^*$, therefore $\hat{\alpha} + c \ge -\tilde{\alpha}$, and hence $E(\tilde{\alpha}) \ge -c$. More detailed comparison requires knowledge about the exact distributions of the MLEs; this can be done by noting that $\tilde{T}_1$ and $\tilde{T}_0$ are distributed as $\chi^2(2N_1)/2N_1$ and $\chi^2(2N_0)/2N_0$ (see Johnston and Kotz 1970, Ch. 18, p. 222), so that $\tilde{T}_0 \tilde{T}_1^{-1}$ is an $F(2N_0, 2N_1)$ since $\tilde{T}_1$ and $\tilde{T}_0$ are independent. Therefore $\tilde{\beta} - \beta_0$ is $\ln F(2N_0, 2N_1)$ whilst $\tilde{\alpha} - \alpha_0$ is $-\ln(\chi^2(2N_0)/2N_0)$. Noting that the distribution of $\ln F(2N_0, 2N_1)$ is the same as $2z$, where z is "Fisher's z" distribution, a simple transformation of results in Johnston and Kotz (1970, Ch. 26, pp. 78–81) gives the pdf, mgf, and cumulants. Corresponding results for $\tilde{\alpha} - \alpha_0$ can be readily obtained from first principles.

Consequently, the pdf expressions for the MLE estimators are given by

$$\text{pdf}(y = \tilde{\alpha} - \alpha_0) = \frac{N_0^{N_0}}{\Gamma(N_0)} e^{-N_0 y} \exp(-N_0 e^{-y}), \quad -\infty < y < \infty \tag{7}$$

and

$$\text{pdf}(g = \tilde{\beta} - \beta_0) = \frac{N_0^{N_0} N_1^{N_1}}{B(N_0, N_1)} \frac{e^{N_0 g}}{(N_1 + N_0 \exp(g))^{N_1 + N_0}}, \quad -\infty < g < \infty \tag{8}$$

and their corresponding mgfs by

$$\tilde{\psi}_1(s) = \frac{N_0^s \Gamma(N_0 - s)}{\Gamma(N_0)} \tag{9}$$

and

$$\tilde{\psi}_2(s) = \left(\frac{N_1}{N_0}\right)^s \frac{\Gamma(N_0 + s)\Gamma(N_1 - s)}{\Gamma(N_0)\Gamma(N_1)} \tag{10}$$

We notice immediately that for $N_1 = N_0$ both the LLSE and MLE estimators of β have symmetric distribution since

$$\psi_2(s) = \psi_2(-s)$$

and

$$\tilde{\psi}_2(s) = \tilde{\psi}_2(-s)$$

Since we wish to see how these estimators may differ distributionally, it is clear that we need to know at least the first few moments.

Thus, turning to moments, and in particular to the cumulant generating functions, we have, by taking logarithms of $\psi_2(s)$ and $\tilde{\psi}_2(s)$ and differentiating,

$$\kappa_r = \left(\frac{1}{N_0^{r-1}} + \frac{(-1)^r}{N_1^{r-1}}\right)\psi^{(r-1)}(1), \ r \geq 2$$

and

$$\kappa_1 = 0$$

for LLSE, and

$$\tilde{\kappa}_r = [\psi^{(r-1)}(N_0) + (-1)^r \psi^{(r-1)}(N_1)] \ , r \geq 2$$

with

$$\tilde{\kappa}_1 = \ln(N_1/N_0) + \psi(N_0) - \psi(N_1)$$

for MLE, which may alternatively be written as

$$\tilde{\kappa}_r = (1 + (-1)^r)\psi^{(r-1)}(1) + (-1)^{r-1}(r-1)!\left(\sum_{j=1}^{N_0-1} 1/j^r + (-1)^r \sum_{j=1}^{N_1-1} 1/j^r\right)$$

$$(11)$$

$\psi^{(m)}(x)$ is the polygamma function defined as the $(m+1)$th derivative of $\ln \Gamma(x)$ (see Abramowitz and Stegum 1972, equations 6.4.1–6.4.3, p. 260).

In particular, the first four cumulants in each case are

$$\kappa_1 = 0$$

$$\kappa_2 = \left(\frac{1}{N_0} + \frac{1}{N_1}\right)\psi^{(1)}(1)$$

$$\kappa_3 = \left(\frac{1}{N_0^2} - \frac{1}{N_1^2}\right)\psi^{(2)}(1)$$

$$\kappa_4 = \left(\frac{1}{N_0^3} + \frac{1}{N_1^3}\right)\psi^{(3)}(1)$$

$$(12)$$

and

$$\tilde{\kappa}_1 = \ln N_1 - \ln N_0 + \sum_{j=1}^{N_0-1} 1/j - \sum_{j=1}^{N_1-1} 1/j$$

$$\tilde{\kappa}_2 = 2\psi^{(1)}(1) - \left(\sum_{j=1}^{N_0-1} 1/j^2 + \sum_{j=1}^{N_1-1} 1/j^2 \right)$$

$$\tilde{\kappa}_3 = 2\left(\sum_{j=1}^{N_0-1} 1/j^3 - \sum_{j=1}^{N_1-1} 1/j^3 \right) \tag{13}$$

$$\tilde{\kappa}_4 = 2\psi^{(3)}(1) - 3!\left(\sum_{j=1}^{N_0-1} 1/j^4 + \sum_{j=1}^{N_1-1} 1/j^4 \right)$$

We shall delay any numerical comparisons for the moment. In the next section we shall look at the asymptotic properties of our rival estimators.

3. ASYMPTOTIC RESULTS

Our exact results in Section 2 allow us to investigate the behavior of the estimators as the sample size increases. There are three possibilities in our model, large N_1, large N_0, and finally the case of N_1 and N_0 both increasing. It may strike the reader that large N_1 and large N_0 asymptotics are rather esoteric; however, in cross-sectional data sets with many exogenous variables, an increase in N may still leave the number of observations, in a given category, fixed or increasing at a lesser rate. There seems to be a widespread feeling that very large samples, $N = 1000+$, are necessary in these models before one can rely on asymptotic normality; our investigation may shed some light on this belief.

Initially, we fix N_0, for the LLS estimators, from (3); $\psi_1(s)$ is invariant to changes in N_1, thus the density of $\hat{\alpha} - \alpha_0$ stays the same for fixed N_0 as N_1 tends to infinity. For $\psi_2(s)$, it can be expressed via an infinite product expansion for the gamma function (see Abramowitz and Stegum 1972, equations. 6.1.3 and 6.1.17, pp. 255–256), as

$$\psi_2(s) = e^{cs}(\Gamma(1 + s/N_0))^{N_0} \prod_{j=1}^{\infty} e^{-s/j}\left(1 - \frac{s}{jN_1} \right)^{-N_1} \tag{14}$$

Therefore $\lim_{N_1 \to \infty} \psi_2(s) = e^{cs}(\Gamma(1 + s/N_0))^{N_0} = \psi_1(-s)$, the characteristic function of $-(\hat{\alpha} - \alpha_0)$; see (3).

Turning now to the MLEs, again $\tilde{\psi}_1(s)$ does not depend upon N_1, so turning to $\tilde{\psi}_2(s)$, see (10), and using the known distribution of $\tilde{\beta} - \beta_0$,

$$\lim_{N_1 \to \infty} P(\tilde{\beta} - \beta_0 < v) = \lim_{N_1 \to \infty} P(F_{2N_0, 2N_1} < e^v) = P(\chi^2_{2N_0} < 2N_0 e^v)$$

$$= 1 - \sum_{j=0}^{N_0=1} \exp(-N_0 e^v) \frac{(N_0 e^v)^j}{j!} \tag{15}$$

The last step follows from Abramowitz and Stegum (1972, equation 26.4.21, p. 941). We note in this case that $(\tilde{\beta} - \beta_0)$ has the same distribution as $-(\tilde{\alpha} - \alpha_0)$. In particular, if $N_0 = 1$,

$$\lim_{N_1 \to \infty} P(\tilde{\beta} - \beta_0 < v) = 1 - \exp(-\exp(v))$$

which can be recognized as a type 1 extreme value distribution; see Johnston and Kotz (1970, p. 272). From (14), for $N_0 = 1$, as $N_1 \to \infty$, we see that $\hat{\beta} - \beta_0$ has a limiting distribution identical to $(\tilde{\beta} - \beta_0 + c)$; i.e., they differ only in location.

We now fix N_1 and let $N_0 \to \infty$; the results are very similar. Both $\hat{\alpha}$ and $\tilde{\alpha}$ converge in probability to α_0, whilst for the estimators of β_0

$$\lim_{N_0 \to \infty} \psi_2(s) = e^{-cs}[\Gamma(1 - s/N_1)]^{N_1}$$

$$\lim_{N_0 \to \infty} P(\tilde{\beta} - \beta_0 < v) = \sum_{j=0}^{N_1-1} \exp(-N_1 e^{-v})(N_1 e^{-v})^j / j! \tag{16}$$

We note that when $N_1 = 1$,

$$\lim_{N_0 \to \infty} P(\tilde{\beta} - \beta_0 < v) = \exp(-\exp(v))$$

Again, $\hat{\beta} - \beta_0$ has the same distribution as the bias-corrected MLE, i.e., as $(\tilde{\beta} - \beta_0 - c)$. This can be checked by putting $N_1 = 1$ in $\lim_{N_0 \to \infty} \psi_0(s)$ in (16), and comparing with the limit as $N_0 \to \infty$ of (10).

In summary, estimators of α are invariant to changes in N_1 and converge in probability to α_0 for large N_0. Estimators of β converge to different distributions for large N_1 and N_0, different for MLE and LLSE; for the cases where $N_1 = 1$ and $N_0 = 1$, however, they differ only in location.

For large N asymptotics, we need to specify how N_0 and N_1 tend to infinity. Let $N_1 = \lambda N_0$, so $N = (1 + \lambda)N_0$. For the MLEs, large sample asymptotics for Fisher's z are well known and, utilizing results from Stuart and Ord (1987, Sect. 16.18), we have

$$(\tilde{\beta} - \beta_0) = \ln F_{2N_0,2N_1} \rightarrow N\left[\frac{1}{2N_1} - \frac{1}{2N_0}, \frac{1}{N_0} + \frac{1}{N_1}\right]$$

as $N \rightarrow \infty$, and thus asymptotically

$$\sqrt{\frac{N_0\lambda}{1+\lambda}}(\tilde{\beta} - \beta_0) = \sqrt{\frac{N_0N_1}{N_0+N_1}}(\tilde{\beta} - \beta_0) \sim N(0, 1) \tag{17}$$

For LLSE, consider

$$\sqrt{\frac{N_0N_1}{N_0+N_1}}(\hat{\beta} - \beta_0) = \sqrt{\frac{N_0\lambda}{1+\lambda}}(\hat{\beta} - \beta_0)$$

and let $M(s)$ be the mgf of this quantity. That is,

$$M(s) = \psi_2\left(s\sqrt{N_0\lambda/(1+\lambda)}\right)$$

From (3) and (14), it follows that

$$M(s) = \prod_{j=1}^{\infty}\left[1 - s\frac{\sqrt{N_0\lambda/(1+\lambda)}}{jN_0\lambda}\right]^{-\lambda N_0}\left[1 + s\frac{\sqrt{N_0\lambda/(1+\lambda)}}{jN_0}\right]^{-N_0} \tag{18}$$

Clearly, the limit of $M(s)$ as $N_0 \rightarrow \infty$ will be $[(s^2/2) \cdot (\pi^2/6)]$, leading to the asymptotic distribution

$$\sqrt{\frac{N_0N_1}{N_0+N_1}}(\hat{\beta} - \beta_0) \sim N\left[0, \frac{\pi^2}{6}\right] \tag{19}$$

One can expand (18) to obtain an Edgeworth expansion, and higher-order terms could be compared with those existing for Fisher's z. However, by merely comparing (17) with (19) we see that MLE is clearly more asymptotically efficient than LLSE.

The conclusion that arises from the analysis so far is that a large sample size is not enough to guarantee that LLSE and MLE will converge in distribution to multivariate distributions as N tends to infinity. Exactly as in the case of linear regression with dummy variables, we require that N_0 and N_1 both become large. The interesting feature of our previous calculations is that we have computed the marginal distributions of the estimator for large $N_0(N_1)$ and fixed $N_1(N_0)$.

Given the above conclusions, it is natural to consider what implications this will have for testing. Rather than answer this question exhaustively, we shall consider a specific example, testing that $\beta = 0$, and use a likelihood ratio (LR) test. Our choice is quite arbitrary, one could do similar calculations for other hypotheses using other test procedures.

After some straightforward calculations, we find that the LR test statistic is given by the following expression:

$$
\text{LR} = (\tilde{T}_0/\tilde{T}_1)^{N_0} \left[\frac{N_1}{N} + \frac{N_0\tilde{T}_0}{N\tilde{T}_1} \right]^{-N}
$$

$$
= \left[\frac{N_0}{N} + \frac{N_1\tilde{T}_1}{N\tilde{T}_0} \right]^{-N_0} \left[\frac{N_1}{N} + \frac{N_0\tilde{T}_0}{N\tilde{T}_1} \right]^{-N_1}
$$

(20)

Since $\tilde{T}_1/\tilde{T}_0 \sim F_{2N_1,2N_0}$ we may write (20) as

$$
\text{LR} = \left[\frac{N}{N_0} \right]^N [F_{2N_1,2N_0}]^{N_1} \left[1 + \frac{N_1}{N_0} F_{2N_1,2N_0} \right]^{-N}
$$

For testing one uses $-2\ln\text{LR}$, and so we now examine the moment-generating function of this statistic. The results are given in the following theorem.

Theorem 2. For the LR test statistic defined in (20) we have the mgf of $-2\ln\text{LR}$ given by

$$
\phi(s) = E[\exp(s(-2\ln\text{LR}))]
$$

$$
= \left[\frac{N}{N_0} \right]^{-2sN} \left[\frac{N_1}{N_0} \right]^{2sN_1} \frac{B(N_0(1-2s), N_1(1-2s))}{B(N_1,N_0)}
$$

(21)

where $B(a,b)$ is the standard Beta function.

Proof. The result follows straightforwardly from the moments of an F distribution.

To investigate the asymptotic behaviour of $-2\ln\text{LR}$ we consider in turn (i) $\lim_{N_0\to\infty}\phi(s)$ for fixed N_1, (ii) $\lim_{N_1\to\infty}\phi(s)$ for fixed N_0, and (iii) $\lim_{N\to\infty}\phi(s)$. In so doing we develop the limiting distribution for large $N = (N_0 + N_1)$ as well as appropriate expansions allowing assessment of the error involved in using the asymptotic distribution for $-2\ln\text{LR}$. The following theorem gives the expansions for ,the mgf of $-2\ln\text{LR}$ and the correction for the distribution.

Theorem 3. For the mgf of $-2\ln\text{LR}$ given in Theorem 2 we have

(i) For fixed N_1, large N_0

$$\lim_{N_0\to\infty} \phi(s) = \phi_0(s)$$

$$= N_1^{2sN_1}(1-2s)^{-N_1(1-2s)}e^{-2sN_1}\Gamma(N_1(1-2s))/\Gamma(N_1) \quad (22)$$

$$= \phi_0(s) = (1-2s)^{-1/2}\left[1 + \frac{1}{6N_1}\left[\frac{s}{1-2s}\right]\right] + O(N_1^{-2})$$

and

$$P\left(\lim_{N_1\to\infty} -2\ln LR > x\right) = P(\chi^2_{(1)} > x) + \frac{1}{6N_1}\mathrm{pdf}(q=\chi^2_{(3)})|_{q=x} \quad (23)$$

(ii) For fixed N_0, large N_1

$$\lim_{N_1\to\infty} \phi(s) = \phi_1(s)$$

$$= N_0^{2sN_0}(1-2s)^{-N_0(1-2s)}e^{-2sN_0}\Gamma(N_0(1-2s))/\Gamma(N_0) \quad (24)$$

$$= \phi_1(s) = (1-2s)^{-1/2}\left[1 + \frac{1}{6N_0}\left[\frac{s}{1-2s}\right]\right] + O(N_0^{-2})$$

and

$$P\left(\lim_{N_1\to\infty} -2\ln LR > x\right) = P(\chi^2_{(1)} > x) + \frac{1}{6N_0}\mathrm{pdf}(q=\chi^2_{(3)})|_{q=x} \quad (25)$$

(iii) For large N

$$\phi(s) = (1-2s)^{-1/2}\left[1 + \frac{1}{6}\left[\frac{1}{N_0} + \frac{1}{N_1} - \frac{1}{N_0+N_1}\right]\left[\frac{s}{1-2s}\right]\right] + O(N^{-2}) \quad (26)$$

and

$$P\left(\lim_{N\to\infty} -2\ln LR > x\right)$$

$$= P(\chi^2_{(1)} > x) + \frac{1}{6}\left[\frac{1}{N_0} + \frac{1}{N_1} - \frac{1}{N_0+N_1}\right]\mathrm{pdf}(1=\chi^2_{(3)})|_{q=x} \quad (27)$$

Proof. The results for $\phi_0(s)$, $\phi_1(s)$, and $\phi(s)$ are readily found by applying Sterling's approximation to the iGamma functions in (21). The approximation to the distribution of $-2\ln LR$ are found by merely inverting the associated characteristic functions.

Using (23), (25), and (27), we can readily compare the tail area of the three distributions defined by (22), (24), and (26) with that of a $\chi^2_{(1)}$. In each case we see that the true size is always greater than the nominal size. Table 1

Table 1. Approximate sizes for LR test[*]

		Nominal sizes of								
		0.050			0.025			0.010		
N_0	N_1	LG N	LG N_0	LG N_1	LG N	LG N_0	LG N_1	LG N	LG N_0	LG N_1
$N = 10$										
1	9	0.0693	0.0521	0.0691	0.0372	0.0263	0.0371	0.0163	0.0107	0.0162
2	8	0.0600	0.0524	0.0595	0.0313	0.0265	0.0310	0.0133	0.0108	0.0131
3	7	0.0572	0.0527	0.0564	0.0295	0.0267	0.0290	0.0123	0.0109	0.0121
4	6	0.0560	0.0532	0.0548	0.0288	0.0270	0.0280	0.0120	0.0110	0.0116
5	5	0.0557	0.0538	0.0538	0.0286	0.0274	0.0274	0.0119	0.0112	0.0112
6	4	0.0560	0.0548	0.0532	0.0288	0.0280	0.0270	0.0120	0.0116	0.0110
7	3	0.0572	0.0564	0.0527	0.0295	0.0290	0.0267	0.0123	0.0121	0.0109
8	2	0.0600	0.0595	0.0524	0.0313	0.0310	0.0265	0.0133	0.0131	0.0108
$N = 50$										
1	49	0.0691	0.0504	0.0691	0.0371	0.0252	0.0371	0.0162	0.0101	0.0162
5	45	0.0539	0.0504	0.0538	0.0274	0.0253	0.0274	0.0113	0.0101	0.0112
10	40	0.0520	0.0505	0.0519	0.0263	0.0253	0.0262	0.0107	0.0102	0.0106
15	35	0.0514	0.0505	0.0513	0.0259	0.0253	0.0258	0.0105	0.0102	0.0104
20	30	0.0512	0.0506	0.0510	0.0258	0.0254	0.0256	0.0104	0.0102	0.0103
25	25	0.0511	0.0508	0.0508	0.0257	0.0255	0.0255	0.0104	0.0102	0.0102
30	20	0.0512	0.0510	0.0506	0.0258	0.0256	0.0254	0.0104	0.0103	0.0102
35	15	0.0514	0.0513	0.0505	0.0259	0.0258	0.0253	0.0105	0.0104	0.0102
40	10	0.0520	0.0519	0.0505	0.0263	0.0262	0.0253	0.0107	0.0106	0.0102
45	5	0.0539	0.0538	0.0504	0.0274	0.0274	0.0253	0.0113	0.0112	0.0101
49	1	0.0691	0.0691	0.0504	0.0371	0.0371	0.0252	0.0162	0.0162	0.0101
$N = 100$										
1	99	0.0691	0.0502	0.0691	0.0371	0.0251	0.0371	0.0162	0.0101	0.0162
2	98	0.0595	0.0502	0.0595	0.0310	0.0251	0.0310	0.0131	0.0101	0.0131
5	95	0.0538	0.0502	0.0538	0.0274	0.0251	0.0274	0.0112	0.0101	0.0112
10	90	0.0519	0.0502	0.0519	0.0262	0.0251	0.0262	0.0106	0.0101	0.0106
20	80	0.0510	0.0502	0.0510	0.0256	0.0252	0.0256	0.0103	0.0101	0.0103
50	50	0.0506	0.0504	0.0504	0.0254	0.0252	0.0252	0.0102	0.0101	0.0101
80	20	0.0510	0.0510	0.0502	0.0256	0.0256	0.0252	0.0103	0.0103	0.0101
90	10	0.0519	0.0519	0.0502	0.0262	0.0262	0.0251	0.0106	0.0106	0.0101
95	5	0.0538	0.0538	0.0502	0.0274	0.0274	0.0251	0.0112	0.0112	0.0101
98	2	0.0595	0.0595	0.0502	0.0310	0.0310	0.0251	0.0131	0.0131	0.0101
99	1	0.0691	0.0691	0.0502	0.0371	0.0371	0.0251	0.0162	0.0162	0.0101

[*]LG N, LG N_0, and LG N_1 refer to the approximations from using equations (32), (30), and (31) respectively. Note that the value given by (32) is very close to the exact probability.

displays the computed sizes for various N and different combinations of N_0 and N_1. The actual sizes of the tests are always larger than the nominal, associated with the asymptotic $\chi^2_{(1)}$. The difference is very pronounced for small N (e.g., $N = 10$) and even for large N when either N_0 or N_1 is less than 10% of the value of N. Consequently, in these situations, which are not uncommon, inference using asymptotic results may lead to incorrect conclusions.

4. EFFICIENCY COMPARISONS AND THE CONSTRUCTION OF MINIMUM VARIANCE UNBIASED ESTIMATORS

Returning to the properties of the estimators, we shall investigate the exact variances. From (2) and (6), this reduces to the problem of comparing the variances of W_1 and $\ln \tilde{T}_1$. In what follows we show that equations (12) and (13) can be compared for all N. Employing the Rao–Blackwell theorem, we show that MLE always has a smaller variance than LLS for $N \geq 2$ (for $N = 1$, they are equal). Since we can adjust the bias of MLE using (13), we can construct an unbiased estimator based on the MLE which always has a smaller variance than LLS.

Explicitly, our bias-adjusted estimators are

$$\tilde{\tilde{\alpha}} = \tilde{\alpha} - \ln N_0 - \sum_{j=1}^{N_0-1} \frac{1}{j} + c$$

(where c is Euler's constant) and

$$\tilde{\tilde{\beta}} = \tilde{\beta} - \ln N_1 + \ln N_0 + \sum_{j=1}^{N_1-1} \frac{1}{j} - \sum_{j=1}^{N_0-1} \frac{1}{j}$$

Since our model is a member of the exponential family, we know that $N_1 \tilde{T}_1$ and $N_0 \tilde{T}_0$ are jointly sufficient for $\exp(\alpha + \beta)$ and $\exp(\alpha)$. In fact, since

$$L = \exp(N_1(\alpha + \beta)) \exp(-N_1 \tilde{T}_1 \exp(\alpha + \beta)) \exp(N_0\alpha) \exp(-N_0 \tilde{T}_0 \exp(\alpha))$$

this follows directly from the factorization criterion for sufficiency (see Lehmann 1983, Theorem 5.2, p. 39).

Now the Rao–Blackwell theorem (Lehmann 1983, p. 291) states that if L is convex and a complete sufficient statistic, T, exists, then any U-estimable g has a unique unbiased estimator depending only on T and this estimator uniformly minimizes the risk among all unbiased estimators. In our context, a U-estimable g means that there exist estimators that are unbiased estimators of α and β; a proof of completeness follows the argument in Lehmann (1983, Example 5.14, p. 47; see also Barndorff-Nielsen 1978, Lemma 8.2). Now, if we take our MLEs $\tilde{\alpha}$ and $\tilde{\beta}$ and subtract $E(\tilde{\alpha} - \alpha_0)$ and $E(\tilde{\beta} - \beta_0)$ (both known quantities), the resulting estimators depend only on the complete sufficient statistics. Thus our bias-corrected maximum likelihood estimators are the unique minimum variance unbiased estimators (MVUE) of α_0 and β_0.

5. EXTENSIONS TO A NON-DUMMY EXOGENOUS VARIABLE

To see whether the results of the previous section can be utilized for a more general set of values of the Xs other than 0 or 1, we consider now that X_j, $j = 1, \ldots, N$, takes values $0, 1, 2, \ldots, r$ with frequencies $N_0, N_1, \ldots, N_r$ so that $\sum_{j=0}^{r} N_j = N$; we shall concentrate on estimating β. Our model now has likelihood given by

$$
L = \prod_{j=1}^{N} \exp(\alpha + X_j\beta) \exp(- \exp(\alpha + X_j\beta)t_j)
$$

$$
= \exp(N\alpha + (N_1 + 2N_2 + \cdots + rN_r)\beta) \exp\left[- \sum_{j=0}^{r} \exp(\alpha + j\beta)T_j \right] \quad (28)
$$

where $T_j = \sum_{s_j} t_j$ with $s_j = \{l; X_l = j\}$. We note that L is still a member of the linear exponential family but that the fundamental parameters $\delta_j = \exp(\alpha + j\beta)$ are connected by the following $(r - 1)$ nonlinear restrictions

$$
\delta_j = \delta_1^j / \delta_0^{j-1}, \quad j = 2, \ldots, r \quad (29)
$$

Combining these restrictions with (28) gives us a member of the curved exponential family (see Barndorff-Nielsen 1978).

An alternative way of viewing this model is to express X_j as a linear combination of 0–1 dummy variables; i.e.,

$$
X_j = Z_{1j} + 2Z_{2j} + 3Z_{3j} + \cdots + rZ_{rj}
$$

where $X_j = \sum_{k=1}^{r} kZ_{kj}$, with $Z_{kj} = 1, 0$ with frequency N_k and $N - N_k$ respectively. Then, defining

$$
\lambda_j = \exp\left(\alpha + \sum_{k=1}^{r} Z_{kj}\gamma_k \right)
$$

imposes $(r-1)$ linear restrictions on the γ_ks, viz.

$$\gamma_1 = \gamma_k/k, \qquad k = 2, 3, \ldots, r$$

The MLE of β is now just the restricted MLE of γ_1 in the above model subject to the $(r-1)$ restrictions

$$R\gamma = 0$$

where R is the $(r-1) \times r$ matrix given by

$$R = (e_{(r-1)} : -W)$$

with $e_{(r-1)}$ an $(r-1)$ vector of 1's and $W = \text{diag}\,(w_2, w_3 \ldots w_r)$, where $w_k = 1/k$, and γ is an $(r \times 1)$ vector with γ_i in the ith position.

Explicitly, we have the Lagrangian associated with the restricted MLE given by

$$H = \ln L(\alpha, \gamma) - \lambda' R\gamma$$

$$= N\alpha + \sum_{k=1}^{r} \gamma_k N_k - e^{\alpha} T_0 - \sum_{k=1}^{r} e^{\alpha + \gamma_k} T_k - \lambda' R\gamma$$

The appropriate first-order conditions are

$$\frac{\partial H}{\partial \alpha} = N - e^{\alpha} T_0 - \sum_{k=1}^{r} e^{\alpha + \gamma_k} T_k = 0$$

$$\frac{\partial H}{\partial \gamma} = \frac{\partial \ln L(\alpha, \gamma)}{\partial \gamma} - R'\lambda = 0$$

$$\frac{\partial H}{\partial \lambda} = R\gamma = 0$$

Defining $Q' = (1, w_2^{-1}, w_3^{-1}, \ldots, w_r^{-1})$, we note immediately that since $Q'R' = 0$, $Q'(\partial \ln L(\alpha, \gamma)/\partial \gamma) = 0$; i.e.,

$$\sum_{j=1}^{r} jN_j - \sum_{j=1}^{r} e^{\alpha + \gamma_k} jT_j = 0$$

Setting $\gamma_j = j\gamma_1$, these are the equations formed from $\partial \ln L(\alpha, \beta)/\partial \beta = 0$.

The unrestricted estimators of the γ_js are those satisfying

$$\frac{\partial \ln L(\alpha, \gamma)}{\partial \gamma} = 0$$

and are given by

$$\hat{\gamma}_j = \ln(T_0/N_0) - \ln(T_j/N_j); \qquad j = 1, 2, \ldots, r$$

In essence, with the restrictions on the γ_j's there is only one γ_j to be estimated, viz. γ_1; however, we generate r estimates of γ_1, that is, $\hat{\gamma}_1$, $\hat{\gamma}_2/2$, $\hat{\gamma}_3/3, \ldots, \hat{\gamma}_r/r$. It may be possible to establish a complete sufficiency argument in this situation. We shall, however, proceed along different lines by exploiting the independence of the $T_j, j = 0, \ldots, r$.

Appealing to ideas from restricted least squares, let us find a γ to minimize $(\hat{\gamma} - \gamma)'\Omega^{-1}(\hat{\gamma} - \gamma)$ subject to $R\gamma = 0$, where $\Omega = \text{var} - \text{cov}(\hat{\gamma})$. This new estimator is given by

$$\tilde{\gamma} = \hat{\gamma} - \Omega R'(R\Omega R')^{-1} R\hat{\gamma}$$

and in particular the estimator of γ_1 is

$$\tilde{\gamma}_1 = \hat{\gamma}_1 - \sum_{k=2}^{r} q_{(k-1)}(\hat{\gamma}_1 - \hat{\gamma}_k/k)$$

$$= 1 - \sum_{k=2}^{r} q_{(k-1)}\hat{\gamma}_1 + \sum_{k=2}^{r} q_{(k-1)}\hat{\gamma}_k/k$$

$$= \sum_{k=1}^{r} g_k\hat{\gamma}_k/k$$

where $g_1 = (1 - \sum_{k=2}^{r} q_{(k-1)})$, $g_k = g_{(k-1)}$, $k = 2, \ldots, r$, and $(q_1, q_2, \ldots, q_{r-1})$ are the elements in $e_1'\Omega R'(R\Omega R)^{-1}$ for $e_1' = (1, 0, 0, \ldots, 0)$. We note in passing that $\sum_{k=1}^{r} g_k = 1$.

Identical results can be obtained by defining a new estimator

$$\bar{\gamma}_1 = \lambda'(W_1\hat{\gamma}) = \sum_{k=1}^{r} \lambda_k\hat{\gamma}_k/k; \qquad W_1 = \text{diag}\left[1, \frac{1}{2}, \frac{1}{3}, \ldots, \frac{1}{r}\right]$$

and finding λ to minimize the var–cov of $\bar{\gamma}$, i.e. $\lambda'W_1\Omega W_1'\lambda$ subject to $\lambda'e = 1$ with e a vector of ones. Letting $\sum = W_1\Omega W_1'$, the appropriate λ is given by

$$\tilde{\lambda} = \sum^{-1} e/e' \sum^{-1} e$$

Although not immediately obvious, $\bar{\gamma}_1 = \tilde{\gamma}_1$.

Since the estimators $\hat{\gamma}_j, j = 1, \ldots, r$, are of a form similar to the estimator $\hat{\beta}$ considered in Section 2, we can readily find their means, variances, and covariances. In particular we have, for any j, that $\hat{\gamma}_j$ are given by identical equations to (6) if we transform to $NE(1)$s, i.e.

$$(\hat{\gamma}_j - \gamma_j) = \ln(T_0/N_0) - \ln(T_j/N_j)$$

and from (13) we have

$$E(\hat{\gamma}_j - \gamma_j) = \tilde{\kappa}_{1j}$$

$$\mathrm{var}(\hat{\gamma}_j - \gamma_j) = \tilde{\kappa}_{2j}$$

where $\tilde{\kappa}_{nj}$ are found from (13) by replacing N_1 by N_j. The covariance of any two $\hat{\gamma}_j$s is given by

$$\mathrm{cov}(\hat{\gamma}_j - \gamma_j, \hat{\gamma}_k - \gamma_k) = \mathrm{cov}(\ln T_0/N_0, \ln T_0/N_0)$$
$$= \mathrm{var}(\ln T_0/N_0)$$
$$= \psi^{(1)}(N_0)$$

The covariance matrix Ω of the $\hat{\gamma}_j$'s is thus

$$\Omega = S + \sigma^2 ee'$$

where $S = \mathrm{diag}(s_1, s_2 \ldots, s_r)$ an $r \times r$ diagonal matrix with $s_j = \psi^{(1)}(N_j)$ and $\sigma^2 = \psi^{(1)}(N_0)$. Consequently, Σ is given by

$$\Sigma = W_1 \Omega W_1' = W_1 S W_1' + \sigma^2 W_1 ee' W_1'$$
$$= V + \sigma^2 M$$

with $V = \mathrm{diag}(v_1, v_2, \ldots v_r)$, an $r \times r$ diagonal matrix with $v_j = s_j/j^2$. Noting that $M = ww'$ where $w' = [1, 1/2, \ldots, 1/r] = e' W_1'$, and that

$$\Sigma^{-1} = V^{-1} - \frac{\alpha^2 V^{-1} ww' V^{-1}}{1 + \sigma^2 w' V^{-1} w}$$

we can now form an unbiased estimator of γ_1 and consequently β by considering

$$\bar{\beta} = \tilde{\lambda}' W_1(\hat{\gamma} - E(\hat{\gamma}))$$

$$= \sum_{k=1}^{r} \tilde{\lambda}_k (\hat{\gamma}_k - \tilde{\kappa}_{1k})/k$$

This estimator will have minimum variance amongst all estimators which are linear combinations of the $\hat{\gamma}_j/j, j = 1, 2, \ldots, r$.

Now

$$w' V^{-1} w = \sum_{j=1}^{r} 1/\psi^{(1)}(N_j)$$

$$e' V^{-1} e = \sum_{j=1}^{r} j^2/\psi^{(1)}(N_j)$$

and

$$e'V^{-1}w = \sum_{j=1}^{r} j/\psi^{(1)}(N_j)$$

so that

$$e'\Sigma^{-1}e = \sum_{j=1}^{r} j^2/\psi^{(1)}(N_j) - \frac{\sigma^2 \left[\displaystyle\sum_{j=1}^{r} j/\psi^{(1)}(N_j)\right]^2}{1 + \sigma^2 \displaystyle\sum_{j=1}^{r} 1/\psi^{(1)}(N_j)} \tag{30}$$

and

$$(\Sigma^{-1}e)_k = \frac{k^2}{\psi^{(1)}(N_k)} \left[1 - \frac{(\sigma^2/k) \displaystyle\sum_{j=1}^{r} j/\psi^{(1)}(N_j)}{1 + \sigma^2 \displaystyle\sum_{j=1}^{r} 1/\psi^{(1)}(N_j)} \right] \tag{31}$$

Thus λ_k is given by the ratio of (31) to (30).

The above procedure seems computationally feasible, and, if $N_j \geq 1$, from (10) we see that all positive integral moments exist; this also follows from the properties of Fisher's z distribution. Therefore our procedure could be applied to the case where the exogenous variable takes distinct values.

Nothing in the above analysis hinges upon X_j being a positive integer; we could work with values $\theta_1, \ldots, \theta_r$. If we redefine $w'_{\sim\theta} = (1/\theta_1, \ldots, 1/\theta_r)$, then similar equations follow with our new definition of $w_{\sim\theta}$. In particular we now find that

$$(\Sigma^{-1}e)_k = \frac{\theta_k^2}{\psi^{(1)}(N_k)} \left[1 - \frac{(\sigma^2/\theta_k) \displaystyle\sum_{j=1}^{r} \theta_j/\psi^{(1)}(N_j)}{1 + \sigma^2 \displaystyle\sum_{j=1}^{r} 1/\psi^{(1)}(N_j)} \right] \tag{32}$$

and

$$e'\Sigma^{-1}e = \sum_{j=1}^{r} \theta_j^2/\psi^{(1)}(N_j) - \frac{\sigma^2 \left[\displaystyle\sum_{j=1}^{r} \theta_j/\psi^{(1)}(N_j)\right]^2}{1 + \sigma^2 \displaystyle\sum_{j=1}^{r} 1/\psi^{(1)}(N_j)} \tag{33}$$

λ_k is now defined as the ratio of (32) to (33).

Finally, note that the variance of the new estimator defined by $\bar{\beta}$ is equal to $(e'\Sigma^{-1}e)^{-1}$ and is given by the inverse of (30) or (33).

It is interesting to note that when $r = 2$ with $X_j = 0$, 1, or 2 and $N_0 = N_2$, the MLE can be solved explicitly. In this case the first-order conditions result in

$$N - e^\alpha T_0 - e^{\alpha+\gamma_1} T_1 - e^{\alpha+\gamma_2} T_2 = 0$$

$$N_1 - e^{\alpha+\gamma_1} T_1 - \lambda = 0$$

$$N_2 - e^{\alpha+\gamma_2} T_2 + \frac{\lambda}{2} = 0$$

and, since $N = N_1 + 2N_2$ when $N_0 = N_2$,

$$e^\alpha T_0 = e^{\alpha+\gamma_2} T_2$$

giving

$$\tilde{\gamma}_2 = \ln(T_0/T_2)$$

Therefore the estimator of β_0, viz. $\tilde{\gamma}_1$, is given by

$$\tilde{\gamma}_1 = \tfrac{1}{2}\tilde{\gamma}_2 = \tfrac{1}{2}\ln(T_0/T_2)$$

Also, from (30) and (31), we have $\lambda_1 = 0$ and $\lambda_2 = 1$, giving

$$\bar{\gamma}_1 = \tfrac{1}{2}\hat{\gamma}_2$$
$$= \tfrac{1}{2}\ln(T_0/T_2), \quad \text{from (38)}$$

It would appear that this is the only situation where $\tilde{\gamma}_1 = \bar{\gamma}_1$.

We know that the smallest variance that any unbiased estimator can attain will be given by the Cramer–Rao lower bound (CRLB). We finish this section by comparing the variance of our unbiased estimator with that of the CRLB.

We should note in passing that the previous bias-adjusted MLE did not achieve the CRLB; see (13) where $\tilde{\kappa}_2 = 2\psi^{(1)}(1) - \left[\sum_{j=1}^{N_0} j^{-2} + \sum_{j=1}^{N_1-1} j^{-2} \right]$ and the CRLB is $[1/N_0 + 1/N_1]$. However, as N_0 and N_1 tend to infinity, the CRLB is reached in the limit. This result, that the unique minimum variance unbiased estimator does not achieve the CRLB in a situation where there is a two-dimensional family of complete sufficient statistics, may seem surprising, but this agrees with existing results on this topic; see Wijsman (1973) and Lehmann (1983, Problem 6.14, p. 142).

For the model in (28), the CRLB for an unbiased estimator of β is given by:

$$\text{CRLB} = N\left[N\sum_{j=1}^{r} j^2 N_j - \left(\sum jN_j\right)^2 \right]^{-1}$$

This should be compared with the inverse of (30). We present our numerical comparison below.

Table 2 gives the values of the variance and the associated CRLB for $\bar{\beta}$ for various sample sizes and different combinations of N_0, N_1, and N_2. We notice immediately that both the variance, given by the inverse of (30), and

Table 2. Variance and C–R lower bound for the unbiased estimator

N	N_0	N_1	N_2	VAR-B	CRLB-B
30	15	9	6	0.5818E − 01	0.5464E − 01
50	25	15	10	0.3404E − 01	0.3279E − 01
100	50	30	20	0.1670E − 01	0.1639E − 01
200	100	60	40	0.8273E − 02	0.8197E − 02
500	250	150	100	0.3291E − 02	0.3279E − 02
1000	500	300	200	0.1642E − 02	0.1639E − 02
30	6	9	15	0.5818E − 01	0.5464E − 01
50	10	15	25	0.3404E − 01	0.3279E − 01
100	20	30	50	0.1670E − 01	0.1639E − 01
200	40	60	100	0.8273E − 02	0.8197E − 02
500	150	250	250	0.3291E − 02	0.3279E − 02
1000	200	300	500	0.1642E − 02	0.1639E − 02
30	9	15	6	0.7294E − 01	0.6803E − 01
50	15	25	10	0.4255E − 01	0.4082E − 01
100	30	50	20	0.2084E − 01	0.2041E − 01
200	60	100	40	0.1031E − 01	0.1020E − 01
500	150	250	100	0.4099E − 02	0.4082E − 02
1000	300	500	200	0.2045E − 02	0.2041E − 02
30	9	6	15	0.4597E − 01	0.4386E − 01
50	15	10	25	0.2706E − 01	0.2632E − 01
100	30	20	50	0.1334E − 01	0.1316E − 01
200	60	40	100	0.6625E − 02	0.6579E − 02
500	150	100	250	0.2639E − 02	0.2632E − 02
1000	300	200	500	0.1318E − 02	0.1316E − 02

the CRLB are symmetric functions of N_0 and N_2. This can also be proved analytically by noting that

$$e'\Sigma^{-1}e = \frac{\psi^{(1)}(N_0) + 4\psi^{(1)}(N_1) + \psi^{(1)}(N_2)}{\psi^{(1)}(N_1)\psi^{(1)}(N_2) + \psi^{(1)}(N_0)\psi^{(1)}(N_1) + \psi^{(1)}(N_0)\psi^{(1)}(N_2)}$$

and

$$(CRLB)^{-1} = \frac{N_1(N_0 + N_2) + 4N_0N_2}{N_0 + N_1 + N_2} = \frac{N_0(N_1 + 2N_2) + N_2(N_1 + 2N_0)}{N_0 + N_1 + N_2}$$

If we now consider values of the ratio

$$R(N_0, N_1, N_2) = CRLB/var(\bar{\beta})$$

we have

$$R(N_0, N_1, N_2) = \frac{(N_0 + N_1 + N_2)}{(N_0(N_1 + 2N_2) + N_2(N_1 + 2N_0))}$$

$$\times \frac{(\psi^{(1)}(N_0) + 4\psi^{(1)}(N_1) + \psi^{(1)}(N_2))}{(\psi^{(1)}(N_0)(\psi^{(1)}(N_1) + \psi^{(1)}(N_2)) + \psi^{(1)}(N_1)\psi^{(1)}(N_2))}$$

By evaluating $R()$ for fixed N_0 with N_1 and N_2 varying and for fixed N_1 with N_0 and N_2 varying, we found that even for quite large sample sizes, e.g., 1002 or 1003, if either N_0, N_1 or N_2 is very small the exact variance is considerably greater than CRLB. Nor does $R()$ increase toward one monotonically as the sample size increases.

When $N_0 = N_2$ we have

$$R(N_0, N_1, N_2) = (N_2\psi^{(1)}(N_2))^{-1}$$

indicating that sample size increases brought about by increasing N_1 have no effect on efficiency. Since in this case the unbiased estimator equals the MLE, the inverse of $e'\Sigma^{-1}e$, viz., $\psi^{(1)}(N_2)/2$, will give the lower bound on the variance for a fixed $N_2(= N_0)$.

Quite accurate approximations to $R()$ could be developed by using the asymptotic expansion of $\psi^{(1)}(N_j)$, $j = 0, 1, 2$ (see Abramowitz and Stegum 1972, equation (6.4.12), p. 260), i.e., $\psi^{(1)}(N_j) \approx 1/N_j + 1/(2N_j^2) + 1/(6N_j^3) - 1/(30N_j^5) + O(N_j^{-7})$. For the case of $N_0 = N_2$ this results in $R(N_0, N_1, N_2) \approx 1 - 1/(2N_2) + 1/(12N_2^2) + O(N_2^{-3})$.

6. CONCLUSION

This paper has examined some efficiency considerations associated with the negative exponential regression model with a constant and one (dummy) regressor. We first showed that in finite samples MLE is more efficient, having smaller variance, than LLSE. This dominance remains asymptotically. Secondly, since the exact bias for MLE can be calculated, the bias-adjusted estimator is shown to be the unique MVUE. Thirdly, the asymptotic distributions associated with suitably standardized $(\hat{\beta} - \beta_0)$ and the likelihood ratio test statistic may not be suitable approximations to the exact distribution when the dummy variable has a small number of zeros or ones. This leads us to conjecture that even in models with larger numbers of regressors, if one of them is a dummy with a small number of zeros or ones in relation to sample size, the use of asymptotic results may lead to incorrect inference. Whilst many of the results discussed are not new, we have provided a unified treatment appropriate to the regression model under consideration.

Finally, we extended our model to allow the regressor to take on $r + 1$ distinct values, say $0, 1, 2, \ldots, r$, with frequencies $N_0, N_1, \ldots, N_r$. We develop a minimum variance unbiased estimator for this case and show that when $r = 2$ with $N_0 = N_2$ this estimator equals the MLE. Further we show that the exact variance and the Cramer–Rao lower bound are symmetric functions of N_0 and N_2 and that the efficiency measure defined as the ratio CRLB/variance is not a monotonic function in N_1 and N_2 for fixed N_0. Consequently, for some configurations of N_0, N_1, and N_2, efficiency cannot be improved by increasing the sample size if this results only in increasing one of the N_j $(j = 0, 1, 2)$. As previously mentioned, in the case of limited distinct values for the regressor care is needed in the application of standard asymptotic results. We have not fully investigated the question of the relative efficiency of our estimator versus the MLE versus the CRLB, instead, for computational reasons, we concentrated on the efficiency of the first compared with the last. Further work on this topic needs to be done.

ACKNOWLEDGMENTS

We wish to thank an anonymous referee for helpful comments on an earlier draft. The first author acknowledges financial support from the Natural Sciences and Engineering Research Council of Canada.

7. REFERENCES

Abramowitz, M., and I. Stegum, 1972, *Handbook of Mathematical Functions*, 9th edn. Dover Publications.

Amemiya, T., 1985, *Advanced Econometrics*. Blackwell.

Barndorff-Nielson, O., 1978, *Information and Exponential Families in Statistical Theory*. Wiley.

Johnston, N. L., and S. Kotz, 1970, *Continuous Univariate Distributions*, Vols. 1 and 2. Houghton Mifflin.

Kalbfleisch, J. D., and R. L. Prentice, 1980, *The Statistical Analysis of Failure Time Data*. Wiley.

——, 1993, Exact critical regions and confidence intervals for MLE in exponential regression model. *Economics Letters* 41, 225–229.

Knight, J. L., and S. E. Satchell, 1996, The exact distribution of the maximum likelihood estimator for the linear regression negative exponential model. *Journal of Statistical Planning and Inference* 50, 91–102.

Lehmann, E. L., 1983, *Theory of Point Estimation*. Wiley.

Stuart, A., and J. K. Ord, 1987, *Kendall's Advanced Theory of Statistics*, Vol. 1. Charles Griffin.

Wijsman, R. A., 1973, On the attainment of the Cramer–Rao lower bound. *Annals of Statistics*, 1, 538–542.

25

On Specifying Double-Hurdle Models

MURRAY D. SMITH   University of Sydney, Sydney, Australia

1. INTRODUCTION

The double-hurdle model (DHM hereafter) has been used in microeconometrics to analyse a wide range of individual and household commodity demand. Indeed, a search of "EconLit" (the American Economic Association's electronic bibliography of economic literature) reveals 30 citations for "double hurdle" dating back to 1989, although the DHM literature dates initially from Cragg's seminal paper [6] published in 1971. Important contributions to the DHM literature include Jones [16], in which the demand for cigarettes is modeled, and Blundell and Meghir [3], which was concerned with the labour supply of married women. Other fields in which the DHM has been applied include finance (e.g., Dionne et al. [9] examine credit-scoring) and sociology (e.g., Zorn [35] examines legislative response to court rulings). An interesting discussion on inferential uses of the DHM in the context of recreational activities appears in Shonkwiler and Shaw [28]. The DHM has been applied to infrequency-of-purchase contexts; e.g., Deaton and Irish [7] report on p-Tobit models of household consumption of tobacco, alcohol, and durable goods—the p-Tobit model is a special case of the DHM, see Maki and Nishiyama [24]. However, the majority of commodity demand applications of the DHM cater for situations in which individual preferences generate zero observations due to corner solutions as well as abstentions from consumption. Generally, the data used in most applications have been cross-sectional on individuals or households; however, recently Labeaga [19] has applied the DHM to estimate reduced form parameters in a structural model for tobacco demand using panel data.

The DHM is designed to explain the mechanism of individual demand whereby an individual's decision process is decomposed into separate components: (i) a market participation decision (whether to buy or not); and (ii) a consumption level decision (how much to buy). Motivating this decomposition is the desire to allow different factors to influence demand; for example, psychological influences may play a prominent role in determining participation, whereas economic considerations are more likely to be important in determining consumption (see DeSarbo and Choi [8] for related discussion). Pudney [26, pp. 160–162] gives a basis for the DHM in consumer choice theory.

The DHM is constructed by assuming the existence of a pair of latent variables designed to represent the utility an individual derives from market participation and the utility the individual derives from consumption. These random variables are then linked to consumer expenditure, the latter being observable. The procedure is this: the utility variables are transformed to a pair of hurdle decision variables (participation decision and consumption decision), then the hurdle variables are linked through a further transformation to consumer expenditure. Constructions like this are not unusual when specifying models of consumer demand.

Consumer expenditure, the observed random variable, is non-negative valued, with the data it generates having the distinctive feature of an excessive number of zeroes; what may be termed zero-inflated data. Accordingly, expenditure would appear distributed according to some member of the family of discrete-continuous mixture distributions, with a single point mass at the origin. While there are many members of this family of distributions, only the DHM will be considered here. Intuitively, the DHM can represent zero-inflated data for it allows for two sources of zeroes: either an individual can elect not to participate in the market, and/or an individual chooses not to consume. It is assumed that neither source of zeroes is separably identifiable in the data, and that the data is drawn from a general population. If further information was available concerning either of these, it may be appropriate to use another model; e.g., a self-selection Heckman-type model arises when preferences are such that participation dominates consumption.

In the DHM literature, it is typical to find the statistical form of the model derived by specifying a reduced-form bivariate model for the utility variables—the specification being linear in its parameters with zero mean bivariate normal disturbances (usually termed Cragg's model). There is, however, no reason why attention should be confined to normal linear models, so in Section 2 of this survey there appears an alternate derivation of the model unencumbered in this respect.

An ongoing concern in the DHM literature is whether or not the two hurdle decisions—participation and consumption—are jointly taken, or else tackled in sequence. Indeed, a number of articles in the DHM literature commit neither way to this debate, simply opting instead to assert that the participation decision and the consumption decision are "separate"; separateness is the consequence of assuming the decomposition of demand into hurdle decisions. Of course, individual behavior is not governed by the DHM; like all statistical models, the DHM is merely attempting to act as a reasonable approximation to reality. Mindful of this, Section 3 inspects the statistical nature of the DHM to show that the model implies the hurdle decisions are jointly taken. The DHM cannot accommodate sequential hurdle decisions, for it is incapable of identifying any particular sequence in which the hurdle decisions are taken.

In Section 4, a number of specifications of DHM that have been applied in practice are examined. Most popular here is Cragg's model, as it is based on specifying bivariate normality for the underlying utility variables. Extensions to Cragg's model are then considered. These include transformed DHM, and non-normal DHM. The former induces aspects of non-normality by suitably transforming normal-based models—the classic method here is to apply the ubiquitous Box–Cox transformation. The second extension—to non-normal DHM—has yet to be be applied in practice. This approach is based on copula theory (see [15] and [25]), which enables construction of multivariate distributions when there is knowledge about only the marginal distributions. This method is adapted to yield non-normal DHM specifications. The copula method is also shown to be useful if model building for the hurdle decision variables is undertaken. Lastly, it is demonstrated how special cases of the DHM result under certain restrictions; these cases include the Tobit model and Heckman's sample selection model.

2. STATISTICAL CONSTRUCTION

Denote individual consumption expenditure by Y, and assume that this random variable is defined on the non-negative portion of the real line; i.e., $Y = y \geq 0$ (assuming a continuous support is convenient, not essential).

The probability density function (pdf) of Y is a discrete-continuous mixture, taking the form of

$$f(y) = \begin{cases} f_+(y) & \text{if } y > 0 \\ f_0 & \text{if } y = 0 \end{cases} \tag{1}$$

The discrete component, f_0, is a probability mass measured at the origin, while the support for the continuous component, $f_+(y)$, is the positive part of the real line. The rationale for a mixture distribution is empirical: in surveys of consumer expenditure there often arises an excessive number of zero-valued observations.

The DHM is a member of the class of statistical models which lead to mixture densities of the form of (1). Other prominent members include censoring models, and zero-inflated component-mix models, such as the zero-inflated Poisson for discrete data (for recent examples see Lambert [20] and Böhning et al. [4]) and Aalen's [1] compound Poisson distribution. At first sight, a censoring model would seem inappropriate for expenditure data for the simple reason that consumer expenditure cannot be left censored at zero. However, as with many econometric models, its applicability is due to assumptions which underpin the economic process of interest. For example, in the case of a censoring model like the Tobit, it is not actual consumer expenditure which is modeled; rather, it is the utility of consumption, a latent variable, which is the object of the modeling exercise, with utility and expenditure linked according to some assumed transformation. In the case of the DHM, its statistical specification is also predicated on modeling latent utility variables.

The economic framework underpinning the DHM begins with a pair (Y_1^{**}, Y_2^{**}) of latent utility random variables: Y_1^{**} represents the utility derived by an individual from participation in the market, and Y_2^{**} represents the utility derived by an individual from consumption. Assume that these variables are continuous and real-valued. Next, assume a parametric bivariate model for (Y_1^{**}, Y_2^{**}) by assigning a joint cumulative distribution function (cdf)

$$F(y_1^{**}, y_2^{**}) \tag{2}$$

for real-valued pairs (y_1^{**}, y_2^{**}). For example, Cragg's model is based on specifying F as the cdf of the bivariate normal distribution. In practice, parameters and covariates are involved in the specification of F.

The relationship between utility and expenditure variables is established by defining two intermediate random variables—the hurdle variables— denoted by

$$Y_1^* = 1\{Y_1^{**} > 0\} \quad \text{and} \quad Y_2^* = 1\{Y_2^{**} > 0\}Y_2^{**} \tag{3}$$

where $1\{A\}$ is an indicator function taking value 1 if event A holds, and 0 otherwise. Y_1^* represents the first hurdle participation decision (if $Y_1^* = 0$ the hurdle is failed, the individual does not participate; if $Y_1^* = 1$ the hurdle is passed, the individual is *potentially* a consumer), and Y_2^* represents the second hurdle consumption decision (if $Y_2^* = 0$ the hurdle is failed, the individual elects not to consume; if $Y_2^* > 0$ the hurdle is passed, the individual is *potentially* a consumer). In general, Y_1^* and Y_2^* are latent. The functional form of the joint distribution of (Y_1^*, Y_2^*) is induced by the specification assumed for F; however, it is easy to see from (3) that the (marginal) distribution of Y_1^* has to be Bernoulli, and that the (marginal) distribution of Y_2^* must take a zero-inflated form as in (1).

To complete the construction of the model, assume

$$Y = Y_1^* Y_2^* \tag{4}$$

or, equivalently,

$$Y = 1\{Y_1^{**} > 0 \cap Y_2^{**} > 0\} Y_2^{**} \tag{5}$$

the latter is the transformation that links the utility variables directly to the expenditure variable.

Due to the decomposition into separate decisions (jointly taken), a zero observation on Y can occur in either of two ways: (i) when the first hurdle is failed, or the first hurdle is passed and second hurdle failed $(Y_1^* = 0) \cup (Y_1^* = 1 \cap Y_2^* = 0)$; and (ii) the second hurdle failed, or the first hurdle failed but the second hurdle passed $(Y_2^* = 0) \cup (Y_1^* = 0 \cap Y_2^* > 0)$. A positive observation on Y is observed only when both hurdles are passed $(Y_1^* = 1 \cap Y_2^* > 0)$. In terms of the utility space (Y_1^{**}, Y_2^{**}), Figure 1 depicts the sample space of Y. For any pair of values of (Y_1^{**}, Y_2^{**}) generated in the L-shaped region given by quadrants II, III, and IV, a zero is generated for Y as shown by the direction arrows, while any pair generated in quadrant I maps to the vertical line (normalized to $Y_1^{**} = 1$) at the value of Y_2^{**}.

For the framework given by (2)–(5), the components $f_+(y)$ and f_0 of the pdf of Y can now be determined. The discrete component f_0 is given by the probability mass contained in quadrants II, III, and IV:

$$f_0 = \Pr(Y = 0)$$
$$= F_1(0) + F_2(0) - F(0, 0) \tag{6}$$

where $F_i(\cdot)$ denotes the marginal cdf of Y_i^{**} $(i = 1, 2)$. For any real $y > 0$, the continuous component $f_+(y)$ may be derived by differentiating, with respect to y, the following probability:

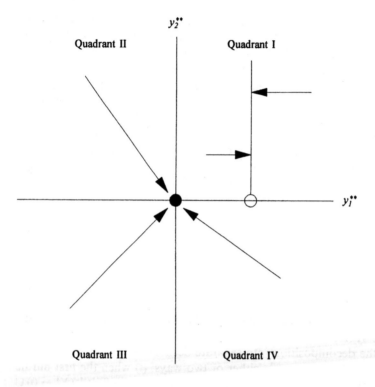

Figure 1. Sample space of Y induced by values assigned to (Y_1^{**}, Y_2^{**}).

$$\Pr(Y \le y) = \Pr(Y = 0) + \Pr(0 < Y \le y)$$
$$= F_1(0) + F_2(y) - F(0, y) \tag{7}$$

Thus,

$$f_+(y) = \frac{\partial}{\partial y}(F_2(y) - F(0, y)) \tag{8}$$

Once a specific functional form is assumed for F, the pdf (6) and (8) may be evaluated.

For a random sample of size n on Y, the log-joint pdf is given by

$$\sum_0 \log f_{0i} + \sum_+ \log f_+(y_i) \tag{9}$$

where $\sum_0$ is the sum over all zero observations, $\sum_+$ is the sum over all positive observations, and i is the observation index. Of course, the log-joint pdf may be viewed as the log-likelihood function.

3. ARE DECISIONS JOINT OR SEQUENTIAL?

The issues surrounding the debate "joint versus sequential decision making" in the DHM may be explored through the statistical structure of the model, leading to the conclusion that the participation and consumption decisions are jointly taken. The reasoning behind this is that the DHM is based on the specification of a *joint* distribution in the space of the utility variables Y_1^{**} and Y_2^{**}. Due to the transformation of (Y_1^{**}, Y_2^{**}) to (Y_1^*, Y_2^*), a *joint* distribution is induced on the space of the hurdle decision variables. The outcomes of the hurdle decisions, even though they cannot be observed, must be generated jointly. Under the assumptions of the DHM, the decision process of an individual involves taking two separate hurdle decisions, both of which are jointly taken. Should individual decision processes truly be sequential (in the population), a statistical model different to that of the DHM would need to be specified.

As obvious as the previous statistically-oriented argument is, it is evident in the DHM literature that considerable confusion reigns on the "joint versus sequential" issue. Particularly notable in this respect is Gao et al. [10, p. 364], in which the authors hedge their bets by asserting that the DHM encompasses both joint and sequential hurdle decision making. Intuitively, the DHM appears to lend itself to a sequential interpretation; namely, the outcome of the first hurdle participation decision precedes the outcome of the second hurdle consumption decision: the individual decides first whether to participate in the market, after that deciding upon the extent of their consumption. This interpretation has been given by Gould [12, p. 453], Roosen et al. [27, p. 378], and Zorn [35, p. 371], amongst others. However, the views expressed by these authors on this point are mistaken, for the commutative rule of multiplication applies in (4); thus the DHM is incapable of identifying any particular sequence in which the hurdles are tackled. Cragg also noted the non-identifiability of the decision sequence [6, p. 832].

Maddala too is adamant that Cragg's model is sequential [23, pp. 4–5]. In fact, the model which Maddala has in mind is one in which the first hurdle decision does indeed precede the second hurdle decision. Put simply, individuals decide first whether or not to participate, and only then, if they have elected to participate, do they determine their amount of consumption. However, this sequential model is *not* Cragg's model. In the DHM all individuals face both hurdle decisions, while in Maddala's sequential model only a sub-population faces both hurdle decisions. In contrast to the definition of the hurdle decision variables in the DHM given in (5), in Maddala's sequential model the hurdle decision variables are given by

$$Y_1^* = 1\{Y_1^{**} > 0\} \quad \text{and} \quad \tilde{Y}_2^* = 1\{\tilde{Y}_2^{**} > 0 | Y_1^* = 1\}\tilde{Y}_2^{**}|(Y_1^* = 1)$$

Note the conditioning event $Y_1^* = 1$ (the individual participates) appearing in the second hurdle decision. The utility space which induces these hurdle decision variables is the pair $(Y_1^{**}, \tilde{Y}_2^{**}|(Y_1^{**} > 0))$. In Maddala's sequential model, expenditure can only be observed on the sub-population of market participants; thus any inferences drawn from the model pertain only to this sub-population. This is not so in the DHM, where model inference applies to all in the population—participants as well as nonparticipants. The sequential model is discussed at length in Lee and Maddala [22].

A further point of confusion in the "joint versus sequential" debate concerns the role of dependence between the utility variables (Y_1^{**}, Y_2^{**}). For example, Blaylock and Blisard [2, pp. 700 and 702] assert that correlated utility variables imply simultaneous decisions, whereas independent utility variables imply that decisions are not simultaneous (Gould [12, p. 453] expresses a similar view). This represents an attempt to create a motive for the existence of dependence in the DHM where none exists in its economic underpinnings. Neither the presence nor the absence of dependency between Y_1^{**} and Y_2^{**} influences the earlier argument that the DHM implies that the hurdle decisions are taken jointly.

4. SPECIFICATIONS FOR DOUBLE-HURDLE MODELS
4.1 Cragg's Model

Selecting F in (2), corresponding to the cdf of the bivariate normal distribution, is by far and away the specification adopted most commonly in practice. The bivariate normal DHM is commonly referred to as "Cragg's model." Occasionally, the terms "independent Cragg model" and "dependent Cragg model" are used. These models are distinguished according to whether correlation between Y_1^{**} and Y_2^{**} is set to zero, or is parameterized. Apart from Cragg himself [6], other prominent examples of the (independent) Cragg model appear in Jones [16] and Blundell and Meghir [3].

Under bivariate normality, specify

$$\begin{bmatrix} Y_1^{**} \\ Y_2^{**} \end{bmatrix} \sim N\left(\begin{bmatrix} x_1'\beta_1 \\ x_2'\beta_2 \end{bmatrix}, \begin{bmatrix} 1 & \sigma\theta \\ \sigma\theta & \sigma^2 \end{bmatrix} \right) \tag{10}$$

The vectors x_1 and x_2 may contain fixed dummy variables and individual-specific covariates, the latter are assumed not determined by Y_1^{**} or Y_2^{**}. Without loss of generality, $\mathrm{Var}(Y_1^{**})$ is normalized to unity because in (3) all scaling information on Y_1^{**} is lost due to the transformation of Y_1^{**} to the hurdle variable Y_1^*. The unknown parameters consist of the elements of the vectors β_1 and β_2, and the scalars σ^2 and θ.

Unlike in the sample selection literature, it is unresolved whether the covariates x_1 and x_2 in the DHM must differ. Pudney [26, pp. 160–162], for example, argues that x_1 should contain psychological covariates influential on participation, whereas x_2 should contain economic covariates pertinent to consumption. In the dependent Cragg model, Jones [17, pp. 69–70] states that covariate exclusion restrictions emphasize identification of the model; however, inspection of the likelihood function (see [17, eq. (1)]) reveals that such restrictions are not necessary for parameter identification. On the other hand, a number of studies take the "kitchen-sink" approach and include all available covariates in participation and consumption ($x_1 = x_2$); the effect then of a given covariate on participation and consumption will differ according to its coefficients.

A further source of covariates may also enter the model if heteroscedasticity is suspected; specifying, for example,

$$\sigma^2 = \exp(x_3\beta_3)$$

Amongst others, Yen and Jones [34] and Yen et al. [32] report on heteroscedastic DHM.

Studies in which θ has been estimated have, on the whole, reported insignificant estimates (e.g., [2, 5, 11, 17, 34]). This is because θ is weakly identified. In fact, θ is identified only through the parametric form assumed for F; otherwise, the (non-parametric) joint distribution of (Y_1^{**}, Y_2^{**}) is not identified. For example, take the distribution of Y_2^{**} as that of Y when $Y = y > 0$, and anything for negative values that leads to $\int_{-\infty}^{0} dF_2(y) = \Pr(Y = 0)$, combined with $Y_1^{**} = Y_2^{**}$. Such a model, with perfect correlation between Y_1^{**} and Y_2^{**}, is always valid. For parametric DHM, such as the dependent Cragg model, evidence in Smith [29] establishes circumstances under which significant correlation estimates may arise, namely, when the data on Y contains excessive numbers of zero observations. To illustrate, [10], [12], [31] report significant correlation estimates, yet these studies analyse data with zeroes amounting to 67%, 59%, 89.9%, respectively, of the overall sample size.

Returning to the bivariate normal specification (10), let $\Phi(\cdot)$ denote the cdf of a $N(0, 1)$ random variable, thus $F_1(0) = \Phi(-x_1'\beta_1)$ and $F_2(0) = \Phi(-x_2'\beta_2/\sigma)$. For the independent Cragg model (i.e., θ set to zero), $F(\cdot, \cdot) = F_1(\cdot)F_2(\cdot)$. The pdf of Y in this case is given by

$$f_0 = \Phi(-x_1'\beta_1) + \Phi(-x_2'\beta_2/\sigma) - \Phi(-x_1'\beta_1)\Phi(-x_2'\beta_2/\sigma)$$
$$= 1 - \Phi(x_1'\beta_1)\Phi(x_2'\beta_2/\sigma)$$

and

$$f_+(y) = \sigma^{-1}\phi\left(\frac{y - x_2'\beta_2}{\sigma}\right) - \sigma^{-1}\Phi(-x_1'\beta_1)\phi\left(\frac{y - x_2'\beta_2}{\sigma}\right)$$

$$= \sigma^{-1}\phi\left(\frac{y - x_2'\beta_2}{\sigma}\right)\Phi(x_1'\beta_1)$$

where $\phi(\cdot)$ denotes the pdf of a $N(0, 1)$ random variable. Substitution into (9) gives the log-likelihood for β_1, β_2, and σ^2. Greene [13, pp. 596–597] gives LIMDEP code for maximum likelihood estimation of the independent Cragg model, and Jones [17] outlines GAUSS code for estimation of the independent and dependent variants of Cragg's model.

4.2 Transformed Models

A more flexible form of DHM may be obtained by specifying a parametric transformation of Y. The transformation serves to alter only the continuous component of the pdf of Y, and in existing studies it has been applied in the context of Cragg's model. Thus, the transformation is motivated by the desire to parameterize features of Y such as any skewness or kurtosis in excess of that which normal-based DHM are able to represent.

A transformed DHM is specified by replacing the left hand side of (4) by a function $T(Y)$, that is

$$T(Y) = Y_1^* Y_2^*$$

The function T is assumed positive-valued, non-decreasing, and differentiable in y for all $Y = y > 0$. Also, because the right-hand side may realize zero, T must be defined such that $T(0) = 0$. Under these conditions on T

$$\Pr(Y \le y) = \Pr(T(Y) \le T(y))$$

Thus, the pdf of Y is given by (6) at the origin, and when $Y = y > 0$ by

$$f_+(y) = \frac{\partial}{\partial y}(F_2(T(y)) - F(0, T(y)))$$

$$= \frac{\partial}{\partial y}(F_2(y) - F(0, y))\bigg|_{y \to T(y)} \frac{\partial T(y)}{\partial y}$$

(11)

which may be evaluated once functional forms for F and T are specified. The second line of (11) shows that the transformed DHM is obtained by scaling an untransformed DHM in which y is replaced by $T(y)$; the scaling factor is simply $\partial T(y)/\partial y$.

Examples of transformed DHM reported in the literature include Yen [30] and Yen and Jones [33], which were based on the Box–Cox transformation, and Gao et al. [10] and Yen and Jones [34], which were based on an inverse hyperbolic sine transformation. All report significant improvements in fit compared to that of the Cragg model. In the case of the Box–Cox transformation:

$$T(y) = \begin{cases} (y^\lambda - 1)/\lambda & \text{if } \lambda \neq 0 \\ \log y & \text{if } \lambda = 0 \end{cases}$$

for which $\partial T(y)/\partial y = y^{\lambda-1}$; usually λ is constrained to lie between 0 and 1. In the case of the inverse hyperbolic sine transformation:

$$T(y) = \begin{cases} \lambda^{-1} \sinh^{-1}(\lambda y) & \text{if } \lambda \neq 0 \\ y & \text{if } \lambda = 0 \end{cases}$$

for which $\partial T(y)/\partial y = (1 + \lambda^2 y^2)^{-1/2}$. In both specifications, λ represents an additional parameter to be estimated.

4.3 Non-Normal Models

The successes reported on fitting transformed Cragg models suggests that further improvements may be sought by specifying non-normal F. This is straightforward if Y_1^{**} and Y_2^{**} are independent, for then their joint cdf factors. Thus, if $F(\cdot, \cdot) = F_1(\cdot)F_2(\cdot)$, the pdf of Y is given by

$$f_0 = F_1(0) + F_2(0) - F_1(0)F_2(0)$$

and

$$f_+(y) = f_2(y)(1 - F_1(0))$$

where $f_2(y) = \partial F_2(y)/\partial y$ is the marginal pdf of Y_2^{**}. In this case, obtaining the DHM requires only specification of the marginal distributions. The flexibility to specify different distributional types for each marginal should be exploited.

More interesting from a practical viewpoint are settings in which Y_1^{**} and Y_2^{**} are dependent.[*] The conventional approach here would be to specify a suitable bivariate distribution from amongst those discussed in, for example, Kotz et al. [18]. However, this approach suffers due to the loss of flexibility in functional form, for typically the marginal distributions are of the same distributional type. In light of the general functional form of the pdf of Y with its interaction between joint and marginal distributions (see (6) and

---

[*]Although it is important to keep in mind the point raised in the previous section concerning the weak identification of the correlation coefficient in the DHM. Potentially, the parameterization of correlation between Y_1^{**} and Y_2^{**} in the DHM may be a statistically spurious generalization.

(8)), an alternative specification method based on the theory of copulas (see Joe [15] and Nelsen [25]) would seem particularly attractive. The basic idea of this approach is to build joint distributions from specific marginals, it is thus the opposite of the conventional method which derives marginals from a specific joint distribution. To illustrate, here is one particular family of copulas—the Farlie–Gumbel–Morgenstern (FGM) copula:

$$C_\theta(u, v) = uv + \theta uv(1 - u)(1 - v)$$

where u and v must be nondecreasing on $[0, 1]$. In terms of the FGM copula (although it applies quite generally for other copulae), if $u = F_1(\cdot)$, $v = F_2(\cdot)$; and provided $-1 \le \theta \le 1$, then by Sklar's theorem (Nelsen [25, p. 15]) C_θ is a bivariate cdf with marginal cdf given by $F_1(\cdot)$ and $F_2(\cdot)$; θ measures dependence,[*] if it is zero, Y_1^{**} and Y_2^{**} are independent. The functional form of the joint cdf of Y_1^{**} and Y_2^{**} induced by this copula is

$$F(y_1^{**}, y_2^{**}) = C_\theta(F_1(y_1^{**}), F_2(y_2^{**}))$$

for real-valued pairs (y_1^{**}, y_2^{**}), while the joint pdf of Y_1^{**} and Y_2^{**} (the following result requires continuity of both random variables) induced by this copula is

$$f(y_1^{**}, y_2^{**}) = \frac{\partial}{\partial y_1^{**}} \frac{\partial}{\partial y_2^{**}} C_\theta(F_1(y_1^{**}), F_2(y_2^{**}))$$

$$= f_1(y_1^{**}) f_2(y_2^{**}) \frac{\partial}{\partial u} \frac{\partial}{\partial v} C_\theta(u, v) \bigg|_{u \to F_1(y_1^{**}), v \to F_2(y_2^{**})}$$

where $f_1(\cdot)$ and $f_2(\cdot)$ are the marginal pdf. For the FGM copula, substitution into (6) and (8) yields the pdf of Y in this case as

$$f_0 = F_1(0) + F_2(0) - F_1(0)F_2(0) - \theta F_1(0)F_2(0)(1 - F_1(0))(1 - F_2(0))$$

and

$$f_+(y) = \frac{\partial}{\partial y}(F_2(y) - F_1(0)F_2(y) - \theta F_1(0)F_2(y)(1 - F_1(0))(1 - F_2(y)))$$

$$= f_2(y)(1 - F_1(0) - \theta F_1(0)(1 - F_1(0))(1 - 2F_2(y)))$$

Given any copula (many are listed in [15] and [25]), specification of a DHM involves selecting from amongst the class of univariate distributions. The attractiveness of the copula method is the flexibility it allows to specify different distributional types for each marginal.

---

[*]In the copula literature, the term "association" is often used instead of "dependence."

4.4 Hurdle Models

Each previous DHM has resulted from a specification assumed for F, the joint cdf of the utility variables (Y_1^{**}, Y_2^{**}). There is, however, a further way to arrive at a DHM; namely, through specifying a joint distribution for the hurdle variables (Y_1^*, Y_2^*). Indeed, due to the intermediate transformation from (Y_1^{**}, Y_2^{**}) to (Y_1^*, Y_2^*), see (3), it is apparent that there is considerable information about the marginal distributions of the hurdle variables: Y_1^* must be Bernoulli distributed, and the distribution of Y_2^* must take a zero-inflated form as in (1).

Let $G(y_1^*, y_2^*)$ denote the joint distribution of (Y_1^*, Y_2^*), for pairs (y_1^*, y_2^*) such that $y_1^* \in \{0, 1\}$ and $y_2^* \geq 0$. Then, the pdf of Y is given by

$$f_0 = G_1(0) + G_2(0) - G(0, 0)$$

when $y = 0$, and, when $y > 0$, by

$$f_+(y) = \frac{\partial}{\partial y}(G_2(y) - G(0, y))$$

It is known that

$$G_1(y_1^*) = p^{y_1^*}(1 - p)^{1 - y_1^*}, \qquad (0 \leq p \leq 1)$$

and

$$G_2(y_2^*) = \begin{cases} g_+(y_2^*) & \text{if } y_2^* > 0 \\ g_0 & \text{if } y_2^* = 0 \end{cases}$$

To arrive at a DHM, the attractiveness of the copula method discussed in Section 4.3 is obvious, for the marginal distributions are of known types.

Given any copula, the DHM is arrived at by specifying the probability p in $G_1(\cdot)$ and the cdf $G_2(\cdot)$. For $p = p(x_1, \beta_1)$, any function whose range is on $[0, 1]$ is suitable. For example, specifying p as the cdf of the standard normal (e.g., $p = \Phi(x_1'\beta_1)$) leads to the familiar probit, whereas another, more flexible alternative would be to specify p as the pdf of a beta random variable. For $G_2(y) = G_2(y; x_2, \beta_2)$, a zero-inflated distribution must be specified; for example, Aalen's [1] compound Poisson distribution.

4.5 Dominance Models

Special cases of the DHM arise if restrictions are placed on the joint distribution of (Y_1^{**}, Y_2^{**}). For example, if $\Pr(Y_1^{**} > 0) = 1$, then all individuals participate and $Y = Y_2^{**}$; furthermore, specifying $Y_2^{**} \sim N(x_2'\beta_2, \sigma^2)$ leads to the well-known Tobit model. Thus, the Tobit model can be viewed as nested within the Cragg model (or at least a DHM in which the marginal distribution of Y_2^{**} is normal), which leads to a simple specification test based on the likelihood ratio principle.

Dominance models, in which the participation decision is said to dominate the consumption decision, provide yet another special case of the DHM. Jones [16] characterizes dominance as a population in which all individuals electing to participate are observed to have positive consumption; hence, $\Pr(Y > 0 | Y_1^* = 1) = 1$. In terms of the underlying utility variables, dominance is equivalent to

$$\Pr(Y_1^{**} > 0, Y_2^{**} \leq 0) = 0$$

So, referring back to Figure 1, dominance implies that quadrant IV is no longer part of the sample space of (Y_1^{**}, Y_2^{**}). Dominance implies that the event $Y = 0$ is equivalent to $Y_1^{**} \leq 0$, for when individuals do not consume then they automatically fail to participate whatever their desired consumption—individuals cannot report zero consumption if they participate in the market. Unlike the standard (non-dominant) DHM in which zeroes can be generated from two unobservable sources, in the dominant DHM there is only one means of generating zero consumption. Thus,

$$Y_1^* = 1\{Y > 0\}$$

meaning that the first hurdle participation decision is observable under dominance. The cdf of Y may be written

$$\Pr(Y \leq y) = \Pr(Y_1^* = 0) + \Pr(0 < Y_1^* Y_2^* \leq y | Y_1^* = 1) \Pr(Y_1^* = 1)$$

for real-valued $y \geq 0$. This form of the cdf has the advantage of demonstrating that the utility associated with participation Y_1^{**} acts as a sample selection mechanism. Indeed, as Y_2^{**} is Y when $Y_1^{**} > 0$, the dominant DHM is the classic sample selection model discussed in Heckman [14]. Accordingly, a dominant DHM requires specification of a joint distribution for $(Y_1^{**}, Y_2^{**} | Y_2^{**} > 0)$.* As argued earlier, specification based on the copula method makes this task relatively straightforward. Lee and Maddala [21] give details of a less flexible method based on a bivariate normal linear specification.

---

*The nomenclature "complete dominance" and "first-hurdle dominance" is sometimes used in the context of a dominant DHM. The former refers to a specification in which Y_1^{**} is independent of $Y_2^{**} | Y_2^{**} > 0$, while the latter is the term used if Y_1^{**} and $Y_2^{**} | Y_2^{**} > 0$ are dependent.

5. CONCLUDING REMARKS

The DHM is one of a number of models that can be used to fit data which is evidently zero-inflated. Such data arise in the context of consumer demand studies, and so it is in this broad area of economics that a number of applications of the DHM will be found. The DHM is motivated by splitting demand into two separate decisions: a market participation decision, and a consumption decision. Zero observations may result if either decision is turned down, giving two sources, or reasons, for generating zero observations. Under the DHM, it was established that individuals tackle the decisions jointly.

The statistical construction of the DHM usually rests on assigning a bivariate distribution, denoted by F, to the utilities derived from participation and consumption. The derivation of the model that was given is general enough to embrace a variety of specifications for F which have appeared throughout the literature. Moreover, it was demonstrated how copula theory was useful in the DHM context, opening numerous possibilities for statistical modeling over and above that of the bivariate normal Cragg model which is the popular standard in this field.

ACKNOWLEDGMENT

The research reported in this chapter was supported by a grant from the Australian Research Council.

REFERENCES

1. Aalen, O. O. (1992), Modelling heterogeneity in survival analysis by the compound Poisson distribution. *The Annals of Applied Probability*, **2**, 951–972.
2. Blaylock, J. R., and Blisard, W. N. (1992), U.S. cigarette consumption: the case of low-income women. *American Journal of Agricultural Economics*, **74**, 698–705.
3. Blundell, R., and Meghir, C. (1987), Bivariate alternatives to the Tobit model. *Journal of Econometrics*, **34**, 179–200.
4. Böhning, D., Dietz, E., Schlattman, P., Mendonça, L., and Kirchner, U. (1999), The zero-inflated Poisson model and the decayed, missing and filled teeth index in dental epidemiology. *Journal of the Royal Statistical Society A*, **162**, 195–209.
5. Burton, M., Tomlinson, M., and Young, T. (1994), Consumers' decisions whether or not to purchase meat: a double hurdle analysis of

single adult households", *Journal of Agricultural Economics*, **45**, 202–212.

6. Cragg, J. G. (1971), Some statistical models for limited dependent variables with applications to the demand for durable goods. *Econometrica*, **39**, 829–844.

7. Deaton, A., and Irish, M. (1984), Statistical models for zero expenditures in household budgets. *Journal of Public Economics*, **23**, 59–80.

8. DeSarbo, W. S., and Choi, J. (1999), A latent structure double hurdle regression model for exploring heterogeneity in consumer search patterns. *Journal of Econometrics*, **89**, 423–455.

9. Dionne, G., Artís, M., and Guillén, M. (1996), Count-data models for a credit scoring system. *Journal of Empirical Finance*, **3**, 303–325.

10. Gao, X. M., Wailes, E. J., and Cramer, G. L. (1995), Double-hurdle model with bivariate normal errors: an application to U.S. rice demand. *Journal of Agricultural and Applied Economics*, **27**, 363–376.

11. Garcia, J., and Labeaga, J. M. (1996), Alternative approaches to modelling zero expenditure: an application to Spanish demand for tobacco. *Oxford Bulletin of Economics and Statistics*, **58**, 489–506.

12. Gould, B. W. (1992), At-home consumption of cheese: a purchase-infrequency model. *American Journal of Agricultural Economics*, **72**, 453–459.

13. Greene, W. H. (1995), LIMDEP, *Version 7.0 User's Manual*. New York: Econometric Software.

14. Heckman, J. J. (1976), The common structure of statistical models of truncation, sample selection and limited dependent variables and a simple estimator for such models. *Annals of Economic and Social Measurement*, **5**, 465–492.

15. Joe, H. (1997), *Multivariate Models and Dependence Concepts*. London: Chapman and Hall.

16. Jones, A. M. (1989), A double-hurdle model of cigarette consumption. *Journal of Applied Econometrics*, **4**, 23–39.

17. Jones, A. M. (1992), A note on computation of the double-hurdle model with dependence with an application to tobacco expenditure. *Bulletin of Economic Research*. **44**, 67–74.

18. Kotz, S., Balakrishnan, N., and Johnson, N. L. (2000), *Continuous Multivariate Distributions*, volume 1, 2nd edition. New York: Wiley.

19. Labeaga, J. M. (1999), A double-hurdle rational addiction model with heterogeneity: estimating the demand for tobacco. *Journal of Econometrics*, **93**, 49–72.

20. Lambert, D. (1992), Zero-inflated Poisson regression, with an application to defects in manufacturing. *Technometrics*, **34**, 1–14.

21. Lee, L.-F., and Maddala, G. S. (1985), The common structure of tests for selectivity bias, serial correlation, heteroscedasticity and non-normality in the Tobit model. *International Economic Review*, **26**, 1–20. Also in: Maddala, G. S. (ed.) (1994), *Econometric Methods and Applications* (Volume II). Aldershot: Edward Elgar, pp. 291–310.

22. Lee, L.-F., and Maddala, G. S. (1994), Sequential selection rules and selectivity in discrete choice econometric models. In: Maddala, G. S. (ed.), *Econometric Methods and Applications* (Volume II). Aldershot: Edward Elgar, pp. 311-329.

23. Maddala, G. S. (1985), A survey of the literature on selectivity bias as it pertains to health care markets. *Advances in Health Economics and Health Services Research*, **6**, 3–18. Also in: Maddala, G. S. (ed.) (1994), *Econometric Methods and Applications* (Volume II), Aldershot: Edward Elgar, pp. 330–345.

24. Maki, A., and Nishiyama, S. (1996), An analysis of under-reporting for micro-data sets: the misreporting or double-hurdle model. *Economics Letters*, **52**, 211–220.

25. Nelsen, R. B. (1999), *An Introduction to Copulas*, New York: Spinger-Verlag.

26. Pudney, S. (1989), *Modelling Individual Choice: the Econometrics of Corners, Kinks, and Holes*. London: Basil Blackwell.

27. Roosen, J., Fox, J. A., Hennessy, D. A., and Schreiber, A. (1998), Consumers' valuation of insecticide use restrictions: an application to apples. *Journal of Agricultural and Resource Economics*, **23**, 367–384.

28. Shonkwiler, J. S., and Shaw, W. D. (1996), Hurdle count-data models in recreation demand analysis. *Journal of Agricultural and Resource Economics*, **21**, 210–219.

29. Smith, M. D. (1999), Should dependency be specified in double-hurdle models? In: Oxley, L., Scrimgeour, F., and McAleer, M. (eds.), *Proceedings of the International Congress on Modelling and Simulation* (Volume 2), University of Waikato, Hamilton, New Zealand, pp. 277–282.

30. Yen, S. T. (1993), Working wives and food away from home: the Box–Cox double hurdle model. *American Journal of Agricultural Economics*, **75**, 884–895.

31. Yen, S. T., Boxall, P. C., and Adamowicz, W. L. (1997), An econometric analysis of donations for environmental conservation in Canada. *Journal of Agricultural and Resource Economics*, **22**, 246–263.

32. Yen, S. T., Jensen, H. H., and Wang, Q. (1996), Cholesterol information and egg consumption in the US: a nonnormal and heteroscedastic

double-hurdle model. *European Review of Agricultural Economics*, **23**, 343–356.

33. Yen, S. T., and Jones, A. M. (1996), Individual cigarette consumption and addiction: a flexible limited dependent variable approach. *Health Economics*, **5**, 105–117.

34. Yen, S. T., and Jones, A. M. (1997), Household consumption of cheese: an inverse hyperbolic sine double-hurdle model with dependent errors. *American Journal of Agricultural Economics*, **79**, 246–251.

35. Zorn, C. J. W. (1998), An analytic and experimental examination of zero-inflated and hurdle Poisson specifications. *Sociological Methods and Research*, **26**, 368–400.

26

Econometric Applications of Generalized Estimating Equations for Panel Data and Extensions to Inference

H. D. VINOD   Fordham University, Bronx, New York

This chapter hopes to encourage greater sharing of research ideas between biostatistics and econometrics. Vinod (1997) starts, and Mittelhammer et al.'s (2000) econometrics text explains in detail and with illustrative computer programs, how the theory of estimating functions (EFs) provides general and flexible framework for much of econometrics. The EF theory can improve over both Gauss–Markov least squares and maximum likelihood (ML) when the variance depends on the mean. In biostatistics literature (Liang and Zeger 1995), one of the most popular applications of EF theory is called generalized estimating equations (GEE). This paper extends GEE by proposing a new pivot function suitable for new bootstrap-based inference. For longitudinal (panel) data, we show that when the dependent variable is a qualitative or dummy variable, variance does depend on the mean. Hence the GEE is shown to be superior to the ML. We also show that the generalized linear model (GLM) viewpoint with its link functions is more flexible for panel data than some fixed/random effect methods in econometrics. We illustrate GEE and GLM methods by studying the effect of monetary policy (interest rates) on turning points in stock market prices. We find that interest rate policy does not have a significant effect on stock price upturns or downturns. Our explanations of GLM and GEE are missing in Mittelhammer et al. (2000) and are designed to improve lines of communication between biostatistics and econometrics.

1. INTRODUCTION AND MOTIVATION

This chapter explains two popular estimation methods in biostatistics literature, called GLM and GEE, which are largely ignored by econometricians. The GEE belongs in the statistical literature dealing with Godambe–Durbin estimating functions (EFs) defined as functions of data and parameters, $g(y, \beta)$. EFs satisfy "unbiasedness, sufficiency, efficiency and robustness" and are more versatile than "moment conditions" in econometrics of generalized method of moments (GMM) models. Unbiased EFs satisfy $E(g) = 0$. Godambe's (1960) optimal EFs minimize $[\mathrm{Var}(g)] [E \partial g / \partial \beta]^{-2}$. Vinod (1997, 1998, 2000) and Mittelhammer et al. (2000, chapters 11 to 17) discuss EF-theory and its applications in econometrics. The EFs are guaranteed to improve over both generalized least-squares (GLS) and maximum likelihood (ML) when the variance depends on the mean, as shown in Godambe and Kale (1991) and Heyde (1997). The GEEs are simply a panel data application of EFs from biostatistics which exploits the dependence of variance on mean, as explained in Dunlop (1994), Diggle et al. (1994), and Liang and Zeger (1995).

The impact of EF-theory on statistical practice is greatest in the context of GEE. Although the underlying statistical results are known in the EF literature, the biometric application to the panel data case clarifies and highlights the advantages of EFs. This chapter uses the same panel data context with an econometric example to show exactly where EFs are superior to both GLS and ML.

A score function is the partial derivative of the log likelihood function. If the partial is specified without any likelihood function, it is called quasi-score function (QSF). The true integral (i.e., likelihood function) can fail to exist when the "integrability condition" that second-order cross partials be symmetric (Young's theorem) is violated. Wedderburn (1974) invented the quasi-likelihood function (QLF) as a hypothetical integral of QSF. Wedderburn was motivated by applications to the generalized linear model (GLM), where one is unwilling to specify any more than mean and variance properties. The quasi-maximum likelihood estimator (QML) is obtained by solving QSF = 0 for β. Godambe (1985) proved that the optimal EF is the quasi-score function (QSF) when QSF exists. The optimal EFs (QSFs) are computed from the means and variances, without assuming further knowledge of higher moments (skewness, kurtosis) or the form of the density. Liang et al. (1992, p. 11) prove that traditional likelihoods require additional restrictions. Thus, EF methods based on QSFs are generally regarded as "more robust," which is a part of unbiasedness, sufficiency, efficiency, and robustness claimed above. Mittelhammer et al.'s (2000, chapter 11) new text gives details on specification-robustness, finite sample optimality, closeness in terms of Kullback–Leibler discrepancy, etc. Two GAUSS programs on a CD accompanying the text provide a graphic demonstration of the advantages of QML.

In econometrics, the situations where variance depends on the mean are not commonly recognized or mentioned. We shall see later that panel data models do satisfy such dependence. However, it is useful to consider such dependence without bringing in the panel data case just yet. We use this simpler case to derive and compare the GLS, ML, and EF estimators.

Consider T real variables y_i $(i = 1, 2, \ldots, T)$:

$$y_i \sim \text{IND}\big(\mu_i(\beta),\ \sigma^2 v_i(\beta)\big), \quad \text{where } \beta \text{ is } p \times 1 \tag{1}$$

where IND denotes an independently (not necessarily identically) distributed random variable (r.v.) with mean $\mu_i(\beta)$, variance $\sigma^2 v_i(\beta)$, and σ^2 does not depend on β. Let $y = \{y_i\}$ be a $T \times 1$ vector and $V = \sigma^2(\text{Var}(y)) = \text{Var}(\mu_i(\beta))$ denote the $T \times T$ covariance matrix. The IND assumption implies that $V(\mu)$ is a diagonal matrix depending only on the ith component μ_i of the $T \times 1$ vector μ. The common parameter vector of interest is β which measures how μ depends on X, a $T \times p$ matrix of covariates. For example, $\mu_i = X\beta$ for the linear regression case. Unlike the usual regression case, however, the heteroscedastic variances $v_i(\beta)$ here are functionally related to the means $\mu_i(\beta)$ through the presence of β. We call this "special" heteroscedasticity, since it is rarely mentioned in econometrics texts.

If y_i are discrete stochastic processes (time series data), then μ_i and v_i are conditional on past data. In any case, the log-likelihood for (1) is

$$\text{LnL} = -(T - 2)\ln 2\pi - (T/2)(\ln \sigma^2) - S_1 - S_2 \tag{2}$$

where $S_1 = (1/2)\sum_{i=1}^{T} \ln v_i$ and $S_2 = \sum_{i=1}^{T}[y_i - \mu_i]^2/[2\sigma^2 v_i]$. The GLS estimator is obtained by minimizing the error sum of squares S_2 with respect to (wrt) β. The first-order condition (FOC) for GLS is simply $\partial(S_2)/\partial\beta = [\partial(S_2)/\partial v_i][\partial v_i/\partial\beta] + [\partial(S_2)/\partial\mu_i][\partial\mu_i/\partial\beta] = T_1 + \text{QSF} = 0$, which explicitly defines T_1 as the first term and QSF as the quasi-score function. The theory of estimating functions suggests that, when available, the optimal EF is the QSF itself. Thus the optimal EF estimator is given by solving $\text{QSF} = 0$ for β. The GLS estimator is obtained by solving $T_1 + \text{QSF} = 0$ for

β. When heteroscedasticity is "special," $[\partial v_i/\partial \beta]$ is nonzero and T_1 is non-zero. Hence, the GLS solution obviously differs from the EF solution.

To obtain the ML estimator one maximies the LnL of (2). The FOC for this is

$$\partial(-S_1 - S_2)/\partial \beta = [\partial(S_1 + S_2)/\partial v_i][\partial v_i/\partial \beta] + [\partial(S_2)/\partial \mu_i][\partial \mu_i/\partial \beta] = 0$$

where we use $\partial(S_1)/\partial \mu_i = 0$. We have

$$\partial LnL/\partial \beta = [\partial(-S_1 - S_2)/\partial v_i][\partial v_i/\partial \beta] + \sum_{i=1}^{T}\{[y_i - \mu_i]/[\sigma^2 v_i]\}[\partial \mu_i/\partial \beta]$$

$$= T_1' + QSF$$

which defines T_1' as a distinct first term. Since special heteroscedasticity leads to nonzero $[\partial v_i/\partial \beta]$, T_1' is also nonzero. Again, the ML solution is obviously distinct from the EF.

The unbiasedness of EF is verified as follows. Since $v_i > 0$, and $E[y_i - \mu_i] = 0$, we do have $E(QSF) = 0$. Now the ML estimator is obtained by solving $T_1' + QSF = 0$ for β, where the presence of T_1' leads to a biased equation since $E(T_1')$ is nonzero. Similarly, since T_1 is nonzero, the equation yielding GLS estimator as solution is biased. The "sufficiency" property is shared by GLS, ML, and EF. Having shown that the EF defined by $QSF = 0$ is unbiased, we now turn to its "efficiency" (see Section 11.3 of Mitttelhammer et al. 2000). To derive the covariance matrix of QSF it is convenient to use matrix notation and write (1) as: $y = \mu + \varepsilon$, $E\varepsilon = 0$, $E\varepsilon\varepsilon' = V = \text{diag}(v_i)$. If $D = \{\partial \mu_i/\partial \beta_j\}$ is a $T \times p$ matrix, McCullagh and Nelder (1989, p. 327) show that the $QSF(\mu, v)$ in matrix notation is

$$QSF(\mu, v) = D'V^{-1}(y - \mu)/\sigma^2 = \sum_{i=1}^{T}\{[y_i - \mu_i]/[\sigma^2 v_i]\}[\partial \mu_i/\partial \beta] \qquad (3)$$

The last summation expression of (3) requires V to be a diagonal matrix. For proving efficiency, V need not be diagonal, but it must be symmetric positive definite. For generalized linear models (GLM) discussed in the next section the V matrix has known functions of β along the diagonal.

In this notation, as before, the optimal EF estimator of the β vector is obtained by solving the (nonlinear) equation $QSF = 0$ for β. The unbiasedness of EF in this notation is obvious. Since $E(y - \mu) = 0$, $E(QSF) = 0$, implying that QSF is an unbiased EF. In order to study the "efficiency" of EF we evaluate its variance–covariance matrix as: $\text{Cov}(QSF) = D'V^{-1}D = I_F$, where I_F denotes the Fisher information matrix. Since $-E(\partial QSF/\partial \beta) = \text{Cov}(QSF) = I_F$, the variance of QSF reaches the Cramer–Rao lower bound. This means EF is minimum variance in the class of unbiased estimators and obviously superior to ML and GLS (if EF is different from ML or GLS) in that class. This does not at all mean that we abandon GLS or ML. Vinod (1997) and Mittelhammer et al. (2000) give examples where the optimal EF estimator coincides with the least-squares (LS) and maximum likelihood (ML) estimators and that EF approach provides a "common tie" among several estimators.

In the 1970s, some biostatisticians simply ignored dependence of $\mathrm{Var}(\varepsilon_t)$ on β for computational convenience. The EF-theory proves the surprising result that it would be suboptimal to incorporate the complicating dependence of $\mathrm{Var}(\varepsilon_t)$ on β by including the chain-rule related extra term(s) in the FOCs of (4). An intuitive explanation seems difficult, although the computer programs in Mittelhammer et al. (2000, chapters 11 and 12) will help. An initial appeal of EF-theory in biostatistics was that it provided a formal justification for the quasi-ML estimator used since the 1970s. The "unbiasedness, sufficiency, efficiency and robustness" of EFs are obvious bonuses. We shall see that GEE goes beyond quasi-ML by offering more flexible correlation structures for panel data.

We summarize our arguments in favor of EFs as follows.

Result 1. Assume that variances $v_i(\beta)$ here are functionally related to the means $\mu_i(\beta)$. Then first-order conditions (FOCs) for GLS and ML imply a superfluous term arising from the chain rule in

$$T_1 + \mathrm{QSF}(\mu, v) = 0 \quad \text{and} \quad T_1' + \mathrm{QSF}(\mu, v) = 0 \qquad (4)$$

respectively, where $\mathrm{QSF}(\mu, v)$ in matrix notation is defined in (3). Under special heteroscedasticity, both T_1 and T_1' are nonzero when variances depend on β.

Only when $[\partial v_i / \partial \beta] = 0$, i.e., when heteroscedasticity does not depend on β, $T_1 = 0$ and $T_1' = 0$, simplifying the FOCs in (4) as QSF $= 0$. This alone is an unbiased EF. Otherwise, normal equations for both the GLS and ML are biased estimating functions and are flawed for various reasons explained in Mittelhammer et al.'s (2000, Section 11.3.2.a) text, which also cites Monte Carlo evidence. The intuitive reason is that a biased EF will produce incorrect β with probability 1 in the limit as variance tends to zero. The technical reasons have nothing to do with a quadratic loss function, but do include inconsistency in asymptotics, laws of large numbers, etc. Thus, FOCs of both GLS and ML lead to biased and inefficient EFs, with a larger variance than that of optimal EF defined by QSF $= 0$.

An interesting lesson of the EF-theory is that biased estimators are acceptable but biased EFs are to be avoided. Indeed, unbiased EFs may well yield biased EF estimators. Thus we have explained why for certain models the quasi-ML estimator, viewed as an EF estimator, is better than the full-blown ML estimator. Although counter-intuitive, this outcome arises whenever we have "special" heteroscedasticity in v_i, which depend on β, in light of Result 1 above. The reason for discussing the desirable properties of QSFs is that the GEE estimator for panel data is obtained by solving the appropriate QSF = 0. The special heteroscedasticity is also crucial and the next section explains that GLMs give rise to it. The GLMs do not need to have panel data, but do need limited (binary) dependent variables.

The special heteroscedasticity variances leading to nonzero $[\partial v_i / \partial \beta]$ are not as rare as one might think. For example, consider the binomial distribution involving n trials with probability p of success; the mean is $\mu = np$ and variance $v = np(1-p) = \mu(1-p)$. Similarly, for the Poisson distribution the mean μ equals the variance. In the following section, we discuss generalized linear models (GLM) where we mention canonical link functions for such familiar distributions.

2. THE GENERALIZED LINEAR MODELS (GLM) AND ESTIMATING FUNCTION METHODS

Recall that besides GEE we also plan to introduce the second popular biostatistics tool, called GLM, in the context of econometrics. This section indicates how the GLM methods are flexible and distinct from the logits or fixed/random effects approach popularized by Balestra, Nerlove, and others and covered in econometrics texts (Greene 2000). Recent econometrics monographs, e.g., Baltagi (1995), dealing with logit, probit and limited dependent variable models, also exclusively rely on the ML methods, ignoring both GLM and GEE models. In statistical literature dealing with designs of experiments and analysis of variance, the fixed/random effects have been around for a long time. They can be cumbersome when the interest is in the effect of covariates or in so-called analysis of covariance. The GLM approach considers a probability distribution of μ_i, usually cross-sectional individual means, and focuses on their relation with the covariates X. Dunlop (1994) derives the estimating functions for the GLM and explains how the introduction of a flexible link function $h(\mu_i)$ linearizes the solution of the EFs or QSF = 0.

We describe the GLM as a generalization of the familiar GLS in the familiar econometric notation of the regression model with T $(t = 1, \ldots, T)$ observations and p regressors, where the subscript t can mean either time series or cross-sectional data. We assume:

$$y = X\beta + \varepsilon, \quad E(\varepsilon) = 0, \quad E\varepsilon\varepsilon' = \sigma^2 \Omega \tag{5}$$

Note that the probability distribution of y with mean $E(y) = X\beta$ is assumed to be normal and linearly related to X. The maximum likelihood (ML) and generalized least-squares (GLS) estimator is obtained by solving the following score functions or normal equations for β. If the existence of the likelihood function is not assumed, it is called a "quasi" score function (QSF):

$$g(y, X, \beta) = X'\Omega^{-1}X\beta - X'\Omega^{-1}y = 0 \tag{6}$$

When a QSF is available, the optimal EF estimator is usually obtained by solving the QSF $= 0$ for β. If Ω is a known (diagonal) matrix which is not a function of β, the EF coincides with the GLS. An explicit (log) likelihood function (not quasi-likelihood function) is needed for defining the score function as the derivative of the log likelihood. The ML estimator is obtained by solving (score function) $= 0$ for β.

Remark 1. The GLS is extended into the generalized linear model (GLM is mostly nonlinear, but linear in parameters) in three steps (McCullagh and Nelder 1989).

(i) Instead of $y \sim N(\mu, \sigma^2\Omega)$ we admit any distribution from the exponential family of distributions with a flexible choice of relations between mean and variance functions. Non-normality permits the expectation $E(y) = \mu$ to take on values only in a meaningful restricted range (e.g., non-negative integer counts or $[0, 1]$ for binary outcomes).

(ii) Define the systematic component $\eta = X\beta = \sum_{j=1}^{p} x_j\beta$; $\eta \in (-\infty, \infty)$, as a linear predictor.

(iii) A monotonic differentiable link function $\eta = h(\mu)$ relates $E(y)$ to the systematic component $X\beta$. The tth observation satisfies $\eta_t = h(\mu_t)$. For GLS, the link function is identity, or $\eta = \mu$, since $y \in (-\infty, \infty)$. When y data are counts of something, we obviously cannot allow negative counts. We need a link function which makes sure that $X\beta = \mu > 0$. Similarly, for y as binary (dummy variable) outcomes, $y \in [0, 1]$, we need a link function $h(\mu)$ which maps the interval $[0,1]$ for y on the $(-\infty, \infty)$ interval for $X\beta$

Remark 2. To obtain generality, the normal distribution is often replaced by a member of the exponential family of distributions, which includes Poisson, binomial, gamma, inverse-Gaussian, etc. It is well known that "sufficient statistics" are available for the exponential family. In our context, $X'y$, which is a $p \times 1$ vector similar to β, is a sufficient statistic. A "canonical" link function is one for which a sufficient statistic of $p \times 1$ dimension exists. Some well-known canonical link functions for distributions in the exponential family are: $h(\mu) = \mu$ for the normal, $h(\mu) = \log\mu$

for the Poisson, $h(\mu) = \log[\mu/(1 - \mu)]$ for the binomial, and $h(\mu) = -1/\mu$ is negative for the gamma distribution. Iterative algorithms based on Fisher scoring are developed for GLM, which also provide the covariance matrix of β as a byproduct. The difference between Fisher scoring and the Newton–Raphson method is that the former uses the expected value of the Hessian matrix. The GLM iterative scheme depends on the distribution of y only through the mean and variance. This is also one of the practical appeals of the estimating function quasi-likelihood viewpoint largely ignored in econometrics.

Now we state and prove the known result that when y is a binary dependent variable, heteroscedasticity measured by $\mathrm{Var}(\varepsilon_t)$, the variance of ε_t, depends on the regression coefficients β. Recall that this is where we have claimed the EF estimator to be better than ML or GLS. This dependence result also holds true for the more general case where y is a categorical variable (e.g., poor, good, and excellent as three categories) and to panel data where we have a time series of cross-sections. The general case is discussed in the EF literature.

Result 2. The "special" heteroscedasticity is present by definition when $\mathrm{Var}(\varepsilon_t)$ is a function of the regression coefficients β. When y_t is a binary (dummy) variable from time series or cross-sectional data (up to a possibly unknown scale parameter) it does possess special heteroscedasticity.

Proof Let P_t denote the probability that $y_t = 1$. Our interest is in relating this probability to various regressors at time t, or X_t. If the binary dependent variable in (5) can assume only two values (1 or 0), then regression errors also can and must assume only two values: $1 - X_t\beta$ or $-X_t\beta$. The corresponding probabilities are P_t and $(1 - P_t)$ respectively, which can be viewed as realizations of a binomial process. Note that

$$E(\varepsilon_t) = P_t(1 - X_t\beta) + (1 - P_t)(-X_t\beta) = P_t - X_t\beta \tag{7}$$

Hence the assumption that $E(\varepsilon_t) = 0$ itself implies that $P_t = X_t\beta$. Thus, we have the result that P_t is a function of the regression parameters β. Since $E(\varepsilon_t) = 0$, then $\mathrm{Var}(\varepsilon_t)$ is simply the square of the two values of ε_t weighted by the corresponding probabilities. After some algebra, thanks to certain cancellations, we have $\mathrm{Var}(\varepsilon_t) = P_t(1 - P_t) = X_t\beta(1 - X_t\beta)$. This proves the key result that both the mean and variance depend on β. This is where EFs have superior properties.

We can extend the above result to other situations involving limited dependent variables. In econometrics, the canonical link function terminology of Remark 2 is rarely used. Econometricians typically replace y_t by unobservable (latent) variables and write the regression model as

$$y_t^* = X_t\beta + \varepsilon_t \tag{8}$$

where the observable $y_t = 1$ if $y_t^* > 0$, and $y_t = 0$ if $y_t^* \leq 0$.

Now write $P_t = Pr(y_t = 1) = Pr(y_t^* > 0) = Pr(X_t\beta + \varepsilon_t > 0)$, which implies that

$$P_t = Pr(\varepsilon_t > -X_t\beta) = 1 - Pr(\varepsilon_t \leq X_t\beta) = 1 - \int_{-\infty}^{X_t\beta} f(\varepsilon_t)d\varepsilon_t \tag{9}$$

where we have used the fact that ε_t is a symmetric random variable defined over an infinite range with density $f(\varepsilon_t)$. In terms of cumulative distribution functions (CDF) we can write the last integral in (9) as $F(X_t\beta) \in [0, 1]$. Hence, $P_t \in [0, 1]$ is guaranteed. It is obvious that if we choose a density, which has an analytic CDF, the expressions will be convenient. For example, $F(X_t\beta) = [1 + \exp(-X_t\beta)]^{-1}$ is the analytic CDF of the standard logistic distribution. From this, econometric texts derive the logit link function $h(P_t) = \log[P_t/(1 - P_t)]$ somewhat arduously. Clearly, as $P_t \in [0, 1]$, the logit is defined by $h(P_t) \in (-\infty, \infty)$. Since $[P_t/(1 - P_t)]$ is the ratio of the odds of $y_t = 1$ to the odds of $y_t = 0$, the practical implication of the logit link function is to regress the log odds ratio on X_t. We shall see that the probit (used for bioassay in 1935) implicitly uses the binomial model. Its link function, $h(P_t) = \Phi^{-1}(P_t)$, not only needs the numerical inverse CDF of the unit normal, it is not "canonical."

The normality assumption is obviously unrealistic when the variable assumes only limited values, or when the researcher is unwilling to assume precise knowledge about skewness, kurtosis, etc. In the present context of a binary dependent variable, $Var(\varepsilon_t)$ is a function of β and minimizing the S_2 with respect to β by the chain rule would have to allow for the dependence of $Var(\varepsilon_t)$ on β. See (4) and Result 1 above. Econometricians allow for this dependence by using a "feasible GLS" estimator, where the heteroscedasticity problem is solved by simply replacing $Var(\varepsilon_t)$ by its sample estimates. By contrast, we shall see that the GEE is based on the QSF of (3), which is the optimum EF and satisfies unbiasedness, sufficiency, efficiency, and robustness.

As in McCullagh and Nelder (1989), we denote the log of the quasi-likelihood by $Q(\mu; y)$ for μ based on the data y. For the normal distribution $Q(\mu; y) = -0.5(y - \mu)^2$, the variance function $Var(\mu) = 1$, and the canonical link is $h(\mu) = \mu$. For the binomial, $Q(\mu; y) = y \log[\mu/(1 - \mu)] + \log(1 - \mu)$, $Var(\mu) = \mu(1 - \mu)$, the canonical link called logit is $h(\mu) = \log[\mu/1(1 - \mu)]$. For the gamma, $Q(\mu; y) = -y/\mu - \log\mu$, $Var(\mu) = \mu^2$, and $h(\mu) = -1/\mu$. Since the link function of the gamma has a negative sign, the signs of all regression coefficients are reversed if the gamma distribution is used. In general, the quasi-score functions (QSFs) become our EFs as in (3):

$$\partial Q / \partial \beta = D' \Omega^{-1}(y - \mu) = 0 \tag{10}$$

where $\mu = h^{-1}(X\beta)$, $D = \{\partial \mu_t / \partial \beta_j\}$ is a $T \times p$ matrix of partials, and Ω is a $T \times T$ diagonal matrix with entries $\text{Var}(\mu_t)$ as noted above. The GLM estimate of β is given by solving (10) for β. Thus, the complication arising from a binary (or limited range) dependent variable is solved by using the GLM method. With the advent of powerful computers, more sophisticated modeling of dispersion has become feasible (Smyth and Verbyla 1999). In biomedical and GLM literature, there is much discussion of the "overdispersion" problem. This simply means that the actual dispersion exceeds the dispersion based on the analytical formulas arising from the assumed binomial, Poisson, etc.

3. GLM FOR PANEL DATA AND SUPERIORITY OF EFS OVER ML AND GLS

A limited dependent variable model for panel data (time series of cross-sections) in econometrics is typically estimated by the logit or probit, both of which have been known in biostatistics since the 1930s. GEE models can be viewed as generalizations of logit and probit models by incorporating time dependence among repeated measurements for an individual subject. When the biometric panel is of laboratory animals having a common heritage, the time dependence is sometimes called the "litter effect." The GEE models incorporate different kinds of litter effects characterized by serial correlation matrices $R(\phi)$, defined later as functions of a parameter vector ϕ.

The panel data involve an additional complication from three possible subscripts i, j, and t. There are $(i = 1, \ldots, N)$ individuals about which cross-sectional data are available in addition to the time series over $(t = 1, \ldots, T)$ on the dependent variable y_{it} and p regressors x_{ijt}, with $j = 1, \ldots, p$. We avoid subscript j by defining x_{it} as a $p \times 1$ vector of p regressors on ith individuals at time t.

Let y_{it} represent a dependent variable; the focus is on finding the relationship between y_{it} and explanatory variables x_{ijt}. Econometrics texts (e.g., Greene 2000, Section 15.2) treat this as a time series of cross-sectional data. To highlight the differences between the two approaches let us consider a binary or multiple-choice dependent variable. Assume that y_{it} equals one of a limited number of values. For the binary case, the values are limited to be simply 1 and 0. The binary case is of practical interest because we can simply assign $y_{it} = 1$ for a "true" value of a logical variable when some condition is satisfied, and $y_{it} = 0$ for its "false" value. For example, death, upward movement, acceptance, hiring, etc. can all be the "true" values of a logical variable. Let $P_{i,t}$ denote the probability of a "true" value for the ith individual ($i = 1, \ldots, N$) in the cross-section at time t ($t = 1, \ldots, T$), and note that

$$E(y_{it}) = 1P_{i,t} + 0(1 - P_{i,t}) = P_{i,t} \tag{11}$$

Now, we remove the time subscript by collecting elements of $P_{i,t}$ into $T \times 1$ vectors and write $E(y_i) = P_i$ as a vector of probabilities of "true" values for the ith individual. Let X_i be a $T \times p$ matrix of data on p regressors for the ith individual. As before, let β be a $p \times 1$ vector of regression parameters. If the method of latent variables is used, the "true or false" is based on a latent unobservable condition, which makes y_{it} be "true." Thus

$$\left.\begin{array}{l} y_{it} = 1, \text{ if the latent } y_{it}^* > 0, \text{ causing the condition to be "true"} \\ = 0, \text{ if } y_{it}^* \leq 0, \text{ where the "false" condition holds} \end{array}\right\} \tag{12}$$

Following the GLM terminology of link functions, we may write the panel data model as

$$h(P_i) = X_i\beta + \varepsilon_i, \quad E(\varepsilon_i) = 0, \quad E\varepsilon_i\varepsilon_i' = \sigma^2\Omega_i \text{ for } i = 1, \ldots, N \tag{13}$$

Now the logit link has $h(P_i) = \log[P_i/(1 - P_i)]$, whereas the probit link has $h(P_i) = \Phi^{-1}(P_i)$.

Econometricians often discuss the advisability of pooling or aggregation of the data for all N individuals and all T time points together. It is recognized that an identical error structure for N individuals and T time points cannot be assumed. Sometimes one splits the errors as $\varepsilon_{it} = M_i + v_{it}$, where v_{it} represents "random effects" and M_i denotes the "individual effects." Using the logit link, the log-odds ratio in a so-called random effects model is written as

$$\log(P_{i,t}/(1 - P_{i,t})) = x_{it}'\beta + M_i + v_{it} \tag{14}$$

The random effects model also assumes that $M_i \sim \text{IID}(0, \sigma^2 M)$ and $v_{it} \sim \text{IID}(0, \sigma^2)$ are independent and identically distributed. They are independent of each other and independent of the regressors x_{it}. It is explained in the panel data literature (Baltagi 1995, p. 178) that these individual effects complicate matters significantly. Note that, under the random effects assumptions in (13), covariance over time is nonzero, $E(\varepsilon_{it}\varepsilon_{is}) = \sigma_M^2$ is nonzero. Hence, independence is lost and the joint likelihood (probability) cannot be rewritten as a product of marginal likelihoods (probabilities).

Since the only feasible maximum likelihood implementation involves numerical integration, we may consider a less realistic "fixed effects" model, where the likelihood function is a product of marginals. Unfortunately, the fixed effects model still faces the so-called "problem of incidental parameters" (the number of parameters M_i increases indefinitely as $N \to \infty$). Some other solutions from the econometrics literature referenced by Baltagi include Chamberlain's (1980) suggestion to maximize a conditional likelihood function. These ML or GLS methods continue to suffer from unnecessary complications arising from the chain-rule-induced extra term in (4), which would make their FOCs in (3) imply biased and inefficient estimating functions.

4. DERIVATION OF THE GEE ESTIMATOR FOR β AND STATISTICAL INFERENCE

This section describes how panel data GEE methods can avoid the difficult and inefficient GLS or ML solutions used in the econometrics literature. We shall write a quasi-score function justified by the EF-theory as our GEE. We achieve a fully flexible choice of error covariance structures by using link functions of the GLM. Since GEE is based on the QSF (see equation (3)), we assume that only the mean and variance are known. The distribution itself can be any member of the exponential family with almost arbitrary skewness and kurtosis, which generalizes the often-assumed normal distribution. Denoting the log likelihood for the ith individual by L_i, we construct a $T \times 1$ vector $\partial L_i/\partial \beta$. Similarly, we construct $\mu_i = h^{-1}(X_i'\beta)$ as $T \times 1$ vectors and suppress the time subscripts. We denote a $T \times p$ matrix of partial derivatives by $D_i = \{\partial \mu_t/\partial \beta_j\}$ for $j = 1, \ldots, p$. In particular, we can often assume that $D_i = X_i$. When there is heteroscedasticity but no autocorrelation $\mathrm{Var}(y_i) = \Omega_i = \mathrm{diag}(\Omega_t)$ is a $T \times T$ diagonal matrix of variances of y_i over time. Using these notations, the ith QSF similar to (10) above is

$$\partial L_i/\partial \beta = D_i'\Omega_i^{-1}(y_i - \mu_i) = 0 \tag{15}$$

When panel data are available with repeated N measurements over T time units, GEE methods view this as an opportunity to allow for both autocorrelation and heteroscedasticity. The sum of QSFs from (15) over i leads to the so-called generalized estimating equation (GEE)

$$\sum_{i=1}^{N} D_i'V_i^{-1}(y_i - \mu_i) = 0, \quad \text{where } V_i = \Omega_i^{0.5} R(\phi)\Omega_i^{0.5} \tag{16}$$

where $R(\phi)$ is a $T \times T$ matrix of serial correlations viewed as a function of a vector of parameters common for each individual i. What econometricians call cross-sectional heteroscedasticity is captured by letting Ω_i be distinct for each i. There is distinct econometric literature on time series models with panel data (see Baltagi 1995 for references including Lee, Hsiao, MaCurdy, etc.). The sandwiching of $R(\phi)$ autocorrelations between two matrices of (heteroscedasticity) standard deviations in (16) makes V_i a proper covariance matrix. The GEE user can simply specify the general nature of the autocorrelations by choosing $R(\phi)$ from the following list, stated in increasing order of flexibility. The list contains common abbreviations used by most authors of the GEE software.

(i)   Independence means that $R(\phi)$ is the identity matrix.
(ii) Exchangeable $R(\phi)$ means that all inter-temporal correlations, $\text{corr}(y_{it}, y_{is})$, are equal to ϕ, which is assumed to be a constant.
(iii) AR(1) means first-order autoregressive model. If $R(\phi)$ is AR(1), its typical (i,j)th element will be equal to the correlation between y_{it} and y_{is}; that is, that it, it will equal $\phi^{|t-s|}$.
(iv) Unstructured correlations in $R(\phi)$ means that it has $T(T-1)/2$ distinct values for all pairwise correlations.

Finally, we choose our $R(\phi)$ and solve (16) iteratively for β, which gives the GEE estimator. Liang and Zeger (1986) suggest a "modified Fisher scoring" algorithm for their iterations (see Remark 2 above). The initial choice of $R(\phi)$ is usually the identity matrix and the standard GLM is first estimated. The GEE algorithm then estimates $R(\phi)$ from the residuals of the GLM and iterates until convergence. We use YAGS (2000) software in S-PLUS (2001) language on an IBM-compatible computer. The theoretical justification for iterations exploits the property that a QML estimate is consistent even if $R(\phi)$ is misspecified (Zeger and Liang 1986, McCullagh and Nelder, 1989, p. 333).

Denoting estimates by hats, the asymptotic covariance matrix of the GEE estimator is

$$\text{Var}(\hat{\beta}_{\text{gee}}) = \sigma^2 A^{-1} B A^{-1} \tag{17}$$

with $A = \sum_{i=1}^{N} D_i' \hat{V}_i^{-1} D_i$ and $B = \sum_{i=1}^{N} D_i' \hat{R}_i^{-1} R_i \hat{R}_i^{-1} D_i$, where R_i refers to the $R_i(\phi)$ value in the previous iteration (see Zeger and Liang 1986). Square roots of diagonal terms yield the "robust" standard errors reported in our numerical work in the next section. Note that the GEE software uses the traditional standard errors of estimated β for inference.

Vinod (1998, 2000) proposes an alternative approach to inference when estimating functions are used for parameter estimation. It is based on Godambe's pivot function (GPF) whose asymptotic distribution is unit normal and does not depend on unknown parameters. The idea is to define a $T \times p$ matrix H_{it} and redefine $\varepsilon_i = (y_i - \mu_i)$ as a $T \times 1$ vector such that we can write the QSF of (15) for the ith individual as a sum of T quasi-score function $p \times 1$ vectors S_{it}, $\sum_{t=1}^{T} H_{it}' \varepsilon_i = \sum_{t=1}^{T} S_{it}$. Under appropriate regularity conditions, this sum of T items is asymptotically normally distributed. In the univariate normal $N(0, \sigma^2)$ case it is customary to divide by σ to make the variance unity. Here we have a multivariate situation, and rescaling S_{it} such that its variance is unity requires pre-multiplication by a square root matrix. Using ($\sim$) to denote rescaled quasi-scores we write for each i the following nonlinear p-dimensional vector equation:

$$\sum_{t=1}^{T} \tilde{S}_{it} = \left[D_i' V_i^{-1} D_i \right]^{-0.05} D_i' V_i^{-1} (y_i - \mu_i) \tag{18}$$

The scaled sum of quasi-scores remains a sum of T items whose variance is scaled to be unity. Hence $\sqrt{T}$ times the sum in (18) is asymptotically unit normal $N(0, 1)$ for each i. Assuming independence of the individuals i from each other, the sum over i remains normal. Note that $\sqrt{N}$ times a sum of N unit normals is unit normal $N(0, 1)$. Thus, we have the desired pivot (GPF), defined as

$$\sum_{i=1}^{N} \sum_{t=1}^{T} \tilde{S}_{it} = \sum_{t=1}^{T} \sum_{i=1}^{N} \tilde{S}_{it} = 0 \tag{19}$$

where we have interchanged the order of summations. For computerized bootstrap inference suggested in Vinod (2000), one should simply shuffle with replacement the T items $[\sum_{i=1}^{N} \tilde{S}_{it}]$ a large number of times ($J = 999$, say). Each shuffle yields a nonlinear equation which should be solved for β to construct $J = 999$ estimates of β. The next step is to use these estimates to construct appropriate confidence intervals upon ordering the coefficient estimates from the smallest to the largest. Vinod (1998) shows that this computer-intensive method has desirable robustness (distribution-free) properties for inference and avoids some pitfalls of the usual Wald-type statistics based on division by standard errors. The derivation of scaled scores for GEE is claimed to be new.

In the following section, we consider applications. First, one must choose the variance-to-mean relation typified by a probability distribution family. For example, for the Poisson family the mean and variance are both λ. By contrast, the binomial family, involving n trials with probability p of success, has its mean np and its variance, $np(1 - p)$, is smaller than the mean. The idea of using a member of the exponential family of distributions to fix the relation between mean and variance is largely absent in econometrics. The lowest residual sum of squares (RSS) can be used to guide this choice. Instead of RSS, McCullagh and Nelder (1989, p. 290) suggest using a "deviance function" defined as the difference between restricted and unrestricted log likelihoods. A proof of consistency of the GEE estimator is given by Li (1997). Heyde (1997, p. 89) gives the necessary and sufficient conditions under which GEE are fully efficient asymptotically. Lipsitz et al. (1994) report simulations showing that GEE are more efficient than ordinary logistic regressions. In conclusion, this section has shown that the GEE estimator is practical, with attractive properties for the typical data available to applied econometricians. Of course, one should pay close attention to economic theory and data generation, and modify the applications to incorporate sample selection, self-selection, and other anomalies when human economic agents rather than animals are involved.

5. GEE ESTIMATION OF STOCK PRICE TURNING POINTS

This section describes an illustrative application of our methods to financial economics. The focus of this study is on the probability of a turn in the direction of monthly price movements of major stocks. Although daily changes in stock prices may be volatile, we expect monthly changes to be less volatile and somewhat sensitive to interest rates. The "Fed effect" discussed in *Wall Street Journal* (2000) refers to a rally in the S&P 500 prices a few days before the meeting of the Federal Reserve Bank and a price decline after the meeting. We use monthly data from May 1993 to July 1997 for all major stocks whose market capitalization exceeds 27 billion. From this list of about 140 companies compiled in the Compustat database, we select the first seven companies in alphabetical order. Our purpose here is to briefly illustrate the statistical methods, not to carry out a comprehensive stock market analysis. The sensitivity of the stock prices to interest rates is studied in Elsendiony's (2000) recent Fordham University dissertation. We study only whether monetary policy expressed through the interest rates has a significant effect on the stock market turning points. We choose the interest on 3-month Treasury bills (TB3) as our monetary policy variable.

Using Compustat's data selection software, we select the companies with the following ticker symbols: ABT, AEG, ATI, ALD, ALL, AOL, and AXP. The selected data are saved as comma delimited text file and read into a workbook. The first workbook task is to use the price data to construct a binary variable, ΔD, to represent the turning points in prices. If S_t denotes the spot price of the stock at time t, S_{t-1} the price at time $t - 1$, if the price is rising (falling) the price difference, $\Delta S_t = S_t - S_{t-1}$, is positive (negative). For the initial date, May 1993, ΔS_t has a data gap coded as "NA." In Excel, all positive ΔS_t numbers are made unity and negatives are made zero. Upon further differencing these ones and zeros, we have ones or negative ones when there are turning points. One more data gap is created in this process for each stock. Finally, replacing the

(−1)s by (+1)s, we create the changing price direction $\Delta D \in [0, 1]$ as our binary dependent variable.

Next, we use statistical software for descriptive statistics for ΔD and TB3. We now report them in parentheses separated by a comma: minimum or min (0, 2.960), first quartile or Q1 (0, 4.430), mean (0.435, 4.699), median (0, 4.990), third quartile or Q3 (1, 5.150), maximum or max (1, 5.810), standard deviation (0.496, 0.767), total data points (523, 523) and NAs (19, 0). Box plots for the seven stocks for other stock characteristics discussed in Elsendiony (2000), including price earnings ratio, log of market capitalization, and earnings per share, etc. (not reproduced, for brevity), show that our seven stocks include a good cross section with distinct box plots for these variables. Simple correlation coefficients $r(\text{TB3}, \Delta S_t)$ over seven stocks separately have min $= -0.146$, mean $= 0.029$, and max $= 0.223$. Similar coefficients $r(\text{TB3}, \Delta D)$ over seven stocks have min $= -0.222$, mean $= -0.042$, and max $= 0.109$. Both sets suggest that the relation between TB3 and changes in individual stock prices do not exhibit any consistent pattern. The coefficient magnitudes are small, with ambiguous signs.

Wall Street specialists study individual companies and make specific "strong buy, buy, accumulate, hold, market outperform, sell" type recommendations privately to clients and influence the market turning points when clients buy or sell. Although monetary policy does not influence these, it may be useful to control for these factors and correct for possible heteroscedasticity caused by the size of the firm. We use a stepwise algorithm to choose the additional regressor for control, which suggests log of market capitalization (logMK).

Our data consist of 7 stocks (cross-sectional units) and 75 time series points including the NAs. One can use dummy variables to study the intercepts for each stock by considering a "fixed effects" model. However, our focus is on the relation of turning points to the TB3 and logMK covariates and not on individual stocks. Hence, we consider the GLM of (13) with $y_i = \Delta S_t$ or ΔD. If $y_i = \Delta S_t \in (-\infty, \infty)$ is a real number, its support suggests the "Gaussian" family and the canonical link function $h(P_i) = P_i$ is identity. The canonical links have the desirable statistical property that they are "minimal sufficient" statistics. Next choice is $y_i = \Delta D \in [0, 1]$, which is 1 when price direction changes (up or down) and zero otherwise. With only two values 0 and 1, it obviously belongs to the binomial family with the choice of a probit or logit link. Of the two, the GLM theory recommends the canonical logit link, defined as $h(P_i) = \log[P_i/(1 - P_i)]$.

We use V. Carey's GEE computer program called YAGS written for S-Plus 2000 and implement it on a home PC having 500 MHz speed. Now, we write $E(y_i) = P_i$ and estimate

$$h(P_i) = \beta_0 + \beta_1[\text{TB3}] + \beta_2[\log(\text{MK})] \tag{20}$$

Ordinary least squares (OLS) regression on pooled data is a special case when $h(P_i) = y_i$. We choose first-order autoregression, AR(1), from among the four choices for the time series dependence of errors listed after (16) above for both $y_i = \Delta S_t$ and ΔD.

Let us omit the details in reporting the $y_i = \Delta S_t$ case, for brevity. The estiamte of β_1 is $-0.073\,24$ with a robust z value of $-0.422\,13$. It is statistically insignificant since these z-values have $N(0, 1)$ distribution. For the Gaussian family, the estimate 24.1862 by YAGS of the scale parameter is interpreted as the usual variance. A review of estimated residuals suggests non-normality and possible outliers. Hence, we estimate (20), when $y_t = \Delta S_t$, a second time by using a robust MM regression method for pooled data. This robust regression algorithm available with S-Plus (not YAGS) lets the software choose the initial S-estimates and final M-estimates with asymptotic efficiency of 0.85. The estimate of β_1 is now 0.2339, which has the opposite sign, while its t-value of 1.95 is almost significant. This sign reversal between robust and YAGS estimates suggests that TB3 cannot predict ΔS_t very well. The estimate of β_2 is 1.6392, with a significant t-value of 2.2107.

Now consider $y_i = \Delta D$, turning points of stock prices in relation to TB3, the monetary policy changes made by the Federal Reserve Bank. Turning points have long been known to be difficult to forecast. Elsendiony (2000) studies a large number of stocks and sensitivity of their spot prices S_t to interest rates. He finds that some groups of stocks, defined in terms of their characteristics such as dividends, market capitalization, product durability, market risk, price earnings ratios, debt equity ratios, etc., are more "interest sensitive" than others. Future research can consider a reliable turning-point forecasting model for interest-sensitive stocks. Our modest objectives here are to estimate the extent to which the turning points of randomly chosen stocks are interest sensitive.

Our methodological interest here is to illustrate estimation of (20) with the binary dependent variable $y_i = \Delta D$, where the superiority of GEE is clearly demonstrated above. Recall that the definition of ΔD leads to two additional NAs for each firm. Since YAGS does not permit missing values (NAs), we first delete all observation rows having NAs for any stock for any month. The AR(1) correlation structure for the time series for each stock has the "working" parameter estimated to be 0.115\,930\,9 by YAGS. We report the estimates of (20) under two regimes, where we force $\beta_2 = 0$ and where $\beta_2 \neq 0$, in two sets of three columns in Table 1. Summary statistics for the raw residuals (lack of fit) for the first regime are: min $= -0.484\,27$, QI $= -0.427\,23$, median $= -0.413\,10$, Q3 $= 0.572\,07$, and max $= 0.596\,33$.

Table 1. GEE estimation results for the binary dependent variable ΔD (turning point)

Regressor	Coefficient	Robust SE	Robust z	Coefficient	Robust SE	Robust z
Intercept	0.276 986 5	0.602 811 4	0.459 491 2	0.002 729 5	0.455 454	0.005 992 9
TB3	−0.114 835 4	0.123 491 9	−0.929 902 2	−0.121 834 6	0.127 234	−0.957 56
logMK				0.031 636 70	0.027 638	1.144 671
Scale est.	1.000 805			1.000 752		
AR(1)	0.115 930 9			0.115 597 2		
Median of residuals	−0.413 10 5			−0.408 351 7		
Family	Binomial			Binomial		

The residuals summary statistics for the second regime are similar. Both suggest a wide variation. The estimated scale parameter is 1.00081, which is fairly close to unity. Whenever this is larger than unity, it captures the overdispersion often present in biostatistics. The intercept 0.27699 has the statistically insignificant z-value of 0.4599 at conventional levels. The slope coefficient β_1 of TB3 is (0.11484, robust SE from (17) is 0.12349, and the z-value (0.92990 is also statistically insignificant. This suggests that that a rise in TB3 does not signal a turning point for the stock prices. The GEE results for (20) when $\beta_2 \neq 0$, that is when it is not forced out of the model, reported in the second set of three columns of Table 1, are seen to be similar to the first regime. The estimated scale parameters in Table 1 slightly exceed unity in both cases where $\beta_2 = 0$ and $\beta_2 \neq 0$, suggesting almost no overdispersion compared to the dispersion assumed by the binomial model. For brevity, we omit similar results for differenced TB3, ΔTB3

6. CONCLUDING REMARKS

Mittelhammer et al.'s (2000) text, which cites Vinod (1997), has brought EFs to mainstream econometrics. However, it has largely ignored two popular estimation methods in biostatistics literature, called GLM and GEE. This chapter reviews related concepts, including some of the estimating function literature, and shows how EFs satisfy "unbiasedness, sufficiency, efficiency, and robustness." It explains why estimation problems involving limited dependent variables are particularly promising for applications of the EF theory. Our Result 1 shows that whenever heteroscedasticity is related to β, the traditional GLS or ML estimators have an unnecessary extra term, leading to biased and inefficient EFs. Result 2 shows why binary dependent variables have such heteroscedasticity. The panel data GEE estimator in (16) is implemented by Liang and Zeger's (1986) "modified Fisher scoring" algorithm with variances given in (17). The flexibility of the GEE estimator arises from its ability to specify the matrix of autocorrelations $R(\phi)$ as a function of a set of parameters . We use the AR(1) structure and a "canonical link" function satisfying "sufficiency" properties available for all distributions from the exponential family. It is well known that this family includes many of the familiar distributions, including normal, binomial, Poisson, exponential, and gamma.

We discuss the generalized linear models (GLM) involving link functions to supplement the strategy of fixed and random effects commonly used in econometrics. When there is no interest in the fixed effects themselves, the GLM methods choose a probability distribution of these effects to control the mean-to-variance relation. For example, the binomial distribution is suited for binary data. Moreover, the binomial model makes it easy to see why its variance (heteroscedasticity) is related to the mean, how its canonical link is logit, and how it is relevant for sufficiency. For improved computer intensive inference, we develop new pivot functions for bootstrap shuffles. It is well known that reliable bootstrap inference needs pivot functions to resample, which do not depend on unknown parameters. Our pivot in (19) is asymptotically unit normal.

We apply these methods to the important problem involving estimation of the "turning points." We use monthly prices of seven large company stocks for 75 months in the 1990s. We then relate the turning points to monetary policy changes by the Federal Reserve Bank's targeting the market interest rates. In light of generally known difficulty in modeling turning points of stock market prices, and the fact that we use only limited data on very large corporations, our results may need further corroboration. Subject to these limitations, we conclude from our various empirical estimates that Federal Reserve's monetary policy targeting market interest rates cannot significantly initiate a stock market downturn or upturn in monthly data.

A change in interest rate can have an effect on the prices of some interest-sensitive stocks. It can signal a turning point for such stocks. Elsendiony (2000) has classified some stocks as more interest sensitive than others. A detailed study of all relevant performance variables for all interest-sensitive stocks and prediction of stock market turning points is left for future research. Vinod and Geddes (2001) provide another example.

Consistent with the theme of the handbook, we hope to encourage more collaborative research between econometricians and statisticians in the areas covered here. We have provided an introduction to GLM and GEE methods in notations familiar to economists. This is not a comprehensive review of latest research on GEE and GLM for expert users of these tools. However, these experts may be interested in our new bootstrap pivot of (19), our reference list and our notational bridge for better communication with economists. Econometrics offers interesting challenges when choice models have to allow for intelligent economic agents maximizing their own utility and sometimes falsifying or cheating in achieving their goals. Researchers in epidemiology and medical economics have begun to recognize that patient behavior can be different from that of passive animal subjects in biostatistics. A better communication between econometrics and biostatistics is obviously worthwhile.

REFERENCES

B. H. Baltagi. *Econometric Analysis of Panel Data*. New York: Wiley, 1995.

G. Chamberlain. Analysis of covariance with qualitative data. *Review of Economic Studies* 47: 225–238, 1980.

P. J. Diggle, K.-Y. Liang, S. L. Zeger. *Analysis of Longitudinal Data*. Oxford: Clarendon Press, 1994.

D. D. Dunlop. Regression for longitudinal data: A bridge from least squares regression. American *Statistician* 48: 299–303, 1994.

M. A. Elsendiony. An empirical study of stock price-sensitivity to interest rate. Doctoral dissertaion, Fordham University, Bronx, New York, 2000.

V. P. Godambe. An optimum property of regular maximum likelihood estimation. *Annals of Mathematical Statistics* 31: 1208–1212, 1960.

V. P. Godambe. The foundations of finite sample estimation in stochastic processes. *Biometrika* 72: 419–428, 1985.

V. P. Godambe, B. K. Kale. Estimating functions: an overview. Chapter 1 In: V. P. Godambe (ed.). *Estimating Functions*. Oxford: Clarendon Press, 1991.

W. H. Greene. *Econometric Analysis*. 4th edn. Upper Saddle River, NJ: Prentice Hall, 2000.

C. C. Heyde. *Quasi-Likelihood and its Applications*. New York: Springer-Verlag, 1997.

B. Li. On consistency of generalized estimating equations. In I. Basawa, V. P. Godambe, R. L. Taylor (eds.). *Selected Proceedings of the Symposium on Estimating Functions*. IMS Lecture Notes—Monograph Series, Vol. 32, pp. 115–136, 1997.

K. Liang, S. L. Zeger. Longitudinal data analysis using generalized linear models. *Biometrika* 73: 13–22, 1986.

K. Liang, S. L. Zeger. Inference based on estimating functions in the presence of nuisance parameters. *Statistical Science* 10: 158–173, 1995.

K. Liang, B. Quaqish, S. L. Zeger. Multivariate regression analysis for categorical data. *Journal of the Royal Statistical Society (Series B)* 54: 3–40 (with discussion), 1992.

S. R. Lipsitz, G. M. Fitzmaurice, E. J. Orav, N. M. Laird. Performance of generalized estimating equations in practical situations. *Biometrics* 50: 270–278, 1994.

P. McCullagh, J. A. Nelder. *Generalized Linear Models*, 2nd edn. New York: Chapman and Hall, 1989.

R. C. Mittelhammer, G. G. Judge, D. J. Miller. *Econometric Foundations*. New York: Cambridge University Press, 2000.

G. K. Smyth, A. P. Verbyla. Adjusted likelihood methods for modelling dispersion in generalized linear models. *Environmetrics* 10: 695–709, 1999.

S-PLUS (2001) Data Analysis Products Division, Insightful Corporation, 1700 Westlake Avenue N, Suite 500, Seattle, WA 98109. E-mail: *info@insightful.com* Web: www.insightful.com.

H. D. Vinod. Using Godambe–Durbin estimating functions in econometrics. In: I. Basawa, V. P. Godambe, R. L. Taylor (eds). *Selected Proceedings of the Symposium on Estimating Functions*. IMS Lecture Notes–Monograph Series, Vol. 32: pp. 215–237, 1997.

H. D. Vinod. Foundations of statistical inference based on numerical roots of robust pivot functions (Fellow's Corner). *Journal of Econometrics* 86: 387–396, 1998.

H. D. Vinod. Foundations of multivariate inference using modern computers. To appear in *Linear Algebra and its Applications*, Vol. 321: pp 365-385, 2000.

H. D. Vinod, R. Geddes. Generalized estimating equations for panel data and managerial monitoring in electric utilities, in N. Balakrishnan (Ed.) *Advances on Methodological and Applied Aspects of Probability and Statistics*. Chapter 33. London, Gordon and Beach Science Publishers: pp. 597-617, 2001.

Wall Street Journal. Fed effect may roil sleepy stocks this week. June 19, 2000, p. C1.

R. W. M. Wedderburn. Quasi-likelihood functions, generalized linear models and the Gaussian method. *Biometrika* 61: 439–447, 1974.

YAGS (yet another GEE solver) 1.5 by V. J. Carey. See *http://biosun1. harvard.edu/carey/gee.html*, 2000.

S. L. Zeger, K.-Y. Liang. Longitudinal data analysis for discrete and continuous outcomes. *Biometrics* 42: 121–130, 1986.

27

Sample Size Requirements for Estimation in SUR Models

WILLIAM E. GRIFFITHS   University of Melbourne, Melbourne, Australia

CHRISTOPHER L. SKEELS   Australian National University, Canberra, Australia

DUANGKAMON CHOTIKAPANICH   Curtin University of Technology, Perth, Australia

1. INTRODUCTION

The seemingly unrelated regressions (SUR) model [17] has become one of the most frequently used models in applied econometrics. The coefficients of individual equations in such models can be consistently estimated by ordinary least squares (OLS) but, except for certain special cases, efficient estimation requires joint estimation of the entire system. System estimators which have been used in practice include two-stage methods based on OLS residuals, maximum likelihood (ML), and, more recently, Bayesian methods; e.g. [11, 2]. Various modifications of these techniques have also been suggested in the literature; e.g. [14, 5]. It is somewhat surprising, therefore, that the sample size requirements for joint estimation of the parameters in this model do not appear to have been correctly stated in the literature. In this paper we seek to correct this situation.

The usual assumed requirement for the estimation of SUR models may be paraphrased as: the sample size must be greater than the number of explanatory variables in each equation and at least as great as the number of equations in the system. Such a statement is flawed in two respects. First, the estimators considered sometimes have more stringent sample size requirements than are implied by this statement. Second, the different estimators may have different sample size requirements. In particular, for a given model the maximum likelihood estimator and the Bayesian estimator with a noninformative prior may require larger sample sizes than does the two-stage estimator.

To gain an appreciation of the different sample size requirements, consider a 4-equation SUR model with three explanatory variables and a constant in each equation. Suppose that the explanatory variables in different equations are distinct. Although one would not contemplate using such a small number of observations, two-stage estimation can proceed if the number of observations (T) is greater than 4. For ML and Bayesian estimation $T > 16$ is required. For a 10-equation model with three distinct explanatory variables and a constant in each equation, two-stage estimation requires $T \geq 11$—the conditions outlined above would suggest that 10 observations should be sufficient—whereas ML and Bayesian estimation need $T > 40$. This last example illustrates not only that the usually stated sample size requirements can be incorrect, as they are for the two-stage estimator, but also just how misleading they can be for likelihood-based estimation.

The structure of the paper is as follows. In the next section we introduce the model and notation, and illustrate how the typically stated sample size requirement can be misleading. Section 3 is composed of two parts: the first part of Section 3 derives a necessary condition on sample size for two-stage estimation and provides some discussion of this result; the second part illustrates the result by considering its application in a variety of different situations. Section 4 derives the analogous result for ML and Bayesian estimators. It is also broken into two parts, the first of which derives and discusses the result, while the second part illustrates the result by examining the model that was the original motivation for this paper. Concluding remarks are presented in Section 5.

2. THE MODEL AND PRELIMINARIES

Consider the SUR model written as

$$y_j = X_j\beta_j + e_j, \qquad j = 1, \ldots, M \tag{1}$$

where y_j is a $(T \times 1)$ vector of observations on the dependent variable for the jth equation, X_j is a $(T \times K_j)$ matrix of observations on K_j explanatory variables in the jth equation. We shall assume that each of the X_j have full column rank. It will also be assumed that the $(TM \times 1)$ continuously distributed random vector $e = [e_1', e_2', \ldots, e_M']'$ has mean vector zero and covariance matrix $\Sigma \otimes I_T$. We are concerned with estimation of the $(K \times 1)$ coefficient vector $\beta = [\beta_1', \ldots, \beta_M']'$, where $K = \sum_{j=1}^m K_j$ is the total number of unknown coefficients in the system.

The OLS estimators for the β_j are

$$b_j = (X_j'X_j)^{-1}X_j'y_j, \qquad j = 1, \ldots, M$$

The corresponding OLS residuals are given by

$$\hat{e}_j = y_j - X_j b_j = M_{X_j} y_j \tag{2}$$

where $M_A = I - A(A'A)^{-1}A' = I - P_A$ for any matrix A of full column rank. We shall define

$$\hat{E} = [\hat{e}_1, \hat{e}_2, \ldots, \hat{e}_M]$$

The two-stage estimator for this system of equations is given by

$$\hat{\beta} = [\hat{X}'(\hat{\Sigma}^{-1} \otimes I_T)\hat{X}]^{-1}\hat{X}'(\hat{\Sigma}^{-1} \otimes I_T)y \tag{3}$$

where the $(TM \times K)$ matrix $\hat{X}$ is block diagonal with the X_j making up the blocks, $y = [y_1', y_2', \ldots, y_M']'$, and*

$$T\hat{\Sigma} = \hat{E}'\hat{E}$$

Clearly, $\hat{\beta}$ is not operational unless $\hat{\Sigma}$ is nonsingular, and $\hat{\Sigma}$ will be singular unless the $(T \times M)$ matrix $\hat{E}$ has full column rank. A standard argument is that $\hat{E}$ will have full column rank with probability one provided that (i) $T \geq M$, and (ii) $T > k_{\max} = \max_{j=1,\ldots,M} K_j$. *Observe that (ii) is stronger than the* $T \geq k_{\max}$ *requirement implicit in assuming that all* X_j *have full column rank*; it is required because for any $K_j = T$ the corresponding $\hat{e}_j$ is identically equal to zero, ensuring that $\hat{\Sigma}$ is singular. Conditions (i) and (ii) can be summarized as[†]

$$T \geq \max(M, k_{\max} + 1) \tag{4}$$

---

*A number of other estimators of Σ have been suggested in the literature; they differ primarily in the scaling applied to the elements of $\hat{E}'\hat{E}$ (see, for example, the discussion in [13, p. 17]), but $\hat{\Sigma}$ is that estimator most commonly used. Importantly, $\hat{\Sigma}$ uses the same scaling as does the ML estimator for Σ, which is the appropriate choice for likelihood-based techniques of inference. Finally, $\hat{\Sigma}$ was also the choice made by Phillips [12] when deriving exact finite-sample distributional results for the two-stage estimator in this model. Our results are consequently complementary to those earlier ones.

[†]Sometimes only part of the argument is presented. For example, Greene [6, p. 627] formally states only (i).

That (4) does not ensure the nonsingularity of $\hat{\Sigma}$ is easily demonstrated by considering the special case of the multivariate regression model, arising when $X_1 = \cdots = X_M = X^*$ (say) and hence $K_1 = \cdots = K_M = k^*$ (say). In this case model (1) reduces to

$$Y = X^*B + E$$

where

$$
\begin{aligned}
Y &= [y_1, y_2, \ldots, y_M] \\
B &= [\beta_1, \beta_2, \ldots, \beta_M] \\
E &= [e_1, e_2, \ldots, e_M]
\end{aligned}
\tag{5}
$$

and

$$T\hat{\Sigma} = E'E = Y'M_{x^*}Y$$

Using a full rank decomposition, we can write

$$M_{X^*} = CC'$$

where C is a $(T \times (T - k^*))$ matrix of rank $T - k^*$. Setting $W = C'Y$, we have

$$T\hat{\Sigma} = W'W$$

whence it follows that, in order for $\hat{\Sigma}$ to have full rank, the $((T - k^*) \times M)$ matrix W must have full column rank, which requires that $T - k^* \geq M$ or, equivalently, that*

$$T \geq M + k^* \tag{6}$$

In the special case of $M = 1$, condition (6) corresponds to condition (4), as $k^* = k_{max}$. For $M > 1$, condition (6) is more stringent in its requirement on sample size than condition (4), which begs the question whether even more stringent requirements on sample size exist for the two-stage estimation of SUR models. It is to this question that we turn next.

---

*It should be noted that condition (6) is not new and is typically assumed in discussion of the multivariate regression model; see, for example, [1, p. 287], or the discussion of [4] in an empirical context.

3. TWO-STAGE ESTIMATION

3.1 A Necessary Condition and Discussion

Our proposition is that, subject to the assumptions given with model (1), two-stage estimation is feasible provided that $\hat{\Sigma}$ is nonsingular. Unfortunately, $\hat{\Sigma}$ is a matrix of random variables and there is nothing in the assumptions of the model that precludes $\hat{\Sigma}$ being singular. However, there are sample sizes that are sufficiently small that $\hat{\Sigma}$ must be singular and in the following theorem we shall characterize these sample sizes. For all larger sample sizes $\hat{\Sigma}$ will be nonsingular with probability one and two-stage estimation feasible. That $\hat{\Sigma}$ is only non-singular *with probability one* implies that our condition is necessary but not sufficient for the two-stage estimator to be feasible; unfortunately, no stronger result is possible.[*]

Theorem 1. In the model (1), a necessary condition for the estimator (3) to be feasible, in the sense that $\hat{\Sigma}$ is nonsingular with probability one, is that

$$T \geq M + \rho - \eta \tag{7}$$

where $\rho = \text{rank}([X_1, X_2, \ldots, X_M])$ and η is the rank of the matrix D defined in equations (9) and (10).

Proof. Let $X = [X_1, X_2, \ldots, X_M]$ be the $(T \times K)$ matrix containing all the explanatory variables in all equations. We shall define $\rho = \text{rank}(X) \leq T$. Next, let the orthogonal columns of the $(T \times \rho)$ matrix V comprise a basis set for the space spanned by the columns of X, so that there exists $(\rho \times k_j)$ matrices F_j such that $X_j = VF_j$ (for all $j = 1, \ldots, M$).[†] Under the assumption that X_j has full column rank, it follows that F_j must also have full column rank and so we see that $k_{\max} \leq \rho \leq T$. It also follows that there exist $(\rho \times (\rho - k_j))$ matrices G_j such that $[F_j, G_j]$ is nonsingular and $F_j'G_j = 0$. Given G_j we can define $(T \times (\rho - k_j))$ matrices

$$Z_j = VG_j, \qquad j = 1, \ldots, M$$

---

[*]A necessary and sufficient condition is available only if one imposes on the support of e restrictions which preclude the possibility of $\hat{\Sigma}$ being singular except when the sample size is sufficiently small. For example, one would have to preclude the possibility of either multicollinearity between the y_j or any $y_j = 0$, both of which would ensure a singular $\hat{\Sigma}$. We have avoided making such assumptions here.

[†]Subsequently, if the columns of a matrix A (say) form a basis set for the space spanned by the columns of another matrix Q (say), we shall simply say that A is a basis for Q.

The columns of each Z_j span that part of the column space of V not spanned by the columns of the corresponding X_j. By Pythagoras' theorem,

$$P_V = P_{X_j} + M_{X_j} Z_j \left(Z_j' M_{X_j} Z_j \right)^{-1} Z_j' M_{X_j}$$

Observe that, because $X_j' Z_j = 0$ by construction, $M_{X_j} Z_j = Z_j$ giving $P_V = P_{X_j} + P_{Z_j}$ or, equivalently,

$$M_{X_j} = M_V + P_{Z_j}$$

From equation (2),

$$\hat{e}_j = [M_V + P_{Z_j}] y_j$$

so that

$$\hat{E} = M_V Y + D \tag{8}$$

where

$$D = [d_1, \ldots, d_M] \tag{9}$$

with

$$d_j = \begin{cases} P_{z_j} y_j, & \text{if } \rho > k_j, \\ 0, & \text{otherwise}, \end{cases} \qquad j = 1, \ldots, M \tag{10}$$

Thus, the OLS residual for each equation can be decomposed into two orthogonal components, $M_V y_j$ and d_j. $M_V y_j$ is the OLS residual from the regression of y_j on V, and d_j is the orthogonal projection of y_j onto that part of the column space of V which is not spanned by the columns of X_j. Noting that $Y' M_V D = 0$, because $M_V Z_j = 0$ ($j = 1, \ldots, M$), equation (8) implies that

$$\hat{E}' \hat{E} = Y' M_V Y + D' D \tag{11}$$

It is well known that if R and S are any two matrices such that $R + S$ is defined, then[*]

$$\text{rank}(R + S) \leq \text{rank}(R) + \text{rank}(S) \tag{12}$$

Defining $\theta = \text{rank}(\hat{E}' \hat{E})$, $\delta = \text{rank}(Y' M_V Y)$, and $\eta = \text{rank}(D)$, equations (11) and (12) give us

$$\theta \leq \delta + \eta$$

---

[*]See, for example, [10, A.6(iv)].

Now, M_V admits a full rank decomposition of the form $M_V = HH'$, where H is a $(T \times (T - \rho))$ matrix. Consequently, $\delta = \text{rank}(Y'M_V Y) = \text{rank} (Y'H) \leq \min(M, T - \rho)$, with probability one, so that

$$\theta \leq \min(M, T - \rho) + \eta$$

Clearly, $\hat{E}'\hat{E}$ has full rank if and only if $\theta = M$, which implies

$$M \leq \min(M, T - \rho) + \eta \tag{13}$$

If $T - \rho \geq M$, equation (13) is clearly satisfied. Thus, the binding inequality for (13) occurs when $\min(M, T - \rho) = T - \rho$. □

As noted in equation (6), and in further discussion below, $T \geq M + \rho$ is the required condition for a class of models that includes the multivariate regression model. When some explanatory variables are omitted from some equations, the sample size requirement is less stringent, with the reduced requirement depending on $\eta = \text{rank}(D)$.

Care must be exercised when applying the result in (7) because of several relationships that exist between M, T, ρ, and η. We have already noted that $\rho \leq T$. Let us examine η more closely. First, because D is a $(T \times M)$ matrix, and $\eta = \text{rank}(D)$, it must be that $\eta \leq \min(M, T)$. Actually, we can write $\eta \leq \min(d, T)$, where $0 \leq d \leq M$ denotes the number of nonzero d_j; that is, d is the number of X_j which do not form a basis for X.

Second, the columns of D are a set of projections onto the space spanned by the columns of $Z = [Z_1, Z_2, \ldots, Z_M]$, a space of possibly lower dimension than ρ, say $\rho - \omega$, where $0 \leq \omega \leq \rho$. In practical terms, Z_j is a basis for that part of V spanned by the explanatory variables excluded from the jth equation; the columns of Z span that part of V spanned by all variables excluded from at least one equation. If there are some variables common to all equations, and hence not excluded from any equations, then Z will not span the complete ρ-dimensional space spanned by V. More formally, we will write $V = [V_1, V_2]$, where the $(T \times (\rho - \omega))$ matrix V_2 is a basis for Z and the $(T \times \omega)$ matrix V_1 is a basis for a subspace of V spanned by the columns of each X_j, $j = 1, \ldots, M$. The most obvious example for which $\omega > 0$ is when each equation contains an intercept. Another example is the multivariate regression model, where $\omega = \rho$, so that V_2 is empty and $D = 0$. Clearly, because $T \geq \rho \geq \rho - \omega$, the binding constraint on η is not $\eta \leq \min(d, T)$ but rather $\eta \leq \min(d, \rho - \omega)$.

Note that $\eta \leq \min(d, \rho - \omega) \leq \rho - \omega \leq \rho$, which implies that $T \geq M + \rho - \eta \geq M$ is a necessary condition for $\hat{\Sigma}$ to be nonsingular. Obviously $T \geq M$ is part of (4). The shortcoming of (4) is its failure to recognize the interactions of the X_j in the estimation of (1) as a system of equations. In particular, $T \geq k^* + 1$ is an attempt to characterize the entire system on the basis of those equations which are most extreme in the sense of having the most regressors. As we shall demonstrate, such a characterization is inadequate.

Finally, it is interesting to note the relationship between the result in Theorem 1 and results in the literature for the existence of the mean of a two-stage estimator. Srivastava and Rao [15] show that sufficient conditions for the existence of the mean of a two-stage estimator that uses an error covariance matrix estimated from the residuals of the corresponding unrestricted multivariate regression model are (i) the errors have finite moments of order 4, and (ii) $T > M + K^* + 1$, where K^* denotes the number of distinct regressors in the system.* They also provide other (alternative) sufficient conditions that are equally relevant when residuals from the restricted SUR model are used. The existence of higher-order moments of a two-step estimator in a two-equation model have also been investigated by Kariya and Maekawa [8]. In every case we see that the result of Theorem 1 for the existence of the estimator is less demanding of sample size than are the results for the existence of moments of the estimator. This is not surprising. The existence of moments requires sufficiently thin tails for the distribution of the estimator. Reduced variability invariably requires increased information which manifests itself in a greater requirement on sample size.[†] This provides even stronger support for our assertion that the usually stated sample size requirements are inadequate because the existence of moments is important to many standard techniques of inference.

3.2 Applications of the Necessary Condition

In what follows we shall exploit the randomness of the d_j to obtain $\eta = \min(d, \rho - \omega)$ with probability one. Consequently, (7) reduces to

$$T \geq \begin{cases} M + \omega, & \text{for } \eta = \rho - \omega \\ M + \rho - d, & \text{for } \eta = d \end{cases} \tag{14}$$

Let us explore these results through a series of examples.

First, $\eta = 0$ requires either $d = 0$ or $\rho = \omega$ (or both). Both of these requirements correspond to the situation where each X_j is a basis for X, so that $\rho = k_{max} = K_j$ for $j = 1, \ldots, M$.[‡] Note that this does not require

---

*Clearly K^* is equivalent to ρ in the notation of this paper.

[†]The arguments underlying these results are summarized by Srivastava and Giles [14, Chapter 4].

[‡]The equality $\rho = k_{max} = K_j$ follows from our assumption of full column rank for each of the X_j. If this assumption is relaxed, condition (15) becomes $T \geq M + \rho$, which is the usual sample size requirement in rank-deficient multivariate regression models; see [9, Section 6.4].

$X_1 = \cdots = X_M$ but does include the multivariate regression model as the most likely special case. In this case, (14) reduces to

$$T \geq M + k^* \tag{15}$$

which is obviously identical to condition (6) and so need not be explored further.

Next consider the model

$$y_1 = x_1\beta_{11} + e_1 \tag{16a}$$

$$y_2 = x_1\beta_{21} + x_2\beta_{22} + x_3\beta_{23} + e_2 \tag{16b}$$

$$y_3 = x_1\beta_{31} + x_2\beta_{32} + x_3\beta_{33} + e_3 \tag{16c}$$

Here $M = 3$, $\rho = 3$, $d = 1$, and $\omega = 1$, so that $\eta = \min(1, 3 - 1) = 1$. Such a system will require a minimum of five observations to estimate. If $\beta_{23} \equiv 0$, so that x_3 no longer appears in equation (16b), then $\eta = d = \rho - \omega = 2$ and the sample size requirement for the system reduces to $T \geq 4$. Suppose now that, in addition to deleting x_3 from equation (16b), we include x_2 in equation (16a). In this case, $\omega = d = 2$ but $\eta = \rho - \omega = 1$ and, once again, the sample size requirement is 5. Finally, if we add x_2 and x_3 to equation (16a) and leave equation (16b) as stated, so that the system becomes a multivariate regression equation, the sample size requirement becomes $T \geq 6$. None of the changes to model (16) that have been suggested above alter the prediction of condition (4), which is that four observations should be sufficient to estimate the model. Hence, condition (4) typically underpredicts the actual sample size requirement for the model and it is unresponsive to certain changes in the composition of the model which do impact upon the sample size requirement of the two-stage estimator.

In the previous example we allowed d and $\rho - \omega$ and η to vary but at no time did the sample size requirement reduce to $T = M$. The next example provides a simple illustration of this situation. A common feature with the previous example will be the increasing requirement on sample size as the commonality of regressors across the system of equations increases, where again ω is the measure of commonality. Heuristically, the increase in sample size is required to compensate for the reduced information available when the system contains fewer distinct explanatory variables. Consider the two equation models

$$\left. \begin{aligned} y_1 &= x_1\beta_1 + e_1 \\ y_2 &= x_2\beta_2 + e_2 \end{aligned} \right\} \tag{17}$$

and

$$
\left.\begin{array}{l}
y_1 = x_1\beta_1 + e_1 \\
y_2 = x_1\beta_2 + e_2
\end{array}\right\} \tag{18}
$$

In model (17) there are no common regressors and so $\omega = 0$, which implies that $\eta = \min(d, \rho) = \min(2, 2) = 2$. Consequently, (14) reduces to $T \geq M + \rho - \eta = 2$. That is, model (17) can be estimated using a sample of only two observations or, more importantly, one can estimate as many equations as one has observations.* Model (18) is a multivariate regression model and so $T \geq M + \rho = 3$.

As a final example, if $\rho = T$ then $Y'M_V Y = 0$ and the estimability of the model is determined solely by η. From condition (14) we see that, in order to estimate M equations, we require $\eta = M$. But $\eta \leq \rho - \omega \leq T - \omega$ and so $M = T - \omega$ is the largest number of equations that can be estimated on the basis of T observations. This is the result observed in the comparison of models (17) and (18); it will be encountered again in Section 4, where each equation in a system contains an intercept, so that $\omega = 1$, and $M = T - 1$ is the largest number of equations that can be estimated for a given sample size. The case of $\rho = T$ would be common in large systems of equations where each equation contributes its own distinct regressors; indeed, this is the context in which it arises in Section 4.

4. MAXIMUM LIKELIHOOD AND BAYESIAN ESTIMATION

4.1 A Necessary Condition and Discussion

Likelihood-based estimation, be it maximum likelihood or Bayesian, requires distributional assumptions and so we will augment our earlier assumptions about e by assuming that the elements of e are jointly normally distributed. Consequently, the log-likelihood function for model (1) is

$$
L = -(TM/2)\log(2\pi) - (T/2)\log|\Sigma| - (1/2)\mathrm{tr}(S\Sigma^{-1}) \tag{19}
$$

where $S = E'E$. The ML estimates for β and Σ are those values which simultaneously satisfy the first-order conditions

$$
\tilde{\beta} = [\hat{X}'(\tilde{\Sigma}^{-1} \otimes I_T)\hat{X}]^{-1}\hat{X}'(\tilde{\Sigma}^{-1} \otimes I_T)y \tag{20}
$$

*This is a theoretical minimum sample size and should not be interpreted as a serious suggestion for empirical work!

and

$$T\tilde{\Sigma} = \tilde{E}'\tilde{E} \tag{21}$$

with $\mathrm{vec}(\tilde{E}) = y - \hat{X}\hat{\beta}$. This is in contrast to the two-stage estimator which obtains estimates of Σ and β sequentially. Rather than trying to simultaneously maximize L with respect to both β and Σ, it is convenient to equivalently maximize L with respect to β subject to the constraint that $\Sigma = S/T$, which will ensure that the ML estimates, $\tilde{\beta}$ and $\tilde{\Sigma}$, satisfy equations (20) and (21). Imposing the constraint by evaluating L at $\Sigma = S/T$ gives the concentrated log-likelihood function*

$$L^*(\beta) = \text{constant} - \frac{T}{2}\log|S|$$

Similarly, using a prior density function $f(\beta, \Sigma) \propto |\Sigma|^{-(M+1)/2}$, it can be shown that the marginal posterior density function for β is[†]

$$f(\beta|y) \propto |S|^{-T/2}$$

Consequently, we see that both the ML and Bayesian estimators are obtained by minimizing the generalized variance $|S|$ with respect to β.

The approach adopted in this section will be to demonstrate that, for sufficiently small samples, there necessarily exist βs such that S is singular (has rank less than M), so that $|S| = 0$. In such cases, ML estimation cannot proceed as the likelihood function is unbounded at these points; similarly, the posterior density for β will be improper at these βs. Since $S = E'E$, S will be singular if and only if E has rank less than M. Consequently, our problem reduces to determining conditions under which there necessarily exist βs such that E is rank deficient.

Theorem 2. In the model (1), augmented by a normality assumption, a necessary condition for $S = E'E$ to be nonsingular, and hence for the likelihood function (19) to be bounded, is that

$$T \ge M + \rho \tag{22}$$

where $\rho = \mathrm{rank}([X_1, X_2, \ldots, X_M])$ and E is defined in equation (5).

Proof. E will have rank less than M if there exists an $(M \times 1)$ vector $c = [c_1, c_2, \ldots, c_M]'$ such that $Ec = 0$. This is equivalent to the equation

$$\Phi\alpha = 0$$

---

*See, for example, [7, p. 553].
[†] See, for example, [18, p. 242].

where the $(T \times (M + K))$ matrix $\Phi = [Y, -X]$ and the $((M + K) \times 1)$ vector $\alpha = [c', c_1\beta_1', c_2\beta_2', \ldots, c_M\beta_M']'$. A nontrivial solution for α, and hence a nontrivial c, requires that Φ be rank deficient. But rank $\Phi = \min(T, M + \rho)$ with probability one and so, in order for E to be rank deficient, it follows that $T < M + \rho$. A necessary condition for E to have full column rank with probability one is then the converse of the condition for E to be rank deficient. □

A number of comments are in order. First, (22) is potentially more stringent in its requirements on sample size than is (14). Theorem 1 essentially provides a spectrum of sample size requirements, $T \geq M + \rho - \eta$, where the actual requirement depends on the specific data set used for estimation of the model. The likelihood-based requirement is the most stringent of those for the two-stage estimator, corresponding to the situation where $\eta = 0$. That η can differ from zero stems from the fact that the OLS residuals used in the construction of $\hat{\Sigma}$ by the two-stage estimator satisfy $\hat{e}_j = M_{X_j}y_j$, whereas the likelihood-based residuals used in the construction of $\tilde{\Sigma}$ need not satisfy the analogous $\tilde{e}_j = M_{X_j}y_j$.

Second, the proof of Theorem 2 is not readily applicable to the two-stage estimator considered in Theorem 1. In both cases we are concerned with determining the requirements for a solution to an equation of the form $Ac = 0$, with $A = \hat{E} = [M_{X_1}y_1, M_{X_2}y_2, \ldots, M_{X_M}y_M]$ in Theorem 1 and $A = E$ in Theorem 2. For the two-stage estimator, the interactions between the various X_j are important and complicate arguments about rank. The decomposition (11) provides fundamental insight into these interactions, making it possible to use arguments of rank to obtain the necessary condition on sample size. The absence of corresponding relationships between the vectors of likelihood-based residuals means that a decomposition similar to equation (11) is not required and that the simpler proof of Theorem 2 is sufficient.

An alternative way of thinking about why the development of the previous section differs from that of this section is to recognize that the problems being addressed in the two sections are different. The difference in the two problems can be seen by comparing the criteria for estimation. For likelihood-based estimators the criterion is to choose β to minimize $|S|$, a polynomial of order $2M$ in the elements of β. For the two-stage estimator a quadratic function in the elements of β is minimized. The former problem is a higher-dimensional one for all $M > 1$ and, consequently, its solution has larger minimal information (or sample size) requirements when the estimators diverge.*

---

*There are special cases where the two-stage estimator and the likelihood-based estimators coincide; see [14]. Obviously, their minimal sample size requirements are the same in these cases.

4.2 Applications of the Necessary Condition

In light of condition (22), it is possible to re-examine some empirical work undertaken by Chotikapanich and Griffiths [3] to investigate the effect of increasing M for fixed T and K_j. This investigation was motivated by the work of Fiebig and Kim [5]. The data set was such that $T = 19$ and $K_j = 3$ for all j. Each X_j contained an intercept and two other regressors that were unique to that equation. Consequently, in this model $\rho = \min(2M + 1, T)$, so that rank$(\Phi) = \min(3M + 1, T)$, and condition (22) predicts that likelihood-based methods should only be able to estimate systems containing up to $M = (T - 1)/3 = 6$ equations.* Conversely, as demonstrated below, condition (14) predicts that two-stage methods should be able to estimate systems containing up to $M = T - 1 = 18$ equations. This is exactly what was found. Although the two-stage estimator for the first two equations gave relatively similar results for M all the way up to 18, the authors had difficulty with the maximum likelihood and Bayesian estimators for $M \geq 7$. With maximum likelihood estimation the software package SHAZAM ([16]) sometimes uncovered singularities and sometimes did not, but, when it did not, the estimates were quite unstable. With Bayesian estimation the Gibbs sampler broke down from singularities, or got stuck in a narrow, nonsensical range of parameter values.

The statement of condition (14) implicitly assumes that the quantity of interest is sample size. It is a useful exercise to illustrate how (14) should be used to determine the maximum number of estimable equations given the sample size; we shall do so in the context of the model discussed in the previous paragraph.[†] To begin, observe that (i) $d = M$, because each equation contains two distinct regressors; (ii) $\omega = 1$, because each equation contains an intercept; and (iii) $\rho = \min(2M + 1, T)$, as before, so that $\rho - \omega = \min(2M, T - 1)$. Unless $M > T - 1$, $d \leq \rho - \omega$, so that $\eta = d$, and condition (14) reduces to $T \geq M + \rho - d$. Clearly, condition (14) is satisfied for all $T \geq M + 1$ in this model, because $d = M$ and $T \geq \rho$ by definition. Next, suppose that $M > T - 1$. Then $\eta = \rho - \omega = T - 1 < M = d$ and (14) becomes $M \leq T - \omega = T - 1$. But this is a contradiction and so, in this model, the necessary condition for $\hat{\Sigma}$ to be nonsingular with prob-

---

*Strictly the prediction is $M = [(T - 1)/3]$ equations, where $[x]$ denotes the integer part of x. Serendipitously, $(T - 1)/3$ is exactly 6 in this case.

[†]The final example of Section 3 is similar to this one except for the assumption that $\rho = T$, which is not made here.

ability one is that $M \leq T - 1.$[*] It should be noted that this is one less equation than is predicted by condition (4), where $T \geq M$ will be the binding constraint, as $19 = T > k_{max} + 1 = 3$ in this case.

5. CONCLUDING REMARKS

This paper has explored sample size requirements for the estimation of SUR models. We found that the sample size requirements presented in standard treatments of SUR models are, at best, incomplete and potentially misleading. We also demonstrated that likelihood-based methods potentially require much larger sample sizes than does the two-stage estimator considered in this paper.

It is worth noting that the nature of the arguments for the likelihood-based estimators is very different from that presented for the two-stage estimator. This reflects the impact of the initial least-squares estimator on the behaviour of the two-stage estimator.[†] In both cases the results presented are necessary but not sufficient conditions. This is because we are discussing the nonsingularity of random matrices and so there exist sets of Y (of measure zero) such that $\hat{\Sigma}$ and $\tilde{\Sigma}$ are singular even when the requirements presented here are satisfied. Alternatively, the results can be thought of as necessary and sufficient with probability one.

Our numerical exploration of the results derived in this paper revealed that standard packages did not always cope well with undersized samples.[‡] For example, it was not uncommon for them to locate local maxima of likelihood functions rather than correctly identify unboundedness. In the

---

[*]An alternative proof of this result comes from working with condition (13) directly and recognizing that the maximum number of equations that can be estimated for a given sample size will be that value at which the inequality is a strict equality. Substituting for $d = M$, $\rho = \min(2M + 1, T)$, and $\eta = \min(d, \rho - \omega)$ in (13) yields

$$M \leq \min(M, T - \min(2M + 1, T)) + \min(M, \min(2M + 1, T) - 1)$$

which will be violated only if

$$T - \min(2M + 1, T) + \min(2M + 1, T) - 1 = T - 1 < M$$

as required.

[†]Although likelihood-based estimators are typically obtained iteratively, and may well use the same initial estimator as the two-stage estimator considered here, the impact of the initial estimator is clearly dissipated as the algorithm converges to the likelihood-based estimate.

[‡]These experiments are not reported in the paper. They served merely to confirm the results derived and to ensure that the examples presented were, in fact, correct.

case of two-stage estimation, singularity of $\hat{\Sigma}$ sometimes resulted in the first stage OLS estimates being reported without meaningful further comment. Consequently, we would strongly urge practitioners to check the minimal sample size requirements and, if their sample size is at all close to the minimum bound, take steps to ensure that the results provided by their computer package are valid.

ACKNOWLEDGMENTS

The authors would like to thank Trevor Breusch, Denzil Fiebig and an anonymous referee for helpful comments over the paper's development. They are, of course, absolved from any responsibility for shortcomings of the final product.

REFERENCES

1. T. W. Anderson. *An Introduction to Multivariate Statistical Analysis*, 2nd edition. John Wiley, New York, 1984.
2. S. Chib and E. Greenberg. Markov chain Monte Carlo simulation methods in econometrics. *Econometric Theory*, 12(3): 409–431, 1996.
3. D Chotikapanich and W. E. Griffiths. Finite sample inference in the SUR model. Working Papers in Econometrics and Applied Statistics No. 103, University of New England, Armidale, 1999.
4. A. Deaton. Demand analysis. In: Z. Griliches and M. D. Intriligator, editors, *Handbook of Econometrics*, volume 3, chapter 30, pages 1767–1839. North-Holland, New York, 1984.
5. D. G. Fiebig and J. Kim. Estimation and inference in SUR models when the number of equations is large. *Econometric Reviews*, 19(1): 105–130, 2000.
6. W. H. Greene. *Econometric Analysis*, 4th edition. Prentice-Hall, Upper Saddle River, New Jersey, 2000.
7. G. G. Judge, R. C. Hill, W. E. Griffiths, H. Lutkepohl, and T.-C. Lee. *Introduction to the Theory and Practice of Econometrics*, 2nd edition. John Wiley, New York, 1988.
8. T. Kariya and K. Maekawa. A method for approximations to the pdf's and cdf's of GLSE's and its application to the seemingly unrelated regression model. *Annals of the Institute of Statistical Mathematics*, 34: 281–297, 1982.
9. K. V. Mardia, J. T. Kent, and J. M. Bibby. *Multivariate Analysis*. Academic Press, London, 1979.

10. R. J. Muirhead. *Aspects of Multivariate Statistical Theory*. John Wiley, New York, 1982.
11. D. Percy. Prediction for seemingly unrelated regressions. *Journal of the Royal Statistical Society*, 44(1): 243–252, 1992.
12. P. C. B. Phillips. The exact distribution of the SUR estimator. *Econometrica*, 53(4): 745–756, 1985.
13. V. K. Srivastava and T. Dwivedi. Estimation of seemingly unrelated regression equations. *Journal of Econometrics*, 10(1): 15–32, 1979.
14. V. K. Srivastava and D. E. A. Giles. *Seemingly Unrelated Regression Equations Models: Estimation and Inference*. Marcel Dekker, New York, 1987.
15. V. K. Srivastava and B. Raj. The existence of the mean of the estimator in seemingly unrelated regressions. *Communications in Statistics A*, 8: 713–717, 1979.
16. K. J. White. *SHAZAM User's Reference Manual Version 8.0*. McGraw-Hill, New York, 1997.
17. A. Zellner. An efficient method of estimating seemingly unrelated regressions and tests of aggregation bias. *Journal of the American Statistical Association*, 57(297): 348–368, 1962.
18. A. Zellner. *Introduction to Bayesian Inference in Econometrics*. John Wiley, New York, 1971.

28

Semiparametric Panel Data Estimation: An Application to Immigrants' Homelink Effect on U.S. Producer Trade Flows

AMAN ULLAH   University of California, Riverside, Riverside, California

KUSUM MUNDRA   San Diego State University, San Diego, California

1. INTRODUCTION

Panel data refers to data where we have observations on the same cross-section unit over multiple periods of time. An important aspect of the panel data econometric analysis is that it allows for cross-section and/or time heterogeneity. Within this framework two types of models are mostly estimated; one is the fixed effect (FE) and the other is the random effect. There is no agreement in the literature as to which one should be used in empirical work; see Maddala (1987) for a good discussion on this subject. For both types of models there is an extensive econometric literature dealing with the estimation of linear parametric models, although some recent works on nonlinear and latent variable models have appeared; see Hsiao (1985), Baltagi (1998), and Mátyás and Sevestre (1996). It is, however, well known that the parametric estimators of linear or nonlinear models may become inconsistent if the model is misspecified. With this in view, in this paper we consider only the FE panel models and propose semiparametric estimators which are robust to the misspecification of the functional forms.

The asymptotic properties of the semiparametric estimators are also established.

An important objective of this paper is to explore the application of the proposed semiparametric estimator to study the effect of immigrants' "home link" hypothesis on the U.S. bilateral trade flows. The idea behind the home link is that when the migrants move to the U.S. they maintain ties with their home countries, which help in reducing transaction costs of trade through better trade negotiations, hence effecting trade positively. In an important recent work, Gould (1994) analyzed the home link hypothesis by considering the well-known gravity equation (Anderson 1979, Bergstrand 1985) in the empirical trade literature which relates the trade flows between two countries with economic factors, one of them being transaction cost. Gould specifies the gravity equation to be linear in all factors except transaction cost, which is assumed to be a nonlinear decreasing function of the immigrant stock in order to capture the home link hypothesis.* The usefulness of our proposed semiparametric estimators stems from the fact that the nonlinear functional form used by Gould (1994) is misspecified, as indicated in Section 3 of this paper. Our findings indicate that the immigrant home link hypothesis holds for producer imports but does not hold for producer exports in the U.S. between 1972 and 1980.

The plan of this paper is as follows. In Section 2 we present the FE model and proposed semiparametric estimators. These semiparametric estimators are then used to analyze the "home link" hypothesis in Section 3. Finally, the Appendix discusses the aysmptotic properties of the semiparametric estimators.

2.   THE MODEL AND ESTIMATORS

Let us consider the parametric FE model as

$$y_{it} = x_{it}'\beta + z_{it}'\gamma + \alpha_i + u_{it} \qquad (i = 1, \cdots, n; \ t = 1, \cdots, T) \tag{2.1}$$

where y_{it} is the dependent variable, x_{it} and z_{it} are the $p \times 1$ and $q \times 1$ vectors, respectively, β, γ, and α_i are the unknown parameters, and u_{it} is the random error with $E(u_{it} \mid x_{it}, z_{it}) = 0$. We consider the usual panel data case of large n and small T. Hence all the asymptotics in this paper are for $n \to \infty$ for a fixed value of T. Thus, as $n \to \infty$, $\sqrt{nT}$ consistency and $\sqrt{n}$ consistency are equivalent.

---

*Transaction costs for obtaining foreign market information about country j in the U.S. used by Gould (1994) in his study is given by $A_e^{-\rho[M_{US,j}/(\theta+MU_{US,j})]}$, $\rho > 0$, $\theta > 0$, $A > 0$, where $M_{US,j} =$ stock of immigrants from country j in the United States.

From (2.1) we can write

$$Y_{it} = X'_{it}\beta + Z'_{it}\gamma + U_{it} \tag{2.2}$$

where $R_{it} = r_{it} - \bar{r}_{i.}$, $\bar{r}_{i.} = \sum_t^T r_{it}/T$. Then the well-known parametric FE estimators of β and γ are obtained by minimizing $\sum_i \sum_t U_{it}^2$ with respect to β and γ or $\sum_i \sum_t u_{it}^2$ with respect to β, γ, and α_i. These are the consistent least-squares (LS) estimators and are given by

$$b_p = \left[\sum_i \sum_t (X_{it} - \tilde{X}_{it})(X_{it} - \tilde{X}_{it})' \right]^{-1} \sum_i \sum_t (X_{it} - \tilde{X}_{it}) Y_{it}$$

$$= (S_{X-\tilde{x}})^{-1} S_{X-\tilde{x},Y}$$

$$= (X'M_Z X)^{-1} X'M_Z Y \tag{2.3}$$

and

$$c_p = S_Z^{-1} (S_{Z,Y} - S_X b_p) \tag{2.4}$$

where p represents parametric, $\tilde{X}'_{it} = Z'_{it}(\sum_i \sum_t Z_{it} Z'_{it})^{-1} \sum_i \sum_t Z_{it} X'_{it}$, $S_{A,B} = A'B/nT = \sum_i \sum_t A_{it} B'_{it}/nT$ for any scalar or column vector sequences A_{it} and B_{it}, $S_A = S_{A,A}$, and $M_Z = I - Z(Z'Z)^{-1}Z'$. The estimator $\hat{\alpha}_i = \bar{y}_{i.} - \bar{x}'_{i.}b_p - \bar{z}'_{i.}c_p$ is not consistent, and this will also be the case with the semiparametric estimators given below.

New semiparametric estimators of β and γ can be obtained as follows. From (2.2) let us write

$$E(Y_{it}|Z_{it}) = E(X'_{it}|Z_{it})\beta + Z'_{it}\gamma \tag{2.5}$$

Then, subtracting (2.5) from (2.2), we get

$$Y_{it}^* = X_{it}^{*'}\beta + U_{it} \tag{2.6}$$

which gives the LS estimator of β as

$$\hat{\beta}_{sp} = \left(\sum_i \sum_t X_{it}^* X_{it}^{*'} \right)^{-1} \sum_i \sum_t X_{it}^* Y_{it}^* = S_{X^*}^{-1} S_{X^*,Y^*} \tag{2.7}$$

where $R_{it}^* = R_{it} - E(R_{it}|Z_{it})$ and sp represents semiparametric. We refer to this estimator as the semiparametric estimator, for the reasons given below.

The estimator $\hat{\beta}_{sp}$ is not operational since it depends on the unknown conditional expectations $E(A_{it}|Z_{it})$, where A_{it} is Y_{it} or X_{it}. Following Robinson (1988), these can however be estimated by the nonparametric kernel estimators

$$\hat{A}_{it} = \sum_j \sum_s A_{js} K_{it,js} \bigg/ \sum_j \sum_s K_{it,js} \tag{2.8}$$

where $K_{it,js} = K((Z_{it} - Z_{js})/a)$, $j = 1, \ldots, n$; $s = 1, \ldots, T$, is the kernel function and a is the window width. We use product kernel $K(Z_{it}) = \prod_{l=1}^{q} k(Z_{it}, l)$, k is the univariate kernel and $Z_{it,l}$ is the ℓth component of Z_{it}. Replacing the unknown conditional expectations in (2.7) by the kernel estimators in (2.8), an operational version of $\hat{\beta}_{sp}$ becomes

$$b_{sp} = \left(\sum_i \sum_t (X_{it} - \hat{X}_{it})(X_{it} - \hat{X}_{it})' \right)^{-1} \sum_i \sum_t (X_{it} - \hat{X}_{it})(Y_{it} - \hat{Y}_{it})$$

$$= S_{X-\hat{X}}^{-1} S_{X-\hat{X}, Y-\hat{Y}} \tag{2.9}$$

Since the unknown conditional expectations have been replaced by their nonparametric estimates we refer to b_{sp} as the semiparametric estimator. After we get b_{sp},

$$c_{sp} = S_Z^{-1} \left(S_{Z,Y} - S_{Z,X} b_{sp} \right) \tag{2.10}$$

The consistency and asymptotic normality of b_{sp} and c_{sp} are discussed in the Appendix.

In a special case where we assume the linear parametric form of the conditional expectation, say $E(A_{it}|Z_{it}) = Z_{it}' \delta$, we can obtain the LS predictor as $\tilde{A}_{it} = Z_{it}' (\sum_i \sum_t Z_{it} Z_{it}')^{-1} \sum_i \sum_t Z_{it} A_{it}$. Using this in (2.7) will give $\hat{\beta}_{sp} = b_p$. It is in this sense that b_{sp} is a generalization of b_p for situations where, for example, X and Z have a nonlinear relationship of unknown form.

Both the parametric estimators b_p, c_p and the semiparametric estimators b_{sp}, c_{sp} described above are the $\sqrt{n}$ consistent global estimators in the sense that the model (2.2) is fitted to the entire data set. Local pointwise estimators of β and γ can be obtained by minimizing the kernel weighted sum of squares

$$\sum_i \sum_t [y_{it} - x_{it}'\beta - z_{it}'\gamma - \alpha_i]^2 K\left(\frac{x_{it} - x}{h}, \frac{z_{it} - z}{h}\right) \tag{2.11}$$

with respect to β, γ, and α; h is the window width. The local pointwise estimators so obtained can be denoted by $b_{sp}(x, z)$ and $c_{sp}(x, z)$, and these are obtained by fitting the parametric model (2.1) to the data close to the points x, z, as determined by the weights $K()$. These estimators are useful for studying the local pointwise behaviors of β and γ, and their expressions are given by

$$\begin{bmatrix} b_{sp}(x, z) \\ c_{sp}(x, z) \end{bmatrix} = \left(\sum_i \sum_t (w_{it} - \hat{w}_{i.})(w_{it} - \hat{w}_{i.})' \right)^{-1} \sum_i \sum_t (w_{it} - \hat{w}_{i.})(y_{it.} - \hat{y}_{i.})$$

$$= S_{w-\hat{w}.}^{-1} S_{w-\hat{w}, y-\hat{y}} \tag{2.12}$$

where $w_{it}' = [\sqrt{K_{it}} x_{it}' \sqrt{K_{it}} z_{it}']$, $K_{it} = K((x_{it} - x)/h, (z_{it} - z)/h)$, $\hat{A}_i = \sum_t A_{it} K_{it} / \sum_t K_{it}$.

While the estimators b_p, c_p and b_{sp}, c_{sp} are the $\sqrt{n}$ consistent global estimators, the estimators $b_{sp}(x, z)$, $c_{sp}(x, z)$ are the $\sqrt{nh^{p+q+2}}$ consistent local estimators (see Appendix). These estimators also provide a consistent estimator of the semiparametric FE model

$$y_{it} = m(x_{it}, z_{it}) + \alpha_i + u_{it} \tag{2.13}$$

where $m()$ is the nonparametric regression. This model is semiparametric because of the presence of the parameters α_i. It is indicated in the Appendix that

$$\hat{m}_{sp}(x_{it}, z_{it}) = x_{it}' b_{sp}(x_{it}, z_{it}) + z_{it}' c_{sp}(x_{it}, z_{it}) \tag{2.14}$$

is a consistent estimator of the unknown function $m(x_{it}, z_{it})$, and hence b_{sp}, c_{sp} are the consistent estimators of its derivatives. In this sense $\hat{m}_{sp}(x_{it}, z_{it})$ is a local linear nonparametric regression estimator which estimates the linear model (2.1) nonparametrically; see Fan (1992, 1993) and Gozalo and Linton (1994). We note however the well-known fact that the parametric estimator $x_{it}' b_p + z_{it}' c_p$ is a consistent estimator only if $m(x_{it}, z_{it}) = x_{it}' \beta + z_{it}' \gamma$ is the true model. The same holds for any nonlinear parametric specification estimated by the global parametric method, such as nonlinear least squares.

In some situations, especially when the model (2.13) is partially linear in x but nonlinear of unknown form in z, as in Robinson (1988), we can estimate β globally but γ locally and vice-versa. In these situations we can first obtain the global $\sqrt{n}$ consistent estimate of β by b_{sp} in (2.9). After this we can write

$$y_{it}^o = y_{it} - x_{it}' b_{sp} = z_{it}' \gamma + \alpha_i + v_{it} \tag{2.15}$$

where $v_{it} = u_{it} + x_{it}'(\beta - b_{sp})$. Then the local estimation of γ can be obtained by minimizing

$$\sum_i \sum_t [y_{it}^o - z_{it}' \gamma - \alpha_i]^2 K\left(\frac{z_{it} - z}{h}\right) \tag{2.16}$$

which gives

$$c_{sp}(z) = \left(\sum_i \sum_t (z_{it} - \hat{z}_{i.})(z_{it} - \hat{z}_{i.})' K_{it} \right)^{-1} \sum_i \sum_t (z_{it} - \hat{z}_{i.})(y^o_{it} - \hat{y}^o_{it.}) K_{it}$$

$$= S^{-1}_{\sqrt{K}(z-\hat{z}.)} S_{\sqrt{K}(z-\hat{z}.)(y^o-\hat{y}.^o)} \tag{2.17}$$

where $K_{it} = K((z_{it} - z)/h)$ and $\hat{A}_i = \sum_t A_{it} K_{it} / \sum_t K_{it}$. Further, $\hat{\alpha}_i(z) = \hat{y}^o_{i.} - \hat{z}_{i.} c_{sp}(z)$. As in (2.14), $\hat{m}_{sp}(z_{it}) = z'_{it} c_{sp}(z)$ is a consistent local linear estimator of the unknown nonparametric regression in the model $y^o_{it} = m(z_{it}) + \alpha_i + u_{it}$. But the parametric estimator $z'_{it} \hat{\gamma}_p$ will be consistent only if $m(z_{it}) = z'_{it} \gamma$ is true. For discussion on the consistency and asymptotic normality of $b_{sp}(z)$, $c_{sp}(z)$, and $\hat{m}_{sp}(z)$, see Appendix.

3. MONTE CARLO RESULTS

In this section we discuss Monte Carlo simulations to examine the small sample properties of the estimator given by (2.9). We use the following data generating process (DGP):

$$y_{it} = \beta x_{it} + \delta z_{it} + \gamma z^2_{it} + u_{it}$$

$$= \beta x_{it} + m(z_{it}) + u_{it} \tag{3.1}$$

where z_{it} is independent and uniformly distributed in the interval $[-\sqrt{3}, \sqrt{3}]$, x_{it} is independent and uniformly distributed in the interval $[-\sqrt{5}, \sqrt{5}]$, u_{it} is i.i.d. $N(0, 5)$. We choose $\beta = 0.7$, $\delta = 1$, and $\gamma = 0.5$. We report estimated bias, standard deviation (Std) and root mean squares errors (Rmse) for the estimators. These are computed via Bias $(\hat{\beta}) = M^{-1} \sum^M (\hat{\beta}_j - \beta_j)$, Std$(\hat{\beta}) = \{M^{-1} \sum^M (\hat{\beta}_j - \text{Mean}(\hat{\beta}))^2\}^{1/2}$, and Rmse$(\hat{\beta}) = \{M^{-1} \sum^M (\hat{\beta}_j - \beta)^2\}^{1/2}$, where $\hat{\beta} = b_{sp}$, M is the number of replications and $\hat{\beta}_j$ is the jth replication. We use $M = 2000$ in all the simulations. We choose $T = 6$ and $n = 50, 100, 200$, and 500. The simulation results are given in Table 1. The results are not dependent on δ and γ, so one can say that the results are not sensitive to different functional forms of $m(z_{it})$. We see that Std and Rmse are falling as n increases.

4. EMPIRICAL RESULTS

Here we present an empirical application of our proposed semiparametric estimators. In this application we look into the effect of the immigrants' "home link" hypothesis on U.S. bilateral producer trade flows. Immigration

Table 1. The case of $\beta = 0.7$, $\delta = 1$, $\gamma = 0.5$

	b_{sp}		
	Bias	Std	Rmse
$n = 50$	−0.116	0.098	0.152
$n = 100$	−0.115	0.069	0.134
$n = 200$	−0.118	0.049	0.128
$n = 500$	−0.117	0.031	0.121

has been an important economic phenomenon for the U.S., with immigrants varying in their origin and magnitude. A crucial force in this home link is that when migrants move to the U.S. they maintain ties with their home countries, which helps in reducing transaction costs of trade through better trade negotiations, removing communication barriers, etc. Migrants also have a preference for home products, which should effect U.S. imports positively. There have been studies to show geographical concentrations of particular country-specific immigrants in the U.S. actively participating in entrepreneurial activities (Light and Bonacich 1988). This is an interesting look at the effect of immigration other than the effect on the labor market, or welfare impacts, and might have strong policy implications for supporting migration into the U.S. from one country over another.

A parametric empirical analysis of the "home link" hypothesis was first done by Gould (1994). His analysis is based on the gravity equation (Anderson 1979, Bergstrand 1985) extensively used in the empirical trade literature, and it relates trade flows between two countries with economic forces, one of them being the transaction cost. Gould's important contribution specifies the transaction cost factor as a nonlinear decreasing function of the immigrant stock to capture the home link hypothesis: decreasing at an increasing rate. Because of this functional form the gravity equation becomes a nonlinear model, which he estimates by nonlinear least squares using an unbalanced panel of 47 U.S. trading partners.

We construct a balance panel of 47 U.S. trading partners over nine years (1972–1980), so here $i = 1, \ldots, 47$ and $t = 1, \ldots, 9$, giving 423 observations. The country specific effects on heterogeneity are captured by the fixed effect. In our case, y_{it} = manufactured U.S. producers' exports and imports, x_{it} includes lagged value of producers' exports and imports, U.S. population, home-country population, U.S. GDP, home-country GDP, U.S. GDP deflator, home-country GDP deflator, U.S. export value index, home-country export value index, U.S. import value index, home-country import value index, immigrant stay, skilled–unskilled ratio of the migrants, and z_{it} is immigrant stock to the U.S. Data on producer-manufactured imports and exports were taken from OECD statistics. Immigrant stock, skill level and length of stay of migrants were taken from INS public-use data on yearly immigration. Data on income, prices, and population were taken from IMF's International Financial Statistics.

We start the analysis by first estimating the immigrants' effect on U.S. producer exports and imports using Gould's (1994) parametric functional form and plot it together with the kernel estimation; see Figures 1 and 2. The kernel estimator is based on the normal kernel given as $K((z_{it} - z)/h) = 1/\sqrt{2\pi} \exp\{-(1/2)((z_{it} - z)/h)^2\}$ and h, the window-width, is taken as $cs(nT)^{-1/5}$, c is a constant, and s is the standard derivation for variable z; for details on the choice of h and K see Härdle (1990) and Pagan and Ullah (1999). Comparing the results with the actual trade flows, we see from Figures 1 and 2 that the functional form assumed in the parametric estima-

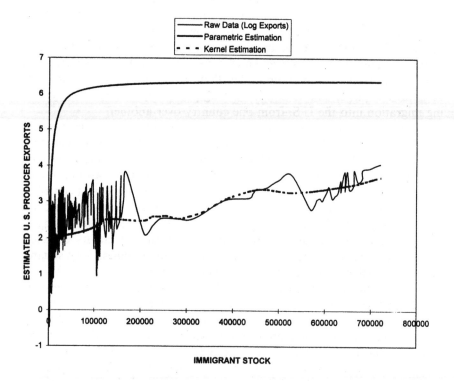

Figure 1. Comparison of U.S. producer exports with parametric functional estimation and kernel estimation.

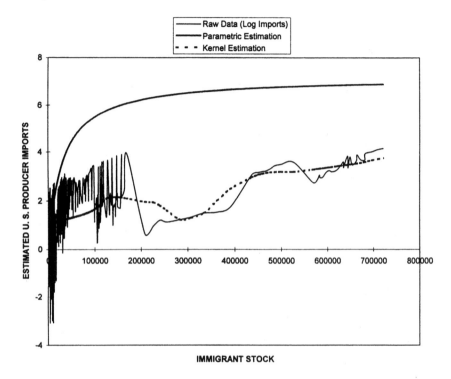

Figure 2. Comparison of U.S. producer imports with parametric functional estimation and kernel estimation

tion is incorrect and hence Gould's nonlinear LS estimates may be inconsistent. In fact the parametric estimates, b_p and c_p, will also be inconsistent. In view of this we use our proposed $\sqrt{n}$ consistent semiparametric estimator of β, b_{sp}, in (2.9) and the consistent semiparametric local linear estimator of γ, $c_{sp}(z)$, in (2.17).

First we look at the semiparametric estimates b_{sp} given in Table 2. The immigrant skilled–unskilled ratio affects exports and imports positively, though it is insignificant. This shows that skilled migrants are bringing better foreign market information. As the number of years the immigrant stays in the U.S. increases, producer exports and producer imports fall at an increasing rate. It can be argued that the migrants change the demand structure of the home country adversely, decreasing U.S. producer exports and supporting imports. But once the home country information, which they carry becomes obsolete and their tastes change, their effect on the trade falls. When the inflation index of a country rises, exports from that

Table 2. Bilateral manufactured producer trade flows between the U.S. and the immigrant home countries

Dependent variable	U.S. producer exports		U.S. producer imports	
	Parametric model	SPFE	Parametric model	SPFE
U.S. GDP deflator	0.52	−9.07	12.42	5.45
	(3.34)	(18.22)	(9.69)	(77.62)
Home-country GDP deflator	−0.25	−0.09	0.29	−0.11
	(0.09)[a]	(0.06)	(0.26)	(0.35)
U.S. GDP	−1.14	−3.29	6.71	5.35
	(2.13)	(11.01)	(6.71)	(53.74)
Home-country GDP	0.60	0.17	0.56	−0.16
	(0.11)[a]	(0.09)[c]	(0.34)[b]	(0.45)[a]
U.S. population	5.09	88.24	6.05	−67.18
	(40.04)	(236.66)	(123.8)	(1097.20)
Home-country population	0.41	0.58	0.58	−5.31
	(0.18)[a]	(0.48)	(0.53)[c]	(2.47)
Immigrant stay	−0.06	0.01	−0.16	−0.13
	(0.05)	(0.25)	(0.01)	(1.18)
Immigrant stay (squared)	0.002	0.001	0.01	0.003
	(0.003)	(0.02)	(0.01)	(0.07)
Immigrant skilled–unskilled ratio	0.01	0.02	0.06	0.02
	(0.02)	(0.02)	(0.06)	(0.06)
U.S. export unit value index	1.61	1.91		
	(0.46)[a]	(0.57)[a]		
Home-country import unit value index	−0.101	0.072		
	(0.04)	(0.09)		
Home-country export unit value index			1.72	0.37
			(0.77)[a]	(1.85)
U.S. import unit value index			−0.10	0.004
			(0.34)	(0.22)

Newey–West corrected standard errors in parentheses. [a]Significant at 1% level. [b]Significant at 5% level. [c]Significant at 10% level.

country may become expensive and are substituted by domestic production in the importing country. Hence, when the home-country GDP deflator is going up, U.S. producer imports fall and the U.S. GDP deflator affects U.S. producer exports negatively. The U.S. GDP deflator has a positive effect on U.S. imports, which might be due to the elasticity of substitution among imports exceeding the overall elasticity between imports and domestic production in the manufactured production sector in the U.S., whereas the opposite holds in the migrants' home country. The U.S. export value index reflects the competitiveness for U.S. exports and has a significant positive effect on producer exports. This may be due to the supply elasticity of transformation among U.S. exports exceeding the overall elasticity between exports and domestic goods, which is true for the home-country export unit value index too. The U.S. and the home country import unit value indexes have a positive effect on producer imports and producer exports respectively. This shows that the elasticity of substitution among imports exceeds the overall elasticity between domestic and imported goods, both in the U.S. and in the home country. The immigrants' home-country GDP affects the producer exports positively and is significant at the 10% level of significance. The U.S. GDP affects producer exports negatively and also the home-country GDP affects producer imports negatively, showing that the demand elasticity of substitution among imports is less than unity both for the U.S. and its trading partners.

To analyze the immigrant "home link" hypothesis, which is an important objective here, we obtain elasticity estimates $c_{sp}(z)$ at different immigrant stock levels for both producer's exports and producer's imports. This shows how much U.S. bilateral trade with the ith country is brought about by an additional immigrant from that country. Based on this, we also calculate in Table 3 the average dollar value change (averaged over nine years) in U.S. bilateral trade flows: $\bar{c}_{isp} \times \bar{z}_i$, where $\bar{c}_{isp} = \sum_t c_{sp}(z_{it})/T$ and $\bar{z}_i = \sum_t z_{it}/T$ is the average immigrant stock into the U.S. from the ith country. When these values are presented in Figures 3 and 4, we can clearly see that the immigrant home link hypothesis supports immigrant stock affecting trade positively for U.S. producer imports but not for U.S. producer exports. These findings suggest that immigrant stock and U.S. producer imports are complements in general, but the immigrants and producer exports are substitutes. In contrast, Gould's (1994) nonlinear parametric framework suggests support for the migrants' "homelink hypothesis" for both exports and imports. The difference in our results for exports with those of Gould may be due to misspecification of the nonlinear transaction cost function in Gould and the fact that he uses unbalanced panel data. All these results however indicate that the "home link" hypothesis alone may not be sufficient to look at the broader effect of immigrant stock on bilateral trade flows. The labor role of migrants and the welfare effects of immigration, both in the receiving and the sending country, need to be taken into account. These results also crucially depend on the sample period; during the 1970s the U.S. was facing huge current account deficits. In any case, the above analysis does open interesting questions as to what should be the U.S. policy on immigration; for example, should it support more immigration from one country over another on the basis of dollar value changes in import or export?

Table 3. Average dollar value change in U.S. producer trade flows from one additional immigrant between 1972 and 1980

Country		Producer exports	Producer imports
1	Australia	−84 447.2	107 852.2
2	Austria	−257 216	332 576.7
3	Brazil	−72 299.9	91 995.54
4	Canada	−1 908 566	2 462 421
5	Colombia	−300 297	381 830.7
6	Cyprus	−11 967.4	15 056.1
7	Denmark	−65 996.3	85 321.2
8	El Salvador	−115 355	146 500.3
9	Ethiopia	−11 396.6	13 098.77
10	Finland	−93 889.6	121 071.7
11	France	−174 535	225 599.7
12	Greece	−557 482	718 292.1
13	Hungary	−172 638	163 015.4
14	Iceland	−13 206.8	17 003.16
15	India	−311 896	383 391.8
16	Ireland	−577 387	742 629.5
17	Israel	−126 694	159 101.8
18	Italy	−2 356 589	3 045 433
19	Japan	−446 486	575 985.8
20	Jordan	−33 074.7	41 427
21	Kenya	−3 604.1	4 044.627
22	Malaysia	−9 761.78	11 766
23	Malta	−23 507.1	30 184.8
24	Morocco	−2 899.56	2 797.519
25	Netherlands	−346 098	447 181.1
26	New Zealand	−23 666.3	30 182.7
27	Nicaragua	−74 061.1	93 930.9
28	Norway	−231 098	298 533.2
29	Pakistan	−35 508.4	42 682.64
30	Philippines	−214 906	258 027.4
31	S. Africa	−29243.3	37247.1
32	S. Korea	−89567.5	109286.9
33	Singapore	−4095.1	4863.85
34	Spain	−161804	207276.4
35	Sri Lanka	−7819.8	9685.5
36	Sweden	−220653	28500.9
37	Switzerland	−91599.2	118259.2

Table 3. Continued

Country		Producer exports	Producer imports
38	Syria	−358 830.3	44 644.6
39	Tanzania	−2 875.3	2 679.2
40	Thailand	−49 734.8	58 071.3
41	Trinidad	−113 210	142 938.1
42	Tunisia	−3 285.2	3 066.1
43	Turkey	−115 192	147 409.5
44	U.K.	0	0
45	W. Germany	−193 8678	2505652
46	Yugoslavia	−468 268	598 664.1
47	Zimbabwe	−2 209.5	1 997.1

ACKNOWLEDGMENTS

The first author gratefully acknowledges the research support from the Academic Senate, UCR. The authors are thankful for useful discussions with Q. Li and the comments by the participants of the Western Economics Association meeting at Seattle. They are also grateful to D. M. Gould for providing the data used in this paper.

APPENDIX

Here we present the asymptotic properties of the estimators in Section 2. First we note the well-known results that, as $n \to \infty$,

$$\left.\begin{array}{l} \sqrt{nT}(b_p - \beta) \sim N\left(0, \sigma^2\left(P\lim S_{X-\hat{x}}\right)^{-1}\right) \\ \sqrt{nT}(c_p - \beta) \sim N\left(0, \sigma^2\left(P\lim S_{Z-\tilde{z}}\right)^{-1}\right) \end{array}\right\} \quad \text{(A.1)}$$

where $\tilde{Z}$ is generated by $\tilde{Z}'_{it} = X'_{it}(\sum_i \sum_t X_{it} X'_{it})^{-1} \sum_i \sum_t X_{it} Z'_{it}$ and $P\lim$ represents probability limit; see the book White (1984).

Next we describe the assumptions that are needed for the consistency and asymptotic normality of b_{sp}, c_{sp}, $b_{sp}(x,z)$, $c_{sp}(x,z)$, and $c_{sp}(z)$ given above. Following Robinson (1988), let G_μ^λ denote the class of functions such that if $g \in G_\mu^\lambda$, then g is μ times differentiable; g and its derivatives (up to order μ) are all bounded by some function that has λth-order finite moments. Also,

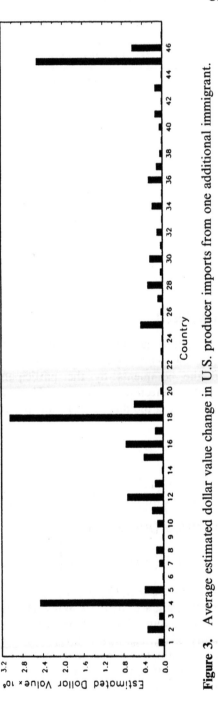

Figure 3. Average estimated dollar value change in U.S. producer imports from one additional immigrant.

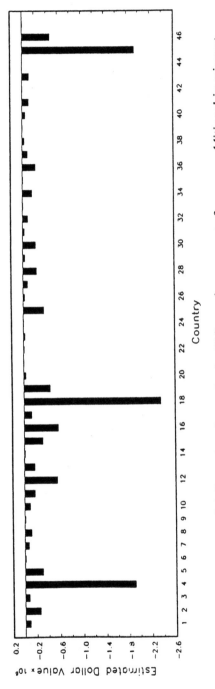

Figure 4. Average estimated dollar value change in the U.S. producer exports from one additional immigrant.

K_2 denotes the class of nonnegative kernel functions k: satisfying $\int k(v)v^m dv = \delta om$ for $m = 0, 1$ (δom is the Kronecker's delta), $\int k(v)vv' dv = C_k I$ ($I > 0$), and $k(u) = O((1 + |u|^{3+\eta})^{-1})$ for some $\eta > 0$. Further, we denote $\int k^2(v)vv' dv = D_k . I$. We now state the following assumptions:

(A1) (i) for all t, (y_{it}, x_{it}, z_{it}) are i.i.d. across i and z_{it} admits a density function $f \in G_{\mu-1}^{\infty}$, $E(x|z)$ and $E(z|x) \in G_{\mu}^4$ for some positive integer $\mu > 2$; (ii) $E(u_{it}|x_{it}, z_{it}) = 0$, $E(u_{it}^2|x_{it}, z_{it}) = \sigma^2(x_{it}, z_{it})$ is continuous in x_{it} and z_{it}, and u_{it}, $\eta_{it} = x_{it} - E(x_{it}|z_{it})$, $\xi_{it} = (z_{it} - E(z_{it}|x_{it})$ have finite $(4 + \delta)$th moment for some $\delta > 0$.

(A2) $\bar{k} \in K_\lambda$; as $n \to \infty$, $a \to 0$, $na^{4\lambda} \to 0$, and $na^{\max(2q-4, q)} \to \infty$.

(A3) $k \in K_2$ and $k(v) \geq 0$; as $n \to \infty$, $h \to 0$, $nh^{q+2} \to \infty$, and $nh^{q+4} \to 0$.

(A1) requires independent observations across i, and gives some moment and smoothness conditions. The condition (A2) ensures b_{sp} and c_{sp} are $\sqrt{n}$ consistent. Finally (A3) is used in the consistency and asymptotic normality of $b_{sp}(x, z)$; $c_{sp}(x, z)$, and $c_{sp}(z)$.

Under the assumptions (A1) and (A2), and taking $\sigma^2(x, z) = \sigma^2$ for simplicity, the asymptotic distributions of the semiparametric estimators b_{sp} and c_{sp} follow from Li and Stengos (1996), Li (1996) and Li and Ullah (1998). This is given by

$$\sqrt{nT}(b_{sp} - \beta) \sim N(0, \sigma^2 \Sigma^{-1}) \text{ and } \sqrt{nT}(c_{sp} - \beta) \sim N(0, \sigma^2 \Omega^{-1}) \quad (A.2)$$

where $\Sigma = E(\eta_1' \eta_1 / T)$ and $\Omega = E(\xi_1' \xi_1 / T)$; $\eta_i' = (\eta_{i1}, \ldots, \eta_{iT})$. Consistent estimators for Σ^{-1} and Ω^{-1} are $\hat{\Sigma}^{-1}$ and $\hat{\Omega}^{-1}$, respectively, where $\hat{\Sigma} = (1/(nT)) \Sigma_i \Sigma_t (X_{it} - \hat{X}_{it})(X_{it} - \hat{X}_{it})' = 1/(nT)) \Sigma_i (X_i - \hat{X}_i)'(X_i - \hat{X}_i)$ and $\hat{\Omega} = (1/(nT)) \Sigma_i \Sigma_t (Z_{it} - \hat{Z}_{it})(Z_{it} - \hat{Z}_{it})'$.

The semiparametric estimators b_{sp} and c_{sp} depend upon the kernel estimators which may have a random denominator problem. This can be avoided by weighting (2.5) by the kernel density estimator $\hat{f}_{it} = \hat{f}(Z_{it}) = (1/(nTa^q)) \sum_j \sum_s K_{it,js}$. This gives $\tilde{b}_{sp} = S_{(X-\hat{X})\hat{f}, (Y-\hat{Y})\hat{f}}^{-1}$. In this case $\hat{\Sigma}$ will be the same as above with $X - \hat{X}$ replaced by $(X - \hat{X})\hat{f}$. Finally, under the assumptions (A1) to (A3) and noting that $(nTh^{q+2})^{1/2}(b_{sp} - \beta) = o_p(1)$, it follows from Kneisner and Li (1996) that for $n \to \infty$

$$(nTh^{q+2})^{-1}\left(c_{sp^{(z)}} - \gamma(z)\right) \sim N(0, \Sigma_1) \quad (A.3)$$

where $\Sigma_1 = (\sigma^2(z)/f(z))C_k^{-1} D_k C_k^{-1}$, C_k and D_k are as defined above. In practice we replace $\sigma^2(z)$ by its consistent estimator $\hat{\sigma}^2(z_{it}) = \Sigma_j \Sigma_s (y_{js}^o$

$-z_{js} c_{sp}(z_{js}))^2 K_{it,js} / \Sigma_j \Sigma_s K_{it,js}$. Further, denoting $m(z) = z'\gamma$ and $\hat{m}_{sp}(z)$ $= z' c_{sp}(z)$, as $n \to \infty$

$$(nTh^q)^{1/2}(\hat{m}_{sp}(z) - m(z)) \sim N(0, \Sigma_2) \tag{A.4}$$

where $\Sigma_2 = (O^2(z)/f(z)) \int K^2(v) dv$; see Gozalo and Linton (1994). Thus the asymptotic variance of $\hat{m}(z)$ is independent of the parametric model $z\gamma$ used to get the estimate $\hat{m}(z)$ and it is the same as the asymptotic variance of Fan's (1992, 1993) nonparametric local linear estimator. In this sense $c_{sp}(z)$ and $\hat{m}_{sp}(z)$ are the local linear estimators.

The asymptotic normality of the vector $[b'_{sp}(x, z), c'_{sp}(x, z)]$ is the same as the result in (A.3) with $q + 2$ replaced by $p + q + 2$ and z replaced by (x, z). As there, these estimators are also the local linear estimators.

REFERENCES

Anderson, J. E. (1979), A theoretical foundation for the gravity equation. *American Economics Review* 69: 106–116.

Baltagi, B. H. (1998), Panel data methods. *Handbook of Applied Economics Statistics* (A. Ullah and D. E. A. Giles, eds.)

Bergstrand, J. H. (1985), The gravity equation in international trade: some microeconomic foundations and empirical evidence. *The Review of Economics and Statistics* 67: 474–481.

Fan, J. (1992), Design adaptive nonparametric regression. *Journal of the American Statistical Association* 87: 998–1024.

Fan, J. (1993), Local linear regression smoothers and their minimax efficiencies. *Annals of Statistics* 21: 196–216.

Gould, D. (1994), Immigrant links to the home-country: Empirical implications for U.S. bilateral trade flows. *The Review of Economics and Statistics* 76: 302–316.

Gozalo, P., and O. Linton (1994), Local nonlinear least square estimation: using parametric information nonparametrically. *Cowler Foundation Discussion paper No. 1075*.

Härdle. W. (1990), *Applied Nonparametric Regression*, New York: Cambridge University Press.

Hsiao, C. (1985), Benefits and limitations of panel data. *Econometric Review* 4: 121–174.

Kneisner, T., and Q. Li (1996), Semiparametric panel data model with dynamic adjustment: Theoretical considerations and an application to labor supply. *Manuscript*, University of Guelph.

Li, Q. (1996), A note on root-N-consistent semiparametric estimation of partially linear models. *Economic Letters* 51: 277–285.

Li, Q., and T. Stengos (1996), Semiparametric estimation of partially linear panel data models. *Journal of Econometrics* 71: 389–397.

Li, Q., and A. Ullah (1998), Estimating partially linear panel data models with one-way error components errors. *Econometric Reviews* 17(2): 145–166.

Light, I., and E. Bonacich (1988), *Immigrant Entrepreneurs*. Berkeley: University of California Press.

Maddala, G. S. (1987), Recent developments in the econometrics of panel data analysis. *Transportation Research* 21: 303–326.

Mátyás, L., and P. Sevestre (1996), *The Econometrics of Panel Data: A Handbook of the Theory with Applications*. Dordrecht: Kluwer Academic Publishers.

Pagan, A., and A. Ullah (1999), *Nonparametric Econometrics*. New York: Cambridge University Press.

Robinson, P. M. (1988), Root-N-consistent semiparametric regression. *Econometrica* 56(4): 931–954.

White, H. (1984), *Asymptotic Theory for Econometricians*. New York: Academic Press.

29
Weighting Socioeconomic Indicators of Human Development: A Latent Variable Approach

A. L. NAGAR and SUDIP RANJAN BASU National Institute of Public Finance and Policy, New Delhi, India

I. INTRODUCTION

Since the national income was found inadequate to measure economic development, several composite measures have been proposed for this purpose [1–7]. They take into account more than one social indicator of economic development.

The United Nations Development Programme (UNDP) has been publishing Human Development Indices (HDIs) in their Human Development Reports (HDRs, 1990–1999) [7]. Although, over the years, some changes have been made in the construction of HDIs, the methodology has remained the same. As stated in the HDR for 1999, the HDI is based on three indicators:

(a) longevity: as measured by life expectancy (LE) at birth;
(b) educational attainment: measured as a weighted average of (i) adult literacy rate (ALR) with two-thirds weight, and (ii) combined gross primary, secondary, and tertiary enrolment ratio (CGER) with one-third weight; and

(c) standard of living: as measured by real gross domestic product (GDP) per capita (PPP$).

Fixed minimum and maximum values have been assigned for each of these indicators. Then, for any component (X) of the HDI, individual indices are computed as

$$I_X = \frac{\text{actual value of } X - \text{minimum value of } X}{\text{maximum value of } X - \text{minimum value of } X}$$

The real GDP (Y) is transformed as

$$I_Y = \frac{\log_e Y_{\text{actual}} - \log_c Y_{\text{min}}}{\log_e Y_{\text{max}} - \log_e Y_{\text{min}}}$$

and HDI is obtained as a simple (unweighted) arithmetic mean of these indices.[*]

The Physical Quality of Life Index (PQLI) was proposed by Morris [4]. The social indicators used in the construction of PQLI are:

(i) life expectancy at age one,
(ii) infant mortality rate, and
(iii) literacy rate.

For each indicator, the performance of individual countries is placed on a scale of 0 to 100, where 0 represents "worst performance" and 100 represents "best performance." Once performance for each indicator is scaled to this common measure, a composite index is calculated by taking the simple (unweighted) arithmetic average of the three indicators.[†]

Partha Dasgupta's [9] international comparison of the quality of life was based on:

(i) Y: per capita income (1980 purchasing power parity);
(ii) LE: life expectancy at birth (years);
(iii) IMR: infant mortality rate (per 1000);
(iv) ALR: adult literacy rate (%);
(v) an index of political rights, 1979;
(vi) an index of civil rights, 1979.

He noted that "The quality of data being what it is for many of the countries, it is unwise to rely on their cardinal magnitudes. We will, therefore,

[*]See UNDP HDR 1999, p. 159.
[†]"The PQLI proved unpopular, among researchers at least, since it had a major technical problem, namely the close correlation between the first two indicators" [8].

base our comparison on ordinal measures" [9]. He proposed using the Borda rank for countries.*

It should be noted that taking the simple unweighted arithmetic average of social indicators is not an appropriate procedure.[†] Similarly, assigning arbitrary weights (two-thirds to ALR and one-third to CGER in measuring educational attainment) is questionable. Moreover, complex phenomena like human development and quality of life are determined by a much larger set of social indicators than only the few considered thus far.[‡] These indicators may also be intercorrelated.

In the analysis that follows, using the principal components method and quantitative data (from HDR 1999), it turns out that the ranks of countries are highly (nearly perfectly) correlated with Borda ranks, as shown in Tables 7 and 13. Thus, Dasgupta's assertion about the use of quantitative data is not entirely upheld.[§]

In Section 2 we discuss the principal components method. The principal components are normalized linear functions of the social indicators, such that the sum of squares of coefficients is unity; and they are mutually orthogonal. The first principal component accounts for the largest proportion of total variance (trace of the covariance matrix) of all causal variables (social indicators). The second principal component accounts for the second largest proportion, and so on.

Although, in practice, it is adequate to replace the whole set of causal variables by only the first few principal components, which account for a substantial proportion of the total variation in all causal variables, if we compute as many principal components as the number of causal variables 100% of the total variation is accounted for by them.

We propose to compute as many principal components as the number of causal variables (social indicators) postulated to determine the human development. An estimator of the human development index is proposed as the

---

*The Borda [10] rule is an ordinal measurement. The rule says "award each alternative (here, country) a point equal to its rank in each criterion of ranking (here, the criteria being, life expectancy at birth, adult literacy rate, etc), adding each alternative's scores to obtain its aggregate score, and then ranking alternatives on the basis of their aggregate scores" [11]. For more discussion, see [12–16].

[†]On this point, see [17].

[‡]See UNDP HDR 1993 "Technical Note 2", pp. 104–112, for a literature review of HDI. See also, for example [8, 18–28]

[§]In several other experiments also we have found that the ranks obtained by the principal component method are highly (nearly perfectly) correlated with Borda ranks.

weighted average of the principal components, where weights are equal to variances of successive principal components. This method helps us in estimating the weights to be attached to different social indicators.

In Section 3 we compute estimates of the human development index by alternative methods and rank the countries accordingly. We use the data for 174 countries on the same variables as were used for computing HDI in HDR 1999. Table 6 provides human development indices for 174 countries by alternative methods, and Table 7 provides ranks of countries by different methods including the Borda ranks.*

In Section 4 we compute human development indices for 51 countries by the principal component method. We use eleven social indicators as determinants of the human development index. The data on the variables were obtained from HDR 1999 and World Development Indicators 1999. Table 14 provides the ranks of different countries according to estimates of the human development index. This table also provides the Borda ranks. The principal component method, outlined in Section 2, helps us in estimating weights to be attached to different social indicators determining the human development index.

As shown in Sections 3 and 4, GDP (measured as $\log_e Y$) is the most dominant factor. If only four variables ($\log_e Y$, LE, CGER, and ALR) are used (as in HDR 1999), $\log_e Y$ has the largest weight, followed by LE and CGER. The adult literacy rate (ALR) has the least weight. This result is at variance with the proposition of assigning higher weight (two-thirds) to ALR and lower weight (one-third) to CGER as suggested by HDR 1999.

The analysis in Section 4 shows that, after the highest weight for $\log_e Y$, we have health services (in terms of hospital beds available) and an environmental variable (measured as average annual deforestation) with the second and third largest weights. LE ranks fourth in order of importance, ASW is the fifth, and CGER sixth. Again, CGER has higher weight than ALR.

*The "best performing" country is awarded rank 1' and rank 174 goes to the "worst performing" country. In Section 4, for 51 countries, rank 51 is awarded to the "worst performing" country.

2. THE PRINCIPAL COMPONENTS METHOD OF ESTIMATING THE HUMAN DEVELOPMENT INDEX*

If human development index H could be measured quantitatively, in the usual manner, we would simply postulate a regression of H on the causal variables (social indicators) and determine optimal estimates of parameters by a suitable statistical method. In that case, we could determine the partial rates of change in H for a small unit change in any one of the causal variables, holding other variables constant, and also obtain the estimated level of H corresponding to given levels of the causal variables. However, human development is, in fact, an abstract conceptual variable, which cannot be directly measured but is supposed to be determined by the interaction of a large number of socioeconomic variables.

Let us postulate that the "latent" variable H is linearly determined by causal variables $x_1, \ldots, x_K$ as

$$H = \alpha + \beta_1 x_1 + \cdots + \beta_K x_K + u$$

so that the total variation in H is composed of two orthogonal parts: (a) variation due to causal variables, and (b) variation due to error.

If the model is well specified, including an adequate number of causal variables, so that the mean of the probability distribution of u is zero ($Eu = 0$) and error variance is small relative to the total variance of the latent variable H, we can reasonably assume that the total variation in H is largely explained by the variation in the causal variables.

We propose to replace the set of causal variables by an equal number of their principal components, so that 100% of variation in causal variables is accounted for by their principal components. In order to compute the principal components we proceed as follows:

Step 1. Transform the causal variables into their standardized form. We consider two alternatives:

$$X_k = \frac{x_k - \bar{x}_k}{S_{x_k}} \tag{1}$$

where $\bar{x}_k$ is the arithmetic mean and S_{x_k} is the standard deviation of observations on x_k; and

$$X_k^* = \frac{x_k - \min x_k}{\max x_k - \min x_k} \tag{2}$$

---

*For a detailed mathematical analysis of principal components, see [29]; see also [30]. For application of principal component methodology in computation of HDI, see HDR 1993, Technical Note 2, pp.109–110; also [22, 24].

where max x_k and min x_k are prespecified maximum and minimum values of x_k, for $k = 1, \ldots, K$.

The correlation coefficient is unaffected by the change of origin and scale. Therefore, the correlation matrix R of $X_1, \ldots, X_K$ will be the same as that of $X_1^*, \ldots, X_K^*$.

Step 2. Solve the determinantal equation

$$|R - \lambda I| = 0$$

for λ. Since R is a $K \times K$ matrix this provides a Kth degree polynomial equation in λ, and hence K roots. These roots are called the characteristic roots or eigenvalues of R. Let us arrange the roots in descending order of magnitude, as

$$\lambda_1 > \lambda_2 > \cdots > \lambda_K$$

Step 3. Corresponding to each value of λ, solve the matrix equation

$$(R - \lambda I)\alpha = 0$$

for the $K \times 1$ characteristic vector α, subject to the condition that

$$\alpha'\alpha = 1$$

Let us write the characteristic vectors as

$$\alpha_1 = \begin{pmatrix} \alpha_{11} \\ \vdots \\ \alpha_{1K} \end{pmatrix}, \ldots, \alpha_K = \begin{pmatrix} \alpha_{K1} \\ \vdots \\ \alpha_{KK} \end{pmatrix}$$

which corresponds to $\lambda = \lambda_1, \ldots, \lambda = \lambda_K$, respectively.

Step 4. The first principal component is obtained as

$$P_1 = \alpha_{11}X_1 + \cdots + \alpha_{1K}X_K$$

using the elements of the characteristic vector α_1 corresponding to the largest root λ_1 of R.

Similarly, the second principal component is

$$P_2 = \alpha_{21}X_1 + \cdots + \alpha_{2K}X_K$$

using the elements of the characteristic vector α_2 corresponding to the second largest root λ_2.

We compute the remaining principal components $P_3, \ldots, P_K$ using elements of successive characteristic vectors corresponding to roots $\lambda_3, \ldots, \lambda_K$, respectively. If we are using the transformation suggested in (2) of Step 1, the principal components are

$$P_1^* = \alpha_{11}X_1^* + \cdots + \alpha_{1K}X_K^*$$

$$\vdots$$

$$P_K^* = \alpha_{K1}X_1^* + \cdots + \alpha_{KK}X_K^*$$

where the coefficients are elements of successive characteristic vectors of R, and remain the same as in $P_1, \ldots, P_K$.

Step 5. The estimator of the human development index is obtained as the weighted average of the principal components: thus

$$\hat{H} = \frac{\lambda_1 P_1 + \cdots + \lambda_K P_K}{\lambda_1 + \cdots + \lambda_K}$$

and

$$\hat{H}^* = \frac{\lambda_1 P_1^* + \cdots + \lambda_K P_K^*}{\lambda_1 + \cdots + \lambda_K}$$

where the weights are the characteristic roots of R and it is known that

$$\lambda_1 = \text{Var } P_1 = \text{Var } P_1^*, \ldots, \lambda_K = \text{Var } P_K = \text{Var } P_K^*$$

We attach the highest weight to the first principal component P_1 or P_1^* because it accounts for the largest proportion of total variation in all causal variables. Similarly, the second principal component P_2 or P_2^* accounts for the second largest proportion of total variation in all causal variables and therefore the second largest weight λ_2 is attached to P_2 or P_2^*; and so on.

3. CONSTRUCTION OF HUMAN DEVELOPMENT INDEX USING THE SAME CAUSAL VARIABLES AND DATA AS IN THE CONSTRUCTION OF HDI IN HDR 1999

In this section we propose to construct the human development index by using the same causal variables as were used by the Human Development Report 1999 of the UNDP. This will help us in comparing the indices obtained by our method with those obtained by the method adopted by UNDP.

The variables are listed below and their definitions and data are as in HDR 1999.

1. LE: Life Expectancy at Birth; for the definition see HDR 1999, p. 254, and data relating to 1997 for 174 countries given in Table 1, pp. 134–137, of HDR 1999.

Table 1. Arithmetic means and standard deviations of observations on the causal variables, 1997

	LE (years)	ALR (%)	CGER (%)	$\log_e Y$
Arithmetic means	65.5868	78.8908	64.7759	8.3317
Standard deviations	10.9498	21.6424	19.8218	1.0845

2. ALR: Adult Literacy Rate; for the definition see HDR 1999, p. 255, and data relating to 1997 for 174 countries given in the Table 1 of HDR 1999.
3. CGER: Combined Gross Enrolment Ratio; for the definition see HDR 1999, p. 254 (also see pp. 255 and 256 for explanations of primary, secondary, and tertiary education). The data relating to 1997 for 174 countries are obtained from Table 1 of HDR 1999.
4. Y: Real GDP per capita (PPP$), defined on p. 255 of HDR 1999; data relating to 1997 for 174 countries are obtained from Table 1 of HDR 1999.

The arithmetic means and standard deviations of observations on the above causal variables are given in Table 1, and product moment correlations between them are given in Table 2.

3.1. Construction of Human Development Index by the Principal Component Method when Causal Variables Have Been Standardized as $(x - \bar{x})/S_x$

In the present case we transform the variables into their standardized form by subtracting the respective arithmetic mean from each observation and dividing by the corresponding standard deviation.

Table 2. Correlation matrix R of causal variables

	LE (years)	ALR (%)	CGER (%)	$\log_e Y$
LE (years)	1.000			
ALR (%)	0.753	1.000		
CGER (%)	0.744	0.833	1.000	
$\log_e Y$	0.810	0.678	0.758	1.000

Table 3. Characteristic roots (λ) of R in descending order

Values of λ			
$\lambda_1 = 3.288\,959$	$\lambda_2 = 0.361\,926$	$\lambda_3 = 0.217\,102$	$\lambda_4 = 0.132\,013$

The matrix R of correlations between pairs of causal variables, for data obtained from HDR 1999, is given in Table 2.* We compute the characteristic roots of the correlation matrix R by solving the determinantal equation $|R - \lambda I| = 0$. Since R is a 4×4 matrix, this will provide a polynomial equation of fourth degree in λ. The roots of R (values of λ), arranged in descending order of magnitude, are given in the Table 3.

Corresponding to each value of λ, we solve the matrix equation $(R - AI)\,\alpha = 0$ for the 4×1 characteristic vector α, such that $\alpha'\alpha = 1$. The characteristic vectors of R are given in Table 4.

The principal components of causal variables are obtained as normalized linear functions of standardized causal variables, where the coefficients are elements of successive characteristic vectors. Thus

$$P_1 = 0.502\,840\left(\frac{LE - 65.59}{10.95}\right) + 0.496\,403\left(\frac{ALR - 78.89}{21.64}\right)$$

$$+ 0.507\,540\left(\frac{CGER - 64.78}{19.82}\right) + 0.493\,092\left(\frac{\log_e Y - 8.33}{1.08}\right)$$

and, similarly, we compute the other principal components P_2, P_3, and P_4 by using successive characteristic vectors corresponding to the roots λ_2, λ_3, and λ_4, respectively. It should be noted that $\log_e Y$ is the natural logarithm of Y.

Table 4. Characteristic vectors (α) corresponding to successive roots (λ) of R

$\lambda = 3.288\,959$	$\lambda_2 = 0.361\,926$	$\lambda_3 = 0.217\,102$	$\lambda_4 = 0.132\,013$
α_1	α_2	α_3	α_4
0.502\,840	-0.372\,013	-0.667\,754	0.403\,562
0.496\,403	0.599\,596	-0.295\,090	-0.554\,067
0.507\,540	0.369\,521	0.524\,116	0.575\,465
0.493\,092	-0.604\,603	0.438\,553	-0.446\,080

*See UNDP HDR 1999, pp. 134–137, Table 1.

Values of the principal components for different countries are obtained by substituting the corresponding values of the causal variables. The variance of successive principal components, proportion of total variation accounted for by them, and cumulative proportion of variation explained, are given in Table 5.

We observe that 82.22% of the total variation in all causal variables is accounted for by the first principal component alone. The first two principal components together account for 91.27% of the total variation, the first three account for 96.70%, and all four account for 100% of the total variation in all causal variables.

The estimates of human development index, $\hat{H}$, for 174 countries are given in column (2) of Table 6, and ranks of countries according to the value of $\hat{H}$ are given in column (2) of Table 7. Since $\hat{H}$ is the weighted average of principal components, we can write, after a little rearrangement of terms,

$$\hat{H} = 0.356\,871\left(\frac{LE - 65.59}{10.95}\right) + 0.428\,112\left(\frac{ALR - 78.89}{21.64}\right)$$

$$+ 0.498\,193\left(\frac{CGER - 64.78}{19.82}\right) + 0.359\,815\left(\frac{\log_e Y - 8.33}{1.08}\right)$$

$$= -8.091 + 0.0326\,LE + 0.0198\,ALR + 0.0251\,CGER + 0.3318\,\log_e Y$$

as a weighted sum of the social indicators.

This result clearly indicates the order of importance of different social indicators in determining the human development index.* GDP is the

Table 5. Proportion of variation accounted for by successive principal components

Variance of P_k, $k = 1, \ldots, 4$	Proportion of variance accounted	Cumulative proportion of variance accounted
λ_k	$\lambda_k / \sum \lambda_k$	
3.288 959	0.822 240	
0.361 926	0.090 482	0.912 722
0.217 102	0.054 276	0.966 998
0.132 013	0.033 003	1

*It is not advisable to interpret the coefficients as partial regression coefficients because the left-hand dependent variable is not observable.

Table 6. Human development index as obtained by principal component method and as obtained by UNDP in HDR 1999

Country	HDI by principal component method, with variables standardized as		HDI as obtained in HDR 1999 (UNDP)
	$(x - \bar{x})/s_x$	$(x_{actual} - x_{min})/$ $(x_{max} - x_{min})$	
	$\hat{H}$	$\hat{H}^*$	HDI
Col (1)	Col (2)	Col (3)	Col (4)
Canada	2.255	1.563	0.932
Norway	2.153	1.543	0.927
United States	2.139	1.540	0.927
Japan	1.959	1.504	0.924
Belgium	2.226	1.558	0.923
Sweden	2.222	1.558	0.923
Australia	2.219	1.557	0.922
Netherlands	2.174	1.548	0.921
Iceland	1.954	1.504	0.919
United Kingdom	2.195	1.553	0.918
France	2.043	1.522	0.918
Switzerland	1.778	1.468	0.914
Finland	2.147	1.544	0.913
Germany	1.902	1.495	0.906
Denmark	1.914	1.497	0.905
Austria	1.857	1.486	0.904
Luxembourg	1.532	1.419	0.902
New Zealand	2.002	1.516	0.901
Italy	1.754	1.465	0.900
Ireland	1.864	1.488	0.900
Spain	1.897	1.494	0.894
Singapore	1.468	1.404	0.888
Israel	1.597	1.433	0.883
Hong Kong, China (SAR)	1.281	1.368	0.880
Brunei Darussalam	1.380	1.387	0.878
Cyprus	1.500	1.416	0.870
Greece	1.488	1.414	0.867
Portugal	1.621	1.439	0.858
Barbados	1.457	1.410	0.857
Korea, Rep. of	1.611	1.441	0.852
Bahamas	1.296	1.376	0.851
Malta	1.336	1.382	0.850
Slovenia	1.314	1.383	0.845
Chile	1.305	1.379	0.844
Kuwait	0.770	1.263	0.833

Table 6. Continued

Country	HDI by principal component method, with variables standardized as		HDI as obtained in HDR 1999 (UNDP)
	$(x - \bar{x})/s_x$	$(x_{actual} - x_{min})/$ $(x_{max} - x_{min})$	
	$\hat{H}$	$\hat{H}^*$	HDI
Col (1)	Col (2)	Col (3)	Col (4)
Czech Republic	1.209	1.363	0.833
Bahrain	1.249	1.364	0.832
Antigua and Barbuda	1.189	1.357	0.828
Argentina	1.246	1.370	0.827
Uruguay	1.210	1.363	0.826
Qatar	0.915	1.295	0.814
Slovakia	1.110	1.345	0.813
United Arab Emirates	0.832	1.276	0.812
Poland	1.080	1.341	0.802
Costa Rica	0.847	1.291	0.801
Trinidad and Tobago	0.838	1.292	0.797
Hungary	0.985	1.322	0.795
Venezuela	0.789	1.279	0.792
Panama	0.890	1.299	0.791
Mexico	0.801	1.281	0.786
Saint Kitts and Nevis	0.914	1.305	0.781
Grenada	0.932	1.312	0.777
Dominica	0.893	1.304	0.776
Estonia	0.894	1.325	0.773
Croatia	0.711	1.269	0.773
Malaysia	0.573	1.234	0.768
Colombia	0.715	1.266	0.768
Cuba	0.751	1.277	0.765
Mauritius	0.494	1.217	0.764
Belarus	0.911	1.311	0.763
Fiji	0.857	1.297	0.763
Lithuania	0.801	1.289	0.761
Bulgaria	0.681	1.265	0.758
Suriname	0.665	1.259	0.757
Libyan Arab Jamahiriya	0.939	1.306	0.756
Seychelles	0.407	1.202	0.755
Thailand	0.430	1.212	0.753
Romania	0.608	1.251	0.752
Lebanon	0.650	1.252	0.749
Samoa (Western)	0.543	1.238	0.747
Russian Federation	0.755	1.282	0.747

Table 6.   Continued

Country	HDI by principal component method, with variables standardized as		HDI as obtained in HDR 1999 (UNDP)
	$(x - \bar{x})/s_x$	$(x_{actual} - x_{min})/$ $(x_{max} - x_{min})$	
	$\hat{H}$	$\hat{H}^*$	HDI
Col (1)	Col (2)	Col (3)	Col (4)
Ecuador	0.625	1.251	0.747
Macedonia, TFYR	0.590	1.246	0.746
Latvia	0.628	1.256	0.744
Saint Vincent and the Grenadines	0.643	1.250	0.744
Kazakhstan	0.694	1.270	0.740
Philippines	0.777	1.285	0.740
Saudi Arabia	0.156	1.146	0.740
Brazil	0.671	1.257	0.739
Peru	0.654	1.257	0.739
Saint Lucia	0.527	1.227	0.737
Jamaica	0.324	1.189	0.734
Belize	0.413	1.201	0.732
Paraguay	0.364	1.201	0.730
Georgia	0.537	1.240	0.729
Turkey	0.242	1.171	0.728
Armenia	0.548	1.242	0.728
Dominican Republic	0.317	1.186	0.726
Oman	0.060	1.126	0.725
Sri Lanka	0.339	1.196	0.721
Ukraine	0.597	1.253	0.721
Uzbekistan	0.578	1.249	0.720
Maldives	0.490	1.230	0.716
Jordan	0.281	1.183	0.715
Iran, Islamic Rep. of	0.300	1.179	0.715
Turkmenistan	0.781	1.291	0.712
Kyrgyzstan	0.327	1.199	0.702
China	0.229	1.172	0.701
Guyana	0.236	1.181	0.701
Albania	0.214	1.170	0.699
South Africa	0.646	1.258	0.695
Tunisia	0.104	1.139	0.695
Azerbaijan	0.314	1.198	0.695
Moldova, Rep. of	0.240	1.185	0.683
Indonesia	0.028	1.135	0.681
Cape Verde	0.150	1.153	0.677

Table 6. Continued

Country	HDI by principal component method, with variables standardized as		HDI as obtained in HDR 1999 (UNDP)
	$(x - \bar{x})/s_x$	$(x_{actual} - x_{min})/(x_{max} - x_{min})$	
	$\hat{H}$	$\hat{H}^*$	HDI
Col (1)	Col (2)	Col (3)	Col (4)
El Salvador	−0.064	1.113	0.674
Tajikistan	0.121	1.164	0.665
Algeria	−0.155	1.086	0.665
Viet Nam	−0.064	1.122	0.664
Syrian Arab Republic	−0.238	1.076	0.663
Bolivia	−0.033	1.125	0.652
Swaziland	−0.068	1.116	0.644
Honduras	−0.416	1.042	0.641
Namibia	0.083	1.148	0.638
Vanuatu	−0.741	0.974	0.627
Guatemala	−0.746	0.974	0.624
Solomon Islands	−0.801	0.961	0.623
Mongolia	−0.521	1.031	0.618
Egypt	−0.416	1.035	0.616
Nicaragua	−0.519	1.020	0.616
Botswana	−0.346	1.061	0.609
São Tomé and Principe	−0.592	1.012	0.609
Gabon	−0.603	1.005	0.607
Iraq	−0.950	0.933	0.586
Morocco	−1.091	0.898	0.582
Lesotho	−0.682	1.001	0.582
Myanmar	−0.744	0.990	0.580
Papua New Guinea	−1.200	0.892	0.570
Zimbabwe	−0.571	1.031	0.560
Equatorial Guinea	−0.782	0.984	0.549
India	−1.148	0.896	0.545
Ghana	−1.310	0.870	0.544
Cameroon	−1.306	0.874	0.536
Congo	−0.824	0.976	0.533
Kenya	−1.221	0.898	0.519
Cambodia	−1.135	0.909	0.514
Pakistan	−1.675	0.786	0.508
Comoros	−1.665	0.796	0.506
Lao People's Dem. Rep.	−1.436	0.847	0.491
Congo, Dem. Rep. of	−1.682	0.808	0.479
Sudan	−1.950	0.741	0.475

Table 6. Continued

Country	HDI by principal component method, with variables standardized as		HDI as obtained in HDR 1999 (UNDP)
	$(x - \bar{x})/s_x$	$(x_{actual} - x_{min})/(x_{max} - x_{min})$	
	$\hat{H}$	$\hat{H}^*$	HDI
Col (1)	Col (2)	Col (3)	Col (4)
Togo	−1.490	0.835	0.469
Nepal	−1.666	0.793	0.463
Bhutan	−2.517	0.623	0.459
Nigeria	−1.659	0.806	0.456
Madagascar	−2.039	0.723	0.453
Yemen	−1.906	0.748	0.449
Mauritania	−2.083	0.709	0.447
Bangladesh	−2.240	0.679	0.440
Zambia	−1.788	0.791	0.431
Haiti	−2.460	0.639	0.430
Senegal	−2.348	0.656	0.426
Côte d'Ivoire	−2.226	0.686	0.422
Benin	−2.253	0.676	0.421
Tanzania, U. Rep. of	−2.172	0.713	0.421
Djibouti	−2.595	0.615	0.412
Uganda	−2.187	0.707	0.404
Malawi	−1.605	0.823	0.399
Angola	−2.596	0.615	0.398
Guinea	−2.620	0.606	0.398
Chad	−2.547	0.628	0.393
Gambia	−2.454	0.638	0.391
Rwanda	−2.290	0.689	0.379
Central African Republic	−2.748	0.585	0.378
Mali	−2.831	0.565	0.375
Eritrea	−3.036	0.521	0.346
Guinea-Bissau	−2.862	0.561	0.343
Mozambique	−2.996	0.538	0.341
Burundi	−3.110	0.520	0.324
Burkina Faso	−3.436	0.443	0.304
Ethiopia	−3.307	0.478	0.298
Niger	−3.612	0.404	0.298
Sierra Leone	−3.469	0.449	0.254

Table 7. Ranks of countries according to human development index obtained by alternative methods

Country	Rank of country according to			
	$\hat{H}$	$\hat{H}^*$	HDI	Borda
Col (1)	Col (2)	Col (3)	Col (4)	Col (5)
Canada	1	1	1	1
Norway	7	8	2	2
United States	9	9	3	10
Japan	12	12	4	6
Belgium	2	2	5	3
Sweden	3	3	6	4
Australia	4	4	7	4
Netherlands	6	6	8	10
Iceland	13	13	9	7
United Kingdom	5	5	10	8
France	10	10	11	8
Switzerland	19	19	12	12
Finland	8	7	13	13
Germany	15	15	14	13
Denmark	14	14	15	17
Austria	18	18	16	15
Luxembourg	24	24	17	21
New Zealand	11	11	18	15
Italy	20	20	19	19
Ireland	17	17	20	18
Spain	16	16	21	20
Singapore	27	28	22	31
Israel	23	23	23	22
Hong Kong, China (SAR)	34	35	24	39
Brunei Darussalam	29	29	25	37
Cyprus	25	25	26	25
Greece	26	26	27	23
Portugal	21	22	28	30
Barbados	28	27	29	26
Korea, Rep. of	22	21	30	28
Bahamas	33	33	31	40
Malta	30	31	32	32
Slovenia	31	30	33	23
Chile	32	32	34	34
Kuwait	60	66	35	61

Table 7. Continued

Country	Rank of country according to			
	$\hat{H}$	$\hat{H}^*$	HDI	Borda
Col (1)	Col (2)	Col (3)	Col (4)	Col (5)
Czech Republic	38	38	36	27
Bahrain	35	36	37	42
Antigua and Barbuda	39	39	38	41
Argentina	36	34	39	35
Uruguay	37	37	40	35
Qatar	46	51	41	58
Slovakia	40	40	42	29
United Arab Emirates	54	61	43	61
Poland	41	41	44	33
Costa Rica	52	53	45	54
Trinidad and Tobago	53	52	46	50
Hungary	42	43	47	38
Venezuela	57	59	48	65
Panama	50	49	49	49
Mexico	55	58	50	63
Saint Kitts and Nevis	47	47	51	53
Grenada	45	44	52	46
Dominica	49	48	53	47
Estonia	43	42	54	43
Croatia	64	63	55	65
Malaysia	79	82	56	82
Colombia	63	64	57	70
Cuba	62	60	58	56
Mauritius	84	85	59	85
Belarus	48	45	60	44
Fiji	51	50	61	50
Lithuania	56	55	62	45
Bulgaria	66	65	63	67
Suriname	68	67	64	71
Libyan Arab Jamahiriya	44	46	65	59
Seychelles	88	87	66	87
Thailand	86	86	67	90
Romania	75	75	68	78
Lebanon	70	73	69	72
Samoa (Western)	81	81	70	79
Russian Federation	61	57	71	48

Table 7. Continued

Country	Rank of country according to			
	$\hat{H}$	$\hat{H}^*$	HDI	Borda
Col (1)	Col (2)	Col (3)	Col (4)	Col (5)
Ecuador	74	74	72	81
Macedonia, TFYR	77	78	73	74
Latvia	73	71	74	59
Saint Vincent and the Grenadines	72	76	75	68
Kazakhstan	65	62	76	52
Philippines	59	56	77	69
Saudi Arabia	102	105	78	98
Brazil	67	69	79	76
Peru	69	70	80	79
Saint Lucia	83	84	81	84
Jamaica	92	93	82	88
Belize	87	88	83	83
Paraguay	89	89	84	93
Georgia	82	80	85	55
Turkey	97	100	86	104
Armenia	80	79	87	73
Dominican Republic	93	94	88	93
Oman	107	108	89	102
Sri Lanka	90	92	90	89
Ukraine	76	72	91	56
Uzbekistan	78	77	92	64
Maldives	85	83	93	86
Jordan	96	96	94	97
Iran, Islamic Rep. of	95	98	95	92
Turkmenistan	58	54	96	74
Kyrgyzstan	91	90	97	98
China	100	99	98	106
Guyana	99	97	99	103
Albania	101	101	100	96
South Africa	71	68	101	76
Tunisia	105	106	102	101
Azerbaijan	94	91	103	91
Moldova, Rep. of	98	95	104	100
Indonesia	108	107	105	110
Cape Verde	103	103	106	104

Table 7. Continued

Country	Rank of country according to			
	$\hat{H}$	$\hat{H}^*$	HDI	Borda
Col (1)	Col (2)	Col (3)	Col (4)	Col (5)
El Salvador	111	112	107	113
Tajikistan	104	102	108	107
Algeria	113	113	109	109
Viet Nam	110	110	110	114
Syrian Arab Republic	114	114	111	116
Bolivia	109	109	112	111
Swaziland	112	111	113	108
Honduras	117	116	114	119
Namibia	106	104	115	95
Vanuatu	124	128	116	123
Guatemala	126	127	117	122
Solomon Islands	128	129	118	120
Mongolia	119	119	119	124
Egypt	116	117	120	115
Nicaragua	118	120	121	121
Botswana	115	115	122	112
São Tomé and Principe	121	121	123	126
Gabon	122	122	124	117
Iraq	130	120	125	130
Morocco	131	132	126	129
Lesotho	123	123	127	125
Myanmar	125	124	128	131
Papua New Guinea	134	135	129	132
Zimbabwe	120	118	130	117
Equatorial Guinea	127	125	131	128
India	133	134	132	134
Ghana	137	137	133	135
Cameroon	136	136	134	133
Congo	129	126	135	127
Kenya	135	133	136	138
Cambodia	132	131	137	135
Pakistan	144	146	138	139
Comoros	142	143	139	142
Lao People's Dem. Rep.	138	138	140	141
Congo, Dem. Rep. of	145	141	141	143
Sudan	148	148	142	144

Table 7. Continued

Country	Rank of country according to			
	$\hat{H}$	$\hat{H}^*$	HDI	Borda
Col (1)	Col (2)	Col (3)	Col (4)	Col (5)
Togo	139	139	143	139
Nepal	143	144	144	144
Bhutan	160	161	145	152
Nigeria	141	142	146	148
Madagascar	149	149	147	151
Yemen	147	147	148	149
Mauritania	150	151	149	146
Bangladesh	154	155	150	154
Zambia	146	145	151	147
Haiti	159	158	152	156
Senegal	157	157	153	153
Côte d'Ivoire	153	154	154	150
Benin	155	156	155	154
Tanzania, U. Rep. of	151	150	156	158
Djibouti	162	162	157	162
Uganda	152	152	158	157
Malawi	140	140	159	137
Angola	163	163	160	162
Guinea	164	164	161	159
Chad	161	160	162	164
Gambia	158	159	163	160
Rwanda	156	153	164	160
Central African Republic	165	165	165	165
Mali	166	166	166	166
Eritrea	169	169	167	167
Guinea-Bissau	167	167	168	168
Mozambique	168	168	169	169
Burundi	170	170	170	170
Burkina Faso	172	173	171	172
Ethiopia	171	171	172	173
Niger	174	174	173	171
Sierra Leone	173	172	174	174

most dominant factor. Next are LE and CGER. ALR has the least weight.

3.2 Construction of Human Development Index $\hat{H}^*$ by the Principal Component Method when Causal Variables Have Been Standardized as $(x_{actual} - x_{min})/(x_{max} - x_{min})$

In this section we adopt UNDP methodology to standardize the causal variables.[*]

Suppose x is one of the causal variables (LE, ALR, CGER, or $\log_e Y$). Firstly, obtain the maximum and minimum attainable values of x, say x_{max} and x_{min}; then define

$$X = \frac{x_{actual} - x_{min}}{x_{max} - x_{min}}$$

as the standardized form of X. This is also called the "achievement index" by the HDR. The correlation matrix of the causal variables, so standardized, will be the same as R given in Table 2. The characteristic roots and vectors will be the same as in Tables 3 and 4, respectively.

We use the same maximum and minimum values for the causal variables as given in HDR.[†] Thus the first principal component is computed as

$$P_1^* = 0.502\,840\left(\frac{\text{LE}_{actual} - 25}{85 - 25}\right) + 0.496\,403\left(\frac{\text{ALR}_{actual} - 0}{100 - 0}\right)$$
$$+ 0.507\,540\left(\frac{\text{CGER}_{actual} - 0}{100 - 0}\right) + 0.493\,092\left(\frac{\log_e Y_{actual} - 4.60}{10.60 - 4.60}\right)$$

where the coefficients are elements of the characteristic vector corresponding to the largest root of R, and $\log_e Y$ is the natural logarithm. Similarly, we obtain P_2^*, P_3^*, and P_4^* by using characteristic vectors corresponding to the successive characteristic roots.

The estimates of human development index $\hat{H}^*$ for 174 countries and the rank of each country according to the value of $\hat{H}^*$ are given in column (3) of Tables 6 and 7, respectively. After a rearrangement of terms we may express

---

[*]See UNDP HDR 1999, p. 159.
[†]See UNDP HDR 1999, p. 159.

$$\hat{H}^* = 0.356\,871\left(\frac{LE_{actual} - 25}{85 - 25}\right) + 0.428\,112\left(\frac{ALR_{actual} - 0}{100 - 0}\right)$$

$$+ 0.498\,193\left(\frac{CGER_{actual} - 0}{100 - 0}\right) + 0.359\,815\left(\frac{\log_e Y_{actual} - 4.60}{10.60 - 4.60}\right)$$

$$= -0.425\,254 + 0.005\,948\,LE + 0.004\,281\,ALR + 0.004\,982\,CGER$$

$$+ 0.060\,055\,\log_e Y$$

We should note that the GDP is again the most dominant factor in determining $\hat{H}^*$; LE has the second largest and CGER the third largest weight. ALR gets the least weight.

3.3 Construction of HDI by the Method Outlined in HDR 1999

The UNDP method of calculating the human development index (HDI) has been outlined in HDR 1999.* HDIs for 1997 data for 174 countries and ranks of different countries according to HDI are reproduced in column (4) of Tables 6 and 7, respectively.† The Borda ranks are given in column 5 of Table 7.

Writing $\hat{H}$ and $\hat{H}^*$ for estimates of human development index as obtained by the principal component method, when the variables have been transformed as $(x - \bar{x})/s_x$ and $(x - x_{min})/(x_{max} - x_{min})$, respectively, and using HDI for the one computed in HDR 1999, we obtain the rank correlations given in Table 8. As we should expect, due to the very construction of $\hat{H}$ and $\hat{H}^*$ the correlation between ranks obtained by these two methods is the

Table 8. Spearman correlations between ranks of countries by different methods

	$\hat{H}$	$\hat{H}^*$	HDI	Borda
$\hat{H}$	1.0000			
$\hat{H}^*$	0.9995	1.0000		
HDI	0.9864	0.9832	1.0000	
Borda	0.9936	0.9940	0.9819	1.0000

*See UNDP HDR 1999, p. 159.
†See UNDP HDR 1999, pp. 134–137, Table 1.

highest. Next highest correlations are $r_{\hat{H},\text{Borda}} = 0.9936$ and $r_{\hat{H}^*,\text{borda}} = 0.9940$. The correlation between ranks obtained by $\hat{H}$ and HDI is equal to 0.9864 and that between $\hat{H}^*$ and HDI is 0.9832. The lowest correlation, 0.9819, is obtained between ranks according to HDI and Borda.

4. CONSTRUCTION OF THE HUMAN DEVELOPMENT INDEX BY THE PRINCIPAL COMPONENT METHOD WHEN ELEVEN SOCIAL INDICATORS ARE USED AS DETERMINANTS OF HUMAN DEVELOPMENT

In this section, we suppose that the latent variable "human development" is determined by several variables.[*] The selected variables, their definitions, and data sources are given below.[†]

(1) Y: Real GDP per capita (PPP$); defined on p. 255 of the Human Development Report (HDR) 1999 of the UNDP; data relating to 1997 are obtained from Table 1, pp. 134–137 of HDR 1999.

(2) LE: Life expectancy at birth; for definition see HDR 1999, p. 254; data relating to 1997 are given in Table 1 of HDR 1999.

(3) IMR: Infant mortality rate (per 1000 live births); defined on p. 113 of the World Development Indicators (WDI) 1999 of the World Bank; data relating to 1997 are available in Table 2.18, pp. 110–112, of WDI 1999.[‡]

(4) HB: Hospital beds (per 1000 persons); for definition see p. 93 and, for data relating to the period 1990–1997, Table 2.13, pp. 90–92, of WDI 1999.[§]

[*]On this point, we would like to mention that there is also a need to take into account the indicator of human freedom, especially an indicator to measure negative freedom, namely political and civil freedom, that is about one's ability to express oneself "without fear of reprisals" [9] (p. 109). However, "should the freedom index be integrated with the human development index? There are some arguments in favor, but the balance of arguments is probably against" [23]. For more discussions, see also, for example, UNDP HDR 1990, Box 1.5, p. 16; UNDP HDR 1991, pp. 18–21 and Technical Note 6, p. 98; references [9] (pp. 129–131), [20, 31].

[†]The data on all eleven variables are available for 51 countries only.

[‡]Infant mortality rate is an indicator of "nutritional deficiency" [32] and "(lack of) sanitation" [33].

[§]This indicator is used to show "health services available to the entire population" [34] (p. 93).

(5) ASA: Access to sanitation (percent of population); for definition see p. 97 and, for data relating to 1995, Table 2.14, pp. 94–96, of WDI 1999.

(6) ASW: Access to safe water (percent of population); definition is given on p. 97 and data relating to 1995 are available in Table 2.14, pp. 94–96, of WDI 1999.[*]

(7) HE: Health expenditure per capita (PPP$); for definition see p. 93 and, for data relating to the period 1990–1997, Table 2.13, pp. 90–92, of WDI 1999.[†]

(8) ALR: Adult literacy rate (percent of population); for definition we refer to p. 255, and data relating to 1997 have been obtained from Table 1, pp. 134–137, of HDR 1999.

(9) CGER: Combined (first, second and third level) gross enrolment ratio (percent of population); for definition see p. 254 and, for data relating to 1997, Table 1, pp. 134–137, of HDR 1999.

(10) CEU: Commercial energy use per capita (kg of oil equivalent); we refer to the explanation on p.147 and data relating to 1996 in Table 3.7, pp. 144–146, of WDI 1999.[‡]

(11) ADE: Average annual deforestation (percent change in km^2); an explanation is given on p. 123 and data relating to 1990–1995 are given in Table 3.1, pp. 120–122, of WDI 1999.[§]

The arithmetic means and standard deviations of observations on these variables for 51 countries are given in Table 9, and pair-wise correlations between them are given in Table 10.

The characteristic roots of the 11×11 correlation matrix R of selected causal variables are given in Table 11, and the corresponding characteristic vectors are given in Table 12.

---

[*]As noted by WDI [34] (p. 97), "People's health is influenced by the environment in which they live. A lack of clean water and basic sanitation is the main reason diseases transmitted by feces are so common in developing countries. Drinking water contaminated by feces deposited near homes and an inadequate supply of water cause diseases that account for 10 percent of the total diseases burden in developing countries."

[†]This indicator takes into account how much the country is "involved in health care financing" [34] (p. 93).

[‡]It is noted that "commercial energy use is closely related to growth in the modern sectors—industry, motorized transport and urban areas" [34] (p. 147).

[§]It appears that "Deforestation, desertification and soil erosion are reduced with poverty reduction" [23].

Table 9. Arithmetic means and standard deviations of observations on selected causal variables

Variable	A.M.	S.D.
$\log_e Y$	8.4553	0.9781
LE	67.4294	9.0830
IMR	37.0392	27.3188
HB	2.8000	3.2117
ASA	68.4314	24.9297
ASW	76.3529	19.4626
HE	418.0588	427.2688
ALR	78.6627	19.5341
CGER	65.6667	17.3962
CEU	1917.1373	2563.5089
ADE	1.2275	1.3800

The principal components of eleven causal variables are obtained by using successive characteristic vectors as discussed before. Variances of principal components are equal to the corresponding characteristic roots of R. The variance of successive principal components, proportion of total variation accounted for by them, and cumulative proportion of variation explained, are given in Table 13.

Table 14 gives the values of human development index for 51 countries when eleven causal variables have been used.* Analytically, we may express

$$\hat{H} = -6.952\,984 + 0.223\,742\log_e Y + 0.028\,863\,\text{LE} - 0.008\,436\,\text{IMR}$$

$$+ 0.046\,214\,\text{HB} + 0.009\,808\,\text{ASA} + 0.012\,754\,\text{ASW} + 0.000\,336\,\text{HE}$$

$$+ 0.007\,370\,\text{ALR} + 0.011\,560\,\text{CGER} + 0.000\,067\,\text{CEU}$$

$$+ 0.037\,238\,\text{ADE}$$

Real GDP per capita (measured as $\log_e Y$) has the highest weight in determining $\hat{H}$. Health services available in terms of hospital beds (HB) has the second largest weight. The environmental variable, as measured by average annual deforestation (ADE), is third in order of importance in determining $\hat{H}$. Then we have LE, ASW, CGER, etc., in that order.

---

*The Spearman correlation coefficient between ranks obtained by $\hat{H}$ and Borda rank is 0.977.

Table 10. Correlation matrix R of causal variables

	$\log_e Y$	LE	IMR	HB	ASA	ASW	HE	ALR	CGER	CEU	ADE
$\log_e Y$	1.000										
LE	0.851	1.000									
IMR	-0.896	-0.908	1.000								
HB	0.663	0.541	-0.571	1.000							
ASA	0.811	0.741	-0.814	0.521	1.000						
ASW	0.727	0.684	-0.661	0.436	0.774	1.000					
HE	0.793	0.634	-0.647	0.839	0.589	0.546	1.000				
ALR	0.756	0.679	-0.820	0.496	0.664	0.493	0.578	1.000			
CGER	0.810	0.735	-0.756	0.612	0.689	0.627	0.763	0.775	1.000		
CEU	0.702	0.511	-0.573	0.491	0.574	0.551	0.591	0.368	0.473	1.000	
ADE	-0.302	0.000	0.044	-0.352	-0.035	-0.111	-0.398	-0.064	-0.288	-0.323	1.000

Table 11. Characteristic roots (λ) of R in descending order

Values of λ
$\lambda_1 = 7.065\,467$
$\lambda_2 = 1.359\,593$
$\lambda_3 = 0.744\,136$
$\lambda_4 = 0.527\,319$
$\lambda_5 = 0.424\,499$
$\lambda_6 = 0.280\,968$
$\lambda_7 = 0.231\,752$
$\lambda_8 = 0.159\,822$
$\lambda_9 = 0.115\,344$
$\lambda_{10} = 0.048\,248$
$\lambda_{11} = 0.042\,852$
$\sum \lambda_k = 11.000\,000$

ACKNOWLEDGMENTS

The first author is National Fellow (ICSSR) at the National Institute of Public Finance and Policy in New Delhi, and is grateful to the Indian Council of Social Science Research, New Delhi, for funding the national fellowship and for providing all assistance in completing this work.

Both the authors express their gratitude to the National Institute of Public Finance and Policy for providing infrastructural and research facilities at the Institute. They also wish to express their thanks to Ms Usha Mathur for excellent typing of the manuscript and for providing other assistance from time to time.

Table 12. Characteristic vectors (α) corresponding to successive roots (λ) of R

$\lambda_1 =$ 7.065 467	$\lambda_2 =$ 1.359 593	$\lambda_3 =$ 0.744 136	$\lambda_4 =$ 0.527 319	$\lambda_5 =$ 0.424 499	$\lambda_6 =$ 0.280 968	$\lambda_7 =$ 0.231 752	$\lambda_8 =$ 0.159 822	$\lambda_9 =$ 0.115 344	$\lambda_{10} =$ 0.048 248	$\lambda_{11} =$ 0.042 852
α_1	α_2	α_3	α_4	α_5	α_6	α_7	α_8	α_9	α_{10}	α_{11}
0.363 021	−0.020 423	0.056 856	−0.097 385	−0.132 687	0.145 080	−0.093 506	−0.004 807	−0.314 855	−0.827 743	−0.152 870
0.328 039	0.243 191	−0.019 339	0.050 356	−0.056 878	0.707 711	−0.010 862	0.044 803	0.131 679	0.114 571	0.541 901
−0.344 504	−0.227 806	0.065 478	0.053 847	0.258 806	−0.215 099	0.263 654	0.060 320	−0.000 200	−0.385 465	0.698 034
0.278 243	−0.317 803	−0.267 873	0.612 959	0.152 481	−0.101 611	−0.314 322	−0.196 894	0.431 698	−0.128 348	−0.011 411
0.322 146	0.217 303	0.221 135	−0.276 675	0.192 332	−0.392 541	−0.451 575	0.598 610	−0.055 853	0.088 983	0.186 985
0.291 544	0.137 001	0.448 905	−0.114 152	0.637 901	−0.037 293	0.127 937	−0.499 217	0.001 064	0.059 960	−0.064 780
0.316 228	−0.297 775	−0.184 005	0.333 261	0.112 374	−0.008 233	0.336 611	0.121 213	−0.666 327	0.284 718	0.025 660
0.299 332	0.195 558	−0.434 331	−0.315 927	−0.266 067	−0.441 525	−0.037 849	−0.440 959	−0.077 715	0.074 038	0.332 520
0.329 021	−0.031 791	−0.282 683	−0.271 463	0.167 448	−0.065 501	0.569 364	0.369 672	0.452 728	−0.086 611	−0.177 499
0.261 503	−0.206 351	0.605 071	0.152 158	−0.575 211	−0.201 654	0.270 732	−0.046 932	0.196 172	0.078 800	0.089 718
−0.097 412	0.742 329	−0.037 569	0.536 516	−0.051 901	−0.161 426	0.300 547	0.003 883	−0.023 787	−0.157 759	−0.075 157

Table 13. Proportion of variation accounted for by successive principal components

Variance of P_k, $k = 1, \ldots, 11$	Proportion of variance accounted for	Cumulative proportion of variance accounted for
λ_k	$\lambda_k / \sum \lambda_k$	
7.065 467	0.642 315	
1.359 593	0.123 599	0.765 915
0.744 136	0.067 649	0.833 563
0.527 319	0.047 938	0.881 501
0.424 499	0.038 591	0.920 092
0.280 968	0.025 543	0.945 635
0.231 752	0.021 068	0.966 703
0.159 822	0.014 529	0.981 232
0.115 344	0.010 486	0.991 718
0.048 248	0.004 386	0.996 104
0.042 852	0.003 896	1

Table 14. Rank of selected countries according to principal component method and Borda rule

Country	$\hat{H}$	Rank($\hat{H}$)	Borda rank
Argentina	0.8703	14	11
Bangladesh	−2.1815	45	47
Bolivia	−1.0286	38	36
Brazil	0.0862	27	21
Cameroon	−2.0089	43	42
Canada	3.0548	3	3
Chile	1.2983	10	9
Colombia	0.1696	26	23
Costa Rica	1.2534	11	13
Côte d'Ivoire	−2.3582	48	46
Croatia	0.3703	22	14
Dominican Republic	0.0040	28	27
Ecuador	−0.3194	30	29
Egypt, Arab Rep.	−0.5217	33	30
El Salvador	−0.4220	32	34
Ethiopia	−4.2114	51	51
Finland	2.8693	5	5
Ghana	−1.8505	42	45
Guatemala	−0.7268	34	39
Haiti	−2.8393	50	50
Honduras	−0.3460	31	35
India	−1.7235	41	41
Indonesia	−0.7779	35	37
Iran, Islamic Rep. of	0.3269	23	26
Jamaica	0.7192	17	25
Japan	3.0815	2	4
Jordan	0.5892	20	20
Kenya	−2.1879	46	43
Malaysia	0.9057	13	18
Mexico	0.6327	19	19
Morocco	−1.2233	39	40
Nepal	−2.4046	49	48
Netherlands	3.0350	4	2
Nicaragua	−0.8209	37	38
Nigeria	−2.2061	47	44
Norway	3.1919	1	1
Pakistan	−2.1051	44	48
Panama	0.8540	15	16

Table 14. Continued

Country	$\hat{H}$	Rank($\hat{H}$)	Borda rank
Paraguay	−0.8110	36	33
Peru	−0.2656	29	28
Philippines	0.2481	24	31
Saudi Arabia	0.7489	16	17
Singapore	2.2751	7	7
Thailand	0.5235	21	24
Trinidad and Tobago	1.3819	9	12
Tunisia	0.1924	25	22
United Arab Emirates	2.0272	8	8
United Kingdom	2.4542	6	6
Uruguay	0.9057	12	10
Venezuela	0.6924	18	15
Zimbabwe	−1.2879	40	32

REFERENCES

1. J. Drewnowski and W. Scott. *The Level of Living Index*. United Nations Research Institute for Social Development, Report No. 4, Geneva, September 1966.
2. I. Adelman and C. T. Morris. *Society, Politics and Economic Development*. Baltimore: Johns Hopkins University Press, 1967.
3. V. McGranahan, C. Richard-Proust, N. V. Sovani, and M. Subramanian. *Contents and Measurement of Socio-economic Development*. New York: Praeger, 1972.
4. Morris D. Morris. *Measuring the Condition of the World's Poor: The Physical Quality of Life Index*. New York: Pergamon, 1979.
5. N. Hicks and P. Streeten. Indicators of development: the search for a basic needs yardstick. *World Development* 7: 567–580, 1979.
6. Morris D. Morris and Michelle B. McAlpin. *Measuring the Condition of India's Poor: The Physical Quality of Life Index*. New Delhi: Promilla, 1982.
7. United Nations Development Programme (UNDP). *Human Development Report*. New York: Oxford University Press (1990–1999).
8. Michael Hopkins. Human development revisited: A new UNDP report. *World Development* 19(10): 1469–1473, 1991.
9. Partha Dasgupta. *An Inquiry into Wellbeing and Destitution*. Oxford: Clarendon Press, 1993.

10. J.C. Borda. *Memoire sur les Elections au Scrutin; Memoires de l'Académie Royale des Sciences;* English translation by A. de Grazia. *Isis* 44: 1953.

11. Partha Dasgupta and Martin Weale. On measuring the quality of life. *World Development* 20(1): 119–131, 1992.

12. L. A. Goodman and H. Markowitz. Social welfare functions based on individual ranking. *American Journal of Sociology* 58: 1952.

13. J. H. Smith. Aggregation of preferences with variable electorate. *Econometrica* 41(6): 1027–1042, 1973.

14. B. Fine and K. Fine. Social choice and individual rankings, I and II. *Review of Economic Studies,* 44(3 & 4): 303–322 and 459–476, 1974.

15. Jonathan Levin and Barry Nalebuff. An introduction to vote-counting schemes. *Journal of Economic Perspectives* 9(1): 3–26, 1995.

16. Mozaffar Qizilbash. Pluralism and well-being indices. *World Development* 25(12): 2009–2026, 1997.

17. A. K. Sen. Poverty: an ordinal approach to measurement. *Econometrica* 44: 219–231, 1976.

18. V. V. Bhanoji Rao. Human development report 1990: review and assessment. *World Development* 19(10): 1451–1460, 1991.

19. Mark McGillivray. The human development index: yet another redundant composite development indicator? *World Development* 19(10): 1461–1468, 1991.

20. Meghnad Desai. Human development: concepts and measurement. *European Economic Review* 35: 350—357, 1991.

21. Mark McGillivray and Howard White. Measuring development: the UNDP's human development index. *Journal International of Development* 5: 183–192, 1993.

22. Tomson Ogwang. The choice of principal variables for computing the human development index. *World Development* 22(12): 2011–2014, 1994.

23. Paul Streeten. Human development: means and ends. *American Economic Review* 84(2): 232–237, 1994.

24. T. N. Srinivasan. Human development: a new paradigm or reinvention of the wheel? *American Economic Review* 84(2): 238–243, 1994.

25. M. V. Haq. *Reflections on Human Development.* New York: Oxford University Press, 1995.

26. Mozaffar Qizilbash. Capabilities, well-being and human development: a survey. *The Journal of Development Studies* 33(2), 143–162, 1996

27. Douglas A. Hicks. The inequality-adjusted human development index: a constructive proposal. *World Development* 25(8): 1283–1298, 1997.

28. Farhad Noorbakhsh. A modified human development index. *World Development* 26(3): 517–528, 1998.

29. T. W. Anderson. *An Introduction to Multivariate Statistical Analysis.* New York: John Wiley, 1958, Chapter II, pp. 272–287.
30. G. H. Dunteman. *Principal Components Analysis.* Nenbury Park, CA: Sage Publications, 1989.
31. A. K. Sen. Development: which way now? *Economic Journal*, 93: 745–762, 1983.
32. R. Martorell. *Nutrition and Health Status Indicators. Suggestions for Surveys of the Standard of Living in Developing Countries.* Living Standard Measurement Study No. 13, The World Bank, Washington, DC, 1982.
33. P. Streeten, with S. J. Burki, M. U. Haq, N. Hicks, and F. J. Stewart. *First Things First: Meeting Basic Human Needs in Developing Countries.* New York: Oxford University Press, 1981.
34. The World Bank. *World Development Indicators.* 1999.

30

A Survey of Recent Work on Identification, Estimation, and Testing of Structural Auction Models

SAMITA SAREEN   Bank of Canada, Ottawa, Ontario, Canada

1. INTRODUCTION

Auctions are a fast, transparent, fair, and economically efficient mechanism for allocating goods. This is demonstrated by the long and impressive list of commodities being bought and sold through auctions. Auctions are used to sell agricultural products like eggplant and flowers. Procurement auctions are being conducted by government agencies to sell the right to fell timber, drill offshore areas for oil, procure crude-oil for refining, sell milk quotas, etc. In the area of finance, auctions have been used by central banks to sell Treasury bills and bonds. The assets of bankrupt firms are being sold through auctions. Transactions between buyers and sellers in a stock exchange and foreign-exchange markets take the form of double auctions. Licenses for radio spectrum were awarded on the basis of "beauty contests" previously; spectrum auctions have now become routine in several countries. Recently, five third-generation mobile telecommunication licenses have been auctioned off in the United Kingdom for an unprecedented and unexpected sum of £22.47 billion, indicating that a carefully designed auction mechanism will provide incentives for the bidders to reveal their true value. Finally, internet sites like eBay, Yahoo and Amazon are being used to bring together buyers and sellers across the world to sell goods ranging from airline tickets to laboratory ventilation hoods (Lucking-Reiley, 2000). The experience with spectrum and internet auctions has demonstrated that the auction mechanism can be designed in a manner to make bidders reveal their "true" value for the auctioned object, not just in laboratories but in real life as well.

Insights into the design of auctions have been obtained by modelling auctions as games of incomplete information. A survey of the theory developed for single-unit auctions can be found in McAfee and McMillan (1987) and Milgrom and Weber (1982) and for multi-unit auctions in Weber (1983). A structural or game-theoretic auction model emerges once a seller pre-commits to a certain auction form, and rules and assumptions are made about the nature of "incomplete" information possessed by the bidders. The latter includes assumptions about the risk attitude of the bidders, the relationship between bidders' valuations, whether the "true" value of the object is known to them, and whether bidders are symmetric up to information differences. The rules and form of the auctions are determined by the seller in a manner that provides the bidders with incentives to reveal their valuation for the auctioned object.

Empirical work in auctions using field data has taken two directions. The first direction, referred to as the structural approach, attempts to recover the structural elements of the game-theoretic model. For example, if bidders are assumed to be risk-neutral, the structural element of the game-theoretic model would be the underlying probability law of valuations of the bidders. Any game-theoretic or structural model makes certain nonparametric predictions. For example, the revenue equivalence theorem is a prediction of a single-unit, independent-private-values auction model with symmetric and risk-neutral players. The second approach, referred to as the reduced-form approach, tests the predictions of the underlying game-theoretic model.

Identification of a private-values, second-price auction is trivial if there is a nonbinding reserve price and each bidder's bid in an auction is available. In a private-values auction a bidder knows the value of the auctioned object to herself. In a second-price auction the bidder pays the value of the object to the second-highest bidder. Hence in a private-values, second-price auction, a bidder submits a bid equal to her value of the auctioned object. If the bids of all bidders in an auction were observed, then in essence the private values of all bidders are observed. Identification then amounts to identifying the distribution of bidders' valuations from data on these valuations.

Outside this scenario identification, estimation and testing of structural models is a difficult exercise. First, outside the private-values, second-price auction, bidders do not bid their valuation of the object. In most cases, an explicit solution is not available for the Bayesian–Nash equilibrium strategy.

Second, since not all bids are observed, identification and estimation now involve recovering the distribution of bidder valuations when one observes a sample of order statistics from this distribution. The properties of these estimators have to be investigated since they do not have standard $\sqrt{T}$ asymptotics, where T is the number of auctions in the sample.

Laffont (1997), Hendricks and Paarsch (1995), and Perrigne and Vuong (1999) provide a survey of issues in empirical work in auctions using field data. Perrigne and Vuong focus on identification and estimation of single-unit, first-price, sealed-bid structural auctions when data on all bids in an auction is available. Bidders are assumed to be risk-neutral and to know the "true" value of the auctioned object. In addition to surveying the empirics of the above-mentioned structural auction model, Laffont and Hendricks and Paarsch discuss several approaches to testing predictions of various single-unit structural auction models through the reduced-form approach. The focus of the former is the work by Hendricks, Porter and their coauthors on drainage sales in the OCS auctions. (Hendricks, Porter and Boudreau; 1987; Hendricks and Porter, 1988; Porter, 1995).

The current survey adds to these surveys in *at least* three ways.

First, Guerre, et al. (2000) have established that the common-value model and private-value model cannot be distinguished if only data on all bids in an auction are observed.[*] Distinguishing between the two models of valuation is important for several reasons. First, the two models of valuations imply different bidding and participation behavior; this may be of interest by itself. Second, depending on the model of valuation, the optimal auction is different.

The need to distinguish the two models of valuation raises the question as to what additional data would be required to distinguish the two models. Several papers have addressed this issue recently: Hendricks et al. (2000), Athey and Haile (2000), and Haile, Hong and Shum (2000). Section 3 integrates these papers with the currently existing body of work by Guerre et al. (2000) on identification of several classes of private-values models from bid data. Exploiting results on competing risks models as in Berman (1963) and the multi-sector Roy model as in Heckman and Honoré (1990), Athey and Haile establish identification of several asymmetric models if additional data in the form of the identity of the winner is observed. Hendricks et al. (2000) and Haile, Hong and Shum (2000) use the ex-post value of the auctioned object and the variation in the number of potential bidders, respectively, to distinguish the common-value model from the private-values model.

---

[*]Parametric assumptions on the joint distribution of private and public signals about the value of the auctioned object will identify the two models.

Second, existing surveys make a cursory mention of the power of Bayesian tools in the estimation and testing of structural auction models. This gap in the literature is remedied in this survey in Section 4.

Third, several field data sets exhibit heterogeneity in the auctioned object, violating one of the assumption made in the theory of auctions that homogeneous objects are being auctioned. The issue is how to model object heterogeneity and still be able to use the results in auction theory that a single homogeneous object is being auctioned. Available approaches to model object heterogeneity are discussed in Section 5.

Section 6 concludes with some directions for future research.

2. DEFINITIONS AND NOTATION

The framework used in this paper follows that of Milgrom and Weber (1982). Unless otherwise mentioned, a single and indivisible object is auctioned by a single seller to n potential bidders. The subscript i will indicate these potential bidders, with $i = 1, \ldots, n$. X_i is bidder i's private signal about the auctioned object with $\mathbf{X} = (X_1, \ldots, X_n)$ indicating the private signals of the n potential bidders. $\mathbf{S} = (S_1, \ldots, S_m)$ are additional m random variables which affect the value of the auctioned object for *all* bidders. These are the *common* components affecting the utility function of all bidders. $U_i(\mathbf{X}, \mathbf{S})$ is the utility function of bidder i; $u_i \geq 0$; it is assumed to be continuous and nondecreasing in its variables. The probability density function and the distribution function of the random variable $\mathbf{X}$ will be indicated by $f_{\mathbf{X}}(\mathbf{x})$ and $F_{\mathbf{X}}(\mathbf{x})$, respectively. The joint distribution of $(\mathbf{X}, \mathbf{S})$ is defined on the support $[\underline{x}, \overline{x}]^n \times [\underline{s}, \overline{s}]^m$. If $\mathbf{X}$ is an n-dimensional vector with elements X_i, then the $n - 1$ dimensional vector *excluding* element X_i will be indicated by $\mathbf{X}_{-i}$. The support of a variable Z, if it depends on θ, will be indicated by $\zeta_Z(\theta)$.

The ith order statistic from a sample of size n will be denoted by $X^{i:n}$, with $X^{1:n}$ and $X^{n:n}$ being the lowest and the highest order statistic, respectively. If $X^{i:n}$ is the ith order statistic from the distribution F_x, its distribution function will be indicated by $F_x^{i:n}$ and its density function by $f_x^{i:n}$.

Once the seller has pre-committed to a set of rules, the genesis of a structural model is in the following optimization exercise performed by each of the $n + 1$ players: each player i chooses his equilibrium bid b_i by maximizing

$$E_{\mathbf{X}_{-i}\mathbf{S}|x_i}[(U_i(\mathbf{X}, \mathbf{S}) - b_i)\Pr(\text{Bidder } i \text{ wins})] \tag{1}$$

The concept of equilibrium employed is the Bayesian–Nash equilibrium. For example, if the seller pre-commits to the rules of a first-price auction, the first-order conditions for the maximization problem are the following set of differential equations

$$v(x_i, x_i; n) = b_i + \frac{\Pr\left(\max_{j \neq n} B_j \leq b_i | B_i = b_i\right)}{\frac{\partial}{\partial \tau} \Pr\left(\max_{j \neq n} B_j \leq \tau | B_i = b_i\right)\big|_{\tau = b_i}} \equiv \xi(b_i, F_{\mathbf{B}}(\mathbf{b}); n) \qquad (2)$$

where $v(x_i, y_i; n) = E\left(U_i | X_i = x_i, X_{-i}^{n-1:n-1} = y_i\right)$. $X_{-i}^{n-1:n-1}$ is the maximum over the variables $X_1, ..., X_n$ *excluding* X_i. This set of n differential equations will simplify to a single equation if bidders are symmetric. The first-order conditions involve variables that are both observed and unobserved to the econometrician.[*] The solution of this set of differential equations gives the equilibrium strategy

$$b_i = e_i(v(x_i, y_i; n), F_{\mathbf{XS}}(\mathbf{x}, \mathbf{s}), n) \qquad (3)$$

The equilibrium strategy of a bidder differs depending on the assumptions made about the auction format or rules and the model of valuation.[†] The structure of the utility function and the relationship between the variables $(\mathbf{X}, \mathbf{S})$ comprises the model of valuation. For example, the simplest scenario is a second-price or an English auction with independent private values; a bidder bids her "true" valuation for the auctioned object in these auctions. Here the auction format is a second-price or English auction. The utility function is $U_i(\mathbf{X}, \mathbf{S}) = X_i$ and the X_is are assumed to be independent.

The *models of valuation* which will be encountered in this paper at various points are now defined. The definitions are from Milgrom and Weber (1982), Laffont and Vuong (1996), Li et al. (2000), and Athey and Haile (2000).

Model 1: Affiliated values (AV)

This model is defined by the pair $[U_i(\mathbf{X}, \mathbf{S}), F_{XS}(\mathbf{x}, \mathbf{s})]$, with variables $(\mathbf{X}, \mathbf{S})$ being affiliated. A formal definition of affiliation is given in Milgrom and Weber (1982, p. 8). Roughly, when the variables $(\mathbf{X}, \mathbf{S})$ are affiliated, large values for some of the variables will make other variables large rather than small. Independent variables are always affiliated; only strict affiliation rules out independence.

The private-values model and common-value model are the two polar cases of the **AV**.

---

[*]Either some bids or bids of all n potential bidders may be observed. If participation is endogenous, then the number of potential bidders, n, is not observed. The private signals of the bidders are unobserved as well.

[†]If bidders are symmetric, $e_i(\bullet) = e(\bullet)$ and $\xi_i(\bullet) = \xi(\bullet)$ for all n bidders.

Model 2: Private values (PV)

Assuming risk neutrality, the private-values model emerges from the affiliated-values model if $U_i(\mathbf{X}, S) = X_i$. Hence, from equation (2),

$$x_i = e^{-1}(b_i) = \xi(b_i, F_\mathbf{B}(\mathbf{b}); n)$$

is also the inverse of the equilibrium bidding rule with respect to x_i. Different variants of this model emerge depending on the assumption made about the relationship between $(\mathbf{X}, S)$.

The *affiliated private values* (AV) model emerges if $(\mathbf{X}, S)$ are affiliated. The structural element of this model is $F_{\mathbf{X}S}$.

It is possible that the private values have a component common to all bidders and an idiosyncratic component. Li et al. (2000) justify that this is the case of the OCS wildcat auctions. The common component would be the unknown value of the tract. The idiosyncratic component could be due to differences in operational cost as well as the interpretation of geological surveys between bidders. Conditional on this common component, the idiosyncratic cost components could be independent. This was also the case in the procurement auctions for crude-oil studied by Sareen (1999); the cost of procuring crude-oil by bidders was independent, conditional on past crude-oil prices. This gives rise to the *conditionally independent private values* (CIPV henceforth) model. In a CIPV model,

(1) $U_i(\mathbf{X}, S) = X_i$ with $m = 1$. S is the auction-specific or common component that affects the utility of all bidders;
(2) X_i are independent, conditional on S.

The structural elements of the CIPV are $(F_S(s), F_{X|S}(x_i))$. In the CIPV model an explicit relationship could be specified for the common and idiosyncratic components of the private signals X_i. Thus

(1) $U_i(\mathbf{X}, S) = X_i = S + A_i$ where A_i, is the component idiosyncratic to bidder i;
(2) A_i are independent, conditional on S.

This is the CIPV model with additive components (CIPV-A); its structural elements are $(F_S(s), F_{A|S}(a_i))$.

If the assumption of conditional independence of A_i in the CIPV-A model is replaced with mutual independence of $(A_1, \ldots, A_n, S)$, the CIPV with independent components (CIPV-I) emerges. The independent private value (IPV) model is a special case of the CIPV-I with $S = 0$, so that $U_i = X_i = A_i$.

Model 3: Common value (CV)

The pure CV model emerges from the AV model if $U_i(\mathbf{X}, \mathbf{S}) = S$. S is the "true" value of the object, which is unknown but the *same* for all bidders; bidders receive private signals X_i about the unknown value of this object. If the OCS auctions, are interpreted as common-value auctions, as in Hendricks and Porter (1988) and Hendricks et al. (1999), then an example of s is the ex-post value of the tract. The structural elements of the pure CV model are $(F_S(s), F_{X|S}(x_i))$. Notice its similarity to the CIPV model with $U_i(\mathbf{X}, \mathbf{S}) = X_i$ and no further structure on X_i.

A limitation of the pure CV model is that it does not allow for variation in utility across the n bidders. This would be the case if a bidder's utility for the auctioned object depended on the private signals of other bidders; thus $U_i = U_i(X_{-i})$. Alternatively, bidders' tastes could vary due to factors idiosyncratic to a bidder, $U_i = U_i(A_i, S)$. Athey and Haile (2000) consider an additively separable variant of the latter. Specifically, $U_i = S + A_i$ with U_i, X_j strictly affiliated but not *perfectly* correlated. Excluding perfect correlation between U_i, X_j rules out *exclusively* private values, where $U_i = X_i = A_i$, and guarantees that the winner's curse exists. This will be referred to as the CV model in this survey.

The *Linear Mineral Rights* (LMR) model puts more structure on the pure CV model. Using the terminology of Li et al. (2000), the LMR model assumes (1) $U_i(\mathbf{X}, \mathbf{S}) = S$; (2) $X_i = SA_i$, with A_i independent, conditional on S; and (3) $V(x_i, x_i)$ is loglinear in $\log x_i$. If in addition $(S, \mathbf{A})$ are mutually independent the *LMR with independent components* (LMR-I) emerges. The LMR model is analogous to the CIPV-A and the LMR-I to the CIPV-I.

All the above-mentioned models of valuation can be either symmetric or asymmetric. A model of valuation is *symmetric* if the distribution $F_{\mathbf{XS}}(\mathbf{x}, \mathbf{s})$ is exchangeable in its first n arguments; otherwise the model is *asymmetric*. Thus a symmetric model says that bidders are ex ante symmetric. That is, the index i which indicates a bidder is exchangeable: it does not matter whether a bidder gets the label $i = 1$ or $i = n$. Note that symmetry of $F_{\mathbf{XS}}(\mathbf{x}, \mathbf{s})$ in it first n arguments $\mathbf{X}$ implies that $F_{\mathbf{X}}(\mathbf{x})$ is exchangeable as well.

The *auction formats* discussed in this survey are the first-price auction, Dutch auction, second-price, and the ascending auction of Milgrom and Weber (1982). In both the first-price and second-price auction the winner is the bidder who submits the highest bid; the transaction price is the highest bid in a first-price auction but the second-highest bid in the second-price auction. In a Dutch auction an auctioneer calls out an initial high price and continuously lowers it till a bidder indicates that she will buy the object for that price and stops the auction. In the ascending auction or the "button" auction the price is raised continuously by the auctioneer. Bidders indicate

to the auctioneer and the other bidders when they are exiting the auction;*
when a single bidder is left, the auction comes to an end. Thus the price level
and the number of active bidders are observed continuously.

Once an auction format is combined with a model of valuation, an equilibrium bidding rule emerges as described in equation (3). The bidding rules
for the AV model with some auction formats are given below.

Example 1: AV model and second-price auction

$$e(x_i) = v(x_i, x_i; n) = E\big(U_i|X_i = x_i, X_{-i}^{n-1:n-1} = x_i\big)\forall i \tag{4}$$

$X_{-i}^{n-1:n-1} = \max_{j\neq i,j=1,...,n} X_j$ is the maximum of the private signals over $n-1$
bidders excluding bidder i; and

$$v(x, y; n) = E\big(U_i|X_i = x_i, X_{-i}^{n-1:n-1} = y_i\big) \tag{5}$$

The subscript i does not appear in the bidding rule $e(x_i)$ since bidders are
symmetric. $e(x_i)$ is increasing in x_i.[†] If $(\mathbf{X}, S)$ are *strictly* affiliated, as in the
CV model, $e(x_i)$ is *strictly* increasing in x_i. For a PV model, $U_i(\mathbf{x}, \mathbf{s}) = X_i$.
Hence $v(x_i, x_i) = x_i$. The bidding rule simplifies to

$$e(x_i) = x_i \tag{6}$$

Example 2: AV model and first-price, sealed-bid/Dutch auction

$$b_i = e(x_i) = v(x_i, x_i) - \int_{\underline{x}}^{x_i} L(\alpha|x_i)dv(\alpha, \alpha) \tag{7}$$

where

$$L(\alpha|x_i) = \exp\left(-\int_\alpha^{x_i} \frac{f_{x_{-i}}^{n-1:n-1}(t|t)}{F_{x_{-i}}^{n-1:n-1}(t|t)}dt\right)$$

Again, from affiliation of $(\mathbf{X}, S)$ and that U_i is nondecreasing in its arguments, it follows that $e(x_i)$ is strictly increasing in x_i.[‡]

---

*For example, bidders could press a button or lower a card to indicate exit.

[†]A reasoning for this follows from Milgrom and Weber (1982). By definition U_i is a
nondecreasing function of its arguments $(\mathbf{X}, S)$. Since $(\mathbf{X}, S)$ are affiliated,
$(X_i, X_{-i}^{n-1:n-1})$ are affiliated as well from Theorem 4 of Milgrom and Weber. Then,
from Theorem 5, $E(U_i|\mathbf{X}, S)|X_i, X_{-i}^{n-1:n-1})$ is strictly increasing in all its arguments.

[‡]Since $(\mathbf{X}, S)$ are affiliated, $(X_i, X_{-i}^{n-1:n-1})$ are affiliated as well. Then it follows that
$(\int_{x_{-i}}^{n-1:n-1}(t|t))/(F_{x_{-i}}^{n-1:n-1}(t|t))$ is increasing in t. Hence $L(\alpha|x_i)$ is decreasing in x_i, which
implies that $b(x_i)$ is increasing in x_i since the first term in the equilibrium bid function
is increasing in x_i, from Example 1.

Again, for the PV model, the bidding rule simplifies to

$$e(x_i) = x_i - \int_{\underline{x}}^{x_i} L(\alpha|x_i)d\alpha \tag{8}$$

3. IDENTIFICATION, ESTIMATION, AND TESTING: AN OVERVIEW

Identification and estimation of a structural model involves recovering the elements $[U(\mathbf{X}, S), F_{\mathbf{XS}}(\mathbf{x}, \mathbf{s})]$ from data on the observables. Identification, estimation, and testing of structural auction models is complicated, for several reasons. First, an explicit solution for the Bayesian–Nash equilibrium strategy $e_i(\bullet)$ is available for few models of valuation. In most instances all an empirical researcher has are a set of differential equations that are first-order conditions for the optimization exercise given by (1). Second, even if an explicit solution for $e_i(\bullet)$ is available, complication is introduced by the fact that the equilibrium bid is not a function of $(\mathbf{x}, \mathbf{s})$ *exclusively*. $(\mathbf{x}, \mathbf{s})$ affects the bidding rule through the distribution function $F_{\mathbf{XS}}(\mathbf{x}, \mathbf{s}|\bullet)$ as well. Sections 3.1–3.3 provide a brief description of the approaches to identification and estimation; these approaches have been discussed at length in the surveys by Hendricks and Paarsch (1995) and Perrigne and Vuong (1999). Testing is discussed in Section 3.4.

3.1 Identification and Estimation

There have been two approaches to identification and estimation of structural auction models when data on all bids is available and the number of potential bidders, n, is known. The *direct approach* starts by making some distributional assumption, $f_{\mathbf{XS}}(\mathbf{x}, \mathbf{s}|\theta)$, about the signals $(\mathbf{X}, S)$ of the auctioned object and the utility function U. Assuming risk-neutrality, the structural element is now θ. From here there are three choices. First, the likelihood function of the observed data can be used. Alternatively, the posterior distribution of the parameters that characterize the distribution of the signals $(\mathbf{X}, S)$ could be used; this is the product of the likelihood function and the prior distribution of the parameters. Both these alternatives are essentially *likelihood-based* methods. These methods are relatively easy as long as an explicit one-to-one transformation relating the unobserved signals with the observed bids is available. The former approach has been used in Paarsch (1992, 1997) and Donald and Paarsch (1996) for identification and estimation assuming an IPV model of valuation. The latter approach has been used in several papers by Bajari (1998b), Bajari and Hortacsu (2000), Sareen (1999), and Van den Berg and Van der Klaauw (2000).

When an explicit solution is not available for the optimization problem in equation (1), it may be preferable to work with simulated features of the distribution specified for the signals $(\mathbf{X}, \mathbf{S})$; centred moments and quantiles are some examples. Simulated features are used since the analytical moments or quantiles of the likelihood function are likely to be intractable. These are called *simulation-based* methods.

For certain structural models the simulation of the moments could be simplified through the predictions made by the underlying game-theoretic models; Laffont et al. (1995) and Donald et al. (1999) provide two excellent examples.

Example 3 (Laffont et al. 1995): In this paper, the revenue equivalence theorem is used to simulate the expected winning bid for a Dutch auction of eggplants with independent private values and risk-neutral bidders. The revenue equivalence theorem is a prediction of the independent-private-values model with risk-neutral bidders. It says that the expected revenue of the seller is identical in Dutch, first-price, second-price, and English auctions. The winning bid in a second-price, and English auction is the second-highest value. If N and $F_X(x)$ were known, this second-highest value could be obtained by making n draws from $F_X(x)$ and retaining the second-highest draw. This process could be repeated to generate a sample of second-highest private values and the average of these would be the expected winning bid. Since N and $F_X(x)$ are not known, an importance function is used instead of $F_X(x)$; the above process is repeated for each value of N in a fixed interval.

Example 4 (Donald et al. 1999): The idea in the previous example is extended to a sequential, ascending-price auction for Siberian timber-export permits assuming independent private values. The key feature of this data set is that bidders demand multiple units in an auction. This complicates single-unit analysis of auctions since bidders may decide to bid strategically across units in an auction. Indeed, the authors demonstrate that winning prices form a sub-martingale. An efficient equilibrium is isolated from a class of equilibria of these auctions. The direct incentive-compatible mechanism which will generate this equilibrium is found. This mechanism is used to simulate the expected winning bid in their simulated nonlinear least-squares objective function. The incentive-compatible mechanism comprises each bidder revealing her true valuation. The T lots in an auction are allocated to bidders with the T highest valuation with the price of the tth lot being the $(t+1)$th valuation. Having specified a functional form for the distribution of private values, they recreate T highest values for each of the N bidders from some importance function. The T highest valuations of these NT draws then identify the winners in an auction and hence the price each winner pays for the lot. This process is repeated several times to get a sample of winning prices for each of the T lots. The average of the sample is an estimate of the expected price in the simulated nonlinear least-squares objective function.

The key issue in using any simulation-based estimator is the choice of the importance function. It is important that the tails of the importance function be fatter than the tails of the target density to ensure that tails of the target density get sampled. For example, Laffont et al. (1995) ensure this by fixing the variance of the lognormal distribution from the data and using this as the variance of the importance function.

In structural models where the bidding rule is a *monotonic* transformation of the unobserved signals, Hong and Shum (2000) suggest the use of simulated quantiles instead of centred moments. Using quantiles has two advantages. First, quantile estimators are more robust to outliers in the data than estimators based on centred moments. Second, quantile estimators dramatically reduce the computational burden associated with simulating the moments of the equilibrium bid distribution when the bidding rule is a monotonic transformation of the unobserved signals. This follows from the quantiles of a distribution being invariant to monotonic transformations so that the quantiles of the unobserved signals and the equilibrium bidding rule are identical. This is not the case for centred moments in most instances.

Instead of recovering θ, the *indirect approach* recovers the distribution of the unobserved signals, $f_{\mathbf{XS}}(x, s)$. It is indirect since it does not work directly off the likelihood function of the observables. Hence it avoids the problem of inverting the equilibrium bidding rule to evaluate the likelihood function. The key element in establishing identification is equation (2). An example illustrates this point.

Example 5: Symmetric PV model. From Guerre et al. (2000) and Li et al. (2000) the first-order condition in equation (2) is

$$x_i = \xi_i(b_i, f_{B_i|b_i}(b_i|b_i), F_{B_i|b_i}(b_i|b_i), n) = b_i + \frac{F_{B_i|b_i}(b_i|b_i)}{f_{B_i|b_i}(b_i|b_i)} \qquad (9)$$

$\xi_i(\bullet)$ is the inverse of the bidding rule given in equation (3) with respect to the private signal x_i in a PV model. Identification of the APV model is based on this equation since $F_{\mathbf{X}}(x)$, the distribution of private signals, is completely determined by n, $f_{B_i|b_i}(b_i|b_i)$, and $F_{B_i|b_i}(b_i|b_i)$. This idea underlies their estimation procedure as well. First, $(f_{B_i|b_i}(\bullet))/(F_{B_i|b_i}(\bullet))$ is nonparametrically estimated through kernel density estimation for each n. Then a sample of nT pseudo private values is recovered using the above relation. Finally this sample of pseudo private values is used to obtain a kernel estimate of the distribution $F_X(x)$. There are several technical and practical issues with this class of estimators. First, the kernel density estimator of bids, $\widehat{f_B}(b)$, is biased at the boundary of the support. This implies that obtaining pseudo private values from relation (9) for observed bids that are close to the boundary will be problematic. The pseudo private values are, as a result, defined by the relation in equation (9) only for that part of the support of $f_B(b)$ where $\widehat{f_B}(b)$ is its unbiased estimator. The second problem concerns the rate of convergence of the estimator $\widehat{f_X}(x)$. Since the density that is being estimated nonparametrically is not the density of observables, standard results on the convergence of nonparametric estimators of density do not apply. Guerre et al. (2000) prove that the best rate of convergence that these estimators can achieve is $(T/\log T)^{r/(2r+3)}$, where r is the number of bounded continuous derivatives of $f_X(x)$.

The *indirect approach* has concentrated on identification and estimation from bid data. Laffont and Vuong (1996) show that several models of valuations described in Section 2 are not identified from bid data. Traditionally, identification of nonidentified models has been obtained through either additional data or dogmatic assumptions about the model or both. Eventually, the data that is available will be a guide as to which course an empirical researcher has to take to achieve identification. For a particular auction format, identification and estimation of more general models of valuation will require more detailed data sets. These issues are discussed in Sections 3.2 and 3.3 for symmetric models; I turn to asymmetric models in Section 3.4. Nonparametric identification implies parametric identification but the reverse is not true; hence the focus is on nonparametric identification.

3.2   Identification from Bids

When bids of all participants are observed, as in this subsection, identification of a PV model, second-price auction is trivial. This follows from the bidders submitting bids identical to their valuation of the auctioned object. Formats of interest are the first-price and the second-price auctions. For a model of valuation other than the PV model, the identification strategy from data on bids for the second-price auction will be similar to that of a first-price auction. The structure of the bidding rules in Examples 1 and 2 makes this obvious. The second-price auction can be viewed as a special case of the first-price auction. The second term in the bidding rule of a first-price auction is the padding a bidder adds to the expected value of the auctioned object. This is zero in a second-price auction since a bidder knows that he will have to pay only the second-highest bid. For example, for the PV model, while the bids are strictly increasing transformations of the private signals under both auction formats, this increasing transformation is the identity function for a second-price auction. A model of valuation identified for a first-price auction will be identified for a second-price auction. Identifying auction formats other than the first-price and second-price auction from data on all bids is not feasible since for both the Dutch and the "button" auctions all bids can never be observed. In a Dutch auction only the transaction price is observed. The drop-out points of all except the highest valuation bidder are observed in the "button" auction. Hence the discussion in this section is confined to first-price auctions.

Within the class of symmetric, risk-neutral, private-values models with a nonbinding reserve price, identification based on the *indirect approach* has been established for models as general as the APV model (Perrigne and Vuong 1999). Since risk neutrality is assumed, the only unknown structural element is the latent distribution, $F_X(x)$.

The CIPV model is a special case of the APV model. The interesting identification question in the CIPV model is whether the structural elements $[F_S(s), F_{X|S}(x)]$ can be determined uniquely from data on bids. From Li et al., (2000), the answer is no; an observationally equivalent CV model can always be found. For example replace the conditioning variable S by an increasing transformation λS in the CIPV model. The utility function is unchanged since $U_i(X, S) = X_i$. The distribution of bids generated by this model would be identical to the CV model where $U_i(X, S) = S$ and the X_i are scaled by $(\lambda)^{1/n}$. Thus $F_{XS}(x, s)$ is identified, but not its individual components. Data could be available on S; this is the case in the OCS auctions where the *ex post* value of the tract is available. Even though $F_S(s)$ can be recovered, this would still not guarantee identification since the relationship between the unknown private signals of the bidders, X_i, and the common component which is affecting all private values S is not observed by the econometrician. Without further structure on the private signals, the CIPV model is not identified *irrespective* of the auction form and the data available.

Both Li et al. (2000) and Athey and Haile (2000) assume $X_i = S + A_i$. Note that $U_i = X_i$ from the PV assumption. With this additional structure, the CIPV-A emerges when $(A_1, \ldots, A_n)$ are independent, conditional on S. This again is not enough to guarantee identification of the structural elements $(F_S(s), F_{A|S}(a))$ from data on all bids in an auction since bids give information about the private valuations but not the individual components of these valuations; hence the observational equivalence result of the CIPV model is valid here as well. At this point identification of the CIPV-A model can be accomplished by either putting further structure on it or through additional data. These are discussed in turn.

In addition to (i) $U_i = X_i = S + A_i$, assume (ii) $(A_1, \ldots, A_n, S)$ are mutually independent, with A_is identically distributed with mean equal to one; and (iii) the characteristic functions of $\log S$ and $\log A_i$ are nonvanishing everywhere. This is the CIPV-I model defined above. Li et al. (2000) and Athey and Haile (2000) draw upon the results in the measurement error literature, as in Kotlarski (1966) and Li and Vuong (1998), to prove that (F_s, F_A) are identified.[*] The importance of the assumptions about the decomposability of the X_i and the mutual independence is that $\log X_i$ can be written as

$$\log X_i = \log c + \log \epsilon_i \qquad (10)$$

where $\log c \equiv [\log s + E(\log A_i)]$ and $\log \epsilon_i \equiv [\log A_i - E(\log \eta_i)]$. Hence the $\log X_i$s are indicators for the unknown $\log c$ which are observed with a measurement error $\log \epsilon_i$. $\log X_i$ or X_i can be recovered from data on bids since the bids are strictly increasing functions of X_i. Identification is then straightforward from results concerning error-in-variable models with multiple indicators, as in Li and Vuong (1998, Lemma 2.1).

S is the common component through which the private signals of the bidders are affiliated. There are several examples where data on S could be available. In the crude-oil auctions studied by Sareen (1999) the past prices of crude-oil would be a candidate for S. In the timber auctions studied by Sareen (2000d) and Athey and Levin (1999), the species of timber, tract location, etc. would be examples of S. In the OCS auctions S could be the ex-post value of the tract. If data were available on S, then the CIPV-A model with $U_i = X_i = S + A_i$ would be identified; refining CIPV-A to CIPV-I is not needed. For each s, $A_i = X_i - s$; hence identification of $F_{A|s}$ (a) follows from the identification of the IPV model. $F_S(s)$ is identified from data on S.[†]

The same argument applies if, instead of S, some auction-specific covariates are observed conditional on which the A_i are independent. If $S = 0$, but auction-specific covariates Z are observed, then again $F_{A|z}(a)$ is identified for each z. It could also be the case that $U_i = g(Z) + A_i$; then $g(Z)$ and $F_{A|z}(a)$ for each z are identified. In fact, when either S or auction-specific character-

---

[*] Li et al. (2000) consider a multiplicative decomposition of X_i with $X_i = SA_i$. This will be a linear decomposition if the logarithm of the variables is taken.

[†] Since more than one realization of S is observed, F_A is identified too. If the auctions are assumed to be independent and *identical*, then *the variation in S* across auctions can be used to identify the conditional distribution $F_{A|S}$.

istics are observed, all bids are not needed in either the first-price or the second-price auctions to identify the CIPV-A model. Any bid, for example the transaction price, will do in a second-price auction (Athey and Haile 2000, Proposition 6).

The picture is not as promising when one comes to *common value* auctions when the data comprises exclusively bids in a first-price auction. The CV model is not identified from data on bids; the bids can be rationalized by an APV model as well. This follows from Proposition 1 of Laffont and Vuong (1996).

The symmetric pure CV model is not identified from all n bids; this is not surprising in view of the observational equivalence of the pure CV and the CIPV model discussed above. The identification strategy is similar to the one followed for the CIPV model: more structure on the model of valuation or additional data.

The LMR and LMR-I models put more structure on the pure CV model. The LMR model is analogous to the CIPV-A and the LMR-I to the CIPV-I; hence identification strategies for the two sets of models are similar. The LMR model is not identified from bids. Assuming mutual independence of (S, A), the LMR-I emerges; Li et al. (2000) prove its identification for the first-price auction and Athey and Haile (2000, Proposition 4) for the second-price auction. Again, if data on the realized value of the object, S, is available, the LMR model is identified from just the transaction price (Athey and Haile 2000, Proposition 15).

3.2.1 Binding reserve price

Once the assumption of a nonbinding reserve price is relaxed there is an additional unknown structural element besides the distribution of the latent private signals. Due to the binding reserve price p_0 the number of players who submit bids, p_j, is less than the number of potential bidders which is now unknown. Bidders with a valuation for the auctioned object greater than p_0 will submit bids in the auction; thus a truncated sample of bids will be observed.

The question now is whether the joint distribution of (N, V) is identified from the truncated sample of p_j bids in an auction. Guerre et al. (2000), in the context of an IPV model, examine a variant of this issue. Using the *indirect approach* they establish the identification of the *number of potential bidders* and the nonparametric identification of the distribution of valuations. A necessary and sufficient condition, in addition to those obtained for the case of the nonbinding reserve price, emerges; the distribution of the number of potential bidders is binomial, with parameters $(n, [1 - F_v(p_0)])$. In essence, though there is an extra parameter here, there is also an additional observable present: the number of potential bidders. This extra observable pins down the parameter n.

3.2.2 Risk-averse models of valuation

Another noteworthy extension here is the relaxation of the assumption of risk neutrality. The structural model is now characterized by $[U, F_X(x)]$. Campo et al. (2000) show that, for nonbinding reserve prices, an IPV model with constant relative risk aversion cannot be distinguished from one with constant absolute risk aversion from data on observed bids. Since the risk-neutral model is a special case of the risk-averse models, this implies that a risk-neutral model is observationally equivalent to a risk-averse model. The redeeming aspect of this exercise is that the converse is not true. This follows from the risk-neutral model imposing a strict monotonicity restriction on the inverse bidding rule $\xi(\bullet)$, a restriction that the risk-averse model does *not* impose.

Identification of $[U, F_X(x)]$ is achieved through two assumptions. First, the utility function is parametrized with some k dimensional parameter vector η; this, on its own, does not deliver identification. Then heterogeneity across the auctioned objects is exploited by identifying η through k distinct data points. Second, the upper bound of $F_X(x)$ is assumed to be known but does not depend on the observed covariates. This seems logical since object heterogeneity can be exploited for identification only once—either to identify η, the parameters of the utility function, or some parts of the distribution $F_X(x)$, but not both.

3.3 Identification with Less than All Bids

In Section 3.2 data was available on the bids of all n agents. As a result, identification of a model of valuation for a first-price and a second-price auction was similar. Outside this scenario, identification of a model of valuation for a first-price auction does not automatically lead to the identification of the same model for a second-price auction. The reason is that, unlike the first-price auction, the transaction price is not the highest bid but the second-highest bid. I start with identification when only the transaction price is recorded in an auction. For models that cannot be identified, the additional data or parametrization needed to identify the models is described. The discussion in Sections 3.3.1 and 3.3.2 is conditional on n, the number of potential bidders; identification of n raises additional issues which are discussed in Section 3.3.3. Thus n is common knowledge, fixed across auctions and observed by the empirical researcher.

3.3.1 First-price, sealed-bid auction

The most general PV model that has been identified from winning bids across auctions is the symmetric IPV model. Guerre et al. (1995) discuss identification and estimation of F_X from data on winning bids. The discussion is analogous to identification from bids in the last subsection with b replaced with w, $f_B(b)$ with $f_W(w)$, and $F_B(b)$ with $F_W(w)$.[*]

It is independence of private values and the symmetry of bidders that make identification of the distribution of private values feasible from only the winning bid. Beyond this scenario, all bids will be needed if the independence assumption is relaxed but symmetry is retained (Athey and Haile 2000, Corollary 3). If the latter is relaxed as well, identity of the bidders will be needed for any further progress; asymmetry is the subject of Section 3.4. Also note that observing s or auction-specific covariates, in addition to the winning bid, does not help; the nature of the additional data should be such that it gives information about either the affiliation structure and/or the asymmetry of the bidders. For example, Athey and Haile give several instances when the top two bids are recorded.[†] Assuming symmetry, the top two bids should give information about the affiliation structure compared to observing just the winning bid. Under the additional assumption of the CIPV-A model, Athey and Haile (2000, Corollary 2) show that observing the top two bids and S identifies only the joint distribution of the idiosyncratic components, $F_A(a)$.

3.3.2 Second-price auctions

Unlike the first-price auction, the winning bid and the transaction price are different in a second-price and an ascending auction. The transaction price is the second-highest bid. Hence, to establish identification, the properties of second-highest order statistic have to be invoked.

The symmetric APV model is not identified from transaction prices; this is not surprising in view of the comments in Section 3.3.1. A formal proof is provided in Proposition 8 of Athey and Haile (2000); linking it to the discussion for first-price auctions, they prove that all n bids are sufficient to identify the symmetric APV model.

Independence of the private signals buys identification. The symmetric IPV model is a special case of the APV with strict affiliation between the X_i replaced by independence. Athey and Haile (2000) prove that the symmetric

---

[*]Since the estimator of $F_w()$ converges at a slow rate of $(T/\log T)^{r/(2r+3)}$, the data requirements and hence estimation could be burdensome. As a result, parametric or direct methods may be necessary.

[†]The difference between the top two bids is the "money left on the table."

IPV model is identified from the transaction price. This follows from the distribution function of any ith order statistic from a distribution function F_X being an increasing function of F_X,

$$F_X^{2:n}(z) = \frac{n!}{(n-2)!} \int_0^{F_X(z)} t(1-t)^{n-2} dt \qquad (11)$$

The CIPV-A puts parametric structure on the APV model in this sense of decomposing the private values into a common and an idiosyncratic component, $X_i = S + A_i$. If s was observed, the identification of the CIPV-A would follow from the identification of the IPV model with transaction prices exclusively since the A_i are independent, conditional on s. Athey and Haile (2000, Propositions 6, 7) prove that the CIPV-A model is identified from the transaction prices and observing the the *ex post* realization of S or auction-specific covariates, Z, conditional on which the A_i are independent. Alternatively, if $X_i = g(Z) + A_i$, then both $g(Z)$ and $F_{A|z}$ are identified up to a locational normalization from the transaction prices and the auction-specific covariate Z.

The prognosis for the CV model can be obtained from the PV model since the CIPV-A is similar to the LMR model and the CIPV-I to the LMR-I model. Athey and Haile (2000, Proposition 15) prove that if the *ex post* realization of the "true" value of the object was observed, then the LMR model (and consequently the LMR-I model) is identified from the transaction price.

3.3.3 Identifying the number of potential bidders

In the previous sections the number of potential bidders is assumed to be given. With n given, the IPV model is identified from transaction prices for both the first-price and the second-price auctions. In many auctions the empirical researcher does not observe the number of potential bidders even though it is common knowledge for the bidders. Inference about N may be of interest since it affects the average revenue of the seller in PV auctions; the larger the number of bidders, the more competitive is the bidding. In CV auctions the number of potential bidders determines the magnitude of the winner's curse.

Issues in identification of n will differ depending on whether it is *endogenous* or *exogenous*. If the number of potential bidders is systematically correlated with the underlying heterogeneity of the auctioned object, then n is endogenous. For example, this is the case in Bajari and Hortacsu (2000) when they model the entry decision of bidders in eBay auctions. If these object characteristics are observed, identification of n involves conditioning on these characteristics. Unobserved object heterogeneity will cause problems for nonparametric identification.* This is similar in spirit to the identification of the CIPV-A in Athey and Haile (2000) that is discussed in Section 3.3.2; nonparametric identification of the CIPV-A model is possible in the case that the *ex post* realization of S or auction-specific covariates is observed.

Exogeneity of n implies that the distribution of unobserved values in an auction is the same for all potential bidders, conditional on the symmetry of the bidders. This accommodates both the case when the number of actual bidders differs from the number of potential bidders and the case of a random n. Donald and Paarsch (1996), Donald et al. (1999), and Laffont et al. (1995) take the former approach. In these papers n is fixed but the actual number of bidders varies across auctions due to either the existence of a reserve price or the costs of preparing and submitting bids. Hendricks et al. (1999) take the latter approach; they make a serious effort to elicit the number of potential bidders on each tract in the OCS wildcat auctions.

Assuming exogeneity of the number of potential bidders, and in view of the IPV model being the most general model of valuation that is identified from the transaction price, I now examine whether the highest bid identifies both the distribution of private values, $F_X(x_i)$, and the number of potential bidders, N. The proposition below formalizes these ideas when the highest bid is observed.

Proposition 1. For a single-unit first-price, Dutch or second-price symmetric APV auction, the distribution of the winning bid can be expressed as

$$F_w(u) = Q\big([F_{X_n}(u)]^{1/n}\big)$$

where $Q(s) = \sum_{n=0}^{\infty} s^n \Pr(N = n)$ and $F_{X_n}(u)$ is the joint distribution of n private values.

Proof.   Let $W = \text{Max}\{B_1, \ldots, B_N\}$. Then

$$F_w(u) = \Pr(W \leq u)$$

$$= \sum_{n=0}^{\infty} \Pr(W \leq u | N = n)\Pr(N = n)$$

*Li (2000) endogenizes entry but makes parametric assumptions about the distribution of private values. Bajari and Hortacsu (2000) make parametric assumptions about both the distribution of unobserved values and the entry process.

$$= \sum_{n=0}^{\infty} \Pr(B_1 \leq u, \ldots, B_{n-1} \leq u, \ W \leq u)\Pr(N = n)$$

$$= \sum_{n=0}^{\infty} \Pr(X_1 \leq u, \ldots, X_{n-1} \leq u, \ W \leq u)\Pr(N = n)$$

$$= \sum_{n=0}^{\infty} F_{\mathbf{X}_n}(u)\Pr(N = n)$$

$F_{\mathbf{X}_n}(u)$ is the joint distribution of $\mathbf{X}$ conditional on $N = n$. The second last step follows from the bids being strictly increasing monotonic transformations of the private signals of the bidders. With $s = [F_{\mathbf{V}_n}(u)]^{1/n}$ in the function $Q(s)$,

$$F_w(u) = Q\big([F_{\mathbf{V}_n}(u)]^{1/n}\big) \qquad\qquad\qquad \square$$

From Proposition 1 it is obvious that the distribution of private signals is not identified as in Athey and Haile (2000, Corollary 3), given n.*

Corollary 1.   For a single-unit first-price, Dutch, or second-price symmetric IPV auction, either $F_X(x_i)$ or N but not both are identified from the highest bid.

Proof.   Suppose n is fixed. For a symmetric IPV auction

$$F_{\mathbf{X}_n}(u) = [F_X(u)]^n$$

Then, from Proposition 1, $F_X(u)$ is uniquely determined,

$$F_X(u) = Q^{-1}(F_w(u))$$

Assume $F_X(u)$ is fixed. A counter-example shows that n is determined uniquely from

$$F_w(u) = Q([F_X(u)]) = \sum_{n=0}^{\infty}(F_X(u))^n \Pr(N = n)$$

*Choose

$$F_{\mathbf{X}_n}(u) = \big[Q^{-1}(F_w(u))\big]^n$$

then, from Proposition 1,

$$F_w(u) = Q\big([F_{\mathbf{X}_n}(u)]^{1/n}\big) = Q\Big([[Q^{-1}(F_w(u))]^n]^{1/n}\Big)$$

$$= F_w(u)$$

Suppose $\Pr(N = n)$ puts point mass on n and n is not uniquely determined. Then there exists n^* with $n \neq n^*$ and

$$F_w(u; n) = F_w(u; n^*)$$

But that cannot be true since

$$F_w(u; n) = (F_X(u))^n \neq (F_X(u))^{n^*} = F_w(u; n^*) \qquad \qquad \square$$

In most cases, if the structural model is not identified nonparametrically, parametric assumptions may identify the model; this is not the case for the result in Corollary 1. Even if parametric assumptions are made about $F_X(x_i|\theta)$, Corollary 1 holds.[*] Corollary 1 forms the basis for identification and estimation when data on just the winning bids is available. Laffont et al. (1995), studying a Dutch auction for eggplants, use a simulated nonlinear least-squares function to obtain $\hat{\theta}$; an estimate of N is obtained, conditional on $\hat{\theta}$. Similarly Guerre et al. (1995) study nonparametric identification and estimation of F_v, the distribution of private values, from winning bids, for a first-price, sealed-bid auction with independent private values, conditional on n. n is obtained by solving

$$\bar{v} = \xi\left(\bar{b}, F_W(w), n\right)$$

where $\bar{v}$ and $\bar{b}$ are the *known* upper bounds of the distributions $F_X(x)$ and $F_B(b)$.

3.4 Asymmetric Models of Valuation

Since symmetric models of valuation are special cases of asymmetric models, if a symmetric model is not identified, the corresponding asymmetric model will not be identified either. Thus the asymmetric pure CV model and the asymmetric CIPV-A model are not identified from n bids. The asymmetric APV model will not be identified from transaction prices.

What about symmetric models of valuation that are identified when all n bids are observed? As long as a complete set of n bids is available, the asymmetric counterparts of the identified symmetric models should be identified. The *indirect method* is used to identify the asymmetric APV model by

---

[*]Guerre et al. (1995, pp. 27–28) give a numerical example of two specifications of the symmetric IPV model which generate the same distribution of winning bids.

Campo et al. (1998) for a first-price auction and by Athey and Haile (2000) for a second-price auction.[*]

The interesting question with regard to asymmetric PV models is their identification from transaction prices. Asymmetric models cannot be identified from transaction prices only. An example clarifies this point. Suppose all auctions observed are won by a single bidder j. Then $F_{X_j}(x_j)$, bidder j's value distribution, will be identified; since bidders are not exchangeable now, $F_{X_j}(x_j)$ will not identify $F_X(x)$.

The example suggests that identification from transaction prices may be possible if the identity of the winner was observed as well. Borrowing the work on competing risk models (Berman 1963, Rao 1992) and the multi-sector Roy model (Heckman and Honoré 1990), Athey and Haile (2000) use the additional data on the identity of the winner to establish identification of several PV models, with a caveat.[†] They assume that the support of the distribution of valuation of each of the n potential bidders is identical. If this is not the case, then the identity of the bidders could be used to establish identification from transaction prices if each bidder won some auctions. It is very rare that a researcher gets to observe all potential bidders winning several auctions in the data set. The identification result of Athey and Haile (2000) is important since in many data sets it is possible to establish symmetry with respect to groups of bidders. This was the case in Sareen (2000d) where one observed three nonfringe firms and several fringe firms, some of whom participated only once; in essence there were two nonexchangeable groups of bidders in this auction.

3.5 Testing of Structural Auction Models

A structural model comprises many elements. For example, a structural model makes assumptions about bidder rationality, that bidders bid accord-

---

[*]Proposition 8 of Athey and Haile shows that the APV move cannot be identified from an incomplete set of bids.

[†]For a first-price and a second-price auction, Athey and Haile (2000, Propositions 16, 2 respectively) establish the identification of the IPV model from the highest bid and the identity of the winner. For ascending and second-price auctions the IPV model is identified from the lowest bid and the identity of the lower as well. Additional data on the identity of the winner can also be used for identifying the APV model for a second-price or ascending auction if further structure, described in Proposition 11, is imposed on the private signals; the similarity of the auction model to the multi-sector Roy model is exploited for this. A similar structure on the private signals for first-price auction does *not* identify the APV model (Corollary 2).

ing to Bayesian–Nash equilibrium strategies, about the correlation between bidders' valuations, common value or private-values components, exchangeability, etc. Testing of structural models has followed three broad directions.

The first direction focuses on whether a structural auction model imposes any testable restrictions on the observables.* For example, for an IPV first-price auction, the structural elements $[U_i(\mathbf{X}, \mathbf{S}), F_{\mathbf{XS}}(\mathbf{x}, \mathbf{s})] = [X_i, F_X(x_i)]$. Laffont and Vuong (1996) show that this structure imposes two testable restrictions on the distribution of bids: the bids in an auction are *iid*, and that $\xi(b)$ is strictly increasing in b.† The former restriction ascertains the independence of bidders' valuations irrespective of whether they are playing Bayesian–Nash strategies; the latter tests whether they are playing Bayesian–Nash strategies irrespective of the model of valuation. Through these restrictions, individual tests can be designed for different aspects of a structural model.

If all bidders were bidding in accordance with a proposed structural model, estimates of this model from different subsets of the data should be identical. If that is the case, the data are consistent with the model; if not, the data are inconsistent with the model. This idea is used by Athey and Haile (2000) for testing several structural models. For example, assuming risk-neutrality, they test a symmetric IPV model for a second-price auction by ascertaining whether $F_X(x)$ recovered from the transaction prices is identical to the $F_X(x)$ recovered from data on the highest bid. In the case that it is not, the null hypothesis that the data is consistent with the IPV model is rejected. From the decision to reject a structural model it is unclear as to which component of the model led to its rejection; rejection could be due to violation of independence or symmetry, or to playing Bayesian–Nash strategies. This problem arises because the alternative model being tested is not explicit.

A third direction specifies explicit structural models under the null and the alternative hypothesis. The focus of this research has been on distinguishing the CV model from the PV model and a symmetric model from its asymmetric counterpart. These are discussed in turn.

Laffont and Vuong (1996) have proved that for first-price auctions, conditional on the number of potential bidders, the CV model cannot be distinguished from the PV model on the basis of bids of the n potential bidders.

---

*According to Laffont and Vuong, a distribution of bids $F_\mathbf{B}$ is rationalized by a structural model $[U(\mathbf{X}, \mathbf{S}), F_{\mathbf{XS}}]$ if $F_\mathbf{B}$ is the equilibrium bid distribution of the corresponding game.

†These two restrictions comprise necessary and sufficient conditions for the bids to be generated by a symmetric IPV first-price auction.

Distinguishing the two models of valuation is important for several reasons, one of which is mechanism design. Hendricks et al. (1999), Haile et al. (2000), and Athey and Haile (2000) examine what additional data would distinguish the two models in a first-price auction. As the discussion below will show, there is no unique way of distinguishing between the two models; eventually which test to use would be decided by the data that is available.

Haile et al. (2000) and Athey and Haile (2000) work with bidding data, as do Laffont and Vuong (1996). However, they do not condition on the number of potential bidders. Rather, the comparative statics implication of varying the number of potential bidders on the equilibrium bidding rule under the two models is exploited for testing. The winner's curse is an adverse selection phenomenon arising in CV but not PV auctions. In a CV auction winning is bad news since the winner is the bidder who has been most optimistic about the unknown value of the auctioned object. A rational bidder will therefore account for the winner's curse by lowering his expected value of the auctioned object and hence his bid. The more competition a bidder expects, the more severe is the curse and hence the larger is the adjustment in expected value of the auctioned object to mitigate the curse. The winner's curse does not arise in the PV setting since the value of the auctioned object to a bidder does not depend on the information that his opponents have. On this basis Haile et al. (2000) suggest the following test:

$$
\left.\begin{array}{l}
\text{PV: } F_{v,2} = F_{v,3} = \cdots = F_{v,N} \\[6pt]
\text{CV: } F_{v,2} < F_{v,3} < \cdots < F_{v,N}
\end{array}\right\} \tag{12}
$$

in the sense of first-order stochastic dominance. $F_{v,n}$ is the distribution of $v(x, x; n)$ induced by the distribution specified for the private signals $\mathbf{X}$; $v(x, x; n)$ is given in equation (4). Instead of testing the stochastic dominance hypothesis, Haile et al. (2000) compare various features, such as quantiles of the distributions, as the number of potential bidders varies. The *key* data requirement for their test is to observe auctions in which the number of potential bidders varies from 1 to N; in an auction with n bidders, all n bids should be observed.

Athey and Haile (2000, Corollary 3) exploit recurrence relations between order statistics and reduce the data requirements of Haile et al. (2000); all they need is the top two bids in an auction with $n \geq 3$ bidders and the top bid in an auction with $n - 1$ bidders.[*]

---

[*]If, in addition, the identity of the bidders corresponding to the observed bids is available, the asymmetric version of this test is possible.

A key feature of the OCS data set is the observation of the *ex post* value of the tract. Therefore Hendricks et al. (1999) can compute the average rent over tracts and average bid markdowns over tracts and bidders under the CV and PV models.* Since entry is determined by a zero expected profit condition, the average rents under the PV and CV models are compared with the total entry cost on a tract.† They find that the average rents under the CV model are comparable with the entry costs. They find no difference in the average bid markdowns under the CV hypothesis between tracts with a small and a large number of potential bidders. In both cases they are comparable to the average rents. Thus bidders seems to have anticipated the winner's curse.

An alternative test suggested by Hendricks et al. (1999) applies to first-price and second-price auctions with nonbinding reserve prices. This involves observing the behavior of $\xi(x, F_B)$ near the reserve price. In a PV model, since $v(x, x) = x$, $\xi(x, F_B)$ will satisfy the boundary condition, $\lim_{b \downarrow r} \xi(x, F_B) = r$ from equation (2). Thus

$$\lim_{b \downarrow r} \frac{\Pr(\max_{j \neq n} B_j \leq b | B_i = b)}{\frac{\partial}{\partial r} \Pr(\max_{j \neq n} B_j \leq b | B_i = b)|_{r=b}} = 0$$

For the CV model, since $v(x, x) \neq x$, $\lim_{b \downarrow r} \xi(x, F_B) = v(x, x) > r$. Hence the limit above approaches a nonzero constant.

Finally, can symmetric models be distinguished from asymmetric models? On the basis of bidding data alone, progress is limited. Laffont and Vuong (1996) give an example to show that asymmetric pure CV models may not be distinguished from their symmetric counterparts for first-price auctions. This result is not surprising since their asymmetric model comprises one informed bidder and $n - 1$ uninformed bidders. The uninformed players

---

*If the sample size is T, the average rents under the PV model are

$$R = T^{-1} \sum_{t=1}^{T} [\hat{x}_t^{n:n} - w_t]$$

where $\hat{x}_t^{n:n} = \hat{\xi}(w_t)$ in the PV model and $\hat{x}_t^{n:n} = s_t$ in the CV model, where s_t is the observed ex post value of tract t. The bid markdowns are given by

$$M = T^{-1} \sum_{t=1}^{T} \sum_{i=1}^{n_t} [\hat{x}_{it}^{n:n} - b_{it}]$$

where $\hat{x}_{it}^{n:n} = \hat{\xi}(b_{it})$ in the PV model and $\hat{x}_t^{n:n} = E(S | w = b_{it})$ in the CV model.
†Entry costs are the price of conducting the seismic survey per acre, hiring engineers to study the survey data and prepare a bid.

will play a mixed strategy whereby their maximum bid mimics the bid distribution of the informed bidder. Thus, if $n = 2$, the observational equivalence noted by Laffont and Vuong becomes obvious.

In Section 3.4 I have explained how Athey and Haile have used bidder identity to identify asymmetric models. It should be possible to exploit bidder identity to test for symmetry versus asymmetry. For example, let I_i indicate the identity of the bidder who submits bid b_i. Then the conditional distribution $F_X(x_i i)$ should be invariant with respect to I_i for a symmetric model but not an asymmetric model.

4. BAYESIAN INFERENCE OF STRUCTURAL AUCTION MODELS

Bayesian methods are *direct methods* in that they require specification of a likelihood function. Parametric assumption about the unobservable $(\mathbf{X}, \mathbf{S})$ and the utility function U and a one-to-one mapping from the unobservable to the observables ensure the specification of the likelihood. The structural element of the auction model is now θ, where θ characterizes the distribution of $(\mathbf{X}, \mathbf{S})$. In *addition* to the likelihood function, Bayesian tools require specification of prior distribution for the parameters θ. Estimation proceeds through the posterior distribution of the parameters, which is a product of the likelihood function and the prior distribution. The posterior odds ratio or the posterior predictive distributions may be used for testing. These concepts will be explained below.

There are several reasons why Bayesian tools may be attractive for inference in structural auction models. Despite progress on nonparametric identification of structural auction models, the results pertain to single-unit auctions. A large number of important auctions are, however, multi-unit auctions; for example, the auctioning of T-bills, spectrum auctions, timber export licenses, auctions to sell the right to fell timber in Southern Ontario, etc. Internet auctions are multi-unit auctions too, since more than one seller could be conducting an auction for the same object at different sites at almost the same time. Further, in most cases identification is achieved conditional on the number of potential bidders. In several auctions for which field data is available, participation is endogenous. In many cases it is linked to the characteristics of the auctioned object, some of which may not be observed or even obvious to an econometrician. The section on identification has illustrated that, even with single-unit auctions and a fixed number of potential bidders, it is difficult to establish identification outside the private-values model without further assumptions about the model of valuation. The key point is that each auction has characteristics that are idiosyncratic to it; establishing general results on identification may not be of much help while doing applied work. Hence, making parametric assumptions may be a necessity to proceed with applied work. A structural model is identified from a Bayesian perspective if the prior and the posterior distribution of the structural elements are different. Equivalently, a likelihood that is not flat in any direction of the parameter space will ensure identification; prior beliefs about the structural elements will always be revised through the likelihood.

Coming to estimation, the *indirect methods* are a comprehensive method for estimating structural auction models nonparametrically. However, proving the properties of these estimators is a difficult exercise; Perrigne and Vuong (1999) point out that the second step of these two-step estimators involves proving properties of the density estimator of unobserved variables. This leads them to comment that it may be more attractive to become parametric in the second stage by specifying a parametric distribution for the pseudo-sample of private values generated in the first stage. An undesirable consequence could be that the parametric distribution for the pseudo private values may not correspond to the distribution of bids obtained in the first step since the equilibrium bidding rule has not been used to obtain this distribution.

A well recognized problem for maximum likelihood estimation is the dependence of the support of the data on the parameters. Standard asymptotic theory breaks down when the support of the data depends on the parameters. Donald and Paarsch (1996) and Hong (1997) have obtained the asymptotic theory of the maximum likelihood estimator for a first-price, sealed-bid auction with independent private values. The key point in these papers is that the sample minimum or maximum is a superconsistent estimator of the support of the data; that is, the sample minimum or maximum converges to the support of the data evaluated at the "true" parameter values at the same rate as the sample size. Since the support of the data is a function of some or all of the parameters, the properties of the maximum likelihood estimator are, as a result, based on the properties of the sample maximum or minimum.

Compared to estimators based on *indirect methods* and maximum likelihood estimators, the properties of the simulation-based estimators are relatively easy to establish. However, simulation-based estimators are method-of-moment estimators; drawbacks of method-of-moment estimators apply to the simulation-based methods as well.

Bayesian estimation proceeds by examining the posterior distribution of parameters. As a result, even though they are likelihood based, Bayesian methods are not affected by the dependence of the support of the data on the parameters. A prior sensitivity analysis can be used to ensure the robustness of the estimates of the elements of the structural model.

Testing of structural auction models involves non-nested hypothesis testing with the support of the data being different under the null and the alternative structural auction model. The key challenge in testing is to find pivotal quantities in this scenario and to prove their properties. I have described testing in Section 3.5. The properties of the test statistics described in Section 3.5 are currently unknown. Further, the tests are extremely sensitive to the data that is available.. Bayesian testing can proceed either through the posterior odds ratio or by comparing the posterior predictive density function of the structural models being considered. Since the emphasis is on comparing two structural models *conditional* on the observed data, the problems that plague testing described in Section 3.5 are not encountered.

Bayesian inference involves specifying the likelihood and the prior; these are discussed in the two subsections that follow. Section 4.3 deals with testing.

4.1   Likelihood Specification

The likelihood for structural auction models is a function of the observed data and the unobserved variables $(\theta, \mathbf{X}, \mathbf{S})$. There are two ways in which the likelihood could be specified. One could work directly with the likelihood of observed data; this has been done in several papers by Bajari (1998a, b) and Bajari and Hortacsu (2000). There are three problems with this approach.

First, to obtain the posterior "probability" of a particular value of θ, the equilibrium bidding rule has to be solved to evaluate the likelihood function. Ordinary differential equation solvers could be used for this. An alternative is to approximate the equilibrium bidding rule in a way to bypass inverting it to obtain the unobserved $(\mathbf{X}, \mathbf{S})$. Bajari (1998a) suggests the use of the quantal response equilibrium (QRE) approach first suggested by McKelvey and Palfrey (1997) to solve normal form games. A Bayesian–Nash equilibrium is a set of probability measures $B_1^*(b_1|x_1), \ldots, B_n^*(b_n|x_n)$ that maximizes expected utility for all i and all x_i. Note that x_i now indicates a player's type.[*] $B_i(b_i|x_i)$ is a probability measure over agent i's strategy set $B_i = (b_i^1, \ldots, b_i^\tau, \ldots, b_i^{J_i})$. If $B_i(b_i|x_i)$ puts a point mass on an element b_i^τ of B_i and zero on the other elements of B_i, then b_i^τ will be interpreted as a probability measure. If the set of possible types of a player and the set of strategies is finite, expected utility is discretized; I will indicate the expected utility of bidder i by $\bar{u}(b_i, \mathbf{B}_{-i}; x_i, \theta)$, where $\mathbf{B}_{-i}$ is an $n-1$ dimensional

---

[*]For example, in a procurement auction the set of possible types would be the possible set of costs for an agent. The private signal x_i for an agent i that has been referred to in Section 2 is just one element of the set of types.

vector each element of which is the set of strategies for the n players excluding i. The basic idea underlying the QRE approach is that bidder i observes this expected utility with an error ϵ_i,

$$\hat{u}_i(x_i) = \bar{u}(b_i, \mathbf{B}_{-i}; x_i, \theta) + \epsilon_i(x_i) \tag{13}$$

Assuming that $\epsilon_i(x_i)$ follows an extreme value distribution, player i's QRE strategy is

$$B_i(b_i|x_i, \lambda) = \frac{\exp(\lambda \bar{u}(b_i, \mathbf{B}_{-i}; x_i, \theta))}{\sum_{\tau=1}^{J_i} \exp(\lambda \bar{u}(b_i^\tau, \mathbf{B}_{-i}; x_i, \theta))} \tag{14}$$

The likelihood function based on the QRE is

$$f_{b_1,\dots,b_n}(b_1^\tau, \dots, b_n^\tau|\theta, \lambda) = \sum_{x_i \in X} \left\{ \prod_{i=1}^{n} B_i(b_i|x_i, \lambda) f_{\mathbf{x}}(\mathbf{x}|\theta) \right\} \tag{15}$$

where X is the Cartesian product of all possible types for all n players in the game. λ is an additional parameter in the QRE approach; as λ becomes large, the QRE approaches the Bayesian–Nash equilibrium.

This approach, like all direct approaches, is implemented by specifying a distribution for $\mathbf{X}|\theta$. For each value of θ, the x_is and b_i are obtained from assumptions about the underlying game-theoretic model; this allows the likelihood to be evaluated. Note that the simulated bids are the QRE bids and not the Bayesian–Nash equilibrium bids. The key is to put a prior on the parameter λ so that the simulated bids mimic the observed Bayesian–Nash bids with a few draws of θ. For example, Bajari (1998a) puts a dogmatic prior of $\lambda = 25$ to obtain QRE bids which are good approximations to the observed bids. The idea is to sample the parameter space where the likelihood puts most of its mass.

Second, methods based on the likelihood of the bids have an additional problem in that the support of the bids is truncated with the truncation point depending on the parameter vector θ. For example, in a first-price, symmetric IPV auction, a bidder's bid will be no lower than the expected value of the second-highest bid. Suppose $l(\theta)$ is the lower bound of the bid distribution and a bid b_j less than $l(\theta)$ is observed; then either the first moment of the distribution of private values will be underestimated or the second moment will be overestimated to accommodate the outlier b_j. Unlike the likelihood function based on the Bayesian–Nash equilibrium bidding rule, the QRE is robust to the presence of outliers in the bidding data since it is based on the full likelihood.

Third, it is rarely the case that the bids of all n potential bidders are observed. Preparing and submitting bids is not a costless activity; there could be other reasons like the characteristics of the auctioned object, which could prevent a bidder from submitting a bid. Hence additional problems of working with censored data arise whether one works with the likelihood based on the QRE or the Bayesian–Nash bidding rule.

Indirect methods shed an important insight into an alternative way of specifying the likelihood. First a pseudo-sample of private values is generated; then this pseudo-sample is used to obtain an estimate of the private signals of the bidders. The key is to work with the unobserved private signals.

Making use of this idea, Sareen (1998) suggests working with the likelihood of the latent data $(\mathbf{X}, \mathbf{S})$ of all n potential bidders; this will be referred to as the *full* likelihood of the latent data. Since the support of the latent data does not depend on parameters θ the kind of problem mentioned with outliers for the Bayesian–Nash likelihood will not arise. Further, the *full* likelihood of the latent data includes all n potential bidders; hence, censoring of the data will not be an issue. Working with the latent structure is also simplified because the likelihood of the latent data does not involve the Jacobian of the transformation from the distribution of the signals to that of the bids.

Assuming that the numbers of participants and potential bidders are identical, estimation is based on the following posterior distribution,

$$f_{\theta|\text{data}}(\theta|\mathbf{X}, \mathbf{S}, \text{data}) \propto f_{\theta}(\theta)f_{\mathbf{XS}|\theta}(\mathbf{x}, \mathbf{s})1_{b=e(\bullet)} \tag{16}$$

$f_{\theta}(\theta)$ is the prior density function for the parameters θ. $f_{\mathbf{XS}|\theta}(\mathbf{x}, \mathbf{s})$ is the specified density function of the signals $(\mathbf{X}, \mathbf{S})$. $1_{(\bullet)}$ is an indicator function. It is equal to one if $(\theta, \mathbf{X}, \mathbf{S}, \text{data})$ solve the differential equations that are the first-order conditions for the optimization exercise given by equation (1); otherwise it is zero. The indicator function is the likelihood function of the bids conditional on realization of $(\mathbf{X}, \mathbf{S})$ that solve the Bayesian–Nash equilibrium bidding rule given by (2). The basic idea of the method is to sample $(\mathbf{X}, \mathbf{S})$ for each draw of θ and retain only those draws of $(\mathbf{X}, \mathbf{S}, \theta)$ which, along with the observed bids, solve the system of differential equations given by (2).

If participation is endogenous, the above scheme can be modified. Indicate by n_* the number of bidders who submit bids. N is the number of potential bidders; it is now a random variable which has to be estimated. First, consider estimation conditional on n; even if n is not observed, a close approximation is available.

$$f_{\theta|\text{data}}(\theta|\mathbf{X}, \text{data}) \propto f_{\theta}(\theta)f_{\mathbf{X}_{n_*}\mathbf{S}|\theta}(\mathbf{x}_{n_*}, \mathbf{s})1_{b_*=e_*(\bullet)}f_{\mathbf{X}_{n-n_*}|\mathbf{X}_{n_*},\mathbf{S},\theta}(\mathbf{x}_{n-n_*}) \tag{17}$$

$\mathbf{X}_{n_*}$ and $\mathbf{X}_{n-n_*}$ are n_* and $n - n_*$ dimensional vectors of private signals, respectively. For each auction only n_* bids, $\mathbf{b}_{n_*}$, are observed. Given θ,

$(\mathbf{x}_{n_*}, \mathbf{s})$ is drawn from the distribution specified for the signals $(\mathbf{X}_{n_*}, \mathbf{S})$;[*] note that, unlike equation (16), n_* instead of n values of $\mathbf{X}$ are drawn. If a draw of $(\mathbf{x}_{n_*}, \mathbf{s}, \theta)$ solves the equilibrium bidding rule in equation (2) it is retained; otherwise it is discarded. This is what the indicator function is doing. Next, $n - n_*$ signals are drawn from the distribution $\mathbf{X}_{n-n_*}|\mathbf{x}_{n_*}, \mathbf{s}, \theta$. Each of these draws will be less than the minimum of $\mathbf{x}_{n_*}$; conditioning on $\mathbf{x}_{n_*}$ ensures this. For example, in the IPV model n_* draws will be made from the distribution $X|\theta$; they will be accepted if they solve the equilibrium bidding rule. Suppose the minimum of these accepted draws is $x_{n_*}^{\min}$. Next, $n - n_*$ draws are made from $X|\theta$ such that each draw is less than $x_{n_*}^{\min}$.

In the case that a model of participation is specified, then sampling can be done iteratively in two blocks. Conditional on n, $(\mathbf{X}, \mathbf{S}, \theta)$ can be sampled in the manner above; then, conditional on $(\mathbf{x}, \mathbf{s}, \theta)$, the specified model of participation can be used to sample N. The problematic scenario of Corollary 1 will emerge here too if N has to be estimated from the high bid data without a model of participation.[†]

The key point is to work with the latent data. The latent data need not be obtained in the manner discussed above. It could be obtained nonparametrically as in the indirect methods. Alternatively, in models like the IPV the equilibrium bidding rule is solved in a few iterations; here ordinary differential equations solvers could be used to obtain the latent data conditional on θ. Van den Berg and Van der Klaauw (2000) also use the *full* likelihood of the latent data to study a Dutch auction for flowers, assuming the IPV model. They observe not just the winning bid but all bids made within 0.2 seconds of the winning bid. Their full likelihood of the latent data is a modified version of that given by equation (17),

$$f_{\theta|\text{data}}(\theta|\mathbf{X}, \text{data}) \propto f_\theta(\theta) f_{\mathbf{X}|\theta}(\xi(\bullet)|\theta) 1_{b^{1:n} \leq E(V_{n-1}|\theta)} \tag{18}$$

$1_{b^{1:n} \leq E(V_{n-1}|\theta)}$ is an indicator function that equals one if the smallest observed bid is less than the expected value of the second-highest private value. To

---

[*]The distribution of $\mathbf{X}_{n_*}, \mathbf{S}|\theta$ is obtained from the distribution specified for $(\mathbf{X}, \mathbf{S})$ by integrating out $\mathbf{X}_{n-n_*}$.

[†]The inability of the winning bid to identify *both* N and θ shows up in the posterior distribution for N. For a simulated Dutch auction, Sareen (1998) observes a spike in the posterior density function of N at the lower boundary for the parameter space for N. Suppose N is defined on the *known* interval $[l_n, u_n]$. To estimate N, for each value of $N \in [l_n, u_n]$, the posterior for $\theta(n)$ can be obtained; features of the posterior will decide the estimate of N. This approach is similar in spirit to the SNLLS method of Laffont et al. (1995).

simplify the exposition, let $n = n_*$ and let n be given.* The basic idea is to set up a Gibbs sampler with conditional distributions $\mathbf{X}|\theta$ and $\theta|\mathbf{X}$. But $\mathbf{X}$ is unobserved; so for some $\tilde{\theta}$ they obtain $x(\mathbf{b}, \tilde{\theta}) = \xi(\mathbf{b}, \tilde{\theta})$, the private signals of the bidders, by numerically solving the equilibrium bidding rule; hence $\mathbf{X}|\theta$ is a spike at $x(\mathbf{b},\tilde{\theta})$. Next, substituting $x(\mathbf{b}, \tilde{\theta})$ in the likelihood for private signals, a draw from the posterior $\theta|x(\mathbf{b},\tilde{\theta})$ given by equation (18) is obtained. Ideally, what is required are $(\theta, \mathbf{X}, \mathbf{S}, \text{data})$ combinations that solve the equilibrium bidding rule; but then this sampling scheme would not work since it will remain at $\tilde{\theta}$. Hence Van den Berg and Van der Klaauw (2000) accept a draw from $\theta|x(\mathbf{b}, \tilde{\theta})$ as long as it satisfies the constraint imposed by the likelihood of bids: the smallest observed bid is less than the expected value of the second highest private value. This will ensure that the posterior samples that part of the parameter space where the likelihood function puts most of its mass.

4.2 Prior Specification

Prior specification for structural auction models has to be done with care, for several reasons.

Structural auction models are nonregular models in that the support of the data depends on θ. A prior on θ implies a prior on the support of the data $\zeta_B(\theta)$ as well. This implicit prior on the support of the data should be consistent with the observed data; that is, it should put nonzero mass on the part of the parameter space within which the observed data falls. An example will help to clarify this issue. Suppose the support of the data $\zeta_B(\theta)$ is the interval $\zeta_B(\theta) = [\underline{b}(\theta), \infty)$. A proper prior, f_θ, is specified for θ such that the implied prior on the function $\underline{b}(\theta)$ is uniform on the interval $[\underline{s}, \overline{s}]$; thus, $\underline{s} \le \underline{b}(\theta) \le \overline{s}$. If the observed bids b_i lie in the interval $[\underline{s}, \overline{s}]$, the prior on θ is consistent with the observed data. Suppose some $b_i < \underline{s}$; then the prior for θ is not consistent with the observed data. f_θ puts a mass of zero on the part of the parameter space where the likelihood function of the observed bid b_i is nonzero.

They key point that emerges from this discussion is that the part of the parameter space where the likelihood and the prior put most of their mass should be similar. This is feasible in many auctions since the parameters

*Van den Berg and Van der Klaauw (2000) assume participation is endogenous. Since the private signals are independent, $\mathbf{X}_{n-n_*}|\mathbf{X}_{n_*}$ is sampled by taking $n - n_*$ draws less than $\min\{\mathbf{x}_{n_*}\}$ from $f_{X|\theta}(\bullet)$. The sampling scheme is in two blocks, as described above.

have a natural interpretation as the limiting form of some observables.*
Bajari (1998b), Bajari and Hortacsu (2000), and Sareen's (1999) work provide many examples of this.

Priors specified in the manner above require a serious prior elicitation effort on the part of the empirical researcher. For example, interpreting parameters in terms of observables to impose reasonable priors may not be obvious. It may also be the case that an empirical researcher may want to be noninformative to conduct a prior sensitivity analysis since the prior specification above is highly informative. I next turn to the specification of noninformative priors.

Jeffreys' prior is used as a standard noninformative prior in many instances. If $\mathbf{J}_Z(\theta)$ is the Fisher information matrix of the likelihood of the data $\mathbf{Z}$, the Jeffreys' prior for θ is

$$f_{\text{jeff}}^z(\theta) \propto \sqrt{\det \mathbf{J}_Z(\theta)} \tag{19}$$

where "det" indicates the determinant of the matrix.

It is attractive to represent "vague" prior beliefs through the Jeffreys' prior, for several reasons. It is easy to understand and implement. Unlike uniform priors, it is *invariant*; inference about θ is identical whether it is based on the Jeffreys' prior for θ or a one-to-one transformation of θ. Jeffreys recommended this prior to represent "noninformativeness" for single-parameter problems from *iid* data. Beyond the single-parameter case it is less certain whether Jeffreys' prior is a candidate for the position of "noninformative" priors. This has led to several modifications and reinterpretations of Jeffreys' prior.†

A modification proposed by Bernardo (1979) to represent "noninformativeness" is the reference prior. The reference prior emerges from maximizing an asymptotic expansion of Lindley's measure of information. Lindley's (1956) measure of information is defined as the expected Kulback–Liebler divergence between the posterior and the prior; the larger the measure, the more informative the data and hence less informative the prior. Reference priors are appealing to both nonsubjective Bayesians and frequentists since the posterior probabilities agree with sampling probabilities to a certain order (Ghosh et al. 1994). They are appealing to subjective Bayesians as well since they serve as a reference point in a prior sensitivity analysis. When

---

*The notion of parameters being a limiting form of observables is from de Finetti's Representation theorem. Bernardo and Smith (1994, pp. 172–181) provide a lucid explanation of this theorem.
†See Kleinberger (1994), Zellner (1971, pp. 216–220), and Phillips (1991) for examples.

there are no nuisance parameters and certain regularity conditions are satisfied, Bernardo's reference prior is the Jeffreys' prior.

In several papers, Ghosal and Samanta (1995) and Ghosal (1997), extending Bernardo's work, have obtained the reference prior when the support of the data depends on the parameters in structural auction models; for an overview of the reference prior idea and its extension to the nonregular case, see Sareen (2000c, pp. 51–57). Suppose $\theta = [\eta, \varphi]$, with η being a scalar. The support of the likelihood of bids $\zeta_B(\theta) = \zeta_B(\eta)$ is strictly monotonic in η; η is referred to as the "nonregular" parameter. Since the support of the likelihood of the bids conditional on η does not depend on a parameter, φ is called the "regular" parameter. The reference prior for $\theta = [\eta, \varphi]$ is

$$\pi_{\text{ref}}^B(\eta, \varphi) \propto |c(\eta, \varphi)| \sqrt{\det \mathbf{J}_B^{\varphi\varphi}(\eta, \varphi)} \tag{20}$$

where $c(\eta, \varphi)$ is the score function of the likelihood of bids

$$c(\eta, \varphi) = E_{B|\eta,\varphi}\left[\frac{\partial}{\partial \eta} \log f_B(b_i|\eta, \varphi)\right]$$

and $\mathbf{J}_B^{\varphi\varphi}(\eta, \varphi)$ is the lower right-hand block of $\mathbf{J}_B(\eta, \varphi)$, the Fisher information from bids. Intuitively, both the regular and the nonregular parameters contribute to the standard deviation of the asymptotic distribution of the relevant estimator. Since the asymptotic distribution of the estimator of the regular and the nonregular parameter are different, the regular parameter φ contributes $\sqrt{\mathbf{J}_Z^{\varphi\varphi}(\eta, \varphi)}$ and the nonregular parameter η contributes $c(\eta, \varphi)$ to the reference prior.*

Like the Jeffreys' prior, the reference prior is invariant. In addition, it provides a way for handling "nuisance" parameters which Jeffreys' prior does not. The distinction between "nuisance parameters" and "parameters of interest" could be important in structural auction models since the support $\zeta_B(\theta)$ is like a nuisance parameter. Further, since the reference prior is based on the sampling density of the data, the nonregularity in the likelihood is taken into account while constructing the prior; inconsistency between the prior and the data of the kind discussed above will not arise with the reference prior.

---

*The MLE of the "regular" parameter φ after "concentrating" out the "nonregular" parameter from the likelihood function converges to a normal distribution; $\sqrt{T}(\hat{\varphi}_{ml} - \varphi_o) \sim N(0, \mathbf{J}_B^{\varphi\varphi}(\hat{\eta}, \varphi))$, where $\hat{\eta} = w_*$. The MLE of the nonregular parameter η is the minimum winning bid, W_*. It converges to an exponential distribution; $T(W_* - \eta_o) \sim \exp(c(\eta, \varphi))$, where $(c(\eta, \varphi)$ is the sampling expectation of the score function of the bids.

When there are no nuisance parameters and certain regularity conditions are satisfied, it is well known that Jeffrey's prior and the reference prior coincide. When the support of the data depends on the parameters, the reference prior is *not* the Jeffreys' prior, in general. Sareen (2000a) proves that the two *coincide*, even when the support of the data depends on the parameters, if the winning bid is sufficient for a scalar parameter θ. The necessary conditions under which a sample maximum or minimum is sufficient for a scalar parameter have been established by Huzurbazar (1976). These conditions restrict the functional form of the support and the density function from which the order statistics are drawn. Specifically, if $B_j^i \geq \zeta_B(\theta)$, then the necessary and sufficient conditions under which the winning bid is sufficient for θ are:

(1) $\zeta_B(\theta)$ is a strictly monotonic, continuous, and differentiable function of θ; and
(2) the form of the density function of bids is

$$f_b(b_j^i|\theta) = m(b_j^i)q(\theta) \tag{21}$$

where $m(b_j^i)$, $q(\theta)$ are strictly positive functions of b_j^i and θ, respectively.

Outside of this scenario the use of Jeffrey's prior to represent "noninformativeness" may lead to pathologies of the kind discussed above.

4.3 Testing

The posterior odds ratio is a means of comparing two models. It gives the odds of one model compared with another conditional on the observed data. Formally, let M_1 and M_2 be two models. Then the posterior odds ratio between model 1 and 2, P_{12}, is

$$P_{12} = \frac{f_{M_1|\text{data}}(\bullet)}{f_{M_2|\text{data}}(\bullet)} = \frac{f_{M_1}(\bullet)f_{\text{data}|M_1}(\bullet)}{f_{M_2}(\bullet)f_{\text{data}|M_2}(\bullet)} \tag{22}$$

where $f_{M_\tau}(\bullet), f_{\text{data}|M_\tau}(\bullet)$, $\tau = 1, 2$, are the prior probability and the marginal likelihood of model τ, respectively. The ratio $f_{M_1}(\bullet)/f_{M_2}(\bullet)$ is a researcher's beliefs, prior to observing the data, about which model is a more probable explanation of the auctions she is studying. These prior beliefs are revised on observing the data through the marginal likelihood

$$f_{\text{data}|M_\tau}(\bullet) = \int_\Theta f_{\theta|M_\tau}(\bullet)l(\theta; \text{data}, M_\tau)d\theta \tag{23}$$

where $l(\theta; \text{data}, M_\tau)$ is the likelihood function under model M_τ; $f_{\theta|M_\tau}(\bullet)$ is the prior for θ under model M_τ.

As long as $f_{\theta|M_\tau}(\bullet)$, the prior for θ under model M_τ, is proper, the marginal likelihood of the data can be evaluated using the draws from the posterior distribution of the parameters (Chib and Jeliazkov 2000). Improper priors for θ cause problems in calculating the marginal likelihood; since improper priors are defined up to a constant of proportionality, the scaling of the marginal likelihood is arbitrary. Using "vague proper priors" does not solve this problem. For example, Bajari (1998b) and Bajari and Hortacsu (2000) use uniform priors on $[l_i, u_i]$, $i = 1, \ldots, k$, for each of the k parameters. The Bayes factor is proportional to a constant so that the resultant answer will again be arbitrary, like the "vague improper prior"; Berger and Pericchi (1997, p. 2) point this out, with several examples.

Several "default" Bayes factors have been suggested to get around this problem; Kass and Raftrey (1995, pp. 773–795) provide a survey of these options. The basic idea underlying these "default" options is to use a subset of the data, called a training sample, to "convert" the improper priors under each model into a proper posterior; this is referred to as an intrinsic prior. As its name suggests, the intrinsic prior is used as a prior to define a Bayes factor for the remaining data. Since the resultant Bayes factor will depend on the size of the training sample, an "average" over all possible training samples is taken.

They key point is that a "default" option cannot be used indiscriminately; what will work best depends on the applied problem at hand. For example, for a scalar nonregular parameter and truncated exponential likelihoods, Berger and Pericchi (1997, pp. 12–13) discuss "default" Bayes factors for a one-sided hypothesis testing. The problem they encounter is that the intrinsic prior is unbounded under the null hypothesis.

Informal means to compare models could be used to circumvent the problem of working with the posterior odds ratio and improper priors. For example, Sareen (1999) compares the posterior predictive distribution under the pure CV and IPV models. The posterior predictive distribution represents beliefs about an out-of-sample bid after observing the data; this is in contrast to the prior predictive distribution which represents beliefs about this out-of-sample bid before observing the data. The extent of "divergence" between the prior and posterior predictive distribution for a model indicates the learning from data about that model. A model for which this "divergence" is large should be favored as there is more to be learnt from the data about this model.

5. UNOBSERVED HETEROGENEITY

Empirical work requires data on several auctions. It is rarely the case that these auctions are identical; either the auctioned object or the environment under which an auction is held, or both, could differ across auctions. Some of this auction/object heterogeneity may be observed by the econometrician. Many characteristics which make the auctioned object different will not be observed; this is termed as unobserved heterogeneity. Since the value of an object determines whether a bidder will participate or not, these unobserved characteristics would affect not just the bidding behavior but participation by bidders as well.

A standard approach to modeling unobserved heterogeneity is to give it a parametric form. This makes direct estimation more complicated since this unobserved heterogeneity enters the Bayesian–Nash equilibrium bid for each auction in addition to the unobserved private values already present in the equilibrium bidding rule. Indirect estimation methods have been unable to address this issue; this is one of the reasons that Perrigne and Vuong (1999) recommend a parametric second stage in their survey.

An alternative approach has been suggested and implemented for the IPV model by Chakraborty and Deltas (2000). It falls within the class of direct methods in that parametric assumptions are made about the distribution of $\mathbf{X}|\theta_j$; it is assumed to belong to the location-scale class either unconditionally or conditional on a shape parameter. The subscript j in θ_j indicates that the estimation of the distribution of valuations is auction specific; *only* bids within an auction are used to obtain $\hat{\theta}_j$ for each auction j. θ_j incorporates both observed and unobserved heterogeneity despite the fact that no auction-specific covariates are used to obtain $\hat{\theta}_j$. Indicating observed heterogeneity in auction j by $\mathbf{Z}_j$ and unobserved heterogeneity by v_j, the following relationship

$$\theta_j = \mathbf{r}\big(\mathbf{Z}_j, v_j | \delta\big) \tag{24}$$

is used to recover estimates of the coefficients of observed heterogeneity δ in the second step. $\mathbf{r}(\bullet)$ is a k-dimensional function of the observed and unobserved heterogeneity in each auction j. The estimates of θ_j obtained in the first step are used instead of θ_j. The estimates of δ obtained in the second step are robust to unobserved heterogeneity in the sense that they do not depend on the distribution of v. In addition, breaking the estimation into two parts simplifies the estimation of δ. The first stage of the estimation which obtains θ is structural as it is based on the Bayesian–Nash equilibrium bidding rule. Reduced-form estimation is done in the second stage to obtain $\hat{\delta}$.

The two-stage procedure of Chakraborty and Deltas (2000) is similar to the following three-stage Bayesian hierarchical model,

$$
\left.\begin{array}{l}
\mathbf{B}_j|\theta_j \\
\theta_j|\mathbf{Z}_j, \delta \\
\delta
\end{array}\right\} \tag{25}
$$

where $\mathbf{b}_j = (b_1, \ldots, b_{n_j})$ are the bids observed for auction j. The first stage is obtained from the parametric assumptions made about the distribution of the signals $\mathbf{X}|\theta_j$ and the Bayesian–Nash equilibrium bidding rule. The second stage is obtained from the relation specified in equation (24). The last stage is the specification of the prior for δ. Sampling the posterior for (θ_j, δ) can be done in blocks. The following blocks fall out naturally from the ideas of Chakraborty and Deltas (2000):

$$
\left.\begin{array}{ll}
\theta_j|bfB_j & \text{for each auction } j \\
\delta|\mathbf{Z}, \theta & \text{across auctions}
\end{array}\right\} \tag{26}
$$

where $\mathbf{Z} = (\mathbf{Z}_1, \ldots, \mathbf{Z}_n)$ and $\theta = (\theta_1, \ldots, \theta_n)$. In the case that participation is endogenous, an iterative sampling scheme could be set up. First, the blocks in equation (25) are sampled, conditional on n. Then, conditional on draws of (θ, δ), the parameters of the process used to model entry could be sampled.

6. CONCLUSION

This survey describes the menu of techniques for identification, estimation, and testing available to an empirical researcher. The choice between structural versus reduced form, direct methods versus indirect methods, will depend on the goal of the empirical exercise, the available data, and the theory developed for the underlying game-theoretic model. Each of these aspects is commented on in turn.

   If the eventual goal of the empirical exercise is mechanism design, the structural approach would be preferred over the reduced-form approach. In many auctions, a seller may be interested in goals other than revenue maximization. For example, in the timber auctions in Southern Ontario, the county of Simcoe wants to encourage a viable local timber industry. Since a when-issued market and a secondary market exists for T-bills and other government debt, one of the aims of a central bank in conducting T-bill auctions would be to promote liquidity in the market for government debt. In these instances reduced-form estimation may be preferred, especially if the theory for the underlying game-theoretic model is not well developed.

In certain instances various features of the distribution of private values could be of interest; for example, Li et al. (2000) were interested in estimating the magnitude of the "money left on the table" for the OCS wildcat auctions. For estimation, indirect methods could be preferred since they do not impose any distributional assumption on the private signals of the bidders. Since the rate of convergence of estimators based on indirect methods is slow, large data sets are required to implement these estimators. For example, if data on winning bids is observed, one may be forced to turn to direct methods to establish the identification of the underlying game-theoretic model in the first instance. In general, when testing a structural model against an explicit alternative, Bayesian methods may be preferred since they allow testing *conditional* on the observed data.

Estimating and testing structural models is a commendable goal; it presupposes that the underlying game-theoretic model exists and that its properties have been established. With few exceptions, our understanding of game-theoretic models is confined to single-unit, symmetric private-values auctions. The challenge in doing empirical work is how to adapt the existing theory to the idiosyncratic features of a specific auction; here the role of conditioning and the details observed about each auction will be the guiding tools. Chakraborty and Deltas (2000) incorporate unobserved heterogeneity without altering the equilibrium bidding rule of a symmetric IPV first-price auction. Bajari and Hortacsu (2000) explain last minute bidding in eBay auctions through a two-stage game; conditional on a stochastic entry process in the first stage, the second stage is similar to a symmetric sealed-bid, second-price auction. Athey and Haile (2000) and Hendricks et al. (1999) identify and estimate standard auction models by observing additional details of that auction Thus *ex post*, forthcoming, 2000value of the auctioned object is used to identify the CV model; bidder identity is used to identify asymmetric models. Similarly, if the component of private values that makes bidders asymmetric was observed, then it may be possible to use the theory of symmetric games by conditioning on this asymmetric component.

ACKNOWLEDGMENTS

I have benefitted from several clarifications by Susan Athey, Patrick Bajari, Phil Haile, and Matthew Shum.

REFERENCES

S. Athey, P. A. Haile. Identification of standard auction models. Working paper, University of Wisconsin–Madison, 2000.

S. Athey, J. Levin. Information and competition in U.S. forest service timber auctions. *Journal of Political Economy* 109(2); 375-417, 2001.

P. Bajari. Econometrics of sealed bid auctions. *American Statistical Association, Proceedings of the Business and Economic Statistics Section,* 1998a.

P. Bajari. Econometrics of first price auctions with asymmetric bidders. Working paper, Stanford University, 1998b.

P. Bajari, A. Hortaçsu. Winner's curse, reserve price and endogenous entry: empirical insights from eBay auctions. Working paper, Stanford University, 2000.

J. O. Berger, L. Pericchi. On criticisms and comparisons of default Bayes factors for model selection and hypothesis testing. In: W Racugno, ed. *Proceedings of the Workshop on Model Selection.* Bologna: Pitagora Editrice, 1997.

S. M. Berman. Note on extreme values, competing risks and semi-Markov processes. *Annals of Mathematical Statistics* 34:1104–1106, 1963.

J. M. Bernardo. Reference posterior distributions for Bayesian inference. *Journal of the Royal Statistical Society B* 41:113–147, 1979.

J. M. Bernardo, A. F. M. Smith. *Bayesian Theory.* New York: Wiley, 1994.

S. Campo, I. Perrigne, Q. Vuong. Asymmetry and joint bidding in OCS wildcat auctions. Working paper, University of Southern California, 1998.

S. Campo, I. Perrigne, Q. Vuong. Semiparametric estimation of first-price auctions with risk-averse bidders. Working paper, University of Southern California, 2000.

I. Chakraborty, G. Deltas. A two-stage approach to structural econometric analysis of first-price auctions. Working paper, University of Illinois, 2000

S. Chib, I. Jeliazkov. Marginal likelihood from the Metropolis–Hastings output. Working paper, Washington University, 2000.

S. Donald, H. J. Paarsch. Identification, estimation and testing in parametric models of auctions within the independent private values paradigm. *Econometric Theory* 12:517–567, 1996.

S. Donald, H. J. Paarsch, J. Robert. Identification, estimation and testing in empirical models of sequential, ascending-price auctions with multi-unit demand: an application to Siberian timber-export permits. Working paper, University of Iowa, 1999.

S. Ghosal. Reference priors in multiparameter nonregular case. *Test* 6:159–186, 1997.

S. Ghosal, T. Samanta. Asymptotic behavior of Bayes estimates and posterior distributions in multiparameter nonregular case. *Mathematical Methods in Statistics* 4:361–388, 1995.

J. K. Ghosh, S. Ghosal, T. Samanta. Stability and convergence of posterior in nonregular problems. In: S. S. Gupta and J. O. Berger, eds. *Statistical Decision Theory and Related Topics V*. New York: Springer-Verlag, 1994, pp. 183–199.

E. Guerre, I. Perrigne, Q. Vuong. Nonparametric estimation of first-price auctions. Working paper, Economie et Sociologie Rurales, Toulouse, no. 95-14D, 1995.

E. Guerre, I. Perrigne, Q. Vuong. Optimal nonparametric estimation of first-price auctions. *Econometrica* 68:525–574, 2000.

J. J. Heckman, B. E. Honoré. The empirical content of the Roy model. *Econometrica* 58:1121–1149, 1990.

K. Hendricks, H. J. Paarsch. A survey of recent empirical work concerning auctions. *Canadian Journal of Economics* 28:403–426, 1995.

K. Hendricks, J. Pinkse, R. H. Porter. Empirical implications of equilibrium bidding in first-price, symmetric, common value auctions. Working paper, North-Western University, 2000.

K. Hendricks, R. H. Porter. An empirical study of an auction with asymmetric information. *American Economic Review* 78:865–883, 1988.

K. Hendricks, R. H. Porter, B. Boudreau. Information, returns, and bidding behavior in OCS auctions: 1954–1969. *The Journal of Industrial Economics* 34:517–542, 1987.

P. A. Haile, H. Hong, M. Shum. Nonparametric tests for common values in first-price auctions. Working paper, University of Wisconsin–Madison, 2000.

H. Hong. Nonregular maximum likelihood estimation in auction, job search and production frontier models. Working paper, Princeton University, 1997.

H. Hong, M. Shum. Increasing competition and the winner's curse: evidence from procurement. Working paper, University of Toronto, 2000.

Vis. Huxurbazar. Sufficient Statistics: Selected Contribution. New York, Marcel Dekker, 1976.

R. E. Kass, A. E. Raftrey. Bayes factors. *Journal of the American Statistical Association* 90:773–795, 1995.

F. Kleiberger. Identifiability and nonstationarity in classical and Bayesian econometrics. PhD dissertation, Erasmus University, Rotterdam, 1994.

I. Kotlarski. On some characterization of probability distributions in Hilbert spaces. *Annali di Matematica Pura ed Applicata* 74:129–134, 1966.

J. J. Laffont. Game theory and empirical economics: the case of auction data. *European Economic Review* 41:1–35, 1997.

J. J. Laffont, H. Ossard, Q. Vuong. Econometrics of first-price auctions. *Econometrica* 63:953–980, 1995.

J. J. Laffont, Q. Vuong. Structural analysis of auction data. *American Economics Review, Papers and Proceedings* 86:414–420, 1996.

T. H. Li, Econometrics of First Price Auctions with Binding Restoration Prices. Working Paper, Indiana University, 2000.

T. Li, I. Perrigne, Q. Vuong. Conditionally independent private information in OCS wildcat auctions. Journal of Econometrics. 98(1): 129-161, 2000.

T. Li, Q. Vuong. Nonparametric estimation of the measurement error model using multiple indicators. *Journal of Multivariate Analysis* 65:139–165, 1998.

D. V. Lindley. On a measure of information provided by an experiment. *Annals of Mathematical Statistics* 27:986–1005, 1956.

D. Lucking-Reiley. Auctions on the internet: what is being auctioned and how? *Journal of Industrial Economics*, forthcoming, 2000.

R. P. McAfee, J. McMillan. Auctions and bidding. *Journal of Economic Literature* 25:699–738, 1987.

R. McKelvey, T. Palfrey. Quantal response equilibria for extensive form games. Working paper, CalTech, 1997.

P. R. Milgrom, J. Weber. A theory of auctions and competitive bidding. *Econometrica* 50:1089–1122, 1982.

H. J. Paarsch. Deciding between the common and private values paradigm in empirical models of auctions. *Journal of Econometrics* 51:191–215, 1992.

H. J. Paarsch. Deciding an estimate of the optimal reserve price: an application to British Columbian timber sales. *Journal of Econometrics* 78:333–357, 1997.

I. Perrigne, Q. Vuong. Structural economics of first-price auctions: a survey of Methods. Canadian Journal of Agricultural Economics 47:203–233, 1999.

P. C. B. Phillips. To criticize the critics: an objective Bayesian analysis of stochastic trends. *Journal of Applied Econometrics* 6:333–364, 1991.

R. Porter. The role of Information in the U.S. Offshore Oil and Gias Lease Auctions. Econometrica 63: 1-27, 1995.

B. L. P. Rao. *Identifiability in Stochastic Models: Characterization of Probability Distributions.* San Diego: Academic Press, 1992.

S. Sareen. Likelihood based estimation of symmetric, parametric structural auction models with independent private values. Unpublished manuscript, 1998.

S. Sareen. Posterior odds comparison of a symmetric low-price, sealed-bid auction within the common-value and the independent-private-values paradigms. *Journal of Applied Econometrics* 14:651–676, 1999.

S. Sareen. Relating Jeffreys' prior and reference prior in nonregular problems: an application to a structural, low-price, sealed-bid auction. Forthcoming in "Proceedings of the ISBA 6th World Congress" 2001

S. Sareen. Evaluating data in structural parametric auctions, job-search and Roy models. Unpublished manuscript, 2000b.

S. Sareen. Estimation and testing of structural parametric sealed-bid auctions. PhD dissertation, University of Toronto, 2000c.

S. Sareen. The economics of timber auctions in Southern Ontario. Unpublished manuscript, 2000d.

G. Van den Berg, B. Van der Klaauw. Structural empirical analysis of Dutch flower auctions. Working paper, Free University Amsterdam, 2000.

R. Weber. Multiple-object auctions. In: R. Engelbrecht-Wiggans, M. Shubik, R. Stark, eds. *Auctions, Bidding and Contracting: Uses and Theory*. New York: New York University Press, 1983, pp. 165–191.

A. Zellner. *An Introduction to Bayesian Inference in Econometrics*. New York: Wiley, 1971.

31

Asymmetry of Business Cycles: The Markov-Switching Approach*

BALDEV RAJ   Wilfrid Laurier University, Waterloo, Ontario, Canada

1. INTRODUCTION

In modern business cycle research there is a growing interest in asymmetric, nonlinear time series models utilizing a Markov regime-switching framework. The influential contribution by Hamilton (1989), who proposed a very tractable approach to modeling regime changes, has been particularly popular. His approach views the parameters of an autoregression as the outcome of a latent discrete-state Markov process. He applied this approach to the identification of "turning points" in the behavior of quarterly GNP from 1951 to 1984. The idea of using such a process in business cycles has a long history in economics that dates back to Keynes (1936). Keynes informally argued that expansions and contractions are different from each other, with the former being long-lived while the latter are more violent.

*The author acknowledges helpful comments from an anonymous referee, James Hamilton, Lonnie Maggie, Alastair Robertson and Danial J. Slottje on an earlier version of this paper. Responsibility for errors rests with the author. Comments are welcome.

The consideration of regime switching raises a variety of interesting econometric issues with respect to specification, estimation and testing (for an overview, see Hamilton 1994, Kim and Nelson 1999a). For instance, testing a two-state regime-switching model of asymmetry against the null of a one-state model using classical hypothesis-testing procedures is difficult since the distribution of the test statistic is nonstandard. Moreover, as pointed out by Boldin (1990), Hansen (1992, 1996), and Garcia (1998), the transition probability parameters that govern the Markov regime-switching model are not identified under the null hypothesis of a one-state model. Finally, if the probability of staying in state 1 is either zero or one, the score of the likelihood function with respect to the mean parameters and the information matrix is zero.

Significant progress has been made in testing such models using parametric classical methods, but a few issues remain to be resolved. The question of interest in testing is whether or not asymmetry of business cycles, as characterized by regime switching, is supported by the data in a particular application. In contrast, the traditional approach to testing for nonlinearity has been to test the null of linearity against a nonspecified nonlinear alternative. We will have more to say about the specification-testing aspects of the regime-switching framework later in the paper. Estimation of the regime-switching models is based on the maximum likelihood method (full or quasi or approximate), depending on a particular model or application (see Hamilton 1994, Kim 1994).

The primary focus of modern business cycle research has been on using a probabilistic parametric model-base approach to account for the asymmetry and/or the comovement attributes of macro variables. Asymmetry and comovement were identified by Burns and Mitchell (1946) and their colleagues at the National Bureau of Economic Research (NBER) as the two key attributes of business cycles.

In the beginning of the modern approach to business cycle research, each of these two features of business cycles was treated separately from the other without an effort to specify a possible link between them. For instance, Hamilton (1989) proposed a model of the asymmetric nonlinear attribute of business cycles in which the mean parameter of the time series of the nonstationary growth rate of output is modeled as a latent discrete-state Markov switching process with two regimes. In the empirical application of this model, the mean of the growth rate of the quarterly postwar output in the U.S. was allowed to switch concurrently between expansions and contractions over the time period. Similarly, Stock and Watson (1991) proposed a dynamic factor framework that is designed to extract a single common factor from many macroeconomic series where the comovement attribute of business cycles is the primary focus of analysis. This framework was used to construct an experimental composite coincidental index of economic activity of the U.S. economy. According to a recent assessment of the modern approach to business cycle research by Diebold and Rudebusch (1996), the focus on modeling one of the two attributes of asymmetry and comovement without considering any link between them may have been convenient but unfortunate. For a recent review of business cycle research, see also Diebold and Rudebusch (1998). It was unfortunate because these two attributes of cycles have been treated jointly for many decades by classical researchers of the business cycles. Diebold and Rudebusch (1996) went on to sketch a prototype model for the two attributes of the business cycles. Their synthesis used a dynamic factor model with a regime-switching framework. It extends the Markov regime-switching autoregressive model to obtain some potential advantages for estimation and testing of the assumption of regime switching of means as a by-product. For one, their approach to a synthesis of the two attributes of the business cycle is intuitively appealing in that it implements a common idea that there may be fewer sources of uncertainty than the number of variables, thereby enabling more precise tracking of cycles. It could also increase the power of the classical likelihood-based test of the null of a one-state model against the regime-switching model because the multivariate data are less likely to be obscured by idiosyncratic variations (see Kim and Nelson 2001). Some recent advances in computing and development of numerical and simulation techniques have aided and advanced the empirical synthesis considerably, and the consideration of other types of multivariate regime-switching models (e.g., see Kim 1994 and references therein).

In retrospect, the classical research program mainly used a model-free approach since its primary focus was on establishing stylized facts or regularities of business cycles. A prominent example of such a research program for business cycles research is Burns and Mitchell (1946). This research program considered comovement among hundreds of series over the business cycles, taking into account possible leads and lags. It helped to create composite leading, coincident, and lagging indexes that served to signal recessions and recovery (see Shishkin 1961). This approach was criticized for lacking a probabilistic underpinning. The research program during the postwar period shifted to using a model-based approach to studying the time series properties of business cycles on the basis of a framework of stochastic linear difference equations. One popular approach has been to specify the time series as a stationary auto-regressive moving average (ARMA) process around a deterministic trend (see Blanchard 1981). Another approach used the ARIMA model that differs from the popular approach by allowing a unit root in the ARMA process (e.g., see Nelson and Plosser 1982, Campbell and Mankiw 1987). Yet another approach used the linear unobserved component modeling framework that specifies the time series as a sum of a random walk and a stationary ARMA process (see Watson 1986). Finally, King et al. (1991) utilized the cointegration approach to business cycle analysis that permits stochastic equilibrium relationships among integrated variables in the sense of Engle and Granger (1987).

The use of the linear model or the associated linear filter theory in empirical models of business cycles had a number of advantages. For one, as mentioned before, it allowed the use of a model-base approach to analysis instead of the classical model-free approach. Also, the use of linear filter theory is easy to understand in comparison to the nonlinear filter theory. Examples of the applications of linear filtering methods to business cycles include the essays by Hodrick and Prescott (1981) and Stock and Watson (1999), among others. However, the use of the linear framework had the unfortunate side effect of sidelining the consideration of the asymmetric nonlinear feature of the business cycle that requires a separate analysis of contractions from expansions.

In the famous paper on business cycles by Lucas (1976), it was emphasized that the outputs of broadly defined sectors tend to move together. This occurs because economic agents tend to coordinate their activities for mutual benefit. The standard way to analyze comovement among macroeconomic variables has been to use the nonparametric method of the autocorrelation function as a tool for modeling multivariate dynamics in variables (e.g., Backus and Kehoe 1992). Another approach has been to use the vector autoregression (VAR) framework that was introduced by Sims (1980). While the use of the VAR has proved to be useful for business cycle analysis when a few variables are used, it has a drawback of running into the degrees of freedom problem when one attempts to capture the pervasive comovement in a large number of macroeconomic variables. One useful way to achieve a crucial reduction in dimensionality is to use a dynamic factor or index structure imbedded among a large set of macrovariables. Sargent and Sims (1977), Geweke (1977), and Watson and Engle (1983) articulated the use of factor models in a dynamic setting. A recent key contribution has been by Stock and Watson (1991), who exploited the idea that comovement among macro-variables is driven in large part by a common shock or economic fundamentals. The use of dynamic-factor models has been influential since it is one of the effective ways to achieve both parsimony and co-movement in business cycle models.

In this paper, we provide a selective survey of the regime-switching literature relating mostly to business cycle research by focusing on some of the new developments since the review by Diebold and Rudebusch (1996). We shall focus not only on models of business cycle asymmetry within the univariate setup but also on those within the multivariate setup. One advantage of using the multivariate setup is that it permits examination of asymmetry using a dynamic one-factor model of output implied by either a common stochastic shock or a common stochastic trend. The multivariate setup can also be used to assess asymmetry, using a dynamic two-factor model of output that incorporates both the common stochastic trend and/or a common cycle among variables. The motivation of the consideration of asymmetry in particular and nonlinearity in dynamic time series models in general is discussed in Section 2 both from the point of view of economic theory and from an empirical perspective. The model of the asymmetry proposed by Hamilton, and some of its key extensions, are reviewed in Section 3. This section also presents some other characterizations of business cycle asymmetry. The dynamics of the one-factor and two-factor models are described in Section 4 along with some selective empirical results. The final section concludes the chapter.

2. MOTIVATION FOR ASYMMETRY
2.1 Empirical Evidence in Support of Asymmetry

It would be instructive to start with a review of the arguments used by Hamilton (1989) to motivate his contribution to the literature on modeling asymmetry. One of the primary objectives of his paper was to explore the consequence of specifying the first differences of the log of output as a nonlinear process rather than a linear process within the Markov-switching framework. He pointed out that abundant evidence had accumulated showing that departures from linearity are an important feature of many macro series in economics. Moreover, researchers had proposed a variety of ways to characterize such nonlinear dynamics. These included papers by Neftci (1984) and Sichel (1987, 1989), who provided direct evidence on business cycle asymmetry, and the time transformation effect by Stock (1987) and the deterministic chaos found by Brock and Sayers (1988) provide indirect evidence on business cycles asymmetry. Other approaches stressed the need to account for nonlinear conditional heterogeneity in asset pricing applications, and found support for it.

The specific form of nonlinearity with which Hamilton's regime-switching model is concerned arises if the dynamic process is subject to discrete shifts such that the dynamic behavior of the variable is markedly different across episodes. It builds on the Markov-switching regression approach of Goldfeld and Quandt (1973) that allowed parameters of the static regression model to switch endogenously. The econometrics of time-varying parameter models that preceded the contribution by Hamilton is surveyed by Raj and Ullah (1981).

The filters and smoothing function used by Hamilton provide nonlinear inference about a discrete-valued unobserved state vector based on observed variables. In contrast, a linear approach is used in the Kalman filter for generating estimates of a continuous unobserved vector based on observations on a series. The application of the regime-switching model to the quarterly postwar U.S. real GNP growth rate found that it produced a best fit when a positive growth rate was associated with a normal state of the economy and a negative growth state associated with recessions in the U.S. economy. The smooth probability plots of the U.S. growth rate were found to have a remarkable correspondence with the NBER dating of business cycles. Another conclusion was that business cycles are associated with a large permanent effect on long-run output. Earlier, Nelson and Plosser (1982) and Campbell and Mankiw (1987) had reached a similar conclusion by using a different inference strategy.

2.2 Economic Theory-Based Arguments in Support of Asymmetry

One possible interpretation of the evidence for asymmetry in output just presented would be that these are examples of measurement ahead of theory. A number of macroeconomics models have since been developed that are consistent with the central idea of the regime-switching model for a presence of multiple equilibria whose dynamics can be adequately approximated by statistical models involving regime switching. Also, a number of recent papers have articulated various mechanisms that may explain why switches between equilibria take place. For example, in the model by Cooper and John (1988), the coordination failure mechanism in an imperfectly competitive environment is used as a source of the existence of multiple equilibrium. As argued by Cooper and John, with no coordination there is a strong possibility of the presence of situations known as "spillover" and "strategic complementarity." The former refers to a situation where one's own payoff is affected by others' strategies. The latter refers to a situation in which one's own optimal strategy is affected by the others' strategies. Chamley (1999) addressed two questions that are central to models of multiple equilibria: why does one equilibrium arise rather than another? How do shifts between them occur? Imposing the twin assumptions of payoff uncertainty and learning from history in a dynamic model he provides a basis for fluctuations and cycles. He shows that even in an imperfect information world there could be a unique equilibrium with phases of high and low economic activity and random switches. He also discussed several applications in macroeconomics and revolutions. Cooper (1994) utilized a conventional approach found in models of multiple equilibria by assuming that economic agents coordinate on the one closest to equilibrium in the previous period where the economy evolved smoothly. In his model when the equilibrium of this 'type' disappears, the economy jumps to a new equilibrium. Specifically, he allowed for the presence of serially correlated shocks that shift between high marginal productivity/high-cost and low marginal productivity technologies for multiple equilibria. Finally, Acemoglu and Scott (1997) utilized high fixed costs and internal increasing returns to scale to generate regime-switching.

In an optimal search model by Diamond (1982), the two situations of spillover and complementarities arise because of the presence of "think-market" externality. The likelihood of one's own success depends on the intensity of search undertaken by others. One's search is more desirable when other economic agents are also searching, both because it is likely to have a larger payoff and also because search would likely be more productive. The presence of both spillovers and complementarities can have important macroeconomic effects, such as increasing returns to scale, because of either high levels of output or high returns to firms' technologies, or a combination of both.

There are other economic theories that can explain why switches between equilibria might take place. For instance, Diamond and Fundberg (1989) built a model to show that rational-expectations sunspot equilibrium would exist if agent beliefs about cycles were self-fulfilling. A recent paper that utilized a similar mechanism is Jeanne and Mason (2000). Employing a model of switching consumer confidence between waves of optimism and pessimism, Howitt and McAfee (1992) have shown that multiple equilibria with statistical properties characterized by a Markov regime-switching process can exist. Finally, the dynamic mechanism of a "learning-by-doing" type externality that drives the "new growth theory" models has been shown to produce endogenous switching between high-growth and low-growth states as a consequence of large shocks which can produce persistence in the selected equilibrium. For instance, Startz (1998) utilized the economic principles for explaining long-run growth over decades for understanding about fluctuation at business cycle frequencies. The model economy considered by him has two goods (or sectors) that share of single factor input, which is in fixed supply. The productivity increases due to both an exogenous technology change of Solowian variety and an endogenous technological change of the type used in the 'new growth theoory' models of Romer (1986, 1990) and Lucas (1988). Each of the two technologies has its own growth rate plus cross-fertilization between them is permitted. Endogenous technology component is a source of positive feedback, because of learning-by-doing or other similar reasons, which could destabilize the economy whereas exogenous technology tends to stabilize the economy. Also, the greater the use of productive input in the process, the faster the growth in the process-specific producing multiple states. In such a model economy shocks to preferences and technologies cause not only endogenous switching of the leading sector between the high-growth and low growth processes but also generate data of the type indentified by Hamilton (1989). The mechanism used here is based more on the endogenous comparative advantage rather than the coordination mechanism used by Cooper 1994. The latter mechanism is more Keynesian as changes in intermediate level of aggregate demand could move the economy into the region where history matters.

2.3 Other Arguments for Asymmetry

There is a variety of other reasons besides those listed above to motivate consideration of asymmetry in particular and nonlinearity in general. First, Markov regime-switching models tend to provide a good fit to aggregate macro data. This is because macroeconomic variables undergo episodes in which behavior of the series dramatically changes and becomes quite different. These dramatic shifts are sometimes caused by such events as wars, economic recessions, and financial panics. Abrupt changes, however, can be the result of deliberate policy actions of the government. For instance, government may choose to bring down inflation after a prolonged inflationary environment, or eliminate previous regulations or reduce taxes to expand the economy. Also, evidence for asymmetry in either, especially, the basic Markov switching or one of its extensions has been generally strong. The reported mixed outcome from the classical likelihood ratio tests of Hansen (1992) and Garcia (1998) which are designed to take care of the nuisance parameters problem, sometimes also known as the Davies' problem (see Davies 1977), is perhaps related to the poor power of these tests in small samples. Finally, the cost of ignoring regime switching, if in fact it occurs, would be large. The benefit of using regime-switching models, even if they were to contribute tiny improvements in forecast accuracy, can be large differences in profits.

3. ASYMMETRY OF THE BUSINESS CYCLE USING A SINGLE TIME SERIES SETUP: THE MARKOV-SWITCHING APPROACH

A typical empirical formulation of a regime-switching model by Hamilton (1989) permitted the mean growth rates of the gross domestic product to switch between two states of positive and negative growth rate in the United States. As mentioned below, he found that such a model corresponds well with the National Bureau of Economic Research business cycle peaks and troughs.

3.1 The Hamilton Empirical Model

In particular, the autoregessive model of order 4 used by Hamilton (1989) had the following specification for the time series, y_t, for the U.S. postwar quarterly-growth rate of real GNP from 1953-Q2 to 1984-Q4:

$$\left.\begin{array}{l} y_t = \mu(S_t) + \phi_1(x_{t-1} - \mu(S_{t-1})) + \phi_2(x_{t-2} - \mu(S_{t-2})) \\[2mm] \quad + \cdots + \phi_4(x_{t-4} - \mu(S_{t-4})) + \varepsilon_t \\[2mm] \varepsilon_t \sim \text{i.i.d. } N(0, \sigma^2) \end{array}\right\} \tag{3.1}$$

We assume that the discrete-valued variable S_t is generated by a two-state first-order Markov process. The true state of the process is treated as unobservable (hidden or imbedded) and it must be inferred on the basis of observations on the series y_t from time 0 to t (the filter probability) or from the series y_t for all t (the smoothed probability). With a two-state, first-order Markov stochastic process, where S_t can take a value 0 or 1, we write:

Prob.$[S_t = 1/S_{t-1} = 1] = p$

Prob.$[S_t = 0/S_{t-1} = 1] = 1 - p$

Prob.$[S_t = 0/S_{t-1} = 0] = q$

Prob.$[S_t = 1/S_{t-1} = 0] = 1 - q$

The state-dependent mean is linearly specified as

$\mu(S_t) = \alpha_0 + \alpha_1 S_t$

such that $\mu(S_t) = \alpha_0$ for $S = 0$ and $\mu(S_t) = \alpha_0 + \alpha_1$ for $S = 1$. The Markov chain is assumed to be ergodic and irreducible so that there does not exist an absorbing state. The transition probabilities are assumed to be time invariant so that the probability of a switch between regions does not depend on how long the process is in a given region. Some of the assumptions of the model can be relaxed, if necessary.

In the basic model the autoregressive coefficients are assumed to be constant so that regime switches shift only the mean growth rates. The density of y_t conditional on S_t is assumed normal with two means around which growth rate y_t moves to episodes of expansion and contraction.

However, since we do not observe S_t but only the sample path of y_t from time 0 to T, we must find a way to make an optimal inference about the current state of the history of observations on y_t. Hamilton proposed the use of a nonlinear filter in a recursive fashion on lines similar to the Kalman filter for linear models. The use of a nonlinear filter gives a likelihood function of the y_ts as a by-product. An algorithm for estimating the parameters of the regime-switching model that uses the limiting unconditional probabilities as initial values for starting it was developed by Hamilton. One powerful feature of the regime-switching model is its ability to generate smooth probabilities that correspond closely to the NBER business-cycle peaks and troughs.

3.1.1 Empirical results

The maximum likelihood estimates of the model are associated in a direct way with the business cycles in the U.S. instead of identifying long-term trends in the U.S. economy. Moreover, recessions corresponding to state 0 are associated with a low negative growth rate of output with an estimate value $a_0 = -0.4\%$ per quarter, and expansions corresponding to state 1 are associated with a positive growth of $(a_0 + a_1) = +1.2\%$ per quarter. Also the probability of remaining in the expansion states $p = 0.90$ is higher than the probability $q = 0.76$ of remaining in the recession state. This evidence is consistent with the business cycle's being asymmetric. These estimates imply that expansions and contractions have an expected duration value of $1/(1 - p) = 10.5$ quarters and $1/(1 - q) = 4.1$ quarters, respectively. In comparison, an average duration of expansion and recession in the NBER dating method is 14.3 quarters and 4.7 quarters, respectively.

The filter and smooth probability estimates of the two-state first-order Markov model of the growth rate, using quarterly data from 1952 through 1984, correspond to the NBER dating of peaks and troughs within three months of each other.

3.2   A Generalization and Other Extensions of the Hamilton Model

The Hamilton approach has been found to be useful in characterizing business cycles by several researchers, although they have sought to augment the basic model in some way, supported by either a theoretical or an empirical argument. Below we give a brief account of some of the important extensions of the Hamilton model.

3.2.1 The Hamilton model with a general autoregressive component

The Hamilton model is attractive because it is simple to estimate and interpret even though it allows for complicated dynamics such as asymmetry and
conditional heteroskedasticity. The time series in this model is composed of
two unobserved components, one following a random walk with drift evol-
ving as a two-state Markov process, and the second following an autoregressive process with a unit root. Lam (1990) generalized the Hamilton
model by allowing its cyclical component to follow a general autoregressive
process without a unit root so that the Markov-switching model may be
specified in the time series in log level rather than in growth rate of real
output. This generalization allows some shocks to have temporary effects
and others to have permanent effects. It is in part supported by the empirical
work of Perron (1989), who found that once a shift in the deterministic trend
is allowed, the evidence for a unit root is weak in most time series.
International evidence casting doubt on a unit root in output is provided
by Raj (1992). The empirical results obtained by Lam for his model for
postwar quarterly data confirmed Hamilton's conclusion that recurrent
shifts between low- and high-growth states constitute an important characteristic of real output. However, the in-sample forecasting performance of
the model proposed by Lam (in terms of the root-mean-square forecast
error criterion) was better than the Hamilton model for long horizons.

3.2.2 The Hamilton model with duration dependence

Another extension of the Hamilton model allowed each of the state transition probabilities to evolve as logistic functions of some exogenous variables
or economic fundamentals (e.g., see Filardo et al. 1998, Diebold et al. 1993,
Filardo 1994 and Diebold et al. 1994) or an integer-valued variable (e.g. see,
Durland and McCurdy 1994). Such an extension is motivated by a view that
the transition probabilities depend on how long the process has been in a
particular regime or is duration dependent rather than constant. This viewpoint is argued to be consistent with the evidence that the longer the contraction persists, the more likely it is to end soon.

3.2.3 The Hamilton model with a three-state Markov process

Another extension has been to add a "third state" to the basic two-state
regime-switching model. This extension is in response to an empirical result
found by Sichel (1994) showing that there are three phases of a business
cycle: recession, high-growth recovery, and the moderate growth period
following recovery. Clements and Krolzig (1998), using the quarterly growth
rate of real GDP from 1948Q2 to 1990Q4 for the U.S., found support for a

third phase of the cycle. They also showed that episodes of expansion and contraction for the three-state model tend to correspond fairly closely to the NBER classification of business-cycle turning points for the U.S. Strong evidence of a third phase also appears to exist for the Canadian GDP over the period 1947Q1 to 1997Q3, as shown by Bodman and Crosby (2000). However, the rapid growth phase observed for previous recessions appears not to have materialized in the 1990s.

3.2.4 The Hamilton model with a two-state Markov process where the mean and the variance follow independent switches

Yet another extension was to permit residual variances of the autoregressive specification to be state dependent instead of constant across states. The augmentation of the basic Hamilton model proposed by McConnell and Perez-Quiros (2000) first allows the mean and the variance to follow independent switching processes. Secondly, it allows for the two-state process for the mean of the business-cycle component of the model to vary according to the state of the variance. The empirical evidence in support of this extension is based on the existence of a structural break in the volatility of the U.S growth rate for a period since 1984. The dramatic reduction in output fluctuations in the U.S. are shown in turn to emanate from a reduction in the volatility of durable goods that is roughly coincident with a break in the proportions of output accounted for by inventories. This increased instability is about quarter-to-quarter fluctuations rather than cyclical fluctuations as considered earlier in the three-state Markov-switching model.

3.2.5 The Hamilton model with two types of asymmetry

Kim and Nelson (1999b) extended the Hamilton model by modeling two types of asymmetry of recessions. One type of asymmetry allowed by them has roots in the work of Friedman (1964, 1993), who argued that there is an optimal level of potential output for the economy that is determined by both resources and the way they are organized, which acts as a ceiling. He argued that occasionally the output is "plucked" downward from the ceiling by recessions caused by a transitory shock in aggregate demand or other disturbances. Output reverts back to the normal phase after the trough of a recession. Some empirical support for this point of view is provided by Beaudry and Koop (1993) and Wynne and Balke (1992, 1996), who found that the deeper the recession, the stronger the ensuing recovery. The empirical support for a high growth phase after recessions, found by Sichel (1994), implies that negative shocks are likely to be less persistent than positive shocks, which is consistent with this type of asymmetry.

The second type of asymmetry of the business cycle considered is about the shift of growth rate of trend that was modeled by Hamilton. This type of asymmetry permits the growth rate of output to shift from boom to recession due to infrequent large shocks, so that the effect is permanent. In contrast, the second type of asymmetry of recessions, a deviation from the output ceiling due to large infrequent negative shocks, has a transitory effect on the growth rate. Kim and Nelson (1999b) model two types of asymmetry by specifying a mixture of two types of shocks. One type consists of a set of discrete asymmetric shocks that depend on an unobserved state. The other type is a set of symmetric shocks with state-dependent variances that can switch during normal and recession times. They show that the hypothesis that potential real GDP provides an upper limit to the output cannot be rejected by the quarterly real GDP data for the U.S.

4. THE MULTIVARIATE MODELS OF ASYMMETRY: DYNAMIC FACTOR WITH MARKOV-SWITCHING MODELS.

4.1 One-Factor Dynamic Model with Markov Regime Switching

The framework of Markov-switching autoregressions for a single time series has been extended to a model of several time series to model the asymmetry attribute of the business cycle. The dynamic one-factor model with regime switching is one prominent example of the multivariate model of asymmetry. In this model the co-movement attribute of business cycles is accounted for through factor structure whereas the asymmetry attribute is specified through regime switching in a dynamic setting. The cycle component C_t in this model is an unobserved component that is common to more than one observed coincident variable in the multivariate case. Suppose we denote observed coincident economic variables such as personal income less transfer payments, an index of industrial production, manufacturing and trade sales, and hours of employees on nonagricultural payrolls, by y_{it} for $i = 1, 2, 3, 4$. In addition, assume that each of the observed variables has a unit root but the variables themselves are not cointegrated, so that the term ΔC_t is treated as a common factor component in a dynamic factor model with regime switching. A prototype model developed by Diebold and Rudebusch (1996) for a one-factor dynamic model with regime switching had the following form:

$$
\left.\begin{aligned}
\Delta y_{it} &= \gamma_i(L)\Delta c_t + e_{it}, & i &= 1, 2, 3, 4 \\
\psi_i(L)e_{it} &= \varepsilon_{it}, & \varepsilon_{it} &\sim \text{i.i.d. } N(0, \sigma_i^2) \\
\varphi(L)(\Delta c - \mu(s_t)) &= v_t, & v_t &\sim \text{i.i.d. } N(0, \sigma^2)
\end{aligned}\right\} \tag{4.1}
$$

where $\mu(s_t)$ is specified as a two-state first-order Markov-switching process; the changes in observed variable $\Delta y_{it} = (\Delta Y_{it} - \Delta Y)$ and unobserved component Δc_t are as deviations from the long-run mean of coincident economic variables, or the (demeaned) long-run growth of the cyclical component. Expressing the data in deviations from means helps to solve the problem of over-parameterization of the observed variables. One also needs to set $\sigma^2 = 1$ for identification. While the above framework was proposed by Diebold and Rudebush, they did not actually estimate it. The implementation was taken by Chauvet (1998), Kim and Yoo (1995) and Kim and Nelson (1998) for business cycle analysis.

The specification (4.1) integrates two features of the business cycles: co-movement and asymmetry. This model collapses to a pure dynamic factor model formulated by Stock and Watson (1991) when no Markov switching is permitted. The empirical test of asymmetry of business cycles for a univariate Markov-switching model, using the classical likelihood ratio testing approach, has generally produced mixed results for the U.S. data. For instance, neither Hansen (1992) nor Garcia (1998) reject the null hypothesis of no Markov switching in quarterly real output. However, Diebold and Rudebusch (1996) found strong evidence of Markov switching in a dynamic factor with a regime-switching model for the composite index of coincident economic indicators published by the Department of Commerce. Several researchers have conjectured that a multivariate framework should provide more reliable and consistent test results than a univariate framework if the dynamic factor model is successful in capturing the co-movement across indicators. Variations on the basic setup of the one-factor model are possible. For instance, Kim and Murray (2001), who incorporated two types of asymmetry in a dynamic factor model with regime switching for estimating to allow high-growth recovery phase during which economy partially reverts to its previous peaks.

4.1.1 One-factor dynamic model with Markov regime switching with time-varing transitional probabilities

Kim and Yoo (1995) have used an extended multivariate Markov switching factor model of Diebold and Rudebusch (1996) that permits the transitional probabilities to vary over time. The use of this type of model is motivated by a desire to achieve a high degree of correlation between the

NBER business cycle dates and the estimated probabilities of the state for the purposes of constructing a new index of coincident indicator for monthly data. Their use of an extended model is consistent with some earlier results obtained by Filardo (1994) and Filardo et al. (1998), who argued that the use of a Markov switching model with fixed transitional probabilities is not as successful for noisy monthly data as it has been for quarterly data.

4.2 The Two-Factor Dynamic Model with Regime Switching

In some multivariate modeling situations, it would be beneficial to consider two common features instead of one common feature to characterize the dynamics, producing a two-factor dynamic model with regime switching for business cycle analysis. For example, for a set of three time series such as log of output, log of fixed investment, and log of consumption taken together, it is not difficult to consider the possibility of the presence of both common stochastic trend and common cycle. Each of the three time series is individually integrated but can be combined with each of the other two to form a linear combination that is stationary. For instance, the common stochastic trend would be the productivity growth shock in the one-sector model of capital accumulation developed by King et al. (1988), where output is produced by two factors, capital and labor, and is subject to an exogenous growth in labor-augmented technology. Kim and Piger (1999), building on the work by Cochrane (1994) and the neoclassical growth model by King et al., identified one common stochastic trend of consumption. Furthermore, they specified an empirical dynamic two-factor model with a regime-switching model (and a second type of asymmetry) in which logarithm of output and logarithm of investment are influenced by a common stochastic trend and a common cycle shock. The common stochastic trend and common cycle are modeled as two dynamic factors plus an idiosyncratic shock affecting the log of output and the log of investment. The logarithm of consumption is influenced by the common stochastic trend only in their two-factor dynamic model. While they go on to consider the possibility of two types of asymmetry (the Hamilton type and the Friedman type), a simpler version of their model might allow for one type of asymmetry or regime switching only. The focus of their paper is to test the marginal significance of each type of asymmetry, and they find support for the Friedman type of asymmetry. However, it would be interesting to assess whether an improvement either in the power of the test for asymmetry or in the correspondence of dating the turning points of cycle performance of the two-factor dynamic model with asymmetry is obtained versus the Hamilton model.

4.3   Other Multivariate Applications

Below we provide a partial list of studies that have used a multivariate setup for the Markov-switching model in econometric analysis. Phillips (1991) used a bivariate Markov-switching model to analyzing the transmission mechanism of business cycles across two countries. The linkage between stock returns and output was studied by Hamilton and Lin (1996), using a bivariate Markov-switching model. Engel and Hamilton (1990) used interest rates and exchange rates to test a hypothesis about the public's behavior to policy changes on inflation. Ruge-Murcia (1992) analyzed three major attempts of the government in Brazil during 1986, 1987, and 1989, using a setup similar to that of Engel and Hamilton. Hamilton and Parez-Quinos (1996) jointly modeled the index of leading indications with the real GNP, and Krolzig and Sensier (2000) used a disaggregated approach for their analysis of the business cycles in UK manufacturing.

5.   CONCLUDING REMARKS

In this paper we have provided a selective interpretive survey of the econometrics of asymmetry of the type focused on by Hamilton. The univariate Markov-switching model of Hamilton (1989), and its extensions to a multivariate setting, to either a dynamic one-factor structure with regime switching or a dynamic two-factor structure with regime switching, have the potential of forming a progressive research program of the modern business cycle. The univariate Markov-switching model focuses only on the asymmetry feature of economic fluctuations. The dynamic one-factor, Markov-switching model allows jointly for common cycles (or co-movement) in variables and asymmetry in economic fluctuations. The dynamic two-factor, Markov-switching model accommodates common stochastic trends, common cycles, and asymmetry features in variables. The last specification may be interpreted as a prototype model that attempts to combine features of movements in trends of variables used in economic growth models with common cycles and symmetry features of their cyclical attributes.

The idea of using Markov-switching models in econometrics was first introduced by Goldfeld and Quandt (1973) for serially uncorrected data. Hamilton (1989, 1990, 1994) used this concept for serially uncorrelated data by modeling booms and recessions as regime switches in the growth rate of the output according to the past history of the dynamic system but treated as an unobservable (latent or imbedded or hidden) variable. Also, the latent state is modeled as a finite-dimensional process which is inferred from the observations of the variable up to time t. This model is often regarded to be more appropriate for business-cycles research than the "threshold" models of Tong (1983), Potter (1995), among others. This model can identify the NBER business cycles dates nonjudgmentally as compared to the judgmental method used by the classical researchers at the NBER. The Markov switching approach has found other applications that are not covered in this chapter. A partial list of such omitted topics is provided below for completeness.

The basic approach of Markov switching has been found to be useful in modeling financial variables. The parallel developments of regime switching in finance are extensive and reviewing this literature requires a separate paper. The question of how economic rational-economic agents identify and respond to major changes in regimes brought about by deliberate economic policy to bring down, say, inflation has also been studied using the Markov-switching model. Hamilton (1988) developed some technical tools for studying this type of question, and illustrated their use in the context of the term structure of interest rates during the monetary experiment of 1979. A sizeable literature followed on this topic, which is surveyed by Hamilton (1995). Finally, the asymmetry of the seasonal cycle has been another area of interest to economists. Two recent contributions in this area are by Ghysels (1994) and Frances et al. (1997).

ACKNOWLEDGEMENTS

The author wishes to thank an anonymous referee, James Hamilton, Lonnie Maggie, Alastair Robertson and Daniel J. Slottje for helpful comments.

REFERENCES

Acemoglu, Daron, and Andrew Scott (1997), Asymmetric Business Cycles: Theory and Time-series Evidence, *Journal of Monetary Economics*, 40, 501-33.

Backus, D. K., and P. J. Kehoe (1992), International evidence on the historical properties of business cycles. American Economic Review, 82(4), 864–888.

Beaudry, P., and G. Koop (1993), Do recessions permanently change output? *Journal of Monetary Economics*, 31, 149–163.

Blanchard, O. J. (1981), What is left of the multiplier accelerator? *American Economic Review*, 71, 150–154.

Bodman, P. M., and M. Crosby (2000), Phases of the Canadian business cycle. *Canadian Journal of Economics*, 33, 618–638.

Boldin, Michael E., Business cycles and stock market volatility: theory and evidence of animal spirits. PhD Dissertation, Department of Economics, University of Pennsylvania, 1990.

Brock, W. A., and Chera L. Sayers (1988), Is the business cycle characterized by deterministic chaos? *Journal of Monetary Economics*, 22, 71–80.

Burns, A. F., and W. C. Mitchell (1946), *Measuring Business Cycles*. New York: National Bureau of Economic Research.

Campbell, J. Y., and N. G. Mankiw (1987), Permanent and transitory components in macroeconomic fluctuations. *American Economic Review Papers and Proceedings*, 77, 111–117.

Chamley, Christophe, Cordinating Regime Switches (1999), *Quarterly Journal of Economics*, 114, 869-905.

Chauvet, Marcelle 1998), An Econometric Characterization of Business Cycle Dyunamics with Factor Structure and Regime Switches, *The International Economic Review*, 39, 969-96.

Clements, M. P., and H.-M. Krolzig (1998), A comparison of the forecast performance of Markov-switching and threshold autoregressive models of US GNP. *The Econometrics Journal*, 1, C47–C75.

Cochrane, J. H. (1994), Permanent and transitory components of GNP and stock prices. *Quarterly Journal of Economics*, 109, 241–263.

Cooper, Russell, and Andrew John (1988), Coordinating coordination failures in Keynesian models. *Quarterly Journal of Economics*, 103 (August), 441–463.

Cooper, Russell (1994), Equilibrium Selection in Imperfectly Competitive Economics with Multiple Equilibria, *Economic Journal*, 1106-1122.

Davies, R. B. (1977), Hypothesis testing when a nuisance parameter is present only under the alternative. *Biometrika*, 64, 247–254.

Diamond, Peter A. (1982), Aggregate demand management in search equilibrium. *Journal of Political Economy*, 90, 881–894.

Diamond, Peter A., and Drew Fudenberg (1989), Rational expectations business cycles in search equilibrium. *Journal of Political Economy*, 97 (June), 606–619.

Diebold, Francis X., and Glenn D. Rudebusch (1996), Measuring business cycles: a modern perspective. *Review of Economics and Statistics*, 78, 67–77.

Diebold, Francis X., and Glenn D. Rudebusch (1998*)*, *Business Cycles: Durations, Dynamics, and Forecasting*. Princeton, NJ: Princeton University Press.

Diebold, F. X., G. D. Rudebusch and D. E. Sichel (1993), Further evidence on business cycle duration dependence. In: J. H. Stock and M. W. Watson (eds.), *Business Cycles, Indicators and Forecasting*. Chicago: University of Chicago Press for NBER, pp. 255–284.

Diebold, Francis., Joon-Haeng Lee, and Gretchen C. Weinbach (1994). Regime Switching with Time-Varying Transition Probabilities, in C. Hargreaves, ed., Nonstationary Times Series Analysis and Cointegration (Oxford University Press).

Durland, J. Michael, and Thomas H. McCurdy (1994), Duration-dependent transitions in a Markov model of U.S. GNP growth. *Journal of Business and Economic Statistics*, 12 (July), 279–288.

Engel, C. M., and J. D. Hamilton (1990), Long swings in the dollar: are they in the data and do markets know it. *American Economic Review*, 80, 689–713.

Engle, R. F., and C. W. J. Granger (1987), Cointegration and error correction: representation, estimation, and testing. *Econometrica*, 55, 251–276.

Filardo, A. (1994), Business-cycle phases and their transitional dynamics. *Journal of Business and Economic Statistics*, 12, 299–308.

Filardo, J. Andrew and Stephen F. Gordon (1998), Business Cycle Durations, *Journal of Econometries*, 85, 99-123.

Frances, P. H., H. Hock, and R. Rapp (1997), Bayesian analysis and seasonal unit roots and seasonal mean shifts. *Journal of Econometrics*, 78, 359–380.

Friedman, M. (1964), *Monetary Studies of the National Bureau, the National Bureau Enters its 45th Year, 44th Annual Report*. 7–25 NBER, New York. Reprinted in: Friedman, M. (1969*)*, *The Optimum Quantity of Money and Other Essays*. Chicago: Aldine.

Friedman, M. (1993), The "plucking model" of business fluctuations revisited. *Economic Inquiry*, 31, 171–177.

Garcia, Rene (1998), Asymptotic null distribution of the likelihood ratio test in Markov switching models. *International Economic Review*, 39(3), 763–788.

Geweke, John (1977), The dynamic factor analysis of economic time-series models. In: D. J. Aigner and A. S. Goldberger (eds.), *Latent Variables in Socioeconomic Models*. Amsterdam: North-Holland, pp. 365–383.

Ghysels, E. (1994), On a periodic structure of the business cycle. *Journal of Business and Economic Studies*, 12, 289–298.

Goldfeld, Stephen M., and Richard E. Quandt (1973), A Markov model for switching regressions. *Journal of Econometrics*, 1, 3–16.

Hamilton, J. D. (1988), Rational-expectations econometric analysis of changes in regime: an investigation of the term structure of interest rates. *Journal of Economic Dynamics and Control*, 12, 385–423.

Hamilton, James D. (1989), A new approach to the economic analysis of nonstationary time series and the business cycle. *Econometrica*, 57(2), 357–384.

Hamilton, James D. (1990), Analysis of time series subject to change in regime. *Journal of Econometrics*, 45, 39–70.

Hamilton, James D. (1994), State-space models. In: R. F. Engle and D. McFadden (eds.), *Handbook of Econometrics*, Vol. 4. Amsterdam: Elsevier Science.

Hamilton, James D. (1995), Rational expectations and the economic consequences of regime change. Chapter 9 in: Kevin D. Hoover (ed.), *Macroeconomics: Development, Tensions and Prospects*, Boston: Kluwer Academic Publishers.

Hamilton, J. D., and G. Lin (1996), Stock market volatility and the business cycles. *Journal of Applied Econometrics*, 11, 411–434.

Hamilton, James D. and Gabriel Perez-Quiros (1996), What Do the Leading Indicators Lead?, Journal of Business, 69, 27-49.

Hansen, Bruce E. (1992), The likelihood ratio test under non-standard conditions: testing the Markov trend model of GNP. *Journal of Applied Econometrics*, 561-582.

Hansen, Bruce E. (1996), Erratum: The likelihood ratio test under non-standard conditions: testing the Markov trend model of GNP. *Journal of Applied Econometrics*, 11, 195-198.

Hodrick, R., and E. Prescott (1981), Post-war U.S. business cycles: an empirical investigation. Working paper, Carnegie-Mellon University, printed in *Journal of Money, Credit and Banking* (1997), 29, 1–16.

Howitt, Peter, and Preston McAfee (1992), Animal spirits. *American Economic Review*, 82 (June), 493–507.

Jeanne, Olivier and Paul Masson (2000), Currency Crises, Sunspots, and Markov-Switching Regimes, *Journal of International Economics*, 50, 327-350.

Keynes, J. M. (1936), *The General Theory of Employment, Interest, and Money*. London: Macmillan.

Kim, Chang-Jin (1994), Dynamic linear models with Markov switching. *Journal of Econometrics*, 60 (January–February), 1–22.

Kim, Chang-Jin, and C. J. Murray (2001), *Permanent and Transitory Components of Recessions*. Empirical Economics, forthcoming.

Kim, C.-J., and C. R. Nelson (2001), A Bayesian approach to testing for Markov-switching in univariate and dynamic factor models. *International Economic Review* (in press)

Kim, C.-J., and C. R. Nelson (1999a), *State-Space Models with Regime Switching: Classical and Gibbs-Sampling Approaches with Applications*. Cambridge, MA: MIT Press.

Kim, C.-J., and C. R. Nelson (1999b), Friedman's plucking model of business fluctuations: tests and estimates of permanent and transitory components. *Journal of Money, Credit and Banking*, 31, 317–334.

Kim, Chang-Jin, and Charles R. Nelson (1998), Business Cycle Turning Points, A New Coincident Index, and Tests of Duration Dependence Based on a Dynamic Factor Model with Regime-Switching, *Review of Economics and Statistics*, 80, 188-201.

Kim, Chang-Jin, and J. Piger (1999), Common stochastic trends, common cycles, and asymmetry in economic fluctuations. Discussion paper, University of Washington, Seattle, WA.

Kim, Myung-Jin, and Ji-Sung Yoo (1995), New index of coincident indicators: a multivariate Markov switching factor model approach. *Journal of Monetary Economics*, 36, 607–630.

King, R. G., C. I. Plosser, and S. T. Rebelo (1988), Production, growth and business cycles: II. New directions. *American Economic Review*, 21, 309–341.

King, R. G., C. I. Plosser, J. H. Stock, and M. W. Watson (1991), Stochastic trends and economic fluctuations. *American Economic Review*, 81, 819–840.

Krolzig, H.-M., and M. Sensier (2000), A disaggregated Makov-switching model of the business cycle in UK manufacturing. *Manchester School*, 68, 442–460.

Lam, P. (1990), The Hamilton model with a general autoregressive component. *Journal of Monetary Economics*, 26, 409–432.

Lucas, Robert E. (1976), Understanding business cycles. In: K. Brunner and A. Meltzer (eds.), *Stabilization of the Domestic and International Economy, Carnegie–Rochester Series on Public Policy*, 5. Amsterdam: North-Holland, pp. 7–29.

Lucas, Robert E. (1988), On the Mechanics of Economic Development, *Journal of Monetary Economics*, XXII, 3-42.

McConnell, M. M., and Perez-Quiros (2000), Output fluctuations in the United States: What has changed since the early 1980s? *American Economic Review*, 90, 1964–1976.

Neftci, Salih N. (1984), Are economic time series asymmetric over the business cycle? *Journal of Political Economy*, 92 (April), 307–328.

Nelson, C. W., and C. I. Plosser (1982), Trends and random walks in economic time series: some evidence and implications. *Journal of Monetary Economics*, 10, 139–162.

Perron, P. (1989), The Great Crash, the oil price shock and the unit root hypothesis. *Econometrica*, 57, 1361–1401.

Phillips, K. L. (1991), A two-country model of stochastic output with changes in regimes. *Journal of International Economics*, 31, 121–142.

Potter, S. M. (1995), "A nonlinear approach to U.S. GNP. *Journal of Applied Econometrics*, 10, 109–125.

Raj, B. (1992), International evidence on persistence in output in the presence of an episodic change. *Journal of Applied Econometrics* 7, 281–193.

Raj, B., and A. Ullah (1981), *Econometrics: A Varying Coefficient Approach*, London: Croom Helm, 1–372.

Romer, P (1996), Increasing Returns to Scale and Long Run Growth, *Journal of Political Economy*, 1(1), 95-124.

Romer, P. (1990), Endogenous Technological Change, *Journal of Political Economy*, Part 2, S71-S102.

Ruge-Murcia, F. (1992), Government credibility in heterodox stabilization program. PhD dissertation, University of Virginia, VA.

Sargent, Thomas J., and Christopher Sims (1977), Business cycle modeling without pretending to have too much *a priori* theory. In: C. Sims (ed.), *New Methods of Business Cycle Research*. Minneapolis, MN: Federal Reserve Bank of Minneapolis.

Shishkin, Julius (1961), *Signals of Recession and Recovery*, NBER Occasional Paper #77. New York: NBER.

Sichel, Daniel E. (1987), Business cycle asymmetry: a deeper look. Mimeograph, Princeton University.

Sichel, D. E. (1989), Are business cycles asymmetric? A correction. *Journal of Political Economy*, 97, 1255–1260.

Sichel, Daniel E. (1994), Inventories and the three phases of the business cycle. *Journal of Business and Economic Studies*, 12 (July), 269–277.

Sims, Christopher A. (1980), Macroeconomics and reality. *Econometrica*, 48, 1–48.

Startz, Richard, Growth States and Shocks, *Journal of Economic Growth* (1998), 3, 203-15.

Stock, J. H. (1987), Measuring business cycle time. *Journal of Political Economy*, 95, 1240–1261.

Stock, James H., and Mark W. Watson (1991), A probability model of the coincident economic indicators. In: K. Lahiri and G. H. Moore (eds.), *Leading Economic Indicators: New Approaches and Forecasting Records*. Cambridge, UK: Cambridge University Press.

Stock, J. H., and M. W. Watson (1999), Business fluctuations in U.S. macroeconomics time series. In: John B. Taylor and M. Woodford (eds.), *Handbook of Macroeconomics*, Vol. 1a, Chapter 1, New York, US: Elsviez Science, B.V.

Tong, H. (1983), *Threshold Models in Non-linear Time-Series Models*. New York: Springer-Verlag.

Watson, M. W. (1986), Univariate detrending methods with stochastic trends. *Journal of Monetary Economics*, 18, 49–75.

Watson, Mark W., and Robert F. Engle (1983), Alternative algorithms for the estimation of dynamic factor, mimic and varying coefficient models. *Journal of Econometrics*, 15 (December), 385–400.

Wynne, M. A., and N. S. Balke (1992), Are deep recessions followed by strong recoveries? *Economic Letters*, 39, 183–189.

Wynne, M. A., and N. S. Balke (1996), Are deep recessions followed by strong recoveries? Results for the G-7 countries. *Applied Economics*, 28, 889–897.

Index

9 780367 578671